# Engineering Electrodynamics

A collection of theorems, principles and field representations

# Engineering Electrodynamics

A collection of theorems, principles and field representations

**Ramakrishna Janaswamy**
*Department of Electrical and Computer Engineering, University of Massachusetts, Amherst, MA, USA*

**IOP** Publishing, Bristol, UK

ISBN 978-0-7503-1716-0 (ebook)
ISBN 978-0-7503-1714-6 (print)
ISBN 978-0-7503-1808-2 (myPrint)
ISBN 978-0-7503-1715-3 (mobi)

DOI 10.1088/978-0-7503-1716-0

Version: 20201201

IOP ebooks

British Library Cataloguing-in-Publication Data: A catalogue record for this book is available from the British Library.

Published by IOP Publishing, wholly owned by The Institute of Physics, London

IOP Publishing, Temple Circus, Temple Way, Bristol, BS1 6HG, UK

US Office: IOP Publishing, Inc., 190 North Independence Mall West, Suite 601, Philadelphia, PA 19106, USA

*To my grandfather Sreeramulu Janaswamy, who always encouraged me to pursue my dreams.*

न चोरहार्यं न च राजहार्यं नभ्रातृभाज्यं न च भारकारि।
व्यये कृते वर्धतैव नित्यं विद्याधनं सर्वधनप्रधानम्।।

*It is not stolen by thieves, nor is it taken away by rulers. It is not divisible between siblings nor is it heavy to be carried.*

*If expended it only grows eternally. Wealth of knowledge is the foremost of all.*

आचार्यात्पादमादत्ते पादं शिष्यस्स्वमेधया।
पादं सब्रह्मचारिभ्यः पादः कालेन पूर्यते।।

*One quarter of knowledge is attained from the Preceptor. One quarter is attained through one's own effort and intelligence.*

*One quarter is attained from fellow students. One quarter of knowledge is completed in time.*

# Contents

# Preface

This book is a collection of key results pertaining to electrodynamics, elucidated though theorems and examples. It arose from my desire to gather and present what I feel are the central results in engineering electrodynamics that aspiring graduate students and researchers in the area may want to be familiar with. Therefore, the choice of contents and style of the book are necessarily subjective. The book contains a number of theorems and principles which afford the reader mathematical justification for many of the methodologies one often takes for granted. For example, when one expresses the solution of fields inside a metallic waveguide in terms of an infinite number of modes, what guarantees that the solution so constructed is sufficiently general? Hopefully the contents in this book will pave the way to bring to rest such troubling queries. Many mathematically oriented theorems are stated, but only those that are not beyond the scope of the book are proved. Our interest here is to explore the relevance of these theorems to the solution of Maxwell's equations and demonstrate their application in solving practical problems.

The order of material does not always transition from elementary topics to advanced topics as is typical of a textbook. For example, in chapter 1 there is already material involving spherical harmonics and Green's functions. The advantage of such a loose structure is that the chapters need not be read in a sequential manner, but can be read more or less independent of each another. However, an attempt has been made to place more advanced topics at the end. I may have deliberately relegated seemingly important physical problems to the status of examples. This is done so that students do not become distracted by the intricate details of the physical problem itself, but rather focus more on the common mathematical theme for treating a similar class of problems. In that sense the book may serve the purpose of answering *how to solve a physical problem* rather than *what the physical problem is.* The book will therefore be satisfactory to students who want to learn and gain proficiency in the foundations of mathematical methods dealing with engineering electromagnetics.

However, this book is not outside the scope of serving as a textbook. If the instructor is already certain of what topics to cover in a course, he/she could potentially tap into the relevant sections and gather the key results available through theorems and examples. For example, in a traditional graduate course on electromagnetic theory one would start off with Maxwell's equations and progress to the topics of boundary conditions, vector potentials, conservation of power, uniqueness of fields, reciprocity, equivalence principles, and the induction theorem, culminating in treating typical problems in Cartesian, cylindrical, and spherical coordinates. There is an ample amount of material dispersed throughout this book that covers all of these topics.

Chapter 1 is comprehensive, where we introduce Maxwell's equations in the time and frequency domains and the associated material boundary conditions. The use of both Lorenz potentials and Coulomb potentials in arriving at causal solutions to

Maxwell's equations is demonstrated. Several important theorems related to radiation conditions and the representation of fields in terms of Lorenz, Coulomb, and Debye potentials are discussed. The derivation of numerical absorbing boundary conditions from these theorems is illustrated.

In chapter 2 we discuss via variational principles the interesting theorem of Thomson which characterizes the behavior of the electric field and potential in electrostatics in the presence of conducting bodies. The principle of virtual displacement, reciprocation of potentials in electrostatics, as well as the maximum principle and mean-value theorem of static fields are all discussed and illustrated through examples.

As is well known, gauge conditions are imposed on potentials to make their definition unique. Apart from the usual Lorenz and Coulomb gauge conditions, there is a whole latitude of gauge functions one can choose from when Hertzian potentials and associated stream potentials are considered. This is treated in detail in chapter 3 for two separate ways of accounting for medium presence: polarization sources and constitutive parameters.

The notion of causality is built into our understanding of the physical world and this is also true with electromagnetic phenomena. Causality manifests itself in the form of relating the real and imaginary parts of electric and magnetic susceptibilities through the Kramers–Krönig relations. The lesser known result of the impact of causality on the scattering amplitude and the interesting ramifications of the latter on the gain-bandwidth performance of antennas is discussed in detail in chapter 4 along with key examples.

The uniqueness and energy aspects of electromagnetic fields are discussed in chapter 5. To gain insight into a physical process it is often desirable to cast the involved energy expressions in a definite quadratic form. The conditions under which this becomes possible and the requirements it imposes on the material properties in a dispersive medium are explored in the chapter. Finally, the momentum of electromagnetic fields and its conservation are also touched upon.

The duality principle results from the symmetry evidenced in Maxwell's equations when both electric and magnetic sources are included. Babinet's principle leads to interesting conclusions on the transmission of fields through apertures in planar conducting screens. A consequence of both of these principles is Booker's relation for the impedance relation between complementary antennas. These topics form the contents of chapter 6.

The notion of reciprocity refers to the similarity in the manifestation of electromagnetic interaction in spacetime between two sources. Various aspects of electromagnetic reciprocity in the frequency and time domains are discussed in chapter 7. A lesser known theorem, known as the compensation theorem, that is useful in accounting for incremental changes in an electromagnetic quantity when the environment or geometry is changed is also discussed at length along with some practical examples.

When an electric circuit is built using passive, lossless elements and energized by a source, the behavior of the imaginary (reactive) part of its driving-point impedance satisfies some fundamental properties that are encapsulated in a reactance theorem.

The field counterpart of this behavior also exists and this is treated in chapter 8. Through a detailed discussion and illustrations through examples, it is shown that a passive lossless antenna does not satisfy the so-called Foster's reactance theorem.

It is not always possible to solve electromagnetic boundary value problems exactly, particularly for general shapes. A high-frequency technique known as the geometrical optics technique has been in existence for several centuries for approximately treating the case of smooth bodies. The rigorous derivation of the geometrical optics field comprised of wavefronts and rays and the various laws pertaining to the latter, such as the eikonal equation, Sommerfeld–Runge law, Snell's law, and intensity law, are all discussed in chapter 9. The more widely applicable variational principle of Fermat is included in the same chapter. With the recent emergence of electromagnetic metasurfaces for the control of reflected and transmitted electromagnetic energies, there has been a renewed interest in geometrical optics that witnessed some generalizations to the classical laws. Gradient metasurfaces and the generalized Snell's law are also treated in this chapter.

It is possible to integrate Maxwell's equations in an environment comprised of material bodies and come up with integral representations of fields that involve integrals of surface fields on the bodies. Popular representations are the Stratton–Chu integral representation and the Franz integral representation. The equivalence of currents (equivalence principle) to produce the same field in a certain region of space follows from these integral representations. The integral representations are also useful in deriving integral equations involving the surface unknowns for general bodies. It is possible to solve the integral equations based on certain high-frequency approximations and come up with useful identities (the so-called Bojarski identity) that have applications in inverse scattering. All of these topics are discussed at length in chapter 10. The closely related topics of induction theorem and the forward scattering theorem are covered in chapter 11.

The separation of variables technique that is so commonly used in solving boundary value problems involving conforming geometries relies heavily on the ingredients of eigenfunctions, Green's functions, and the notion of completeness in Hilbert space. Chapter 12 discusses the basic foundations of eigenfunction expansions, Green's functions, functions of operators, spectral representation of regular and singular operators, and sums of operators. Some well known theorems such as Floquet's theorem, the Hilbert–Schmidt theorem, the completeness theorem for Green's functions, the commutative operators theorem, etc, are all discussed in the chapter with ample examples.

The notion of electromagnetic degrees of freedom is important in multiple input–multiple output (MIMO) communication systems, inverse scattering, and in the design of algorithms in computational electromagnetics, among others. The fundamental result that the degrees of freedom between communicating volumes depends on the electrical size of transmitting and receiving antennas as well as on the distance between them is derived in chapter 13. An example involving broadside coupled rectangular apertures and the associated eigenfunctions consisting of prolate spheroidal wave functions is introduced. Finally the fundamental result on the upper bound on the maximum gain of any finite-sized antenna with limited degrees of freedom is derived.

In computed x-ray tomography, the object shape is determined by illuminating it with x-rays and measuring the transmitted intensity at the exit plane of the object for multiple scan angles. The mathematical foundation of tomography including the topics of radon transform and projection slice theorem are discussed in detail in chapter 14.

In chapter 15, various integral/discrete representations of the free-space Green's function are given and their application in solving several canonical problems in Cartesian, cylindrical, and spherical coordinates is considered through a comprehensive list of examples.

Asymptotic evaluation of integrals is central in antenna and scattering theories. In this regard, the complications introduced by the presence of branch points or poles in the integrand is a common source of frustration for students. Asymptotic evaluation of integrals by the steepest descent technique, the stationary phase method, and the modified saddle point technique in the presence of pole and branch-point singularities is discussed in detail in chapter 16. Several examples are considered.

Transformational electromagnetics, wherein it becomes possible to design material shapes and their constituent properties, is emerging as a viable means of controlling wave–material interactions. Transformational electromagnetic techniques are based on the form invariance of Maxwell's equations under coordinate transformations and central to this is tensor calculus. In chapter 17, we discuss the preliminaries of tensor calculus including the topics of Riemannian space, geodesics, absolute and covariant derivatives, tensor gradients, divergence, and curl operations. The covariant form of Maxwell's equations is then presented. The transformational equations necessary to understand the operation of an electromagnetic cloak are then derived and an example cloak design is presented.

Traditionally boundary conditions are derived by applying an integral form of Maxwell's equations to limiting geometries such as pill-boxes, etc. A more powerful and general approach that can handle higher order source densities residing on the interface between two media is to cast Maxwell's equations at the outset in the sense of distributions. The emergence of exotic man-made surfaces such as metasurfaces and their analysis makes the latter methodology even more relevant. In chapter 18, we discuss the mathematical background required to treat Maxwell's equations in the sense of distributions. The theory is culminated in the derivation of classical boundary conditions at interfaces as well as the derivation of generalized sheet transition conditions (GSTC) at metasurfaces.

For some selected problems, such as those encountered in plasmonics and low-frequency geometries, efficient computational techniques can be devised based on stochastic representations of the solution of wave and Helmholtz equations. In chapter 19, we discuss the theory of stochastic calculus and stochastic processes and derive several interesting results including the Itô–Doeblin formula and the Feynman–Kac formula of representing solutions of wave and Helmholtz equations.

Apart from the main chapters, there is a sizable amount of material in the appendices on vector analysis, complex variable theory, Bessel functions, and associated Legendre functions, familiarity with and knowledge of which are valuable in tackling practical electromagnetic problems.

I must thank all my past students, whose rightful suspicions of the paucity of my knowledge made me amass more of it over the years. I would like to thank the numerous professional colleagues whom I meet at conferences and from whom I constantly learn the topics I am ignorant of. I would like to thank my past teachers: Professor Kishan Rao (National Institute of Technology, Warangal, India), Professor J Das and Professor B N Das (Indian Institute of Technology, Kharagpur, India), Professor R E McIntosh, Professor D M Pozar, and Professor D H Schaubert (University of Massachusetts, Amherst, MA), Professor S W Lee (University of Illinois, Urbana-Champaign), and Professor J-B Andersen (Aalborg University, Denmark), who have all contributed to my learning and inspired me in my career as an electromagnetician. Thanks are due to Dr K Zekios (Florida International University, Miami) and Dr C Miloslav (Czech Technical University, Prague) for reading certain sections of this book. I would like to thank the developer of the Akshāntara keyboard, Dr Rolf Noyer (Department of Linguistics, University of Pennsylvania, Philadelphia) for providing me with the Mac-compatible Devanāgari keyboard for typing up the Sanskrit quotations in the book. Finally, I would like to acknowledge Ashley Gasque, Senior Commissioning Editor at the Institute of Physics, for her constant encouragement and for granting me several extensions to the ever-rolling deadlines during the preparation of the manuscript.

I may have inadvertently omitted some theorems from this book what some readers may regard as important. I attribute this entirely to my own ignorance and to writer's fatigue. If brought to my attention, I will be happy to accommodate them in a future edition.

July 2020

Amherst, MA

# Author biography

## Ramakrishna Janaswamy

Ramakrishna Janaswamy is a Professor in the Department of Electrical and Computer Engineering, University of Massachusetts, Amherst, MA, USA. His teaching and research interests include analytical and computational electromagnetics, deterministic and stochastic radiowave propagation modeling, antenna theory and design, system theory, mathematical physics, and wireless communications. He is a Fellow of IEEE, was the recipient of the 1995 R W P King Prize Paper Award, IEEE Transactions on Antennas and Propagation, and of the IEEE 3rd Millennium Medal in 2000. He received the Distinguished Alumni Professional Achievement Award from the National Institute of Technology, Warangal, India in 2017 and the College of Engineering Outstanding Teacher Award from the University of Massachusetts, Amherst in 2014. He is an elected member of the US National Committee of International Union of Radio Science, Commissions B and F. He served/is serving as an Associate Editor of *Radio Science*, *IEEE Transactions on Vehicular Technology*, *IEEE Transactions on Antennas & Propagation*, and *IET Electronics Letters*, and as a contributing member of the IEEE Antennas & Propagation Standards Committee. He is the author of the book *Radiowave Propagation and Smart Antennas for Wireless Communications* (Kluwer, 2000) and a contributing author to *Handbook of Antennas in Wireless Communications* (L Godara (ed), CRC Press, 2001) and *Encyclopedia of RF and Microwave Engineering* (Wiley, 2005). During his spare time he indulges in nature photography, particularly of birds.

IOP Publishing

Engineering Electrodynamics
A collection of theorems, principles and field representations
Ramakrishna Janaswamy

# Chapter 1

# Maxwell's equations, potentials, and boundary conditions

In this chapter we introduce Maxwell's equations in the time and frequency domains, examine the representation of fields by the Lorenz and Debye potentials, and look at the boundary conditions that fields need to satisfy across material interfaces and at infinity. The Lorenz and Coulomb gauge conditions are introduced and it is shown that the Coulomb gauge condition also leads to retarded fields despite the intermediate quantities becoming non-causal. A basic existence theorem for fields related to Maxwell's equations is discussed. Several theorems involving radiation conditions and multipole expansion of fields by spherical harmonics are presented. The pertinent theorems we discuss in this chapter with regards to the latter are the Wilcox theorems, the Whittaker theorem, the theorem of Bouwkamp and Casimir, and spherical harmonics expansion theorems. All of these theorems have a bearing on the field representations carried out external to the sphere enclosing all sources.

## 1.1 The time-domain Maxwell's equations

Maxwell's equations for linear and simple media ($\mathscr{D} = \epsilon\mathscr{E}$, $\mathscr{B} = \mu\mathscr{H}$, and $\epsilon$ and $\mu$ are scalars independent of the field[1]) can be written as

$$\nabla \times \mathscr{E} = -\frac{\partial \mathscr{B}}{\partial t} - \mathscr{M}; \qquad \nabla \cdot \mathscr{D} = q_{\mathrm{ev}}, \tag{1.1}$$

$$\nabla \times \mathscr{H} = \frac{\partial \mathscr{D}}{\partial t} + \mathscr{J}; \qquad \nabla \cdot \mathscr{B} = q_{\mathrm{mv}}, \tag{1.2}$$

[1] In more complicated media we may have the bianisotropic relations $\mathscr{D} = \bar{\bar{\epsilon}}\mathscr{E} + \bar{\bar{\xi}}\mathscr{H}$; $\mathscr{B} = \bar{\bar{\mu}}\mathscr{H} + \bar{\bar{\zeta}}\mathscr{E}$, where $\bar{\bar{\epsilon}}$, $\bar{\bar{\mu}}$, $\bar{\bar{\xi}}$, and $\bar{\bar{\zeta}}$ are matrices.

 

where $(\mathscr{E}, \mathscr{H})$ are the real-valued, time-instantaneous electric (V m$^{-1}$) and magnetic (A m$^{-1}$) field strengths, $(\mathscr{D}, \mathscr{B})$ are the real-valued, time-instantaneous electric flux (C m$^{-2}$) and magnetic flux (Wb m$^{-2}$) densities, $(\epsilon, \mu)$ are the permittivity (F m$^{-1}$) and permeability (H m$^{-1}$) of the medium, $(\mathscr{J}, \mathscr{M})$ are the real-valued, time-instantaneous electric (A m$^{-2}$) and magnetic (V m$^{-2}$) volume current densities, and $(q_{\mathrm{ev}}, q_{\mathrm{mv}})$ are the real-valued, time-instantaneous electric (C m$^{-3}$) and magnetic (V sm$^{-3}$) volume charge densities, respectively. The electric current density is assumed to comprise both the conduction current density and the source current density. The equations of continuity relate the current and charge densities:

$$\nabla \cdot \mathscr{J} + \frac{\partial q_{\mathrm{ev}}}{\partial t} = 0; \quad \nabla \cdot \mathscr{M} + \frac{\partial q_{\mathrm{mv}}}{\partial t} = 0, \tag{1.3}$$

and are implied by Maxwell's equations. For field computations it is very advantageous to introduce intermediate field quantities called the scalar and vector potentials. The four relevant quantities are the electric and magnetic scalar potentials $\psi_{\mathrm{e}}$ (V) and $\psi_{\mathrm{m}}$ (A), respectively, and the electric and magnetic vector potentials $\mathscr{F}$ (V m$^{-1}$) and $\mathscr{A}$ (Wb m$^{-1}$), respectively. For example, in the absence of magnetic current density, the fields can be related to the magnetic vector potential $\mathscr{A}$ and the scalar electric potential $\psi_{\mathrm{e}}$ via

$$\mathscr{B} = \nabla \times \mathscr{A}; \quad \mathscr{E} = -\frac{\partial \mathscr{A}}{\partial t} - \nabla \psi_{\mathrm{e}}. \tag{1.4}$$

In a similar fashion the fields in the absence of electric currents may be related to the electric vector potential $\mathscr{F}$ and magnetic scalar potential $\psi_{\mathrm{m}}$ via

$$\mathscr{D} = -\nabla \times \mathscr{F}; \quad \mathscr{H} = -\frac{\partial \mathscr{F}}{\partial t} - \nabla \psi_{\mathrm{m}}. \tag{1.5}$$

### 1.1.1 The Lorenz gauge, Coulomb gauge, and causality

We explore equation (1.4) further in the determination of fields from potentials. Writing $\mathscr{D} = \epsilon_0 \mathscr{E} + \mathscr{P}_{\mathrm{e}}$ and $\mathscr{B} = \mu_0(\mathscr{H} + \mathscr{P}_{\mathrm{m}})$, where $\mathscr{P}_{\mathrm{e}}$ (C m$^{-2}$) and $\mathscr{P}_{\mathrm{m}}$ (A m$^{-1}$) represent electric and magnetic polarization vectors, respectively, it is easy to see from Maxwell's equations that the potentials satisfy the equations

$$\nabla^2 \psi_{\mathrm{e}} + \frac{\partial}{\partial t}(\nabla \cdot \mathscr{A}) = -\frac{q_{\mathrm{ev}}}{\epsilon_0} + \frac{1}{\epsilon_0} \nabla \cdot \mathscr{P}_{\mathrm{e}}, \tag{1.6}$$

$$\nabla \times \nabla \times \mathscr{A} + \frac{1}{c^2}\frac{\partial^2 \mathscr{A}}{\partial t^2} = -\frac{1}{c^2}\frac{\partial \nabla \psi_{\mathrm{e}}}{\partial t} + \mu_0\left(\mathscr{J} + \frac{\partial \mathscr{P}_e}{\partial t} + \nabla \times \mathscr{P}_{\mathrm{m}}\right), \tag{1.7}$$

where $c = 1/\sqrt{\mu_0 \epsilon_0}$ is the speed of the wave in a vacuum. Additional constraints must be placed on the potentials $\mathscr{A}$ and/or $\psi_{\mathrm{e}}$ through any one of the gauge conditions to decouple and determine them uniquely.

In the *Lorenz gauge*

$$\nabla \cdot \mathscr{A} = -\frac{1}{c^2}\frac{\partial \psi_e}{\partial t}, \tag{1.8}$$

and the potentials satisfy the non-homogeneous *wave equations*

$$\Box^2 \psi_e = \frac{q_{ev}}{\epsilon_0} - \frac{1}{\epsilon_0}\nabla \cdot \mathscr{P}_e, \tag{1.9}$$

$$\Box^2 \mathscr{A} = \mu_0 \left( \mathscr{J} + \frac{\partial \mathscr{P}_e}{\partial t} + \nabla \times \mathscr{P}_m \right), \tag{1.10}$$

where $\nabla \times \nabla \times \mathscr{A} = \nabla\nabla \cdot \mathscr{A} - \nabla^2 \mathscr{A}$ and $\Box^2 = \frac{1}{c^2}\frac{\partial^2}{\partial t^2} - \nabla^2$ is the d'Alembertian operator. The sources for the potentials are impressed and the polarization charges and currents, and the d'Alembertian operator carries the signal speed $c$ at which the source fluctuations are transmitted. Since the solution of a wave equation is causal[2], the potentials $\psi_e$ and $\mathscr{A}$ (as will be shown in example 1.1.1.1 below) are both retarded functions[3] of the sources, which, in turn, generate retarded electric and magnetic fields. Under the Lorenz gauge condition, the complete electromagnetic field in any region, including the source region, is completely determined in terms of three scalar functions out of the four functions ($\mathscr{A}\ \psi_e$).

In the *Coulomb gauge*

$$\nabla \cdot \mathscr{A} = 0, \tag{1.11}$$

and the potentials satisfy

$$\nabla^2 \psi_e = -\frac{q_{ev}}{\epsilon_0} + \frac{1}{\epsilon_0}\nabla \cdot \mathscr{P}_e, \tag{1.12}$$

$$\Box^2 \mathscr{A} = \mu_0 \left( \mathscr{J} + \frac{\partial \mathscr{P}_e}{\partial t} - \epsilon_0 \frac{\partial \nabla \psi_e}{\partial t} \right) + \mu_0 \nabla \times \mathscr{P}_m. \tag{1.13}$$

The scalar potential satisfies Poisson's equation, wherein it responds instantaneously to the charge density fluctuations at any distance as evidenced by the presence of the purely spatial Laplacian operator in equation (1.12). The presence of the time-instantaneous term proportional to $\frac{\partial \psi_e}{\partial t}$ on the right-hand side (rhs) of equation (1.13) raises the question of whether the magnetic vector potential is also non-causal. If

[2] According to the causality principle, all present phenomena are exclusively determined by past events and equations depicting causal relations between physical phenomena must, in general, be equations where a present time quantity (the effect) relates to one or more quantities (causes) that existed at some previous time [1]. Thus an equation between two or more quantities simultaneous in time but separated in space cannot represent a causal relation between these quantities.

[3] Meaning, the effect of disturbance at a location is retarded (delayed) by the time it takes for the wave to reach it from the source of the disturbance. Mathematically, if $t$ describes time for causes, the effects must contain temporal terms of the form $t - t_d$, where $t_d$ is the delay.

that were true then the entire electromagnetic field would be non-causal. If the total current density (impressed plus polarized) is decomposed[4] into longitudinal and solenoidal parts as

$$\mathscr{J} + \frac{\partial \mathscr{P}_e}{\partial t} = \mathscr{J}_\ell + \mathscr{J}_{so}, \tag{1.14}$$

such that $\nabla \times \mathscr{J}_\ell = 0$ and $\nabla \cdot \mathscr{J}_{so} = 0$, then the equation of continuity together with equation (1.12) yields

$$\nabla \cdot \left(\mathscr{J}_\ell - \epsilon_0 \frac{\partial \nabla \psi_e}{\partial t}\right) = 0; \quad \nabla \times \left(\mathscr{J}_\ell - \epsilon_0 \frac{\partial \nabla \psi_e}{\partial t}\right) = 0, \tag{1.15}$$

where the latter is due to the definition of $\mathscr{J}_\ell$ and the identity $\nabla \times \nabla \psi_e = 0$. It is seen that both the divergence and the curl of the quantity within parenthesis is zero, suggesting that the quantity is a constant vector. Since the only constant function that vanishes at infinity is the constant function zero we have

$$\mathscr{J}_\ell = \epsilon_0 \frac{\partial \nabla \psi_e}{\partial t}, \tag{1.16}$$

which reduces equation (1.13) to the non-homogeneous wave equation

$$\Box^2 \mathscr{A} = \mu_0 \mathscr{J}_{so} + \mu_0 \nabla \times \mathscr{P}_m. \tag{1.17}$$

Thus the Coulomb gauge once again results in a retarded solution for the quantities $\mathscr{A}$, $\mathscr{B}$, $\mathscr{H}$ in terms of the solenoidal current $\mathscr{J}_{so}$ and polarization current $\nabla \times \mathscr{P}_m$. Since the electric field is proportional to the gradient of $\psi_e$, but not $\psi_e$ itself, we will demonstrate in the example below that the electric field also remains retarded relative to the sources.

**Example 1.1.1.1** Causality in the Coulomb gauge.
Show that the electric field generated under the Coulomb gauge in an unbounded medium remains causal and that it coincides with the field produced under the Lorenz gauge [2].

The starting point is the expression for the causal Green's function, $G_a$, in spacetime for the d'Alembert's equation

$$G_a = \frac{\delta(t - \tau - R/c)}{4\pi R}, \; R = |\mathbf{r} - \mathbf{r}'|, \; t - \tau > 0; \quad \Box^2 G_a = \delta(t - \tau)\delta(\mathbf{r} - \mathbf{r}'). \tag{1.18}$$

It is easy to derive this expression by solving the d'Alembert differential equation either in the spacetime [3, p 838] or in wavenumber-time [4, p 267] domains. For example, the steps in the spacetime approach are as follows:

(i) Use the simpler representation $\delta(\mathbf{r} - \mathbf{r}') = \delta(R)/2\pi R^2$ from equation (18.29) for the spatial delta function. Note that $\int_0^\infty \delta(R)\,\mathrm{d}R = 1/2$.

[4] This is possible by the Helmholtz theorem (see theorem B.2.6) assuming that the total current density has finite support. See also example 1.2.0.1.

(ii) Write $G_a = g/R$ and use spherical coordinates $(R, \theta, \phi)$ to reduce the d'Alembert equation to

$$\frac{1}{c^2}\frac{\partial^2 g}{\partial u^2} - \frac{\partial^2 g}{\partial R^2} = \frac{\delta(R)\delta(u)}{2\pi R}, \tag{1.19}$$

where $u = t - \tau$. This is now a simple 1D wave equation in spacetime $(R, u)$.

(iii) Employ the change of variables $\xi = u - R/c$, $\eta = u + R/c$ and the corresponding Jacobian relation $\mathrm{d}t\mathrm{d}R = \mathrm{d}u\mathrm{d}R = (c/2)\mathrm{d}\xi\mathrm{d}\eta$.

(iv) Using equation (18.118), the effect of distribution $\delta(R)\delta(u)/2\pi R$ (whose mapping into the $\xi\eta$ domain is simply denoted by $\{\cdot\}$ in the ensuing material) on a testing function $\varphi(R, u)$ is

$$I = \int_{R=0}^{\infty}\int_{u=-\infty}^{\infty}\frac{\delta(R)}{2\pi R}\delta(u)\varphi(R, u)\,\mathrm{d}R\mathrm{d}u = \frac{1}{4\pi}\frac{\partial\varphi(0, 0)}{\partial R}$$
$$= \frac{1}{4\pi c}\left(\frac{\partial\varphi}{\partial\eta} - \frac{\partial\varphi}{\partial\xi}\right)_{\xi=0,\eta=0}.$$

However, since the singularity of the distribution is all concentrated at the origin

$$I = \frac{c}{2}\int_{\xi=-\infty}^{\infty}\int_{\eta=\xi}^{\infty}\{\cdot\}\varphi(\xi, \eta)\,\mathrm{d}\xi\mathrm{d}\eta = \frac{c}{4}\iint_{-\infty}^{\infty}\{\cdot\}\varphi(\xi, \eta)\,\mathrm{d}\xi\mathrm{d}\eta.$$

Therefore

$$\iint_{-\infty}^{\infty}\{\cdot\}\varphi(\xi, \eta)\,\mathrm{d}\xi\mathrm{d}\eta = \frac{1}{\pi c^2}\left(\frac{\partial\varphi}{\partial\eta} - \frac{\partial\varphi}{\partial\xi}\right)_{\xi=0,\eta=0},$$

which implies that the distribution $\delta(R)\delta(u)/2\pi R$ is transformed as

$$\frac{\delta(R)\delta(u)}{2\pi R} \to \frac{1}{\pi c^2}(\delta'(\xi)\delta(\eta) - \delta'(\eta)\delta(\xi)).$$

(v) The operator in equation (1.19) is transformed to

$$\frac{1}{c^2}\frac{\partial^2}{\partial u^2} - \frac{\partial^2}{\partial R^2} = \left(\frac{1}{c}\frac{\partial}{\partial u} - \frac{\partial}{\partial R}\right)\left(\frac{1}{c}\frac{\partial}{\partial u} + \frac{\partial}{\partial R}\right) = \frac{4}{c^2}\frac{\partial^2}{\partial\xi\partial\eta},$$

resulting in the equation

$$\frac{\partial^2 g}{\partial\xi\partial\eta} = \frac{1}{4\pi}(\delta'(\xi)\delta(\eta) - \delta'(\eta)\delta(\xi)).$$

(vi) Finally, integrating this equation and retaining only the non-zero terms yields the desired Green's function

$$g = \frac{\delta(\xi)\Theta(\eta)}{4\pi} \Longrightarrow G_a = \Theta(t-\tau)\frac{\delta(t-\tau-R/c)}{4\pi R}, \tag{1.20}$$

where $\Theta(\cdot)$ is the unit step function.

Clearly, in the case of d'Alembert's equation, an impulse excited at the point $\mathbf{r}'$ at epoch $t = \tau$ is only felt at a later time $t = \tau + R/c > \tau$ at an $\mathbf{r}$ that is located at a distance $R$ from $\mathbf{r}'$. The solution is thus causal and is a retarded function. By linearity the solution of equation (1.17) with an arbitrary rhs leads to the solution

$$\begin{aligned}\mathscr{A}_{\mathrm{c}}(\mathbf{r};\, t) &= \frac{\mu_0}{4\pi}\iiint\limits_V \int_{\tau=0}^{\infty} \frac{\delta(t-\tau-R/c)}{R}\left\{\mathscr{J}_{\mathrm{so}}(\mathbf{r}';\, \tau) + \nabla' \times \mathscr{P}_{\mathrm{m}}(\mathbf{r}';\, \tau)\right\} \mathrm{d}\tau \mathrm{d}v' \\ &= \frac{\mu_0}{4\pi}\iiint\limits_V \frac{1}{R}\left\{\mathscr{J}_{\mathrm{so}}(\mathbf{r}';\, t-R/c) + \nabla' \times \mathscr{P}_{\mathrm{m}}(\mathbf{r}';\, t-R/c)\right\} \mathrm{d}v',\end{aligned} \tag{1.21}$$

where the prime on the curl indicates that it is w.r.t. the primed coordinates and the subscript 'c' on $\mathscr{A}$ qualifies that the expression is for the Coulomb gauge. It is seen that vector potential remains causal under the Coulomb gauge condition.

When specialized to $c \to \infty$, the expression for Green's function in equation (1.18) also gives the Green's function, $G_\ell$, in spacetime for the Laplace equation involved in equation (1.12):

$$\nabla^2 G_\ell = -\delta(t-\tau)\delta(\mathbf{r}-\mathbf{r}'); \quad G_\ell = \Theta(t-\tau)\frac{\delta(t-\tau)}{4\pi R}, \; R = |\mathbf{r}-\mathbf{r}'|. \tag{1.22}$$

In this case an impulse excited at $(\mathbf{r}', \tau)$ is felt instantly at any $\mathbf{r}$ no matter how far its distance from $\mathbf{r}'$ is. By linearity this leads to the non-causal solution for the scalar potential in equation (1.12):

$$\psi_{\mathrm{ec}}(\mathbf{r};\, t) = \frac{1}{4\pi\epsilon_0}\iiint\limits_V \frac{1}{R}\left\{q_{\mathrm{ev}}(\mathbf{r}';\, t) - \nabla' \cdot \mathscr{P}_{\mathrm{e}}(\mathbf{r}',\, t)\right\} \mathrm{d}v', \tag{1.23}$$

where the subscript 'c' on $\psi_{\mathrm{e}}$ once again qualifies that the expression is for the Coulomb gauge. For ease we use the notation $[\cdot]$ such that the quantity within the square brackets is evaluated at a retarded time $\tau = t - R/c$. According to this notation, $[\mathscr{J}_{\mathrm{so}}] = \mathscr{J}_{\mathrm{so}}(\mathbf{r}';\, \tau = t - R/c)$. Note also that the time derivatives $\partial/\partial t = \partial/\partial\tau$. Under this notation, equation (1.21) may be rewritten as

$$\mathscr{A}_{\mathrm{c}}(\mathbf{r};\, t) = \frac{\mu_0}{4\pi}\iiint\limits_V \frac{1}{R}\left[\mathscr{J} + \frac{\partial}{\partial t}\mathscr{P}_{\mathrm{e}} - \epsilon_0\frac{\partial}{\partial t}\nabla'\psi_{\mathrm{ec}} + \nabla' \times \mathscr{P}_{\mathrm{m}}\right] \mathrm{d}v', \tag{1.24}$$

where the definition of $\mathscr{J}_{\mathrm{so}}$ given in equation (1.14) and the expression for $\mathscr{J}_\ell$ given in equation (1.16) were used in obtaining the rhs above. The time derivative of this expression is

$$\frac{\partial \mathscr{A}_c}{\partial t} = \frac{\mu_0}{4\pi}\frac{\partial}{\partial t}\iiint\limits_V \frac{1}{R}\left[\mathscr{J} + \frac{\partial \mathscr{P}_e}{\partial t} + \nabla' \times \mathscr{P}_m\right] dv' \\ - \frac{1}{4\pi c^2}\iiint\limits_V \frac{1}{R}\frac{\partial^2[\nabla'\psi_{ec}]}{\partial t^2} dv'. \tag{1.25}$$

Since the electric field $\mathscr{E}$ is equal to $-\nabla\psi_{ec} - \partial\mathscr{A}_c/\partial t$ from equation (1.4), let us explore the behavior of the gradient of the scalar potential term. By adding a term $-(1/c^2)\partial^2\psi_{ec}/\partial t^2$ to the left-hand side (lhs) of equation (1.12) and then taking its negative gradient we obtain the non-homogeneous d'Alembert's equation,

$$\Box^2\nabla\psi_{ec} = \frac{1}{\epsilon_0}\nabla\left(q_{ev} - \nabla\cdot\mathscr{P}_e\right) + \frac{1}{c^2}\frac{\partial^2\nabla\psi_{ec}}{\partial t^2}, \tag{1.26}$$

for $\nabla\psi_{ec}$. Solving this formally using the Green's function (1.18) leads to the *integral equation* for the unknown $\nabla\psi_{ec}$

$$\nabla\psi_{ec} = \frac{1}{4\pi\epsilon_0}\iiint\limits_V \frac{1}{R}\nabla'[q_{ev} - \nabla'\cdot\mathscr{P}_e]\, dv' \\ + \frac{1}{4\pi c^2}\iiint\limits_V \frac{1}{R}\frac{\partial^2[\nabla'\psi_{ec}]}{\partial t^2} dv'. \tag{1.27}$$

The last integrals on the rhs of equations (1.25) and (1.27) are equal and opposite and will cancel with each other when added. Hence the expression for electric field under Coulomb gauge is

$$-\mathscr{E} = \frac{\mu_0}{4\pi}\frac{\partial}{\partial t}\iiint\limits_V \frac{1}{R}\left[\mathscr{J} + \frac{\partial \mathscr{P}_e}{\partial t} + \nabla' \times \mathscr{P}_m\right] dv' \\ + \frac{1}{4\pi\epsilon_0}\iiint\limits_V \frac{1}{R}\nabla'[q_{ev} - \nabla'\cdot\mathscr{P}_e]\, dv', \tag{1.28}$$

clearly indicating that the electric field is a retarded function of the sources.

To show that this is identical to the electric field derived under the Lorenz gauge condition, note that the first integral on the rhs of equation (1.28) is identical to what would be obtained from the vector potential equation (1.10). The explicit solution for the vector potential equation (1.10) under the Lorenz gauge condition is

$$\mathscr{A}_l = \frac{\mu_0}{4\pi}\iiint\limits_V \frac{1}{R}\left[\mathscr{J} + \frac{\partial \mathscr{P}_e}{\partial t} + \nabla' \times \mathscr{P}_m\right] dv', \tag{1.29}$$

where the subscript '$l$' is a qualifier for the Lorenz gauge. The solution of the scalar potential under the Lorenz gauge condition in equation (1.9) is

$$\psi_{el} = \frac{1}{4\pi\epsilon_0} \iiint\limits_V \frac{1}{R}\left[q_{ev} - \nabla' \cdot \mathscr{P}_e\right] dv'$$

with a gradient

$$\begin{aligned}
\nabla\psi_{el} &= \frac{1}{4\pi\epsilon_0} \iiint\limits_V \nabla\left(\frac{1}{R}\right)[q_{ev} - \nabla' \cdot \mathscr{P}_e] \, dv' \\
&= -\frac{1}{4\pi\epsilon_0} \iiint\limits_V \nabla'\left(\frac{1}{R}\right)[q_{ev} - \nabla' \cdot \mathscr{P}_e] \, dv' \\
&= \frac{1}{4\pi\epsilon_0} \iiint\limits_V \frac{1}{R}\nabla'[q_{ev} - \nabla' \cdot \mathscr{P}_e] \, dv' - \frac{1}{4\pi\epsilon_0} \iiint\limits_V \nabla'\left(\frac{1}{R}[q_{ev} - \nabla' \cdot \mathscr{P}_e]\right) dv' \\
&= \frac{1}{4\pi\epsilon_0} \iiint\limits_V \frac{1}{R}\nabla'[q_{ev} - \nabla' \cdot \mathscr{P}_e] \, dv' - \frac{1}{4\pi\epsilon_0} \oiint\limits_S \hat{\mathbf{n}}\left(\frac{1}{R}[q_{ev} - \nabla' \cdot \mathscr{P}_e]\right) ds' \\
&= \frac{1}{4\pi\epsilon_0} \iiint\limits_V \frac{1}{R}\nabla'[q_{ev} - \nabla' \cdot \mathscr{P}_e] \, dv',
\end{aligned} \tag{1.30}$$

assuming that the charge density and electric polarization have finite support to make the surface integral at infinity vanish. The conversion from the volume to surface integral of the last integral on the rhs above is due to the gradient theorem (B.7). The final expression in equation (1.30) is identical to the last integral on the rhs in equation (1.28), both of which are causal in nature. Clearly, the electromagnetic fields derived under the Coulomb gauge coincide with those obtained under the Lorenz gauge despite the fact that the scalar potential under the former fails to be causal. Thus use of the Coulomb gauge does not violate finite speed of signal propagation for the electromagnetic fields. ■■

### 1.1.2 Existence theorem for fields

At this point it is appropriate to state without proof an existence theorem for fields from given charge conservation.

**Theorem 1.1.1.** Existence theorem for fields [5].
*Given localized sources* $q_e(\mathbf{r}; t)$ *and* $\mathscr{J}(\mathbf{r}; t)$ *which satisfy the equation of continuity*

$$\nabla \cdot \mathscr{J} + \frac{\partial q_e}{\partial t} = 0, \tag{1.31}$$

*there exist retarded fields* $\mathscr{F}(\mathbf{r}; t)$ *and* $\mathscr{G}(\mathbf{r}: t)$ *defined by*

$$\mathscr{F} = \frac{\alpha}{4\pi} \iiint\limits_V \left(\frac{\hat{\mathbf{R}}}{R^2}[q_e] + \frac{\hat{\mathbf{R}}}{cR}\left[\frac{\partial q_e}{\partial t}\right] - \frac{1}{c^2 R}\left[\frac{\partial \mathscr{J}}{\partial t}\right]\right) dv' \tag{1.32}$$

$$\mathscr{G} = \frac{\beta}{4\pi} \iiint_V \left( [\mathscr{J}] \times \frac{\hat{\mathbf{R}}}{R^2} + \left[ \frac{\partial \mathscr{J}}{\partial t} \right] \times \frac{\hat{\mathbf{R}}}{cR} \right) \mathrm{d}v', \tag{1.33}$$

*that satisfy the field equations*

$$\nabla \cdot \mathscr{F} = \alpha q_{\mathrm{e}}; \quad \nabla \cdot \mathscr{G} = 0 \tag{1.34}$$

$$\nabla \times \mathscr{F} = -\gamma \frac{\partial \mathscr{G}}{\partial t}; \quad \nabla \times \mathscr{G} = \frac{\beta}{\alpha} \frac{\partial \mathscr{F}}{\partial t} + \beta \mathscr{J}, \tag{1.35}$$

*where the constants* $\alpha$, $\beta$, $\gamma$, *and* $c$ *(not necessarily the speed of light) are related by* $\alpha = \beta\gamma c^2$, $\hat{\mathbf{R}} = \mathbf{R}/R = (\mathbf{r} - \mathbf{r}')/|\mathbf{r} - \mathbf{r}'|$, *and the square brackets* $[\cdot]$ *indicate that the quantity enclosed is to be evaluated at retarded time* $t' = t - R/c$.

The proof is straightforward but lengthy and the reader is referred to [5] for details. Identifying $\mathscr{J}$ with electric current density, $q_{\mathrm{e}}$ with electric charge density, $c$ with the speed of light in vacuum, $\mathscr{F}$ with electric field, $\mathscr{G}$ with magnetic flux density, $\gamma$ with 1, $\alpha$ with $\epsilon_0^{-1}$, $\beta$ with $\mu_0$, we arrive at Maxwell's equations in free-space given the equation of continuity. In other words, Maxwell's equations can be obtained by postulating charge conservation at the outset. Thus the equation of continuity assumes a fundamental status as a precursor to Maxwell's equations, rather than as an induced result from Maxwell's equations.

## 1.2 Frequency domain Maxwell's equations

Assuming an $\mathrm{e}^{j\omega t}$ convention in the time variable $t$, where $\omega$ is the radian frequency, Maxwell's equations for simple media can be written as

$$\nabla \times \mathbf{E} = -j\omega\mu\mathbf{H} - \mathbf{M}; \qquad \nabla \cdot \mathbf{D} = q_{\mathrm{ev}}, \tag{1.36}$$

$$\nabla \times \mathbf{H} = j\omega\epsilon\mathbf{E} + \mathbf{J}; \qquad \nabla \cdot \mathbf{B} = q_{\mathrm{mv}}, \tag{1.37}$$

where $(\mathbf{E},\ \mathbf{H})$ are the complex-valued, phasor electric, and magnetic field strengths, $(\mathbf{D},\ \mathbf{B})$ are the complex-valued, phasor electric, and magnetic flux densities, $(\epsilon(\omega),\ \mu(\omega))$ are the permittivity and permeability of the medium, respectively, $(\mathbf{J},\ \mathbf{M})$ are the complex-valued, phasor electric, and magnetic current densities, $(q_{\mathrm{ev}},\ q_{\mathrm{mv}})$ are the complex-valued, phasor electric, and magnetic charge densities and $j = \sqrt{-1}$. The relation between a phasor field quantity and the original time-dependent field quantity, illustrated below for the electric field, is

$$\mathscr{E}(\mathbf{r};\, t) = \Re[\mathbf{E}(\mathbf{r})\mathrm{e}^{j\omega t}], \tag{1.38}$$

where $\Re$ stands for 'real part of'.

The media constants can change from point to point in space in general. To accommodate lossy media, the permittivity is assumed to be complex of the form $\epsilon = \epsilon' - j\epsilon''$ with $\epsilon',\ \epsilon'' > 0$. Similarly, the permeability can be made complex by assuming $\mu = \mu' - j\mu''$ with $\mu', \mu'' > 0$. For simple media, Maxwell's equations also imply the equations of continuity:

$$\nabla \cdot \mathbf{J} = -j\omega q_{\mathrm{ev}}, \quad \nabla \cdot \mathbf{M} = -j\omega q_{\mathrm{mv}}. \tag{1.39}$$

By taking $\nabla\times$ of a Maxwell curl equation in (1.36) or (1.37) and substituting from the other, we obtain the non-homogeneous, vector Helmholtz equations

$$\nabla \times \nabla \times \mathbf{E} - k^2\mathbf{E} = -j\omega\mu\mathbf{J} - \nabla \times \mathbf{M}, \tag{1.40}$$

$$\nabla \times \nabla \times \mathbf{H} - k^2\mathbf{H} = -j\omega\epsilon\mathbf{M} + \nabla \times \mathbf{J}, \tag{1.41}$$

satisfied by the electric and magnetic fields, where $k = \omega\sqrt{\mu\epsilon} = k_r - jk_i$, $k_r > 0$, $k_i > 0$ is the wavenumber at the frequency $\omega$ in the medium. By observing the rhs of equations (1.40) and (1.41) it is clear that $\mathbf{J}$ and $\nabla \times \mathbf{M}/j\omega\mu$ generate the same electric field in the sense that in an expression relating $\mathbf{E}$ to $\mathbf{J}$ one can replace $\mathbf{J}$ with $\nabla \times \mathbf{M}/j\omega\mu$ and obtain the expression for the electric field generated by the magnetic current $\mathbf{M}$. Hence $\mathbf{J}$ and $\nabla \times \mathbf{M}/j\omega\mu$ are equivalent. Likewise the current densities $-\mathbf{M}$ and $\nabla \times \mathbf{J}/j\omega\epsilon$ are equivalent. For instance, an azimuthal electric current density of the form $\mathbf{J} = j\omega\epsilon\hat{\boldsymbol{\phi}}J_\phi(\rho)$ in cylindrical coordinates is equivalent to a vertical magnetic current density $\mathbf{M} = -\nabla \times (\hat{\boldsymbol{\phi}}J_\phi) = \hat{\mathbf{z}}\left(\frac{\partial J_\phi}{\partial\rho} + \frac{1}{\rho}J_\phi\right)$.

As with time-domain equations, magnetic vector potential $\mathbf{A}$ and electric vector potential $\mathbf{F}$ (the so-called Lorenz potentials) are introduced such that $\mathbf{B} = \nabla \times \mathbf{A}$, $\mathbf{D} = -\nabla \times \mathbf{F}$. The general relationship between the field quantities and the Lorenz potentials are

$$\mathbf{E} = \frac{\epsilon^{-1}}{j\omega}(\nabla \times \mu^{-1}\nabla \times \mathbf{A} - \mathbf{J}) - \epsilon^{-1}\nabla \times \mathbf{F}, \tag{1.42}$$

$$\mathbf{H} = \frac{\mu^{-1}}{j\omega}(\nabla \times \epsilon^{-1}\nabla \times \mathbf{F} - \mathbf{M}) + \mu^{-1}\nabla \times \mathbf{A}. \tag{1.43}$$

There is complete duality between the electric and magnetic quantities, which will be elaborated upon in the duality principle (see section 6.1.1).

In an infinite homogeneous medium carrying electric and magnetic currents $(\mathbf{J}, \mathbf{M})$, the magnetic vector potential $\mathbf{A}$ is generally associated with the electric current density and the electric vector potential $\mathbf{F}$ is associated with the magnetic current density. Consequently, the complete field can be expressed in terms of three scalar functions (the three components of $\mathbf{A}$ or the three components of $\mathbf{F}$) when only one of these two sources is present. However, the freedom provided by a particular choice of gauge can potentially reduce the required scalar functions to two. For instance, under Coulomb gauge $\nabla \cdot \mathbf{A} = 0$ and one component of $\mathbf{A}$ can be determined from the other two. Additional gauge conditions under which the complete electromagnetic field can be expressed in terms of only two scalar functions is discussed in section 3.1.2. Under the Lorenz gauge condition $\nabla \cdot \mathbf{A} = -j\omega\mu\epsilon\psi_{\mathrm{e}}$, $\nabla \cdot \mathbf{F} = -j\omega\mu\epsilon\psi_{\mathrm{m}}$, and the vector potential satisfies $\nabla^2\mathbf{A} + k^2\mathbf{A} = -\mu\mathbf{J}$, where $\psi_e$ and $\psi_m$ are time-harmonic electric and magnetic scalar potentials. One consequently obtains

$$\mathbf{E} = -j\omega\mathbf{A} - \nabla\psi_e - \epsilon^{-1}\nabla \times \mathbf{F}, \tag{1.44}$$

$$\mathbf{H} = -j\omega\mathbf{F} - \nabla\psi_m + \mu^{-1}\nabla \times \mathbf{A}. \tag{1.45}$$

These expressions are valid at all points in space including the source region. On adapting the solution (1.18) for each component of $\mathbf{A}$ in the time-harmonic case it is easy to see that the solution of the vector Helmholtz equation

$$\nabla^2\mathbf{A} + k^2\mathbf{A} = -\mu\hat{\mathbf{q}}p\delta(\mathbf{r} - \mathbf{r}'), \tag{1.46}$$

with the dipole source, where the unit vector $\hat{\mathbf{q}}$ is an arbitrary constant vector representing the dipole orientation and the quantity $p$ is the dipole moment (units (A m)), is

$$\mathbf{A} = \hat{\mathbf{q}}p\mu\frac{\mathrm{e}^{-jkR}}{4\pi R} =: \hat{\mathbf{q}}p\mu\psi, \tag{1.47}$$

where $R = |\mathbf{r} - \mathbf{r}'|$ is the distance between the source and field points. Using equation (1.29), an expression for $\mathbf{A}$ for an arbitrary electric current distribution radiating in a homogeneous medium with parameters $(\epsilon, \ \mu)$ can be written as

$$\mathbf{A}(\mathbf{r}) = \frac{\mu}{4\pi}\iiint_V \mathbf{J}(\mathbf{r}')\frac{\mathrm{e}^{-jkR}}{R}\,\mathrm{d}v'. \tag{1.48}$$

Thus causality in the time-domain with a delay $t_d = R/c$ is manifested in the frequency domain by the appearance of the phase lag factor $\mathrm{e}^{-jkR}$.

In the far-zone where the approximations $R \approx r - \mathbf{r}' \cdot \hat{\mathbf{r}}$ in the phase term and $R \approx r$ in the denominator of the integrand hold, the expression for the vector potential reduces to

$$\begin{aligned}\mathbf{A}(\mathbf{r}) &\sim \frac{\mu\mathrm{e}^{-jkr}}{4\pi r}\iiint_V \mathbf{J}(\mathbf{r}')\mathrm{e}^{jk\hat{\mathbf{r}}\cdot\mathbf{r}'}\,\mathrm{d}v', \\ &= \frac{\mu\mathrm{e}^{-jkr}}{4\pi r}\tilde{\mathbf{J}}(\boldsymbol{\kappa} = -\hat{\mathbf{r}}k),\end{aligned} \tag{1.49}$$

where $\tilde{\mathbf{J}}(\boldsymbol{\kappa})$ is the 3D vector Fourier transform of the current distribution as defined in equation (1.149). Thus the vector potential in the far-zone is proportional to the Fourier transform of the current distribution and the uncertainty relations associated with Fourier transforms hold in the space and wavenumber domains. For the $\theta$ and $\phi$ components of the electric field in the far-zone, only the first term on the rhs of equation (1.44) remains. The dominant electric field components in the far-zone are then

$$E_\theta \sim -jk\eta \frac{e^{-jkr}}{4\pi r} \hat{\boldsymbol{\theta}} \cdot \tilde{\mathbf{J}}(-\hat{\mathbf{r}}k); \qquad E_\phi \sim -jk\eta \frac{e^{-jkr}}{4\pi r} \hat{\boldsymbol{\phi}} \cdot \tilde{\mathbf{J}}(-\hat{\mathbf{r}}k). \tag{1.50}$$

*Thus the transverse components of* $\tilde{\mathbf{J}}(\boldsymbol{\kappa})$ *on the circle* $|\boldsymbol{\kappa}| = k$ *are completely determined by the far-zone electric field components and vice versa.*

**Example 1.2.0.1.** Decomposition of current distribution by the Helmholtz theorem. In this example we apply the Helmholtz theorem (see theorem B.2.6) and show how to decompose a compact-support current density into a solenoidal part and an irrotational part. We do this for electric current density by assuming that there are no magnetic sources (i.e. the magnetic current and charge densities are both zero, $\mathbf{M} = 0$, $q_{\text{mv}} = 0$). The goal is to see how the decomposition of current by the Helmholtz theorem translates to the properties of the electromagnetic field [6]. Furthermore, as already seen in section 1.1.1, the proof that the Coulomb gauge gives rise to retarded fields relies on this decomposition.

By the Helmholtz theorem, the electric current density $\mathbf{J}$ can be expressed as

$$\mathbf{J} = \nabla \boldsymbol{\theta} + \nabla \times \mathbf{w} =: \mathbf{J}_\ell + \mathbf{J}_\text{t}, \tag{1.51}$$

where $\mathbf{J}_\ell = \nabla \boldsymbol{\theta}$ is the longitudinal part and $\mathbf{J}_\text{t} = \nabla \times \mathbf{w}$ is the transverse part. Here $\mathbf{J}$ consists of both the impressed currents $\mathbf{J}_\text{i}$ (charge density $q_{\text{ev}} = -\nabla \cdot \mathbf{J}_\text{i}/j\omega$) and polarization currents $\mathbf{J}_\text{p} = j\omega\mathbf{P}$, $\mathbf{P}$ being electric polarization, so that the divergence equation becomes

$$\begin{aligned} \epsilon_0 \nabla \cdot \mathbf{E} &= \nabla \cdot \mathbf{D} - \nabla \cdot \mathbf{P} \\ &= -\frac{1}{j\omega} \nabla \cdot \mathbf{J}_\text{i} - \frac{1}{j\omega} \nabla \cdot \mathbf{J}_\text{p} \\ &= -\frac{1}{j\omega} \nabla \cdot \mathbf{J} = -\frac{1}{j\omega} \nabla \cdot \mathbf{J}_\ell, \end{aligned} \tag{1.52}$$

by virtue of the equation of continuity. Hence it is possible to express the electric field in a material medium as

$$\mathbf{E} = -\frac{1}{j\omega\epsilon_0} \mathbf{J}_\ell + \mathbf{E}_\text{t} =: \mathbf{E}_\ell + \mathbf{E}_\text{t}; \quad \nabla \cdot \mathbf{E}_\text{t} = 0, \tag{1.53}$$

and

$$\nabla \times \mathbf{E} = \nabla \times \mathbf{E}_\text{t} = -j\omega \mathbf{B} = -j\omega \nabla \times \mathbf{A}_\text{c} \Longrightarrow \mathbf{E}_\text{t} = -j\omega \mathbf{A}_\text{c}, \tag{1.54}$$

where $\mathbf{A}_\text{c}$ is the magnetic vector potential and $\mathbf{B} = \nabla \times \mathbf{A}_\text{c}$ is the magnetic flux density. Since $\nabla \cdot \mathbf{E}_\text{t} = 0$, the Coulomb gauge condition $\nabla \cdot \mathbf{A}_\text{c} = 0$ is implied in this decomposition. The longitudinal current generates an electric field, $\mathbf{E}_\ell$, whose curl vanishes, thereby resembling an electrostatic field.

Expressing $\mathbf{B} = \mu_0(\mathbf{H} + \mathbf{P}_{\rm m})$, where $\mathbf{P}_{\rm m}$ is the magnetization vector, and using Maxwell's equation $\nabla \times \mathbf{H} = \mu_0^{-1}\nabla \times \mathbf{B} - \nabla \times \mathbf{P}_{\rm m} = j\omega(\epsilon_0\mathbf{E} + \mathbf{P}) + \mathbf{J}_{\rm i} = j\omega\epsilon_0\mathbf{E} + \mathbf{J}$ we obtain

$$\begin{aligned} &\nabla \times \nabla \times \mathbf{A}_{\rm c} - k_0^2\mathbf{A}_{\rm c} = \mu_0\nabla \times \mathbf{P}_{\rm m} + \mu_0\mathbf{J}_{\rm t} \\ &\Longrightarrow \nabla^2\mathbf{A}_{\rm c} + k_0^2\mathbf{A}_{\rm c} = -\mu_0\nabla \times \mathbf{P}_{\rm m} - \mu_0\mathbf{J}_{\rm t}. \end{aligned} \tag{1.55}$$

Thus the transverse current together with the magnetization vector generates the vector potential $\mathbf{A}_{\rm c}$ and, consequently, the magnetic field. Clearly, the magnetic vector potential still satisfies the Helmholtz equation under the Coulomb gauge condition and, thereby, generates a retarded solution as also shown in section 1.1.1. The magnetic vector potential under the Coulomb gauge condition contributes to the transverse electric field and is responsible for the radiated field. If the transverse current is zero, both $\mathbf{A}_{\rm c}$ and the radiated field vanish. ■■

### 1.2.1 Classification of media

It is most convenient to classify media in the frequency domain because the relations involving them turn out to be algebraic rather than of convolutional type. The most general linear medium is described in terms of four matrices or dyadics, $\bar{\bar{\epsilon}}$, $\bar{\bar{\xi}}$, $\bar{\bar{\zeta}}$, and $\bar{\bar{\mu}}$. They relate the flux densities to the field quantities [7]

$$\mathbf{D} = \bar{\bar{\epsilon}}\mathbf{E} + \bar{\bar{\xi}}\mathbf{H} \tag{1.56}$$

$$\mathbf{B} = \bar{\bar{\zeta}}\mathbf{E} + \bar{\bar{\mu}}\mathbf{H}. \tag{1.57}$$

Whether one regards them as matrices or dyadics, note that the medium parameters have to be placed anterior to the field variables. The most general linear medium above is termed *bianisotropic* or *magnetoelectric*. If the medium parameters are all scalars, then the medium is called *bi-isotropic*. Special cases of bi-isotropic media are the *isotropic chiral medium* with $\xi = -\zeta$ and the *Tellegen medium* $\xi = \zeta$. A chiral medium can be produced by inserting suitable base material particles with specific handedness, i.e. particles whose mirror image cannot be brought into coincidence with the original particles; for example, helical particles (for a detailed discussion see [8, 9]). A Tellegen medium can be produced by combining permanent electric and magnetic dipoles in similar parallel pairs and making a mixture with such particles. If $\bar{\bar{\xi}} = 0 = \bar{\bar{\zeta}}$ and $\bar{\bar{\epsilon}}$ and $\bar{\bar{\mu}}$ are scalars, then the medium is simply *isotropic*. If the medium dyadics are functions of frequency, the medium is said to be *time-dispersive* or simply dispersive.

For a non-dispersive Tellegen medium, writing $\zeta = \xi = \chi/c$, $\mu = \mu_0\mu_{\rm r}$, $\epsilon = \epsilon_0\epsilon_{\rm r}$, and $\alpha_{\rm T} = (\mu_{\rm r}\epsilon_{\rm r} - \chi^2)$, where $c$ is the speed of light in a vacuum, it can be shown that with the *modified Lorenz condition*

$$\nabla \cdot \mathscr{A} = -\frac{\alpha_{\rm T}}{c^2}\frac{\partial \psi_{\rm e}}{\partial t}, \tag{1.58}$$

the Lorenz potentials satisfy the decoupled d'Alembertian equations

$$\frac{\alpha_T}{c^2}\frac{\partial^2 \mathscr{A}}{\partial t^2} - \nabla^2 \mathscr{A} = \mu \mathscr{J}, \tag{1.59}$$

$$\frac{\alpha_T}{c^2}\frac{\partial^2 \psi_e}{\partial t^2} - \nabla^2 \psi_e = \frac{\mu c^2}{\alpha_T} q_{ev}, \tag{1.60}$$

in the presence of electric sources ($\mathscr{J}$, $q_{ev}$).

*1.2.1.1 The plasmonic medium and Drude model*

Metals behave as good conductors at radio frequencies, but have other interesting properties at optical frequencies in that the real part of permittivity turns negative at sufficiently high frequencies. A simple approach to describe the effect of the free electrons in a metal is to approximate the metal as a homogeneous domain with a complex dielectric permittivity described by the *Drude model* [10]. The equation of motion of a bound electron under the action of an applied electromagnetic field is

$$m\left(\frac{\mathrm{d}^2\mathbf{r}}{\mathrm{d}t^2} + \gamma\frac{\mathrm{d}\mathbf{r}}{\mathrm{d}t} + \omega_0^2\mathbf{r}\right) = -q_e\mathbf{E}_{\mathrm{eff}}, \tag{1.61}$$

where $\mathbf{E}_{\mathrm{eff}}$ is the local electric field intensity that acts on the electron of charge $q_e$ and effective mass $m$, $\gamma$ is the damping constant, and $\omega_0$ is the undamped resonance frequency of the transverse vibrational mode of the ionic lattice structure. The Drude model is derived based on setting the oscillating force term $m\omega_0^2\mathbf{r}$ to zero for time-harmonic excitation:

$$\epsilon_s = \epsilon_\infty - \frac{\omega_p^2}{\omega(\omega - j\gamma)} = \epsilon_\infty - \frac{\omega_p^2}{\omega^2 + \gamma^2} - j\frac{\gamma\omega_p^2}{\omega(\omega^2 + \gamma^2)} =: \epsilon_{rs}(\omega) - j\sigma_s(\omega), \tag{1.62}$$

where $\omega_p = \sqrt{Nq_e^2/m\epsilon_0}$ is known as the *plasma frequency*, $N$ is the number of free electrons per unit volume, and $\epsilon_\infty$ is the relative permittivity at infinite frequency. Commonly used values are $\epsilon_\infty = 5$, $\omega_p = 1.443\,3 \times 10^{16}$ rad s$^{-1}$, and $\gamma = 10^{14}$ s$^{-1}$. Figure 1.1 shows a plot of the real and imaginary parts of the complex relative permittivity of the metal as a function of free-space wavelength $\lambda_0$. Note that the real part $\epsilon_{rs}$ becomes negative for wavelengths above 300 nm. The upper horizontal axis is the normalized wavelength $\lambda_0/\lambda_p$, where $\lambda_p$ is the wavelength corresponding to the plasma frequency.

### 1.2.2 Boundary conditions

Maxwell's equations must be supplemented by boundary conditions that must be satisfied by the electric and magnetic fields at material interfaces or at infinity for open domains.

### 1.2.2.1 Material interface conditions

If $(\epsilon_1, \ \mu_1)$ and $(\epsilon_2, \ \mu_2)$ are the material parameters on two sides of an interface, and $\hat{\mathbf{n}}$ is a unit normal from medium 1 to medium 2, figure 1.2, the interface conditions are[5]:

$$\hat{\mathbf{n}} \times (\mathbf{E}_2 - \mathbf{E}_1) = -\mathbf{M}_s; \qquad \hat{\mathbf{n}} \cdot (\mathbf{D}_2 - \mathbf{D}_1) = q_{es}, \tag{1.63}$$

$$\hat{\mathbf{n}} \times (\mathbf{H}_2 - \mathbf{H}_1) = \mathbf{J}_s; \qquad \hat{\mathbf{n}} \cdot (\mathbf{B}_2 - \mathbf{B}_1) = q_{ms}, \tag{1.64}$$

$$\hat{\mathbf{n}} \cdot (\mathbf{J}_2 - \mathbf{J}_1) = -j\omega q_{es}; \qquad \hat{\mathbf{n}} \cdot (\mathbf{M}_2 - \mathbf{M}_1) = -j\omega q_{ms}, \tag{1.65}$$

where the subscripts '1' and '2' denote the interface fields in regions 1 and 2, respectively, and the subscript 's' on the current and charge densities denotes surface values. The boundary condition (1.65) follows from applying the Gauss divergence theorem to the equation of continuity (1.31). If medium 1 is a perfect electric/magnetic conductor (PEC/PMC), then the boundary conditions reduce to

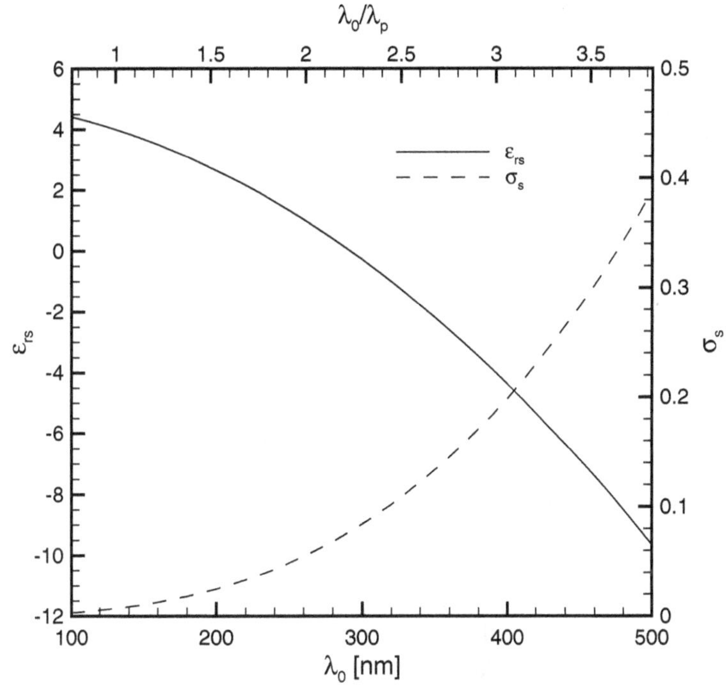


**Figure 1.1.** Complex permittivity of a typical metal at optical wavelengths.

---

[5] The usual boundary conditions of continuity of tangential electric fields and continuity of normal magnetic flux densities in the absence of surface magnetic sources are under the assumption that the fields are finite at the interface [11]. In particular, they preclude the presence of the delta-function-like behavior along the normals. For example, if there is a double layer of *electric* charge density $\tau$ at an interface (which implies that the normal electric field behaves like a delta function on the interface) and if $\hat{\mathbf{u}}_\ell$ in the unit tangent vector on the interface, then $\hat{\mathbf{u}}_\ell \cdot (\mathbf{E}_2 - \mathbf{E}_1) = -\frac{1}{\epsilon_0}\hat{\mathbf{u}}_\ell \cdot \nabla_s \tau \neq 0$. Derivation of more complicated boundary conditions can be accomplished by treating Maxwell's equations in the sense of distributions (see section 18.1).

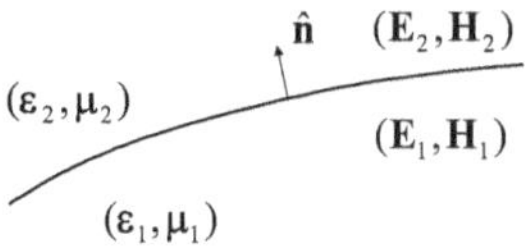


**Figure 1.2.** Material boundary conditions at an interface.

$$\text{PEC: } \hat{\mathbf{n}} \times \mathbf{E}_2 = 0;\ \hat{\mathbf{n}} \times \mathbf{H}_2 = \mathbf{J}_s;\ \hat{\mathbf{n}} \cdot \mathbf{D}_2 = q_{es};\ \hat{\mathbf{n}} \cdot \mathbf{B}_2 = 0, \tag{1.66}$$

$$\text{PMC: } \hat{\mathbf{n}} \times \mathbf{H}_2 = 0;\ \hat{\mathbf{n}} \times \mathbf{E}_2 = -\mathbf{M}_s;\ \hat{\mathbf{n}} \cdot \mathbf{D}_2 = 0;\ \hat{\mathbf{n}} \cdot \mathbf{B}_2 = q_{ms}. \tag{1.67}$$

Another useful interface boundary condition is an impedance boundary condition in which the tangential electric and magnetic fields in medium 2 are related through a surface impedance $Z_s$ (complex-valued):

$$\hat{\mathbf{n}} \times (\hat{\mathbf{n}} \times \mathbf{E}_2) = -Z_s \hat{\mathbf{n}} \times \mathbf{H}_2. \tag{1.68}$$

**Definition 1.2.1.** *Scalar wave function.* A complex-valued scalar function $u$ is called a scalar wave function for a domain $\mathcal{D}$ if it is defined and of class $C^2$ in $\mathcal{D}$ and satisfies the scalar homogeneous Helmholtz equation $\nabla^2 u + k^2 u = 0$ at every point in $\mathcal{D}$.

**Definition 1.2.2.** *Vector wave function.* A complex-valued vector function $\mathbf{A}(\mathbf{r})$ is called a vector wave function for a domain $\mathcal{D}$ if it is defined and of class $C^2$ in $\mathcal{D}$ and satisfies the vector homogeneous Helmholtz equation $\nabla \times \nabla \times \mathbf{A} - k^2 \mathbf{A} = 0$.

The electric and magnetic fields $\mathbf{E}$ and $\mathbf{H}$ in source-free regions in the frequency domain given by equations (1.40) and (1.41) are examples of vector wave functions.

*1.2.2.2 Radiation, absorbing boundary conditions, and related theorems*

Radiation conditions are needed to yield a unique solution to Maxwell's equations when finite extent sources radiate in an infinite medium. Radiation conditions to time-harmonic waves are what causality conditions are for time-instantaneous waves. If $r$ denotes the radial distance from a point within the sources and if $\eta = \sqrt{\mu/\epsilon}$ is the intrinsic impedance of the infinite medium, *Silver–Müller radiation conditions* state that[6] for $r \to \infty$

$$|r\mathbf{E}|, |r\mathbf{H}| = \mathcal{O}(1), \quad r[\hat{r} \times \mathbf{H}\eta + \mathbf{E}] = o(1), \quad r[\hat{r} \times \mathbf{E} - \eta\mathbf{H}] = o(1), \tag{1.69}$$

where $\hat{r}$ is the unit vector in the radial direction.

[6] The statement $u(\mathbf{r}; \epsilon) = \mathcal{O}[v(\mathbf{r}; \epsilon)]$ for $\epsilon$ in some interval $I$ means that for each point $\mathbf{r}$ in some domain $D$ there exists some positive number $\kappa(\mathbf{r})$ such that $|u(\mathbf{r}; \epsilon)| \leqslant \kappa(\mathbf{r})|v(\mathbf{r}; \epsilon)|$ for all $\epsilon$ in $I$. The statement $u(\mathbf{r}; \epsilon) = o[v(\mathbf{r}; \epsilon)]$ as $\epsilon \to a$ for each point $\mathbf{r}$ in some domain $D$ means that $|u|$ becomes arbitrarily small compared to $|v|$ as $\epsilon \to a$. Note that $u = o(v)$ always implies $u = \mathcal{O}(v)$, but not conversely.

For a scalar function $\psi$ satisfying the Helmholtz equation, the radiation conditions are referred to as the *Sommerfeld's radiation conditions*:

$$|r\psi| = \mathcal{O}(1), \ r\left[\frac{\partial\psi}{\partial r} + jk\psi\right] = o(1) \text{ as } r \to \infty. \tag{1.70}$$

If $\psi$ satisfies the non-homogeneous scalar Helmholtz equation

$$\nabla^2\psi + k^2\psi = -s(\mathbf{r}), \tag{1.71}$$

with the source function $s(\mathbf{r})$, then the only solution that satisfies the Sommerfeld's radiation condition is

$$\psi(\mathbf{r}) = \iiint\limits_V s(\mathbf{r}')\frac{\mathrm{e}^{-jkR}}{4\pi R}\,\mathrm{d}v', \tag{1.72}$$

where $R = |\mathbf{r} - \mathbf{r}'|$ is the distance between the observation and source points. Wilcox [12] has shown that the radiation condition can be expressed in a variety of equivalent statements, which we quote here without proof. For the scalar case the relevant theorem is the following.

**Theorem 1.2.1.** Wilcox theorem (equivalent forms of radiation condition) [12]. *Let $\psi(\mathbf{p})$ satisfy the Helmholtz equation in an exterior domain $\mathcal{D}$. Let $\mathcal{R}(\mathbf{p}, r)$ denote a sphere with a center at $\mathbf{p}$ and radius r. Let $r_0(\mathbf{p})$ be the least radius r for which $\mathcal{R}(\mathbf{p}, r)$ contains the sources and other material boundaries in the problem. Then the following radiation conditions (Sommerfeld radiation conditions) are equivalent for $\psi$; i.e. each implies all others:*

*RC 1:* $\displaystyle\oiint\limits_{\mathcal{R}(\mathbf{p},r)}\left[\frac{\partial(r\psi)}{\partial r} + jkr\psi\right]\mathrm{d}s = 0$ *for all* $\mathbf{p}$ *and* $r > r_0(\mathbf{p})$.

*RC 2:* $\displaystyle\lim_{r\to\infty}\frac{1}{r}\oiint\limits_{\mathcal{R}(\mathbf{p},r)}\left(\frac{\partial\psi}{\partial r} + jk\psi\right)\mathrm{d}s = 0$ *for all* $\mathbf{p}$.

*RC 3:* $\displaystyle\lim_{r\to\infty}\oiint\limits_{\mathcal{R}(\mathbf{p},r)}\left|\frac{\partial\psi}{\partial r} + jk\psi\right|^2\mathrm{d}s = 0$ *for all* $\mathbf{p}$.

*RC 4:* $\displaystyle\lim_{r\to\infty} r\left(\frac{\partial\psi}{\partial r} + jk\psi\right) = 0$ *for all* $\mathbf{p}$, *uniformly in* $\hat{\mathbf{r}}$.

*RC 5:* $\displaystyle\lim_{r\to\infty}\oiint\limits_{\mathcal{R}(0,r)}\left|\frac{\partial\psi}{\partial r} + jkr\right|^2\mathrm{d}s = 0$ *for a single center* $\mathbf{0}$.

**Definition 1.2.3.** *Scalar radiation function.* Let $\mathscr{D}$ be an exterior domain. A scalar function $u$ in the domain $\mathcal{D}$ is called a scalar radiation function if it is a scalar wave function for $\mathscr{D}$ and, in addition, satisfies the Sommerfeld radiation condition.

**Theorem 1.2.2.** Whittaker theorem [13, 14].

*Every scalar radiation function $\psi(\mathbf{r})$ can be represented in a plane wave expansion of the form*

$$\psi(\mathbf{r}) = \frac{k}{4\pi} \oiint \hat{\psi}(\hat{\mathbf{s}}) e^{-jk\hat{\mathbf{s}}\cdot\mathbf{r}} \sin\alpha \, d\alpha d\beta, \tag{1.73}$$

*where the unit vector* $\hat{\mathbf{s}} = \hat{\mathbf{x}} \sin\alpha \cos\beta + \hat{\mathbf{y}} \sin\alpha \sin\beta + \hat{\mathbf{z}} \cos\alpha$, *with* $\alpha \in (0, \pi)$, $\beta \in (0, 2\pi)$, *and* $\hat{\psi}(\mathbf{s})$ *is the spectrum of the representation.*

For a proof see [13]. Equation (1.73) may also be regarded as a Fourier transform relation between the space function $\psi(\mathbf{r})$ and its spectrum $\hat{\psi}(\hat{\mathbf{s}})$. The Fourier transform is performed on a unit sphere in the wavenumber domain. For the special case of a spherical harmonic $\psi_n^m(\mathbf{r}) = j_n(kr) Y_n^m(\theta, \phi)$, where $j_n(kr)$ is the spherical Bessel function of order $n$ and argument $kr$, and $Y_n^m(\theta, \phi)$ is the surface spherical harmonic as defined in equation (1.103), we have in view of identity (D.127) that $\hat{\psi}_n^m(\hat{\mathbf{s}}) = k^{-1} j^n Y_n^m(\alpha, \beta)$ so that

$$\psi_n^m(\mathbf{r}) = j_n(kr) Y_n^m(\theta, \phi) = \frac{(-j)^n}{4\pi} \oiint Y_n^m(\alpha, \beta) e^{jk\hat{\mathbf{s}}\cdot\mathbf{r}} \sin\alpha \, d\alpha d\beta. \tag{1.74}$$

Thus the Fourier transform of a spherical harmonic yields another surface spherical harmonic in the space domain whose strength is governed by the spherical Bessel function along the radial direction.

**Example 1.2.2.1.** Radial components of EM fields as scalar radiating functions. Determine the equations satisfied by $u = \mathbf{r} \cdot \mathbf{E}$ and $v = \mathbf{r} \cdot \mathbf{H}$ in a homogeneous medium containing electric sources only, where $\mathbf{r} = \hat{\mathbf{r}} r$, $r$ being the radial coordinate in a spherical coordinate system and $\hat{\mathbf{r}}$ being the unit vector along $r$. Obtain a solution for $\mathbf{r} \cdot \mathbf{E}$ and $\mathbf{r} \cdot \mathbf{H}$ in terms of the given current density $\mathbf{J}$.

Expressing the various quantities in rectangular coordinates, it is straightforward to see that for an arbitrary vector $\mathbf{w}$ belonging to class $C^2$,

$$\begin{aligned} \nabla^2(\mathbf{r} \cdot \mathbf{w}) &= \mathbf{r} \cdot \nabla^2 \mathbf{w} + 2\nabla \cdot \mathbf{w} = -\mathbf{r} \cdot (\nabla \times \nabla \times \mathbf{w} - \nabla\nabla \cdot \mathbf{w}) + 2\nabla \cdot \mathbf{w} \\ &= -\mathbf{r} \cdot \nabla \times \nabla \times \mathbf{w}, \quad \text{if } \nabla \cdot \mathbf{w} = 0. \end{aligned} \tag{1.75}$$

Applying the identity (1.75) to the divergenceless vector $\mathbf{w} = \mathbf{E} + \mathbf{J}/j\omega\epsilon$ and noting that $\nabla \times \nabla \times \mathbf{E} - k^2\mathbf{E} = -j\omega\mu\mathbf{J}$ from equation (1.40), we arrive at the non-homogeneous, scalar Helmholtz equation

$$(\nabla^2 + k^2)\mathbf{r} \cdot \left[\mathbf{E} + \frac{1}{j\omega\epsilon}\mathbf{J}\right] = -\frac{1}{j\omega\epsilon}\mathbf{r} \cdot \nabla \times \nabla \times \mathbf{J}, \tag{1.76}$$

for the radial component of the electric field. Using equation (1.72), the solution of this equation that satisfies the Sommerfeld's radiation condition is

$$\mathbf{r}\cdot\mathbf{E} = -\frac{\mathbf{r}\cdot\mathbf{J}}{j\omega\epsilon} + \frac{1}{j\omega\epsilon}\iiint_V \mathbf{r}'\cdot\nabla'\times\nabla'\times\mathbf{J}(\mathbf{r}')\frac{\mathrm{e}^{-jk|\mathbf{r}-\mathbf{r}'|}}{4\pi|\mathbf{r}-\mathbf{r}'|}\,\mathrm{d}v'. \tag{1.77}$$

Similarly, on applying the identity (1.75) to the divergenceless vector $\mathbf{w} = \mathbf{H}$ and noting that $\nabla\times\nabla\times\mathbf{H} - k^2\mathbf{H} = \nabla\times\mathbf{J}$ from equation (1.41), we obtain the non-homogeneous Helmholtz equation

$$(\nabla^2 + k^2)\mathbf{r}\cdot\mathbf{H} = -\mathbf{r}\cdot\nabla\times\mathbf{J}, \tag{1.78}$$

for the radial component of the magnetic field, whose only solution satisfying the Sommerfeld's radiation condition is

$$\mathbf{r}\cdot\mathbf{H} = \iiint_V \mathbf{r}'\cdot\nabla'\times\mathbf{J}(\mathbf{r}')\frac{\mathrm{e}^{-jk|\mathbf{r}-\mathbf{r}'|}}{4\pi|\mathbf{r}-\mathbf{r}'|}\,\mathrm{d}v'. \tag{1.79}$$

Several remarks are in order at this point.

(i) Equations (1.77) and (1.79) are valid everywhere in space including the source region.

(ii) If all sources are contained within a sphere of radius $r_0$ such that the current density and all its derivatives vanish outside the sphere, then it is clear from equations (1.76), (1.77), (1.78), and (1.79) that for $r > r_0$ the radial components $\mathbf{r}\cdot\mathbf{E} = rE_r$ and $\mathbf{r}\cdot\mathbf{H} = rH_r$ are *scalar radiation functions.* Since the spherical harmonics of the type $h_n^{(2)}(kr)P_n^m(\cos\theta)\mathrm{e}^{jm\phi}$ are also scalar radiation functions, it should be possible to expand the radial fields $(\mathbf{r}\cdot\mathbf{E})$ and $(\mathbf{r}\cdot\mathbf{H})$ in terms of these harmonics for $r > r_0$.

(iii) The current density $\mathbf{J}$ could represent either impressed sources or equivalent sources arising from material scatterers embedded inside $S$.

(iv) Discontinuous currents can be accommodated in the current formulation by treating the integrands in equations (1.77) and (1.79) in the sense of distributions. In such a case additional terms will arise in equations (1.77) and (1.79) that are consistent with distribution theory as outlined in section 18.1 (see theorem 18.1.4).

■■

For the vector case the Wilcox theorem for radiation conditions may be stated as follows.

**Theorem 1.2.3.** Wilcox theorem (equivalent forms of Silver–Müller radiation condition) [12].
*Let* $\mathbf{A}(\mathbf{p})$ *satisfy the vector Helmholtz equation*

$$\nabla\times\nabla\times\mathbf{A} - k^2\mathbf{A} = 0, \tag{1.80}$$

*in an exterior domain $\mathcal{D}$ (namely* **A** *is a vector wave function in $\mathcal{D}$). Let $\mathcal{R}(\mathbf{p}, r)$ denote a sphere with a center at* **p** *and radius r. Let $r_0(\mathbf{p})$ be the least radius r for which $\mathcal{R}(\mathbf{p}, r)$ contains the sources and other material boundaries in the problem. Then the following radiation conditions are equivalent; i.e. for* **A***, each implies all others:*

*RC 1:* $\oiint_{\mathcal{R}(\mathbf{p},r)} \left[\frac{\partial(r\mathbf{A})}{\partial r} + jkr\mathbf{A}\right] \mathrm{d}s = 0$ *for all* **p** *and* $r > r_0(\mathbf{p})$.

*RC 2:* $\lim_{r\to\infty}\frac{1}{r} \oiint_{\mathcal{R}(\mathbf{p},r)} [\hat{\mathbf{r}} \times (\nabla \times \mathbf{A}) - jk\mathbf{A}]\, \mathrm{d}s = 0$ *for all* **p**.

*RC 3:* $\lim_{r\to\infty} \oiint_{\mathcal{R}(0,r)} |\hat{\mathbf{r}} \times (\nabla \times \mathbf{A}) - jk\mathbf{A}|^2\, \mathrm{d}s = 0$ *for a single center* **0**.

Based on the radiation condition a vector radiation function is defined as follows.

**Definition 1.2.4.** *Vector radiation function.* Let $\mathcal{D}$ be an exterior domain. A vector field $\mathbf{A}(\mathbf{r})$ is a vector radiation function for $\mathcal{D}$ provided it is vector wave function for $\mathcal{D}$ and satisfies the Silver–Müller radiation condition

$$\lim_{r\to\infty} \oiint_{\mathcal{R}(0,r)} |\hat{\mathbf{r}} \times (\nabla \times \mathbf{A}) - jk\mathbf{A}|^2\, \mathrm{d}s = 0. \tag{1.81}$$

**Example 1.2.2.2** Examples of vector radiation functions.
Show that the vector function **A** defined by $\mathbf{A} = \nabla \times (\mathbf{r}u)$, where $u$ is a scalar radiation function and $\mathbf{r} = \hat{\mathbf{r}}r$, with $r$ being the radial coordinate in the spherical coordinate system, is a vector radiation function.

Using the definition of the curl operator in spherical coordinates we have

$$\mathbf{A} = \nabla \times (\mathbf{r}u) = -\hat{\boldsymbol{\phi}}\frac{\partial u}{\partial \theta} + \frac{\hat{\boldsymbol{\theta}}}{\sin\theta}\frac{\partial u}{\partial \phi}, \tag{1.82}$$

$$\nabla \times \mathbf{A} = \nabla \times \nabla \times (\mathbf{r}u) = -\frac{\hat{\mathbf{r}}}{r\sin\theta}Du + \frac{\hat{\boldsymbol{\theta}}}{r}\frac{\partial}{\partial r}\left(r\frac{\partial u}{\partial \theta}\right) + \frac{\hat{\boldsymbol{\phi}}}{r\sin\theta}\frac{\partial}{\partial r}\left(r\frac{\partial u}{\partial \phi}\right), \tag{1.83}$$

$$\nabla \times \nabla \times \mathbf{A} = -\frac{\hat{\boldsymbol{\theta}}}{\sin\theta}\frac{\partial}{\partial \phi}\nabla^2 u + \hat{\boldsymbol{\phi}}\nabla^2 u, \tag{1.84}$$

where $D$ is the Beltrami operator given in equation (1.94). Therefore

$$\nabla \times \nabla \times \mathbf{A} - k^2\mathbf{A} = -\hat{\boldsymbol{\theta}}\frac{1}{\sin\theta}\frac{\partial}{\partial \phi}(\nabla^2 u + k^2 u) + \hat{\boldsymbol{\phi}}\frac{\partial}{\partial \theta}(\nabla^2 u + k^2 u) = 0, \tag{1.85}$$

since $u$ is a scalar radiation function. Furthermore,

$$\lim_{r\to\infty} r(\hat{\mathbf{r}} \times \nabla \times \mathbf{A} - jk\mathbf{A}) = \left(\hat{\boldsymbol{\phi}}\frac{\partial}{\partial\theta} - \hat{\boldsymbol{\theta}}\frac{1}{\sin\theta}\frac{\partial}{\partial\phi}\right)\left[\lim_{r\to\infty} r\left(\frac{\partial u}{\partial r} + jku + \frac{u}{r}\right)\right] \\ = 0, \tag{1.86}$$

since $u$ satisfies Sommerfeld's radiation condition. Therefore $\mathbf{A} = \nabla \times (\mathbf{r}u)$ is a vector radiation function. Similarly, the associated function $\mathbf{B} = \nabla \times \mathbf{A}$ is also a vector radiation function. ■■

**Theorem 1.2.4** Wilcox representation theorem [15].
*Let* $\mathbf{A}(\mathbf{r})$ *be a vector radiation function for an exterior domain* $\mathcal{D}$ *bounded by a regular surface S (namely comprised of a finite number of smooth sections) and let* $\mathbf{A}(\mathbf{r})$ *be of class* $C^2$ *in the closure of* $\mathcal{D}$. *Then*

$$\mathbf{A}(\mathbf{r}) = \oiint_S [\hat{\mathbf{n}}' \times (\nabla' \times \mathbf{A}) + (\hat{\mathbf{n}}' \times \mathbf{A}) \times \nabla' + (\hat{\mathbf{n}} \cdot \mathbf{A})\nabla']\psi(\mathbf{r}, \mathbf{r}')\, \mathrm{d}s', \quad \mathbf{r} \text{ in } \mathcal{D}, \tag{1.87}$$

*where* $\psi(\mathbf{r}, \mathbf{r}') = \frac{\mathrm{e}^{-jkR}}{4\pi R}$, $R = |\mathbf{r} - \mathbf{r}'|$, $\nabla'$ *is the gradient operator with respect to the point* $\mathbf{r}'$, *and* $\hat{\mathbf{n}}$ *is a unit outward normal on S.*

The field representation here is a special case of the Stratton–Chu integral representation (see equations (10.12) and (10.13)) in a source-free region, although Wilcox proves the theorem without resorting to the stronger assumption of $|\mathbf{A}(\mathbf{r})| = \mathcal{O}\left(\frac{1}{r}\right)$, $r \to \infty$. A direct consequence of the representation theorem is the expansion theorem.

**Theorem 1.2.5.** Expansion theorem [15].
*Let* $\mathbf{A}(\mathbf{r})$ *be a vector radiation function for a region* $r > r_0$ *where* $(r, \theta, \phi)$ *are spherical coordinates. Then* $\mathbf{A}(\mathbf{r})$ *has an expansion*

$$\mathbf{A}(\mathbf{r}) = \frac{e^{-jkr}}{r}\sum_{n=0}^{\infty}\frac{\mathbf{A}_n(\theta, \phi)}{r^n}, \tag{1.88}$$

*which is valid for* $r > r_0$ *and which converges absolutely and uniformly in the parameters* $(r, \theta, \phi)$ *in any region* $r > r_0$. *The series can be differentiated term by term with respect to* $r$, $\theta$, *and* $\phi$ *any number of times and the resulting series all converge absolutely and uniformly.*

**Corollary 1.2.5.1** Corollary 1 to expansion theorem.
*Every vector radiation function* $\mathbf{A}(\mathbf{r})$ *has the asymptotic form*

$$\mathbf{A}(\mathbf{r}) \sim \frac{\mathrm{e}^{-jkr}}{r}\mathbf{A}_0(\theta, \phi), \quad r \to \infty, \tag{1.89}$$

*where the vector field* $\mathbf{A}_0(\theta, \phi)$*, called the radiation pattern of* $\mathbf{A}(\mathbf{r})$*, is tangential to the sphere* $r = constant$*, i.e.* $\hat{\mathbf{r}} \cdot \mathbf{A}_0 = 0$ *or* $\mathbf{A}_0 = A_0^\theta \hat{\boldsymbol{\theta}} + A_0^\phi \hat{\boldsymbol{\phi}}$.

**Corollary 1.2.5.2.** Corollary 2 to the expansion theorem.
*The vector coefficients* $\mathbf{A}_n(\theta, \phi) = A_n^r \hat{\mathbf{r}} + A_n^\theta \hat{\boldsymbol{\theta}} + A_n^\phi \hat{\boldsymbol{\phi}}$, $n > 0$ *of the expansion theorem are completely determined by the radiation pattern (1.88) through the recursion formulas*

$$jkA_1^r = \frac{1}{\sin\theta}\left[\frac{\partial}{\partial\theta}(\sin\theta A_0^\theta) + \frac{\partial A_0^\phi}{\partial\phi}\right] = r\nabla \cdot \mathbf{A}_0, \tag{1.90}$$

$$2jknA_{n+1}^r = -n(n-1)A_n^r - DA_n^r, \quad n = 1, 2, \ldots \tag{1.91}$$

*and*

$$2jknA_n^\theta = -n(n-1)A_{n-1}^\theta - DA_{n-1}^\theta - D_\theta \mathbf{A}_{n-1}, \quad n = 1, 2, \ldots \tag{1.92}$$

$$2jknA_n^\phi = -n(n-1)A_{n-1}^\phi - DA_{n-1}^\phi - D_\phi \mathbf{A}_{n-1}, \quad n = 1, 2, \ldots, \tag{1.93}$$

*where*

$$Df = \frac{1}{\sin\theta}\frac{\partial}{\partial\theta}\left(\sin\theta\frac{\partial f}{\partial\theta}\right) + \frac{1}{\sin^2\theta}\frac{\partial^2 f}{\partial^2\phi}, \tag{1.94}$$

*is the Beltrami operator for the sphere, while* $D_\theta$ *and* $D_\phi$ *are first linear operators (taking vector fields* $\mathbf{F} = \hat{\mathbf{r}}F^r + \hat{\boldsymbol{\theta}}F^\theta + \hat{\boldsymbol{\phi}}F^\phi$ *into scalar functions) defined by*

$$D_\theta \mathbf{F} = 2\frac{\partial F^r}{\partial\theta} - \frac{1}{\sin^2\theta}F^\theta - \frac{2\cos\theta}{\sin^2\theta}\frac{\partial F^\phi}{\partial\phi} \tag{1.95}$$

$$D_\phi \mathbf{F} = \frac{2}{\sin\theta}\frac{\partial F^r}{\partial\phi} + \frac{2\cos\theta}{\sin^2\theta}\frac{\partial F^\theta}{\partial\phi} - \frac{1}{\sin^2\theta}F^\phi. \tag{1.96}$$

**Example 1.2.2.3.** Numerical absorbing boundary conditions.
As an example of the utility of the Wilcox expansion theorem, consider the development of absorbing boundary conditions on a sphere $S_r$ of radius $r$ circumscribing a scatterer. Such boundary conditions are useful in terminating computational domains of numerical algorithms and the scalar case is discussed in detail in [16]. We discuss here the vector case [17]. Let $\hat{\mathbf{r}}$ represent the unit outward normal on $S_r$. Exterior to $S_r$ the scattered electromagnetic fields satisfy the vector Helmholtz equation. Let $\mathbf{F}$ denote a field vector, which could be the scattered electric field, or the scattered magnetic field, or the scattered vector potential, that takes the form given in equation (1.88) exterior to $S_r$. Then

$$\hat{\mathbf{r}} \times \nabla \times \mathbf{F} = jk\mathbf{F} + \frac{e^{-jkr}}{r}\sum_{n=1}^{\infty}\left[-jk\frac{\hat{\mathbf{r}}F_n^r}{r^n} + \frac{r\nabla F_n^r + n(\mathbf{F}_n - \hat{\mathbf{r}}F_n^r)}{r^{n+1}}\right], \tag{1.97}$$

since $\hat{\mathbf{r}} \cdot \mathbf{F}_0 = 0$. If the Silver–Müller type boundary condition $B_{abc1}$: $\hat{\mathbf{r}} \times \nabla \times \mathbf{F} - jk\mathbf{F} = 0$ is employed *on* $S_r$, then the residual error as seen from equation (1.97) will be of order $\mathcal{O}(r^{-2})$. Judging from the form of equation (1.88) a better absorbing boundary condition is the one that removes the first term from the series expression. This results in the absorbing boundary condition $B_{abc2}$: $\hat{\mathbf{r}} \times \nabla \times \mathbf{F} - jk\mathbf{F} + jk\hat{\mathbf{r}}F^r = 0$ on $S_r$. From equation (1.97) this will result in a lower residual error[7] of order $\mathcal{O}(r^{-3})$. In fact a whole hierarchy of absorbing boundary conditions $B_{abc\ell}$ may be constructed on $S_r$ using the Wilcox expansion, which will yield successively lower residual errors of order $\mathcal{O}(r^{-(\ell+1)})$, $\ell \geqslant 1$. ■■

## 1.3 Field determination by radial components

In many applications it is desirable to express the full vector fields in terms of a minimal set of scalar components. For example, in formulations involving Cartesian and cylindrical coordinates, the complete electromagnetic field in a source-free region can be expressed in terms of the $z$-components of the magnetic and electric vector potentials [18]. The fields produced separately by these $z$-components of potentials are labeled as transverse magnetic to $z$ ($TM_z$) and transverse electric to $z$ ($TE_z$), respectively, in the literature. In example 1.2.2.1, it was shown that the radial components of electric and magnetic fields are scalar radiation functions in an unbounded medium. These scalar functions are determined everywhere, including the source region, by the specification of electric current density. This raises the question whether the radial components of electric and magnetic fields are sufficient to describe the complete electromagnetic field in an unbounded medium. The answer to this is affirmative when one recognizes that the radial components of fields are related to the *Debye potentials*, as has been shown in [19]. The field generated separately by the radial component of the electric or magnetic field is labeled as transverse magnetic to $r$ ($TM_r$) or transverse electric to $r$ ($TE_r$), respectively. The methodology of $TM_r$ and $TE_r$ proves to be extremely useful in solving various problems in 3D in unbounded media. The radial components are expanded in terms of spherical harmonics and the coefficients of expansion are known as multipole strengths. The multipole strengths are completely determined by the current density specified in 3D space. In an alternative formulation involving the wavenumber space, the multipole strengths are determined by the transverse components of the spectral components of the current density. Important theorems in this regard are the theorem of Bouwkamp and Casimir [19] and the theorem of Devaney and Wolf [20].

[7] Note that $F_1^r$ will be independent of $r$. Consequently, $r\nabla F_1^r$ will be of order $\mathcal{O}(1)$.

### 1.3.1 Multipole expansion, Debye potentials, and related theorems

We begin by relating the radial components of fields to current density via spherical harmonics.

**Example 1.3.1.1.** Multipole expansion of radial electromagnetic fields.
Using the spherical harmonic expansion of the free-space Green's function, determine the multipole expansions of radial electromagnetic fields in a homogeneous medium *outside the sphere* containing all electric currents. (See also [19].)

Equations (1.77) and (1.79) of example 1.2.2.1 give the expressions for the radial electromagnetic fields in terms of the given electric current density **J**. Let all electric currents be contained within a sphere $S$ of radius $r_0$ so that **J** and all its derivatives vanish on $S$. We now employ the representation (15.21) for the free-space Green's function in the expression (1.77) for a radial electric field to obtain for $r > r_0$ that

$$\begin{aligned}\mathbf{r}\cdot\mathbf{E} = &-\frac{\eta}{4\pi}\sum_{n=0}^{\infty}\sum_{m=-n}^{n}(2n+1)\frac{(n-m)!}{(n+m)!}h_n^{(2)}(kr)P_n^m(\cos\theta)e^{jm\phi}\\ &\times\iiint_V \underbrace{j_n(kr')P_n^m(\cos\theta')e^{-jm\phi'}\mathbf{r}'}_{\mathbf{P}(\mathbf{r}')}\cdot\nabla'\times\nabla'\times\mathbf{J}(\mathbf{r}')\,dv'.\end{aligned} \tag{1.98}$$

Applying the vector Green's second identity (B.15)

$$\begin{aligned}\iiint_V (\mathbf{P}\cdot\nabla\times\nabla\times\mathbf{J} - \mathbf{J}\cdot\nabla\times\nabla\times\mathbf{P})\,dv &= \oiint_S \hat{\mathbf{n}}\times[\mathbf{J}\cdot\nabla\times\mathbf{P} - \mathbf{P}\cdot\nabla\times\mathbf{J}]\,ds\\ &= 0, \text{ since } \mathbf{J} \text{ vanishes identically on } S,\end{aligned}$$

the radial electric field could also be expressed for $r > r_0$ as

$$\begin{aligned}\mathbf{r}\cdot\mathbf{E} = &-\frac{\eta}{4\pi}\sum_{n=0}^{\infty}\sum_{m=-n}^{n}(2n+1)\frac{(n-m)!}{(n+m)!}h_n^{(2)}(kr)P_n^m(\cos\theta)e^{jm\phi}\\ &\times\iiint_V \mathbf{J}(\mathbf{r}')\cdot\nabla'\times\nabla'\times\left[j_n(kr')P_n^m(\cos\theta')e^{-jm\phi'}\mathbf{r}'\right]dv'.\end{aligned} \tag{1.99}$$

In a like manner, starting from equation (1.79) and the vector Green's first identity (B.14)

$$\iiint_V (\mathbf{P}\cdot\nabla\times\mathbf{J} - \mathbf{J}\cdot\nabla\times\mathbf{P})\,dv = \iiint_V \nabla\cdot(\mathbf{J}\times\mathbf{P})\,dv = \oiint_S \hat{\mathbf{n}}\cdot\mathbf{J}\times\mathbf{P}\,ds = 0, \tag{1.100}$$

it is easy to show that the radial magnetic field for $r > r_0$ can be expressed as

$$\mathbf{r}\cdot\mathbf{H} = -\frac{jk}{4\pi}\sum_{n=0}^{\infty}\sum_{m=-n}^{n}(2n+1)\frac{(n-m)!}{(n+m)!}h_n^{(2)}(kr)P_n^m(\cos\theta)\mathrm{e}^{jm\phi} \times \iiint_V \left[j_n(kr')P_n^m(\cos\theta')\mathrm{e}^{-jm\phi'}\mathbf{r}'\right]\cdot\nabla'\times\mathbf{J}(\mathbf{r}')\,\mathrm{d}v', \tag{1.101}$$

$$= -\frac{jk}{4\pi}\sum_{n=0}^{\infty}\sum_{m=-n}^{n}(2n+1)\frac{(n-m)!}{(n+m)!}h_n^{(2)}(kr)P_n^m(\cos\theta)\mathrm{e}^{jm\phi} \times \iiint_V \mathbf{J}(\mathbf{r}')\cdot\nabla'\times\left[j_n(kr')P_n^m(\cos\theta')\mathrm{e}^{-jm\phi'}\mathbf{r}'\right]\mathrm{d}v'. \tag{1.102}$$

We now define the complex-valued surface spherical harmonic of order $m$ and degree $n$ as

$$Y_n^m(\theta,\phi) = \left[\frac{2n+1}{4\pi}\frac{(n-m)!}{(n+m)!}\right]^{\frac{1}{2}} P_n^m(\cos\theta)\mathrm{e}^{jm\phi}, \quad n = 0, 1, 2, \ldots, \; m = (-n, n), \tag{1.103}$$

which implies on using equation (D.92) that

$$\begin{aligned} Y_n^{-m}(\theta,\phi) &= \left[\frac{2n+1}{4\pi}\frac{(n+m)!}{(n-m)!}\right]^{\frac{1}{2}} P_n^{-m}(\cos\theta)\mathrm{e}^{-jm\phi} \\ &= (-1)^m\left[\frac{2n+1}{4\pi}\frac{(n-m)!}{(n+m)!}\right]^{\frac{1}{2}} P_n^m(\cos\theta)\mathrm{e}^{-jm\phi} \\ &= (-1)^m\bar{Y}_n^m(\theta,\phi), \end{aligned} \tag{1.104}$$

where the overhead bar denotes complex conjugation. Using equation (D.116) we can then deduce the orthogonality relationship

$$\int_{\theta=0}^{\pi}\int_{\phi=0}^{2\pi} Y_n^m(\theta,\phi)\bar{Y}_{n'}^{m'}(\theta,\phi)\sin\theta\,\mathrm{d}\phi\mathrm{d}\theta = \delta_n^{n'}\delta_m^{m'}. \tag{1.105}$$

Defining multipole strengths $I_n^m$, $L_n^m$ in terms of these surface harmonics as

$$I_n^m = \iiint_V \mathbf{J}(\mathbf{r}')\cdot\nabla'\times\nabla'\times\left[\mathbf{r}'j_n(kr')\bar{Y}_n^m(\theta',\phi')\right]\mathrm{d}v', \quad \text{(A)} \tag{1.106}$$

$$= \iiint_V \left[j_n(kr')\bar{Y}_n^m(\theta',\phi')\right]\mathbf{r}'\cdot\nabla'\times\nabla'\times\mathbf{J}(\mathbf{r}')\,\mathrm{d}v', \quad \text{(A)} \tag{1.107}$$

$$= \iiint_V \nabla'\times\mathbf{J}(\mathbf{r}')\cdot\nabla'\times\left[\mathbf{r}'j_n(kr')\bar{Y}_n^m(\theta',\phi')\right]\mathrm{d}v', \quad \text{(A)} \tag{1.108}$$

$$L_n^m = jk\eta \iiint_V \mathbf{J}(\mathbf{r}') \cdot \nabla' \times \left[\mathbf{r}' j_n(kr') \bar{Y}_n^m(\theta', \phi')\right] \mathrm{d}v', \quad \text{(V)} \tag{1.109}$$

$$= jk\eta \iiint_V \left[j_n(kr') \bar{Y}_n^m(\theta', \phi')\right] \mathbf{r}' \cdot \nabla' \times \mathbf{J}(\mathbf{r}') \, \mathrm{d}v', \quad \text{(V)} \tag{1.110}$$

the radial fields in equations (1.99) and (1.102) can then be re-expressed as

$$rE_r = \mathbf{r} \cdot \mathbf{E} = -\eta \sum_{n=0}^{\infty} \sum_{m=-n}^{n} I_n^m \, h_n^{(2)}(kr) Y_n^m(\theta, \phi), \tag{1.111}$$

$$rH_r = \mathbf{r} \cdot \mathbf{H} = -\eta^{-1} \sum_{n=0}^{\infty} \sum_{m=-n}^{n} L_n^m \, h_n^{(2)}(kr) Y_n^m(\theta, \phi). \tag{1.112}$$

The last form (1.108) for $I_n^m$ follows from the vector Green's first identity (B.14). Equations (1.111) and (1.112) provide a discrete representation of the radial electromagnetic fields outside $S$ in terms of outgoing spherical harmonics whose strengths, $I_n^m$, $L_n^m$ are governed by the currents flowing within the volume $V$. These currents could be a combination of impressed sources and induced sources on scattering objects residing in $V$. The radial fields $rE_r$ and $rH_r$ are both scalar radiation functions outside $S$. Since the surface spherical harmonics are complete for functions of class $C^2$ (see theorem 1.3.2), and the Hankel functions are bounded for arguments different from zero, the above series converge uniformly and absolutely outside $S$ and the terms can be differentiated any number of times.

Note that for $\mathbf{P}(\mathbf{r}) = \mathbf{r} j_n(kr) \bar{Y}_n^m(\theta, \phi)$, we obtain

$$\nabla \times \mathbf{P} = j_n(kr) \left[ \frac{\hat{\boldsymbol{\theta}}}{\sin\theta} \frac{\partial \bar{Y}_n^m}{\partial \phi} - \hat{\boldsymbol{\phi}} \frac{\partial \bar{Y}_n^m}{\partial \theta} \right], \tag{1.113}$$

$$\nabla \times \nabla \times \mathbf{P} = \frac{1}{r} \left[ \hat{\mathbf{r}} n(n+1) j_n(kr) \bar{Y}_n^m + \hat{\boldsymbol{\theta}} \hat{J}_n'(kr) \frac{\partial \bar{Y}_n^m}{\partial \theta} + \frac{\hat{\boldsymbol{\phi}}}{\sin\theta} \hat{J}_n'(kr) \frac{\partial \bar{Y}_n^m}{\partial \phi} \right], \tag{1.114}$$

where $\hat{J}_n(z) = z j_n(z)$ is the spherical Ricatti–Bessel function (see subsection C.3.8).

If the current density is all radial so that $\mathbf{J}(\mathbf{r}) = \hat{\mathbf{r}} J_r(\mathbf{r})$, then we obtain from equations (1.106) and (1.109) that

$$I_n^m = n(n+1) \iiint_V r' J_r(\mathbf{r}') j_n(kr') \bar{Y}_n^m(\theta', \phi') \sin\theta' \, \mathrm{d}r' \mathrm{d}\theta' \mathrm{d}\phi'; \quad L_n^m = 0. \tag{1.115}$$

For a radial electric dipole of moment $I_0\ell$ and located at $(b, \theta_0, \phi_0)$, for instance, $J_r = I_0\ell\delta(r-b)\delta(\theta-\theta_0)\delta(\phi-\phi_0)/r^2 \sin\theta$. In that case

$$I_n^m = I_0\ell \frac{n(n+1)}{b} j_n(kb) \bar{Y}_n^m(\theta_0, \phi_0); \quad L_n^m = 0, \tag{1.116}$$

and the radial components of fields are

$$E_r = -\frac{I_0\ell\eta}{4\pi rb}\sum_{n=0}^{\infty}\sum_{m=-n}^{n} n(n+1)(2n+1)\frac{(n+m)!}{(n-m)!}j_n(kb)h_n^{(2)}(kr) \\ P_n^m(\cos\theta)P_n^m(\cos\theta_0)\mathrm{e}^{jm(\phi-\phi_0)}, \quad r > b. \tag{1.117}$$

and

$$H_r = 0, \quad r > b. \tag{1.118}$$

Obviously, the $n = 0$ terms will not influence the radial fields and may be dropped. As a check, for a radial dipole located at the origin and oriented along the $z$-axis, $b = 0$ and azimuthal symmetry implies $m = 0$. We then obtain upon using (i) $P_n^0(\cos\theta) = P_n(\cos\theta)$, (ii) $P_1(\cos\theta) = \cos\theta$, (iii) the limit

$$\lim_{z\to 0}\frac{j_n(z)}{z} = \lim_{z\to 0}\frac{2^n n! z^{n-1}}{(2n+1)!} = \begin{cases}\frac{1}{3}, & n = 1\\ 0, & n > 1,\end{cases} \tag{1.119}$$

and (iv)

$$h_1^{(2)}(z) = -\frac{\mathrm{e}^{-jz}}{z}\left(\frac{1}{jz} + 1\right), \tag{1.120}$$

that

$$E_r = \frac{\eta I_0\ell}{2\pi r^2}\left(1 + \frac{1}{jkr}\right)\mathrm{e}^{-jkr}\cos\theta, \quad r > 0, \tag{1.121}$$

for the radial component of an electric field of a $z$-directed infinitesimal electric dipole located at the origin.

For a radial magnetic dipole of moment $V_0\ell$ located at $(b, \theta_0, \phi_0)$, dual radial fields are generated, namely

$$H_r = -\frac{V_0\ell\eta^{-1}}{4\pi rb}\sum_{n=0}^{\infty}\sum_{m=-n}^{n} n(n+1)(2n+1)\frac{(n+m)!}{(n-m)!}j_n(kb)h_n^{(2)}(kr) \\ P_n^m(\cos\theta)P_n^m(\cos\theta_0)\mathrm{e}^{jm(\phi-\phi_0)}, \quad r > b. \tag{1.122}$$

and

$$E_r = 0, \quad r > b. \tag{1.123}$$

■■

**Theorem 1.3.1.** Bouwkamp and Casimir theorem [19].
*Let S be a sphere containing all sources in a homogeneous medium characterized by (ϵ(ω), μ(ω)). The time-harmonic electromagnetic field at the frequency ω outside S is identically zero if the radial components of the electromagnetic field there are zero.*

*Proof.* Let $\mathbf{E} = \hat{\boldsymbol{\theta}}E_\theta + \hat{\boldsymbol{\phi}}E_\phi$ and $\mathbf{H} = \hat{\boldsymbol{\theta}}H_\theta + \hat{\boldsymbol{\phi}}H_\phi$ outside $S$. We need to show that $\mathbf{E} \equiv 0$ and $\mathbf{H} \equiv 0$. From source-free Maxwell's curl equations we have

$$\frac{\partial}{\partial\theta}(\sin\theta E_\phi) - \frac{\partial E_\theta}{\partial\phi} = 0; \ \frac{\partial}{\partial\theta}(\sin\theta H_\phi) - \frac{\partial H_\theta}{\partial\phi} = 0, \tag{1.124}$$

$$\frac{\partial(rE_\phi)}{\partial r} = jk\eta rH_\theta; \ \eta\frac{\partial(rH_\phi)}{\partial r} = -jkrE_\theta, \tag{1.125}$$

$$\frac{\partial}{\partial r}(rE_\theta) = -jk\eta rH_\phi; \ \eta\frac{\partial}{\partial r}(rH_\theta) = jkrE_\phi. \tag{1.126}$$

Introduce a change of variable such that $\sin\theta\partial/\partial\theta = \partial/\partial\xi$, which implies that $d\xi/d\theta = 1/\sin\theta$ or $\xi = \ln[(1 - \cos\theta)/\sin\theta] = \ln[\tan(\theta/2)]$. Define $u(r, \xi, \phi) = r\sin\theta E_\theta$, $v(r, \xi, \phi) = r\sin\theta E_\phi$, $g(r, \xi, \phi) = \eta r\sin\theta H_\phi$, $h(r, \xi, \phi) = -\eta r\sin\theta H_\theta$. Then equations (1.124)–(1.126) can be rewritten as

$$\frac{\partial v}{\partial\xi} - \frac{\partial u}{\partial\phi} = 0; \ \frac{\partial g}{\partial\xi} + \frac{\partial h}{\partial\phi} = 0, \tag{1.127}$$

$$\frac{\partial v}{\partial r} = -jkh; \ \frac{\partial g}{\partial r} = -jku, \tag{1.128}$$

$$\frac{\partial u}{\partial r} = -jkg; \ \frac{\partial h}{\partial r} = -jkv, \tag{1.129}$$

Upon writing $u = u_0(\xi, \phi)e^{-jkr}$, $v = v_0(\xi, \phi)e^{-jkr}$, $g = g_0(\xi, \phi)e^{-jkr}$, $h = h_0(\xi, \phi)e^{-jkr}$ for waves outgoing in the exterior of $S$ and substituting into equations (1.128)–(1.129), we obtain $g_0 = u_0$ and $h_0 = v_0$. Letting $w_0 = u_0 + jh_0 = g_0 + jv_0$ and substituting into equation (1.127) yields

$$\frac{\partial w_0}{\partial\xi} = j\frac{\partial w_0}{\partial\phi}, \tag{1.130}$$

which is the complex form of Cauchy–Riemann equations (A.4) in the complex variable $z = \phi + j\xi$ of the function $w_0(z)$. The function $w_0$ must be analytic in the whole complex plane of $z = \phi + j\xi$; it is bounded and periodic in $\phi$ with a period $2\pi$. The fields are related to $w_0(z)$ through $(E_\theta - j\eta H_\theta) = j(E_\phi - j\eta H_\phi) = w_0(\phi + j\xi)e^{-jkr}/(r\sin\theta)$. Finiteness of the fields at $\theta = 0, \pi$ requires that $|w_0(\phi \pm j\infty)| \to 0$. By Liouville's theorem [21, p 85], every bounded entire function such as $w_0(z)$ must be a constant. Since the constant has to vanish at infinity along the imaginary axis, it must be identically zero. Hence $w_0(z) \equiv 0 \Longrightarrow E_\phi = E_\theta = H_\phi = H_\theta \equiv 0$ and we arrive at the proof that $\mathbf{E} \equiv 0$, $\mathbf{H} \equiv 0$ outside $S$ if $\hat{\mathbf{r}} \cdot \mathbf{E}$ and $\hat{\mathbf{r}} \cdot \mathbf{H}$ are zero there.

We can draw the following conclusions from this theorem:

(i) The theorem indicates that it is not possible to have a purely spherical TEM wave ($E_r = 0$, $H_r = 0$) outside the circumscribing sphere of any radiator. In particular, *it is not possible to design an antenna that generates zero radial fields*. The quest to design antennas with minimal radial components forms the subject of electrically small, wideband antennas.

(ii) From equations (1.107) and (1.110) it is clear that if the current density satisfies $\hat{\mathbf{r}} \cdot \nabla \times \nabla \times \mathbf{J} = 0$ and $\hat{\mathbf{r}} \cdot \nabla \times \mathbf{J} = 0$ in the interior of $S$, then both $I_n^m$ and $L_n^m$, and thereby, $(\mathbf{r} \cdot \mathbf{E})$ and $(\mathbf{r} \cdot \mathbf{H})$ are identically zero. Consequently, the field outside $S$ is identically zero per this theorem. Currents such as these are referred to as *non-radiating* currents.

(iii) Note that the theorem says nothing about the radial and other components of fields *inside S*.

**Corollary 1.3.1.1** Field determination by radial components [19].
*Any electromagnetic field* $(\mathbf{E}, \mathbf{H})$ *satisfying a source-free Maxwell's equation exterior to a sphere S is completely determined by its radial components* $E_r$ *and* $H_r$. *In particular, the field* $(\mathbf{E}, \mathbf{H})$ *due to currents contained inside S is fully characterized by the radial parts* $\mathbf{r} \cdot \mathbf{E}$ *and* $\mathbf{r} \cdot \mathbf{H}$. *Equality of the radial parts for two fields exterior to S implies the equality of other components.*

**Example 1.3.1.2.** Construction of total field from radial fields, Debye potentials. Construct two independent radial potentials $\mathbf{r}\Pi_1$ and $\mathbf{r}\Pi_2$, with $\Pi_1$ and $\Pi_2$ being scalar radiation functions, such that the electromagnetic field in the source-free region exterior to a sphere, $S$, enclosing all sources and defined by the expressions

$$\mathbf{E} = \nabla \times \nabla \times (\mathbf{r}\Pi_1) - jk\eta\nabla \times (\mathbf{r}\Pi_2), \tag{1.131}$$

$$\mathbf{H} = jk\eta^{-1}\nabla \times (\mathbf{r}\Pi_1) + \nabla \times \nabla \times (\mathbf{r}\Pi_2), \tag{1.132}$$

recovers the radial field $(\mathbf{r} \cdot \mathbf{E})$ and $(\mathbf{r} \cdot \mathbf{H})$ specified in equations (1.111) and (1.112).

We shall construct the solution using linear superposition by treating $\Pi_1$ and $\Pi_2$ separately. We assume a spherical harmonic representation and let

$$\Pi_1 = -\eta \sum_{n=0}^{\infty} \sum_{m=-n}^{n} A_{mn} I_n^m \, h_n^{(2)}(kr) Y_n^m(\theta, \phi); \quad \Pi_2 = 0, \tag{1.133}$$

where the coefficients $A_{mn}$ are to be determined from the equality of the radial parts. Substituting equation (1.133) into equations (1.131) and (1.132) and carrying out the curl operations in spherical coordinates using the identity (1.83) and noting $DY_n^m = -n(n+1)\sin\theta Y_n^m$, we obtain

$$rE_r = -\eta \sum_{n=0}^{\infty} \sum_{m=-n}^{n} n(n+1) A_{mn} I_n^m \, h_n^{(2)}(kr) Y_n^m(\theta, \phi), \tag{1.134}$$

$$E_\theta = -\eta \sum_{n=0}^{\infty} \sum_{m=-n}^{n} A_{mn} I_n^m \frac{1}{r} \frac{\mathrm{d}}{\mathrm{d}r}[r h_n^{(2)}(kr)] \frac{\partial}{\partial \theta} Y_n^m(\theta, \phi), \tag{1.135}$$

$$E_\phi = -\eta \sum_{n=0}^{\infty} \sum_{m=-n}^{n} A_{mn} I_n^m \frac{1}{r \sin\theta} \frac{\mathrm{d}}{\mathrm{d}r}[r h_n^{(2)}(kr)] \frac{\partial}{\partial \phi} Y_n^m(\theta, \phi), \tag{1.136}$$

$$rH_r = 0, \tag{1.137}$$

$$H_\theta = -jk \sum_{n=0}^{\infty} \sum_{m=-n}^{n} A_{mn} I_n^m \, h_n^{(2)}(kr) \frac{1}{\sin\theta} \frac{\partial}{\partial \phi} Y_n^m(\theta, \phi), \tag{1.138}$$

$$H_\phi = jk \sum_{n=0}^{\infty} \sum_{m=-n}^{n} A_{mn} I_n^m \, h_n^{(2)}(kr) \frac{\partial}{\partial \theta} Y_n^m(\theta, \phi). \tag{1.139}$$

By comparing equation (1.134) with equation (1.111), we conclude that $A_{mn} = 1/n(n+1)$, $n \neq 0$. Because the field generated by $\Pi_1$ lacks $H_r$, it is appropriate to label this field as being *transverse magnetic* to $r$ or simply $\mathrm{TM}_r$. Note that the $n = 0$ term does not affect the electromagnetic field as determined by the above equations because $Y_0^0(\theta, \phi)$ is a constant. By theorem 1.3.1, except for the addition of terms proportional to $h_0^{(2)}(kr)$, the potential (1.133) is uniquely determined by the radial part $(\mathbf{r} \cdot \mathbf{E})$. With the choice $A_{mn} = 1/n(n+1)$, it generates the complete field (1.134)–(1.139) consistent with the specification of this $(\mathbf{r} \cdot \mathbf{E})$.

In a like manner letting

$$\Pi_2 = -\eta^{-1} \sum_{n=0}^{\infty} \sum_{m=-n}^{n} B_{mn} L_n^m \, h_n^{(2)}(kr) Y_n^m(\theta, \phi); \quad \Pi_1 = 0, \tag{1.140}$$

where the coefficients $B_{mn}$ are to be determined, we obtain upon substituting into equations (1.131) and (1.132)

$$rH_r = -\eta^{-1} \sum_{n=0}^{\infty} \sum_{m=-n}^{n} n(n+1) B_{mn} L_n^m \, h_n^{(2)}(kr) Y_n^m(\theta, \phi), \tag{1.141}$$

$$H_\theta = -\eta^{-1} \sum_{n=0}^{\infty} \sum_{m=-n}^{n} B_{mn} L_n^m \frac{1}{r} \frac{\mathrm{d}}{\mathrm{d}r}[r h_n^{(2)}(kr)] \frac{\partial}{\partial \theta} Y_n^m(\theta, \phi), \tag{1.142}$$

$$H_\phi = -\eta^{-1} \sum_{n=0}^{\infty} \sum_{m=-n}^{n} B_{mn} L_n^m \frac{1}{r \sin\theta} \frac{\mathrm{d}}{\mathrm{d}r}[r h_n^{(2)}(kr)] \frac{\partial}{\partial \phi} Y_n^m(\theta, \phi), \tag{1.143}$$

$$rE_r = 0, \tag{1.144}$$

$$E_\theta = jk \sum_{n=0}^{\infty} \sum_{m=-n}^{n} B_{mn} L_n^m \, h_n^{(2)}(kr) \frac{1}{\sin\theta} \frac{\partial}{\partial\phi} Y_n^m(\theta, \phi), \tag{1.145}$$

$$E_\phi = -jk \sum_{n=0}^{\infty} \sum_{m=-n}^{n} B_{mn} L_n^m \, h_n^{(2)}(kr) \frac{\partial}{\partial\theta} Y_n^m(\theta, \phi). \tag{1.146}$$

By comparing equation (1.141) to equation (1.112), we conclude that $B_{mn} = 1/n(n+1)$, $n \neq 0$. Because the field generated by $\Pi_2$ lacks $E_r$, it is appropriate to label this field as being *transverse electric* to $r$ or simply $\mathrm{TE}_r$. As before, per theorem 1.3.1, except for the addition of terms proportional to $h_0^{(2)}(kr)$, the potential (1.140) is uniquely determined by the radial part $(\mathbf{r} \cdot \mathbf{H})$. With the choice $B_{mn} = 1/n(n+1)$, it generates the complete field (1.141)–(1.146) consistent with the specification of this $(\mathbf{r} \cdot \mathbf{H})$. The potentials $\Pi_1$ and $\Pi_2$ are referred to in the literature as *Debye potentials* [22] and are given explicitly by

$$\Pi_1 = \sum_{n=1}^{\infty} \sum_{m=-n}^{n} \frac{-\eta I_n^m}{n(n+1)} h_n^{(2)}(kr) Y_n^m(\theta, \phi), \tag{1.147}$$

$$\Pi_2 = \sum_{n=1}^{\infty} \sum_{m=-n}^{n} \frac{-\eta^{-1} L_n^m}{n(n+1)} h_n^{(2)}(kr) Y_n^m(\theta, \phi). \tag{1.148}$$

As expected, $\Pi_1$ and $\Pi_2$ are dual quantities (see section 6.1.1). ■■

**Example 1.3.1.3** Multipole strengths via spectral current components.
Using equation (1.74) express the multipole strengths $I_n^m$ and $L_n^m$ of the Debye potentials in terms of the vector Fourier transform of the current distribution:

$$\tilde{\mathbf{J}}(\boldsymbol{\kappa}) = \iiint_V \mathbf{J}(\mathbf{r}) e^{-j\boldsymbol{\kappa}\cdot\mathbf{r}} \, dv, \tag{1.149}$$

where $\boldsymbol{\kappa} = \hat{\mathbf{x}}\kappa_x + \hat{\mathbf{y}}\kappa_y + \hat{\mathbf{z}}\kappa_z$ is the variable conjugate to $\mathbf{r}$, i.e. is the transformed variable.

Denoting the interior spherical harmonic as $\psi_n^m = j_n(kr) Y_n^m(\theta, \phi)$, defining the gradient operator in the wavenumber domain with respect to a 'position' vector $\hat{\mathbf{s}} = (\hat{\mathbf{x}}s_x, \hat{\mathbf{y}}s_y, \hat{\mathbf{z}}s_z)$ with three independent spectral coordinates $(s_x, s_y, s_z)$ as $\tilde{\nabla} = \hat{\mathbf{x}}\partial/\partial s_x + \hat{\mathbf{y}}\partial/\partial s_y + \hat{\mathbf{z}}\partial/\partial s_z$, and a related transverse operator as $\mathscr{L} = -j\hat{\mathbf{s}} \times \tilde{\nabla}$, and employing the vector identity $\nabla \times (\mathbf{r}\psi_n^m) = \nabla\psi_n^m \times \mathbf{r}$ and the result (1.74) in the defining equation (1.109), the multipole strength $L_n^m$ can be obtained as

$$
\begin{aligned}
L_n^m &= jk\eta \iiint\limits_V \mathbf{J}(\mathbf{r}') \cdot \nabla'\bar{\psi}_n^m \times \mathbf{r}' \, \mathrm{d}v' \\
&= jk\eta \frac{j^n}{4\pi} \iiint\limits_V \oiint \mathrm{d}\alpha \mathrm{d}\beta \sin\alpha \bar{Y}_n^m(\alpha, \beta) \mathbf{J}(\mathbf{r}') \cdot \nabla'(\mathrm{e}^{-jk\hat{s}\cdot\mathbf{r}'}) \times \mathbf{r}' \, \mathrm{d}v' \\
&= jk\eta \frac{j^n}{4\pi} \iiint\limits_V \oiint \mathrm{d}\alpha \mathrm{d}\beta \sin\alpha \bar{Y}_n^m(\alpha, \beta) \mathbf{J}(\mathbf{r}') \cdot \hat{\mathbf{s}} \times \widetilde{\nabla}(\mathrm{e}^{-jk\hat{s}\cdot\mathbf{r}'}) \, \mathrm{d}v' \\
&= jk\eta \frac{j^n}{4\pi} \iiint\limits_V \oiint \mathrm{d}\alpha \mathrm{d}\beta \sin\alpha \bar{Y}_n^m(\alpha, \beta) \hat{\mathbf{s}} \cdot \widetilde{\nabla} \times (\mathbf{J}(\mathbf{r}') \mathrm{e}^{-jk\hat{s}\cdot\mathbf{r}'}) \, \mathrm{d}v' \\
&= jk\eta \frac{j^n}{4\pi} \oiint \bar{Y}_n^m(\alpha, \beta) \, \hat{\mathbf{s}} \cdot \widetilde{\nabla} \times \widetilde{\mathbf{J}}(k\hat{\mathbf{s}}) \sin\alpha \, \mathrm{d}\alpha \mathrm{d}\beta \\
&= -jk\eta \frac{j^n}{4\pi} \oiint \hat{\mathbf{s}} \cdot [\widetilde{\nabla}\bar{Y}_n^m(\alpha, \beta) \times \widetilde{\mathbf{J}}(k\hat{\mathbf{s}})] \sin\alpha \, \mathrm{d}\alpha \mathrm{d}\beta \\
&= k\eta \frac{j^n}{4\pi} \oiint \widetilde{\mathbf{J}}(k\hat{\mathbf{s}}) \cdot \mathscr{L}[\bar{Y}_n^m(\alpha, \beta)] \sin\alpha \, \mathrm{d}\alpha \mathrm{d}\beta \\
&= k\eta \frac{j^n}{4\pi} \oiint \widetilde{\mathbf{J}}^T(k\hat{\mathbf{s}}) \cdot \mathscr{L}[\bar{Y}_n^m(\alpha, \beta)] \sin\alpha \, \mathrm{d}\alpha \mathrm{d}\beta,
\end{aligned}
\tag{1.150}
$$

where the Fourier transformed current is decomposed into a transverse part and a longitudinal part as $\widetilde{\mathbf{J}} = \widetilde{\mathbf{J}}^T + \hat{\mathbf{s}}(\hat{\mathbf{s}} \cdot \widetilde{\mathbf{J}})$, $\widetilde{\mathbf{J}}^T = (\hat{\mathbf{s}} \times \widetilde{\mathbf{J}}) \times \hat{\mathbf{s}}$ being the transverse part of $\widetilde{\mathbf{J}}$. The following simplification was used in arriving at the final step

$$
\hat{\mathbf{s}} \cdot [\widetilde{\nabla}\bar{Y}_n^m \times \widetilde{\mathbf{J}}] = [\widetilde{\mathbf{J}} \times \hat{\mathbf{s}}] \cdot \widetilde{\nabla}\bar{Y}_n^m = \left[\widetilde{\mathbf{J}}^T \times \hat{\mathbf{s}}\right] \cdot \widetilde{\nabla}\bar{Y}_n^m = \widetilde{\mathbf{J}}^T \cdot [\hat{\mathbf{s}} \times \widetilde{\nabla}\bar{Y}_n^m]. \tag{1.151}
$$

Equation (1.150) expresses the multipole strength $L_n^m$ of a $\mathrm{TE}_r$ mode in terms of (i) the transverse part of the Fourier transform of the current distribution evaluated at $\boldsymbol{\kappa} = k\hat{\mathbf{s}}$, with $\hat{\mathbf{s}} = \hat{\mathbf{x}} \sin\alpha \cos\beta + \hat{\mathbf{y}} \sin\alpha \sin\beta + \hat{\mathbf{z}} \cos\alpha$, and (ii) the vector $\mathscr{L}\bar{Y}_n^m$ obtained by operating the *orbital angular momentum operator* $\mathscr{L} = -j\hat{\mathbf{s}} \times \widetilde{\nabla}$ on the surface harmonic $\bar{Y}_n^m(\alpha, \beta)$ in the $\boldsymbol{\kappa}$-space. Its main advantage over the alternative spatial form in equation (1.109) arises from the fact that the far-zone radiated fields of an antenna and the transverse component of $\widetilde{\mathbf{J}}$ in the visible part of the spectrum (namely $|\boldsymbol{\kappa}| = k$) are proportional to each other. Note that the substitution $\cos\alpha = s_z$, $\cos\beta = s_x/\sqrt{1 - s_z^2}$, $\sin\beta = s_y/\sqrt{1 - s_z^2}$ may be used in $\bar{Y}_n^m$ before implementing the term $\mathscr{L}\bar{Y}_n^m$ in the wavenumber domain.

In an identical manner, defining $\mathbf{M} = \nabla \times \mathbf{J}$, the multipole strength $I_m^n$ can be obtained as

$$
I_n^m = -j\frac{j^n}{4\pi} \oiint \widetilde{\mathbf{M}}^T(k\hat{\mathbf{s}}) \cdot \mathscr{L}[\bar{Y}_n^m(\alpha, \beta)] \sin\alpha \, \mathrm{d}\alpha \mathrm{d}\beta. \tag{1.152}
$$

But the Fourier transform of $\mathbf{M} = \nabla \times \mathbf{J}$ is

$$\begin{aligned}\widetilde{\mathbf{M}}(\boldsymbol{\kappa}) &= \iiint\limits_V [\nabla \times \mathbf{J}(\mathbf{r})]e^{-j\boldsymbol{\kappa}\cdot\mathbf{r}}\, dv \\ &= \iiint\limits_V [\nabla \times (\mathbf{J}(\mathbf{r})e^{-j\boldsymbol{\kappa}\cdot\mathbf{r}}) + \mathbf{J}(\mathbf{r}) \times \nabla e^{-j\boldsymbol{\kappa}\cdot\mathbf{r}}]\, dv \\ &= \oiint\limits_S \hat{\mathbf{n}} \times \mathbf{J}(\mathbf{r})e^{-j\boldsymbol{\kappa}\cdot\mathbf{r}}\, ds + j\boldsymbol{\kappa} \times \iiint\limits_V \mathbf{J}(\mathbf{r})e^{-j\boldsymbol{\kappa}\cdot\mathbf{r}}\, dv \\ &= j\boldsymbol{\kappa} \times \widetilde{\mathbf{J}},\end{aligned} \tag{1.153}$$

since the localized current density $\mathbf{J}$ vanishes on the bounding surface $S$. The conversion of the volume integral to a surface integral above follows from the curl theorem (B.5). Inserting this result into equation (1.152) gives

$$\begin{aligned}I_n^m &= -j(jk)\frac{j^n}{4\pi}\oiint \hat{\mathbf{s}} \times \widetilde{\mathbf{J}}^T(k\hat{\mathbf{s}}) \cdot \mathscr{L}[\bar{Y}_n^m(\alpha, \beta)] \sin\alpha \, d\alpha d\beta \\ &= -\frac{j^n k}{4\pi}\oiint \widetilde{\mathbf{J}}^T(k\hat{\mathbf{s}}) \cdot (\hat{\mathbf{s}} \times \mathscr{L}[\bar{Y}_n^m(\alpha, \beta)]) \sin\alpha \, d\alpha d\beta,\end{aligned} \tag{1.154}$$

for the multipole strength of a $\text{TM}_r$ mode. Expressions for the multipole strengths in terms of the Fourier transform of the current have been given in [20], whose detailed derivation is completed in this example. Thus it is seen that the multipole strengths of both $\text{TM}_r$ and $\text{TE}_r$ modes depend solely on the transverse part of the current distribution evaluated on the circle $|\boldsymbol{\kappa}| = k$. The latter is completely determined by the transverse components of the electric field in the far-zone (see equation (1.50)). *Thus the transverse components of the far-zone electric field completely determine the electromagnetic field radiated in a homogeneous medium outside the sphere enclosing all electric sources.* Furthermore, it is also clear from equations (1.150), (1.154), (1.131), (1.132), (1.133), and (1.140) that *if the current is such that* $\widetilde{\mathbf{J}}^T(k\hat{\mathbf{s}}) \equiv 0$, *then the electromagnetic field outside the spherical surface enclosing the electric currents is identically zero.* This is an alternative definition of a non-radiating current. ■■

### 1.3.2 Additional theorems related to spherical harmonics

A theorem related to the Wilcox theorem 1.2.4 is the expansion theorem of an arbitrary function in spherical surface harmonics.

**Theorem 1.3.2.** Spherical surface harmonics expansion theorem [23, p 513].
*Let $g(\theta, \phi)$ be an arbitrary function on the surface of a unit sphere, which together with all its first and second derivatives is continuous. Then $g(\theta, \phi)$ may be expanded in an absolutely and uniformly convergent series of surface spherical harmonics*

$$g(\theta, \phi) = \sum_{n=0}^{\infty}\left[a_{n0}P_n(\cos\theta) + \sum_{m=1}^{n}(a_{nm}\cos m\phi + b_{nm}\sin m\phi)P_n^m(\cos\theta)\right], \tag{1.155}$$

*whose coefficients are determined from*

$$a_{n0} = \frac{2n+1}{4\pi} \int_0^{2\pi} \int_0^{\pi} g(\theta, \phi) P_n(\cos\theta) \sin\theta \, d\theta d\phi, \tag{1.156}$$

$$a_{nm} = \frac{2n+1}{2\pi} \frac{(n-m)!}{(n+m)!} \int_0^{2\pi} \int_0^{\pi} g(\theta, \phi) P_n^m(\cos\theta) \cos m\phi \sin\theta \, d\theta d\phi, \quad m \geqslant 1, \tag{1.157}$$

$$b_{nm} = \frac{2n+1}{2\pi} \frac{(n-m)!}{(n+m)!} \int_0^{2\pi} \int_0^{\pi} g(\theta, \phi) P_n^m(\cos\theta) \sin m\phi \sin\theta \, d\theta d\phi. \tag{1.158}$$

*An equivalent representation using the normalized surface harmonic* $Y_n^m(\theta, \phi)$ *of equation (1.103) is*

$$g(\theta, \phi) = \sum_{n=0}^{\infty} \sum_{m=-n}^{n} C_n^m Y_n^m(\theta, \phi), \tag{1.159}$$

*where*

$$C_n^m = \int_0^{2\pi} \int_0^{\pi} g(\theta, \phi) \bar{Y}_n^m(\theta, \phi) \sin\theta \, d\theta d\phi. \tag{1.160}$$

A direct consequence of the above theorem are the two following corollaries.

**Corollary 1.3.2.1.** Closure of surface spherical harmonics [24, p 62].
*Let* $g(\theta, \phi)$ *be an arbitrary function on the surface of a unit sphere, which together with all its first and second derivatives is continuous. If* $a_{n0}$, $a_{nm}$, $b_{nm}$, *or equivalently,* $C_n^m$ *are all zero, then* $g(\theta, \phi)$ *is identically zero.*

**Corollary 1.3.2.2.** Completeness of surface spherical harmonics [24, p 62].
*Let* $g(\theta, \phi)$ *be an arbitrary function on the surface of a unit sphere* $\Omega$, *which together with all its first and second derivatives is continuous. Then*

$$\begin{aligned} \iint_{\Omega} |g(\theta, \phi)|^2 \sin\theta \, d\theta d\phi &= \sum_{n=0}^{\infty} \frac{2}{2n+1} \left[ |a_{n0}|^2 + \sum_{m=1}^{n} \frac{(n+m)!}{(n-m)!} (|a_{nm}|^2 + |b_{nm}|^2) \right] \\ &= \sum_{n=0}^{\infty} \sum_{m=-n}^{n} |C_n^m|^2. \end{aligned} \tag{1.161}$$

**Theorem 1.3.3.** Spherical harmonics representation interior to a sphere [24, p 83].
*If the twice continuously differentiable function* $U(\mathbf{r})$, $r \leqslant r_0$, *satisfies the scalar Helmholtz equation* $\nabla^2 U + k^2 U = 0$, *then for* $0 \leqslant r < r_0$, *this function can be represented in a uniformly convergent series of the form*

$$U(\mathbf{r}) = \sum_{n=0}^{\infty} \sum_{m=-n}^{n} C_n^m j_n(kr) Y_n^m(\theta, \phi), \tag{1.162}$$

*where*

$$j_n(kr_0) C_n^m = \int_0^{2\pi} \int_0^{\pi} U(r_0, \theta, \phi) \bar{Y}_n^m(\theta, \phi) \sin\theta \, \mathrm{d}\theta \mathrm{d}\phi,$$

*and* $j_n(z)$ *is the spherical Bessel function of argument* $z$ *and order* $n$.

**Theorem 1.3.4.** Spherical harmonics representation exterior to a sphere [24, p 84]. *If the twice continuously differentiable function* $U(\mathbf{r})$, $r \geqslant r_0$, *satisfies the scalar Helmholtz equation* $\nabla^2 U + k^2 U = 0$, *and the Sommerfeld's radiation condition*

$$\frac{\partial U}{\partial r} + jkU = \mathcal{O}\left(\frac{1}{r^2}\right), \quad r \to \infty,$$

*i.e. is a scalar radiation function, then for* $r_0 < r < \infty$, *this function can be represented in a uniformly convergent series of the form*

$$U(\mathbf{r}) = \sum_{n=0}^{\infty} \sum_{m=-n}^{n} C_n^m h_n^{(2)}(kr) Y_n^m(\theta, \phi), \tag{1.163}$$

*where*

$$h_n^{(2)}(kr_0) C_n^m = \int_0^{2\pi} \int_0^{\pi} U(r_0, \theta, \phi) \bar{Y}_n^m(\theta, \phi) \sin\theta \, \mathrm{d}\theta \mathrm{d}\phi,$$

*and* $h_n^{(2)}(z)$ *is the spherical Hankel function of the second kind of argument* $z$ *and order* $n$.

## References


[1] Jefimenko O D 2004 Presenting electromagnetic theory in accordance with the principle of causality *Eur. J. Phys.* **25** 287–96

[2] Heras J A 2011 A short proof that the Coulomb-gauge potentials yield the retarded fields *Eur. J. Phys.* **32** 213–16

[3] Morse P M and Feshbach H 1953 *Methods of Theoretical Physics* (New York: McGraw-Hill)

[4] Barton G 1989 *Elements of Green's Function and Propagation: Potentials Difusion and Waves* (New York: Oxford University Press)

[5] Heras J A 2007 Can Maxwell's equations be obtained from the continuity equation? *Am. J. Phys.* **75** 652–7

[6] van Bladel J 1993 A discussion of Helmholtz' theorem *Electromagnetics* **13** 95–110

[7] Lindell I V 1995 *Methods for Electromagnetic Field Analysis* (IEEE Press Series on Electromagnetic Wave Theory) (New York: IEEE)

[8] Caloz C and Sihvola A 2020 Electromagnetic chirality. Part 2: The microscopic perspective *IEEE Antennas Propag. Mag.* **62** 58–71

[9] Caloz C and Sihvola A Electromagnetic chirality. Part 2: The microscopic perspective *IEEE Antennas Propag. Mag.* **62** 82–98

[10] Ougshtun K E 2006 *Electromagnetic and Optical Pulse Propagation* vol 1 (New York: Springer)

[11] van Bladel J 1991 Unusual boundary conditions at an interface *IEEE Antennas Propag. Mag.* **33** 57–8

[12] Wilcox C H 1959 Spherical means and radiation conditions *Arch. Rat. Mech. Anal.* **3** 133–48

[13] Whittaker E T 1903 On the partial differential equations of mathematical physics *Math. Ann.* **57** 333–55

[14] Devaney A J and Wolf E 1974 Multipole expansions and plane wave representations of the electromagnetic field *J. Math. Phys.* **15** 234–44

[15] Wilcox C H 1956 An expansion theorem for electromagnetic fields *Commun. Pure Appl. Math.* **9** 115–34

[16] Bayliss A, Gunzburger M and Turkel E 1982 Boundary conditions for the numerical solution of elliptic equations in exterior regions *SIAM J. Appl. Math.* **42** 430–51

[17] Peterson A F 1988 Absorbing boundary conditions for the vector wave equation *Microw. Opt. Technol. Lett.* **1** 62–4

[18] Harrington R F 1961 *Time-Harmonic Electromagnetic Fields* (New York: McGraw-Hill)

[19] Bouwkamp C J and Casimir H B G 1954 On multipole expansions in the theory of electromagnetic radiation *Physica* **20** 539–54

[20] Devaney A J and Wolf E 1973 Radiating and nonradiating classical current distributions and the fields they generate *Phys. Rev.* D **8** 1044–7

[21] Titchmarsh E C 1939 *The Theory of Functions* 2nd edn (New York: Oxford University Press)

[22] Wilcox C H 1957 Debye potentials *J. Math. Mech.* **6** 167–201

[23] Courant R and Hilbert D 1953 *Methods of Mathematical Physics* vol 1 (New York: Wiley)

[24] Müller C 1969 *Foundations of the Mathematical Theory of Electromagnetic Waves* (New York: Springer)

# Chapter 2

## Electrostatics and magnetostatics

Even though static fields can be thought of as a limiting case of dynamic fields as the frequency tends to zero, there are certain theorems that are unique to static fields that deserve special mention. Important theorems in this regard are Thomson's theorem, Earnshaw's theorem, Green's reciprocation theorem, the principle of virtual displacement, mean value theorem, and the maximum principle for harmonic functions, which will all be discussed in the present chapter.

### 2.1 Energy related theorems in electrostatics

**Theorem 2.1.1.** Thomson's theorem [1, 2].
*Energy of the actual electrostatic field is an extremum relative to the energies of fields which could be produced by any other distribution of the charges on or in perfect conductors.*

*Stated otherwise, if a given total charge $Q$ is distributed in a volume $V_i$ bounded by the surface $S$, then the electrostatic energy is an absolute minimum when charges are placed so that the surface $S$ is an equipotential surface and the interior field is zero as happens when $V_i$ is filled with a perfect conductor.*

*Proof.* Our proof is adapted from the one given in [3, p 267] for a capacitor problem. We will formulate a variational expression involving electrostatic energy under the given constraints and show that the properties follow from the *necessary conditions* for the extremal solution. With reference to figure 2.1 we will, in particular, show (i) that the electrostatic field in the interior region $V_\mathrm{i}$ is identically zero, (ii) the surface $S$ corresponds to an equipotential surface, and (iii) the solution in the exterior region $V_\mathrm{e}$ is governed by the Poisson's equation. In the following we consider only one *finite-extent* surface $S$. The solution in the presence of multiple disjoint surfaces is a trivial extension of the results shown here.

doi:10.1088/978-0-7503-1716-0ch2

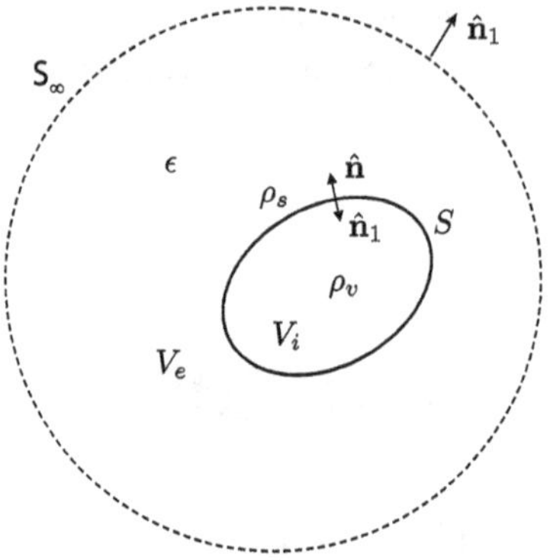


**Figure 2.1.** A charged body of volume $V_i$ carrying a volume charge density $\rho_v$ and a surface charge density $\rho_s$. The permittivity of the medium is $\varepsilon$ and the volume exterior to the body is denoted by $V_e$. Note that $\hat{\mathbf{n}}_1 = -\hat{\mathbf{n}}$ on $S$ and that $S_\infty$ is a large outer sphere of radius $r \to \infty$.

Consider the energy $W_e$ associated with an electrostatic field in the presence of charged body embedded in a dielectric medium [1, p 46], [4, p 98]

$$W_e = \frac{1}{2} \iiint\limits_{V_e + V_i} \mathbf{E} \cdot \mathbf{D} \, dv. \tag{2.1}$$

We use the Cartesian coordinate representation of the electric field intensity $\mathbf{E} = \hat{\mathbf{x}}u + \hat{\mathbf{y}}v + \hat{\mathbf{z}}w =: (u, v, w)$ and express the electric flux density as $\mathbf{D} = \varepsilon \mathbf{E}$. We assume that the permittivity $\varepsilon$ of the medium does not change with the field and is a function only of position. Thus the process of changing the field in the following is regarded as an isothermal process, since the permittivity will, in general, depend on the absolute temperature $T$. To assure isothermal behavior, the dielectric material in question must be in contact with a heat bath which can exchange heat with it to maintain a constant temperature[1].

Let $\rho_v$ and $\rho_s$ denote the volume charge density in $V_i$ and the surface charge density on $S$, respectively. We impose the following constraints on the problem:

$$\nabla \cdot \mathbf{D} = \begin{cases} 0, & \mathbf{r} \in V_e \\ \rho_v, & \mathbf{r} \in V_i \end{cases}, \tag{2.2}$$

$$\hat{\mathbf{n}} \cdot (\mathbf{D}_+(\mathbf{r}_s) - \mathbf{D}_-(\mathbf{r}_s)) = \rho_s(\mathbf{r}_s), \quad \mathbf{r} = \mathbf{r}_s \in S, \tag{2.3}$$

$$\text{The total charge } Q = \iiint\limits_{V_i} \rho_v \, dv + \oint\!\!\oint\limits_{S} \rho_s \, ds, \tag{2.4}$$

[1] Thermodynamically speaking, the maximum work which can be extracted from a system under isothermal conditions is the free energy, $F$, of the system, not the total energy $U$. The relation between these two quantities at constant temperature is $F = U - TS$, where $S$ is the entropy. For arbitrary changes in field and temperature $\delta F = -S\delta T + \delta W_e$. Hence in the presence of dielectrics at constant temperature, $W_e$ as given in equation (2.1) is to be identified as the thermodynamic free energy $F$.

where $\mathbf{D}_{\pm}(\mathbf{r}_s) = \lim_{\mathbf{r}\to\mathbf{r}_s\pm\hat{\mathbf{n}}0} \mathbf{D}(\mathbf{r})$ are, respectively, the electric flux densities just above and below the surface $S$ at an arbitrary point $\mathbf{r}_s \in S$.

The unknown functions for a variational treatment are $u$, $v$, $w$, $\rho_s$, and $\rho_v$. Note that the permittivity of the medium, even though spatially inhomogeneous, is considered to be given and will not have any variations. Constraints (2.2) and (2.3) are valid point-wise and call for spatially varying Lagrangian multipliers $\psi_e(\mathbf{r})$, $\mathbf{r} \in V_e$, $\psi_i(\mathbf{r})$, $\mathbf{r} \in V_i$, and $\lambda(\mathbf{r}_s)$, $\mathbf{r}_s \in S$, respectively, that are functions of spatial coordinates in their respective domains, whereas the constraint (2.4) is of isoperimetric type with an associated Lagrangian multiplier that is a constant $\psi_0$ [3]. Accordingly we consider the Lagrangian (that is, the cost function)

$$\begin{aligned} L = {} & \frac{1}{2} \iiint\limits_{V_e+V_i} \varepsilon(u^2 + v^2 + w^2)\,\mathrm{d}v - \iiint\limits_{V_e} \psi_e \nabla \cdot \mathbf{D}\,\mathrm{d}v \\ & - \iiint\limits_{V_i} \psi_i [\nabla \cdot \mathbf{D} - \rho_v]\,\mathrm{d}v \\ & + \psi_0 \left[ Q - \iiint\limits_{V_i} \rho_v\,\mathrm{d}v - \oint\!\!\oint\limits_{S} \rho_s\,\mathrm{d}s \right] \\ & + \oint\!\!\oint\limits_{S} \lambda[\rho_s - \hat{\mathbf{n}} \cdot (\mathbf{D}_+ - \mathbf{D}_-]\,\mathrm{d}s. \end{aligned} \tag{2.5}$$

Taking the variation with respect to the unknown quantities we obtain[2]

$$\begin{aligned} \delta L = {} & \iiint\limits_{V_e+V_i} \varepsilon(u\delta u + v\delta v + w\delta w)\,\mathrm{d}v \\ & - \iiint\limits_{V_e} \psi_e \nabla \cdot \delta\mathbf{D}\,\mathrm{d}v - \iiint\limits_{V_i} \psi_i \nabla \cdot \delta\mathbf{D}\,\mathrm{d}v \\ & + \iiint\limits_{V_i} (\psi_i - \psi_0)\delta\rho_v\,\mathrm{d}v + \oint\!\!\oint\limits_{S} (\lambda - \psi_0)\delta\rho_s\,\mathrm{d}s \\ & - \oint\!\!\oint\limits_{S} \lambda[\hat{\mathbf{n}} \cdot (\delta\mathbf{D}_+ - \delta\mathbf{D}_-]\,\mathrm{d}s. \end{aligned} \tag{2.6}$$

Making use of the vector identity $\psi\nabla \cdot \mathbf{A} = \nabla \cdot (\psi\mathbf{A}) - \mathbf{A} \cdot \nabla\psi$ and using Gauss's theorem (B.4) in the above leads to

[2] The variations with respect to the Lagrangian parameters when set to zero will yield back the constraints (2.2)–(2.4) for the extremal trajectory and are therefore ignored.

$$\begin{aligned}\delta L = &\iiint_{V_i} \varepsilon\left[\left(u + \frac{\partial \psi_i}{\partial x}\right)\delta u + \left(v + \frac{\partial \psi_i}{\partial y}\right)\delta v + \left(u + \frac{\partial \psi_i}{\partial z}\right)\delta w\right] \mathrm{d}v \\ &+ \iiint_{V_e} \varepsilon\left[\left(u + \frac{\partial \psi_e}{\partial x}\right)\delta u + \left(v + \frac{\partial \psi_e}{\partial y}\right)\delta v + \left(u + \frac{\partial \psi_e}{\partial z}\right)\delta w\right] \mathrm{d}v \\ &+ \iiint_{V_i} (\psi_i - \psi_0)\delta\rho_v \, \mathrm{d}v + \oiint_S (\lambda - \psi_0)\delta\rho_s \, \mathrm{d}s \\ &+ \oiint_S (\psi_e - \lambda)\hat{\mathbf{n}} \cdot \delta\mathbf{D}_+ \, \mathrm{d}s - \oiint_S (\psi_i - \lambda)\hat{\mathbf{n}} \cdot \delta\mathbf{D}_- \, \mathrm{d}s \\ &- \oiint_{S_\infty} \psi_e \, \hat{\mathbf{n}}_1 \cdot \delta\mathbf{D}_+ \, \mathrm{d}s.\end{aligned} \tag{2.7}$$

The surface integral over $S_\infty$ can be taken to be zero on assuming that $|\psi_e\mathbf{D}| = O(r^{-(2+\alpha)})$, $r \to \infty$, $\alpha > 0$. Note that $\hat{\mathbf{n}} \cdot \delta\mathbf{D} = \varepsilon(n_x\delta u + n_y\delta v + n_z\delta w)$, where $(n_x, n_y, n_z)$ are the direction cosines of $\hat{\mathbf{n}}$. Equating the coefficients of $\delta u$, $\delta v$, $\delta w$, $\delta\rho_v$, $\delta\rho_s$ in equation (2.7) to zero for the extremum trajectory we obtain the following necessary conditions (also known as the Euler equations)

$$\mathbf{E} = (u, v, w) = \begin{cases} -\nabla\psi_e, \ \mathbf{r} \in V_e \\ -\nabla\psi_i, \ \mathbf{r} \in V_i, \end{cases} \tag{2.8}$$

$$\psi_i = \psi_0, \quad \mathbf{r} \in V_i, \tag{2.9}$$

$$\lambda(\mathbf{r}_s) = \psi_0 \tag{2.10}$$

$$\psi_e = \lambda = \psi_i, \quad \mathbf{r} = \mathbf{r}_s \in S. \tag{2.11}$$

Hence by equation (2.8), electric field in the interior and exterior regions is expressed as the negative gradient of Lagrangian multipliers, $\psi_i$ and $\psi_e$, respectively, which may be identified as interior and exterior electric potentials. By equation (2.9), the potential inside $S$ is a constant and hence the fields $\mathbf{E}$ and $\mathbf{D}$ in $V_i$ are identically zero. Because $\rho_v = \nabla \cdot \mathbf{D}$, the volume charge density in $V_i$ for generating this minimum electrostatic energy is also zero. All of the given charge $Q$ then resides on the surface whose density is $\rho_s = \hat{\mathbf{n}} \cdot (\mathbf{D}_+ - \mathbf{D}_-) = \hat{\mathbf{n}} \cdot \mathbf{D}_+ = -\varepsilon\partial\psi_e/\partial n$. By equations (2.10) and (2.11) the external potential $\psi_e$ is constant on the surface $S$, being equal to $\psi_0$. Hence $S$ is an equipotential surface having a potential $\psi_0$ and there is no component of the $\mathbf{E}$ field tangential to $S$. All of the electric field just outside $S$ points normal to it. Finally, on utilizing equation (2.2) the equation for determining the potential in the exterior region is given by

$$\nabla \cdot (\varepsilon\nabla\psi_e) = 0 \tag{2.12}$$

subject to

$$\psi_e = \psi_0 \quad \text{on } S, \tag{2.13}$$

$$\oint\!\!\!\oint_S \varepsilon \frac{\partial \psi_e}{\partial n}\, ds = -Q, \tag{2.14}$$

and

$$\psi_e(\mathbf{r}) = O(r^{-\beta}) \text{ as } r = |\mathbf{r}| \to \infty, \ \beta > 0.5. \tag{2.15}$$

We may now calculate the minimum energy $W_{em}$ by inserting the Euler equations into (2.1). Denoting the potential in space as $\psi$ where $\psi = \psi_i$ in $V_i$ and $\psi = \psi_e$ in $V_e$ and writing $\mathbf{E} = -\nabla\psi$, we obtain on using $\nabla \cdot \mathbf{D} = 0$ in $V_i \cup V_e$ and $\psi_e = \psi_i = \psi_0$ on $S$ that

$$\begin{aligned} W_{em} &= -\frac{1}{2} \iiint_{V_i+V_e} \nabla\psi \cdot \mathbf{D}\, dv \\ &= -\frac{1}{2} \iiint_{V_i+V_e} [\nabla \cdot (\psi\mathbf{D}) - \psi\nabla \cdot \mathbf{D}]\, dv \\ &= -\frac{1}{2} \oint\!\!\!\oint_S \psi_i\, \hat{\mathbf{n}} \cdot \mathbf{D}_-\, ds + \frac{1}{2} \oint\!\!\!\oint_S \psi_e\, \hat{\mathbf{n}} \cdot \mathbf{D}_+\, ds \\ &\quad -\frac{1}{2} \oint\!\!\!\oint_{S_\infty} \psi_e\, \hat{\mathbf{n}}_1 \cdot \mathbf{D}\, ds \\ &= \frac{1}{2}\psi_0 \oint\!\!\!\oint_S \rho_s\, ds = \frac{1}{2} Q\psi_0, \end{aligned} \tag{2.16}$$

in view of equations (2.3) and (2.15). Equation (2.16) can easily be generalized to the multiple conductor case. If the conductor $n$ has a total charge $Q^{(n)}$ and is at a potential $\psi_0^{(n)}$, $n = 1, 2, \ldots, N$, then the total electrostatic energy is

$$W_{em} = \frac{1}{2} \sum_{n=1}^{N} Q^{(n)} \psi_0^{(n)}. \tag{2.17}$$

■

As a corollary to Thomson's theorem, if a second *uncharged* conductor with a bounding surface $S_2$ is introduced into the field of the given charge $Q$, it can only lower the total electrostatic energy. To prove this we note that the potential $\psi_0$ on $S$ will be modified to $\psi_0'$ due to the introduction of *induced* charges on $S_2$. The induced charges together with the original charge $Q$ on $S$ will also cause the field in the interior of $S_2$ to vanish. The net charge on $S_2$ will, however, remain zero and the new minimum energy is $W_{em}' = \psi_0' Q/2$. The energy of this two-body problem is for a configuration of field which is identically zero in the interior of $S$ and $S_2$ and in which the electric field lines will be normal to both $S$ and $S_2$. Any other distribution of field with these two surfaces in place will only have energy higher than $W'_{\mathrm{em}}$ by Thomson's theorem. In particular, the energy of an electric field distribution that is (i) null inside $S$ and is normal to $S$ and (ii) penetrable inside $S_2$ and in which there are no induced charges on $S_2$ will only have an energy higher than $W'_{\mathrm{em}}$. But the latter corresponds to the original problem with only the surface $S$ present, the second surface $S_2$ being fictitious. Therefore $W'_{\mathrm{em}} < W_{\mathrm{em}}$.

**Theorem 2.1.2.** Earnshaw's theorem [5, p 116], [6, p 167].
*A charged body placed in an electrostatic field cannot be maintained in stable equilibrium under the influence of electric forces alone.*

*Proof.* For a stable equilibrium the total energy of the system must be a minimum. We prove the theorem by showing that a state of minimum energy is not attained when a charged body is introduced in an electrostatic field. Suppose that the initial field is generated by a set of charges $q_i$, $i = 1, 2, \ldots, n$, distributed on $n$ fixed conductors whose bounding surfaces are denoted by $S_i$. The conductors are embedded in a medium with permittivity $\varepsilon$, which may be a continuous function of position. Given that the potential $\psi$ at any point in the dielectric is governed by the Laplace's equation

$$\nabla^2\psi = \frac{\partial^2\psi}{\partial x^2} + \frac{\partial^2\psi}{\partial y^2} + \frac{\partial^2\psi}{\partial z^2} = 0, \tag{2.18}$$

it is noted that neither the electric potential nor any of its derivatives can assume an extremum value at that point. This is because the three partial derivatives $\frac{\partial^2\psi}{\partial x^2}, \frac{\partial^2\psi}{\partial y^2}, \frac{\partial^2\psi}{\partial z^2}$ must be either strictly negative or strictly positive at the extremum point. However this is incompatible with equation (2.18).

Let us suppose that a charge $q_0$ is now placed on a conducting surface $S_0$. The distributions on all the conductors are momentarily assumed to be fixed and $S_0$ is introduced into the field of the other $n$ charges. If $\rho_{s0}$ is the surface charge distribution on $S_0$, the energy of this conductor is

$$U_0 = \frac{1}{2}\oiint_{S_0} \psi\rho_{s0}\, \mathrm{d}s, \tag{2.19}$$

where $\psi$ is the potential of the initial field. Let $P_0(\xi, \eta, \zeta)$ be an arbitrary point on $S_0$ and let $P(x, y, z)$ be any point in the dielectric in the neighborhood of $P_0$. Then by Taylor's series

$$\psi(P_0) = \psi(P) + (\xi - x)\frac{\partial\psi(P)}{\partial x} + (\eta - y)\frac{\partial\psi(P)}{\partial y} + (\zeta - z)\frac{\partial\psi(P)}{\partial z} + \cdots. \tag{2.20}$$

The energy too may be referred to the potential and its derivatives at this point $P$:

$$U_0 = \frac{1}{2}q_0\psi(P) + \frac{1}{2}\oint\!\!\oint_{S_0}\left[(\xi - x)\frac{\partial\psi(P)}{\partial x} + (\eta - y)\frac{\partial\psi(P)}{\partial y} + (\zeta - z)\frac{\partial\psi(P)}{\partial z}\right] \times \rho_{s0}\,\mathrm{d}s + \cdots. \tag{2.21}$$

Since $\psi$ cannot be a minimum at $P$, it is always possible to displace the conductor $S_0$ in such a way that the energy $U_0$ is decreased. If after this displacement the charges, which thus far have been chosen to be frozen on the surfaces $S_0$, $S_i$ are released, these surfaces will again become equipotentials and by Thomson's theorem the energy of the field will further be diminished. A minimum value of the energy function $U_0(P)$ does not exist in the electrostatic field and consequently the conductor $S_0$ is never in static equilibrium. ■

### 2.1.1 Reciprocity theorem in electrostatics

**Theorem 2.1.3** Green's reciprocation theorem [4, p 43].
*Consider a set of n point charges $q_j$, $j = 1, 2, \ldots, n$ at positions where the potentials in free-space due to the other charges are given by a set of numbers $\psi_j$. Similarly let the set of point charges $q_j'$ at the same positions produce potentials $\psi_j'$. Then*

$$\sum_{i=1}^{n}\psi_j q_j' = \sum_{i=1}^{n} q_j\psi_j'. \tag{2.22}$$

*Proof.* The potential at the point $j$ is related to the charges at the other points by

$$\psi_j = \frac{1}{4\pi\varepsilon_0}\sum_{i=1}^{n}{}' \frac{q_i}{r_{ij}}, \tag{2.23}$$

where a prime on the summation means that the term $i = j$ is excluded and $r_{ij}$ denotes the distance between points $i$ and $j$. If the other hand, a different set of charges $q_j'$ is placed at the same points, giving rise to the corresponding potentials $\psi_j'$, a similar relation holds

$$\psi_j' = \frac{1}{4\pi\varepsilon_0}\sum_{i=1}^{n}{}' \frac{q_i'}{r_{ij}}. \tag{2.24}$$

Multiply equation (2.23) by $q_j'$ and (2.24) by $q_j$ and sum each expression over the index *j*. Interchanging the summation indices *i* and *j* in one of the summations yields the desired result (2.22). A continuum version of this result is that if a charge distribution $\rho_1(\mathbf{r})$ produces the potential $\psi_1(\mathbf{r})$ and a charge distribution $\rho_2(\mathbf{r})$ produces a potential $\psi_2(\mathbf{r})$, then

$$\iiint_V \psi_1(\mathbf{r})\rho_2(\mathbf{r})\,\mathrm{d}v = \iiint_V \psi_2(\mathbf{r})\rho_1(\mathbf{r})\,\mathrm{d}v. \tag{2.25}$$

■

## 2.2 Principle of virtual displacement for static fields

Consider an electrostatic system comprised of charged bodies (both conducting and dielectric) or a magnetostatic system comprised of direct current carrying conductors. The charged bodies (respectively, the current carrying conductors) experience electric (respectively, magnetic) forces as a result of the electric (respectively, magnetic) field existing in space. Let the electric energy in the system be $W_\mathrm{e}$ (respectively, the magnetic energy in the system be $W_\mathrm{m}$). Since the system is assumed to be in static equilibrium with no time-varying occurrences of events, all bodies are stationary and the mutual forces between them are all in balance. Furthermore the total charge (total flux) in the static system is assumed to be constant. The goal here is to relate the electric force $\mathbf{F}_\mathrm{e}$ (magnetic force $\mathbf{F}_\mathrm{m}$) acting on a given body to the stored energy $W_\mathrm{e}$ ($W_\mathrm{m}$) in the system. By imagining a slight displacement $\delta\ell$ (a 'virtual' displacement as the system being considered is static) of the body under consideration and equating the work done by the system on the body to the corresponding decrease in energy stored we write

$$\mathbf{F}_\mathrm{e}\cdot\delta\ell = -\delta W_\mathrm{e} = -\nabla W_\mathrm{e}\cdot\delta\ell;\quad \mathbf{F}_\mathrm{m}\cdot\delta\ell = -\delta W_\mathrm{m} = -\nabla W_\mathrm{m}\cdot\delta\ell.$$

Since the orientation of the displacement is arbitrary, this implies that

$$\mathbf{F}_\mathrm{e} = -\nabla W_\mathrm{e};\quad \mathbf{F}_\mathrm{m} = -\nabla W_\mathrm{m}. \tag{2.26}$$

In other words the energy and forces are such that the negative gradient of energy at a point is equal to the force existing at that point. The forces are also conservative since $\nabla\times\mathbf{F}_\mathrm{e} = -\nabla\times\nabla W_\mathrm{e} = 0$.

**Example 2.2.0.1.** Force exerted on a plate of a parallel plate capacitor.
Consider a parallel plate capacitor whose plates have an area $A$ and whose separation is $z$, see figure 2.2. We know that for fixed charges $\pm Q$ on the plates and ignoring fringing of fields, the total electric energy stored, $W_\mathrm{e}$ and the electric field, $\mathbf{E}_\mathrm{e}$, can be expressed as

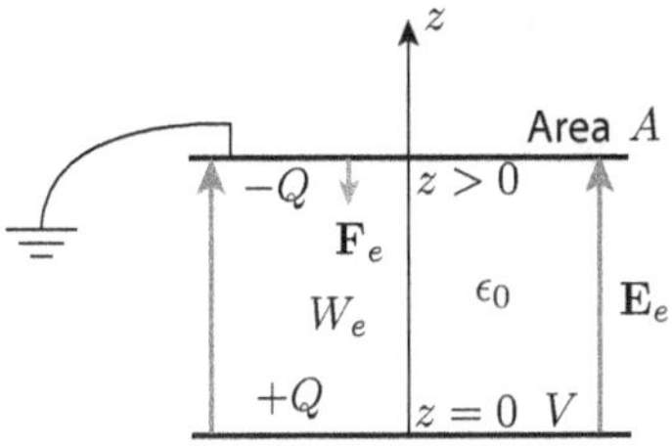


**Figure 2.2.** Parallel plate capacitor in vacuum having a plate separation $z$ and charges $\pm Q$.

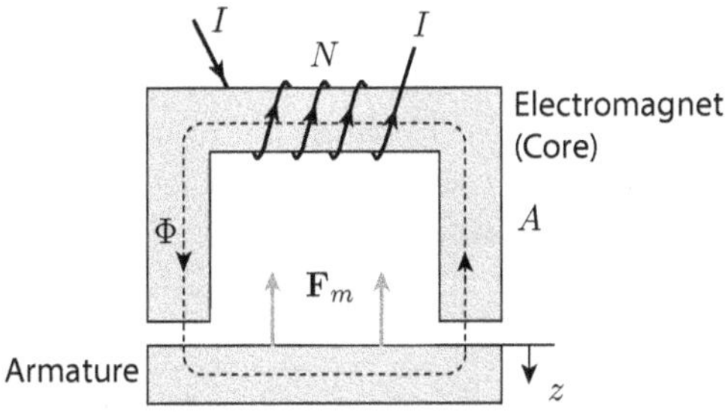


**Figure 2.3.** An electromagnetic consisting of a core and an armature.

$$W_e = \frac{1}{2}CV^2 = \frac{Q^2}{2C} = \frac{Q^2 z}{2\varepsilon_0 A} > 0; \quad \mathbf{E}_e = \hat{\mathbf{z}}\frac{V}{z} = \hat{\mathbf{z}}\frac{Q}{\varepsilon_0 A}.$$

For a given charge, the energy depends on the separation $z$ and increases with $z$. On the other hand, the electric field remains constant between the upper and lower plates. The force on the top plate (with a total charge $-Q$) by equation (2.26) is

$$\mathbf{F}_e = -\nabla W_e = -\hat{\mathbf{z}}\frac{Q^2}{2\varepsilon_a A} = -\frac{1}{2}Q\mathbf{E}_e.$$

It is directed downward because the top plate is attracted towards the bottom one. ■■

**Example 2.2.0.2** Force exerted on an armature.
Consider an electromagnet as shown in figure 2.3. A current $I$ flows in an $n$-turn coil wound around the core having a cross-sectional area $A$. The current produces a magnetic flux $\Phi$ in the magnetic circuit comprised of the core and armature. The virtual displacement of the armature in the $z$-direction (downward) changes the length of the air gap between the core and armature. The magnetic flux density $B$ in the core is related to the flux $\Phi$ via $BA = \Phi$. The magnetic energy density $w_m$ at any point is equal to $B^2/2\mu$, where $\mu$ is the permeability at that point. The change in energy $\delta W_m$ in the air gap of total cross-sectional area $2A$ due to virtual movement of armature by $\delta z$ is equal to

$$\delta W_m = (2A)\frac{B^2}{2\mu_0}\delta z = \frac{\Phi^2}{\mu_0 A}\delta z.$$

Therefore the force on the armature $\mathbf{F}_m$ for a constant flux $\Phi$ from equation (2.26) is

$$\mathbf{F}_m = -\hat{\mathbf{z}}\frac{\delta W_m}{\delta z} = -\hat{\mathbf{z}}\frac{\Phi^2}{\mu_0 A},$$

and is consequently attractive in nature. ■■

Other examples involving the principle of virtual displacement are considered in the introductory book by Cheng [7].

## 2.3 Theorems related to harmonic functions

**Theorem 2.3.1** Mean value theorem for harmonic function [8, p 254].
*The value of a harmonic function (i.e. solution of Laplace equation) at a point P is equal to the arithmetic mean of its values on any sphere with center at P provided that the function is regular inside and continuous in the closed region bounded by the sphere.*

*Proof.* From Green's second identity (B.11) we obtain on choosing $\nabla^2 u = 0$, $g = 1$

$$\oiint_S \frac{\partial u}{\partial n}\,\mathrm{d}s = 0, \tag{2.27}$$

for any harmonic function. Let $G$, $G_0$ be concentric spheres of radii $R$, $R_0$, respectively, with $R_0 < R$ and bounded by respective surfaces $\Gamma$, $\Gamma_0$. Let $r$ denote the distance from the center of the spheres. Take $v = 1/r$ in the Green's second identity (B.11) when applied to the region bounded by the two spheres. Then for any harmonic function $u$ we obtain on using equation (2.27) that

$$\frac{1}{4\pi R_0^2}\oiint_{\Gamma_0} u\,\mathrm{d}s = \frac{1}{4\pi R^2}\oiint_{\Gamma} u\,\mathrm{d}s. \tag{2.28}$$

The left-hand integral gives $4\pi R_0^2\,u(P)$ in the limit as $R_0 \to 0$. Therefore

$$u(P) = \frac{1}{4\pi R^2}\oiint_{\Gamma} u\,\mathrm{d}s. \tag{2.29}$$

■

**Theorem 2.3.2** Maximum principle for the harmonic function [9, p 32].
*A harmonic function u (i.e. solution of the Laplace's equation) in an arbitrary region cannot assume an extremum value at an interior point of the region. Maximum and minimum are assumed in the interior iff the function is constant.*

*Proof.* The proof here is shown by contradiction for the case of a maximum. There is no loss of generality in doing so since if a function $\psi$ assumes a minimum at an interior point, then the function $-\psi$ assumes a maximum at that point.

Consider a subset $F$ of the closed domain $V$ plus $S$ of figure B.1 consisting of points at which $u$ assumes its maximum value $M$ in $V + S$. Since $u$ is continuous in $V + S$, $F$ is a closed set. Now if $F$ should contain an interior point $P_0$ of $V$, then by equation (2.29) of the *mean value theorem* 2.3.1, there would exist a family of spheres with a center $P_0$ of $V$ and the mean value of the function $u$ on each one of these spheres would be $u(P_0) = M$. But since $u \leqslant M$, this is possible only if $u = M$ throughout every sphere centered at $P_0$ and contained entirely in $V$. Thus, $F$ contains at least *all points* inside the largest sphere around $P_0$ which lies entirely in $V + S$. The same argument may now be repeated for any other interior point of $F$, hence $F$ must coincide with $V + S$. But this would mean that $u$ is a constant equal to $M$. Hence for any $u$ which is not a constant, $F$ can consist only of boundary points. ■

As a consequence of the maximum principle we have the following corollary.

**Corollary 2.3.2.1.** *If a regular harmonic function in V, which is continuous in V + S is constant on the boundary, then it is constant in the whole region.*

Furthermore, we have the following theorem.

**Theorem 2.3.3.** Uniqueness theorem for harmonic functions [8, p 255].
*Two harmonic functions in V which are continuous in V + S and coincide on the boundary are identical throughout.*

## References


[1] Landau L D, Lifshitz E M and Pitaevskii L P 1984 *Electrodynamics of Continuous Media* 2nd edn (New York: Pergamon)

[2] Jackson J D 1975 *Classical Electrodynamics* 2nd edn (New York: Wiley)

[3] Courant R and Hilbert D 1953 *Methods of Mathematical Physics* vol 1 (New York: Wiley)

[4] Panofsky W K H and Phillips M 1962 *Classical Electricity and Magnetism* 2nd edn (Reading, MA: Addison-Wesley)

[5] Stratton J A 1941 *Electromagnetic Theory* (New York: McGraw-Hill)

[6] Jeans J 1951 *The Mathematical Theory of Electricity and Magnetism* 5th edn (New York: Cambridge University Press)

[7] Cheng D K 1992 *Field and Wave Electromagnetics* 2nd edn (Reading, MA: Addison-Wesley)

[8] Courant R and Hilbert D 1962 *Methods of Mathematical Physics* vol 2 (New York: McGraw-Hill)

[9] Günter N M 1967 *Potential Theory and its Applications to Basic Problems in Mathematical Physics* (New York: Frederick Ungar)

# Chapter 3

# Gauge invariance for electromagnetic fields

Electromagnetic fields are uniquely determined from the specification of sources and material properties. However, in many problems it is possible to determine the electromagnetic field in terms of intermediate quantities known as potentials, which are themselves related to the sources. There is a considerable amount of latitude in finding potentials that produce a unique electromagnetic field. Commonly used potentials are the magnetic vector potential, the electric scalar potential, and their dual quantities. These are known as the Lorenz potentials and have been employed in chapter 1. To make the definition of potentials unique, additional conditions known as gauge conditions are enforced on these potentials. Some commonly used gauge conditions are the Lorenz gauge and the Coulomb gauge, which were discussed in chapter 1. Other types of potentials are the Debye potentials and the Hertzian potentials, the former of which have also been discussed in chapter 1. In this chapter we consider important theorems that allow us to construct or transform Hertzian potentials to determine the electromagnetic field using a wide variety of gauge functions. Most of the material presented here is adapted from [1].

## 3.1 Gauge invariance for general material media

We will first consider the case where medium properties are expressed in terms of electric and magnetic polarization vectors and their derivatives. These induced sources together with the impressed sources radiate in a vacuum. The parallel case of a homogenized medium, where medium permittivity and permeability are defined to take the place of polarization vectors, is treated in the next section.

### 3.1.1 Stream potentials and Hertzian potentials

Consider Maxwell's equations (1.1) and (1.2) which are repeated below with a slightly different notation

doi:10.1088/978-0-7503-1716-0ch3

$$\nabla \times \mathscr{E} = -\frac{\partial \mathscr{B}}{\partial t} - \mathscr{J}_{\mathrm{m}}; \qquad \nabla \cdot \mathscr{D} = q_{\mathrm{ev}} =: -\nabla \cdot \mathscr{Q}_{\mathrm{e}}, \tag{3.1}$$

$$\nabla \times \mathscr{H} = \frac{\partial \mathscr{D}}{\partial t} + \mathscr{J}_{\mathrm{e}}; \qquad \nabla \cdot \mathscr{B} = q_{\mathrm{mv}} =: -\nabla \cdot \mathscr{R}_{\mathrm{m}}. \tag{3.2}$$

The electric and magnetic charge densities ($q_{\mathrm{ev}}$, $q_{\mathrm{mv}}$) above are defined in terms of the *stream potentials* $\mathscr{Q}_{\mathrm{e}}$ and $\mathscr{R}_{\mathrm{m}}$ respectively, which are *particular solutions* of their defining divergence equations $\nabla \cdot \mathscr{Q}_{\mathrm{e}} = -q_{\mathrm{ev}}$ and $\nabla \cdot \mathscr{R}_{\mathrm{m}} = -q_{\mathrm{mv}}$ in equations (3.1) and (3.2). From their defining equations, the stream potentials $\mathscr{Q}_{\mathrm{e}}$ and $\mathscr{R}_{\mathrm{m}}$ may be interpreted as the *flux density of source charges*. Unlike $\mathscr{D}$ and $\mathscr{B}$ they do not have to be electromagnetic fields, also satisfying the respective modified (modified by the medium constitutive parameters) curl equations. In other words, they can be derived entirely from the source functions without regard to electromagnetic fields. One particular solution in the time-harmonic situation is $\mathbf{Q}_{\mathrm{e}} = \mathbf{J}_{\mathrm{e}}/j\omega$ and $\mathbf{R}_{\mathrm{m}} = \mathbf{J}_{\mathrm{m}}/j\omega$. Thus they vanish outside the source regions as opposed to $\mathbf{D}$ and $\mathbf{B}$ which can pervade all space.

We express the *field flux densities* ($\mathscr{D}$, $\mathscr{B}$) in terms of the electric polarization vector $\mathscr{P}$ and magnetic polarization vector[1] $\mathscr{M}$ via

$$\mathscr{D} = \varepsilon_0 \mathscr{E} + \mathscr{P}; \qquad \mathscr{B} = \mu_0(\mathscr{H} + \mathscr{M}). \tag{3.3}$$

The equations of continuity relate the electric and magnetic current densities ($\mathscr{J}_{\mathrm{e}}$, $\mathscr{J}_{\mathrm{m}}$) to the respective charge densities ($q_{\mathrm{ev}}$, $q_{\mathrm{em}}$) in the usual manner

$$\nabla \cdot \mathscr{J}_{\mathrm{e}} + \frac{\partial q_{\mathrm{ev}}}{\partial t} = 0; \quad \nabla \cdot \mathscr{J}_{\mathrm{m}} + \frac{\partial q_{\mathrm{mv}}}{\partial t} = 0. \tag{3.4}$$

Using the standard identity $\nabla \cdot \nabla \times \mathbf{F} = 0$ in the divergence equation in (3.2) we introduce the magnetic vector potential $\mathscr{A}$ such that

$$\mathscr{B} + \mathscr{R}_{\mathrm{m}} = \nabla \times \mathscr{A} \Longrightarrow \nabla \times \mathscr{E} = -\frac{\partial}{\partial t}\nabla \times \mathscr{A} + \frac{\partial \mathscr{R}_{\mathrm{m}}}{\partial t} - \mathscr{J}_{\mathrm{m}}. \tag{3.5}$$

In the latter part of equation (3.5) we set

$$\frac{\partial \mathscr{R}_{\mathrm{m}}}{\partial t} - \mathscr{J}_{\mathrm{m}} =: \frac{1}{\varepsilon_0}\nabla \times \mathscr{R}_{\mathrm{e}} \Longrightarrow \nabla \cdot \mathscr{J}_{\mathrm{m}} = -\frac{\partial q_{\mathrm{mv}}}{\partial t}. \tag{3.6}$$

Similarly using the first part of equation (3.4) and the second part of equation (3.1) we can also set

$$\frac{\partial \mathscr{Q}_{\mathrm{e}}}{\partial t} - \mathscr{J}_{\mathrm{e}} =: -\frac{1}{\mu_0}\nabla \times \mathscr{Q}_{\mathrm{m}} \Longrightarrow \nabla \cdot \mathscr{J}_{\mathrm{e}} = -\frac{\partial q_{\mathrm{ev}}}{\partial t}. \tag{3.7}$$

[1] Throughout this section the symbol $\mathscr{M}$ refers to the magnetization vector and not the magnetic current density.

Thus the equations of continuity stay intact in the presence of these stream potentials. For given current densities $\mathscr{J}_e$ and $\mathscr{J}_m$ and the stream functions $\mathscr{R}_m$ and $\mathscr{Q}_e$ (these are determined from the given electric and magnetic charge distributions via equations (3.1) and (3.2)), the additional stream potentials $\mathscr{R}_e$ and $\mathscr{Q}_m$ are once again *particular solutions* of their defining equations (3.6) and (3.7). They are dual quantities[2] as are $\mathscr{Q}_e$ and $\mathscr{R}_m$. As with $\mathscr{R}_m$ and $\mathscr{Q}_e$, they can be determined completely from the source functions without regarding them as electromagnetic fields. One particular solution that is consistent with the choice of $\mathbf{R}_m = \mathbf{J}_m/j\omega$, $\mathbf{Q}_e = \mathbf{J}_e/j\omega$ in the time-harmonic case is $\mathbf{R}_e = 0 = \mathbf{Q}_m$. Note that the functions $\mathscr{R}_m$ and $\mathscr{R}_e$ are dependent on each other as are $\mathscr{Q}_m$ and $\mathscr{Q}_e$. The given sources specify only $\nabla \cdot \mathscr{R}_m$, $\nabla \cdot \mathscr{Q}_e$, $\nabla \times \mathscr{R}_e$, and $\nabla \times \mathscr{Q}_m$. The remaining parts, namely, $\nabla \times \mathscr{R}_m$, $\nabla \times \mathscr{Q}_e$, $\nabla \cdot \mathscr{R}_e$, and $\nabla \cdot \mathscr{Q}_m$ are completely arbitrary. Using the standard vector relation that $\nabla \times \nabla\psi = 0$ in equations (3.5) and (3.6), the electric field $\mathscr{E}$ is expressed in terms of the magnetic vector potential $\mathscr{A}$ and an electric scalar potential $\psi_e$ as

$$\mathscr{E} = \frac{1}{\varepsilon_0}\mathscr{R}_e - \frac{\partial \mathscr{A}}{\partial t} - \nabla\psi_e. \tag{3.8}$$

Substituting equations (3.3), (3.5), (3.6), and (3.8) and the definition $\nabla^2\mathbf{F} = -\nabla \times \nabla \times \mathbf{F} + \nabla\nabla \cdot \mathbf{F}$, $c^2 = 1/\mu_0\varepsilon_0$ into (3.2) we obtain

$$\frac{1}{c^2}\frac{\partial^2 \mathscr{A}}{\partial t^2} - \nabla^2\mathscr{A} + \nabla\left[\nabla \cdot \mathscr{A} + \frac{1}{c^2}\dot{\psi}_e\right] = \nabla \times \mathscr{R}_m + \mu_0(\nabla \times \mathscr{M} + \dot{\mathscr{R}}_e + \dot{\mathscr{P}} + \mathscr{J}_e), \tag{3.9}$$

where a dot above a field variable is used to indicate a time derivative. Using the Lorenz gauge condition

$$\nabla \cdot \mathscr{A} = -\frac{1}{c^2}\dot{\psi}_e, \tag{3.10}$$

the definitions $\Box^2 = \frac{1}{c^2}\frac{\partial^2}{\partial t^2} - \nabla^2$ of the d'Alembertian operators and equation (3.3), the following uncoupled differential equations are obtained for the vector and scalar potentials

$$\Box^2\mathscr{A} = \mu_0\left[\mathscr{J}_e + \dot{\mathscr{P}} + \dot{\mathscr{R}}_e + \nabla \times \mathscr{M} + \frac{1}{\mu_0}\nabla \times \mathscr{R}_m\right], \tag{3.11}$$

$$\Box^2\psi_e = -\frac{1}{\varepsilon_0}\nabla \cdot (\mathscr{P} + \mathscr{Q}_e + \mathscr{R}_e). \tag{3.12}$$

Note that the Lorenz condition (3.10) involves the parameters of free space and the electric polarization and magnetic polarization vectors appear as additional sources in the differential equations for $\mathscr{A}$ and $\psi_e$. So one can continue to use the same Lorenz condition in a vacuum and as well in material media, with the caveat that in material

[2] The principle of duality is discussed in chapter 6.

media extra terms will appear in the differential equation for $\mathscr{A}$ and $\psi_e$. Similar equations can be developed for the dual potentials $\mathcal{F}$ and $\psi_m$. For field computations it is very advantageous to introduce super-potentials known as Hertzian vectors, $\mathbf{\Pi}_e$, $\mathbf{\Pi}_m$ in which the Lorenz gauge condition is built in. To this end we define

$$\psi_e = -\nabla \cdot \mathbf{\Pi}_e, \tag{3.13}$$

$$\mathscr{A} = \frac{1}{c^2}\dot{\mathbf{\Pi}}_e + \mu_0 \nabla \times \mathbf{\Pi}_m. \tag{3.14}$$

Using these in equations (3.11) and (3.12) results in the wave equations

$$\Box^2 \mathbf{\Pi}_e = \frac{1}{\varepsilon_0}(\mathscr{P} + \mathscr{Q}_e + \mathscr{R}_e) =: \frac{1}{\varepsilon_0}\boldsymbol{p}, \tag{3.15}$$

$$\Box^2 \mathbf{\Pi}_m = \frac{1}{\mu_0}(\mu_0 \mathscr{M} + \mathscr{R}_m + \mathscr{Q}_m) =: \frac{1}{\mu_0}\boldsymbol{m}. \tag{3.16}$$

The quantities $\boldsymbol{p}$ and $\boldsymbol{m}$ include both the polarization and free-charge sources. The electric and magnetic polarization vectors as well as the various stream functions appear as source functions in the differential equations for the Hertzian potentials. Hence the definitions of Hertzian potentials are valid for any arbitrary medium. In cases where the polarization vectors are dependent on the field intensities (3.15) and (3.16) reduce to integro-differential equations. The Hertzian potentials are considered 'super-potentials' because they are less discontinuous than the Lorenz potentials (the magnetic vector potential $\mathscr{A}$ and the scalar electric potential $\psi_e$) as the source functions that appear on the right-hand sides of equations (3.15) and (3.16) are time integrals of current and charge densities rather than the current and derivatives of the charge densities that appear in the differential equations for the determination of $\mathscr{A}$ and $\psi_e$. The Hertzian vectors $\mathbf{\Pi}_e$ and $\mathbf{\Pi}_m$ are dual quantities. The retarded potential solution of the potential equations are

$$\mathbf{\Pi}_e(\mathbf{r}; t) = \frac{1}{\varepsilon_0} \iiint\limits_V \frac{\boldsymbol{p}(\mathbf{r}'; t')}{|\mathbf{r} - \mathbf{r}'|} \, \mathrm{d}\mathbf{r} \tag{3.17}$$

$$\mathbf{\Pi}_m(\mathbf{r}; t) = \frac{1}{\mu_0} \iiint\limits_V \frac{\boldsymbol{m}(\mathbf{r}'; t')}{|\mathbf{r} - \mathbf{r}'|} \, \mathrm{d}\mathbf{r}', \tag{3.18}$$

where the retarded time $t' = t - |\mathbf{r} - \mathbf{r}'|/c$ and integration is performed over the volume extent, $V$, of the sources. For the time harmonic case using equation (15.12) together with equation (15.19) gives the following general expressions for the Hertzian potentials in the spherical coordinates:

$$\begin{aligned} \mathbf{\Pi}_e(\mathbf{r}) = {} & \frac{k}{4\pi j \varepsilon_0} \sum_{n=0}^{\infty} \sum_{m=-n}^{n} (2n+1)(-1)^m h_n^{(2)}(kr) P_n^m(\cos\theta) \mathrm{e}^{jm\phi} \\ & \times \iiint\limits_V \boldsymbol{p}(\mathbf{r}') P_n^{-m}(\cos\theta') j_n(kr') \mathrm{e}^{-jm\phi'} \, d\mathbf{r}', \end{aligned} \tag{3.19}$$

$$\begin{aligned}\mathbf{\Pi}_m(\mathbf{r}) = \frac{k}{4\pi j\mu_0}\sum_{n=0}^{\infty}\sum_{m=-n}^{n}(2n+1)(-1)^m h_n^{(2)}(kr)P_n^m(\cos\theta)\mathrm{e}^{jm\phi} \\ \times \iiint_V m(\mathbf{r}')P_n^{-m}(\cos\theta')j_n(kr')\mathrm{e}^{-jm\phi'}\,d\mathbf{r}'.\end{aligned} \tag{3.20}$$

Once the Hertzian potentials are obtained from the given source specifications, the field quantities are obtained from

$$\mathscr{E} = \frac{1}{\varepsilon_0}\mathscr{R}_\mathrm{e} + \nabla\nabla\cdot\mathbf{\Pi}_\mathrm{e} - \frac{1}{c^2}\ddot{\mathbf{\Pi}}_\mathrm{e} - \mu_0\nabla\times\dot{\mathbf{\Pi}}_\mathrm{m}, \tag{3.21}$$

$$= \nabla\times\nabla\times\mathbf{\Pi}_\mathrm{e} - \frac{1}{\varepsilon_0}(\mathscr{P} + \mathscr{Q}_\mathrm{e}) - \mu_0\nabla\times\dot{\mathbf{\Pi}}_\mathrm{m}, \tag{3.22}$$

$$\mathscr{B} = \mu_0\nabla\times\nabla\times\mathbf{\Pi}_\mathrm{m} - \mathscr{R}_\mathrm{m} + \frac{1}{c^2}\nabla\times\dot{\mathbf{\Pi}}_\mathrm{e} \tag{3.23}$$

$$\mathscr{D} = \varepsilon_0\nabla\times\nabla\times\mathbf{\Pi}_\mathrm{e} - \mathscr{Q}_\mathrm{e} - \frac{1}{c^2}\nabla\times\dot{\mathbf{\Pi}}_\mathrm{m} \tag{3.24}$$

$$\mathscr{H} = \frac{1}{\mu_0}\mathscr{Q}_\mathrm{m} + \nabla\nabla\cdot\mathbf{\Pi}_\mathrm{m} - \frac{1}{c^2}\ddot{\mathbf{\Pi}}_\mathrm{m} + \varepsilon_0\nabla\times\dot{\mathbf{\Pi}}_\mathrm{e} \tag{3.25}$$

$$= \nabla\times\nabla\times\mathbf{\Pi}_\mathrm{m} - \frac{1}{\mu_0}\left(\mu_0\mathscr{M} + \mathscr{R}_\mathrm{m}\right) + \varepsilon_0\nabla\times\dot{\mathbf{\Pi}}_\mathrm{e}. \tag{3.26}$$

The scalar and vector potentials, the stream potentials, and the Hertzian potentials are not unique for a given electromagnetic field. Additional constraints could be placed on the potentials $\mathscr{A}$ and/or $\psi_\mathrm{e}$ through any one of the gauge conditions. If the scalar and vector potentials are transformed according to

$$\psi_\mathrm{e} = \psi_\mathrm{e}^0 + \dot{\chi}; \qquad \mathscr{A} = \mathscr{A}^0 - \nabla\chi, \tag{3.27}$$

where $\psi_\mathrm{e}^0$, $\mathscr{A}^0$ are other possible potentials and the scalar gauge function $\chi$ is constrained to satisfy $\Box^2\chi = 0$, then the wave equations in (3.11), (3.12), the Lorenz condition (3.10), and the various fields in (3.21)–(3.26) are invariant to the transformation[3]. If the electric stream potentials $\mathscr{Q}_\mathrm{e}$, $\mathscr{Q}_\mathrm{m}$ are transformed according to

$$\mathscr{Q}_\mathrm{e} = \mathscr{Q}_\mathrm{e}^0 + \nabla\times\mathscr{G}; \qquad \mathscr{Q}_\mathrm{m} = \mathscr{Q}_\mathrm{m}^0 - \mu_0\dot{\mathscr{G}} - \nabla g, \tag{3.28}$$

where $\mathscr{Q}_\mathrm{e}^0$ and $\mathscr{Q}_\mathrm{m}^0$ are other possible potentials and the gauge functions $\mathscr{G}$ and $g$ are arbitrary, then it is easy to verify from equations (3.1) and (3.7) that the electric

[3] Even for arbitrary $\chi$, i.e. even without the constraint $\Box^2\chi = 0$, the electromagnetic fields $\mathscr{E}$, $\mathscr{B}$ remain invariant. The constraint is only required to preserve the Lorenz condition (3.10) and the wave equation character (3.11), (3.12) of $\psi_\mathrm{e}$ and $\mathscr{A}$.

charge density $q_{ev}$ and the current density $\mathscr{J}_e$ remain invariant. Similarly, if the magnetic stream potentials $\mathscr{R}_e$, $\mathscr{R}_m$ are transformed according to

$$\mathscr{R}_m = \mathscr{R}_m^0 - \nabla \times \mathscr{L}; \qquad \mathscr{R}_e = \mathscr{R}_e^0 - \varepsilon_0 \dot{\mathscr{L}} - \nabla \ell, \tag{3.29}$$

where and $\mathscr{R}_e^0$ and $\mathscr{R}_m^0$ are other possible potentials and the gauge functions $\mathscr{L}$ and $\ell$ are arbitrary, then the magnetic charge density $q_{mv}$ and the current density $\mathscr{J}_m$ remain invariant.

Since $q_{mv}$ and $\mathscr{J}_m$ are zero in reality, it is always possible to set the magnetic stream potentials to zero. Thus $\mathscr{R}_e^0$ and $\mathscr{R}_m^0$ can be chosen to be zero. If $\mathscr{Q}_m^0$ is also chosen to be zero, then the remaining stream potential is determined according to $\mathscr{Q}_e^0 = \int \mathscr{J}_e \, dt$ (from equation (3.7)). This constitutes one valid choice of stream potentials.

Since any set of stream potentials can be used, the solution of Maxwell's equation contained in equations (3.21)–(3.26) can be written in the general form as

$$\mathscr{E} = \frac{1}{\varepsilon_0}\mathscr{R}_e^0 - \dot{\mathscr{L}} - \frac{1}{\varepsilon_0}\nabla \ell + \nabla\nabla \cdot \mathbf{\Pi}_e - \frac{1}{c^2}\ddot{\mathbf{\Pi}}_e - \mu_0 \nabla \times \dot{\mathbf{\Pi}}_m, \tag{3.30}$$

$$= \nabla \times \nabla \times \mathbf{\Pi}_e - \frac{1}{\varepsilon_0}(\mathscr{P} + \mathscr{Q}_e^0) - \frac{1}{\varepsilon_0}\nabla \times \mathscr{G} - \mu_0 \nabla \times \dot{\mathbf{\Pi}}_m \tag{3.31}$$

$$\mathscr{B} = \mu_0 \nabla \times \nabla \times \mathbf{\Pi}_m - \mathscr{R}_m^0 + \nabla \times \mathscr{L} + \frac{1}{c^2}\nabla \times \dot{\mathbf{\Pi}}_e, \tag{3.32}$$

$$\mathscr{D} = \varepsilon_0 \nabla \times \nabla \times \mathbf{\Pi}_e - \mathscr{Q}_e^0 - \nabla \times \mathscr{G} - \frac{1}{c^2}\nabla \times \dot{\mathbf{\Pi}}_m \tag{3.33}$$

$$\mathscr{H} = \frac{1}{\mu_0}\mathscr{Q}_m^0 - \dot{\mathscr{G}} - \frac{1}{\mu_0}\nabla g + \nabla\nabla \cdot \mathbf{\Pi}_m - \frac{1}{c^2}\ddot{\mathbf{\Pi}}_m + \varepsilon_0 \nabla \times \dot{\mathbf{\Pi}}_e \tag{3.34}$$

$$= \nabla \times \nabla \times \mathbf{\Pi}_m - \frac{1}{\mu_0}\left(\mu_0 \mathscr{M} + \mathscr{R}_m^0\right) + \frac{1}{\mu_0}\nabla \times \mathscr{L} + \varepsilon_0 \nabla \times \dot{\mathbf{\Pi}}_e, \tag{3.35}$$

with the Hertzian potentials satisfying the general equations

$$\Box^2 \mathbf{\Pi}_e = \frac{1}{\varepsilon_0}(\mathscr{P} + \mathscr{Q}_e^0 + \mathscr{R}_e^0) + \frac{1}{\varepsilon_0}\nabla \times \mathscr{G} - \dot{\mathscr{L}} - \frac{1}{\varepsilon_0}\nabla \ell, \tag{3.36}$$

$$\Box^2 \mathbf{\Pi}_m = \frac{1}{\mu_0}\left(\mu_0 \mathscr{M} + \mathscr{R}_m^0 + \mathscr{Q}_m^0\right) - \frac{1}{\mu_0}\nabla \times \mathscr{L} - \dot{\mathscr{G}} - \frac{1}{\mu_0}\nabla g. \tag{3.37}$$

The gauge functions $\mathscr{L}$, $\mathscr{G}$, $g$, and $\ell$ do not satisfy the wave equations but remain arbitrary and serve as sources to the Hertzian potentials. The pair ($\mathscr{L}$, $\ell$) is dual with the pair ($\mathscr{G}$, $g$).

*3.1.1.1 Nisbet's theorem on gauge invariance and Hertzian potentials*

It is clear from the definition of Hertzian potentials in equations (3.13) and (3.14) that $\psi_e$ and $\mathscr{A}$ are invariant if a gauge transformation of the kind in equation (3.28) is applied to the potentials $\mathbf{\Pi}_e$ and $\mathbf{\Pi}_m$. However, $\psi_e$ and $\mathscr{A}$ are themselves not unique. Hence a great deal of arbitrariness is afforded to $\mathbf{\Pi}_e$ and $\mathbf{\Pi}_m$. It can be verified that if the Hertzian potentials are transformed according to

$$\mathbf{\Pi}_e = \mathbf{\Pi}_e^0 + \nabla \times \mathbf{\Gamma} - \mu_0 \dot{\mathbf{\Lambda}} - \nabla \lambda, \tag{3.38}$$

$$\mathbf{\Pi}_m = \mathbf{\Pi}_m^0 - \nabla \times \mathbf{\Lambda} - \varepsilon_0 \dot{\mathbf{\Gamma}} + \nabla \gamma, \tag{3.39}$$

where the gauge functions $\gamma$, $\mathbf{\Gamma}$ are arbitrary but $\lambda$, $\mathbf{\Lambda}$ satisfy

$$\Box^2 \lambda = -\mu_0 \dot{\zeta}; \qquad \Box^2 \mathbf{\Lambda} = \nabla \zeta, \tag{3.40}$$

where $\zeta$ is an arbitrary function (which can be zero), then $\psi_e$ and $\mathscr{A}$ undergo a gauge transformation of the kind in equation (3.27) with the gauge function

$$\chi = \mu_0 \nabla \cdot \mathbf{\Lambda} + \frac{1}{c^2} \dot{\lambda} + \mu_0 \zeta, \tag{3.41}$$

satisfying $\Box^2 \chi = 0$.

One can completely bypass the scalar and vector potentials and focus on the direct relationship between the electromagnetic field with the Hertzian potentials. Substituting equations (3.38) and (3.39) into equations (3.21)–(3.26) along with equations (3.28) and (3.29) it can be seen that the expressions for the electromagnetic fields will be invariant to the transformations (3.38) and (3.39) of the Hertzian potentials with *arbitrary* $\gamma$, $\mathbf{\Lambda}$, $\lambda$, and $\mathbf{\Gamma}$ provided that the stream potentials are simultaneously transformed according to equations (3.28) and (3.29) with the various gauge functions related by

$$\begin{aligned} \mu_0 \Box^2 \mathbf{\Lambda} &= \mu_0 \nabla \zeta + \mathscr{L}; \qquad \Box^2 \lambda = -\mu_0 \dot{\zeta} + \frac{1}{\varepsilon_0} \ell \\ \varepsilon_0 \Box^2 \mathbf{\Gamma} &= \varepsilon_0 \nabla \xi + \mathscr{G}; \qquad \Box^2 \gamma = \varepsilon_0 \dot{\xi} - \frac{1}{\mu_0} g. \end{aligned} \tag{3.42}$$

Thus we arrive at the following theorem proposed by Nisbet [1].

**Theorem 3.1.1** Gauge invariance and Hertzian potentials [1].

*If a given electromagnetic field can be represented by Hertzian potentials* $\mathbf{\Pi}_e^0$ *and* $\mathbf{\Pi}_m^0$ *for a given choice of stream potentials* $\mathscr{R}_e^0$, $\mathscr{R}_m^0$, $\mathscr{Q}_e^0$, *and* $\mathscr{Q}_m^0$ *as in equations (3.30)–(3.35) together with equations (3.36) and (3.37), then other stream potentials exist for which the representation of the same field is given by potentials* $\mathbf{\Pi}_e$ *and* $\mathbf{\Pi}_m$ *of equations (3.38) and (3.39) where the gauge functions* $\gamma$, $\mathbf{\Gamma}$, $\lambda$, $\mathbf{\Lambda}$ *are related to the gauge functions* $\mathscr{L}$, $\ell$, $\mathscr{G}$, $g$ *corresponding to the simultaneously transformed stream potentials* $\mathscr{Q}_e$, $\mathscr{Q}_m$, $\mathscr{R}_e$, $\mathscr{R}_m$ *of equations (3.28) and (3.29)* *through the relations (3.42). The gauge functions $\zeta$ and $\xi$ appearing in equation (3.42) are completely arbitrary.*

Note that the gauge functions are arbitrary, being connected only by equation (3.42). In particular, it is not required to impose the condition (3.40) which arises from the requirement that the electric scalar and magnetic vector potentials undergo a transformation of the kind indicated in equation (3.27). If $\lambda$ and $\mathbf{\Lambda}$ are not arbitrary but satisfy equation (3.40), then equation (3.42) shows that $\ell$ and $\mathscr{L}$ are zero so that the electromagnetic field ($\mathscr{E}$, $\mathscr{B}$) is invariant under equations (3.38) and (3.39) without any change in the stream potentials $\mathscr{R}_e$ and $\mathscr{R}_m$ (which can therefore remain zero). Similarly, if $\gamma$ and $\mathbf{\Gamma}$ are not arbitrary but satisfy $\Box^2\gamma = \varepsilon_0\dot{\xi}$ and $\Box^2\mathbf{\Gamma} = \nabla\xi$, then $g$ and $\mathscr{G}$ are zero and the field ($\mathscr{D}$, $\mathscr{H}$) is invariant under equations (3.38) and (3.39) without any change in the stream potentials $\mathscr{Q}_e$ and $\mathscr{Q}_m$.

The field expressions presented in this theorem are equally valid in two situations:

1. In an arbitrary *heterogeneous* medium wherein the impressed sources (which do not depend on the field in the medium) as well as position dependent polarization sources (which depend on the fields in the medium) appear as excitations in the non-homogeneous wave equation satisfied by the Hertzian potentials. The flexibility provided by the gauge potentials enables one to force certain components of the Hertzian potentials to zero. The expressions for the fields are valid everywhere, including in regions occupied the sources.
2. In a *homogeneous* magneto-dielectric medium (where the permittivity $\epsilon$ and the permittivity $\mu$ are constant throughout) the expressions can still be used by replacing $\varepsilon_0$ and $\mu_0$ by $\epsilon$ and $\mu$, respectively. The various sources that appear in the non-homogeneous wave equation are then only the impressed sources.

We now provide some examples where the Hertzian and gauge potentials can be chosen at will (with some restrictions) to represent the same electromagnetic field by a number of different means.

**Example 3.1.1.1** Different potential representations of the same field.
We show in this example that the same $\mathrm{TM}_z$ field can be obtained in terms of a $z$-directed electric Hertzian potential or a $\phi$-directed magnetic Hertzian potential. Consider a $z$-directed time-harmonic, infinitesimal electric dipole located at the origin with a dipole moment $p_0$ (units A m). Let the surrounding medium be free space. We work directly with phasor quantities and set $\partial/\partial t = j\omega$, $\Box^2 = -(\nabla^2 + k^2)$, where $k = \omega/c$ is the free-space wavenumber. As usual script letters in the time domain are replaced with uppercase letters in the phasor domain for the various field quantities. The electric source current density is $\mathbf{J}_e = \hat{\mathbf{z}}p_0\delta(\mathbf{r})$. All other source densities are zero. Hence we take $\mathbf{Q}_m^0 = 0 = \mathbf{R}_m^0 = \mathbf{R}_e^0$ and $\mathbf{Q}_e^0 = \hat{\mathbf{z}}p_0\delta(\mathbf{r})/j\omega$. The solution of equations (3.15) and (3.16) with $\mathscr{P} = \mathbf{0} = \mathscr{M}$ is

$$\mathbf{\Pi}_{\mathrm{e}}^{0} = \hat{\mathbf{z}}\frac{p_0}{j\omega\varepsilon_0}\frac{\mathrm{e}^{-jkr}}{4\pi r} =: \hat{\mathbf{z}}N_{\mathrm{e}}; \qquad \mathbf{\Pi}_{\mathrm{m}}^{0} = 0. \tag{3.43}$$

The electric field as given by equation (3.21) or (3.22) is

$$\mathbf{E} = k^2\mathbf{\Pi}_{\mathrm{e}}^{0} + \nabla\nabla\cdot\mathbf{\Pi}_{\mathrm{e}}^{0} = -\frac{\mathbf{Q}_{\mathrm{e}}^{0}}{\varepsilon_0} + \nabla\times\nabla\times\mathbf{\Pi}_{\mathrm{e}}^{0}. \tag{3.44}$$

We will now show that the same electric field is obtained by a magnetic Hertzian potential together with non-zero gauge functions. Taking the gauge functions $\mathbf{\Gamma} = 0 = \lambda = \gamma$ and $\mathbf{\Lambda} = \mathbf{\Pi}_{\mathrm{e}}^{0}/j\omega\mu_0$ in equations (3.38) and (3.39) results in $\mathbf{\Pi}_{\mathrm{e}} = 0$ and

$$\mathbf{\Pi}_{\mathrm{m}} = -\nabla\times\mathbf{\Lambda} = \frac{-1}{j\omega\mu_0}\nabla\times\mathbf{\Pi}_{\mathrm{e}}^{0} = \hat{\boldsymbol{\phi}}\frac{\sin\theta}{j\omega\mu_0}\frac{\partial N_{\mathrm{e}}}{\partial r}.$$

We also choose $\mathbf{G} = 0 = \xi = \zeta = \ell$ and determine the remaining gauge function $\mathbf{L}$ using equation (3.42) to yield

$$\mathbf{L} = -\mu_0(\nabla^2 + k^2)\mathbf{\Lambda} = \frac{-1}{j\omega}(\nabla^2 + k^2)\mathbf{\Pi}_{\mathrm{e}}^{0} = \frac{\mathbf{Q}_{\mathrm{e}}^{0}}{j\omega\varepsilon_0}.$$

Using these in equation (3.30) gives

$$\mathbf{E} = -j\omega\mathbf{L} - j\omega\mu_0\nabla\times\mathbf{\Pi}_{\mathrm{m}} = -\frac{\mathbf{Q}_{\mathrm{e}}^{0}}{\varepsilon_0} + \nabla\times\nabla\times\mathbf{\Pi}_{\mathrm{e}}^{0},$$

which is the same as equation (3.44). ■■

**Example 3.1.1.2.** $\mathrm{TE}_k/\mathrm{TM}_k$ decomposition of field.
We would like to show here that the electromagnetic field can be completely represented in terms of two scalar functions (Cartesian components along a fixed direction $\hat{\mathbf{k}}$) in regions including the source. Let $\mathbf{\Pi}_{\mathrm{e}} = \hat{\mathbf{k}}\Pi_{\mathrm{e}}$, $\mathbf{\Pi}_{\mathrm{m}} = \hat{\mathbf{k}}\Pi_{\mathrm{m}}$ in the presence of sources $\mathscr{P}$, $\mathscr{J}$, and $\mathscr{M}$. For now we will assume $\hat{\mathbf{k}} = \hat{\mathbf{z}}$, but other choices of unit vectors are also possible such as $\hat{\mathbf{k}} = \hat{\mathbf{x}}$ or $\hat{\mathbf{k}} = \hat{\mathbf{y}}$. The functions $\Pi_{\mathrm{e}}$ and $\Pi_{\mathrm{m}}$ are one-component electric and magnetic Hertzian potentials. We choose $\mathscr{R}_{\mathrm{e}}^{0} = 0 = \mathscr{R}_{\mathrm{m}}^{0} = \mathscr{Q}_{\mathrm{m}}^{0}$, $\mathscr{G}_x = 0 = \mathscr{G}_y = \mathscr{L}_x = \mathscr{L}_y$, and $\mathscr{Q}_{\mathrm{e}}^{0} = \int\mathscr{J}_{\mathrm{e}}\,\mathrm{d}t$. The non-homogeneous wave equations satisfied by the components of Hertzian potentials are from equations (3.36) and (3.37)

$$\Box^2\Pi_{\mathrm{e}} = \frac{1}{\varepsilon_0}(\mathscr{P}_z + \mathscr{Q}_{\mathrm{e}z}^{0}) - \dot{\mathscr{L}}_z - \frac{1}{\varepsilon_0}\frac{\partial\ell}{\partial z}, \tag{3.45}$$

$$\Box^2\Pi_{\mathrm{m}} = \mathscr{M}_z - \dot{\mathscr{G}}_z - \frac{1}{\mu_0}\frac{\partial g}{\partial z}, \tag{3.46}$$

together with

$$\frac{\partial \ell}{\partial x} - \frac{\partial \mathcal{G}_z}{\partial y} = \mathcal{Q}^0_{\mathrm{e}x} + \mathcal{P}_x, \tag{3.47}$$

$$\frac{\partial \ell}{\partial y} + \frac{\partial \mathcal{G}_z}{\partial x} = \mathcal{Q}^0_{\mathrm{e}y} + \mathcal{P}_y, \tag{3.48}$$

$$\frac{\partial g}{\partial x} + \frac{\partial \mathcal{L}_z}{\partial y} = \mu_0 \mathcal{M}_x, \tag{3.49}$$

$$\frac{\partial g}{\partial y} - \frac{\partial \mathcal{L}_z}{\partial x} = \mu_0 \mathcal{M}_y. \tag{3.50}$$

Defining the transverse del operator $\nabla_{\mathrm{t}} = \nabla - \hat{\mathbf{z}}\partial/\partial z$ and the transverse Laplacian $\nabla_{\mathrm{t}}^2 = \nabla^2 - \partial^2/\partial z^2$, equations (3.47)–(3.50) may be combined to yield an equivalent set

$$\nabla_{\mathrm{t}}^2 \mathcal{G}_z = \hat{\mathbf{z}} \cdot \nabla \times \left(\mathcal{P} + \mathcal{Q}^0_{\mathrm{e}}\right), \tag{3.51}$$

$$\nabla_{\mathrm{t}}^2 \mathcal{L}_z = -\mu_0 \hat{\mathbf{z}} \cdot \nabla \times \mathcal{M}, \tag{3.52}$$

$$\nabla_{\mathrm{t}}^2 \ell = \nabla_{\mathrm{t}} \cdot \left(\mathcal{P} + \mathcal{Q}^0_{\mathrm{e}}\right), \tag{3.53}$$

$$\nabla_{\mathrm{t}}^2 g = \mu_0 \nabla_{\mathrm{t}} \cdot \mathcal{M}. \tag{3.54}$$

After the gauge potentials $\mathcal{G}_z$, $\mathcal{L}_z$, $\ell$, and $g$ are determined from the *particular solutions* of equations (3.51), (3.52), (3.53), and (3.54), respectively, they are substituted into equations (3.45) and (3.46) to find $\Pi_{\mathrm{e}}$ and $\Pi_{\mathrm{m}}$. The electromagnetic field is then given by

$$\mathcal{E} = -\hat{\mathbf{z}}\dot{\mathcal{L}}_z - \frac{1}{\varepsilon_0}\nabla \ell + \nabla\nabla \cdot (\hat{\mathbf{z}}\Pi_{\mathrm{e}}) - \frac{1}{c^2}\hat{\mathbf{z}}\ddot{\Pi}_{\mathrm{e}} - \mu_0 \nabla \times (\hat{\mathbf{z}}\Pi_{\mathrm{m}}), \tag{3.55}$$

$$\mathcal{B} = \mu_0 \nabla \times \nabla \times (\hat{\mathbf{z}}\Pi_{\mathrm{m}}) + \nabla \times (\hat{\mathbf{z}}\mathcal{L}_z) + \frac{1}{c^2}\nabla \times (\hat{\mathbf{z}}\dot{\Pi}_{\mathrm{e}}). \tag{3.56}$$

The purely $\mathrm{TM}_z$ field is obtained when $\Pi_{\mathrm{m}} \equiv 0$. This can happen when $\mathcal{M} \equiv 0$ and if $\mathcal{P}$ and $\mathcal{J}$ have only $z$-components. However, other situations can arise when $\Pi_{\mathrm{m}} \equiv 0$ even when $\mathcal{M} \neq 0$. Likewise the purely $\mathrm{TE}_z$ field is obtained when $\Pi_{\mathrm{e}} \equiv 0$. ■■

**Example 3.1.1.3** Radial Hertzian potentials.
In this example we show that the electromagnetic field can also be completely represented in terms of two radial electric and magnetic Hertzian vector potentials. Let $\mathbf{\Pi}_{\mathrm{e}} = \hat{\mathbf{r}}\Pi_1$, $\mathbf{\Pi}_{\mathrm{m}} = \hat{\mathbf{r}}\Pi_2$. Using elementary vector analysis we can show that $-\nabla^2(\mathbf{r}\Pi_{1/2}) = -\mathbf{r}\nabla^2\Pi_{1/2} - 2\nabla\Pi_{1/2}$. Substituting these into equations (3.36) and

(3.37) and performing a scalar product with $\mathbf{r}$ we obtain the non-homogeneous wave equations for the scalar functions $\Pi_1$ and $\Pi_2$ as

$$\Box^2\Pi_1 = \frac{1}{\varepsilon_0 r}\left(\mathscr{P}_r + \mathscr{Q}^0_{er} + \mathscr{R}^0_{er}\right) - \frac{\dot{\mathscr{L}}_r}{r} - \frac{1}{\varepsilon_0 r}\frac{\partial \ell}{\partial r} + \frac{2}{r}\frac{\partial \Pi_1}{\partial r}, \tag{3.57}$$

$$\Box^2\Pi_2 = \frac{1}{\mu_0 r}\left(\mu_0\mathscr{M}_r + \mathscr{R}^0_{mr} + \mathscr{Q}^0_{mr}\right) - \frac{\dot{\mathscr{G}}_r}{r} - \frac{1}{\mu_0 r}\frac{\partial g}{\partial r} + \frac{2}{r}\frac{\partial \Pi_2}{\partial r}, \tag{3.58}$$

where we have assumed the choice $\mathscr{G}_\theta = 0 = \mathscr{G}_\phi = \mathscr{L}_\theta = \mathscr{L}_\phi$. The last two terms on the rhs of equations (3.57) and (3.58) may be combined and expressed as $-(1/\varepsilon_0 r)\partial\ell'/\partial r$ and $-(1/\mu r)\partial g'/\partial r$, respectively, where the modified gauge functions are $\ell' = \ell - 2\varepsilon_0\Pi_1$, $g' = g - 2\mu_0\Pi_2$. With the additional choice of $\mathscr{Q}^0_{\rm m} = 0 = \mathscr{R}^0_{\rm e} = \mathscr{R}^0_{\rm m}$, the other gauge functions are particular solutions of

$$-\frac{1}{r\sin\theta}\frac{\partial \mathscr{G}_r}{\partial \phi} + \frac{1}{r}\frac{\partial \ell'}{\partial \theta} = \mathscr{P}_\theta + \mathscr{Q}^0_{e\theta}, \tag{3.59}$$

$$\frac{1}{r\sin\theta}\frac{\partial \ell'}{\partial \phi} + \frac{1}{r}\frac{\partial \mathscr{G}_r}{\partial \theta} = \mathscr{P}_\phi + \mathscr{Q}^0_{e\phi}, \tag{3.60}$$

$$\frac{1}{r\sin\theta}\frac{\partial \mathscr{L}_r}{\partial \phi} + \frac{1}{r}\frac{\partial g'}{\partial \theta} = \mu_0\mathscr{M}_\theta, \tag{3.61}$$

$$\frac{1}{r\sin\theta}\frac{\partial g'}{\partial \phi} - \frac{1}{r}\frac{\partial \mathscr{L}_r}{\partial \theta} = \mu_0\mathscr{M}_\phi. \tag{3.62}$$

As in the previous example an equivalent set of equations can be obtained for the gauge functions $\mathscr{G}_r$, $\mathscr{L}_r$, $\ell'$, $g$ by combining equations (3.59)–(3.62):

$$\nabla^2_{\rm t}\mathscr{G}_r = \hat{\mathbf{r}}\cdot\nabla\times\left(\mathscr{P} + \mathscr{Q}^0_{\rm e}\right), \tag{3.63}$$

$$\nabla^2_{\rm t}\mathscr{L}_r = -\mu_0\hat{\mathbf{r}}\cdot\nabla\times\mathscr{M}, \tag{3.64}$$

$$\nabla^2_{\rm t}\ell' = \nabla_{\rm t}\cdot\left(\mathscr{P} + \mathscr{Q}^0_{\rm e}\right), \tag{3.65}$$

$$\nabla^2_{\rm t}g' = \mu_0\nabla_{\rm t}\cdot\mathscr{M}, \tag{3.66}$$

where the transverse Laplacian $\nabla^2_{\rm t}A = \nabla^2 A - (1/r^2)\partial/\partial r[r^2\partial A/\partial r]$ and the transverse divergence $\nabla_{\rm t}\cdot\mathbf{A} = \nabla\cdot\mathbf{A} - (1/r^2)\partial(r^2A_r)/\partial r$. Once the Hertzian potentials and the gauge functions are determined, the electric and magnetic fields are given by

$$\mathscr{E} = \nabla\left(\frac{\partial U}{\partial r}\right) - \frac{\hat{\mathbf{r}}}{c^2}\ddot{U} + \mu_0\hat{\mathbf{r}}\times\nabla\dot{V} - \hat{\mathbf{r}}\dot{\mathscr{L}}_r - \frac{\nabla\ell'}{\varepsilon_0}, \tag{3.67}$$

$$\mathscr{B} = \mu_0 \nabla \times \nabla \times (\hat{\mathbf{r}} V) - \frac{1}{c^2} \hat{\mathbf{r}} \times \nabla \dot{U} - \hat{\mathbf{r}} \times \nabla \mathscr{L}_r, \tag{3.68}$$

where $U = r\Pi_1$ and $V = r\Pi_2$ are termed the *Debye* potentials. When $U = 0$ the resultant field is $\mathrm{TE}_r$. Likewise, when $V = 0$, the corresponding field is $\mathrm{TM}_r$. ■■

**Example 3.1.1.4** Transformation of $\mathrm{TM}_z$ potentials to $\mathrm{TM}_r$ potentials.
In this example we show how gauge transformations can be used to convert a representation of the potential given in one coordinate system to that in the other [2]. Consider a $z$-directed time-harmonic, infinitesimal electric dipole located on the $z$-axis at some height $r'$ and having a dipole moment $p_0$. Let the surrounding medium be free space. As in example 3.1.1.1, we work directly with phasor quantities and set $\partial/\partial t = j\omega$, $\Box^2 = -(\nabla^2 + k^2)$, where $k = \omega/c$ is the free-space wavenumber. As usual script letters in the time domain are replaced with uppercase letters in the phasor domain for the various field quantities. The electric source current density is $\mathbf{J}_\mathrm{e} = \hat{\mathbf{z}}\, p_0 \delta(\mathbf{r} - \hat{\mathbf{z}}r')$. All other source densities are zero. Under the Lorenz condition the magnetic vector potential is equal to

$$\mathbf{A} = \hat{\mathbf{z}} \frac{\mu k p_0}{4\pi j} \sum_{n=0}^{\infty} (2n + 1) h_n^{(2)}(kr_>) j_n(kr_<) P_n(\cos\theta) =: \hat{\mathbf{z}} \mu \psi, \tag{3.69}$$

where $(r, \theta, \phi)$ are the usual spherical coordinates, $r_< = \min(r, r')$, $r_> = \max(r, r')$, $j_n(\cdot)$, $h_n^{(2)}(\cdot)$ are, respectively, the spherical Bessel function of the first kind and Hankel function of the second kind of, respectively, order $n$ and $P_n(\cdot)$ is the Legendre polynomial of degree $n$. Note that the vector potential is independent of the azimuth angle $\phi$, suggesting $\partial/\partial\phi = 0$. We take $\mathbf{Q}_\mathrm{m}^0 = 0 = \mathbf{R}_\mathrm{m}^0 = \mathbf{R}_\mathrm{e}^0$ and $\mathbf{Q}_\mathrm{e}^0 = \hat{\mathbf{z}} p_0 \delta(\mathbf{r} - \hat{\mathbf{z}}r')/j\omega$. The solution of equations (3.15) and (3.16) with $\mathscr{P} = \mathbf{0} = \mathscr{M}$ (or directly from equation (3.14)) is

$$\mathbf{\Pi}_\mathrm{e}^0 = \hat{\mathbf{z}} \frac{1}{j\omega\varepsilon} \psi = (\hat{\mathbf{r}} \cos\theta - \hat{\boldsymbol{\theta}} \sin\theta) \frac{1}{j\omega\varepsilon} \psi; \qquad \mathbf{\Pi}_\mathrm{m}^0 = 0. \tag{3.70}$$

Note that the electric Hertzian potential $\mathbf{\Pi}_\mathrm{e}^0$ has both a radial component and a $\theta$-component. The magnetic field intensity of this $\mathrm{TM}_z$ source in the cylindrical coordinates $(\rho, \phi, z)$ is

$$\mathbf{H} = \frac{1}{\mu} \nabla \times \mathbf{A} = -\hat{\boldsymbol{\phi}} \frac{\partial \psi}{\partial \rho} = -\hat{\boldsymbol{\phi}} \frac{1}{r} \left( r \sin\theta \frac{\partial \psi}{\partial r} + \cos\theta \frac{\partial \psi}{\partial \theta} \right), \tag{3.71}$$

where the latter follows on implementing the cylindrical coordinate radial derivative $\partial/\partial\rho$ in spherical coordinates. An explicit expression for the magnetic field for $r > r'$ is

$$\begin{aligned} H_\phi = {} & \frac{jkp_0}{4\pi r} \sum_{n=0}^{\infty} (2n + 1) j_n(kr') \\ & \times \left[ kr \sin\theta P_n(\cos\theta) h_n^{(2)\prime}(kr) + h_n^{(2)}(kr) \cos\theta \frac{d}{d\theta} P_n(\cos\theta) \right]. \end{aligned} \tag{3.72}$$

We would now like to construct a new Hertzian electric potential $\mathbf{\Pi}_e$ that has only a radial component such that the field is classified as $TM_r$ in spherical coordinates. To this end we choose $\mathbf{\Lambda} = 0 = \mathbf{\Gamma}$ and $\gamma = 0$ in equations (3.38) and (3.39) so that the magnetic Hertzian potential continues to be zero and $\mathbf{\Pi}_e = \mathbf{\Pi}_e^0 - \nabla\lambda$. We also choose $\zeta = 0$ and $\ell = 0$ in equations (3.42). The gauge function $\lambda$ is chosen so that the $\theta$-component of $\mathbf{\Pi}_e$ becomes zero. Hence

$$-\frac{\sin\theta\psi}{j\omega\varepsilon} - \frac{1}{r}\frac{\partial\lambda}{\partial\theta} = 0 \Longrightarrow \frac{\partial\lambda}{\partial\theta} = -\frac{r\sin\theta\psi}{j\omega\varepsilon}. \tag{3.73}$$

The transformed Hertzian electric potential is

$$\mathbf{\Pi}_e = \hat{\mathbf{r}}\left[\frac{\cos\theta\psi}{j\omega\varepsilon} - \frac{\partial\lambda}{\partial r}\right] = \hat{\mathbf{r}}\Pi_e, \tag{3.74}$$

which picks up an additional radial component from the gauge function $\lambda$. The $\phi$-component of $\mathbf{\Pi}_e$ is zero because equation (3.73) also suggests that $\partial\lambda/\partial\phi = 0$. The transformed Hertzian potentials will generate a $TM_r$ field. Inserting all of these into equation (3.35) the magnetic field is

$$\begin{aligned} \mathbf{H} = j\omega\varepsilon\nabla\times\mathbf{\Pi}_e &= -\hat{\boldsymbol{\phi}}\frac{j\omega\varepsilon}{r}\frac{\partial\Pi_e}{\partial\theta} \\ &= -\frac{\hat{\boldsymbol{\phi}}}{r}\left[\frac{\partial(\cos\theta\psi)}{\partial\theta} - j\omega\varepsilon\frac{\partial^2\lambda}{\partial\theta\partial r}\right] \\ &= -\frac{\hat{\boldsymbol{\phi}}}{r}\left[\frac{\partial(\cos\theta\psi)}{\partial\theta} + \frac{\partial(r\sin\theta\psi)}{\partial r}\right] \\ &= -\frac{\hat{\boldsymbol{\phi}}}{r}\left[\cos\theta\frac{\partial\psi}{\partial\theta} + r\sin\theta\frac{\partial\psi}{\partial r}\right], \end{aligned} \tag{3.75}$$

which is the same as equation (3.71). From equation (3.42) the gauge function $\lambda$ satisfies the homogeneous Helmholtz equation subject to equation (3.73). We now demonstrate the determination of $\lambda$ for $r > r'$. The case for $r < r'$ is similar. Letting

$$\lambda = \frac{1}{j\omega\varepsilon}\frac{kp_0}{4\pi j}\sum_{\nu=0}^{\infty} a_\nu(2\nu+1)j_\nu(kr')h_\nu^{(2)}(kr)P_\nu(\cos\theta) =: \frac{1}{j\omega\varepsilon}\frac{1}{4\pi j}\Phi,$$

and substituting in equation (3.73) we arrive at

$$\begin{aligned} &\sum_{\nu=1}^{\infty} a_\nu(2\nu+1)j_\nu(kr')h_\nu^{(2)}(kr)\frac{d}{d\theta}P_\nu(\cos\theta) \\ &\qquad = -r\sum_{n=0}^{\infty}(2n+1)j_n(kr')h_n^{(2)}(kr)\sin\theta P_n(\cos\theta), \end{aligned}$$

for the determination of the constants $a_\nu$. Multiplying both sides by $\sin\theta \mathrm{d}P_m(\cos\theta)/\mathrm{d}\theta$, $m = 1, 2, \ldots$ and integrating over $\theta \in (0, \pi)$ and using the orthogonality relation

$$\int_0^\pi \sin\theta \frac{\mathrm{d}}{\mathrm{d}\theta}P_m(\cos\theta)\frac{\mathrm{d}}{\mathrm{d}\theta}P_\nu(\cos\theta)\,\mathrm{d}\theta = \frac{2m(m+1)}{2m+1}\delta_\nu^m, \tag{3.76}$$

together with [3, p 8.914–2]

$$\sin\theta\frac{\mathrm{d}}{\mathrm{d}\theta}P_m(\cos\theta) = \frac{m(m+1)}{2m+1}[P_{m+1}(\cos\theta) - P_{m-1}(\cos\theta)], \tag{3.77}$$

and

$$\int_0^\pi \sin\theta P_\nu(\cos\theta)P_m(\cos\theta)\,\mathrm{d}\theta = \frac{2\delta_\nu^m}{2m+1}, \tag{3.78}$$

we obtain

$$(2m+1)a_m j_m(kr')h_m^{(2)}(kr) = -r\left[j_{m+1}(kr')h_{m+1}^{(2)}(kr) - j_{m-1}(kr')h_{m-1}^{(2)}(kr)\right]. \tag{3.79}$$

Therefore the gauge function reduces to

$$\begin{aligned}\frac{1}{kp_0}\Phi &= a_0 j_0(kr')h_0^{(2)}(kr)\\ &\quad + r\sum_{m=1}^{\infty}\left[j_{m-1}(kr')h_{m-1}^{(2)}(kr) - j_{m+1}(kr')h_{m+1}^{(2)}(kr)\right]P_m(\cos\theta)\\ &= a_0 j_0(kr')h_0^{(2)}(kr) + r\sum_{m=0}^{\infty} j_m(kr')h_m^{(2)}(kr)P_{m+1}(\cos\theta)\\ &\quad - r\sum_{m=2}^{\infty} j_m(kr')h_m^{(2)}(kr)P_{m-1}(\cos\theta).\end{aligned} \tag{3.80}$$

The coefficient $a_0$ is yet undefined and is immaterial as far the electromagnetic field is concerned. But we will determine it shortly to result in a compact expression for $\Pi_e$. Using the identities $(2m+1)\cos\theta P_m(\cos\theta) = (m+1)P_{m+1}(\cos\theta) + mP_{m-1}(\cos\theta)$, $\hat{B}_n(z) = zb_n(z)$ relating the Ricatti spherical Bessel function $\hat{B}_n(z)$ and the spherical Bessel function $b_n(z)$, equation (3.80) in equation (3.74) we arrive at

$$
\begin{aligned}
\frac{-4\pi\omega\varepsilon}{kp_0}\Pi_e &= ka_0 j_0(kr')h_1^{(2)}(kr) - j_1(kr')\hat{H}_1^{(2)'}(kr) \\
&\quad - \sum_{m=0}^{\infty} j_m(kr')P_{m+1}(\cos\theta)\left[\hat{H}_m^{(2)'}(kr) - \frac{(m+1)}{kr}\hat{H}_m^{(2)}(kr)\right] \\
&\quad + \sum_{m=1}^{\infty} j_m(kr')P_{m-1}(\cos\theta)\left[\frac{m}{kr}\hat{H}_m^{(2)}(kr) + \hat{H}_m^{(2)'}(kr)\right] \\
&= ka_0 j_0(kr')h_1^{(2)}(kr) - j_1(kr')\hat{H}_1^{(2)'}(kr) \\
&\quad + \sum_{m=0}^{\infty} j_m(kr')P_{m+1}(\cos\theta)\hat{H}_{m+1}^{(2)}(kr) \\
&\quad + \sum_{m=1}^{\infty} j_m(kr')P_{m-1}(\cos\theta)\hat{H}_{m-1}^{(2)}(kr) \\
&= ka_0 j_0(kr')h_1^{(2)}(kr) - j_1(kr')\hat{H}_1^{(2)'}(kr) + j_1(kr')\hat{H}_0^{(2)}(kr) \\
&\quad + \sum_{m=1}^{\infty}\left[j_{m+1}(kr') + j_{m-1}(kr')\right]\hat{H}_m^{(2)}(kr)P_m(\cos\theta) \\
&= ka_0 j_0(kr')h_1^{(2)}(kr) + j_1(kr')h_1^{(2)}(kr) \\
&\quad + \sum_{m=1}^{\infty}(2m+1)\frac{j_m(kr')}{kr'}\hat{H}_m^{(2)}(kr)P_m(\cos\theta) \\
&= \frac{1}{(kr')^2}\sum_{m=1}^{\infty}(2m+1)\hat{J}_m(kr')\hat{H}_m^{(2)}(kr)P_m(\cos\theta),
\end{aligned}
\tag{3.81}
$$

on choosing $ka_0 = -j_1(kr')/j_0(kr')$. In arriving at the final formula (3.81) we have repeatedly made use of the identities $\hat{B}_n'(z) = (n+1)\hat{B}_n(z)/z - \hat{B}_{n+1}(z) = \hat{B}_{n-1}(z) - n\hat{B}_n(z)/z$ valid for any Ricatti spherical Bessel function $\hat{B}_n(z)$. The radial Hertzian electric potential of a $z$-directed electric dipole of moment $p_0$ located at a height $r'$ along the $z$-axis is then

$$
\begin{aligned}
\Pi_e &= \frac{1}{4\pi j}\frac{1}{j\omega\varepsilon}\frac{kp_0}{(kr')^2}\sum_{m=1}^{\infty}(2m+1)\hat{J}_m(kr')\hat{H}_m^{(2)}(kr)P_m(\cos\theta), \qquad r > r' \\
&= \frac{1}{4\pi j}\frac{1}{j\omega\varepsilon}\frac{krp_0}{r'}\sum_{m=1}^{\infty}(2m+1)j_m(kr_<)h_m^{(2)}(kr_>)P_m(\cos\theta).
\end{aligned}
\tag{3.82}
$$

From the first of equation (3.75) the magnetic field generated by this TM$_r$ source for $r > r'$ is

$$
H_\phi = \frac{jkp_0}{4\pi r'}\sum_{m=1}^{\infty}(2m+1)j_m(kr')h_m^{(2)}(kr)\frac{\mathrm{d}}{\mathrm{d}\theta}P_m(\cos\theta). \tag{3.83}
$$

The two expressions (3.72) and (3.83), although seemingly different, represent the same magnetic field as guaranteed by the gauge transformations. ■■

### 3.1.2 Summary for linear media

Abstracting from examples 3.1.1.2 through 3.1.1.4 we can state the following.

*The stream potentials, $\mathscr{Q}_e$, $\mathscr{Q}_m$, $\mathscr{R}_m$, and $\mathscr{R}_e$ of the non-homogeneous Maxwell's equations may be chosen such that the fields ($\mathscr{E}$, $\mathscr{B}$) at any arbitrary point in space can be expressed in terms of only two scalar components of the Hertzian potentials. Representations such as the following can be obtained:*

(i) *Two-components of $\mathbf{\Pi}_e$: A choice $\gamma = 0$, $\mathbf{\Lambda} = 0$, and $\varepsilon\dot{\mathbf{\Gamma}} = \mathbf{\Pi}_m^0$ gives $\mathbf{\Pi}_m = 0$; gauge function $\lambda$ can be chosen to reduce one component of $\mathbf{\Pi}_e$ to zero.*

(ii) *Similar remarks apply to two-component $\mathbf{\Pi}_m$.*

(iii) *One component each of $\mathbf{\Pi}_e$ and $\mathbf{\Pi}_m$: This can be accomplished by using two radial components as in example 3.1.1.3, or two parallel Cartesian components of $\mathbf{\Pi}_e$ and $\mathbf{\Pi}_m$ as in example 2, or choose $\lambda$, $\gamma$, $\Lambda_x$, $\Gamma_y$ to give a representation with $\mathbf{\Pi}_e$ having an x-component only and with $\mathbf{\Pi}_m$ having a y-component only. In the last case the gauge functions are determined from $\partial\lambda/\partial y = \Pi_{ey}^0$; $\partial\Gamma_y/\partial x = -\Pi_{ez}^0 + \partial\lambda/\partial z$; $\partial\gamma/\partial x = -\Pi_{mx}^0$; and $\partial\Lambda_x/\partial y = -\Pi_{mz}^0 - \partial\gamma/\partial z$. Other such combinations are possible.*

## 3.2 Gauge invariance in homogenized media

Even though the gauge invariance theory presented in the previous section is general enough to apply to any material medium, the disadvantage is that when solving for the Hertzian potentials via equations (3.15) and (3.16), the unknown appears both on the left-hand and right-hand sides. This is because the electric polarization and magnetic polarization vectors are expressed in terms of the fields, which are themselves expressed in terms of the Hertzian potentials. Hence one encounters a *system of coupled second order equations* involving both $\mathbf{\Pi}_e$ and $\mathbf{\Pi}_m$, which is undesirable. It is possible to develop the invariance theory for heterogeneous, non-conducting media directly along the lines of the previous section and we show the theory here following Nisbet [4].

We assume that the permittivity and permeability tensors, $\varepsilon = \varepsilon_0\varepsilon_r$ and $\mu = \mu_0\mu_r$, are specified as a function of position $\mathbf{r}$, where $\varepsilon_0$ and $\mu_0$ are the free-space values. The tensors are further assumed to be non-singular so that the inverses $\varepsilon_r^{-1}$ and $\mu_r^{-1}$ exist. For isotropic media $\varepsilon_r$ and $\mu_r$ reduce to scalar quantities. We write $\mathscr{D} = \varepsilon_0\varepsilon_r\mathscr{E}$, $\mathscr{B} = \mu_0\mu_r\mathscr{H}$, $c^2 = 1/\varepsilon_0\mu_0$ and introduce the stream functions via

$$\begin{aligned} \mathscr{R}_m - \mathscr{J}_m &= \frac{1}{\varepsilon_0}\nabla \times \varepsilon_r^{-1}\mathscr{R}_e; \quad q_{ev} = -\nabla \cdot \mathscr{Q}_e \\ \dot{\mathscr{Q}}_e - \mathscr{J}_e &= -\frac{1}{\mu_0}\nabla \times \mu_r^{-1}\mathscr{Q}_m; \; q_{mv} = -\nabla \cdot \mathscr{R}_m. \end{aligned} \tag{3.84}$$

Note that the stream potentials $\mathscr{Q}_e$, $\mathscr{Q}_m$, $\mathscr{R}_e$, $\mathscr{R}_m$ are all particular solutions of the governing source equations and that the given sources specify only $\nabla \cdot \mathscr{Q}_e$, $\nabla \cdot \mathscr{R}_m$, $\nabla \times (\varepsilon_r^{-1}\mathscr{R}_e)$, and $\nabla \times (\mu_r^{-1}\mathscr{Q}_m)$. The remaining parts of these stream functions are completely arbitrary. The quantities $\mathscr{R}_m$ and $\varepsilon_r^{-1}\mathscr{R}_e$ are dependent on each other via $\mathscr{J}_m$ as are $\mathscr{Q}_e$ and $\mu_r^{-1}\mathscr{Q}_m$ via $\mathscr{J}_e$.[4] For example, if $\mathscr{R}_e$ and $\mathscr{Q}_m$ are chosen to be zero in a non-static case, then $\dot{\mathscr{R}}_m = \mathscr{J}_m$ and $\dot{\mathscr{Q}}_e = \mathscr{J}_e$. When $\varepsilon_r = \mathbf{I} = \mu_r$, where $\mathbf{I}$ is a unit dyad, then the definitions in equation (3.84) reduce to equations (3.6) and (3.7). The magnetic and electric fields are expressed through the vector potential $\mathscr{A}$ and the scalar potential $\psi_e$ analogous to equation (3.5):

$$\mathscr{B} + \mathscr{R}_m = \nabla \times \mathscr{A} \Longrightarrow \mathscr{E} = -\nabla\psi_e - \dot{\mathscr{A}} + \frac{1}{\varepsilon_0}\varepsilon_r^{-1}\mathscr{R}_e. \tag{3.85}$$

On imposing a *generalized Lorenz condition*[5]

$$\nabla \cdot (\varepsilon_r \mathscr{A}) = -\frac{1}{c^2}\dot{\psi}_e, \tag{3.86}$$

the differential equations for the potentials uncouple as in equations (3.11) and (3.12):

$$\frac{1}{c^2}\varepsilon_r\ddot{\mathscr{A}} + \nabla \times \mu_r^{-1}\nabla \times \mathscr{A} - \varepsilon_r\nabla\nabla \cdot (\varepsilon_r\mathscr{A}) = \mu_0(\dot{\mathscr{R}}_e + \mathscr{J}_e) + \nabla \times \mu_r^{-1}\mathscr{R}_m, \tag{3.87}$$

$$\frac{1}{c^2}\ddot{\psi}_e - \nabla \cdot (\varepsilon_r\nabla\psi_e) = -\frac{1}{\varepsilon_0}\nabla \cdot (\mathscr{Q}_e + \mathscr{R}_e). \tag{3.88}$$

On comparing the set of equations (3.87) and (3.88) with the set (3.11) and (3.12) it is seen that only quantities related to the impressed sources appear on the rhs when the generalized Lorenz condition is used. In contrast both impressed and induced polarization sources appear on the rhs when the normal Lorenz condition is used. The differential operators acting on $\mathscr{A}$ and $\psi_e$ will correspondingly change depending on the type of Lorenz condition used, as is evident by observing the lhs of these sets. Both sets are, however, equally valid in material media.

The scalar and vector potentials are not unique but can be transformed according to $\psi_e = \psi_e^0 + \dot{\chi}$ and $\mathscr{A} = \mathscr{A}^0 - \nabla\chi$, where $\chi$ is an arbitrary gauge function. This transformation preserves the Lorenz condition as well as the electromagnetic field if $\chi$ satisfies

$$\frac{1}{c^2}\ddot{\chi} - \nabla \cdot \varepsilon_r\nabla\chi = 0. \tag{3.89}$$

If the Hertzian vector potentials $\mathbf{\Pi}_e$ and $\mathbf{\Pi}_m$ are defined in terms of the scalar and vector potentials as

[4] However, the addition of a term $\nabla\phi_e$ to $\varepsilon_r^{-1}\mathscr{R}_e$ or a term $\nabla\phi_m$ to $\mu_r^{-1}\mathscr{Q}_m$, where $\phi_e$ and $\phi_m$ are completely arbitrary, does not modify the mutual relation between $\mathscr{R}_e$ and $\mathscr{R}_m$ or that between $\mathscr{Q}_e$ and $\mathscr{Q}_m$.

[5] Note that the generalized Lorenz condition reduces to the Lorenz condition when $\varepsilon_r$ is unity.

$$\psi_e = -\nabla \cdot \varepsilon_r \mathbf{\Pi}_e, \tag{3.90}$$

$$\mathscr{A} = \frac{1}{c^2}\dot{\mathbf{\Pi}}_e + \mu_0 \varepsilon_r^{-1} \nabla \times \mathbf{\Pi}_m, \tag{3.91}$$

so that the generalized Lorenz condition is automatically met, then the Hertzian potentials satisfy

$$\frac{1}{c^2}\varepsilon_r \ddot{\mathbf{\Pi}}_e + \nabla \times \mu_r^{-1} \nabla \times \mathbf{\Pi}_e = \frac{1}{\varepsilon_0}\left(\mathscr{Q}_e + \tilde{\mathscr{R}}_e\right), \tag{3.92}$$

$$\frac{1}{c^2}\mu_r \ddot{\mathbf{\Pi}}_m + \nabla \times \varepsilon_r^{-1} \nabla \times \mathbf{\Pi}_m = \frac{1}{\mu_0}(\mathscr{R}_m + \mathscr{Q}_m). \tag{3.93}$$

The electromagnetic field is given by

$$\mathscr{E} = \frac{1}{\varepsilon_0}\varepsilon_r^{-1}\tilde{\mathscr{R}}_e - \frac{1}{c^2}\ddot{\mathbf{\Pi}}_e - \mu_0 \varepsilon_r^{-1} \nabla \times \dot{\mathbf{\Pi}}_m, \tag{3.94}$$

$$\mathscr{B} = -\mathscr{R}_m + \frac{1}{c^2}\nabla \times \dot{\mathbf{\Pi}}_e + \mu_0 \nabla \times \varepsilon_r^{-1} \nabla \times \mathbf{\Pi}_m, \tag{3.95}$$

$$\mathscr{D} = -\mathscr{Q}_e + \varepsilon_0 \nabla \times \mu_r^{-1} \nabla \times \mathbf{\Pi}_e - \frac{1}{c^2}\nabla \times \dot{\mathbf{\Pi}}_m, \tag{3.96}$$

$$\mathscr{H} = \frac{1}{\mu_0}\mu_r^{-1}\mathscr{Q}_m + \varepsilon_0 \mu_r^{-1} \nabla \times \dot{\mathbf{\Pi}}_e - \frac{1}{c^2}\ddot{\mathbf{\Pi}}_m, \tag{3.97}$$

where a term of the form $\nabla\nabla \cdot (\varepsilon_r \mathbf{\Pi}_e)$ has been absorbed into the definition of a revised $\tilde{\mathscr{R}}_e$ in expressions (3.92) and (3.94)[6]. We shall drop the tilde in $\tilde{\mathscr{R}}_e$ in further development because we will totally bypass the Lorentz potentials ($\mathscr{A}$, $\psi_e$) (which are not invariant to changes in $\mathscr{R}_e$) and focus on the direct relationship between the Hertzian potentials and the electromagnetic fields as given in equation (3.92)–(3.97). Note that the duality between $\mathscr{E}$ and $\mathscr{H}$ as well as that between $\mathscr{B}$ and $\mathscr{D}$ is maintained by the dual variables $\mathbf{\Pi}_e$ and $\mathbf{\Pi}_m$.

Note that in the formulation here, the right-hand sides of equations (3.92) and (3.93) include only free charges and free currents since polarization charges and currents have already been accounted for in the definitions of $\varepsilon_r$ and $\mu_r$. In contrast

[6] Namely, $\varepsilon_r^{-1}\mathscr{R}_e + \varepsilon_0 \nabla\nabla \cdot (\varepsilon_r \mathbf{\Pi}_e) = \varepsilon_r^{-1}\mathscr{R}_e - \varepsilon_0 \nabla \psi^e =: \varepsilon_r^{-1}\tilde{\mathscr{R}}_e$. This can be done because a transformation of $\varepsilon_r^{-1}\mathscr{R}_e$ by the addition of the gradient of an arbitrary scalar function does not change the relationship between $\mathscr{R}_m$ and $\mathscr{R}_e$ in equation (3.84) as already mentioned. For this reason the equation for $\mathscr{E}$ in equation (3.94) as well as the differential equation for $\mathbf{\Pi}_e$ (3.92) are different from the traditional definitions found in other books (see [5, p 24], [6, p 195]). The relationship between the other field components and $\mathbf{\Pi}_e$ remains the same as traditional ones because they all depend on $\nabla \times \mathbf{\Pi}_e$, but not directly on $\mathbf{\Pi}_e$. This is true because if we set $\tilde{\mathscr{R}}_e = \mathscr{R}_e - \varepsilon_r \nabla \ell$ for an arbitrary function $\ell$ and simultaneously set $\ddot{\tilde{\mathbf{\Pi}}}_e = \ddot{\mathbf{\Pi}}_e - c^2 \nabla(\varepsilon_0^{-1}\ell + \nabla \cdot (\varepsilon_r \mathbf{\Pi}_e))$, then $\nabla \times \tilde{\mathbf{\Pi}}_e = \nabla \times \mathbf{\Pi}_e$. Note that in electrostatics, the expression for electric field reduces to the first term, i.e. $\mathscr{E} = \frac{1}{\varepsilon_0}\varepsilon_r^{-1}\tilde{\mathscr{R}}_e$ under this new definition.

the right-hand sides of equations (3.15) and (3.16) contain both the impressed and induced sources with a corresponding simpler lhs.

If the stream functions are transformed according to

$$\mathcal{Q}_{\mathrm{e}} = \mathcal{Q}_{\mathrm{e}}^{0} + \nabla \times \mu_r^{-1}\mathcal{G}; \qquad \mathcal{Q}_{\mathrm{m}} = \mathcal{Q}_{\mathrm{m}}^{0} - \mu_0\dot{\mathcal{G}} - \mu_r\nabla g, \tag{3.98}$$

$$\mathcal{R}_{\mathrm{m}} = \mathcal{R}_{\mathrm{m}}^{0} - \nabla \times \varepsilon_r^{-1}\mathcal{L}; \qquad \mathcal{R}_{\mathrm{e}} = \mathcal{R}_{\mathrm{e}}^{0} - \varepsilon_0\dot{\mathcal{L}} - \varepsilon_r\nabla\ell, \tag{3.99}$$

where $\mathcal{R}_{\mathrm{e}}^{0}$, $\mathcal{R}_{\mathrm{m}}^{0}$, $\mathcal{Q}_{\mathrm{e}}^{0}$, and $\mathcal{Q}_{\mathrm{m}}^{0}$ are other possible potentials and the gauge functions $\mathcal{G}$, $g$, $\mathcal{L}$, and $\ell$ are arbitrary, and the Hertzian potentials are simultaneously transformed according to

$$\mathbf{\Pi}_{\mathrm{e}} = \mathbf{\Pi}_{\mathrm{e}}^{0} + \varepsilon_r^{-1}\nabla \times \mathbf{\Gamma} - \mu_0\dot{\mathbf{\Lambda}} - \nabla\lambda, \tag{3.100}$$

$$\mathbf{\Pi}_{\mathrm{m}} = \mathbf{\Pi}_{\mathrm{m}}^{0} - \mu_r^{-1}\nabla \times \mathbf{\Lambda} - \varepsilon_0\dot{\mathbf{\Gamma}} + \nabla\gamma, \tag{3.101}$$

where the gauge functions $\gamma$, $\mathbf{\Gamma}$, $\lambda$, $\mathbf{\Lambda}$ are arbitrary but satisfy

$$\begin{aligned} \frac{1}{\mu_0}\mathcal{L} &= \nabla \times \mu_r^{-1}\nabla \times \mathbf{\Lambda} + \frac{1}{c^2}\varepsilon_r\ddot{\mathbf{\Lambda}} - \varepsilon_r\nabla\zeta; \quad \frac{1}{c^2}\ddot{\lambda} = -\mu_0\dot{\zeta} + \frac{1}{\varepsilon_0}\ell \\ \frac{1}{\varepsilon_0}\mathcal{G} &= \nabla \times \varepsilon_r^{-1}\nabla \times \mathbf{\Gamma} + \frac{1}{c^2}\mu_r\ddot{\mathbf{\Gamma}} - \mu_r\nabla\xi; \quad \frac{1}{c^2}\ddot{\gamma} = \varepsilon_0\dot{\xi} - \frac{1}{\mu_0}g, \end{aligned} \tag{3.102}$$

with $\zeta$, $\xi$ being completely arbitrary, then it is straightforward to show that the electromagnetic field given in equations (3.94)–(3.97) is preserved. The arbitrariness in equations (3.100) and (3.101) allows the Hertzian potentials to be chosen which have *four of their six* components identically zero. The two remaining non-zero components satisfy non-homogeneous differential equations which can be obtained from equations (3.92) and (3.93).

## References


[1] Nisbet A 1955 Hertzian electromagnetic potentials and associated gauge transformations *Proc. R. Soc.* A **231** 250–63

[2] Janaswamy R 2019 Gauge transformation for recasting potential representations *IEEE Antennas and Propagation Int. Sympos.* vol 1088 (Atlanta, GA: IEEE) pp 1–2

[3] Gradshteyn I S and Ryzhik I M 2007 *Tables of Integrals, Series and Products* 7th edn (New York: Academic)

[4] Nisbet A 1957 Electromagnetic potentials in a heterogeneous non-conducting medium *Proc. R. Soc. Lond.* **240** 375–81

[5] Ishimaru A 2017 *Electromagnetic Wave Propagation, Radiation, and Scattering* (Hoboken, NJ: Wiley)

[6] van Bladel J 1985 *Electromagnetic Fields* (New York: Hemisphere)

# Chapter 4

# Causality and dispersion

As alluded to in chapter 1, causality is a central concept in our understanding of the physical world and this includes the wave propagation phenomenon. Several important theorems exist that characterize the behavior of causal systems, and it is of interest to see how those theorems translate to describing the electrical properties of materials and passive systems in the time and frequency domains. Some of the important results pertaining to electromagnetic systems in this regard are Titchmarsh's theorem, the Kramers–Krönig relations, the Bode–Fano criterion, properties of the scattering amplitude of antennas and scatterers, and their implications for the fundamental limitations on the antenna gain–bandwidth product.

## 4.1 Causal systems

In this section we look at the general requirements of causality on the impulse response and transfer function of a linear system. An important theorem in this regard is Titchmarsh's theorem. An immediate consequence of this theorem is the Kramers–Krönig relation[1], which provides a relationship between the real and imaginary parts of a complex permittivity function.

### 4.1.1 Titchmarsh's theorem and the Kramers–Krönig relations

Consider a *linear, time-invariant* physical system with a real-valued input $x(t)$, a real-valued response function $y(t)$, a real-valued causal impulse response function $h(t)$: $h(t) = 0$ for $t < 0$, and a corresponding transfer function $H(\omega)$. We have the input–output relationship

$$y(t) = \int_0^\infty h(\tau)x(t - \tau)\,\mathrm{d}\tau = \int_{-\infty}^{t} h(t - \tau)x(\tau)\,\mathrm{d}\tau, \tag{4.1}$$

[1] Note that the Kramers–Krönig relations were anterior to the Titchmarsh's theorem.

doi:10.1088/978-0-7503-1716-0ch4

which implies that the effect (i.e. $y(t)$) cannot precede the cause (i.e. $x(t)$). We suppose further that the system is unconditionally stable, meaning that if $x(t) = 0$ for $t > t_0$, then $y(t) \to 0$ as $t \to \infty$. The impulse response is related to the transfer function $H(\omega)$ through the inverse Fourier transform:

$$h(t) = \frac{1}{2\pi} \int_{-\infty}^{\infty} H(\omega) e^{j\omega t} \, d\omega. \tag{4.2}$$

Since $h(t) = 0$ for $t < 0$, $H(\omega)$ must not have any singularities (poles, branch points, etc) in the lower half of the complex $\omega$-plane, namely for $\mathfrak{J}(\omega) < 0$, where $\mathfrak{J}(\cdot)$ indicates the imaginary part. A system is considered passive if it consumes energy but does not produce energy (thermodynamic passivity) or if it is incapable of power gain (incremental passivity). There is a direct link between passivity and causality for a linear system as elaborated in the following example.

**Example 4.1.1.1.** Consider a series *RLC* circuit fed by a time-dependent source. We would like to show in this example that the passivity of a *linear, time-invariant* system implies causality.
Let the circuit consisting of a series connection of a resistor $R > 0$, inductor $L > 0$, and capacitor $C > 0$ be at rest at $t \to -\infty$ (i.e. the inductor current is zero and the capacitor voltage is zero at the remote past), see figure 4.1. The circuit is driven by a voltage source $v_s(t)$ (the cause), resulting in the circuit current $i(t) = C\dot{v}_c(t)$ and hence the capacitor voltage $v_c(t)$ (the effects).

From Kirchhoff's voltage law we have

$$v_s(t) = LC\ddot{v}_c(t) + RC\dot{v}_c(t) + v_c(t) = v_s(t),$$

where an overhead dot means a derivative with respect to time. Multiplying both sides by $i(t) = C\dot{v}_c(t)$ and integrating we obtain

$$C\int_{-\infty}^{t} v_s(\tau)\dot{v}_c(\tau) \, d\tau = R\int_{-\infty}^{t} i^2(\tau) \, d\tau + \frac{1}{2}Cv_c^2(t) + \frac{1}{2}Li^2(t) > 0.$$

The resulting inequality

$$\int_{-\infty}^{t} v_s(\tau)\dot{v}_c(\tau) \, d\tau > 0, \tag{4.3}$$

for arbitrary $t$ is known as the passivity condition. The lhs is the integral of the product of the cause and time rate of change of the effect. Passivity states that this

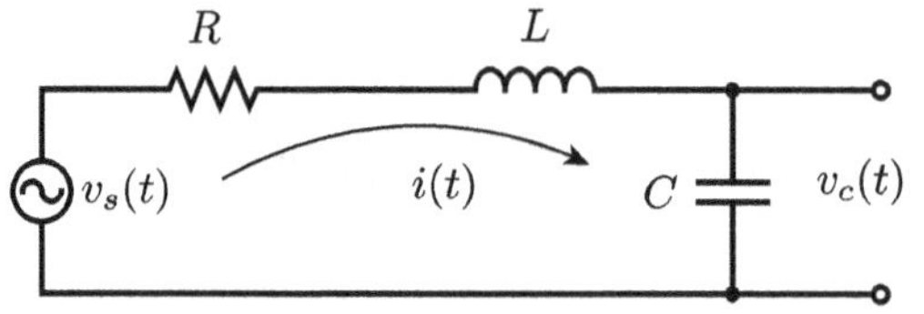


**Figure 4.1.** A series *RLC* linear, time-invariant system.

quantity is always positive for a system that absorbs, but not generates, energy. We now wish to show that if $v_s(t) = 0,\ t < 0$, then passivity implies causality, meaning the effect is also non-existent for negative times: $v_c(t) = 0$ for $t < 0$. Let $u_s(t)$ be an arbitrary source voltage generating the capacitor voltage $u_c(t)$. Then passivity dictates that

$$\int_{-\infty}^{t} u_s(\tau)\dot{u}_c(\tau)\,\mathrm{d}\tau =: k_1 > 0. \tag{4.4}$$

Consider now the sum source $u_s(t) + \lambda v_s(t)$, where $\lambda$ is an arbitrary real number. Then by linearity, the corresponding capacitor voltage it generates is $u_c(t) + \lambda v_c(t)$. Applying the passivity condition to this sum source and the corresponding capacitor voltage, we obtain on using $v_s(t) = 0,\ t < 0$ that

$$\int_{-\infty}^{t} u_s(\tau)[\dot{u}_c(\tau) + \lambda\dot{v}_c(\tau)]\,\mathrm{d}\tau > 0, \quad \text{for } t < 0,$$

or

$$k_1 + \lambda\int_{-\infty}^{t} u_s(\tau)\dot{v}_c(\tau)\,\mathrm{d}\tau > 0, \quad \text{for } t < 0.$$

Since $k_1 > 0$ by equation (4.4) and $\lambda$ (can be negative) and $u_s(t)$ (not necessarily positive) are both arbitrary, the above implies that $\dot{v}_c(t) = 0,\ t < 0$. This, in turn, implies that $v_c(t) = 0,\ t < 0$ since $v_c(-\infty) = 0$. Thus linearity, time-invariance, and passivity imply causality. ■■

**Theorem 4.1.1.** Titchmarsh's theorem [1, chapter 5], [2, p 27], [3, p 183].
*Any square integrable function $H(\omega)$ with inverse Fourier transform $h(t)$ that satisfies one of the following four conditions satisfies them all:*

1. $h(t) = 0$ *for all* $t < 0$.
2. $H(\omega') = \lim_{\omega''\to 0^-} H(\omega' + j\omega'')$ *for almost all* $\omega'$, *where* $H(\omega)$ *is holomorphic in the lower half of the complex* $\omega = \omega' + j\omega''$ *plane and is square integrable over any parallel line to the* $\omega'$*-axis in the lower half plane:*

$$\int_{-\infty}^{\infty} |H(\omega' - j\omega'')|^2\,\mathrm{d}\omega' < \infty, \quad \omega'' > 0.$$

3. *The real part* $\mathfrak{R}\{H(\omega)\}$ *and the imaginary part* $\mathfrak{I}\{H(\omega)\}$ *satisfy the first Plemelj formula*

$$\mathfrak{I}[H(\omega)] = -\frac{1}{\pi}\mathscr{P}\int_{-\infty}^{\infty}\frac{\mathfrak{R}[H(\omega')]}{\omega - \omega'}\,\mathrm{d}\omega', \tag{4.5}$$

*where* $\mathscr{P}$ *stands for the Cauchy's principal part.*

4. *The real part* $\Re\{H(\omega)\}$ *and the imaginary part* $\Im\{H(\omega)\}$ *satisfy the second Plemelj formula*

$$\Re[H(\omega)] = \frac{1}{\pi}\mathscr{P}\int_{-\infty}^{\infty}\frac{\Im[H(\omega')]}{\omega - \omega'}\,\mathrm{d}\omega'. \tag{4.6}$$

*Proof.* We will show the equivalence of items 1, 3, and 4. To this end let us write

$$h(t) = \theta(t)z(t), \tag{4.7}$$

where $\theta(t)$ is the unit step function and $z(t) = h(t)$ for $t > 0$. We are free to choose $z(t)$ for $t < 0$. By treating the step function as a limiting case of $\mathrm{e}^{-at}$ as $a \to 0^+$, we have for its Fourier transform

$$\Theta(\omega) = \lim_{a\to 0^+}\int_0^{\infty}\mathrm{e}^{-at}\mathrm{e}^{-j\omega t}\,\mathrm{d}t = \lim_{a\to 0^+}\frac{1}{j\omega + a} = \pi\delta(\omega) + \mathscr{P}\frac{1}{j\omega}. \tag{4.8}$$

The significance of the latter part can be deciphered when evaluating inverse Fourier transform type integrals along the real axis in the complex $\omega$-plane; the semi-circular indentation around the pole *at the origin* is counter-clockwise ($\Gamma_-$) for $t < 0$ and clockwise ($\Gamma_+$) for $t > 0$, see figure 4.2. Assume $F(\omega)$ to be regular at $\omega = 0$ and $\omega_1,\ \omega_2 > 0$. Then

$$\mathscr{P}\int_{-\omega_1}^{\omega_2}\frac{F(\omega)}{\omega}\mathrm{e}^{j\omega t}\,\mathrm{d}\omega = \lim_{\varepsilon\to 0^+}\left[\int_{-\omega_1}^{-\varepsilon} + \int_{\varepsilon}^{\omega_2}\right]\frac{F(\omega)}{\omega}\mathrm{e}^{j\omega t}\,\mathrm{d}\omega. \tag{4.9}$$

As an example, if $F(\omega)$ is an even function of $\omega$ on the real axis, $F(-\omega) = F(\omega) =: F_e(\omega)$ and we have

$$\begin{aligned}\mathscr{P}\int_{-\infty}^{\infty}\frac{F_e(\omega)}{\omega}\mathrm{e}^{j\omega t}\,\mathrm{d}\omega &= 2j\lim_{\varepsilon\to 0^+}\int_{\varepsilon}^{\infty}\frac{\sin(\omega t)}{\omega}F_e(\omega)\,\mathrm{d}\omega\\ &= 2j\int_0^{\infty}\frac{\sin(\omega t)}{\omega}F_e(\omega)\,\mathrm{d}\omega.\end{aligned} \tag{4.10}$$

In particular, evaluating the above for $t = 0$ gives

$$\mathscr{P}\int_{-\infty}^{\infty}\frac{F_e(\omega)}{\omega}\,\mathrm{d}\omega = 0. \tag{4.11}$$

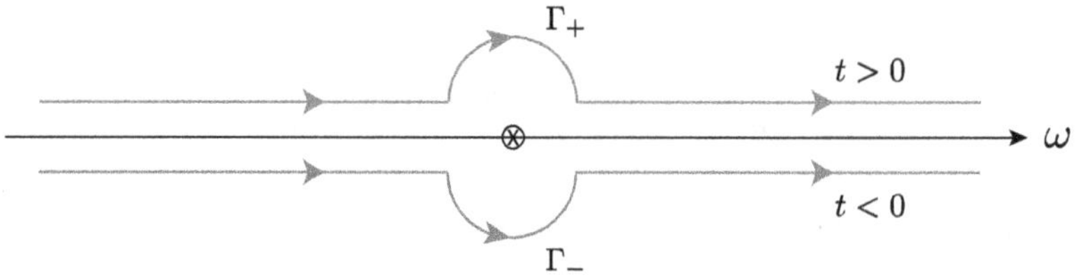


**Figure 4.2.** Contours for evaluating the Cauchy's principal value integrals in the complex $\omega$-plane.

Also, using $F_e = 1/j$ in equation (4.10) and the identity $\int_0^\infty \sin\theta/\theta \, \mathrm{d}\theta = \pi/2$ gives

$$\mathscr{P}\int_{-\infty}^{\infty} \frac{e^{j\omega t}}{j\omega} \, \mathrm{d}\omega = \pi \, \mathrm{sign}(t), \quad t \neq 0. \tag{4.12}$$

Cauchy's integral theorem will give the same result. For instance, for $t < 0$ we consider the closed contour consisting of the punctured real axis, the counter-clockwise, semi-circular indent of radius $\varepsilon$ at the origin and the clockwise, large semi-circular arc, $C_R$, in the lower half plane of radius $R$. The closed contour excludes the singularities and the integrand vanishes on $R$. We then have from Cauchy's integral theorem upon using $\omega = \varepsilon e^{j\theta}$ on $\Gamma_-$ that

$$0 = \mathscr{P}\int_{-\infty}^{\infty} \frac{e^{j\omega t}}{j\omega} \mathrm{d}\omega + \lim_{\varepsilon \to 0} \oint_{\Gamma_-} \frac{e^{j\omega t}}{j\omega} \mathrm{d}\omega + \lim_{R \to \infty} \oint_{C_R} \frac{e^{j\omega t}}{j\omega} \mathrm{d}\omega$$

$$\Longrightarrow \mathscr{P}\int_{-\infty}^{\infty} \frac{e^{j\omega t}}{j\omega} \mathrm{d}\omega = -\lim_{\varepsilon \to 0} \oint_{\Gamma_-} \frac{e^{j\omega t}}{j\omega} \mathrm{d}\omega = -\lim_{\varepsilon \to 0} \int_{\theta=\pi}^{2\pi} \frac{e^{j0} \varepsilon j e^{j\theta}}{j\varepsilon e^{j\theta}} \mathrm{d}\theta = -\pi, \quad t < 0.$$

If $F(\omega)$ is an odd function of $\omega$ on the real axis, $F(-\omega) = -F(\omega) =: -F_o(\omega)$ and we have

$$\begin{aligned} \mathscr{P}\int_{-\infty}^{\infty} \frac{F_o(\omega)}{\omega} e^{j\omega t} \, \mathrm{d}\omega &= 2 \lim_{\varepsilon \to 0^+} \int_{\varepsilon}^{\infty} \frac{F_o(\omega)}{\omega} \cos(\omega t) \, \mathrm{d}\omega \\ &= 2 \int_0^{\infty} \frac{F_o(\omega)}{\omega} \cos(\omega t) \, \mathrm{d}\omega. \end{aligned} \tag{4.13}$$

An arbitrary function can be written as the superposition of even and odd parts $F(\omega) = [F_e(\omega) + F_o(\omega)]/2$, where $F_e(\omega) = F(\omega) + F(-\omega)$, $F_o(\omega) = F(\omega) - F(-\omega)$. Hence

$$\mathscr{P}\int_{-\infty}^{\infty} \frac{F(\omega)}{\omega} e^{j\omega t} \, \mathrm{d}\omega = j \int_0^{\infty} \frac{\sin(\omega t)}{\omega} F_e(\omega) \, \mathrm{d}\omega + \int_0^{\infty} \frac{F_o(\omega)}{\omega} \cos(\omega t) \, \mathrm{d}\omega. \tag{4.14}$$

Recall that if $F(\omega)$ and $G(\omega)$ are the Fourier transforms of the functions $f(t)$ and $g(t)$, respectively, then

$$\int_{-\infty}^{\infty} f(t) g(t) e^{-j\omega t} \, \mathrm{d}t = \frac{1}{2\pi} \int_{-\infty}^{\infty} F(\omega') G(\omega - \omega') \, \mathrm{d}\omega'. \tag{4.15}$$

Using equation (4.15) for the product $h(t) = \theta(t) z(t)$ and utilizing equation (4.8), we have

$$H(\omega) = \frac{1}{2\pi} \int_{-\infty}^{\infty} Z(\omega') \Theta(\omega - \omega') \, \mathrm{d}\omega' = \frac{Z(\omega)}{2} - \frac{j}{2\pi} \mathscr{P}\int_{-\infty}^{\infty} \frac{Z(\omega')}{\omega - \omega'} \, \mathrm{d}\omega'. \tag{4.16}$$

**Even extension of the impulse response**

If we choose $z(-t) = h(t),\ t > 0 \Longrightarrow Z(\omega) = 2\Re[H(\omega)]$, where $\Re(\cdot)$ indicates real part, then we have from equation (4.16) that

$$H(\omega) = \Re[H(\omega)] - \frac{j}{\pi}\mathscr{P}\int_{-\infty}^{\infty}\frac{\Re[H(\omega')]}{\omega - \omega'}\,\mathrm{d}\omega', \tag{4.17}$$

or

$$\Im[H(\omega)] = -\frac{1}{\pi}\mathscr{P}\int_{-\infty}^{\infty}\frac{\Re[H(\omega')]}{\omega - \omega'}\,\mathrm{d}\omega', \tag{4.18}$$

where $\Im(\cdot)$ indicates imaginary part.

**Odd extension of impulse response:**

If we choose $z(-t) = -h(t),\ t > 0 \Longrightarrow Z(\omega) = 2j\Im[H(\omega)]$ then we have from equation (4.16) that

$$H(\omega) = j\Im[H(\omega)] + \frac{1}{\pi}\mathscr{P}\int_{-\infty}^{\infty}\frac{\Im[H(\omega')]}{\omega - \omega'}\,\mathrm{d}\omega', \tag{4.19}$$

or

$$\Re[H(\omega)] = \frac{1}{\pi}\mathscr{P}\int_{-\infty}^{\infty}\frac{\Im[H(\omega')]}{\omega - \omega'}\,\mathrm{d}\omega', \tag{4.20}$$

Equations (4.18) and (4.20) together constitute the *Kramers–Krönig* relations. They state that the real and imaginary parts of the transfer function of a causal impulse response function are not independent but are related to each other. Equation (4.20) that expresses the real part in terms of an integral of the imaginary part is known as the *Hilbert transform* and equation (4.18) that expresses the imaginary part in terms of an integral of the real part is known as the *inverse Hilbert transform.* ■

## 4.2 Dispersive systems

Systems are considered dispersive if their properties change with time or frequency. We now apply the above relations to the electric susceptibility function, $\bar{\bar{\chi}}_e(\mathbf{r}; t)$, which relates the electric flux density in a dielectric medium to the electric field intensity and shows that the real and imaginary parts of the susceptibility functions are dependent on each other. In particular, they cannot be chosen independently.

### 4.2.1 Linear dielectrics

Recall that the electric field intensity $\mathscr{E}(\mathbf{r}; t)$ is the primitive field (or the excitation field) and the flux density $\mathscr{D}(\mathbf{r}; t)$ is the induced field (or the response field). The relationship between the electric flux density and the electric field intensity in the frequency and time domains in a *linear* medium is [4]

$$\mathbf{D}(\mathbf{r};\,\omega) = \varepsilon_0(\bar{\bar{I}} + \bar{\bar{\chi}}_e(\mathbf{r};\,\omega))\mathbf{E}(\mathbf{r};\,\omega) =: \varepsilon_0\bar{\bar{\varepsilon}}_{rc}(\mathbf{r};\,\omega)\mathbf{E}(\mathbf{r};\,\omega), \tag{4.21}$$

$$\mathscr{D}(\mathbf{r};\,t) = \varepsilon_0\mathscr{E}(\mathbf{r};\,t) + \varepsilon_0\int_{-\infty}^{t} \bar{\bar{\chi}}_e(\mathbf{r};\,t-\tau)\mathscr{E}(\mathbf{r};\,\tau)\,\mathrm{d}\tau, \tag{4.22}$$

where $\bar{\bar{I}}$ is the identity tensor and $\varepsilon_0$ is the permittivity of the vacuum. In the most general case the susceptibility function $\bar{\bar{\chi}}_e(\mathbf{r};\,\omega)$ is a tensor of second rank (a 3 × 3 matrix) and the complex relative permittivity $\bar{\bar{\varepsilon}}_{rc}(\mathbf{r};\,\omega) = \bar{\bar{\varepsilon}}_r'(\mathbf{r};\,\omega) - j\bar{\bar{\varepsilon}}_r''(\mathbf{r};\,\omega)$ has both real and imaginary parts. This corresponds to an *anisotropic, spatially dispersive, and temporally dispersive medium.*

As is evident in equation (4.22) the susceptibility function takes the place of the impulse response of a linear system and is hence assumed to be causal. In the following we shall drop the argument **r** from the various quantities as our focus here is on the frequency and time variables. In most cases the susceptibility function vanishes at $|\omega| \to \infty$ and the complex relative permittivity approaches 1 as $|\omega| \to \infty$. In the more general case where the value at infinity is not unity, we write $\bar{\bar{\chi}}_e(\omega) = \bar{\bar{\varepsilon}}_{rc}(\omega) - \bar{\bar{I}}\varepsilon_{r\infty} = \bar{\bar{\varepsilon}}_r'(\omega) - \bar{\bar{I}}\varepsilon_{r\infty} - j\bar{\bar{\varepsilon}}_r''(\omega)$, where $\varepsilon_{r\infty}$ is the value of the real part of relative permittivity at infinite frequency. Applying equations (4.18) and (4.20) to the electric susceptibility tensor we have

$$\bar{\bar{\varepsilon}}_r''(\omega) = \frac{1}{\pi}\mathscr{P}\int_{-\infty}^{\infty}\frac{\bar{\bar{\varepsilon}}_r'(\omega')}{\omega-\omega'}\,\mathrm{d}\omega' = \frac{2\omega}{\pi}\mathscr{P}\int_{0}^{\infty}\frac{\bar{\bar{\varepsilon}}_r'(\omega')}{\omega^2-\omega'^2}\,\mathrm{d}\omega', \tag{4.23}$$

$$\bar{\bar{\varepsilon}}_r'(\omega) = \bar{\bar{I}}\varepsilon_{r\infty} - \frac{1}{\pi}\mathscr{P}\int_{-\infty}^{\infty}\frac{\bar{\bar{\varepsilon}}_r''(\omega')}{\omega-\omega'}\,\mathrm{d}\omega' = \bar{\bar{I}}\varepsilon_{r\infty} - \frac{2}{\pi}\mathscr{P}\int_{0}^{\infty}\frac{\omega'\bar{\bar{\varepsilon}}_r''(\omega')}{\omega^2-\omega'^2}\,\mathrm{d}\omega', \tag{4.24}$$

where we have utilized equation (4.11) for the constant function $\varepsilon_{r\infty}$. We have further assumed that $\bar{\bar{\varepsilon}}_r(\omega)$ is regular on the real axis in arriving at equations (4.23) and (4.24). The last integrals on the rhs of equations (4.23) and (4.24) are true because $\bar{\bar{\varepsilon}}_{rc}^*(-\omega) = \bar{\bar{\varepsilon}}_{rc}(\omega)$, which implies that the real part $\bar{\bar{\varepsilon}}_r'(\omega)$ is an even function of $\omega$ and the imaginary part $\bar{\bar{\varepsilon}}_r''(\omega)$ is an odd function of $\omega$ on the real axis of the complex $\omega$-plane. (More generally $\bar{\bar{\varepsilon}}_{rc}^*(-\omega) = \bar{\bar{\varepsilon}}_{rc}(\omega^*)$.) Note from equations (4.23) and (4.24) that the value of the imaginary part of $\bar{\bar{\varepsilon}}_{rc}$ at any single frequency depends on the values of the real part on the entire real axis and vice versa. Hence the relations are anything but local. Identical Kramers–Krönig relations apply to the magnetic susceptibility tensor $\bar{\bar{\chi}}_m$ that relates the magnetic flux density, **B**, and the magnetic field intensity, **H**,

$$\mathbf{B} = \mu_0(\bar{\bar{I}} + \bar{\bar{\chi}}_m)\mathbf{H}. \tag{4.25}$$

In some cases, the complex relative permittivity will have poles on the real axis. These poles must be extracted before applying the Kramers–Krönig relations.

An example of a valid causal function is $\chi_e(\omega) = \chi_0/(\omega - \gamma)$, $\mathfrak{I}(\gamma) > 0$. However, $\chi_e(\omega) = \chi_0 e^{j\omega\tau}/(\omega - \gamma)$, $\tau > 0$, $\mathfrak{I}(\gamma) > 0$ is not causal as the exponential factor blows up in the lower half plane. Its Fourier inverse $\chi_e(t)$ vanishes only for $t < -\tau$, rather than for $t < 0$ as required by causality.

**Example 4.2.1.1** Drude medium.
Consider the scalar relative permittivity of metals as described by the Drude model [4] (see section 1.2.1.1)

$$\varepsilon_{rc}(\omega) = \varepsilon_r'(\omega) - j\varepsilon_r''(\omega) = 1 - \frac{j\sigma_0}{\omega\varepsilon_0(1 + j\omega\tau)}, \tag{4.26}$$

where $\sigma_0$ and $\tau$ are positive constants representing the dc conductivity and mean time between collisions of electron gas, respectively. The permittivity has a simple pole at $\omega = 0$ with the residue $-j\sigma_0/\varepsilon_0$. In the general case with a pole at the origin let us express the complex permittivity as

$$\varepsilon_{rc}(\omega) = \varepsilon_r'(\omega) - j\varepsilon_r''(\omega) = 1 - \frac{j\sigma_0}{\omega\varepsilon_0} + \chi_e(\omega), \tag{4.27}$$

and apply the Kramers–Krönig relations to

$$\chi_e(\omega) = \varepsilon_{rc}(\omega) - 1 + \frac{j\sigma_0}{\omega\varepsilon_0}. \tag{4.28}$$

Writing $\chi_e(\omega) = \chi_r(\omega) - j\chi_i(\omega)$, it is clear that $\Re[\chi_e(\omega)] = \chi_r(\omega) = \varepsilon_r'(\omega) - 1$ and $-\Im[\chi_e(\omega)] = \chi_i(\omega) = \varepsilon_r''(\omega) - \sigma_0/\omega\varepsilon_0$. For a causal function $\chi_e(\omega)$ is analytic in the lower half of the complex $\omega$-plane and all singularities lie in the upper half-space. Furthermore we suppose that $|\chi_e(\omega)| \sim \mathcal{O}(1/|\omega|)$ as $|\omega| \to \infty$. We then have from the Kramers–Krönig relations that

$$\begin{aligned}
\chi_r(\omega) = \varepsilon_r'(\omega) - 1 &= -\frac{1}{\pi}\mathcal{P}\int_{-\infty}^{\infty} \frac{\chi_i(\omega')}{\omega - \omega'}\,d\omega' \\
&= -\frac{1}{\pi}\mathcal{P}\int_{-\infty}^{\infty} \frac{\varepsilon_r''(\omega') - \sigma_0/(\omega'\varepsilon_0)}{\omega - \omega'}\,d\omega' \\
&= -\frac{1}{\pi}\mathcal{P}\int_{-\infty}^{\infty} \frac{\varepsilon_r''(\omega')}{\omega - \omega'}\,d\omega' + \frac{\sigma_0}{\varepsilon_0\pi}\mathcal{P}\int_{-\infty}^{\infty} \frac{1}{\omega'(\omega - \omega')}\,d\omega' \\
&= -\frac{1}{\pi}\mathcal{P}\int_{-\infty}^{\infty} \frac{\varepsilon_r''(\omega')}{\omega - \omega'}\,d\omega'
\end{aligned}$$

and

$$\chi_i(\omega) = \frac{1}{\pi}\mathcal{P}\int_{-\infty}^{\infty} \frac{\chi_r(\omega')}{\omega - \omega'}\,d\omega'.$$

Note that we have made repeated use of equation (4.11) to conclude

$$\mathcal{P}\int_{-\infty}^{\infty} \frac{1}{\omega'(\omega - \omega')}\,d\omega' = \frac{1}{\omega}\left[\mathcal{P}\int_{-\infty}^{\infty} \frac{1}{\omega'}\,d\omega' + \mathcal{P}\int_{-\infty}^{\infty} \frac{1}{\omega - \omega'}\,d\omega'\right] = 0. \tag{4.29}$$

In terms of the original variables we have the modified Kramers–Krönig relations

$$\varepsilon_r''(\omega) = \frac{\sigma_0}{\omega\varepsilon_0} + \frac{1}{\pi}\mathcal{P}\int_{-\infty}^{\infty} \frac{\varepsilon_r'(\omega')}{\omega - \omega'}\,d\omega', \tag{4.30}$$

$$\varepsilon_r'(\omega) = 1 - \frac{1}{\pi}\mathscr{P}\int_{-\infty}^{\infty} \frac{\varepsilon_r''(\omega')}{\omega - \omega'}\, \mathrm{d}\omega'. \tag{4.31}$$

that are valid for any complex permittivity whose imaginary part has a simple pole at the origin. Obviously equations (4.30) and (4.31) reduce to equations (4.23) and (4.24), respectively, with $\varepsilon_r(\infty) = 1$ when the dc conductivity $\sigma_0 = 0$. As an application of these equations, suppose that $\varepsilon_r'(\omega) = \varepsilon_r$, a constant on the entire real axis. Then equation (4.30) suggests that the imaginary part must be of the form $\varepsilon_r''(\omega) = \sigma_0/\omega\varepsilon_0$, which corresponds to either a lossless medium when $\sigma_0 = 0$ or a lossy medium with constant conductivity equal to the dc conductivity $\sigma_0$ when $\sigma_0 \neq 0$. ■■

**Example 4.2.1.2.** Bandpass medium.
As another example, let us prescribe a bandpass characteristic for the real part of a dielectric material:

$$\varepsilon_r'(\omega) = \begin{cases} K, \text{ if } 0 < \omega_1 \leqslant \omega \leqslant \omega_2 < \infty \\ 0, \text{ otherwise,} \end{cases} \tag{4.32}$$

along the positive real axis. In this example, $\varepsilon_r(\infty) = 0$. Then from the right-most part of equation (4.23) the imaginary part can be shown to be

$$\varepsilon_r''(\omega) = \frac{K}{\pi}\ln\left(\frac{(\omega_2 + \omega)|\omega_1 - \omega|}{(\omega_1 + \omega)|\omega_2 - \omega|}\right), \quad 0 \leqslant \omega < \infty. \tag{4.33}$$

Note that $\varepsilon_r''(\omega)$ must be an odd function of $\omega$ with the property $\varepsilon_r''(0) = 0$. Figure 4.3 shows a plot of the real and imaginary parts of $\varepsilon_{rc}(\omega)$ for $K = \pi$, $\omega_1 = 30$, $\omega_2 = 50$. It is clear that a bandpass characteristic of the real part must be associated with negative values of the imaginary part, which is non-physical. Note that the imaginary part has zero crossings at $\omega = 0$, $\sqrt{\omega_1\omega_2}$ and that the negative part of the spectrum lies in the interval $\omega \in (0, \sqrt{\omega_1\omega_2})$. Therefore if we choose $\omega_1 = 0$, we obtain $\varepsilon_r''(\omega) = K/\pi \ln[(\omega_2 + \omega)/|\omega_2 - \omega|]$, which will be positive for all $\omega > 0$. Hence a lowpass assumption for the real part of $\varepsilon_{rc}(\omega)$ is physical. Figure 4.4 shows the permittivity for this case. If $\omega_1 = 0$ and $\omega_2 \to \infty$, then equation (4.33) rightly gives $\varepsilon_r''(\omega) = 0$. ■■

## 4.3 Causal properties of scattering amplitude

In this section we see how causality can be used to derive some fundamental properties of the scattering amplitude in relation to antennas and scatterers. The results were first established in [5, 6] and the development here parallels those works. The theory here demonstrates the combined power of the causality, complex integration, and low frequency asymptotic analysis in arriving at some useful results.

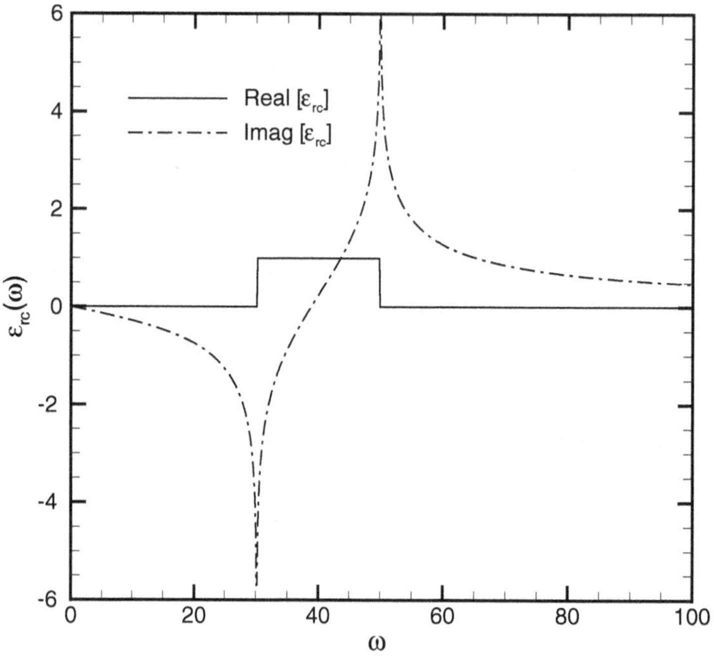


**Figure 4.3.** Real and imaginary parts of a complex permittivity satisfying Kramers–Krönig relations with a bandpass real part.

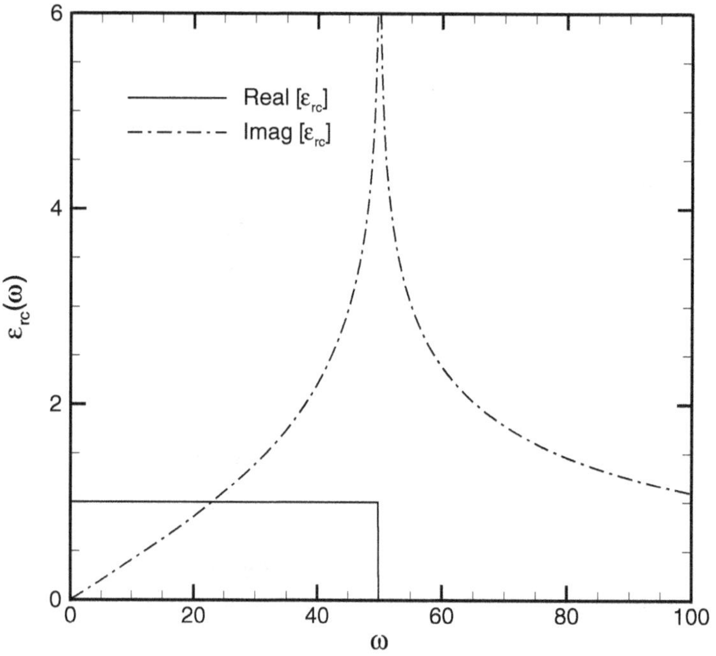


**Figure 4.4.** Real and imaginary parts of a complex permittivity satisfying Kramers–Krönig relations with a lowpass real part.

### 4.3.1 Properties of scattering amplitude

We deduce here the properties of scattering amplitude in the complex wavenumber domain arising from a finite sized, but otherwise arbitrary scatterer, from the considerations of temporal causality (Gustafsson *et al* [5]).

Let $\mathscr{E}^{\mathrm{i}}(t, \mathbf{r}; \hat{\mathbf{k}}_{\mathrm{i}}) = \hat{\mathbf{e}}^{\mathrm{i}} V_0 \delta(ct - \hat{\mathbf{k}}_{\mathrm{i}} \cdot \mathbf{r})$ denote the time-instantaneous electric field of a plane wavefront propagating in the direction $\hat{\mathbf{k}}_{\mathrm{i}}$ with polarization $\hat{\mathbf{e}}^{\mathrm{i}}$, where $c$ is its speed of propagation. The wavefront has been defined such that it reaches the origin of the coordinate system at $t = 0$, see figure 4.5. The plane wave has an amplitude of $V_0$ (units of V) volts per meter. The origin is chosen to be some arbitrary point within the support of the scatterer, whose surface is denoted by $S$. The scatterer can both absorb and redirect incident energy.

The electric field $\mathscr{E}^{\mathrm{s}}$ scattered by an object in a certain direction $\hat{\mathbf{k}}_{\mathrm{s}}$ behaves in the far-zone ($r \gg a$, $a$ being the characteristic dimension of the scattering object) as

$$\mathscr{E}^{\mathrm{s}}(t, \mathbf{r}; \hat{\mathbf{k}}_{\mathrm{i}}) \sim \frac{\mathscr{G}^{\mathrm{s}}(ct - r; \hat{\mathbf{k}}_{\mathrm{s}}, \hat{\mathbf{k}}_{\mathrm{i}})}{r},$$

where $\mathbf{r} = r\hat{\mathbf{k}}_{\mathrm{s}}$. Note that both the incident field $\mathscr{E}^{\mathrm{i}}$ and the time-instantaneous far-zone amplitude $\mathscr{G}^{\mathrm{s}}$ (units of V) are dependent only on the combined space–time argument $ct - \hat{\mathbf{k}} \cdot \mathbf{r}$. Scattering starts to emanate in all directions beginning from the instant when the incident wavefront first strikes the scatterer and as such the scattered field is causally related to the incident field in any direction. However, since we are referencing the far-zone distance relative to an origin chosen inside the scatterer and the time $t = 0$ has been defined as the instant at which the incident field reaches this origin, the time-instantaneous scattered field in the backward sector (i.e. $\hat{\mathbf{k}}_{\mathrm{s}}$ between $\hat{\mathbf{k}}_{\mathrm{s}1}$ and $\hat{\mathbf{k}}_{\mathrm{s}2}$ in the figure) can start to appear *before* $t = 0$. This does not violate causality in any way. In the forward sector (complement of the backward sector), the scattered field in any direction in the far-zone can only appear after $t = 0$.

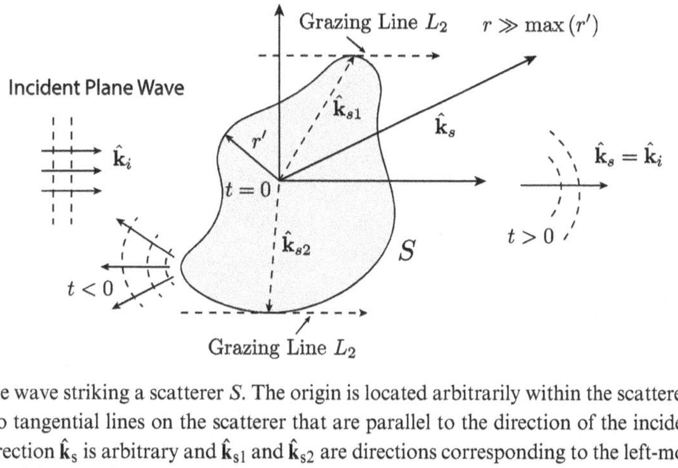


**Figure 4.5.** Plane wave striking a scatterer $S$. The origin is located arbitrarily within the scatterer. Lines $L_1$ and $L_2$ correspond to tangential lines on the scatterer that are parallel to the direction of the incident wave vector $\hat{\mathbf{k}}_{\mathrm{i}}$. Scattering direction $\hat{\mathbf{k}}_{\mathrm{s}}$ is arbitrary and $\hat{\mathbf{k}}_{\mathrm{s}1}$ and $\hat{\mathbf{k}}_{\mathrm{s}2}$ are directions corresponding to the left-most points of the intersection of $S$ with $L_1$ and $L_2$, respectively.

In particular, the scattered field in the forward direction can only appear after $t = 0$ irrespective of the choice of the origin within the scatterer. Henceforth we shall focus our attention on the forward scatter direction $\hat{\mathbf{k}}_s = \hat{\mathbf{k}}_i$. The far-zone amplitude $\mathscr{G}^s$ in the forward direction is causally related to the incident field through a temporal scattering dyadic (a 3 × 3 matrix) $\mathbf{S}_t$ (unitless) and the convolutional integral as

$$\mathscr{G}^s(\tau = ct; \hat{\mathbf{k}}_i, \hat{\mathbf{k}}_i) = \int_{-\infty}^{\tau} \mathbf{S}_t(\tau - \tau'; \hat{\mathbf{k}}_s = \hat{\mathbf{k}}_i, \hat{\mathbf{k}}_i) \cdot \mathscr{E}^i(t', \mathbf{r}' = 0; \hat{\mathbf{k}}_i)\, \mathrm{d}\tau', \tag{4.34}$$

where $\tau' = ct'$. Note that $\mathbf{S}_t$ is independent of the strength of the incident electric field. If $\mathbf{S}(k; \hat{\mathbf{k}}_i, \hat{\mathbf{k}}_i)$ (units of m) is the Fourier transform of $\mathbf{S}_t(\tau; \hat{\mathbf{k}}_i, \hat{\mathbf{k}}_i)$ then

$$\mathbf{S}(k; \hat{\mathbf{k}}_i, \hat{\mathbf{k}}_i) = \int_0^{\infty} \mathbf{S}_t(\tau; \hat{\mathbf{k}}_i, \hat{\mathbf{k}}_i) \mathrm{e}^{-jk\tau}\, \mathrm{d}\tau, \tag{4.35}$$

where $k = \omega/c$ is the complex wavenumber at the complex-valued radian frequency $\omega$. The frequency domain representation of the incident field at the origin is

$$\mathbf{E}^i(k, \mathbf{r}' = 0; \hat{\mathbf{k}}_i) = \hat{\mathbf{e}}_i V_0 \int_{-\infty}^{\infty} \delta(\tau') \mathrm{e}^{-jk\tau'}\, \mathrm{d}\tau' = \hat{\mathbf{e}}_i V_0. \tag{4.36}$$

Note that the frequency domain representation of the electric field here has dimensions of V and is dimensionally not the same as the phasor of a time-harmonic field. This distinction arises because of the extra dimensions introduced by the Fourier transformation process and will not materially change the physics of the problem. If $\mathbf{G}^s(k; \hat{\mathbf{k}}_i, \hat{\mathbf{k}}_i)$ (units of V m) is the Fourier transform of $\mathscr{G}^s(\tau; \hat{\mathbf{k}}_i, \hat{\mathbf{k}}_i)$, then equation (4.34) and result (4.36) imply that

$$\mathbf{G}^s(k; \hat{\mathbf{k}}_i, \hat{\mathbf{k}}_i) = V_0 \mathbf{S}(k; \hat{\mathbf{k}}_i, \hat{\mathbf{k}}_i) \cdot \hat{\mathbf{e}}_i, \tag{4.37}$$

If $k = p$ is real, the scattering dyadic must satisfy the property (i) $\mathbf{S}(-p; \hat{\mathbf{k}}_i; \hat{\mathbf{k}}_i) = \mathbf{S}^*(p; \hat{\mathbf{k}}_i; \hat{\mathbf{k}}_i)$ in order that $\mathscr{G}^s$ is real-valued. By causality, the scattered field at a distance $r$ in the forward direction must be zero until a time $t = r/c$ has elapsed, that is, the scattering amplitude $\mathscr{G}^s(\tau; \hat{\mathbf{k}}_i, \hat{\mathbf{k}}_i)$ must be zero for $\tau = ct - r < 0$. This places a requirement that its Fourier transform $\mathbf{G}^s(k; \hat{\mathbf{k}}_s, \hat{\mathbf{k}}_s)$ be analytic in the lower half of the complex $k$-plane. It follows from equation (4.37) that the forward scattering dyadic $\mathbf{S}(k; \hat{\mathbf{k}}_i, \hat{\mathbf{k}}_i)$ is analytic in the lower half of the complex $k$-plane. From equation (4.35) it is clear that since $\mathbf{S}_t(\tau; \hat{\mathbf{k}}_i, \hat{\mathbf{k}}_i)$ is real-valued, $\mathbf{S}(k = -jq; \hat{\mathbf{k}}_i, \hat{\mathbf{k}}_i)$ is also real-valued for $q$ real. Combining this with the property (i) we have (ii) $\mathbf{S}^*(k = p - jq; \hat{\mathbf{k}}_i, \hat{\mathbf{k}}_i) = \mathbf{S}(-p - jq; \hat{\mathbf{k}}_i, \hat{\mathbf{k}}_i)$. The two properties can be combined to yield the crossing symmetry property that

$$\mathbf{S}^*(k; \hat{\mathbf{k}}_i, \hat{\mathbf{k}}_i) = \mathbf{S}(-k^*; \hat{\mathbf{k}}_i, \hat{\mathbf{k}}_i), \quad k = (p - jq), \text{ complex}. \tag{4.38}$$

Sometimes a more convenient way of expressing this property is $\mathbf{S}(-k; \hat{\mathbf{k}}_i, \hat{\mathbf{k}}_i) = \mathbf{S}^*(k^*; \hat{\mathbf{k}}_i, \hat{\mathbf{k}}_i)$. A direct consequence of this is that $\mathbf{S}(0; \hat{\mathbf{k}}_i, \hat{\mathbf{k}}_i)$ is purely real. The frequency domain far-zone scattered field in the forward direction is

$$
\begin{aligned}
\mathbf{E}^{\mathrm{s}}(k, \hat{\mathbf{k}}_{\mathrm{i}}, \hat{\mathbf{k}}_{\mathrm{i}}) &= \int_{-\infty}^{\infty} \mathscr{E}^{\mathrm{s}}(t, \mathbf{r}; \hat{\mathbf{k}}_{\mathrm{i}}) \mathrm{e}^{-jk\tau}\, \mathrm{d}\tau \\
&= \frac{1}{r} \mathbf{G}^{\mathrm{s}}(k; \hat{\mathbf{k}}_{\mathrm{i}}, \hat{\mathbf{k}}_{\mathrm{i}}) \mathrm{e}^{-jkr} \\
&= V_0 \frac{\mathrm{e}^{-jkr}}{r} \mathbf{S}(k; \hat{\mathbf{k}}_{\mathrm{i}}, \hat{\mathbf{k}}_{\mathrm{i}}) \cdot \hat{\mathbf{e}}_{\mathrm{i}} =: \frac{\mathrm{e}^{-jkr}}{r} \mathbf{F}^{\mathrm{s}}(k; \hat{\mathbf{k}}_{\mathrm{s}} = \hat{\mathbf{k}}_{\mathrm{i}}, \hat{\mathbf{k}}_{\mathrm{i}}),
\end{aligned}
\tag{4.39}
$$

where $\mathbf{F}^{\mathrm{s}}(k; \hat{\mathbf{k}}_{\mathrm{s}} = \hat{\mathbf{k}}_{\mathrm{i}}, \hat{\mathbf{k}}_{\mathrm{i}}) = V_0 \mathbf{S}(k; \hat{\mathbf{k}}_{\mathrm{i}}, \hat{\mathbf{k}}_{\mathrm{i}}) \cdot \hat{\mathbf{e}}_{\mathrm{i}}$ is the *scattering amplitude* defined in theorem 11.1.2 (except that the units of $\mathbf{F}^{\mathrm{s}}$ here are V m). Here we explicitly show the dependence of $\mathbf{F}^{\mathrm{s}}$ on the wavenumber $k$ by including it in the argument. As opposed to the scattering dyadic $\mathbf{S}$, the scattering amplitude is directly proportional to the strength of the incident electric field. The scattering amplitude has the representations given in equations (11.24) and (11.25), which are repeated below

$$
\begin{aligned}
\mathbf{F}^{\mathrm{s}}(k; \hat{\mathbf{k}}_{\mathrm{s}}, \hat{\mathbf{k}}_{\mathrm{i}}) &= \frac{-jk}{4\pi} \hat{\mathbf{k}}_{\mathrm{s}} \times \left[ \oiint_S (\hat{\mathbf{n}}' \times \mathbf{E}^{\mathrm{s}}) \mathrm{e}^{jk\hat{\mathbf{k}}_{\mathrm{s}} \cdot \mathbf{r}'}\, \mathrm{d}s' - \hat{\mathbf{k}}_{\mathrm{s}} \times \oiint_S (\hat{\mathbf{n}}' \times \mathbf{H}^{\mathrm{s}}) \mathrm{e}^{jk\hat{\mathbf{k}}_{\mathrm{s}} \cdot \mathbf{r}'}\, \mathrm{d}s' \right] \\
&= \frac{k^2}{4\pi} \hat{\mathbf{k}}_{\mathrm{s}} \times \left[ \hat{\mathbf{k}}_{\mathrm{s}} \times \oiint_S \mathbf{r}' \left( \eta \hat{\mathbf{k}}_{\mathrm{s}} \cdot \hat{\mathbf{n}}' \times \mathbf{H}^{\mathrm{s}} - \hat{\mathbf{n}}' \cdot \mathbf{E}^{\mathrm{s}} \right) \mathrm{e}^{jk\hat{\mathbf{k}}_{\mathrm{s}} \cdot \mathbf{r}'}\, \mathrm{d}s' \right. \\
&\qquad \left. - \oiint_S \mathbf{r}' \left( \hat{\mathbf{k}}_{\mathrm{s}} \cdot \hat{\mathbf{n}}' \times \mathbf{E}^{\mathrm{s}} + \eta \hat{\mathbf{n}}' \cdot \mathbf{H}^{\mathrm{s}} \right) \mathrm{e}^{jk\hat{\mathbf{k}}_{\mathrm{s}} \cdot \mathbf{r}'}\, \mathrm{d}s' \right],
\end{aligned}
\tag{4.40}
$$

$$
=: V_0 k^2 \rho(k; \hat{\mathbf{k}}_{\mathrm{s}}, \hat{\mathbf{k}}_{\mathrm{i}}),
\tag{4.41}
$$

where the vector

$$
\begin{aligned}
\rho(k; \hat{\mathbf{k}}_{\mathrm{i}}, \hat{\mathbf{k}}_{\mathrm{i}}) &= \mathbf{S}(k; \hat{\mathbf{k}}_{\mathrm{i}}, \hat{\mathbf{k}}_{\mathrm{i}}) \cdot \hat{\mathbf{e}}_{\mathrm{i}} / k^2 \\
&= \frac{1}{4\pi V_0} \hat{\mathbf{k}}_{\mathrm{i}} \times \left[ \hat{\mathbf{k}}_{\mathrm{i}} \times \oiint_S \mathbf{r}' \left( \eta \hat{\mathbf{k}}_{\mathrm{i}} \cdot \hat{\mathbf{n}}' \times \mathbf{H}^{\mathrm{s}} - \hat{\mathbf{n}}' \cdot \mathbf{E}^{\mathrm{s}} \right) \mathrm{e}^{jk\hat{\mathbf{k}}_{\mathrm{i}} \cdot \mathbf{r}'}\, \mathrm{d}s' \right. \\
&\qquad \left. - \oiint_S \mathbf{r}' \left( \hat{\mathbf{k}}_{\mathrm{i}} \cdot \hat{\mathbf{n}}' \times \mathbf{E}^{\mathrm{s}} + \eta \hat{\mathbf{n}}' \cdot \mathbf{H}^{\mathrm{s}} \right) \mathrm{e}^{jk\hat{\mathbf{k}}_{\mathrm{i}} \cdot \mathbf{r}'}\, \mathrm{d}s' \right],
\end{aligned}
\tag{4.42}
$$

has units of volume m$^3$ and is well behaved at the origin in the complex $k$-plane. As remarked in the proof of theorem 11.1.2, it is perfectly valid to change the scattered fields inside the integrals above to total fields. This is because the integrals with incident fields only on the rhs would produce a zero scattered field and, consequently, zero scattering amplitude. The property (4.38) directly translates to that on the scattering amplitude

$$
[\mathbf{F}^{\mathrm{s}}(k; \hat{\mathbf{k}}_{\mathrm{s}} = \hat{\mathbf{k}}_{\mathrm{i}}, \hat{\mathbf{k}}_{\mathrm{i}})]^* = \mathbf{F}^{\mathrm{s}}(-k^*; \hat{\mathbf{k}}_{\mathrm{s}} = \hat{\mathbf{k}}_{\mathrm{i}}, \hat{\mathbf{k}}_{\mathrm{i}}).
\tag{4.43}
$$

Furthermore, the scattering amplitude in the forward direction is analytic in the lower half of the complex $k$-plane. The function $\rho(k;\hat{\mathbf{k}}_i,\hat{\mathbf{k}}_i)$ defined in equation (4.42) is also analytic in the lower half of the complex $k$-plane. It is also obvious from property (4.38) and the relations between $\mathbf{S}(k;\hat{\mathbf{k}}_i,\hat{\mathbf{k}}_i)$, the scattering amplitude $\mathbf{F}^s(k;\hat{\mathbf{k}}_i,\hat{\mathbf{k}}_i)$, and the function $\rho(k;\hat{\mathbf{k}}_i,\hat{\mathbf{k}}_i)$, that *they are all purely real at* $k = 0$.

### 4.3.2 Fundamental limits on the antenna gain–bandwidth product

We shall first derive a summation rule for the extinction cross section of a scatterer and use that result in a following example to establish an upper bound for the gain–bandwidth product of any antenna. Interestingly, the sum rule and the upper bound are dependent only on the low frequency characteristics of the object comprising the scatterer or the antenna.

**Example 4.3.2.1** Summation rule for extinction cross section of scatterer.
Using the properties of the scattering amplitude as elucidated in subsection 4.3.1, the results of the optical theorem (see theorem 11.1.2), and assuming that the extinction cross section is an even function of $k$ and of order $\mathcal{O}(k^2)$ near the origin [7], we shall establish a summation rule for the extinction cross section of an arbitrary scattering object (Gustafsson *et al* [5]).

From the optical theorem it is known that the extinction cross section $\sigma_{\text{ext}}(k)$ is related to the scattering amplitude via equation (11.17)

$$\sigma_{\text{ext}}(k) = \sigma_s(k) + \sigma_a(k) = -\frac{4\pi}{k}\Im\left[\hat{\mathbf{e}}_i^* \cdot \frac{\mathbf{F}^s(k;\hat{\mathbf{k}}_i,\hat{\mathbf{k}}_i)}{V_0}\right], \tag{4.44}$$

where we have used the notation of subsection 4.3.1. In view of the definition given in equation (4.42), the extinction cross section can also be written as

$$\sigma_{\text{ext}}(k) = -4\pi k\Im[\hat{\mathbf{e}}_i^* \cdot \rho(k;\hat{\mathbf{k}}_i,\hat{\mathbf{k}}_i)] = -4\pi k\Im[\rho_{\text{ext}}(k)], \tag{4.45}$$

where $\rho_{\text{ext}}(k) = \hat{\mathbf{e}}_i^* \cdot \rho(k;\hat{\mathbf{k}}_i,\hat{\mathbf{k}}_i)$ can be regarded as *extinction volume* (units of $\text{m}^3$). Now we have from the result (4) of Titchmarsh's theorem (see theorem 4.1.1) that since $\rho_{\text{ext}}(k)$ is analytic in the lower half of the complex $k$-plane,

$$\Re[\rho_{\text{ext}}(k)] = \frac{1}{\pi}\mathscr{P}\int_{k'=-\infty}^{\infty}\frac{\Im[\rho_{\text{ext}}(k')]}{k-k'}\,\mathrm{d}k'. \tag{4.46}$$

Evaluating this at $k = 0$ and using (i) the property that $\rho_{\text{ext}}(0)$ is purely real and (ii) the relation (4.45), and replacing the dummy variable $k'$ in the integrand by $k$, we obtain

$$\begin{aligned}\rho_{\text{ext}}(k=0) &= \frac{1}{4\pi^2}\mathscr{P}\int_{k=-\infty}^{\infty}\frac{\sigma_{\text{ext}}(k)}{k^2}\,\mathrm{d}k = \frac{1}{2\pi^2}\int_{k=0}^{\infty}\frac{\sigma_{\text{ext}}(k)}{k^2}\,\mathrm{d}k\\ &= \frac{1}{4\pi^3}\int_{\lambda=0}^{\infty}\sigma_{\text{ext}}(\lambda)\,\mathrm{d}\lambda,\end{aligned} \tag{4.47}$$

where $\lambda = 2\pi/k$ is the wavelength. The Cauchy's principle value integral has been converted to a regular integral since $\sigma_{\text{ext}}(k)$ is of the order $\mathcal{O}(k^2)$ near the origin and, thereby, $\sigma_{\text{ext}}(k)/k^2$ is finite near the origin. Equation (4.47) expresses a remarkable result in that the lhs is dependent only on the static field behavior (i.e. behavior at zero frequency) of the scatterer, whereas the rhs is the integrated effect of the dynamic response (scattering and absorption by the scatterer) over all wavenumbers. The result is valid for any passive scatterer irrespective of its shape, electrical size, and composition.

If a time-harmonic incident wave with electric field strength $E_0 = 1$ V m$^{-1}$, polarization $\hat{\mathbf{e}}_{\text{i}}$ and magnetic field polarization $\hat{\mathbf{h}}_{\text{i}}$ $(= \hat{\mathbf{k}}_{\text{i}} \times \hat{\mathbf{e}}_{\text{i}})$ induces normalized electric dipole moment $\mathbf{p}$ (units of F m$^2$) and normalized magnetic dipole moment $\mathbf{m}$ (units of Ohms$^{-1}$ m$^3$) in the scatterer at zero frequency, then Kleinman and Senior [7, equation (2.89), p 18] show that

$$\rho_{\text{ext}}(k) = \hat{\mathbf{e}}_{\text{i}}^* \cdot \rho(k; \hat{\mathbf{k}}_{\text{i}}, \hat{\mathbf{k}}_{\text{i}}) \sim \frac{1}{4\pi}\left(\frac{1}{\varepsilon_0}\hat{\mathbf{e}}_{\text{i}}^* \cdot \mathbf{p} + \eta\hat{\mathbf{h}}_{\text{i}}^* \cdot \mathbf{m}\right) + \mathcal{O}(k), \quad k \to 0. \tag{4.48}$$

Note that this becomes an exact relation for $k = 0$. Combining this with equation (4.47) we arrive at

$$\int_0^\infty \frac{\sigma_{\text{ext}}(k)}{k^2}\,\mathrm{d}k = \frac{1}{2\pi}\int_0^\infty \sigma_{\text{ext}}(\lambda)\,\mathrm{d}\lambda = \frac{\pi}{2}\left(\frac{1}{\varepsilon_0}\hat{\mathbf{e}}_{\text{i}}^* \cdot \mathbf{p} + \eta\hat{\mathbf{h}}_{\text{i}}^* \cdot \mathbf{m}\right), \tag{4.49}$$

which is an *exact relation valid for any generic scatterer*. For linear media, it is possible to relate the normalized induced dipole moments $(\mathbf{p}, \mathbf{m})$ to the incident wave polarizations $(\hat{\mathbf{e}}_{\text{i}}, \hat{\mathbf{h}}_{\text{i}})$ through a normalized polarizability matrix $\boldsymbol{\alpha}$ as

$$\begin{pmatrix} \dfrac{\mathbf{p}}{\varepsilon_0} \\ \eta\mathbf{m} \end{pmatrix} = \boldsymbol{\alpha}\begin{pmatrix} \hat{\mathbf{e}}_{\text{i}} \\ \hat{\mathbf{h}}_{\text{i}} \end{pmatrix} = \begin{pmatrix} \alpha_{\text{ee}} & \alpha_{\text{em}} \\ \alpha_{\text{me}} & \alpha_{\text{mm}} \end{pmatrix}\begin{pmatrix} \hat{\mathbf{e}}_{\text{i}} \\ \hat{\mathbf{h}}_{\text{i}} \end{pmatrix} =: \begin{pmatrix} \gamma_{\text{e}} \cdot \hat{\mathbf{e}}_{\text{i}} \\ \gamma_{\text{m}} \cdot \hat{\mathbf{h}}_{\text{i}} \end{pmatrix}, \tag{4.50}$$

where $\gamma_{\text{e}} = (\alpha_{\text{ee}}\hat{\mathbf{e}}_{\text{i}}\hat{\mathbf{e}}_{\text{i}} + \alpha_{\text{em}}\hat{\mathbf{h}}_{\text{i}}\hat{\mathbf{e}}_{\text{i}})$ m$^3$ and $\gamma_{\text{m}} = (\alpha_{\text{me}}\hat{\mathbf{e}}_{\text{i}}\hat{\mathbf{h}}_{\text{i}} + \alpha_{\text{mm}}\hat{\mathbf{h}}_{\text{i}}\hat{\mathbf{h}}_{\text{i}})$ m$^3$ denote the electric and magnetic polarizability dyadics of the scatterer. In terms of these polarizability dyadics, the summation rule reads

$$\int_0^\infty \frac{\sigma_{\text{ext}}(k)}{k^2}\,\mathrm{d}k = \frac{1}{2\pi}\int_0^\infty \sigma_{\text{ext}}(\lambda)\,\mathrm{d}\lambda = \frac{\pi}{2}\left(\hat{\mathbf{e}}_{\text{i}}^* \cdot \gamma_{\text{e}} \cdot \hat{\mathbf{e}}_{\text{i}} + \hat{\mathbf{h}}_{\text{i}}^* \cdot \gamma_{\text{m}} \cdot \hat{\mathbf{h}}_{\text{i}}\right). \tag{4.51}$$

This summation rule remains exactly valid for any generic scatterer and forms the basis for characterizing the behavior of electrically small antennas in [6]. ■■

**Example 4.3.2.2** Upper bound to antenna gain–bandwidth product.
Using the results of example 4.3.2.1, establish an upper bound on the gain–bandwidth product of an arbitrary receiving antenna (Gustafsson *et al* [6]).

Let the receiving antenna have gain $G(\lambda)$, effective aperture $A_{\text{e}}(\lambda)$, load reflection coefficient $\Gamma(\lambda)$, and an absorption efficiency $\eta_{\text{a}}(\lambda)$, each of which is dependent on the operating wavelength $\lambda$. The absorption efficiency relates the absorption cross section to the extinction cross section via $\eta_a = \sigma_a/\sigma_{ext}$. When a plane wave is incident on a receiving antenna, a portion of the incident power is scattered, a portion is dissipated due to material losses and the remaining is delivered to a load connected across the antenna terminals. The absorption cross section accounts for the portion delivered to the load and dissipated and is non-zero even for an antenna with no dissipative losses. If the load is not matched to the antenna input impedance, the power delivered to the load is reduced by a factor $0 \leqslant (1 - |\Gamma|^2) \leqslant 1$. The absorption cross section of an antenna under mismatch loss is taken as $\sigma_a = (1 - |\Gamma|^2)A_e$. The extinction cross section $\sigma_{ext} = \sigma_a/\eta_a = (1 - |\Gamma|^2)A_e/\eta_a$ depends on the polarization of the incoming wave, its direction of propagation, and the wavelength of operation. In general, $\sigma_{ext}$ has bandpass characteristics centered around a nominal wavelength $\lambda_0$ and the two edges defined by $\lambda_1 \leqslant \lambda \leqslant \lambda_2$, see figure 4.6.

The center wavelength is often taken to be the arithmetic mean of $\lambda_1$ and $\lambda_2$, namely $2\lambda_0 = (\lambda_2 + \lambda_1)$. The fractional bandwidth $B$ of the antenna is defined as $B = (\lambda_2 - \lambda_1)/\lambda_0 = 2(\lambda_2 - \lambda_1)/(\lambda_2 + \lambda_1)$. Clearly, $0 \leqslant B \leqslant 2$. The gain and effective area of an antenna are related by $A_e = G\lambda^2/4\pi$. Therefore $\sigma_{ext} = (1 - |\Gamma|^2)G\lambda^2/(4\pi\eta_a) \geqslant (1 - |\Gamma|^2)G\lambda^2/4\pi \geqslant G_\Lambda\lambda^2/4\pi$, where $G_\Lambda = \min[(1 - |\Gamma|^2)G]$ is the minimum value of the imperfectly matched gain over the wavelength interval $\Lambda = (\lambda_1, \lambda_2)$ of interest. Using this we have

$$\int_0^\infty \sigma_{ext}(\lambda)\,d\lambda \geqslant \frac{G_\Lambda}{4\pi}\int_{\lambda_1}^{\lambda_2} \lambda^2\,d\lambda = \frac{\lambda_0^3 G_\Lambda B}{4\pi}\left(1 + \frac{B^2}{12}\right) \geqslant \frac{\lambda_0^3 G_\Lambda B}{4\pi}. \tag{4.52}$$

Therefore, denoting $k_0 = 2\pi/\lambda_0$ and inserting this result into equation (4.51), we arrive at the upper bound

$$G_\Lambda B \leqslant \frac{1}{2}k_0^3\left(\hat{\mathbf{e}}_i^* \cdot \gamma_e \cdot \hat{\mathbf{e}}_i + \hat{\mathbf{h}}_i^* \cdot \gamma_m \cdot \hat{\mathbf{h}}_i\right) = \frac{2\pi}{3}\left(\frac{\hat{\mathbf{e}}_i^* \cdot \gamma_e \cdot \hat{\mathbf{e}}_i + \hat{\mathbf{h}}_i^* \cdot \gamma_m \cdot \hat{\mathbf{h}}_i}{V_s}\right)(k_0 a)^3, \tag{4.53}$$

where $a$ is the radius of the sphere circumscribing the antenna and $V_s = 4\pi a^3/3$ its volume. The rhs gives a fundamental upper bound to the realizable gain–bandwidth product of any passive antenna. The result is reminiscent of the Bode–Fano

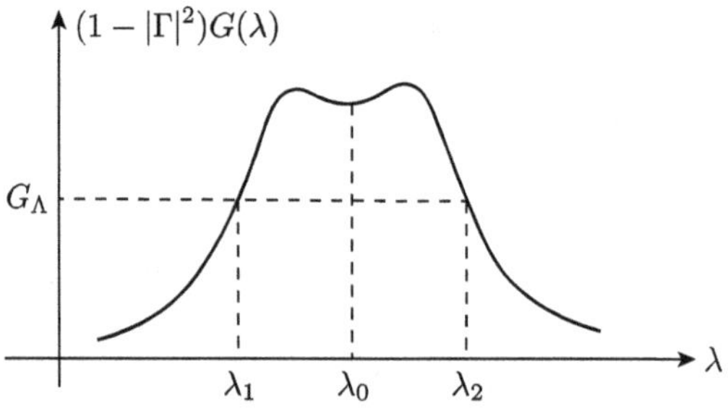


**Figure 4.6.** Bandpass characteristics of a typical antenna as a function of wavelength $\lambda$.

criterion[2] encountered in the design of matching networks. First it is seen that the gain–bandwidth product of an antenna cannot be made arbitrarily large if its electrical size is limited. Further, it is seen that the upper bound varies as frequency raised to the third power; if the frequency is halved, the gain–bandwidth product is expected to decrease by a factor of eight. The quantity within the parenthesis is independent of wavelength and depends only on the material properties, shape, and size of the antenna structure and the polarization and direction of propagation of the incident wave. ■■

**Example 4.3.2.3.** Numerical example of prolate spheroidal dipole.
As an example, consider a perfectly conducting dipole antenna in the shape of a prolate spheroid, figure 4.7. The semi-minor axis has dimension $b$ and the semi-major axis has dimension $h$. When used as a transmitter, the antenna is fed by a delta-gap generator to result in a $\mathrm{TM}_r$ mode ($H_r = 0$). The dominant far-zone field components are ($E_\theta$, $H_\phi$). Let a plane wave having such a polarization be incident from an angle $\theta$ as shown in figure 4.7. In such a case, the electric and magnetic polarizability dyadics have closed form expressions. Let $V_p = 4\pi b^2 h/3$ be the volume of the spheroidal structure and let $\xi = \sqrt{1 - (b/h)^2}$ be the eccentricity of the ellipse in a meridian plane. Then it can be shown upon using the static solutions available for ellipsoids of symmetry [7, pp 22–23, 27–28]: $\gamma_e = V_p[\mathbf{I}P_{11} + \hat{\mathbf{z}}\hat{\mathbf{z}}(P_{33} - P_{11})]$ and $-\gamma_m = V_p[\mathbf{I}M_{11} + \hat{\mathbf{z}}\hat{\mathbf{z}}(\frac{1}{2}P_{11} - M_{11})]$ that the quantity $A$ proportional to the rhs of equation (4.53), namely

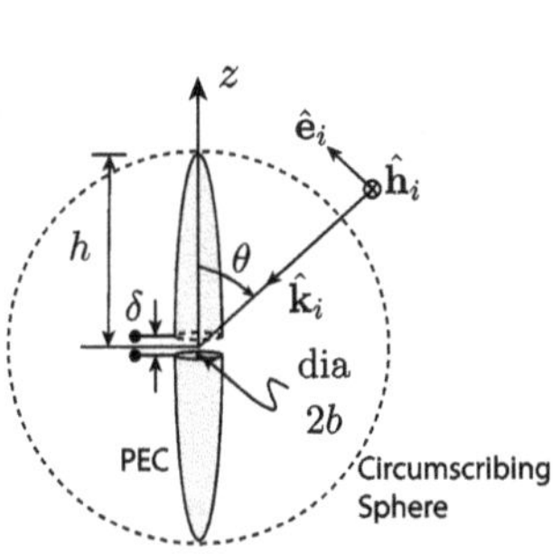


**Figure 4.7.** A $\mathrm{TM}_r$ plane wave incident on a center-fed receiving dipole antenna having the shape of a prolate spheroid. The major component of current flows in the axial direction when $b/h \ll 1$.

[2] The Bode–Fano criterion places an upper limit on the achievable bandwidth of a passive matching network designed to match a source impedance $R_s$ to a first-order load impedance such as a parallel combination of a load resistance $R$ and a capacitance $C$. It states that the frequency-dependent, complex-valued reflection coefficient $\Gamma(\omega)$ seen at the input of the matching network satisfies the relation [8]

$$\int_0^\infty \ln \frac{1}{|\Gamma(\omega)|}\, d\omega \leqslant \frac{\pi}{RC}.$$

$$\begin{aligned} A &= \left(\hat{\mathbf{e}}_{\mathrm{i}}^{*} \cdot \gamma_{\mathrm{e}} \cdot \hat{\mathbf{e}}_{\mathrm{i}} + \hat{\mathbf{h}}_{\mathrm{i}}^{*} \cdot \gamma_{\mathrm{m}} \cdot \hat{\mathbf{h}}_{\mathrm{i}}\right) \\ &= V_p\left[(P_{11} - M_{11}) + |\hat{\mathbf{e}}_{\mathrm{i}} \cdot \hat{\mathbf{z}}|^2(P_{33} - P_{11}) - |\hat{\mathbf{h}}_{\mathrm{i}} \cdot \hat{\mathbf{z}}|^2\left(\frac{1}{2}P_{11} - M_{11}\right)\right], \end{aligned} \tag{4.54}$$

where

$$P_{11} = \frac{1}{L_1}, \quad P_{33} = \frac{1}{L_3}, \quad M_{11} = -\frac{1}{1 - L_1}, \quad M_{33} = \frac{1}{2}P_{11}, \tag{4.55}$$

and

$$L_1 = \frac{1}{2\xi^2}\left[1 - \frac{1 - \xi^2}{2\xi}\ln\left(\frac{1 + \xi}{1 - \xi}\right)\right], \quad L_3 = 1 - 2L_1. \tag{4.56}$$

For the assumed $\mathrm{TM}_r$ incident wave polarization $\hat{\mathbf{e}}_{\mathrm{i}} = -\hat{\boldsymbol{\theta}}$, $\hat{\mathbf{h}}_{\mathrm{i}} = \hat{\boldsymbol{\phi}}$. Therefore

$$A = V_p\left[\frac{1}{L_1(1 - L_1)} + \left(\frac{1}{L_3} - \frac{1}{L_1}\right)\sin^2\theta\right]. \tag{4.57}$$

Note that this is a function of the direction of arrival $\theta$ of the incident wave. It is easy to verify that

$$\frac{1}{L_1} + \frac{1}{(1 - L_1)} \leqslant \frac{A}{V_p} \leqslant \frac{1}{L_3} + \frac{1}{(1 - L_1)}. \tag{4.58}$$

Since the rhs of equation (4.53) must be valid for all $\theta$, we may use its least upper bound obtainable over all $\theta$. This corresponds to the lower bound of $A$. Writing $V_s = 4\pi h^3/3$ with $a = h$ for the volume of the circumscribing sphere, we obtain the upper bound for the gain–bandwidth product of a thin prolate spheroidal dipole antenna to be

$$G_\Lambda B \leqslant \frac{2\pi}{3}\left(\frac{b}{h}\right)^2 \frac{(k_0h)^3}{L_1(1 - L_1)}. \tag{4.59}$$

Figure 4.8 shows a plot of the gain–bandwidth limit of the dipole antenna as a function of $h/\lambda_0$. It is seen that gain–bandwidth product steadily increases in the form of a power law having an exponent of 3 as the dipole electrical length increases. Other interesting parametric studies such as the variation of gain–bandwidth product with respect to the aspect ratio $b/h$ can be carried out from equation (4.59). Figure 4.9 shows a variation of the gain–bandwidth product as the ratio $b/h$ of the dipole is varied over one decade. The electrical length of the dipole is held fixed at $2h/\lambda_0 = 0.25$. The gain of the dipole depends primarily on $k_0h$ and to a lesser extent on $(b/h)$ and as such the curve in figure 4.9 may be treated as variation of bandwidth with the dipole waist size. It is seen that the performance metric improves by two orders of magnitude for every one order magnitude of increase of the ratio $(b/h)$. Thus it is consistent with the appearance of the factor $(b/h)^2$ in equation (4.59)

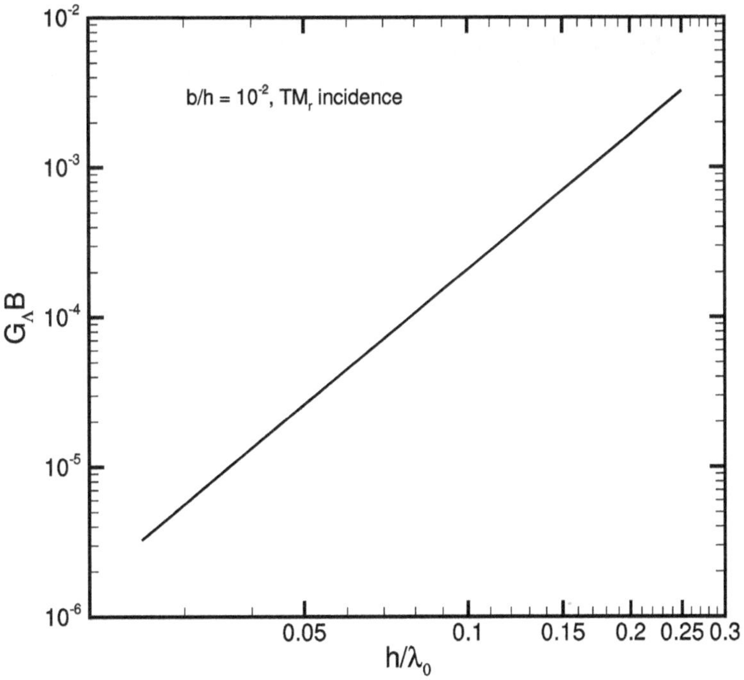


**Figure 4.8.** Maximum gain–bandwidth product of a center-fed prolate spheroidal dipole antenna as a function of its electrical length. Ratio of radius to half-height remains fixed at $b/h = 10^{-2}$. Incident wave is $TM_r$ with magnetic field having only a $\phi$-component.

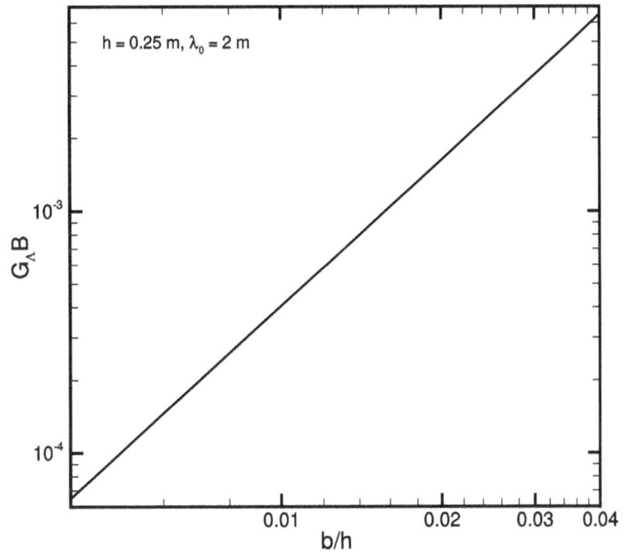


**Figure 4.9.** Variation of maximum possible gain–bandwidth product as a function of dipole waist size. Electrical length of dipole $2h/\lambda_0$ remains fixed at 0.25. Incident wave is $TM_r$ with magnetic field having only a $\phi$-component.

as long as the change in $(b/h)$ is such that it does not substantially influence the other factors. In fact for $(b/h) \to 0$, we have $\xi \sim 1$ and $L_1 \sim 1/2$ so that $G_\Lambda B \lesssim (b/h)^2(k_0h)^3 8\pi/3$, thus showing that the performance improves as $(b/h)^2$. Figure 4.9 indicates clearly that fatter dipoles give larger bandwidths than thinner ones. ■■

## References


[1] Titchmarsh E C 1939 *Introduction to the Theory of Fourier Integrals* (London: Oxford University Press)

[2] Nussenzveig H M 1972 *Causality and Dispersion Relations* (New York: Academic)

[3] Ougshtun K E 2006 *Electromagnetic and Optical Pulse Propagation 1* (New York: Springer)

[4] Landau L D, Lifshitz E M and Pitaevskii L P 1984 *Electrodynamics of Continuous Media* 2nd edn (New York: Pergamon)

[5] Gustafsson M, Sohl C and Kristensson G 2007 Physical limitations on antennas of arbitrary shape *Proc. R. Soc.* A **463** 2589–607

[6] Gustafsson M, Sohl C and Kristensson G 2007 Physical limitations on antennas of arbitrary shape *Technical Report LUTEDX/(TEAT-7153)/1-37/(2007)* Department of Electrical and Information Technology, Lund University

[7] Kleinman R E and Senior T B A 1986 Rayleigh scattering *Low and High Frequency Asymptotics* (Amsterdam: Elsevier) ch 1, pp 1–70

[8] Matthaei G, Young L and Jones E M T 1980 *Microwave Filters, Impedance Matching Networks, and Coupling Structures* (Dedham, MA: Artech House)

# Chapter 5

# Uniqueness, energy, and momentum

In this chapter we shall explore the important theorems related to the uniqueness of fields with respect to the given sources, conservation of electromagnetic energy, and conservation of momentum of electromagnetic fields. To gain additional insight into the aspect of electromagnetic energy, it is desirable to cast the energy expressions in a definite quadratic form. The conditions under which this becomes possible and the requirements it imposes on the material properties in a dispersive medium are also explored in the chapter.

## 5.1 Uniqueness theorem

Uniqueness theorem guarantees that a given set of sources can only generate one electromagnetic field in a lossy medium. We will only consider the time-harmonic situation here, but are wary of the fact that the result is true for arbitrary time dependence owing to a unique relationship between a time-instantaneous field component and its Fourier transform.

**Theorem 5.1.1** Uniqueness theorem [1, p 486].
*Consider time-harmonic sources* $(\mathbf{J}, \mathbf{M})$ *generating the fields* $(\mathbf{E}, \mathbf{H})$ *in a linear, isotropic medium with constitutive parameters* $(\varepsilon, \mu, \sigma)$ *in the presence of a volume* $V_0$ *bounded by the surface S. Let V denote the volume exterior to S and bounded at infinity by* $S_\infty$ *as shown in figure 5.1.*

*Then the electromagnetic field is unique in V when any one of the following three conditions is satisfied:*

(a) *the tangential electric field* $\hat{\mathbf{n}} \times \mathbf{E}$ *is specified on S, or*
(b) *the tangential magnetic field* $\hat{\mathbf{n}} \times \mathbf{H}$ *is specified on S, or*
(c) *the tangential electric field* $\hat{\mathbf{n}} \times \mathbf{E}$ *is specified over a portion,* $S_1$*, of S and the tangential magnetic field* $\hat{\mathbf{n}} \times \mathbf{H}$ *is specified over the remaining portion,* $S_2$*, of S.*

doi:10.1088/978-0-7503-1716-0ch5  

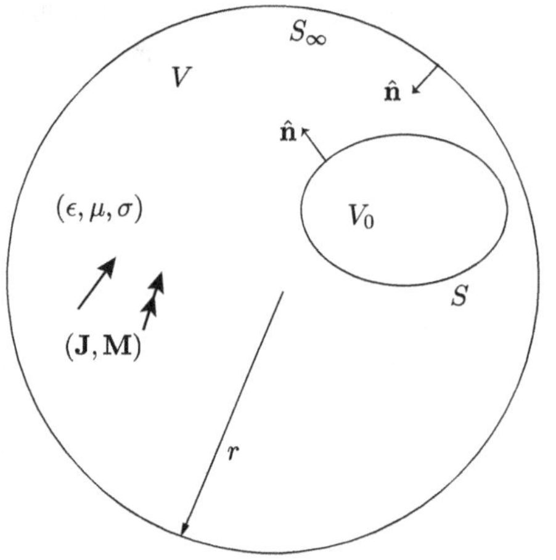


**Figure 5.1.** Geometry under consideration for uniqueness theorem.

*Proof.* We will assume that the source excites a second field $(\mathbf{E} + \delta\mathbf{E}, \mathbf{H} + \delta\mathbf{H})$ and subsequently show that the difference field $(\delta\mathbf{E}, \delta\mathbf{H}) \equiv 0$ everywhere. The difference field, being source-free, satisfies the power conservation result (see equation (5.18))

$$- \oiint\limits_{S+S_\infty} \delta\mathbf{E} \times \delta\mathbf{H}^* \cdot \hat{\mathbf{n}} \, \mathrm{d}s + j\omega \iiint\limits_V (\mu|\delta\mathbf{H}|^2 - \varepsilon|\delta\mathbf{E}|^2) \, \mathrm{d}v + \iiint\limits_V \sigma|\delta\mathbf{E}|^2 \, \mathrm{d}v = 0. \tag{5.1}$$

The negative sign in front of the surface integral is due to the fact that the normal $\hat{\mathbf{n}}$ in power conservation should point away from the volume $V$. The integral over $S_\infty$ in equation (5.1) vanishes because the fields on that surface vanish exponentially due to the loss in the medium. Since $\delta\mathbf{E} \times \delta\mathbf{H}^* \cdot \hat{\mathbf{n}} = \hat{\mathbf{n}} \times \delta\mathbf{E} \cdot \delta\mathbf{H}^* = -(\hat{\mathbf{n}} \times \delta\mathbf{H} \cdot \delta\mathbf{E}^*)^*$, the surface integral over $S$ also vanishes because the difference field has a zero tangential electric field on $S$, or a zero tangential magnetic field on $S$, or zero tangential electric field on a part, $S_1$, of $S$ and zero tangential magnetic field on the remaining portion, $S_2$, of $S$. Equating the real and imaginary parts of equation (5.1) separately to zero we obtain

$$\iiint\limits_V \sigma|\delta\mathbf{E}|^2 \, \mathrm{d}v = 0 \Longrightarrow \delta\mathbf{E} \equiv 0 \text{ in } V \text{ since } \sigma > 0, \tag{5.2}$$

$$\omega \iiint\limits_V (\mu|\delta\mathbf{H}|^2 - \varepsilon|\delta\mathbf{E}|^2) \, \mathrm{d}v = 0 \Longrightarrow \delta\mathbf{H} \equiv 0 \text{ in } V, \ \omega \neq 0 \text{ since } \mu > 0. \tag{5.3}$$

■

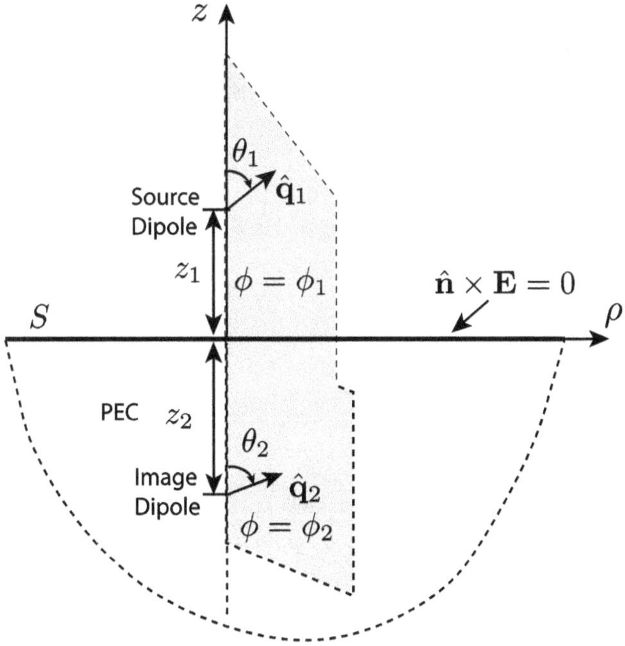


**Figure 5.2.** An infinitesimal dipole radiating in the presence of a perfectly conducting plane.

**Example 5.1.0.1.** Establishing image theory using uniqueness theorem.
Using the uniqueness theorem arrive at the image strengths, location, and orientation for (i) a vertical electric dipole (VED) and (ii) a horizontal electric dipole (HED) placed over a perfectly conducting plane.

Figure 5.2 shows a source dipole of strength (i.e. its dipole moment) $p_1$ and orientation $\hat{\mathbf{q}}_1$ placed over a PEC ground plane at a height $z_1$. If the dipole makes an angle $\theta_1$ with the $z$-axis and lies in the meridian plane $\phi = \phi_1$, then the unit vector $\hat{\mathbf{q}}_1 = \hat{\mathbf{x}} \sin\theta_1 \cos\phi_1 + \hat{\mathbf{y}} \sin\theta_1 \sin\phi_1 + \hat{\mathbf{z}} \cos\theta_1 = \hat{\boldsymbol{\rho}} \sin\theta_1 \cos(\phi - \phi_1) - \hat{\boldsymbol{\phi}} \sin\theta_1 \sin(\phi - \phi_1) + \hat{\mathbf{z}} \cos\theta_1$. The dipole together with the induced currents on the ground plane generates an electric field in the upper half-space such that $\hat{\mathbf{n}} \times \mathbf{E} = 0$ on the surface $z = 0$. For clarity we imagine the ground plane to be closed off at infinity by the curved line in the figure to constitute the closed surface $S$. One can visualize a second problem with the PEC removed and additional sources placed in its place inside $S$ such that the tangential electric field satisfies the same boundary condition on $S$ as the original problem, i.e. $\hat{\mathbf{n}} \times \mathbf{E} = 0$. By the uniqueness theorem the two problems must generate the same field outside $S$, namely in the upper half-space. We shall now construct such an equivalent problem.

Let the image be located at $z = z_2$, $z_2 < 0$, have a strength $p_2$, and an orientation $\hat{\mathbf{q}}_2 = \hat{\mathbf{x}} \sin\theta_2 \cos\phi_2 + \hat{\mathbf{y}} \sin\theta_2 \sin\phi_2 + \hat{\mathbf{z}} \cos\theta_2 = \hat{\boldsymbol{\rho}} \sin\theta_2 \cos(\phi - \phi_2) - \hat{\boldsymbol{\phi}} \sin\theta_2 \sin(\phi - \phi_2) + \hat{\mathbf{z}} \cos\theta_2$. We are allowing for the possibility that the source and image dipoles could lie in different meridian planes and that they could be non-equidistant from the PEC plane. Furthermore, the orientation of the image dipole is kept arbitrary at this point.

Let $R_i = \sqrt{\rho^2 + (z - z_i)^2}$, $i = 1, 2$ be the distance of a field point from the source and image dipoles and let $\psi_i = \mathrm{e}^{-jkR_i}/4\pi R_i$ be the scalar Green's function at the wavenumber $k = \omega\sqrt{\mu\varepsilon}$. It is independent of the azimuthal coordinate $\phi$. Then from equation (1.47) the magnetic vector potential $\mathbf{A}_i = \hat{\mathbf{q}}_i \mu p_i \psi_i$. Using the identity $\nabla \cdot (\mathbf{q}_i\psi_i) = \hat{\mathbf{q}} \cdot \nabla\psi_i = \cos(\phi - \phi_i)\sin\theta_i \partial\psi_i/\partial\rho + \cos\theta_i \partial\psi_i/\partial z$, the electric field due to dipole $i$ is given by equation (1.42)

$$\begin{aligned}\mathbf{E}_i &= -j\omega\left(\mathbf{A}_i + \frac{1}{k^2}\nabla\nabla\cdot\mathbf{A}_i\right)\\ &= -j\omega\mu p_i\left\{\hat{\boldsymbol{\rho}}\left[\sin\theta_i\cos(\phi - \phi_i)\left(\psi_i + \frac{1}{k^2}\frac{\partial^2\psi_i}{\partial\rho^2}\right) + \cos\theta_i\frac{1}{k^2}\frac{\partial^2\psi_i}{\partial z\partial\rho}\right]\right.\\ &\quad + \hat{\boldsymbol{\phi}}\left[\sin\theta_i\sin(\phi_i - \phi)\left(\psi_i + \frac{1}{k^2\rho}\frac{\partial\psi_i}{\partial\rho}\right)\right]\\ &\quad \left.+ \hat{\mathbf{z}}\left[\cos\theta_i\left(\psi_i + \frac{1}{k^2}\frac{\partial^2\psi_i}{\partial z^2}\right) + \cos(\phi - \phi_i)\sin\theta_i\frac{1}{k^2}\frac{\partial^2\psi}{\partial z\partial\rho}\right]\right\},\quad i = 1, 2.\end{aligned} \tag{5.4}$$

Note that

$$\frac{\partial\psi_i}{\partial z} = \left(\frac{z - z_i}{R_i}\right)\frac{\mathrm{d}\psi_i}{\mathrm{d}R_i} = -\left(\frac{z_i}{R_i}\right)\frac{\mathrm{d}\psi_i}{\mathrm{d}R_i}\ \text{at } z = 0. \tag{5.5}$$

The total field in the upper half-space is the sum of $\mathbf{E}_1$ and $\mathbf{E}_2$. Requiring $\hat{\mathbf{n}} \times (\mathbf{E}_1 + \mathbf{E}_2) = 0$ at $z = 0$ with $\hat{\mathbf{n}} = \hat{\mathbf{z}}$ gives

$$p_1\sin\theta_1\cos(\phi - \phi_1)\psi_1 = -p_2\sin\theta_2\cos(\phi - \phi_2)\psi_2, \tag{5.6}$$

$$p_1\cos\theta_1\left(\frac{z_1}{R_1}\right)\frac{\mathrm{d}\psi_1}{\mathrm{d}R_1} = -p_2\cos\theta_2\left(\frac{z_2}{R_2}\right)\frac{\mathrm{d}\psi_2}{\mathrm{d}R_2}, \tag{5.7}$$

$$p_1\sin\theta_1\sin(\phi - \phi_1)\psi_1 = -p_2\sin\theta_2\sin(\phi - \phi_2)\psi_2. \tag{5.8}$$

*Case 1.* In the case of a VED $\theta_1 = 0$ and equation (5.6) gives $\theta_2 = 0$ or $\pi$. Equation (5.7) suggests the possible solution $p_2 = p_1$, $z_2^2 = z_1^2$, $z_2\cos\theta_2 = -z_1$. Since $z_2$ is negative this results in the solution $\theta_2 = 0$ and $z_2 = -z_1$. Thus the image of a vertical electric dipole is oriented in the same direction as the source dipole and it is located at the same distance below the ground plane as the source is above the ground plane. Furthermore, it has the same strength as the source dipole.

*Case 2.* In the case of a HED we take $\theta_1 = \pi/2$, $\phi_1 = \pi/2$ assuming that the dipole lies in a plane parallel to the $yz$-plane. Equations (5.6) and (5.8) then suggest $p_1 = p_2$, $z_2^2 = z_1^2$, $\theta_2 = \pi/2$, and $\cos(\phi - \phi_2) = -\cos(\phi - \phi_1)$, $\sin(\phi - \phi_2) = -\sin(\phi - \phi_1)$. The latter two equations imply $\sin(\phi - \phi_1)\cos(\phi - \phi_2) - \cos(\phi - \phi_1)\sin(\phi - \phi_2)$

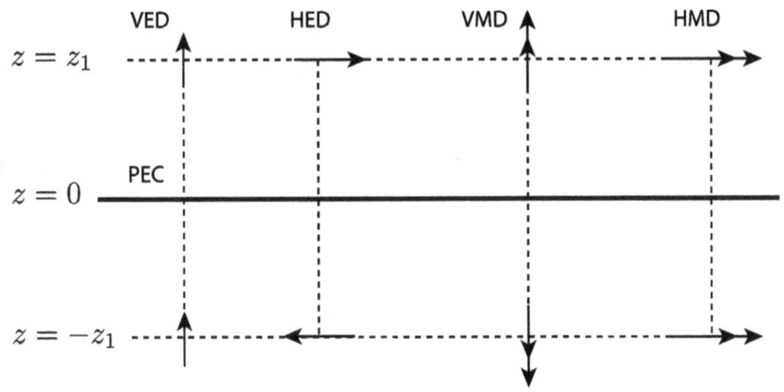


**Figure 5.3.** Image theory for electric and magnetic dipoles radiating in the presence of a perfectly conducting plane.

$=\sin(\phi_2 - \phi_1) = 0$. Therefore $z_2 = -z_1$, $\theta_2 = \pi/2$, and $\phi_2 = \phi_1 + \pi$ satisfy all of the equations (5.6)–(5.8). Thus the image of a horizontal dipole has the same strength as the original dipole, but has an orientation opposite to that of the original dipole. Furthermore, it lies at the same distance below the PEC plane as the source dipole is above the plane.

The two cases along with analogous results for vertical magnetic dipole (VMD) and horizontal magnetic dipole (HMD) are summarized in figure 5.3 below.■■

## 5.2 Energy and momentum

### 5.2.1 Electromagnetic energy and its conservation

**Theorem 5.2.1.** Conservation of energy [1, p 131].
*Let the electromagnetic field* $(\mathscr{E}, \mathscr{D}, \mathscr{B}, \mathscr{H})$ *satisfy Maxwell's equations in a linear medium with conductivity* $\sigma$ *and sources* $(\mathscr{J}_i, \mathscr{M}_i)$. *Define the following quantities*

$$\begin{aligned}
&\mathscr{S} = \mathscr{E} \times \mathscr{H} \quad \textit{Poynting vector, power flux density } (W\ m^{-2}) \\
&k_e = \mathscr{E} \cdot \frac{\partial \mathscr{D}}{\partial t} \quad \textit{rate of change of electric energy density } (W\ m^{-3}) \\
&k_m = \mathscr{H} \cdot \frac{\partial \mathscr{B}}{\partial t} \quad \textit{rate of change of magnetic energy density } (W\ m^{-3}) \\
&p_s = -(\mathscr{E} \cdot \mathscr{J}_i + \mathscr{H} \cdot \mathscr{M}_i) \quad \textit{power supplied by sources per unit volume } (W\ m^{-3}) \\
&p_d = \sigma \mathscr{E} \cdot \mathscr{E} \quad \textit{power dissipated per unit volume } (W\ m^{-3}).
\end{aligned} \tag{5.9}$$

*If S is a surface enclosing the volume V and* $\hat{\boldsymbol{n}}$ *is an outward unit normal on S (figure 5.4), then*

$$p_s = \nabla \cdot \mathscr{S} + k_e + k_m + p_d, \tag{5.10}$$

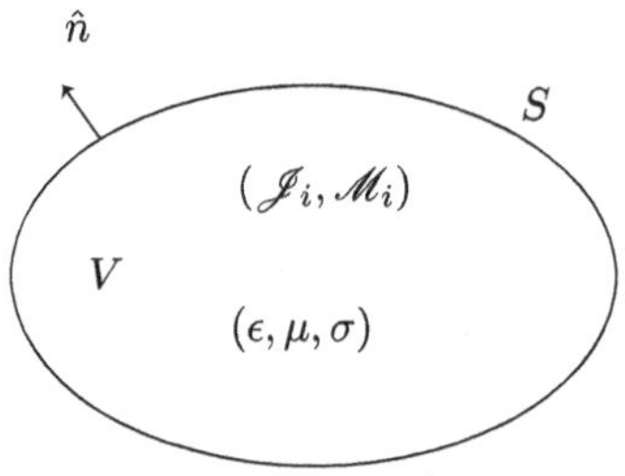


**Figure 5.4.** Power conservation over a volume $V$ bounded by the surface $S$.

$$\iiint_V p_s \, \mathrm{d}v = \oiint_S \mathscr{S} \cdot \hat{\boldsymbol{n}} \, \mathrm{d}s + \frac{\mathrm{d}W}{\mathrm{d}t} + \iiint_V p_d \, \mathrm{d}v, \tag{5.11}$$

*where*

$$\frac{\mathrm{d}W}{\mathrm{d}t} = \iiint_V (k_e + k_m) \, \mathrm{d}v. \tag{5.12}$$

The theorem is valid for linear isotropic, anisotropic, or bianisotropic lossy media where it is possible to write [2, pp 261–2]

$$k_e = \frac{\partial}{\partial t}\left(\frac{1}{2}\mathscr{E} \cdot \mathscr{D}\right), \quad k_m = \frac{\partial}{\partial t}\left(\frac{1}{2}\mathscr{H} \cdot \mathscr{B}\right). \tag{5.13}$$

It is possible to express the rate of change of energy density in terms of scalar and vector potentials. We show this for the case of $\mathscr{M}_i \equiv 0$ using electric scalar potential $\psi_e$ and magnetic vector potential $\mathscr{A}$. Using $\mathscr{B} = \nabla \times \mathscr{A}$ and $\mathscr{E} = -\nabla\psi_e - \partial\mathscr{A}/\partial t$ and Maxwell's equations we obtain

$$\frac{\mathrm{d}W}{\mathrm{d}t} = \iiint_V \left(\psi_e \frac{\partial \rho_e}{\partial t} + \mathscr{J}_i \cdot \frac{\partial \mathscr{A}}{\partial t}\right) \mathrm{d}v - \oiint_S \left(\psi_e \frac{\partial \mathscr{D}}{\partial t} + \mathscr{H} \times \frac{\partial \mathscr{A}}{\partial t}\right) \cdot \hat{\boldsymbol{n}} \, \mathrm{d}s. \tag{5.14}$$

If at instant $t$ after which the sources have been initiated, the size of the surface $S$ is chosen sufficiently large so that the field has not arrived on it yet, the surface integral will be zero. In that case

$$\frac{\mathrm{d}W}{\mathrm{d}t} = \iiint_V \left(\psi_e \frac{\partial \rho_e}{\partial t} + \mathscr{J}_i \cdot \frac{\partial \mathscr{A}}{\partial t}\right) \mathrm{d}v. \tag{5.15}$$

A time-harmonic counterpart to the energy conservation theorem is available. Considering a linear medium we define the following quantities (an over-bar represents quantities averaged over one complete period of the time-harmonic variation)

$$\begin{aligned}
\mathbf{S} &= \frac{1}{2}\mathbf{E} \times \mathbf{H}^* \quad \text{complex Poynting vector} \\
\bar{w}_\mathrm{e} &= \frac{1}{4}\mathbf{E} \cdot \mathbf{D}^* \quad \text{time-averaged electric energy density} \\
\bar{w}_\mathrm{m} &= \frac{1}{4}\mathbf{B} \cdot \mathbf{H}^* \quad \text{time-averaged magnetic energy density} \\
p_\mathrm{s} &= -\frac{1}{2}(\mathbf{E} \cdot \mathbf{J}_i^* + \mathbf{H}^* \cdot \mathbf{M}_i) \quad \text{complex source power density} \\
\bar{p}_\mathrm{d} &= \frac{1}{2}\sigma\mathbf{E} \cdot \mathbf{E}^* \quad \text{time-averaged dissipated power density.}
\end{aligned} \tag{5.16}$$

We then have

$$p_\mathrm{s} = \nabla \cdot \mathbf{S} + 2j\omega(\bar{w}_\mathrm{m} - \bar{w}_\mathrm{e}) + \bar{p}_\mathrm{d}, \tag{5.17}$$

$$\iiint_V p_\mathrm{s}\,\mathrm{d}v = \oiint_S \mathbf{S} \cdot \hat{\boldsymbol{n}}\,\mathrm{d}s + 2j\omega \iiint_V (\bar{w}_\mathrm{m} - \bar{w}_\mathrm{e})\,\mathrm{d}v + \iiint_V \bar{p}_\mathrm{d}\,\mathrm{d}v. \tag{5.18}$$

Note that $\bar{w}_\mathrm{e}$, $\bar{w}_\mathrm{m}$, $\bar{p}_\mathrm{d}$ are all real-valued if $\mathbf{E}$ and $\mathbf{D}^*$ are in phase and $\mathbf{H}^*$ and $\mathbf{B}$ are in phase. This happens when all of the dissipation present in the medium is contained in the conductivity $\sigma$. However, if material losses are separately accounted for in dielectric and magnetic properties then $\bar{w}_\mathrm{e}$ and $\bar{w}_\mathrm{m}$ can also be complex. Further, only the real parts of $\mathbf{S}$ and $p_\mathrm{s}$ correspond to time-averaged Poynting vector and time-averaged source densities, respectively. If the volume $V$ does not enclose sources then the left-hand side of equation (5.18) is zero. For a volume not enclosing any sources and when all losses are accounted for in $\sigma$, we can equate the real, $\mathfrak{R}$, and imaginary, $\mathfrak{J}$, parts of equation (5.18) to result in

$$\iiint_V \bar{p}_\mathrm{d}\,\mathrm{d}v = -\mathfrak{R} \oiint_S \mathbf{S} \cdot \hat{\mathbf{n}}\,\mathrm{d}s, \tag{5.19}$$

$$2\omega \iiint_V (\bar{w}_\mathrm{e} - \bar{w}_\mathrm{m})\,\mathrm{d}v = \mathfrak{J} \oiint_S \mathbf{S} \cdot \hat{\mathbf{n}}\,\mathrm{d}s. \tag{5.20}$$

So the average power dissipated in a volume and the difference between the averaged electric and magnetic energies stored in a volume can be determined solely from the complex Poynting vector specified on the surface enclosing the volume as long as the volume does not contain sources and when all losses are contained in the description of the medium conductivity.

#### *5.2.1.1 Conditions under which energy is a quadratic function of a primitive field in a dispersive medium*

We shall investigate the conditions under which the expression for energy density can be written as a quadratic function of primitive fields. The following results are adapted from [3]. We will limit here the discussion to electric energy but identical results follow for the magnetic energy. Let $\bar{\bar{\varepsilon}}_{\rm rc} = \bar{\bar{\varepsilon}}' - j\bar{\bar{\varepsilon}}''$ be the complex relative permittivity tensor of a linear, dispersive, passive, and possibly inhomogeneous dielectric medium. The following physical conditions are imposed on $\bar{\bar{\varepsilon}}_{\rm rc}$ in additional to the obvious

$$\bar{\bar{\varepsilon}}_{\rm rc}(-\omega^*) = \bar{\bar{\varepsilon}}^*_{\rm rc}(\omega), \tag{5.21}$$

1. The *causality* condition (4.24).
2. *Kinetic symmetry*. In the absence of strong external, static magnetic field such that there is no optical activity [4, p 332] the permittivity tensor is symmetrical, i.e. $\bar{\bar{\varepsilon}}^T_{\rm rc} = \bar{\bar{\varepsilon}}_{\rm rc}$, where the superscript $T$ stands for transpose. This symmetry condition also follows from electromagnetic field reciprocity condition that is considered in section 7.1.
3. *Passivity*. We assume that the spectra $\rho$ (the collection of eigenvalues) of the imaginary part of the complex permittivity tensor $\bar{\bar{\varepsilon}}''_{\rm rc}$ are positive for positive frequencies $\omega$:

$$\rho\{\bar{\bar{\varepsilon}}''_{\rm rc}\} > 0. \tag{5.22}$$

When condition 2 is combined with condition 3 it can be shown that the dissipation condition

$$\mathfrak{J}[\rho\{\bar{\bar{\varepsilon}}_{\rm rc}\}] > 0, \tag{5.23}$$

is automatically met. Since the imaginary part of the relative permittivity is an odd function of real frequency $\omega$ by equation (5.21), equation (5.22) also implies that

$$\rho\{\omega\bar{\bar{\varepsilon}}''_{\rm rc}\} \geqslant 0, \tag{5.24}$$

where equality is only true at $\omega = 0$. Because of the non-negative nature in equation (5.24), it is possible to factorize $\omega\bar{\bar{\varepsilon}}''_{\rm rc}$ such that

$$\omega\bar{\bar{\varepsilon}}''_{\rm rc} = \bar{\bar{\alpha}}^\dagger_{\rm e}\bar{\bar{\alpha}}_{\rm e}, \tag{5.25}$$

for all real frequencies $\omega$, where superscript $\dagger$ denotes Hermitian conjugation. Consider the time rate of change of electric energy density[1]

$$\begin{aligned} k_{\rm e} = \mathscr{E}\cdot\frac{\partial\mathscr{D}}{\partial t} &= \frac{1}{2}\varepsilon_0\frac{\partial\mathscr{E}\cdot\mathscr{E}}{\partial t} + \varepsilon_0\mathscr{E}\cdot\frac{\partial}{\partial t}\int_{-\infty}^{t}\bar{\bar{\chi}}_{\rm e}(t-\tau)\mathscr{E}(\tau)\,{\rm d}\tau \\ &= \frac{1}{2}\varepsilon_0\frac{\partial\mathscr{E}\cdot\mathscr{E}}{\partial t} + \varepsilon_0\mathscr{E}\cdot\int_{-\infty}^{t}\frac{\partial}{\partial t}\bar{\bar{\chi}}_{\rm e}(t-\tau)\mathscr{E}(\tau)\,{\rm d}\tau. \end{aligned} \tag{5.26}$$

In moving the time derivative inside the integral in equation (5.26) we have made use of the fact that $\bar{\bar{\chi}}_{\rm e}(0) = \bar{\bar{\chi}}_{\rm e}(0-) = 0$ due to causality. Our goal is to reduce the

[1] For ease of notation we have suppressed the spatial dependence from all terms.

expression on the right-hand side to the time derivative of a quadratic form. Assuming that the real part of the permittivity tensor approaches an identity tensor at infinite frequency we have

$$
\begin{aligned}
\bar{\bar{\chi}}_e(t) &= \frac{1}{2\pi}\int_{-\infty}^{\infty}\left[\bar{\bar{\varepsilon}}'(\omega) - \bar{\bar{I}} - j\bar{\bar{\varepsilon}}''(\omega)\right]e^{j\omega t}\,d\omega \\
&= -\frac{1}{2\pi^2}\int_{-\infty}^{\infty} d\omega\, e^{j\omega t}\mathscr{P}\int_{-\infty}^{\infty}\frac{\bar{\bar{\varepsilon}}''(\omega')}{\omega-\omega'}\,d\omega' - \frac{j}{2\pi}\int_{-\infty}^{\infty}\bar{\bar{\varepsilon}}''(\omega)e^{j\omega t}\,d\omega \\
&= -\frac{1}{2\pi^2}\int_{-\infty}^{\infty} d\omega'\left(\mathscr{P}\int_{-\infty}^{\infty}\frac{e^{j\omega t}}{\omega-\omega'}\,d\omega\right)\bar{\bar{\varepsilon}}''(\omega') \\
&\quad - \frac{j}{2\pi}\int_{-\infty}^{\infty}\bar{\bar{\varepsilon}}''(\omega)e^{j\omega t}\,d\omega \\
&= -\frac{1}{2\pi^2}\int_{-\infty}^{\infty} d\omega' j\pi e^{j\omega' t}\bar{\bar{\varepsilon}}''(\omega') - \frac{j}{2\pi}\int_{-\infty}^{\infty}\bar{\bar{\varepsilon}}''(\omega)e^{j\omega t}\,d\omega,\ t>0 \\
&= -\frac{j}{\pi}\int_{-\infty}^{\infty}\bar{\bar{\varepsilon}}''(\omega)e^{j\omega t}\,d\omega,\ t>0.
\end{aligned}
\tag{5.27}
$$

Hence the electric susceptibility function in the time domain can be described entirely in terms of the imaginary part of the complex permittivity along the real frequency $\omega$. This is true for all susceptibility functions that are causal in nature. The interchange of the orders of integration with respect to $\omega$ and $\omega'$ and the transfer of $\mathscr{P}$ from the $\omega'$-axis to the $\omega$-axis can be justified by first writing the $\mathscr{P}$ operator as a limiting process and investigating the convergence process. Inserting this into equation (5.26) and using equation (5.25) we obtain

$$
\begin{aligned}
\mathscr{E}(t)\cdot\int_{-\infty}^{t}\frac{\partial}{\partial t}\bar{\bar{\chi}}_e(t-\tau)\mathscr{E}(\tau)\,d\tau &= \frac{1}{\pi}\int_{-\infty}^{t}\int_{-\infty}^{\infty} e^{j\omega t}\mathscr{E}(t)\cdot\bar{\bar{\alpha}}_e^{\dagger}(\omega)\bar{\bar{\alpha}}_e(\omega)\mathscr{E}(\tau)e^{-j\omega\tau}\,d\omega\,d\tau \\
&= \frac{1}{\pi}\int_{-\infty}^{\infty} d\omega\left[\frac{\partial}{\partial t}\int_{-\infty}^{t} d\tau\, e^{j\omega\tau}\mathscr{E}(\tau)\right]\cdot\bar{\bar{\alpha}}_e^{\dagger}(\omega)\bar{\bar{\alpha}}_e(\omega) \\
&\quad\cdot\left[\int_{-\infty}^{t}\mathscr{E}(\tau)e^{-j\omega\tau}\,d\tau\right] \\
&= \frac{1}{\pi}\int_{-\infty}^{\infty}\left[\frac{\partial}{\partial t}\bar{\bar{\alpha}}_e(\omega)\int_{-\infty}^{t}\mathscr{E}(\tau)e^{-j\omega\tau}\,d\tau\right]^{\dagger} \\
&\quad\cdot\left[\bar{\bar{\alpha}}_e(\omega)\int_{-\infty}^{t}\mathscr{E}(\tau)e^{-j\omega\tau}\,d\tau\right]d\omega \\
&= \frac{1}{2\pi}\frac{\partial}{\partial t}\int_{-\infty}^{\infty}\left[\bar{\bar{\alpha}}_e(\omega)\int_{-\infty}^{t}\mathscr{E}(\tau)e^{-j\omega\tau}\,d\tau\right]^{\dagger} \\
&\quad\cdot\left[\bar{\bar{\alpha}}_e(\omega)\int_{-\infty}^{t}\mathscr{E}(\tau)e^{-j\omega\tau}\,d\tau\right]d\omega \\
&= \frac{\partial}{\partial t}\frac{1}{2\pi}\int_{-\infty}^{\infty}\left\|\bar{\bar{\alpha}}_e(\omega)\int_{-\infty}^{t}\mathscr{E}(\tau)e^{-j\omega\tau}\,d\tau\right\|^2 d\omega,
\end{aligned}
\tag{5.28}
$$

where $\|\mathbf{x}\|^2$ indicates the magnitude square $\mathbf{x}^{\dagger}\mathbf{x}$ of a complex column vector $\mathbf{x}$. Using equation (5.28) in equation (5.26) we obtain

$$k_{\mathrm{e}} =: \frac{\partial w_{\mathrm{e}}}{\partial t} = \frac{\partial}{\partial t}\left[\frac{1}{2}\varepsilon_0|\mathscr{E}(t)|^2 + \frac{1}{2\pi}\varepsilon_0 \int_{-\infty}^{\infty} \left\| \bar{\bar{\alpha}}_{\mathrm{e}}(\omega) \int_{-\infty}^{t} \mathscr{E}(\tau)\mathrm{e}^{-j\omega\tau}\, \mathrm{d}\tau \right\|^2 \mathrm{d}\omega\right], \quad (5.29)$$

where $w_{\mathrm{e}}$ can be labeled as the electric energy density. With appropriate initial condition, the electric energy density is then the positive definite quadratic form

$$w_{\mathrm{e}}(t) = \frac{1}{2}\varepsilon_0\left[|\mathscr{E}(t)|^2 + \frac{1}{\pi}\int_{-\infty}^{\infty} \left\| \bar{\bar{\alpha}}_{\mathrm{e}}(\omega) \int_{-\infty}^{t} \mathscr{E}(\tau)\mathrm{e}^{-j\omega\tau}\, \mathrm{d}\tau \right\|^2 \mathrm{d}\omega\right] > 0, \quad (5.30)$$

$$\neq \frac{1}{2}\mathscr{E}(t)\cdot\mathscr{D}(t). \quad (5.31)$$

In contrast to equation (5.30), the rhs of equation (5.31) can alternate sign. However, if time-averaged electric energy, $\bar{w}_{\mathrm{e}}$, at a single frequency is of interest, then one can express $\bar{w}_{\mathrm{e}} = \frac{1}{4}\Re\{\mathbf{E}\cdot\mathbf{D}^*\}$ in terms of the phasors $\mathbf{E}$ and $\mathbf{D}$. But the latter need not be positive definite.

In a similar fashion using the complex relative permeability $\bar{\bar{\mu}}_{\mathrm{rc}} = \bar{\bar{\mu}}' - j\bar{\bar{\mu}}''$ and its decomposition $\omega\bar{\bar{\mu}}'' = \bar{\bar{\alpha}}_{\mathrm{m}}^{\dagger}\bar{\bar{\alpha}}_{\mathrm{m}}$, the magnetic energy density, $w_{\mathrm{m}}(t)$, can be written as

$$w_{\mathrm{m}}(t) = \frac{1}{2}\mu_0\left[|\mathscr{H}(t)|^2 + \frac{1}{\pi}\int_{-\infty}^{\infty} \left\| \bar{\bar{\alpha}}_{\mathrm{m}}(\omega) \int_{-\infty}^{t} \mathscr{H}(\tau)\mathrm{e}^{-j\omega\tau}\, \mathrm{d}\tau \right\|^2 \mathrm{d}\omega\right] > 0, \quad (5.32)$$

$$\neq \frac{1}{2}\mathscr{H}(t)\cdot\mathscr{B}(t). \quad (5.33)$$

### *5.2.1.2 Energy balance equations in a non-conductive, dispersive medium*

The results in this subsection are valid for a general non-conductive, anisotropic dispersive medium that is allowed to have material losses accounted for in the dielectric and magnetic constitutive properties. Below we adapt from the material given in [5].

**Theorem 5.2.2.** *Let the phasor electromagnetic field* ($\mathbf{E}$, $\mathbf{D}$, $\mathbf{B}$, $\mathbf{H}$) *exist due to an electric source* $\mathbf{J}_i$ *operating at a frequency* $\omega$ *in a linear dispersive medium. Define*

$\boldsymbol{S} = \frac{1}{2}\mathbf{E} \times \mathbf{H}^*$ *Poynting vector, power flux density; complex*

$\boldsymbol{K} = \frac{1}{4}\left[\mathbf{E} \times \frac{\partial \mathbf{H}^*}{\partial \omega} - \frac{\partial \mathbf{E}}{\partial \omega} \times \mathbf{H}^*\right]$ *radiated energy flux density vector; complex*

$\bar{w}_\mathrm{e} = \frac{1}{4}\mathbf{E} \cdot \mathbf{D}^*$ *time-averaged electric energy density; complex*

$\bar{k}_\mathrm{e} = \frac{1}{4}\left[\mathbf{E} \cdot \frac{\partial \mathbf{D}^*}{\partial \omega} - \mathbf{D}^* \cdot \frac{\partial \mathbf{E}}{\partial \omega}\right]$ *time-averaged excess stored electric energy density; complex*

$\bar{w}_\mathrm{m} = \frac{1}{4}\mathbf{B} \cdot \mathbf{H}^*$ *time-averaged magnetic energy density; complex*

$\bar{k}_\mathrm{m} = \frac{1}{4}\left[\mathbf{B} \cdot \frac{\partial \mathbf{H}^*}{\partial \omega} - \mathbf{H}^* \cdot \frac{\partial \mathbf{B}}{\partial \omega}\right]$ *time-averaged excess stored magnetic energy density; complex*

$p_\mathrm{s} = -\frac{1}{2}\mathbf{E} \cdot \mathbf{J}_i^*$ *source power density; complex*

$g_\mathrm{s} = \frac{1}{4}\left[\mathbf{E} \cdot \frac{\partial \mathbf{J}_i^*}{\partial \omega} - \frac{\partial \mathbf{E}}{\partial \omega} \cdot \mathbf{J}_i^*\right]$ *complex.*

*Then*

$$p_\mathrm{s} = \nabla \cdot \mathbf{S} + 2j\omega(\bar{w}_\mathrm{m} - \bar{w}_\mathrm{e}), \tag{5.34}$$

$$g_\mathrm{s} = -\nabla \cdot \mathbf{K} + j(\bar{w}_\mathrm{m} + \bar{w}_\mathrm{e}) + j\omega(\bar{k}_\mathrm{e} - \bar{k}_\mathrm{m}). \tag{5.35}$$

*Proof.* We carry out the proof in a complex frequency domain $s = \alpha + j\omega$ and handle the time-harmonic case of interest here by taking the limit as $\alpha \to 0$. We represent a typical time-instantaneous field quantity, $\mathscr{F}$, by means of its phasor, $\widetilde{\boldsymbol{F}}$, in the complex frequency domain via the bilateral Laplace transform (see theorem A.3.2):

$$\widetilde{\boldsymbol{F}}(\mathbf{r}; s) = \int_{-\infty}^{\infty} \mathscr{F}(\mathbf{r}; t)\mathrm{e}^{-st}\,\mathrm{d}t, \tag{5.36}$$

$$\mathscr{F}(\mathbf{r}; t) = \frac{1}{2\pi j}\int_{\alpha_0 - j\infty}^{\alpha_0 + j\infty} \widetilde{\boldsymbol{F}}(\mathbf{r}; s)\mathrm{e}^{st}\,\mathrm{d}s, \quad \alpha_- < \alpha_0 < \alpha_+, \tag{5.37}$$

where the contour of integration in equation (5.37) is as shown in figure A.3 and it has been assumed in equation (5.36) that $|\mathscr{F}(\mathbf{r}; t)|_{t\to\infty} \leqslant A(\mathbf{r})\mathrm{e}^{\alpha_- t}$ and $|\mathscr{F}(\mathbf{r}; t)|_{t\to-\infty} \leqslant B(\mathbf{r})\mathrm{e}^{\alpha_+ t}$. Subject to these latter assumptions it can also be verified that $\partial\mathscr{F}/\partial t$ could be represented in terms of the phasor $s\widetilde{\boldsymbol{F}}$. By theorem A.3.1 the phasor $\widetilde{\boldsymbol{F}}$ is an analytic function of the complex variable $s$ in the strip of convergence $\alpha_- < \Re(s) < \alpha_+$. Because the initial fields are finite and the final fields have to decay to zero in a lossy medium, the limits satisfy $\alpha_- < 0$ and $\alpha_+ > 0$. Thus $\alpha = 0$ is included in the strip of convergence. For an analytic function we have from the Cauchy–Riemann conditions (A.21) and Taylor series that to a first order in $\alpha$

$$\begin{aligned}\widetilde{\boldsymbol{F}}(\mathbf{r}; \alpha + j\omega) &\sim \widetilde{\boldsymbol{F}}(\mathbf{r}; 0 + j\omega) - j\alpha\frac{\partial \widetilde{\boldsymbol{F}}(\mathbf{r}; 0 + j\omega)}{\partial \omega} \\ &= \mathbf{F}(\mathbf{r}; \omega) - j\alpha\frac{\partial \mathbf{F}(\mathbf{r}; \omega)}{\partial \omega},\end{aligned} \tag{5.38}$$

where $\mathbf{F}(\mathbf{r};\omega) = \widetilde{\boldsymbol{F}}(\mathbf{r}; 0 + j\omega)$ is the usual phasor in the time-harmonic case. Applying the identity $\nabla \cdot (\boldsymbol{a} \times \boldsymbol{b}^*) = \boldsymbol{b}^* \cdot \nabla \times \boldsymbol{a} - \boldsymbol{a} \cdot \nabla \times \boldsymbol{b}^*$ to Maxwell's equations $\nabla \times \widetilde{\boldsymbol{E}} = -s\widetilde{\boldsymbol{B}}, \nabla \times \widetilde{\boldsymbol{H}} = \widetilde{\boldsymbol{J}} + s\widetilde{\boldsymbol{D}}$ in the complex frequency domain with $\boldsymbol{a} = \widetilde{\boldsymbol{E}}$ and $\boldsymbol{b} = \widetilde{\boldsymbol{H}}$ it is easy to see that

$$\nabla \cdot (\widetilde{\boldsymbol{E}} \times \widetilde{\boldsymbol{H}}^*) = -\widetilde{\boldsymbol{E}} \cdot \widetilde{\boldsymbol{J}}^* - \alpha(\widetilde{\boldsymbol{E}} \cdot \widetilde{\boldsymbol{D}}^* + \widetilde{\boldsymbol{B}} \cdot \widetilde{\boldsymbol{H}}^*) + j\omega(\widetilde{\boldsymbol{E}} \cdot \widetilde{\boldsymbol{D}}^* - \widetilde{\boldsymbol{B}} \cdot \widetilde{\boldsymbol{H}}^*). \quad (5.39)$$

Using the expansion (5.38) for each field component and retaining terms only up to order $\alpha$ we have

$$\frac{1}{2}\widetilde{\boldsymbol{E}} \times \widetilde{\boldsymbol{H}}^* \sim \mathbf{S} + 2j\alpha\mathbf{K}, \quad (5.40)$$

$$-\frac{1}{2}\widetilde{\boldsymbol{E}} \cdot \widetilde{\boldsymbol{J}}^* \sim p_s - 2j\alpha g_s, \quad (5.41)$$

$$\frac{1}{4}\left(\widetilde{\boldsymbol{E}} \cdot \widetilde{\boldsymbol{D}}^* \pm \widetilde{\boldsymbol{B}} \cdot \widetilde{\boldsymbol{H}}^*\right) \sim (w_e \pm w_m) + j\alpha(\bar{k}_e \pm \bar{k}_m). \quad (5.42)$$

Finally, substituting equations (5.40)–(5.42) into equation (5.39) and equating terms of like power in $\alpha$ we arrive at equations (5.34) and (5.35). ■

The first result (5.34) is the special case of equation (5.10) with $\bar{p}_d = 0$ and describes conservation of power. Let us now gain some insight into the second result (5.35). From theorem 5.2.1 we have for time-instantaneous fields that $-\mathscr{E} \cdot \mathscr{J}_i = \nabla \cdot \mathscr{S} + \mathscr{E} \cdot \frac{\partial \mathscr{D}}{\partial t} + \mathscr{H} \cdot \frac{\partial \mathscr{B}}{\partial t}$. We first write

$$\mathscr{E} \cdot \frac{\partial \mathscr{D}}{\partial t} = \frac{1}{2}\frac{\partial}{\partial t}(\mathscr{E} \cdot \mathscr{D}) + \frac{1}{2}\left(\mathscr{E} \cdot \frac{\partial \mathscr{D}}{\partial t} - \mathscr{D} \cdot \frac{\partial \mathscr{E}}{\partial t}\right), \quad (5.43)$$

and use the phasor representation $\mathscr{F}(\mathbf{r};t) = \Re[\widetilde{\boldsymbol{F}}(\mathbf{r};s)e^{st}] = \frac{1}{2}[\widetilde{\boldsymbol{F}}e^{st} + \widetilde{\boldsymbol{F}}^*e^{s^*t}]$ for each field component. Now

$$\begin{aligned}\left(\mathscr{E} \cdot \frac{\partial \mathscr{D}}{\partial t} - \mathscr{D} \cdot \frac{\partial \mathscr{E}}{\partial t}\right) &= \omega e^{2\alpha t}\Im[\widetilde{\boldsymbol{E}} \cdot \widetilde{\boldsymbol{D}}^*] \\ &\sim \omega e^{2\alpha t}\Im[\mathbf{E} \cdot \mathbf{D}^* + 4j\alpha\bar{k}_e + \mathcal{O}(\alpha^2)] \\ &= \omega e^{2\alpha t}\Im(\mathbf{E} \cdot \mathbf{D}^*) + 2\omega\frac{\partial}{\partial t}\Re\left(e^{2\alpha t}\bar{k}_e\right) + \mathcal{O}(\alpha^2),\end{aligned} \quad (5.44)$$

and

$$\begin{aligned}\mathscr{E} \cdot \mathscr{D} &= \frac{1}{2}e^{2\alpha t}\Re(\widetilde{\boldsymbol{E}} \cdot \widetilde{\boldsymbol{D}}^*) + \frac{1}{2}e^{2\alpha t}\Re(\widetilde{\boldsymbol{E}} \cdot \widetilde{\boldsymbol{D}}e^{2j\omega t}) \\ &\sim \frac{1}{2}e^{2\alpha t}\Re(\mathbf{E} \cdot \mathbf{D}^*) - \frac{\partial}{\partial t}\Im\left(\bar{k}_e e^{2\alpha t}\right) + \mathcal{O}(\alpha^2) \\ &\quad + \frac{1}{2}e^{2\alpha t}\Re(\widetilde{\boldsymbol{E}} \cdot \widetilde{\boldsymbol{D}}e^{2j\omega t}),\end{aligned} \quad (5.45)$$

where the expansion (5.38) was utilized for each field component. The last term in equation (5.45) will drop out on taking the average over one time period $T = 2\pi/\omega$ in the limit of $\alpha \to 0$ and will be ignored in the further development because we will be concerned with only time averaged quantities at the end. Therefore

$$\begin{aligned}\mathscr{E} \cdot \frac{\partial \mathscr{D}}{\partial t} &= 2\omega e^{2\alpha t}\Im(\bar{w}_e) + \omega\frac{\partial}{\partial t}\Re(e^{2\alpha t}\bar{k}_e) \\ &+ \frac{\partial}{\partial t}\Re(\bar{w}_e e^{2\alpha t}) + \mathcal{O}(\alpha^2).\end{aligned} \tag{5.46}$$

Note that the second term on the right-hand side of equation (5.45) will result in a term of order $\mathcal{O}(\alpha^2)$ in equation (5.46). Expressing

$$\mathscr{E} \cdot \frac{\partial \mathscr{D}}{\partial t} = p_{ed} + \frac{\partial w_{es}}{\partial t}, \tag{5.47}$$

where $p_{ed}$ denotes the rate of dissipated electric energy density and $w_{es}$ denotes the stored electric energy density and comparing with equation (5.46) gives

$$\begin{aligned}p_{ed} &= 2\omega e^{2\alpha t}\Im(\bar{w}_e) \\ &\to 2\omega\Im(\bar{w}_e) \text{ as } \alpha \to 0,\end{aligned} \tag{5.48}$$

and

$$\begin{aligned}w_{es} &= \omega\Re(e^{2\alpha t}\bar{k}_e) + \Re(\bar{w}_e e^{2\alpha t}) \\ &\to \Re[\omega\bar{k}_e + \bar{w}_e] \text{ as } \alpha \to 0.\end{aligned} \tag{5.49}$$

For example in an isotropic medium with a constitutive relation $\mathbf{D} = (\varepsilon' - j\varepsilon'')\mathbf{E}$, the rate of dissipated electric energy density is $p_{ed} = \frac{1}{2}\omega\varepsilon''|\mathbf{E}|^2$. Similarly, expressing

$$\mathscr{H} \cdot \frac{\partial \mathscr{B}}{\partial t} = p_{md} + \frac{\partial w_{ms}}{\partial t}, \tag{5.50}$$

where $p_{md}$ denotes the rate of dissipated magnetic field energy density and $w_{ms}$ denotes the stored magnetic field energy density we find

$$\begin{aligned}p_{md} &= -2\omega e^{2\alpha t}\Im(\bar{w}_m) = 2\omega e^{2\alpha t}\Im(\bar{w}_m^*) \\ &\to -2\omega\Im(\bar{w}_m) \text{ as } \alpha \to 0,\end{aligned} \tag{5.51}$$

and

$$\begin{aligned}w_{ms} &= -\omega\Re(e^{2\alpha t}\bar{k}_m) + \Re(\bar{w}_m e^{2\alpha t}) \\ &\to \Re[-\omega\bar{k}_m + \bar{w}_m] \text{ as } \alpha \to 0.\end{aligned} \tag{5.52}$$

For example, in an isotropic medium with a constitutive relation $\mathbf{B} = (\mu' - j\mu'')\mathbf{H}$, the rate of dissipated magnetic energy density is $p_{md} = \frac{1}{2}\omega\mu''|\mathbf{H}|^2$. Because there is no temporal variation at the radian frequency of $\omega$ in the expressions for $w_{es}$ and $w_{ms}$, these are also the time-averaged quantities. Taking the imaginary part of equation

(5.35) and utilizing equations (5.49) and (5.52) we obtain an expression for the total time-average energy stored in the electromagnetic field as

$$w_{es} + w_{ms} = \mathfrak{J}(g_s + \nabla \cdot \mathbf{K}). \tag{5.53}$$

Taking the real part of equation (5.35) and utilizing equations (5.48) and (5.51) we obtain an expression for the difference between the dissipated magnetic and electric energy densities in the field as

$$\frac{1}{2\omega}(p_{md} - p_{ed}) = \mathfrak{R}(g_s + \nabla \cdot \mathbf{K}) + \omega\mathfrak{J}\left(\bar{k}_e - \bar{k}_m\right). \tag{5.54}$$

Integrating equation (5.53) over a volume $V$ bounded by a surface $S$ with an outward normal $\hat{\mathbf{n}}$ and applying the divergence theorem we obtain

$$\begin{aligned}\bar{R} &:= \iiint_V \mathfrak{J}[g_s(\mathbf{r};\,\omega)]\,\mathrm{d}v \\ &= \iiint_V [w_{es}(\mathbf{r};\,\omega) + w_{ms}(\mathbf{r};\,\omega)]\,\mathrm{d}v - \oiint_S \mathfrak{J}[\mathbf{K}(\mathbf{r};\,\omega)]\cdot\hat{\mathbf{n}}\,\mathrm{d}s, \\ &= \iiint_V \mathfrak{R}[\bar{w}_e(\mathbf{r};\,\omega) + \bar{w}_m(\mathbf{r};\,\omega)]\,\mathrm{d}v - \mathfrak{J}\oiint_S[\mathbf{K}(\mathbf{r};\,\omega)]\cdot\hat{\mathbf{n}}\,\mathrm{d}s \\ &\quad - \omega\iiint_V \mathfrak{R}[\bar{k}_m(\mathbf{r};\,\omega) - \bar{k}_e(\mathbf{r};\,\omega)]\,\mathrm{d}v,\end{aligned} \tag{5.55}$$

The first volume integral on the right-hand side of equation (5.55) denotes the total electromagnetic energy stored inside the volume $V$. If the volume $V$ excludes the source, the left-hand side of equation (5.55) is zero resulting in

$$\iiint_V [w_{es}(\mathbf{r};\,\omega) + w_{ms}(\mathbf{r};\,\omega)]\,\mathrm{d}v = \oiint_S \mathfrak{J}[\mathbf{K}(\mathbf{r};\,\omega)]\cdot\hat{\mathbf{n}}. \tag{5.56}$$

Equation (5.56) provides an alternative means of calculating the energy stored in a source-free volume $V$ in terms of surface integrals only. It is particularly useful in a lossless medium as we shall now demonstrate. When the operating frequency falls in a transparent band of an isotropic medium having constitutive relations $\mathbf{D} = \varepsilon(\omega)\mathbf{E}$ and $\mathbf{B} = \mu(\omega)\mathbf{H}$, the imaginary parts of $\varepsilon(\omega)$ and $\mu(\omega)$ are negligible compared to the real part over the band of frequencies of interest and we may regard the latter parameters as being purely real. In such a case it is easy to see that

$$\bar{k}_e = \frac{1}{4}|\mathbf{E}|^2\frac{\partial\varepsilon}{\partial\omega} + j\frac{\varepsilon}{2}\mathfrak{J}\left(\mathbf{E}\cdot\frac{\partial\mathbf{E}^*}{\partial\omega}\right),$$

$$\bar{k}_m = -\frac{1}{4}|\mathbf{H}|^2\frac{\partial\mu}{\partial\omega} + j\frac{\mu}{2}\mathfrak{J}\left(\mathbf{H}\cdot\frac{\partial\mathbf{H}^*}{\partial\omega}\right).$$

Substituting these into equation (5.55) and simplifying we obtain

$$\bar{R} = \frac{1}{4} \iiint_V \left[ \frac{\partial(\omega\varepsilon)}{\partial\omega} |\mathbf{E}|^2 + \frac{\partial(\omega\mu)}{\partial\omega} |\mathbf{H}|^2 \right] \mathrm{d}v - \Im \oiint_S \mathbf{K} \cdot \hat{\mathbf{n}} \, \mathrm{d}s, \tag{5.57}$$

in a transparent medium. As an example of its usefulness, equation (5.57) provides an expression for the total energy stored in a source-free volume in free-space by means of the following surface integral

$$\frac{1}{4} \iiint_V \left[ \varepsilon_0 |\mathbf{E}|^2 + \mu_0 |\mathbf{H}|^2 \right] \mathrm{d}v = \Im \oiint_S \mathbf{K} \cdot \hat{\mathbf{n}} \, \mathrm{d}s. \tag{5.58}$$

### 5.2.2 Electromagnetic momentum and its conservation

Relativistically speaking, conservation of energy and conservation of momentum are not independent principles; one demands the other for a covariant formulation[2]. In this subsection we shall see how to associate a valid momentum with electromagnetic waves. If we consider that the electromagnetic force acting on a given volume containing charges and currents is transmitted across the elements of surface bounding that volume, then the transmitting force can formulated in terms of a quantity called the *stress tensor* ${}^2\mathscr{T}$. However, this stress is an abstract concept and does not depend on the presence of matter in and around the volume. The $\alpha\beta$ component $\mathscr{T}_{\alpha\beta}$ of the second rank stress tensor is so constituted that the $\alpha$th component $\mathrm{d}F_\alpha$ of the force $\mathbf{dF}$ transmitted across a surface $\mathbf{dS}$ whose component in the $\beta$th direction is $\mathrm{d}S_\beta$, is given by

$$\mathrm{d}F_\alpha = \sum_{\beta=1}^{3} \mathscr{T}_{\alpha\beta} \mathrm{d}S_\beta =: \mathscr{T}_{\alpha\beta} \mathrm{d}S_\beta. \tag{5.59}$$

where the latter is Einstein's summation convention for a repeated index (see chapter 17). Thus the components of stress tensor are the tension (or pressure) and tangential shears which represent the forces exerted by fields outside $S$ on a unit area of $S$. By imposing rotational equilibrium of the volume about any point $\mathbf{r}$ under the given surface stresses it can be shown that the stress tensor ${}^2\mathscr{T}$ for a linear medium must be symmetric, i.e. $\mathscr{T}_{\alpha\beta} = \mathscr{T}_{\beta\alpha}$.

The $\alpha$th component of the total force acting on a volume $V$ is (figure 5.5)

$$F_\alpha = \oiint_S \mathscr{T}_{\alpha\beta} \, \mathrm{d}S_\beta =: \iiint_V F_{V_\alpha} \, \mathrm{d}v, \tag{5.60}$$

[2] The components of the energy momentum four-vector in a covariant are $p^i = (c\mathbf{p}, E)$, where the first three components are the three components of the linear momentum vector $\mathbf{p}$ and the fourth component is the energy $E$ in that frame of reference.

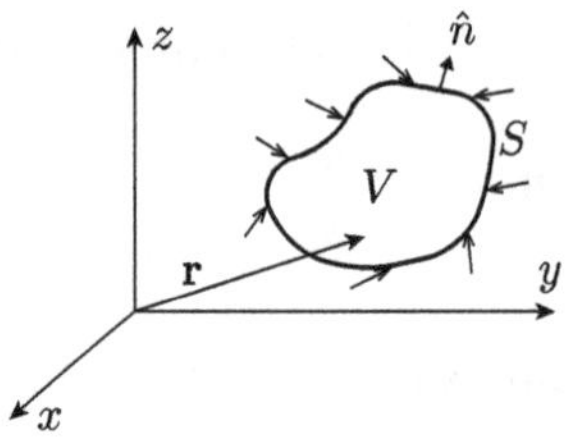


**Figure 5.5.** A region $V$ bounded by $S$ in an elastic-like medium subject to stress.

where we have expressed this force in terms of a volume force $\mathbf{F}_v$ whose $\alpha$th component is $F_{v_\alpha}$. Applying Gauss' divergence theorem to the surface integral it is easy to see that

$$F_{v_\alpha} = \frac{\partial \mathscr{T}_{\alpha\beta}}{\partial x_\beta}. \tag{5.61}$$

Thus if we express a volume force as the tensor divergence of a certain quantity $\mathscr{Q}$, then the quantity $\mathscr{Q}$ can be identified with the surface stress tensor ${}^2\mathscr{T}$ that gave the stress transmitted by the field across the surface of the volume.

**Theorem 5.2.3** Conservation of momentum [6, p 181].
*Consider a volume $V$ containing electric sources $(\rho_e, \mathscr{J})$ and bounded by the surface S. Define the stress tensor ${}^2\mathscr{T}$ with components*

$$\mathscr{T}_{\alpha\beta} = \mathscr{E}_\alpha \mathscr{D}_\beta - \frac{1}{2}\delta_\alpha^\beta \mathscr{E}_\gamma \mathscr{D}_\gamma + \mathscr{H}_\alpha \mathscr{B}_\beta - \frac{1}{2}\delta_\alpha^\beta \mathscr{H}_\gamma \mathscr{B}_\gamma, \tag{5.62}$$

$$= \begin{bmatrix} \mathscr{E}_x\mathscr{D}_x - \frac{1}{2}\mathscr{E}\cdot\mathscr{D} & \mathscr{E}_x\mathscr{D}_y & \mathscr{E}_x\mathscr{D}_z \\ \mathscr{E}_y\mathscr{D}_x & \mathscr{E}_y\mathscr{D}_y - \frac{1}{2}\mathscr{E}\cdot\mathscr{D} & \mathscr{E}_y\mathscr{D}_z \\ \mathscr{E}_z\mathscr{D}_x & \mathscr{E}_z\mathscr{D}_y & \mathscr{E}_z\mathscr{D}_z - \frac{1}{2}\mathscr{E}\cdot\mathscr{D} \end{bmatrix} + \begin{bmatrix} \mathscr{H}_x\mathscr{B}_x - \frac{1}{2}\mathscr{H}\cdot\mathscr{B} & \mathscr{H}_x\mathscr{B}_y & \mathscr{H}_x\mathscr{B}_z \\ \mathscr{H}_y\mathscr{B}_x & \mathscr{H}_y\mathscr{B}_y - \frac{1}{2}\mathscr{H}\cdot\mathscr{B} & \mathscr{H}_y\mathscr{B}_z \\ \mathscr{H}_z\mathscr{B}_x & \mathscr{H}_z\mathscr{B}_y & \mathscr{H}_z\mathscr{B}_z - \frac{1}{2}\mathscr{H}\cdot\mathscr{B} \end{bmatrix}, \tag{5.63}$$

*and the electromagnetic momentum density per unit volume $\mathbf{g} = \mathscr{D} \times \mathscr{B}$. Then we have*

$$\nabla \cdot {}^2\mathscr{T} = \rho_e \mathscr{E} + \mathscr{J} \times \mathscr{B} + \frac{\partial \mathbf{g}}{\partial t}, \tag{5.64}$$

$$\oiint_S {}^2\mathscr{T} \cdot \boldsymbol{n}\, \mathrm{d}s = \iiint_V (\rho_e \mathscr{E} + \mathscr{J} \times \mathscr{B})\, \mathrm{d}v + \frac{\mathrm{d}}{\mathrm{d}t} \iiint_V \boldsymbol{g}\, \mathrm{d}v, \tag{5.65}$$

*where the tensor divergence is defined as*

$$\nabla \cdot {}^2\mathscr{T} = \sum_{\alpha=1}^{3} \mathbf{i}_\alpha \sum_{\beta=1}^{3} \frac{\partial \mathscr{T}_{\alpha\beta}}{\partial x_\beta} =: \mathbf{i}_\alpha \frac{\partial \mathscr{T}_{\alpha\beta}}{\partial x_\beta}. \tag{5.66}$$

Note that the divergence of a second rank tensor is a true vector [1, p 68] and that $\mathbf{i}_\alpha$ in equation (5.66) are unit vectors along the principal Cartesian coordinates. Also note that equations (5.64) and (5.65) are both vector relations. Equations (5.64) and (5.65) follow from imposing the conditions that the charges and currents inside $S$ are in translational and rotational equilibrium. The first integral on the right-hand side of equation (5.65) is the net mechanical force experienced by material charges and currents inside $S$. It may be expressed as the total rate of change of mechanical momentum $\mathbf{p}$. The second term on the right-hand side is proportional to the time rate of change of Poynting vector integrated in $V$ and represents the momentum $\mathscr{G}$ associated with the electromagnetic field contained within $V$. The left-hand side of equation (5.65) is the force exerted on the electromagnetic system within the volume $V$ as a consequence of the electromagnetic stress acting across the bounding surface $S$. Equation (5.65) states that

$$\frac{\mathrm{d}}{\mathrm{d}t}(\mathbf{p} + \mathscr{G}) = \oiint_S {}^2\mathscr{T} \cdot \boldsymbol{n}\, \mathrm{d}s. \tag{5.67}$$

In linear isotropic media the momentum density is proportional to the Poynting vector through $\boldsymbol{g} = \mathscr{E} \times \mathscr{H}/v^2 = \mathscr{S}/v^2$, where $v$ is the wave speed. In general, if energy is being absorbed by a body at a given rate the momentum of the body increases, and to conserve overall momentum we must associate a momentum density per unit volume, $\boldsymbol{g}$, with any agent that transmits energy at the rate $\mathscr{S}$ per unit area in a given direction. Consequently, the Poynting vector assumes a dual role, as carrying energy and also as carrying momentum.

If the surface $S$ is sufficiently large so that the time-instantaneous fields have not reached it yet, then the surface integral term on the right-hand side of equation (5.67) would vanish. In that case the sum of mechanical and electromagnetic momentum is conserved:

$$\mathbf{p} + \mathscr{G} = \text{constant}. \tag{5.68}$$

# References


[1] Stratton J A 1941 *Electromagnetic Theory* (New York: McGraw-Hill)

[2] Page L and Adams N I 1940 *Electrodynamics* (New York: Van Nostrand)

[3] Glasgow S, Ware M and Peatross J 2001 Poynting's theorem and luminal total energy transport in passive dielectric media *Phys. Rev.* E **64** 1–11

[4] Landau L D, Lifshitz E M and Pitaevskii L P 1984 *Electrodynamics of Continuous Media* 2nd edn (New York: Pergamon)

[5] Geyi W 2019 Stored electromagnetic field energies in general media *J. Opt. Soc. Am.* **36** 917–25

[6] Panofsky W K H and Phillips M 1962 *Classical Electricity and Magnetism* 2nd edn (Reading, MA: Addison-Wesley)

# Chapter 6

# Duality principle and Babinet's principle

The duality principle is based on the observation that Maxwell's equations exhibit symmetry when both electric and magnetic sources are included. Babinet's principle is a form of Huygen's principle valid when fields from one half-space are transmitted to the other half-space by means of perforations in a planar, conducting screen. An interesting consequence of the combination of duality principle and Babinet's principle is that the product of the impedance of a flat metallic antenna and a complementarily shaped slot antenna is a constant independent of frequency in a vacuum. All of these results are elaborated upon in the present chapter.

## 6.1 Duality principle and Babinet's principle

### 6.1.1 Duality principle

When magnetic sources are included, Maxwell's equations are symmetric with respect to the electric and magnetic fields as well as with respect to the sources. This symmetry permits one to write solutions for problems involving magnetic sources in terms of solutions already available for electric sources and vice versa with an appropriate change of field variables. This observation is encapsulated in the duality principle.

**Theorem 6.1.1.** Duality principle [1, p 98].
*The principle of duality states that the field quantities in a dual formulation can be obtained as the dual of the field quantities in the original formulation as expounded in table 6.1. Similar correspondence exists for the time-dependent field quantities.*

The duality principle is based on the observation that if the equations describing two different phenomena are of the same mathematical form, solutions to them will take the same mathematical form. Two equations of the same mathematical form are called *dual equations*. Quantities that occupy the same position in dual equations are called *dual quantities*. Note that equations (1.1) and (1.2) are dual equations.

**Table 6.1.** Dual quantities.

| Original formulation | Dual formulation |
|---|---|
| **J** | **M** |
| **M** | −**J** |
| $q_{ev}$ | $q_{mv}$ |
| $q_{mv}$ | $-q_{ev}$ |
| **E** | **H** |
| **H** | −**E** |
| **D** | **B** |
| **B** | −**D** |
| **A** | **F** |
| **F** | −**A** |
| $\psi_e$ | $\psi_m$ |
| $\psi_m$ | $-\psi_e$ |
| $\varepsilon$ | $\mu$ |
| $\mu$ | $\varepsilon$ |

Likewise, equations (1.36) and (1.37) are dual equations. As an application, the duality principle enables one to express the fields radiated by an electrically small loop antenna in terms of the fields available for an electrically small dipole antenna [1]. Another application is to express the fields generated by a thin slot antenna in terms of the fields generated by a thin strip dipole.

### 6.1.2 Babinet's principle

Babinet's principle is useful for relating fields in one half-space to those in the other half-space when the two regions are separated by perforated planar, conducting screens.

**Theorem 6.1.2.** Babinet's principle [1, p 365].
*Consider three cases of a given source (i) radiating in free-space, (ii) radiating in the presence of an electrically conducting planar screen having an aperture* $S_a$ *and conducting portion* $S_c$*, and (iii) magnetically conducting planar screen of obstacle shape* $S_a$*. The electric and magnetic screens are complementary in the sense that* $S_a \cap S_c = \varnothing$, $S_a \cup S_c =$ *complete* $z = 0$ *plane, see figure 6.1. Let the fields for* $z > 0$ *be designated* $(\mathbf{E}^i, \mathbf{H}^i)$, $(\mathbf{E}^e, \mathbf{H}^e)$, *and* $(\mathbf{E}^m, \mathbf{H}^m)$ *for the cases (i), (ii), and (iii), respectively. Then Babinet's principle for complementary screens states that*

$$\mathbf{E}^e + \mathbf{E}^m = \mathbf{E}^i, \quad \mathbf{H}^e + \mathbf{H}^m = \mathbf{H}^i \quad \forall z > 0. \tag{6.1}$$

*Proof.* The total field in cases (ii) and (iii) is the sum of the incident field $\mathbf{E}^i$ and the scattered field $\mathbf{E}^s$ produced by the induced currents on the respective screens. Thus

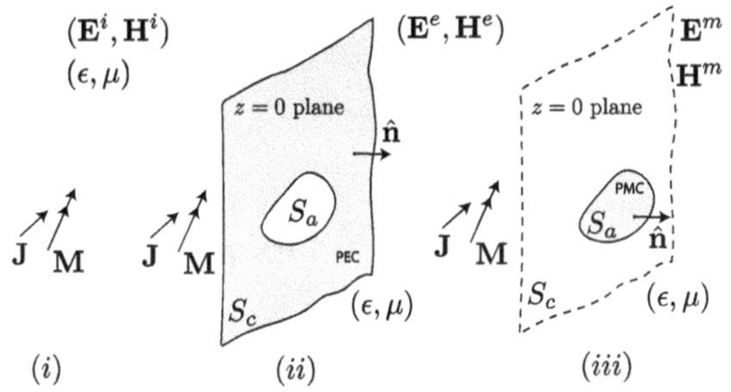


**Figure 6.1.** Sources (**J**, **M**) radiating in the three situations.

$(\mathbf{E}^e, \mathbf{H}^e) = (\mathbf{E}^i + \mathbf{E}^{es}, \mathbf{H}^i + \mathbf{H}^{es})$ and $(\mathbf{E}^m, \mathbf{H}^m) = (\mathbf{E}^i + \mathbf{E}^{ms}, \mathbf{H}^i + \mathbf{H}^{ms})$. Note that an element of electric current in homogeneous space produces no components of **H** tangential to any plane containing the element. (For example a $z$-directed current element in the $xz$-plane produces no $H_z$ and $H_x$ in the $xz$-plane.) The currents induced on the screen thus produce no tangential magnetic field, $\hat{\mathbf{n}} \times \mathbf{H}^{es}$, over the $z = 0$ plane. Hence

$$\hat{\mathbf{n}} \times \mathbf{H}^e = \hat{\mathbf{n}} \times \mathbf{H}^i \quad \text{over } S_a. \tag{6.2}$$

On the electrically conducting screen itself we have the boundary condition

$$\hat{\mathbf{n}} \times \mathbf{E}^e = \mathbf{0} \quad \text{over } S_c. \tag{6.3}$$

For the complementary magnetic screen, we similarly have

$$\hat{\mathbf{n}} \times \mathbf{E}^m = \hat{\mathbf{n}} \times \mathbf{E}^i \quad \text{over } S_c, \tag{6.4}$$

$$\hat{\mathbf{n}} \times \mathbf{H}^m = \mathbf{0} \quad \text{over } S_a. \tag{6.5}$$

Adding equations (6.3) and (6.4) and equations (6.2) and (6.5) together we have

$$\begin{aligned} \hat{\mathbf{n}} \times (\mathbf{E}^e + \mathbf{E}^m) &= \hat{\mathbf{n}} \times \mathbf{E}^i \quad \text{over } S_c \\ \hat{\mathbf{n}} \times (\mathbf{H}^e + \mathbf{H}^m) &= \hat{\mathbf{n}} \times \mathbf{H}^i \quad \text{over } S_a. \end{aligned} \tag{6.6}$$

Hence (ii) plus (iii) field has the same $\hat{\mathbf{n}} \times \mathbf{E}$ as the incident field (i) over part of the $z = 0$ plane and the same $\hat{\mathbf{n}} \times \mathbf{H}$ over the rest of the $z = 0$ plane. By the uniqueness theorem the fields of (ii) plus (iii) are then identical to (i) in all of $z > 0$. Thus equation (6.1) follows. ■

An alternative statement of Babinet's principle can be given in terms of the dual problem to figure 6.1, shown in figure 6.2. Replacing the electric current **J** with the magnetic current **K** of the same size and functional dependence and the magnetic current **M** with electric current −**L** of the same size and functional dependence, the magnetic conductor over $S_a$ with an electric conductor, and the medium parameters $(\varepsilon, \mu)$ with $(\mu, \varepsilon)$ such that the intrinsic impedance $\eta$ is replaced with its inverse $\eta^{-1}$ and denoting the fields for dual problem as $(\mathbf{E}^d, \mathbf{H}^d)$, we have by duality that $\mathbf{H}^d = \mathbf{E}^m$, $\mathbf{E}^d = -\mathbf{H}^m$. Note that because a reciprocal medium ($\eta$ is replaced by $1/\eta$) is

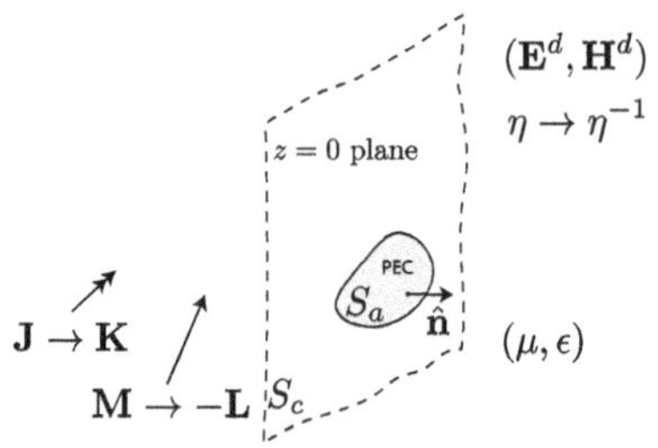


**Figure 6.2.** Electric, magnetic sources ($-\mathbf{L}$, $\mathbf{K}$) radiating in the presence of a PEC obstacle $S_a$ in a reciprocal medium ($\mu$, $\varepsilon$).

considered in the dual problem the units of $\mathbf{E}^m$ and $\mathbf{H}^d$ will also match. The Babinet's principle can be recast as

$$\mathbf{E}^e + \mathbf{H}^d = \mathbf{E}^i, \quad \mathbf{H}^e - \mathbf{E}^d = \mathbf{H}^i, \quad \forall z > 0. \tag{6.7}$$

### 6.1.3 Booker's relation

**Example 6.1.3.1.** Impedance relation between complementary antennas.
Figure 6.3(a) shows a perfectly conducting antenna-A cut from a plane conductor and fed across the slot *ba* by a voltage source. Figure 6.3(b) shows the aperture formed by the remainder of the conducting plane after antenna-A is cut. The aperture antenna, labeled as antenna-B, and fed by a voltage source connected across *dc*, is said to be complementary to antenna-A. Let $Z_d$ be the input impedance of antenna-A and $Z_s$ be the input impedance of antenna-B. Let the intrinsic impedance of the homogeneous medium surrounding the plane be $\eta$. We show using the duality and Babinet's principles that

$$Z_s Z_d = \frac{\eta^2}{4}, \tag{6.8}$$

a result known as *Booker's relation*. Let ($\mathbf{E}^d$, $\mathbf{H}^d$) be the fields radiated by antenna-A due to the currents $\mathbf{J}_s$ flowing on its conducting fins and let ($\mathbf{E}^s$, $\mathbf{H}^s$) be the fields generated by antenna-B due the magnetic current $\mathbf{M}_s$ flowing in its aperture. Because the currents ($\mathbf{J}_s$, $\mathbf{M}_s$) are dual quantities, we have by duality the dimensional relations $\mathbf{H}^s = \eta^{-1}\mathbf{E}^d$ and $\mathbf{E}^s = -\eta\mathbf{H}^d$ upon affecting the change $\mathbf{J}_s \to \eta^{-1}\mathbf{M}_s$ and $\mathbf{M}_s \to -\eta\mathbf{J}_s$ in the sources.

With reference to figure 6.3(a), we define the input impedance of antenna-A as

$$\begin{aligned} Z_d &= \frac{V_b - V_a}{I_d} = \frac{-\int_a^b \mathbf{E}^d \cdot d\ell}{\oint \mathbf{H}^d \cdot d\ell} \\ &= -\frac{\int_a^b \mathbf{E}^d \cdot d\ell}{2\int_c^d \mathbf{H}^d \cdot d\ell}. \end{aligned} \tag{6.9}$$

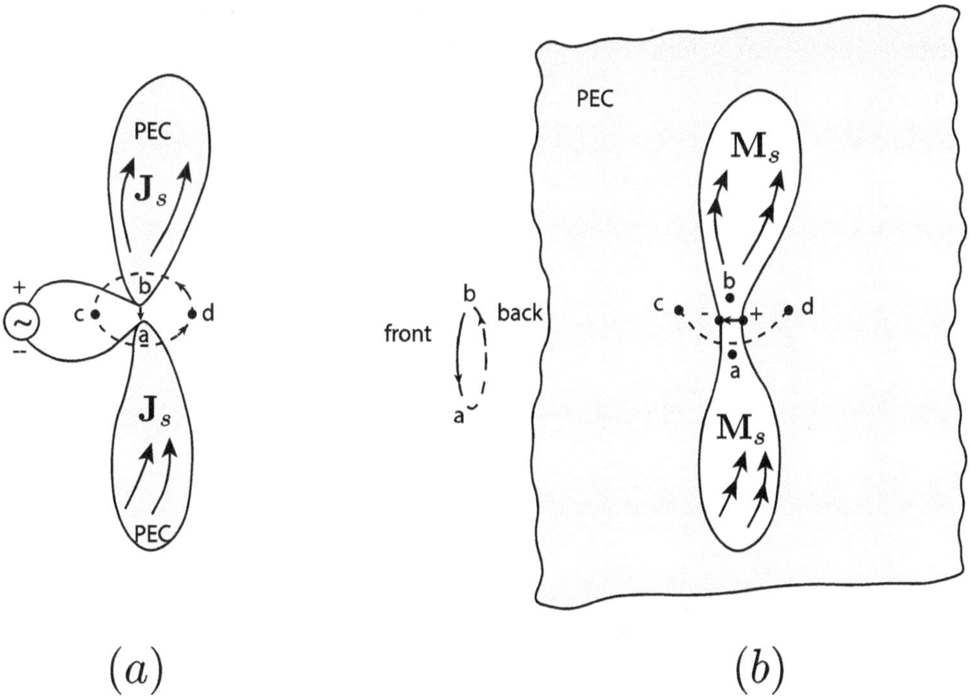


**Figure 6.3.** (a) Conducting antenna-A and (b) its complementary antenna-B carved out of a perfectly conducting plane.

Similarly, with reference to figure 6.3(b), the input impedance of the aperture antenna is defined as

$$Z_s = \frac{V_d - V_c}{I_s} = \frac{-\int_c^d \mathbf{E}^s \cdot d\ell}{2\int_b^a \mathbf{H}^s \cdot d\ell} = \frac{\int_c^d \eta \mathbf{H}^d \cdot d\ell}{-2\int_a^b \eta^{-1}\mathbf{E}^d \cdot d\ell}. \tag{6.10}$$

Multiplying the left-hand sides of equations (6.9) and (6.10) gives the result (6.8). Therefore, the product of the input impedance of two complementary antennas is proportional to the square of the intrinsic impedance of the medium. Note that $Z_d$ and $Z_s$ are frequency dependent and that their product has a frequency dependence of the intrinsic impedance square of the medium.

If the antennas are *self-complementary* (meaning the metal portion and the air portion of the screen are of the same shape), then their input impedances are the same, each equal to half the intrinsic impedance of the surrounding medium:

$$Z_s = Z_d = \frac{1}{2}\eta.$$

In free-space (where $\eta$ is independent of frequency) such antennas will have extremely wide bandwidth performance. ■■

## Reference

[1] Harrington R F 1961 *Time-Harmonic Electromagnetic Fields* (New York: McGraw-Hill)

# Chapter 7

# Electromagnetic reciprocity

The notion of reciprocity refers to the symmetry in the transfer of electromagnetic quantities in spacetime through an intervening medium. For instance, the strength of the electromagnetic field generated by an antenna in the far-zone in a certain direction is related to the strength of the signal it captures from a plane wave arriving from the same direction. That relationship is an example of reciprocity. Another area where reciprocity manifests is in the electromagnetic interaction that exists between two different sources operating at the same frequency and separated in space. The conditions on the medium constituent parameters under which electromagnetic reciprocity holds in the frequency and time domains are captured in the various versions of the reciprocity theorem and discussed in this chapter. A lesser known theorem known as the compensation theorem that is useful in accounting for incremental changes in an electromagnetic quantity when the environment and/or the geometry is changed is also discussed at length along with some interesting practical examples.

## 7.1 Reciprocity theorems in the frequency and time domains

In the frequency domain, the reciprocity theorem relates the cross-coupling relations that must exist between two independent electromagnetic sources, both operating at the same frequency in a given simple medium. In more complex media, certain restrictions are placed on the medium parameters in order for these relations to be valid and we discuss these also in the current section.

### 7.1.1 Reciprocity theorem for fields

**Theorem 7.1.1.** Frequency domain reciprocity theorem [1, p 205].
*Consider time-harmonic sources* $(\mathbf{J}_1, \mathbf{M}_1)$ *generating fields* $(\mathbf{E}_1, \mathbf{H}_1)$ *and sources* $(\mathbf{J}_2, \mathbf{M}_2)$ *generating fields* $(\mathbf{E}_2, \mathbf{H}_2)$ *in a medium with isotropic constitutive parameters*

doi:10.1088/978-0-7503-1716-0ch7

*$(\varepsilon, \mu)$. The sources are all contained within a volume $V_0$ bounded by a surface $S_0$. A spherical surface $S_\infty$ surrounds $S_0$ and forms the outer boundary of the annular volume $V_\infty$. The volume $V_\infty$ is bounded by the surface $S_0$ on the inside (figure 7.1). On a sufficiently large surface $S_\infty$ with an outward normal $\hat{\mathbf{n}}_\infty$, let the fields satisfy Sommerfeld's radiation condition:*

$$\begin{aligned} \hat{\mathbf{n}}_\infty \times \mathbf{E}_i &= \mathcal{O}\left(\frac{1}{r}\right), \ \hat{\mathbf{n}}_\infty \times \mathbf{H}_i = \mathcal{O}\left(\frac{1}{r}\right), \\ \mathbf{E}_i + \eta \hat{\mathbf{n}}_\infty \times \mathbf{H}_i &= o\left(\frac{1}{r}\right), \quad i = 1, 2, \end{aligned} \tag{7.1}$$

*where $\eta = \sqrt{\mu/\varepsilon}$. Then*

$$\oiint_{S_0} \mathbf{E}_1 \times \mathbf{H}_2 \cdot \hat{\mathbf{n}}\, \mathrm{d}s = \oiint_{S_0} \mathbf{E}_2 \times \mathbf{H}_1 \cdot \hat{\mathbf{n}}\, \mathrm{d}s, \tag{7.2}$$

$$\begin{aligned} \langle 2, 1 \rangle &:= \iiint_{V_0} (\mathbf{E}_2 \cdot \mathbf{J}_1 - \mathbf{H}_2 \cdot \mathbf{M}_1)\, \mathrm{d}v \\ &= \iiint_{V_0} (\mathbf{E}_1 \cdot \mathbf{J}_2 - \mathbf{H}_1 \cdot \mathbf{M}_2)\, \mathrm{d}v =: \langle 1, 2 \rangle, \end{aligned} \tag{7.3}$$

*where $\langle a, b \rangle$ denotes reaction of field a on source b.*

*Proof.* We have from Maxwell's equations

$$\begin{aligned} \nabla \times \mathbf{E}_1 &= -j\omega\mu \mathbf{H}_1 - \mathbf{M}_1, \\ \nabla \times \mathbf{H}_1 &= j\omega\varepsilon \mathbf{E}_1 + \mathbf{J}_1, \end{aligned} \tag{7.4}$$

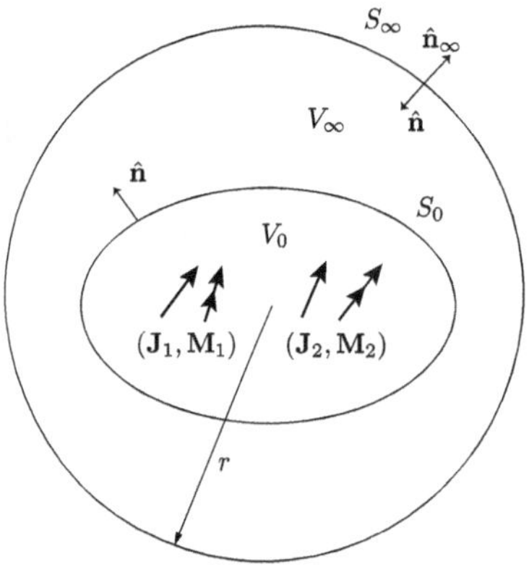


**Figure 7.1.** Volume $V_0$ bounded by $S_0$ and $V_\infty$ bounded by $S_0$ and $S_\infty$.

$$\begin{aligned}\nabla \times \mathbf{E}_2 &= -j\omega\mu\mathbf{H}_2 - \mathbf{M}_2,\\ \nabla \times \mathbf{H}_2 &= j\omega\varepsilon\mathbf{E}_2 + \mathbf{J}_2.\end{aligned} \tag{7.5}$$

Dot multiply the first of equation (7.4) by $\mathbf{H}_2$ and the first of equation (7.5) by $\mathbf{H}_1$ and subtract. Likewise, dot multiply the second of equation (7.4) by $\mathbf{E}_2$ and the second of equation (7.5) by $\mathbf{E}_1$ and subtract. The resulting equations are

$$\mathbf{H}_2 \cdot \nabla \times \mathbf{E}_1 - \mathbf{H}_1 \cdot \nabla \times \mathbf{E}_2 = \mathbf{H}_1 \cdot \mathbf{M}_2 - \mathbf{H}_2 \cdot \mathbf{M}_1, \tag{7.6}$$

$$\mathbf{E}_2 \cdot \nabla \times \mathbf{H}_1 - \mathbf{E}_1 \cdot \nabla \times \mathbf{H}_2 = \mathbf{E}_2 \cdot \mathbf{J}_1 - \mathbf{E}_1 \cdot \mathbf{J}_2. \tag{7.7}$$

Adding equations (7.6) and (7.7) and employing the identity $\nabla \cdot (\mathbf{a} \times \mathbf{b}) = \mathbf{b} \cdot \nabla \times \mathbf{a} - \mathbf{a} \cdot \nabla \times \mathbf{b}$ we arrive at

$$\begin{aligned}\nabla \cdot (\mathbf{E}_1 \times \mathbf{H}_2 - \mathbf{E}_2 \times \mathbf{H}_1) = (&\mathbf{E}_2 \cdot \mathbf{J}_1 - \mathbf{H}_2 \cdot \mathbf{M}_1)\\ &- (\mathbf{E}_1 \cdot \mathbf{J}_2 - \mathbf{H}_1 \cdot \mathbf{M}_2).\end{aligned} \tag{7.8}$$

Equation (7.8) is the differential form of the reciprocity theorem valid point by point in space. Integrating this over any volume $V$ bounded by a surface $S$ with outward normal $\hat{\mathbf{n}}$ and applying the Gauss' divergence theorem (B.4) results in

$$\begin{aligned}&\oiint_S (\mathbf{E}_1 \times \mathbf{H}_2 - \mathbf{E}_2 \times \mathbf{H}_1) \cdot \hat{\mathbf{n}}\, \mathrm{d}s\\ &= \iiint_V [(\mathbf{E}_2 \cdot \mathbf{J}_1 - \mathbf{H}_2 \cdot \mathbf{M}_1) - (\mathbf{E}_1 \cdot \mathbf{J}_2 - \mathbf{H}_1 \cdot \mathbf{M}_2)]\, \mathrm{d}v.\end{aligned} \tag{7.9}$$

In particular, applying this relation to the volume $V_0$ bounded by $S_0$ we obtain

$$\begin{aligned}&\iiint_{V_0} [(\mathbf{E}_2 \cdot \mathbf{J}_1 - \mathbf{H}_2 \cdot \mathbf{M}_1) - (\mathbf{E}_1 \cdot \mathbf{J}_2 - \mathbf{H}_1 \cdot \mathbf{M}_2)]\, \mathrm{d}v\\ &\quad = \oiint_{S_0} (\mathbf{E}_1 \times \mathbf{H}_2 - \mathbf{E}_2 \times \mathbf{H}_1) \cdot \hat{\mathbf{n}}\, \mathrm{d}s.\end{aligned} \tag{7.10}$$

We will now show that that the right-hand side of equation (7.10) is actually equal to zero. Applying equation (7.9) to the annular volume $V_\infty$ that is free from sources and having boundaries $S_0 + S_\infty$ we obtain

$$\oiint_{S_0+S_\infty} (\mathbf{E}_1 \times \mathbf{H}_2 - \mathbf{E}_2 \times \mathbf{H}_1) \cdot \hat{\mathbf{n}}\, \mathrm{d}s = 0, \tag{7.11}$$

where $\hat{\mathbf{n}}$ is chosen to be the unit normal pointing *into* the volume $V_\infty$. But

$$
\begin{aligned}
-\oiint_{S_\infty}(\mathbf{E}_1\times\mathbf{H}_2-\mathbf{E}_2\times\mathbf{H}_1)\cdot\hat{\mathbf{n}}\,\mathrm{d}s &= \oiint_{S_\infty}(\mathbf{E}_1\times\mathbf{H}_2-\mathbf{E}_2\times\mathbf{H}_1)\cdot\hat{\mathbf{n}}_\infty\,\mathrm{d}s \\
&= \oiint_{S_\infty}(\mathbf{E}_1\cdot\mathbf{H}_2\times\hat{\mathbf{n}}_\infty-\mathbf{E}_2\cdot\mathbf{H}_1\times\hat{\mathbf{n}}_\infty)\,\mathrm{d}s \\
&= \frac{1}{\eta}\oiint_{S_\infty}[\mathbf{E}_1\cdot\mathbf{E}_2-\mathbf{E}_2\cdot\mathbf{E}_1]\,\mathrm{d}s \qquad (7.12)\\
&\quad+\oiint_{S_\infty} o\left(\frac{1}{r}\right)\mathcal{O}\left(\frac{1}{r}\right)\mathrm{d}s \\
&= \oiint_{S_\infty} o\left(\frac{1}{r}\right)\mathcal{O}\left(\frac{1}{r}\right) r^2\sin\theta\,\mathrm{d}\theta\mathrm{d}\phi \\
&= 0 \quad \text{as } r\to\infty \text{ on } S_\infty. \qquad (7.13)
\end{aligned}
$$

Using this in equation (7.11) yields

$$
\begin{aligned}
\oiint_{S_0}(\mathbf{E}_1\times\mathbf{H}_2-\mathbf{E}_2\times\mathbf{H}_1)\cdot\hat{\mathbf{n}}\,\mathrm{d}s = 0 \Longrightarrow \oiint_{S_0}\mathbf{E}_1\times\mathbf{H}_2\cdot\hat{\mathbf{n}}\,\mathrm{d}s \\
= \oiint_{S_0}\mathbf{E}_2\times\mathbf{H}_1\cdot\hat{\mathbf{n}}\,\mathrm{d}s,
\end{aligned} \qquad (7.14)
$$

for *any* surface $S_0$ as long as it encloses *all* sources. Inserting this in equation (7.10) we arrive at

$$
\begin{aligned}
\langle 2, 1\rangle =: \iiint_{R_1}(\mathbf{E}_2\cdot\mathbf{J}_1-\mathbf{H}_2\cdot\mathbf{M}_1)\,\mathrm{d}v \\
= \iiint_{R_2}(\mathbf{E}_1\cdot\mathbf{J}_2-\mathbf{H}_1\cdot\mathbf{M}_2)\,\mathrm{d}v := \langle 1, 2\rangle,
\end{aligned} \qquad (7.15)
$$

where $R_i$ is the volume enclosing source $i$, $i = 1, 2$ only. So the reciprocity theorem says that the reaction, $\langle 2, 1\rangle$, of field 2 on source 1 is the same as the reaction, $\langle 1, 2\rangle$, of field 1 on source 2. ■

On observing that the surface integral in equation (7.9) vanishes when the surface $S$ excludes both sources and that the surface integral in equation (7.14) vanishes since $S_0$ encloses both sources, we are led to the following corollary of the reciprocity theorem.

**Corollary 7.1.1.1** Corollary I to reciprocity theorem [2].
*Let two systems of sources (1) and (2) operate in a linear and passive medium. Let (1) alone give rise to fields* $(\mathbf{E}_1, \mathbf{H}_1)$ *and let (2) alone give rise to fields* $(\mathbf{E}_2, \mathbf{H}_2)$. *Let a*

*flux density vector be defined as* $\mathbf{P} = \mathbf{E}_1 \times \mathbf{H}_2 - \mathbf{E}_2 \times \mathbf{H}_1$. *Then the flux across a closed surface T*

$$\oint\!\!\!\oint_T \mathbf{P} \cdot \hat{\mathbf{n}}\, \mathrm{d}s = 0, \tag{7.16}$$

*provided that either all sources are inside T or all sources are outside T.*

**Example 7.1.1.1** Antennas tangential to PEC.

Using reciprocity theorem we prove an important result that an electric current density $\mathbf{J}_s$ placed tangentially and in close contact with a perfect electric conductor (PEC) produces zero overall electric field $\mathbf{E}_1$ external to the conductor. The field $\mathbf{E}_1$ is the sum total of the field produced by $\mathbf{J}_s$ in the absence of the PEC and the scattered field produced by the additional currents $\mathbf{J}_{\text{ind}}$ induced on the PEC. Accordingly we take $\mathbf{J}_1 = \mathbf{J}_s + \mathbf{J}_{\text{ind}}$. Note that the total current $\mathbf{J}_1$ is all tangential to the PEC.

Consider an electric dipole $\mathbf{J}_2 = \hat{\mathbf{p}} I_0 \ell \delta(\mathbf{r} - \mathbf{r}_0)$ oriented in a direction $\hat{\mathbf{p}}$ and located at an arbitrary point $\mathbf{r}_0$ exterior to the PEC. The dipole radiates in the presence of the PEC object. It produces a field $\mathbf{E}_2$ whose tangential component vanishes on the PEC, i.e. $\hat{\mathbf{n}} \times \mathbf{E}_2 = 0$, where $\hat{\mathbf{n}}$ is the outward normal on the PEC. Applying equation (7.15) to the pairs $(\mathbf{J}_1, \mathbf{E}_1)$ and $(\mathbf{J}_2, \mathbf{E}_2)$ and noting that the lhs is zero since $\mathbf{J}_1 \cdot \mathbf{E}_2$ picks up the tangential part of $\mathbf{E}_2$ on the PEC, we obtain $I_0 \ell \hat{\mathbf{p}} \cdot \mathbf{E}_1(\mathbf{r}_0) = 0$. But since $\hat{\mathbf{p}}$ is arbitrary, this implies that $\mathbf{E}_1(\mathbf{r}_0)$ is identically zero. ■■

**Example 7.1.1.2.** Mutual impedance between antennas.

Using reciprocity theorem we can derive an expression for the mutual impedance between two antennas in terms of their radiated fields. Let an infinitesimal current $I_0$ (current density $\mathbf{J} = I_0 \hat{\ell} \delta(x - 0)\delta(y - 0)$) impressed between the terminals of antenna-$i$ generate the field $(\mathbf{E}_i, \mathbf{H}_i)$, $i = 1, 2$ when the antenna-$j$ is open-circuited. Then the mutual impedance, $Z_{12}$, is defined as [3, p 119]

$$Z_{12} = -\frac{\langle 2, 1\rangle}{I_0^2} = -\frac{\langle 1, 2\rangle}{I_0^2}, \tag{7.17}$$

where $\langle 2, 1\rangle = \iiint \mathbf{E}_2 \cdot I_0 \hat{\ell}\, \mathrm{d}z \delta(x - 0)\delta(y - 0)\, \mathrm{d}x\mathrm{d}y = I_0 \int \mathbf{E}_2 \cdot \mathrm{d}\ell = -V_2 I_0$, $V_2$ being the open-circuited voltage induced at port-1 due to the incident field $\mathbf{E}_2$. Similarly, for $\langle 1, 2\rangle$. Applying the result (7.10) to a volume $V$ bounded by an arbitrary surface $T$ that *encloses only source 2* (such as in figure 7.2) gives

$$-\langle 1, 2\rangle = \oint\!\!\!\oint_T (\mathbf{E}_1 \times \mathbf{H}_2 - \mathbf{E}_2 \times \mathbf{H}_1) \cdot \hat{\mathbf{n}}\, \mathrm{d}s. \tag{7.18}$$

Hence

$$Z_{12} = \frac{1}{I_0^2} \oint\!\!\!\oint_T (\mathbf{E}_1 \times \mathbf{H}_2 - \mathbf{E}_2 \times \mathbf{H}_1) \cdot \hat{\mathbf{n}}\, \mathrm{d}s, \tag{7.19}$$

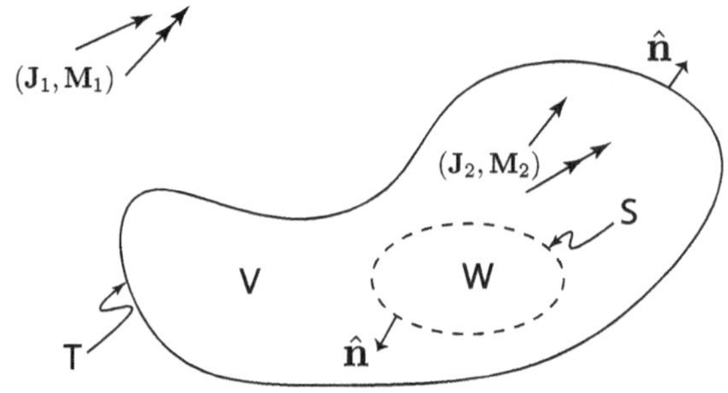


**Figure 7.2.** Volume $V$ exterior to $(\mathbf{J}_1, \mathbf{M}_1)$ and bounded by $T$ and sub-volume $W$ bounded by $S$.

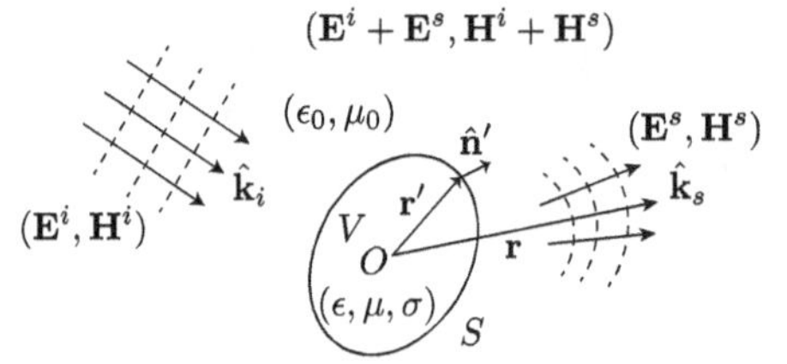


**Figure 7.3.** Plane wave striking a lossy scatterer bounded by surface $S$ and immersed in an ambient medium with constitutive parameters $(\epsilon_0, \mu_0)$.

where the unit normal $\hat{\mathbf{n}}$ points out of the volume $V$ on $T$ (as indicated in figure 7.2). In view of corollary I the expression for mutual impedance remains invariant if the surface $T$ is modified to include a second closed surface $S$ that either contains both sources or excludes both sources. ■■

### 7.1.2 Reciprocity theorem for scattering amplitude

Consider scattering of plane waves by an arbitrary scatterer bounded by a surface $S$ as shown in figure 7.3. The incident plane wave propagates along $\hat{\mathbf{k}}_\mathrm{i}$ and the scattered wave propagates in the direction $\hat{\mathbf{k}}_\mathrm{s}$. In the reciprocity theorem, let the $a$-field, corresponding to field 1, be a plane wave *arriving towards the scatterer* from a direction $\hat{\mathbf{k}}_\mathrm{i} = \hat{\mathbf{k}}_a$. Likewise let the $b$-field, corresponding to field 2, be a plane wave arriving towards the scatterer from a direction $\hat{\mathbf{k}}_\mathrm{i} = \hat{\mathbf{k}}_b$. The electric and magnetic field for the incident plane waves are

$$\mathbf{E}_a^\mathrm{i} = \mathbf{e}_a\, \mathrm{e}^{jk\hat{\mathbf{k}}_a \cdot \mathbf{r}}; \quad \eta \mathbf{H}_a^\mathrm{i} = \mathbf{E}_a^\mathrm{i} \times \hat{\mathbf{k}}_a; \quad \hat{\mathbf{k}}_a \cdot \mathbf{e}_a = 0, \tag{7.20}$$

$$\mathbf{E}_b^\mathrm{i} = \mathbf{e}_b\, \mathrm{e}^{jk\hat{\mathbf{k}}_b \cdot \mathbf{r}}; \quad \eta \mathbf{H}_b^\mathrm{i} = \mathbf{E}_b^\mathrm{i} \times \hat{\mathbf{k}}_b; \quad \hat{\mathbf{k}}_b \cdot \mathbf{e}_b = 0, \tag{7.21}$$

where $\mathbf{e}_a$ and $\mathbf{e}_b$ are complex quantities that specify the polarizations of the $a$- and $b$-plane waves, respectively. The total field in each case is written as the sum of incident and scattered fields. For instance, the total $a$ electric field is written as

$\mathbf{E}_a = \mathbf{E}_a^{\mathrm{i}} + \mathbf{E}_a^{\mathrm{s}}$. Because the sources for incident and scattered fields are either entirely outside or entirely inside $S$, respectively, we have from corollary 7.1.1.1 that

$$\oiint_S \left(\mathbf{E}_a^{\mathrm{i}} \times \mathbf{H}_b^{\mathrm{i}} - \mathbf{E}_b^{\mathrm{i}} \times \mathbf{H}_a^{\mathrm{i}}\right) \cdot \hat{\mathbf{n}}\, \mathrm{d}s = 0 \tag{7.22}$$

and

$$\oiint_S \left(\mathbf{E}_a^{\mathrm{s}} \times \mathbf{H}_b^{\mathrm{s}} - \mathbf{E}_b^{\mathrm{s}} \times \mathbf{H}_a^{\mathrm{s}}\right) \cdot \hat{\mathbf{n}}\, \mathrm{d}s = 0. \tag{7.23}$$

Hence from the reciprocity theorem (7.14) we have

$$\oiint_S \left(\mathbf{E}_a^{\mathrm{i}} \times \mathbf{H}_b^{\mathrm{s}} - \mathbf{E}_b^{\mathrm{i}} \times \mathbf{H}_a^{\mathrm{s}}\right) \cdot \hat{\mathbf{n}}\, \mathrm{d}s = \oiint_S \left(\mathbf{E}_a^{\mathrm{s}} \times \mathbf{H}_b^{\mathrm{i}} - \mathbf{E}_b^{\mathrm{s}} \times \mathbf{H}_a^{\mathrm{i}}\right) \cdot \hat{\mathbf{n}}\, \mathrm{d}s. \tag{7.24}$$

Using equations (11.22) and (11.23) to represent the scattered electric field for the $a$-field propagating in the direction $\hat{\mathbf{k}}_{\mathrm{s}} = \hat{\mathbf{k}}_b$, taking a dot product with $\mathbf{e}_b$ and drawing upon the results from equation (7.21) we obtain

$$\begin{aligned}
\mathbf{e}_b \cdot \mathbf{E}_a^{\mathrm{s}}(\hat{\mathbf{k}}_b) &= \frac{\mathrm{e}^{-jkr}}{r}\frac{jk}{4\pi} \oiint_S \left\{ -\eta\left(\hat{\mathbf{n}}' \times \mathbf{H}_a^{\mathrm{s}}\right) \cdot \mathbf{E}_b^{\mathrm{i}} + \left(\hat{\mathbf{n}}' \times \mathbf{E}_a^{\mathrm{s}}\right) \cdot \left(\hat{\mathbf{k}}_b \times \mathbf{E}_b^{\mathrm{i}}\right)\right\} \mathrm{d}s' \\
&= \frac{\mathrm{e}^{-jkr}}{r}\frac{jk\eta}{4\pi} \oiint_S \left\{\mathbf{E}_b^{\mathrm{i}} \times \mathbf{H}_a^{\mathrm{s}} - \mathbf{E}_a^{\mathrm{s}} \times \mathbf{H}_b^{\mathrm{i}}\right\} \cdot \hat{\mathbf{n}}'\, \mathrm{d}s' \\
&= \frac{\mathrm{e}^{-jkr}}{r}\mathbf{e}_b \cdot \mathbf{F}_a^{\mathrm{s}}\left(\hat{\mathbf{k}}_b, \hat{\mathbf{k}}_a\right),
\end{aligned} \tag{7.25}$$

where $\mathbf{F}^{\mathrm{s}}(\hat{\mathbf{k}}_b, \hat{\mathbf{k}}_a)$ is the scattering amplitude in the direction $\hat{\mathbf{k}}_b$ due to a plane wave arriving from direction $\hat{\mathbf{k}}_a$. The term proportional to $\mathbf{e}_b \cdot \mathbf{k}_b$ drops out because of the orthogonality condition in equation (7.21) between the plane wave polarization and its direction of propagation. Similarly,

$$\begin{aligned}
\mathbf{e}_a \cdot \mathbf{E}_b^{\mathrm{s}}(\hat{\mathbf{k}}_a) &= \frac{\mathrm{e}^{-jkr}}{r}\mathbf{e}_a \cdot \mathbf{F}_b^{\mathrm{s}}(\hat{\mathbf{k}}_a, \hat{\mathbf{k}}_b) \\
&= \frac{\mathrm{e}^{-jkr}}{r}\frac{jk\eta}{4\pi} \oiint_S \left\{\mathbf{E}_a^{\mathrm{i}} \times \mathbf{H}_b^{\mathrm{s}} - \mathbf{E}_b^{\mathrm{s}} \times \mathbf{H}_a^{\mathrm{i}}\right\} \cdot \hat{\mathbf{n}}'\, \mathrm{d}s'.
\end{aligned} \tag{7.26}$$

Therefore by equation (7.24) we obtain the reciprocity result for the scattering amplitudes

$$\mathbf{e}_a \cdot \mathbf{F}_b^{\mathrm{s}}(\hat{\mathbf{k}}_a, \hat{\mathbf{k}}_b) = \mathbf{e}_b \cdot \mathbf{F}_a^{\mathrm{s}}(\hat{\mathbf{k}}_b, \hat{\mathbf{k}}_a). \tag{7.27}$$

A result of such kind was first obtained in [4] and later in [5].

### 7.1.3 Extended reciprocity theorem

The reciprocity theorem may be extended to a bianisotropic system with certain conditions placed on the permittivity and permeability tensors. In a bianisotropic medium one has

$$\nabla \times \mathbf{E} = -j\omega(\bar{\bar{\mu}}\mathbf{H} + \bar{\bar{\zeta}}\mathbf{E}) - \mathbf{M}, \tag{7.28}$$

$$\nabla \times \mathbf{H} = j\omega(\bar{\bar{\varepsilon}}\mathbf{E} + \bar{\bar{\xi}}\mathbf{H}) + \mathbf{J}. \tag{7.29}$$

Using the same manipulations as above with two different sources and the corresponding fields yields

$$\begin{aligned}\langle 1, 2\rangle - \langle 2, 1\rangle = -j\omega \iiint_{V_0} \Big[&\mathbf{E}_2 \cdot (\bar{\bar{\varepsilon}} - \bar{\bar{\varepsilon}}^{\mathrm{T}})\mathbf{E}_1 + \mathbf{H}_1 \cdot (\bar{\bar{\mu}} - \bar{\bar{\mu}}^{\mathrm{T}})\mathbf{H}_2 \\ &+ \mathbf{E}_2 \cdot (\bar{\bar{\xi}} + \bar{\bar{\zeta}}^{\mathrm{T}})\mathbf{H}_1 - \mathbf{H}_1 \cdot (\bar{\bar{\zeta}} + \bar{\bar{\xi}}^{\mathrm{T}})\mathbf{E}_2\Big] \mathrm{d}v,\end{aligned} \tag{7.30}$$

where superscript 'T' stands for transpose. Thus the fields will be reciprocal if

$$\bar{\bar{\varepsilon}}^{\mathrm{T}} = \bar{\bar{\varepsilon}}, \tag{7.31}$$

$$\bar{\bar{\mu}}^{\mathrm{T}} = \bar{\bar{\mu}}, \tag{7.32}$$

$$\bar{\bar{\xi}}^{\mathrm{T}} = -\bar{\bar{\zeta}}, \tag{7.33}$$

in other words the fields satisfy reciprocity condition if the permittivity and permeability tensors are symmetric and if the cross tensors satisfy equation (7.33).

### 7.1.4 Modified reciprocity theorem

Proceeding with an analysis along the same lines as in the previous subsection we arrive at the modified reciprocity theorem that is stated next.

**Theorem 7.1.2.** Modified reciprocity theorem [6, p 411].
*Consider a bianisotropic medium with permittivity, permeability and cross tensors* $\bar{\bar{\varepsilon}}, \bar{\bar{\mu}}, \bar{\bar{\xi}}, \bar{\bar{\zeta}}$ *as defined in equations (7.28) and (7.29). Let sources* $(\mathbf{J}, \mathbf{M})$ *produce fields* $(\mathbf{E}, \mathbf{H})$ *in it. Consider a complementary medium with permittivity, permeability, and cross tensors* $(\bar{\bar{\varepsilon}}^{\mathrm{C}}, \bar{\bar{\mu}}^{\mathrm{C}}, \bar{\bar{\xi}}^{\mathrm{C}}, \bar{\bar{\zeta}}^{\mathrm{C}})$ *with sources* $(\mathbf{J}, \mathbf{M})$ *producing fields* $(\mathbf{E}^{\mathrm{C}}, \mathbf{H}^{\mathrm{C}})$*. Then the reaction of sources 1 on fields 2 in the original bianisotropic medium equals the reaction of sources 2 on fields 1 in the complementary medium, i.e.*

$$\iiint_{V_0} (\mathbf{J}_1 \cdot \mathbf{E}_2 - \mathbf{M}_2 \cdot \mathbf{H}_2)\, \mathrm{d}v = \iiint_{V_0} (\mathbf{J}_2 \cdot \mathbf{E}_1^{\mathrm{C}} - \mathbf{M}_2 \cdot \mathbf{H}_1^{\mathrm{C}})\, \mathrm{d}v, \tag{7.34}$$

*where* $(\mathbf{J}_2, \mathbf{M}_2)$ *produce fields* $(\mathbf{E}_2, \mathbf{H}_2)$ *in the original medium and* $(\mathbf{J}_1, \mathbf{M}_1)$ *produce fields* $(\mathbf{E}_1^{\mathrm{C}}, \mathbf{H}_1^{\mathrm{C}})$ *in the complementary medium, provided the material tensors satisfy*

$$\bar{\bar{\varepsilon}}^{\mathrm{T}} = \bar{\bar{\varepsilon}}^{\mathrm{C}}, \tag{7.35}$$

$$\bar{\bar{\mu}}^{\mathrm{T}} = \bar{\bar{\mu}}^{\mathrm{C}}, \tag{7.36}$$

$$\bar{\bar{\xi}}^{\mathrm{T}} = -\bar{\bar{\zeta}}^{\mathrm{C}}. \tag{7.37}$$

The results of this theorem reduce to equations (7.31)–(7.33) if the complementary medium is identical to the original medium.

### 7.1.5 Time domain reciprocity theorem

**Theorem 7.1.3.** Time domain reciprocity theorem [1, p 193].
*Let* $(\mathscr{E}_1, \mathscr{H}_1)$ *be the retarded fields produced by* $(\mathscr{J}_1, \mathscr{M}_1)$ *and* $(\widetilde{\mathscr{E}_2}, \widetilde{\mathscr{H}_2})$ *be the advanced fields produced by sources* $(\mathscr{J}_2, \mathscr{M}_2)$. *Let* $t_1$ *and* $t_2$, *respectively, be anterior and posterior to the period of activation of the sources. Then*

$$\int_{t_1}^{t_2} \mathrm{d}t \iiint_V \left(\mathscr{E}_1 \cdot \mathscr{J}_2 + \mathscr{H}_1 \cdot \mathscr{M}_2\right) = -\int_{t_1}^{t_2} \mathrm{d}t \iiint_V \left(\widetilde{\mathscr{E}_2} \cdot \mathscr{J}_1 + \widetilde{\mathscr{H}_2} \cdot \mathscr{M}_1\right) \mathrm{d}v. \tag{7.38}$$

*Proof.* From time-instantaneous Maxwell's equations and on following similar steps as in the frequency domain case, we obtain

$$\begin{aligned} -\nabla \cdot (\mathscr{E}_1 \times \widetilde{\mathscr{H}_2} + \widetilde{\mathscr{E}_2} \times \mathscr{H}_1) = \Big[ & \mu \frac{\partial}{\partial t}(\widetilde{\mathscr{H}_2} \cdot \mathscr{H}_1) + \varepsilon \frac{\partial}{\partial t}(\widetilde{\mathscr{E}_2} \cdot \mathscr{E}_1) \\ & + (\mathscr{E}_1 \cdot \mathscr{J}_2 + \mathscr{H}_1 \cdot \mathscr{M}_2 + \widetilde{\mathscr{H}_2} \cdot \mathscr{M}_1 + \widetilde{\mathscr{E}_2} \cdot \mathscr{J}_1)\Big]. \end{aligned} \tag{7.39}$$

Integrating both sides over a time interval $(t_1, t_2)$ and over a volume containing all sources gives

$$\begin{aligned} -\int_{t_1}^{t_2} \mathrm{d}t \iiint_V & (\mathscr{E}_1 \cdot \mathscr{J}_2 + \mathscr{H}_1 \cdot \mathscr{M}_2 + \widetilde{\mathscr{H}_2} \cdot \mathscr{M}_1 + \widetilde{\mathscr{E}_2} \cdot \mathscr{J}_1)\, \mathrm{d}v \\ & = \int_{t_1}^{t_2} \mathrm{d}t \oiint_S (\mathscr{E}_1 \times \widetilde{\mathscr{H}_2} + \widetilde{\mathscr{E}_2} \times \mathscr{H}_1) \cdot \hat{\mathbf{n}}\, \mathrm{d}s \\ & \quad + \left[ \iiint_V \mu \widetilde{\mathscr{H}_2} \cdot \mathscr{H}_1 + \varepsilon \widetilde{\mathscr{E}_2} \cdot \mathscr{E}_1\, \mathrm{d}v \right]_{t_1}^{t_2}. \end{aligned}$$

At $t = t_1$ the retarded fields are zero everywhere; at $t = t_2$ the same is true for the advanced fields. Hence the second integral term in the right-hand side member is zero. If the surface $S$ is taken large enough such that the retarded field has not reached it any point, then the surface integral term on the right-hand side is also zero, thus producing the required result in equation (7.38). ■

## 7.2 Compensation theorem

A theorem closely related to the reciprocity theorem is the compensation theorem. The compensation theorem allows for the calculation of fields or quantities derived from the fields of a modified problem in terms of the fields of the original problem. The formulation itself is exact, but the compensation theorem is often used as a perturbation approach when materials or geometries with a specific region are altered. The time-harmonic field version of the compensation theorem was put forth by Monteath [7] in analogy with its original version for electrical networks. A more general field version was provided by Mittra [8], which we presented below.

**Theorem 7.2.1** Compensation theorem [8].
*Let* $(\mathbf{E}_i, \mathbf{H}_i)$ *be the fields generated in an arbitrary medium in volume V by sources* $(\mathbf{J}_i, \mathbf{M}_i)$, $i = 1, 2$. *The volume V contains sources* $(\mathbf{J}_2, \mathbf{M}_2)$, *but excludes sources* $(\mathbf{J}_1, \mathbf{M}_1)$ *(see figure 7.2). Let* $(\mathbf{E}_1', \mathbf{H}_1')$ *be the fields generated by* $(\mathbf{J}_1, \mathbf{M}_1)$ *when material and/or geometric changes are introduced inside a volume W enclosed by a surface S, which is a subregion of V, but does not enclose the '2' sources. Let* $\delta\mathbf{E}_1 = (\mathbf{E}_1' - \mathbf{E}_1)$, $\delta\mathbf{H}_1 = (\mathbf{H}_1' - \mathbf{H}_1)$ *and* $\hat{\mathbf{n}}$ *be a normal on S pointing out of volume W as indicated in figure 7.2. Then*

$$\oiint_S \left(\mathbf{E}_1' \times \mathbf{H}_2 - \mathbf{E}_2 \times \mathbf{H}_1'\right) \cdot \hat{\mathbf{n}}\, \mathrm{d}s = \iiint_{R_2} [\mathbf{J}_2 \cdot \delta\mathbf{E}_1 - \mathbf{M}_2 \cdot \delta\mathbf{H}_1]\, \mathrm{d}v, \tag{7.40}$$

*where* $R_2$ *is the region occupied by the sources* $(\mathbf{J}_2, \mathbf{M}_2)$.

*Proof.* Let $\hat{\mathbf{n}}_o$ denote the *outward* unit normal on the surfaces $T$ and $S$, chosen to point out of the volume $V - W$. With reference to figure 7.2, $\hat{\mathbf{n}}_o = \hat{\mathbf{n}}$ on $T$ and $\hat{\mathbf{n}}_o = -\hat{\mathbf{n}}$ on $S$. Applying the result (7.10) to the annular volume $V - W$ that is bounded by $T + S$, one obtains

$$\begin{aligned} &\oiint_T [\delta\mathbf{E}_1 \times \mathbf{H}_2 - \mathbf{E}_2 \times \delta\mathbf{H}_1] \cdot \hat{\mathbf{n}}\, \mathrm{d}s \\ &- \oiint_S [\delta\mathbf{E}_1 \times \mathbf{H}_2 - \mathbf{E}_2 \times \delta\mathbf{H}_1] \cdot \hat{\mathbf{n}}\, \mathrm{d}s \\ &= - \iiint_{V-W} [\mathbf{J}_2 \cdot \delta\mathbf{E}_1 - \mathbf{M}_2 \cdot \delta\mathbf{H}_1]\, \mathrm{d}v. \end{aligned} \tag{7.41}$$

The difference field $(\delta\mathbf{E}_1, \delta\mathbf{H}_1)$ has no sources outside $S$ and the field $(\mathbf{E}_2, \mathbf{H}_2)$ has no sources outside $T$. Hence applying the result (7.14) to the volume $V$ and surface $T$ we obtain

$$\oiint_T [\delta\mathbf{E}_1 \times \mathbf{H}_2 - \mathbf{E}_2 \times \delta\mathbf{H}_1] \cdot \hat{\mathbf{n}}\, \mathrm{d}s = 0. \tag{7.42}$$

Both the fields ($\mathbf{E}_1$, $\mathbf{H}_1$) and ($\mathbf{E}_2$, $\mathbf{H}_2$) have no sources inside $S$. Thus application of equation (7.10) yields

$$\oint\!\!\oint_S [\mathbf{E}_1 \times \mathbf{H}_2 - \mathbf{E}_2 \times \mathbf{H}_1] \cdot \hat{\mathbf{n}}\, \mathrm{d}s = 0. \tag{7.43}$$

Subtracting equation (7.43) from equation (7.41) and substituting equation (7.42) yields the desired result (7.40) upon collapsing the volume $V - W$ to $R_2$ in the volume integral. ■

A useful corollary to this theorem gives an expression for the incremental change in mutual impedance between two antennas when the medium inside $S$ is perturbed.

**Corollary 7.2.1.1.** Mutual impedance change between antennas [7, corollary I]. *Consider two pairs of antenna terminals: 1 and 2 outside a closed surface S (see figure 7.4) between which a constant current $I_0$ is impressed separately. Let the current $I_0$ impressed between terminal i result in a field ($\mathbf{E}_i$, $\mathbf{H}_i$), $i = 1, 2$ when the other terminal is open-circuited. Let the mutual impedance between the antenna terminals be $Z_{12} = Z_{21}$ as defined in equation (7.19). Let the medium enclosed by S be perturbed by introducing certain changes in its electrical characteristics. Let the corresponding fields and impedance change to ($\mathbf{E}_1'$, $\mathbf{H}_1'$), ($\mathbf{E}_2'$, $\mathbf{H}_2'$), and $Z_{12}' = Z_{21}'$. Then*

$$Z_{12}' - Z_{12} = \frac{1}{I_0^2} \oint\!\!\oint_S [\mathbf{E}_1 \times \mathbf{H}_2' - \mathbf{E}_2' \times \mathbf{H}_1] \cdot \hat{\mathbf{n}}\, \mathrm{d}s, \tag{7.44}$$

$$= \frac{1}{I_0^2} \oint\!\!\oint_S [\mathbf{E}_2 \times \mathbf{H}_1' - \mathbf{E}_1' \times \mathbf{H}_2] \cdot \hat{\mathbf{n}}\, \mathrm{d}s, \tag{7.45}$$

*where $\hat{\mathbf{n}}$ is the outward normal on S.*

*Proof.* From the equivalence theorem, equivalent surface currents $\mathbf{J}_{s1} = \hat{\mathbf{n}} \times \mathbf{H}_1'$ and $\mathbf{M}_{s1} = -\hat{\mathbf{n}} \times \mathbf{E}_1'$ placed on $S$ and radiating in an unperturbed medium generate the perturbed fields $\delta\mathbf{E}_1 = \mathbf{E}_1' - \mathbf{E}_1$ and $\delta\mathbf{H}_1 = \mathbf{H}_1' - \mathbf{H}_1$ everywhere outside $S$. Application of corollary 7.1.1.1 to the fields ($\delta\mathbf{E}_1$, $\delta\mathbf{H}_1$) and ($\mathbf{E}_2$, $\mathbf{H}_2$) and to a surface

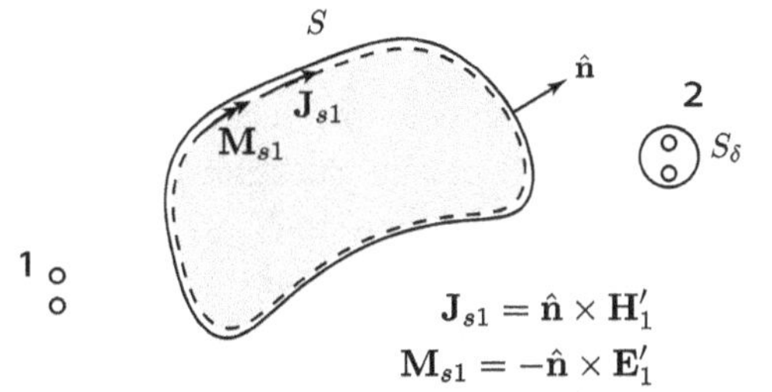


**Figure 7.4.** Mutual impedance between antenna terminals when medium inside $S$ is perturbed.

$T$ consisting of $S$ plus an infinitesimal surface $S_\delta$ enclosing a volume $V_\delta$ around the terminal 2 (see figure 7.4) gives

$$\oiint_{S+S_\delta} [\delta \mathbf{E}_1 \times \mathbf{H}_2 - \mathbf{E}_2 \times \delta \mathbf{H}_1] \cdot \hat{\mathbf{n}}\, \mathrm{d}s = 0. \tag{7.46}$$

But

$$\oiint_{S} [\mathbf{E}_1 \times \mathbf{H}_2 - \mathbf{E}_2 \times \mathbf{H}_1] \cdot \hat{\mathbf{n}}\, \mathrm{d}s = 0, \tag{7.47}$$

as $S$ does not enclose either of the pairs of terminals 1 or 2. And from equation (7.9)

$$\begin{aligned} \oiint_{S_\delta} [\delta \mathbf{E}_1 \times \mathbf{H}_2 - \mathbf{E}_2 \times \delta \mathbf{H}_1] \cdot \hat{\mathbf{n}}\, \mathrm{d}s &= - \iiint_{V_\delta} \delta \mathbf{E}_1 \cdot \mathbf{J}_2\, \mathrm{d}v \\ &= - \int \delta \mathbf{E}_1 \cdot I_0\, \mathrm{d}\ell \\ &= I_0 I_0 (Z_{21}' - Z_{21}) = I_0^2 (Z_{12}' - Z_{12}). \end{aligned} \tag{7.48}$$

Substitution of equations (7.47) and (7.49) into equation (7.46) yields

$$\begin{aligned} Z_{12}' - Z_{12} &= \frac{1}{I_0^2} \oiint_S [\mathbf{E}_2 \times \mathbf{H}_1' - \mathbf{E}_1' \times \mathbf{H}_2] \cdot \hat{\mathbf{n}}\, \mathrm{d}s \\ &= \frac{1}{I_0^2} \oiint_S [\mathbf{M}_{s1} \cdot \mathbf{H}_2 - \mathbf{J}_{s1} \cdot \mathbf{E}_2]\, \mathrm{d}s, \end{aligned} \tag{7.49}$$

which is identical to equation (7.45). The alternative form in equation (7.44) can be obtained by interchanging '1' and '2' and repeating the steps. ■

An immediate application of the compensation theorem is in determining expressions for the incremental electric or magnetic field when changes are made within a certain region $S$. Indeed by choosing $\mathbf{J}_2 = I_0 \ell \hat{\mathbf{u}} \delta(\mathbf{r} - \mathbf{r}_p)$ and $\mathbf{M}_2 = 0$ at an arbitrary point $\mathbf{r}_\mathrm{p}$ outside $S$ and denoting the position and unit normal on $S$ as $\mathbf{r}_\mathrm{q}$ and $\hat{\mathbf{n}}_\mathrm{q}$, respectively, in equation (7.40) one sees that

$$\hat{\mathbf{u}} \cdot \delta \mathbf{E}_1(\mathbf{r}_\mathrm{p}) = \frac{1}{I_0 \ell} \oiint_S \left[ -\mathbf{H}_2^u(\mathbf{r}_\mathrm{q}) \cdot \mathbf{M}_s(\mathbf{r}_\mathrm{q}) + \mathbf{E}_2^u(\mathbf{r}_\mathrm{q}) \cdot \mathbf{J}_s(\mathbf{r}_\mathrm{q}) \right] \mathrm{d}s_\mathrm{q}, \tag{7.50}$$

where $(\mathbf{E}_2^u, \mathbf{H}_2^u)$ is the unperturbed field produced on $S$ by a $u$-directed $\mathbf{J}_2$ and $\mathbf{J}_s(\mathbf{r}_\mathrm{q}) = \hat{\mathbf{n}}_\mathrm{q} \times \mathbf{H}_1'(\mathbf{r}_\mathrm{q})$, $\mathbf{M}_s = -\hat{\mathbf{n}}_\mathrm{q} \times \mathbf{E}_1'(\mathbf{r}_\mathrm{q})$ are the equivalent surface currents on $S$ that are determined from the changes made within $S$. By successively orienting $\hat{\mathbf{u}}$ along the three coordinate directions in an orthogonal coordinate system one obtains the three independent components of $\delta \mathbf{E}_1$ at the point $\mathbf{r}_\mathrm{p}$. Similarly, by choosing $\mathbf{J}_2 = 0$, $\mathbf{M}_2 = -\hat{\mathbf{u}} V_0 \ell \delta(\mathbf{r} - \mathbf{r}_\mathrm{p})$ one obtains the three independent components of $\delta \mathbf{H}_1$ at the point $\mathbf{r}_p$ in terms of the equivalent surface currents $\mathbf{J}_s$, $\mathbf{M}_s$ and the unperturbed fields $(\mathbf{E}_2^u, \mathbf{H}_2^u)$ produced on $S$ by $\mathbf{M}_2$. Denoting the three independent coordinate axes in a three-dimensional space as $\hat{\mathbf{u}}$, $\hat{\mathbf{v}}$, $\hat{\mathbf{w}}$ and defining the electric dyadic Green's functions as $\tilde{\mathscr{G}}_E(\mathbf{r}_p; \mathbf{r}_q) = \hat{\mathbf{u}}\mathbf{E}_2^u + \hat{\mathbf{v}}\mathbf{E}_2^v + \hat{\mathbf{w}}\mathbf{E}_2^w$ and the magnetic dyadic Green's function as $\tilde{\mathscr{G}}_H(\mathbf{r}_p; \mathbf{r}_q) = \hat{\mathbf{u}}\mathbf{H}_2^u + \hat{\mathbf{v}}\mathbf{H}_2^v + \hat{\mathbf{w}}\mathbf{H}_2^w$, the complete vector incremental electric field in equation (7.50) and the corresponding expression for magnetic field may be expressed as

$$\delta\mathbf{E}_1(\mathbf{r}_p) = \frac{1}{I_0\ell} \oiint_S \left[-\tilde{\mathscr{G}}_H(\mathbf{r}_p; \mathbf{r}_q) \cdot \mathbf{M}_s(\mathbf{r}_q) + \tilde{\mathscr{G}}_E(\mathbf{r}_p; \mathbf{r}_q) \cdot \mathbf{J}_s(\mathbf{r}_q)\right] \mathrm{d}s_q, \tag{7.51}$$

$$\delta\mathbf{H}_1(\mathbf{r}_p) = \frac{1}{V_0\ell} \oiint_S \left[\tilde{\mathscr{G}}_H(\mathbf{r}_p; \mathbf{r}_q) \cdot \mathbf{M}_s(\mathbf{r}_q) - \tilde{\mathscr{G}}_E(\mathbf{r}_p; \mathbf{r}_q) \cdot \mathbf{J}_s(\mathbf{r}_q)\right] \mathrm{d}s_q, \tag{7.52}$$

where the dyadic Green's functions use fields $(\mathbf{E}_2^i, \mathbf{H}_2^i)$, $i = u, v, w$ produced at $\mathbf{r}_q$ due to an $i$-directed electric dipole of moment $I_0\ell$ located at $\mathbf{r}_p$ in the case of equation (7.51) and due to a $i$-directed magnetic dipole of moment $V_0\ell$ in the case of equation (7.52).

**Example 7.2.0.1** Setting up integral equations.
Consider an antenna placed over a conducting half-plane as shown in figure 7.5. The unperturbed system is regarded as free space. For the perturbed system we choose the surface $S$ to coincide with the surface of the half-plane. Since the tangential electric field on a PEC is zero one notes that $\hat{\mathbf{n}}_q \times \mathbf{E}(\mathbf{r}_q) = 0 \Longrightarrow \mathbf{M}_s = 0$. Setting the position vector $\mathbf{r}_p$ to a point on $S$ in equation (7.51) and cross-multiplying both sides with $\hat{\mathbf{r}}_p$ yields the electric-field integral equation

$$\begin{aligned} -\hat{\mathbf{n}}_p \times \mathbf{E}_1(\mathbf{r}_p) &= \frac{1}{I_0\ell}\hat{\mathbf{n}}_p \times \oiint_S \tilde{\mathscr{G}}_E(\mathbf{r}_p; \mathbf{r}_q) \cdot \mathbf{J}_s(\mathbf{r}_q)\, \mathrm{d}s_q \\ &= \frac{1}{I_0\ell}\hat{\mathbf{n}}_p \times \int_{x=-\infty}^{0} \int_{z=-\infty}^{\infty} \tilde{\mathscr{G}}_E(\mathbf{r}_p; \mathbf{r}_q) \cdot \mathbf{J}_\Sigma(\mathbf{r}_q)\, \mathrm{d}z\mathrm{d}x, \end{aligned} \tag{7.53}$$

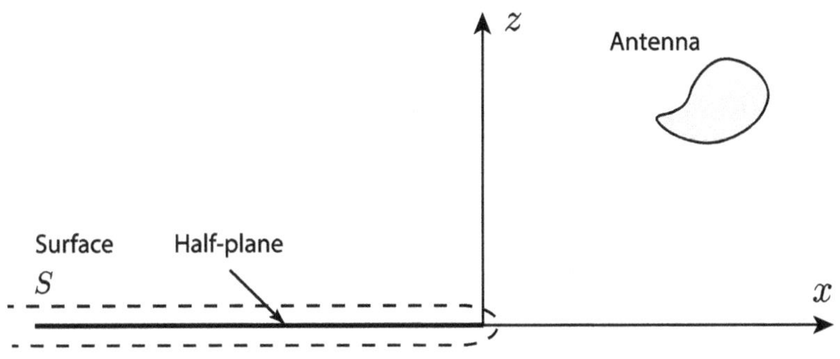


**Figure 7.5.** Antenna placed over a PEC half-plane that is housed inside $S$.

for the total unknown surface current $\mathbf{J}_\Sigma(x, z) = \mathbf{J}_s(x, y = 0+, z) + \mathbf{J}_s(x, y = 0-, z)$ on the half-plane. Here $\mathbf{E}_1$ is the known field produced by the antenna in the unperturbed system, i.e. in free space. ■■

Additional examples on the application of compensation theorem can be found in [7, 9, 10]. We now provide an example for the corollary to the compensation theorem.

**Example 7.2.0.2.** Change in mutual impedance.
Suppose one wishes to determine how the mutual impedance between the elements of a linear array changes when the array is placed in front of a PEC ground plane. We illustrate the case with a linear array of broadside-coupled, short, identical dipole antennas as shown in figure 7.6. Let us assume that the dipoles are short enough $kH < \pi/2$ so that a sinusoidal current distribution $I(z) = I_0 \sin k(H - |z|)$, $-H \leqslant z \leqslant H$ remains valid, where $k = \omega\sqrt{\varepsilon\mu}$ is the wavenumber.

Expressions are available for the mutual impedance $Z_{12}$ between the dipoles as a function of the separation $d$ and length $2H$ in a number of references including [11, p 472]. For a $z$-directed dipole with its center along the $y$-axis at $y = y_0$ and carrying a sinusoidal current distribution, the fields are [12, p 334]

$$E_z(x, y, z; y_0) = -\frac{jI_0\eta}{4\pi}\left(\frac{e^{-jkR_1}}{R_1} + \frac{e^{-jkR_2}}{R_2} - 2\cos(kH)\frac{e^{-jkR_0}}{R_0}\right), \quad (7.54)$$

$$H_x(x, y, z; y_0) = -\frac{jI_0}{4\pi}\frac{1}{\rho_0}(e^{-jkR_1} + e^{-jkR_2} - 2\cos(kH)e^{-jkR_0}), \quad (7.55)$$

where $\rho_0^2 = x^2 + (y - y_0)^2$, $R_0^2 = \rho_0^2 + z^2$, $R_1^2 = \rho_0^2 + (z - H)^2$, $R_2^2 = \rho^2 +(z + H)^2$, and $\eta = \sqrt{\mu/\varepsilon}$. The mutual impedance $Z_{12}$ between two identical dipoles in a side-by-side configuration is

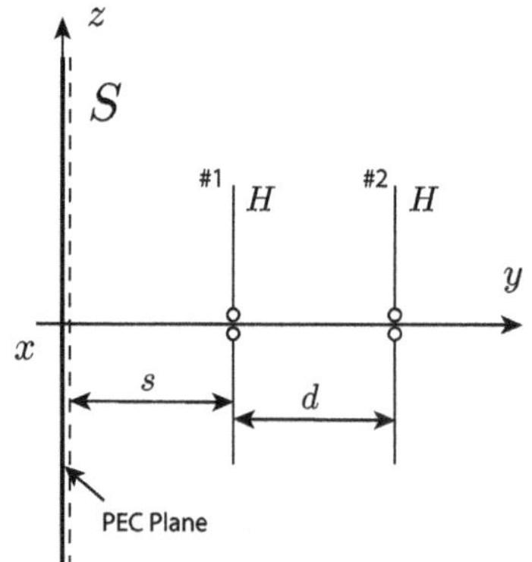

**Figure 7.6.** Dipole array placed in front of a PEC plane.

$$Z_{12} = \frac{\eta}{4\pi}[2Ei(u_0) - Ei(u_1) - Ei(u_2)], \tag{7.56}$$

where $u_0 = kd$, $u_1 = k(\sqrt{d^2 + 4H^2} - 2H)$, $u_2 = k(\sqrt{d^2 + 4H^2} + 2H)$, $Ei(x) = Ci(x) - jSi(x)$, where $Si(x)$ and $Ci(x)$ are sine and cosine integrals, respectively. If the array is now placed in front of a PEC ground plane at a distance $s$ to the first element, the mutual impedance becomes modified according to equation (7.45) by the amount

$$\begin{aligned} Z_{12}' - Z_{12} &= -\frac{1}{I_0^2} \oiint_S \mathbf{n} \times \mathbf{H}_1' \cdot \mathbf{E}_2 \, \mathrm{d}s \\ &= \frac{1}{I_0^2} \iint_{-\infty}^{\infty} E_{z2}(x, 0, z) H_{x1}'(x, 0, z) \, \mathrm{d}x\mathrm{d}z, \\ &= \frac{2}{I_0^2} \iint_{-\infty}^{\infty} E_z(x, 0, z; s + d) H_x(x, 0, z; s) \, \mathrm{d}x\mathrm{d}z, \end{aligned} \tag{7.57}$$

where the surface $S$ corresponds to the front part of the PEC plane on which $\hat{\mathbf{n}} = \hat{\mathbf{y}}$ and $E_z$ and $H_x$ are given in equations (7.54) and (7.55). In the above expressions image theory was used in coming up with $H_{x1}'(x, 0, z) = 2H_x(x, 0, z; s)$. It should be possible to express the integral in equation (7.57) also in terms of sine and cosine integrals in the manner similar to King [13].

Another elegant way to evaluate the fields of a dipole is to take the Fourier transform of the sinusoidal current density distribution $\mathbf{J}_\mathrm{d} = \hat{\mathbf{z}}\delta(x)I(z)$ in the $(k_x, k_z)$ plane and use the Weyl representation (15.2) of the free-space Green's function to express the potential $A_z$ and the electric and magnetic fields as

$$A_z = \frac{k\mu I_0}{4\pi j} \iint_{-\infty}^{\infty} \frac{\cos(k_z H) - \cos(kH)}{k^2 - k_z^2} \frac{\mathrm{e}^{-jk_y|y-y_0|}}{k_y} \mathrm{e}^{j(k_x x + k_z z)} \, \mathrm{d}k_x \mathrm{d}k_z, \tag{7.58}$$

$$\begin{aligned} E_z &= \frac{\eta}{jk}\left(\frac{\partial^2}{\partial z^2} + k^2\right) A_z \\ &= \frac{-\eta I_0}{4\pi^2} \iint_{-\infty}^{\infty} [\cos(k_z H) - \cos(kH)] \\ &\quad \times \frac{\mathrm{e}^{-jk_y|y-y_0|}}{k_y} \mathrm{e}^{j(k_x x + k_z z)} \, \mathrm{d}k_x \mathrm{d}k_z, \end{aligned} \tag{7.59}$$

$$\begin{aligned} H_x &= \frac{1}{\mu}\frac{\partial A_z}{\partial y} \\ &= \mp \frac{kI_0}{4\pi^2} \iint_{-\infty}^{\infty} \frac{\cos(k_z H) - \cos(kH)}{k^2 - k_z^2} \mathrm{e}^{j(k_x x + k_z z - k_y|y-y_0|)} \mathrm{d}k_x \mathrm{d}k_z, \\ & y - y_0 \gtrless 0, \end{aligned} \tag{7.60}$$

where $k_y = \sqrt{k^2 - k_x^2 - k_z^2} = -j\sqrt{k_x^2 + k_z^2 - k^2}$. When these expressions are substituted in equation (7.57) and the spatial integrals carried out, the following alternative expression is obtained for the incremental change in impedance of the dipole array

$$Z_{12}^{'} - Z_{12} = -\frac{k\eta}{2\pi^2}\iint_{-\infty}^{\infty} \frac{[\cos(k_z H) - \cos(kH)]^2}{k^2 - k_z^2} \frac{e^{-jk_y(2s+d)}}{k_y} \, dk_x dk_z. \quad (7.61)$$

The contribution of the integrals in the visible part of the spectrum, $k_x^2 + k_z^2 < k^2$, determines the incremental change of the real part, while the contribution from the invisible part of the spectrum, $k_x^2 + k_z^2 > k^2$, determines the incremental change in the imaginary part of the mutual impedance. Because of the presence of the exponential factor in the integrand, the integrals converge rapid in the invisible region. Note that the expressions (7.57) and (7.61) are both exact, subject only to the assumption that the current distribution along the dipole remains sinusoidal in shape. ■■

## References


[1] van Bladel J 1985 *Electromagnetic Fields* (New York: Hemisphere)

[2] Ballantine S 1929 Reciprocity in electromagnetic, mechanical, acoustical, and interconnected systems *Proc. IRE* **17** 929–51

[3] Harrington R F 1961 *Time-Harmonic Electromagnetic Fields* (New York: McGraw-Hill)

[4] Saxon D S 1955 Tensor scattering matrix for the electromagnetic field *Phys. Rev.* **100** 1771–5

[5] de Hoop A T 1960 A reciprocity theorem for the electromagnetic field scattered by an obstacle *Appl. Sci. Res. Section* B **8** 135–40

[6] Kong J A 1990 *Electromagnetic Wave Theory* (New York: Wiley)

[7] Monteath G D 1951 Application of the compensation theorem to certain radiation and propagation problems *Proc. IEE-Pt IV: Inst. Monogr.* **98** 23–30

[8] Mittra R 1964 A vector form of compensation theorem and its application to boundary value problems *Appl. Sci. Res.* B **11** 26–42

[9] Wait J R 1998 The ancient and modern history of EM ground-wave propagation *IEEE Antennas Propag. Mag.* **40** 7–24

[10] Green H E 2007 Derivation of the Norton surface wave using compensation theorem *IEEE Antennas Propag. Mag.* **49** 47–57

[11] Balanis C A 2005 *Antenna Theory and Design* 3rd edn (New York: Wiley)

[12] Jordan E C and Balmain K G 1968 *Electromagnetic Waves and Radiating Systems* 2nd edn (Englewood Cliffs, NJ: Prentice-Hall)

[13] King H E 1957 Mutual impedance of unequal length antennas in echelon *IRE Trans. Antennas Propag.* **AP-5** 306–13

# Chapter 8

# Reactance theorems

Half-power bandwidth is an important consideration in the design of many linear, passive, time-invariant electromagnetic systems. A typical application is the design of a matching network system to match an electrically small antenna to a receiver or to a given source. The half-power bandwidth seen at the input of a linear, passive, and time-invariant electromagnetic system is known to approach a value that depends on the frequency derivative of the driving point input reactance. In the ensuing problem of matching the electromagnetic system to a source or receiver for maximum power transfer, the matching network is often realized by means of passive resistors, inductors, and capacitors. The question then arises as to what frequency dependences of input impedance can be realized by a matching network comprised of resistors, inductors, and capacitors. One then looks for a similarity between an antenna input port and the input port of a high-frequency network comprised of passive elements insofar as the behavior of the input impedance is concerned. The theorems related to the impedance behavior of these electromagnetic and network systems are collectively known as reactance theorems, which is the subject of discussion of the present chapter.

## 8.1 Reactance theorems for networks and antennas

We first discuss the reactance theorem connected with a network comprised of lossless elements known as Foster's reactance theorem in the circuits community [1, p 97], [2, chapter 6].

### 8.1.1 Foster's reactance theorem for passive lossless networks

**Theorem 8.1.1.** Foster's reactance theorem [3].
*The most general driving point impedance $Z(\omega)$ obtainable by means of a finite resistance-less network is a pure reactance which is an odd rational function of the frequency $\omega/2\pi$ and which is completely determined, except for a constant factor $H$, by assigning the resonant and anti-resonant frequencies, subject to the condition that they*

doi:10.1088/978-0-7503-1716-0ch8

*alternate and include both zero and infinity. Any such impedance may be physically constructed either by combining, in parallel, resonant circuits having impedances in the form* $j\omega L + (j\omega C)^{-1}$, *or by combining, in series, anti-resonant circuits having impedances of the form* $[j\omega C + (j\omega L)^{-1}]^{-1}$. *In more precise form,*

$$Z(\omega) = jX(\omega) = jH\frac{\left(\omega^2 - \omega_1^2\right)\left(\omega^2 - \omega_3^2\right)\cdots\left(\omega - \omega_{2n-1}^2\right)}{\omega\left(\omega^2 - \omega_2^2\right)\left(\omega^2 - \omega_4^2\right)\cdots\left(\omega - \omega_{2n-2}^2\right)}, \tag{8.1}$$

*where* $H \geqslant 0$ *and* $0 = \omega_0 \leqslant \omega_1 \leqslant \omega_2 \leqslant \cdots \leqslant \omega_{2n-1} \leqslant \omega_{2n} = \infty$. *The inductance and capacitance for the n resonant circuits are given by the formula,*

$$L_k = \frac{1}{C_k\omega_k^2} = \lim_{\omega\to\omega_k}\left(\frac{j\omega Z(\omega)}{\omega_k^2 - \omega^2}\right), \quad k = 1, 3,.., 2n - 1, \tag{8.2}$$

*and the inductances and capacitances of the* $(n + 1)$ *anti-resonant circuits are given by the formula,*

$$C_k = \frac{1}{L_k\omega_k^2} = \lim_{\omega\to\omega_k}\left(\frac{j\omega}{\left(\omega_k^2 - \omega^2\right)Z(\omega)}\right), \quad k = 0, 2, \ldots, 2n, \tag{8.3}$$

*which includes the limiting values*

$$C_0 = \frac{\omega_2^2\omega_4^2\cdots\omega_{2n-2}^2}{H\omega_1^2\omega_3^2\cdots\omega_{2n-1}^2}, \; L_0 = \infty, \; C_{2n} = 0, \; L_{2n} = H. \tag{8.4}$$

Clearly, the poles and zeros of the impedance function alternate along the real frequency axis. Furthermore, $Z(-\omega) = -Z(\omega)$ and $Z(\omega)$ is a strictly proper rational function of frequency[1]. An immediate consequence of this theorem is that the frequency derivative of the reactance $\mathrm{d}X(\omega)/\mathrm{d}\omega > 0$, that is, *the reactance function always increases with frequency, except for the discontinuities which carry it back from a positive infinite value to a negative infinite value at the anti-resonant points.* The theorem only applies to a lossless network for the frequency derivative of impedance will no longer be positive if the elements are made lossy.

For passive antennas the fractional half-power bandwidth, $B$, of the antenna tuned to a frequency $\omega_0$ is [4, 5]

$$B \propto \frac{R_0(\omega_0)}{\omega_0\left|\frac{\mathrm{d}X_0(\omega_0)}{\mathrm{d}\omega}\right|}, \tag{8.5}$$

where $Z(\omega_0) = R_0(\omega_0) + jX_0(\omega_0)$ is the antenna input impedance at $\omega = \omega_0$. So, the bandwidth tends to be small if the frequency derivative of reactance is large. An

[1] The degree of the numerator polynomial is less than the degree of the denominator polynomial.

electromagnetic version of Foster's reactance theorem valid for antennas radiating in an inhomogeneous, isotropic, and dispersive media has been worked out by Levis [6] and to a fuller extent by Rhodes [7] and is presented below.

### 8.1.2 Theorem of Levis and Rhodes for antennas

**Theorem 8.1.2** Reactance theorem of Levis [6] and Rhodes [7].
*For any electromagnetic system comprised of complex electric permittivity* $\varepsilon(\mathbf{r};\,\omega) = \varepsilon_r(\mathbf{r};\,\omega) + j\varepsilon_i(\mathbf{r};\,\omega)$, *where* $\omega$ *is the radian frequency, and magnetic permittivity* $\mu(\mathbf{r};\,\omega) = \mu_r(\mathbf{r};\,\omega) + j\mu_i(\mathbf{r};\,\omega)$, *denoted simply by* $\varepsilon = \varepsilon_r + j\varepsilon_i$ *and* $\mu = \mu_r + j\mu_i$, *respectively, the material medium of which is contained wholly within any region of finite dimensions inside the infinite volume* $V$ *enclosed by the sphere* $S_\infty$ *at infinity, the frequency derivative of the input reactance* $X(\omega)$ *with respect to the real variable* $\omega$ *is given by*

$$\begin{aligned}\frac{1}{2}I_0^2\frac{\mathrm{d}X}{\mathrm{d}\omega} &= 2\iiint\limits_V \frac{1}{4}\left(\mu_r\,|\mathbf{H}|^2 + \varepsilon_r\,|\mathbf{E}|^2\right)\mathrm{d}v - \frac{1}{\eta_0}\Im\left[\oint\!\!\!\oint\limits_{S_\infty}\mathbf{E}\cdot\frac{\partial\mathbf{E}^*}{\partial\omega}\,\mathrm{d}s\right]\\ &+\frac{1}{2}\omega\iiint\limits_V\left(\frac{\partial\mu_r}{\partial\omega}|\mathbf{H}|^2 + \frac{\partial\varepsilon_r}{\partial\omega}|\mathbf{E}|^2\right)\mathrm{d}v\\ &+\omega\,\Im\iiint\limits_V\left(\mu_i\mathbf{H}\cdot\frac{\partial\mathbf{H}^*}{\partial\omega} + \varepsilon_i\mathbf{E}\cdot\frac{\partial\mathbf{E}^*}{\partial\omega}\right)\mathrm{d}v,\end{aligned} \tag{8.6}$$

*where* $\eta_0$ *is the intrinsic impedance* $(\mu_0/\varepsilon_0)^{\frac{1}{2}}$ *of free space, the constant* $I_0$ *is the Fourier transform of an impulsive current* $I_0\delta(t)$ *applied at a coaxial input port located at the origin* $\mathbf{r} = 0$, *and* $\mathbf{E}$, $\mathbf{H}$ *are the Fourier transforms of the time-instantaneous electric and magnetic fields* $\mathscr{E}$, $\mathscr{H}$, *respectively, produced at the point* $\mathbf{r}$ *by that current impulse.*

*Proof.* After Rhodes [7]. Note that the medium is allowed to be dispersive ($\varepsilon$, $\mu$ are functions of frequency $\omega$) and dissipative ($\varepsilon_i$, $\mu_i \neq 0$) and $I_0$ is a real constant. For passive media $-\infty < \varepsilon_i,\ \mu_i < 0$. The media parameters are such that $\varepsilon(\mathbf{r};\,\omega) \to \varepsilon_{r\infty}(\mathbf{r})$ as $|\omega| \to \infty$, where $\varepsilon_{r\infty}(\mathbf{r})$ is a real number. Similarly, $\mu(\mathbf{r};\,\omega) \to \mu_{r\infty}(\mathbf{r})$ as $|\omega| \to \infty$. Since the medium is considered lossy only in a region of finite support $\varepsilon(\mathbf{r};\,\omega) \to \varepsilon_0$, $\mu(\mathbf{r};\,\omega) \to \mu_0$ as $|\mathbf{r}| \to \infty$. The fields $\mathscr{E}$, $\mathscr{H}$ are related to their frequency domain counterparts through

$$\mathscr{E}(\mathbf{r};\,t) = \frac{1}{2\pi}\int_\Gamma \mathbf{E}(\mathbf{r};\,\omega)\mathrm{e}^{j\omega t}\,\mathrm{d}\omega;\qquad \mathscr{H}(\mathbf{r};\,t) = \frac{1}{2\pi}\int_\Gamma \mathbf{H}(\mathbf{r};\,\omega)\mathrm{e}^{j\omega t}\,\mathrm{d}\omega, \tag{8.7}$$

where $\Gamma$ is a contour parallel to the real axis in the complex frequency $\omega = \omega_r + j\omega_i$ domain. Because the time-instantaneous fields are real-valued and causal in nature, their frequency domain counterparts satisfy $\mathbf{E}(\mathbf{r};\,-\omega) = \mathbf{E}^*(\mathbf{r};\,\omega^*)$, $\mathbf{H}(\mathbf{r};\,-\omega) = \mathbf{H}^*(\mathbf{r};\,\omega^*)$ and remain analytic in the lower half of the complex $\omega$-plane: $\omega_i < 0$. At a distance $r = |\mathbf{r}|$ from the origin, causality demands that the fields are zero until a time $t = t_0 = r/c$, where $c$ is the speed of light in free space, has elapsed since the impulse has been excited. From equation (8.7) this implies that $e^{j\omega t_0}\mathbf{E}(\mathbf{r};\,\omega)$ must be analytic for $\omega_i < 0$. But this is already true from above. The question remains whether any singularities can lie on the real axis $\omega_i = 0$. In a lossy medium the fields decay exponentially in time at a rate determined by the dissipation in the medium [8]. For a pole or a branch-point located in the upper half-plane at $\omega = \omega_0 + j\alpha_0$, the time-instantaneous field at a distance $r$ will have a decaying envelope of the form $e^{-\alpha_0(t-t_0)}$. So in order to support decaying field, the singularities must all lie above some positive imaginary part in the complex frequency plane dictated by the losses in the medium, leaving the real axis free of singularities. We then conclude that the frequency domain fields must be analytic in the half-plane $\omega_i \leqslant 0$. Hence the contour $\Gamma$ can lie anywhere in the region $-\infty < \mathfrak{J}(\omega) \leqslant 0$.

From causality considerations (see section 4.2), we also know that the material parameters $\varepsilon$, $\mu$ are analytic in the lower half of the complex $\omega$-plane. In fact we have from the Kramers–Krönig relations (4.23), (4.24) (keeping in mind that the notation here differs from that in section 4.2 in that $\varepsilon' = \varepsilon_r$ and $-\varepsilon'' = \varepsilon_i$) that

$$\varepsilon_i(\mathbf{r};\,\omega) = -\frac{1}{\pi}\mathscr{P}\int_{-\infty}^{\infty}\frac{\varepsilon_r(\mathbf{r};\,\omega')}{\omega-\omega'}\,d\omega', \tag{8.8}$$

$$\varepsilon_r(\mathbf{r};\,\omega) = \varepsilon_{r\infty}(\mathbf{r}) + \frac{1}{\pi}\mathscr{P}\int_{-\infty}^{\infty}\frac{\varepsilon_i(\mathbf{r};\,\omega')}{\omega-\omega'}\,d\omega'. \tag{8.9}$$

Similar relations hold the real and imaginary parts of the permeability $\mu$. Equations (8.8) and (8.9) clearly demonstrate that zero dissipation ($\varepsilon_i(\mathbf{r};\,\omega) = 0$) implies zero dispersion ($\varepsilon_r(\mathbf{r};\,\omega) = \varepsilon_{r\infty}(\mathbf{r})$, independent of frequency). Conversely, zero dispersion ($\varepsilon_r(\mathbf{r};\,\omega) = \varepsilon_{r\infty}(\mathbf{r})$) implies zero dissipation ($\varepsilon_i(\mathbf{r};\,\omega) = 0$ upon noting that $\mathscr{P}\int_{-\infty}^{\infty}\frac{1}{\omega-\omega'}\,d\omega' = 0$). Hence dispersion and dissipation go hand in hand.

Figure 8.1 shows an arbitrary antenna fed through a conceptual coaxial cable at a port present at $z = 0$. The inner and outer radii of the cable are $a$ and $b$, respectively,

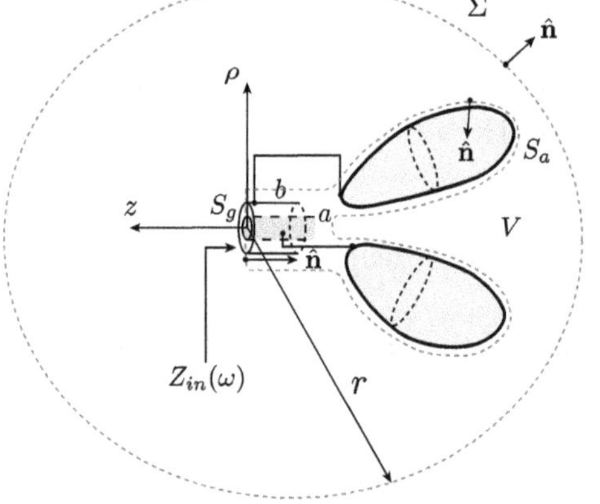


**Figure 8.1.** An arbitrary antenna fed by an impulsive current via a coaxial cable.

such that $0 < a < b < \Delta$, where $\Delta$ is a vanishingly small number. The antenna may consist of both metallic (PEC) and penetrable materials. The volume exterior to the antenna is denoted by $V$ whose surface is bounded by $S$. The unit normal $\hat{\mathbf{n}}$ on $S$ points out of the volume $V$. The surface $S$ consists of the aperture $S_g$ of the coaxial cable at $z = 0$, the surface $S_a$ comprised of the metallic portions of the antenna and a large surface $\sum$ that encloses $S_g$ plus $S_a$ and all material inhomogeneities of the medium. The impulsive current has a Fourier transform $I_0$, which is the total current that flows on the inner conductor of the coaxial cable. The complex-valued input impedance of the antenna looking into the antenna port is denoted by $Z_{\text{in}}(\omega)$. The aperture fields $(\mathbf{E}_g, \mathbf{H}_g)$ in the coaxial cable at $z = 0$ are the radial electric field and the circumferential magnetic fields expressed in terms of a local coordinate system and are

$$\mathbf{E}_g = \frac{\hat{\rho} Z_{\text{in}}(\omega) I_0}{\rho \ln(b/a)}; \quad \mathbf{H}_g = \hat{\boldsymbol{\phi}} \frac{I_0}{2\pi\rho}; \ \hat{\mathbf{n}} = -\hat{\mathbf{z}}. \tag{8.10}$$

Note that the $Z_{\text{in}}(\omega)$ is also the characteristic impedance of the conceptual coaxial cable and is thus complex here. Moreover, it is a function of frequency, which implies that the conceptual coaxial cable is also dispersive. This also implies that the current on the inner conductor and the voltage difference between the inner and outer conductors cannot both be independent of frequency. For the impulsive current excitation assumed here, it is the voltage that becomes frequency dependent. The complex power supplied by the source into the volume $V$ by the surface $S_a + S_g$ is

$$P_s = -\frac{1}{2} \oint\!\!\!\oint_{S_g} \left(\mathbf{E}_g \times \mathbf{H}_g^*\right) \cdot \hat{\mathbf{n}} \, \mathrm{d}s = \frac{I_0^2 Z_{\text{in}}(\omega)}{2\pi \ln(b/a)} \int_0^{2\pi} \int_a^b \frac{1}{\rho} \, \mathrm{d}\rho \mathrm{d}\phi = \frac{1}{2} I_0^2 Z_{\text{in}}(\omega), \tag{8.11}$$

on noting that $\hat{\mathbf{n}} \times \mathbf{E} = 0$ on the portion $S_a$. Applying power conservation (5.18) to volume $V$ with $\omega$ in the lower half-plane we obtain

$$P_s = \frac{1}{2} \oint\!\!\!\oint_{\Sigma} \mathbf{E} \times \mathbf{H}^* \cdot \hat{\mathbf{n}} \, \mathrm{d}s + 2j \left[ \omega \iiint_V \frac{\mu}{4} \mathbf{H} \cdot \mathbf{H}^* \, \mathrm{d}v - \omega^* \iiint_V \frac{\varepsilon^*}{4} \mathbf{E} \cdot \mathbf{E}^* \, \mathrm{d}v \right]. \tag{8.12}$$

Note that the factor $(\omega\varepsilon)^*$ arises from the time-average of $\partial \mathscr{D} / \partial t$ and the relation $\mathbf{D} = \varepsilon \mathbf{E}$. Using equations (8.11) and (8.12) and expression for the input impedance is obtained as

$$\begin{aligned} \frac{1}{2} I_0^2 Z_{\text{in}}(\omega) = {} & \frac{1}{2} \oint\!\!\!\oint_{\Sigma} \mathbf{E} \times \mathbf{H}^* \cdot \hat{\mathbf{n}} \, \mathrm{d}s \\ & + 2j \left[ \omega \iiint_V \frac{\mu}{4} \mathbf{H} \cdot \mathbf{H}^* \, \mathrm{d}v - \omega^* \iiint_V \frac{\varepsilon^*}{4} \mathbf{E} \cdot \mathbf{E}^* \, \mathrm{d}v \right]. \end{aligned} \tag{8.13}$$

The surface and the volume integrals will have both real and imaginary parts resulting in a complex-valued input impedance. The fields $\mathbf{E}(\mathbf{r}; \omega)$, $\mathbf{H}(\mathbf{r}; \omega)$, the material parameters $\varepsilon(\mathbf{r}; \omega)$, $\mu(\mathbf{r}; \omega)$ and the input impedance $Z_{\text{in}}(\omega) = R_{\text{in}}(\omega) + jX_{\text{in}}(\omega)$, where $R_{in}$ and $X_{in}$ are the real and imaginary parts of $Z_{in}$, are all analytic function in the lower half of the complex $\omega$-plane. Each of these functions satisfies the Cauchy–Riemann conditions (CR-conditions for brevity)[2]. In particular we are interested in the frequency derivative of the reactance $X_{in}(\omega)$ on the real axis. From CR-conditions

$$\left.\frac{\partial X_{in}(\omega)}{\partial \omega_r}\right|_{\omega_i=0} = -\left.\frac{\partial R_{in}(\omega)}{\partial \omega_i}\right|_{\omega_i=0}. \tag{8.14}$$

From equation (8.13), the real part of the input impedance is given by

$$I_0^2 R_{in}(\omega) = \Re \oint\!\!\!\oint_\Sigma \mathbf{E} \times \mathbf{H}^* \cdot \hat{\mathbf{n}}\, ds - \iiint_V [(\omega_r\mu_i + \omega_i\mu_r)\mathbf{H} \cdot \mathbf{H}^* + (\omega_r\varepsilon_i + \omega_i\varepsilon_r)\mathbf{E} \cdot \mathbf{E}^*]\, dv.$$

Since the real part of the impedance is non-negative for a passive linear system at any frequency, choosing $\omega_i = 0$ we can assert that

$$\Re \oint\!\!\!\oint_\Sigma \mathbf{E} \times \mathbf{H}^* \cdot \hat{\mathbf{n}}\, ds \geqslant \omega \iiint_V \left(\mu_i \mathbf{H} \cdot \mathbf{H}^* + \varepsilon_i \mathbf{E} \cdot \mathbf{E}^*\right) dv, \ \omega \text{ real}. \tag{8.15}$$

Now

$$\begin{aligned} -I_0^2 \left.\frac{\partial R}{\partial \omega_i}\right|_{\omega_i=0} &= \iiint_V \left(\mu_r \mathbf{H} \cdot \mathbf{H}^* + \varepsilon_r \mathbf{E} \cdot \mathbf{E}^*\right) dv \\ &+ \omega_r \iiint_V \frac{\partial}{\partial \omega_i}\left(\mu_i \mathbf{H} \cdot \mathbf{H}^* + \varepsilon_i \mathbf{E} \cdot \mathbf{E}^*\right) dv \\ &- \Re \oint\!\!\!\oint_\Sigma \frac{\partial}{\partial \omega_i}[\mathbf{E} \times \mathbf{H}^*]_{\omega_i=0} \cdot \hat{\mathbf{n}}\, ds. \end{aligned} \tag{8.16}$$

Using CR-conditions again

$$\begin{aligned} \frac{\partial}{\partial \omega_i}\left(\mu_i \mathbf{H} \cdot \mathbf{H}^*\right) &= \frac{\partial \mu_i}{\partial \omega_i}\mathbf{H} \cdot \mathbf{H}^* + \mu_i \frac{\partial \mathbf{H}}{\partial \omega_i} \cdot \mathbf{H}^* + \mu_i \frac{\partial \mathbf{H}^*}{\partial \omega_i} \cdot \mathbf{H} \\ &= \frac{\partial \mu_r}{\partial \omega_r}\mathbf{H} \cdot \mathbf{H}^* + j\mu_i\left(\frac{\partial \mathbf{H}}{\partial \omega_r} \cdot \mathbf{H}^* - \mathbf{H} \cdot \frac{\partial \mathbf{H}^*}{\partial \omega_r}\right) \\ &= \frac{\partial \mu_r}{\partial \omega_r}|\mathbf{H}|^2 + 2\mu_i \Im\left(\mathbf{H} \cdot \frac{\partial \mathbf{H}^*}{\partial \omega_r}\right), \end{aligned} \tag{8.17}$$

[2] If $W(z) = u(z) + jv(z)$ is an analytic function of the complex variable $z = x + jy$, Cauchy–Riemann conditions state that

$$\frac{\partial u}{\partial x} = \frac{\partial v}{\partial y}; \quad \frac{\partial u}{\partial y} = -\frac{\partial v}{\partial x}, \quad \text{or collectively, } \frac{\partial W}{\partial y} = j\frac{\partial W}{\partial x}.$$

and similarly for the term involving electric field. Furthermore,

$$\Re\left[\frac{\partial}{\partial\omega_i}(\mathbf{E}\times\mathbf{H}^*)\right] = \Re\left[\frac{\partial\mathbf{E}}{\partial\omega_i}\times\mathbf{H}^* + \mathbf{E}\times\frac{\partial\mathbf{H}^*}{\partial\omega_i}\right]$$

$$= \Re\left[j\frac{\partial\mathbf{E}}{\partial\omega_r}\times\mathbf{H}^* - j\mathbf{E}\times\frac{\partial\mathbf{H}^*}{\partial\omega_r}\right]$$

$$= \Im\left[\mathbf{E}\times\frac{\partial\mathbf{H}^*}{\partial\omega_r} - \frac{\partial\mathbf{E}}{\partial\omega_r}\times\mathbf{H}^*\right].$$

If $r \to \infty$ the surface $\Sigma$ recedes to $S_\infty$, the electric and magnetic fields behave as spherical waves and $\eta_0\mathbf{H}\times\hat{\mathbf{n}} \sim \mathbf{E} = \mathbf{F}(\theta, \phi; \omega)e^{-jkr}/r$, $\hat{\mathbf{n}} = \hat{\mathbf{r}}$, $k = \omega_r/c$, where $\mathbf{F}$ is the frequency dependent, vector radiation pattern of the antenna. Using this it is easy to see that

$$\frac{1}{2}\Im\left[\mathbf{E}\times\frac{\partial\mathbf{H}^*}{\partial\omega_r} - \frac{\partial\mathbf{E}}{\partial\omega_r}\times\mathbf{H}^*\right]\cdot\hat{\mathbf{n}} = \frac{1}{\eta_0}\Im\left(\mathbf{E}\cdot\frac{\partial\mathbf{E}^*}{\partial\omega_r}\right), \quad r\to\infty, \tag{8.18}$$

$$=2\frac{r}{c}S_r + \frac{1}{\eta_0}\frac{1}{r^2}\Im\left(\mathbf{F}\cdot\frac{\partial\mathbf{F}^*}{\partial\omega}\right), \quad \omega_r = \omega, \tag{8.19}$$

where $S_r = |\mathbf{F}|^2/2r^2\eta_0 > 0$ is the power density of the radiated fields in the far-zone. Substituting equations (8.17) and (8.18) into equation (8.16) and noting that $\omega_r = \omega$ when $\omega_i = 0$, we arrive at the desired result in equation (8.6). Note that the factor $r$ in front of the power density arises from the frequency derivative of the spherical wavefront in the far-zone. Since the integral of $S_r$ over a sphere will yield a constant positive value, the presence of this term in the frequency derivative of reactance will nullify any possible linear increase of the other terms with $r$. That this is indeed the case will be shown next. The second term on the rhs of equation (8.19) is due to the intrinsic variation of the antenna pattern with frequency. For electrically small antennas (only) this term should be small as the pattern is rather insensitive to frequency. ■

### *8.1.2.1 Further discussion of the theorem*

Let $\bar{w}_m = \mu_r|\mathbf{H}|^2/4$, $\bar{w}_e = \varepsilon_r|\mathbf{E}|^2/4$, $P_{rad} = \oiint_\Omega r^2 S_r\, d\Omega$, where $\Omega$ is the unit sphere and $d\Omega = \sin\theta\, d\theta d\phi$, be the time-average magnetic energy density, time-average electric energy density, time-average radiated power, respectively. The energy densities $\bar{w}_m(r, \theta, \phi)$ and $\bar{w}_e(r, \theta, \phi)$ are defined for all $(r, \theta, \phi)$ and taken to be zero at points *within* the conducting portions of the antenna. By combining half of the first term on the rhs of equation (8.19) with the magnetic and electric stored energies, we define an effective stored energy as

$$W_{eff}(\omega) = \int_0^\infty\left[r^2\oiint_\Omega(\bar{w}_m + \bar{w}_e)\, d\Omega - \frac{P_{rad}}{c}\right]dr. \tag{8.20}$$

The individual terms on the rhs of equation (8.20) will all diverge in an infinite volume, but the combined quantity $W_{\text{eff}}$ is expected to converge. The quantity defined by the surface integral part within the square brackets in equation (8.20) is the total electromagnetic field energy per unit length contained within a sphere of radius $r$. This term will contain the energy contributed by both the near fields and the radiated fields. Integration of this quantity w.r.t. $r$ in an infinite volume gives the total energy contained in the electromagnetic field in all space, which will diverge to infinity. The divergence at infinity will be dictated by the behavior of the radiated fields alone since the near fields will decay faster than the former at large distances. The fixed quantity $P_{\text{rad}}/c$ (this is the lower bound of $P_{\text{rad}}/v(\mathbf{r})$, where $v(\mathbf{r}) < c$ is the signal speed in the inhomogeneous medium) is the energy per unit length contained in the radiated part of the fields and is independent of distance. This quantity, when integrated with respect to $r$ will give rise to the electromagnetic energy contained in the far-field alone in all space, which will diverge to infinity linearly in $r$. Subtraction of the latter from the former gives the net energy in all space that is contributed by the near fields as well as that contributed by the interaction of the near fields with the far fields. (Note that the near fields and far fields are not orthogonal; the latter cannot exist without the former.) Thus $W_{\text{eff}}$ is expected to be non-negative, $W_{\text{eff}} \geqslant 0$.

Using the expression for effective stored energy and equation (8.19) we may express equation (8.6) as

$$\begin{aligned}\frac{1}{2}I_0^2\frac{\mathrm{d}X_{\text{in}}}{\mathrm{d}\omega} &= 2W_{\text{eff}}(\omega) + 2\omega\iiint\limits_V \left(\bar{w}_{\text{m}}\frac{\partial\ln(\mu_{\text{r}})}{\partial\omega} + \bar{w}_{\text{e}}\frac{\partial\ln(\varepsilon_{\text{r}})}{\partial\omega}\right)\mathrm{d}v\\ &+ \omega\iiint\limits_V \Im\left(\mu_{\text{i}}\mathbf{H}\cdot\frac{\partial\mathbf{H}^*}{\partial\omega} + \varepsilon_{\text{i}}\mathbf{E}\cdot\frac{\partial\mathbf{E}^*}{\partial\omega}\right)\mathrm{d}v\\ &- \frac{1}{\eta_0}\oiint\limits_\Omega \Im\left(\mathbf{F}\cdot\frac{\mathrm{d}\mathbf{F}^*}{\mathrm{d}\omega}\right)\mathrm{d}\Omega.\end{aligned} \tag{8.21}$$

If the medium is non-dispersive (equivalently, non-dissipative), the last two volume integrals disappear and one obtains the simpler relation

$$\frac{1}{2}I_0^2\frac{\mathrm{d}X_{\text{in}}}{\mathrm{d}\omega} = 2W_{\text{eff}}(\omega) - \frac{1}{\eta_0}\oiint\limits_\Omega \Im\left(\mathbf{F}\cdot\frac{\mathrm{d}\mathbf{F}^*}{\mathrm{d}\omega}\right)\mathrm{d}\Omega =: 2W_{\text{eff}} - W_\delta. \tag{8.22}$$

Even for this special case it is difficult to infer whether the rhs is non-negative or not at all frequencies due to the presence of the second term. Any negative value occurring can be understood from the fact that radiation plays the same role as dissipation in the equivalent circuit representation of an antenna as seen from its input port. On the other hand Foster's reactance theorem asserts that the slope of $\mathrm{d}X/\mathrm{d}\omega$ is positive at all frequencies for the driving point reactance $X$ of a lossless, passive network. Whether or not such an assertion holds for a radiative system, which is akin to a lossy network, is still debatable [9, 10]. An extension of the Levis and Rhodes' theorem to anisotropic materials is given in [5].

### 8.1.3 Susceptance theorem for an impulsive voltage source

In contrast to the current impulse assumed in theorem (8.6), suppose the voltage between the inner and outer conductors of the feeding coaxial cable is an impulse function $V_0\delta(t)$, where $V_0$ is a real constant. Then the electric and magnetic fields in the cable change to

$$\mathbf{E}_g = \frac{\hat{\rho} V_0}{\rho \ln(b/a)}; \quad \mathbf{H}_g = \hat{\phi}\frac{V_0 Y_{\text{in}}(\omega)}{2\pi\rho}; \ \hat{\mathbf{n}} = -\hat{\mathbf{z}}, \tag{8.23}$$

where $Y_{\text{in}}(\omega) = G_{\text{in}}(\omega) - jB_{\text{in}}(\omega)$ is the input admittance seen into the antenna terminals, $G_{\text{in}}$ is the input conductance, and $B_{\text{in}}$ is the input susceptance[3]. The total current on the inner conductor of the feeding cable is no longer independent of frequency but becomes equal to $V_0 Y_{\text{in}}(\omega)$. As with the impulsive current excitation of the previous subsection, the conceptual coaxial cable here is also dispersive.

The power supplied to the volume $V$ becomes

$$P_s = \frac{1}{2} V_0^2 Y_{\text{in}}^*(\omega),$$

instead of the expression shown in equation (8.11). Note that the complex power is proportional to the complex conjugate of the input admittance. Because of the causal relation between the voltage and current, it is admittance $Y_{\text{in}}(\omega)$ rather than its conjugate that is analytic in the lower half-plane. In place of equation (8.13) we now have

$$\begin{aligned}\frac{1}{2} V_0^2 Y_{\text{in}}^*(\omega) &= \frac{1}{2} \oiint_{\Sigma} \mathbf{E} \times \mathbf{H}^* \cdot \hat{\mathbf{n}}\, \mathrm{d}s \\ &+ 2j\left[\omega \iiint_V \frac{\mu}{4} \mathbf{H} \cdot \mathbf{H}^*\, \mathrm{d}v - \omega^* \iiint_V \frac{\varepsilon^*}{4} \mathbf{E} \cdot \mathbf{E}^*\, \mathrm{d}v\right].\end{aligned} \tag{8.24}$$

The rest of the development proceeds in an identical manner and one obtains the results (keeping in mind that the imaginary part of $Y_{\text{in}}(\omega)$ is $-B_{\text{in}}(\omega)$)

$$-\frac{V_0^2}{2}\frac{\mathrm{d}B_{\text{in}}}{\mathrm{d}\omega} = 2W_{\text{eff}}(\omega) + 2\omega \iiint_V \left(\bar{w}_{\text{m}} \frac{\partial \ln(\mu_{\text{r}})}{\partial \omega} + \bar{w}_{\text{e}} \frac{\partial \ln(\varepsilon_{\text{r}})}{\partial \omega}\right) \mathrm{d}v, \tag{8.25}$$

$$\begin{aligned}&+ \omega \iiint_V \Im\left(\mu_{\text{i}} \mathbf{H} \cdot \frac{\partial \mathbf{H}^*}{\partial \omega} + \varepsilon_{\text{i}} \mathbf{E} \cdot \frac{\partial \mathbf{E}^*}{\partial \omega}\right) \mathrm{d}v - \frac{1}{\eta_0} \oiint_{\Omega} \Im\left(\mathbf{F} \cdot \frac{\mathrm{d}\mathbf{F}^*}{\mathrm{d}\omega}\right) \mathrm{d}\Omega \\ &= 2W_{\text{eff}}(\omega) - \frac{1}{\eta_0} \oiint_{\Omega} \Im\left(\mathbf{F} \cdot \frac{\mathrm{d}\mathbf{F}^*}{\mathrm{d}\omega}\right) \mathrm{d}\Omega \ \text{(non-dispersive medium)},\end{aligned} \tag{8.26}$$

$$=: 2W_{\text{eff}} - W_\delta, \tag{8.27}$$

[3] Note that for the sake of algebraic convenience we have included a negative sign in front of the susceptance function so that $Y_{\text{in}}^*(\omega) = G_{\text{in}}(\omega) + jB_{\text{in}}(\omega)$.

in place of equations (8.21) and (8.22), respectively. Once again due to the presence of the second term on the rhs of equation (8.26), the negative derivative of the susceptance function can no longer be assumed to be positive semi-definite.

*8.1.3.1 Invalidity of Foster's reactance theorem for radiating systems*
We shall now take up a concrete example to investigate the behavior of the susceptance function and its frequency derivative. The goals are to see (i) whether the effective energy as defined in equation (8.20) is positive and (ii) whether Foster's theorem is applicable to a radiating system.

**Example 8.1.3.1** Equatorial aperture antenna on a sphere.
Consider a thin circumferential slot cut symmetrically on the equator of a perfectly conducting sphere of radius $b$, figure 8.2. The only source of loss here is the radiation loss into the open space exterior to the antenna. The external medium is assumed to be a vacuum. The outer conductor of the feeding coaxial cable exciting the sphere is assumed to be along the $z$-axis and in contact with the lower half of the spherical antenna. The inner conductor crosses over and connects to the other half of the spherical antenna at the origin. The two halves of the sphere are separated by a small gap of width $2b\Delta \ll \lambda_s$ and the aperture field is assumed to be azimuthally invariant. We take the aperture field $\mathbf{E}_a$ to be

$$\mathbf{E}_a(\theta, \phi) = \hat{\boldsymbol{\theta}}\frac{V_0}{2b\Delta}[u(\theta - (\pi/2 - \Delta)) - u(\theta - (\pi/2 + \Delta))], \tag{8.28}$$

where $u(\tau)$ is the unit step function in $\tau$ and $\lambda_s$ is the wavelength at the highest frequency of interest.

Adopting the standard procedure, the electric and magnetic fields generated by this antenna can be obtained in a closed form using spherical harmonics. The excitation produces a $\text{TM}_r$ field and the electric and magnetic fields in the region $r \geqslant b$ can be written using equations (1.134), (1.135), and (1.139) as

$$E_r = -\frac{1}{R^2}\sum_{n=1}^{\infty} n(n+1)A_n\hat{H}_n^{(2)}(R)P_n(\cos\theta), \tag{8.29}$$

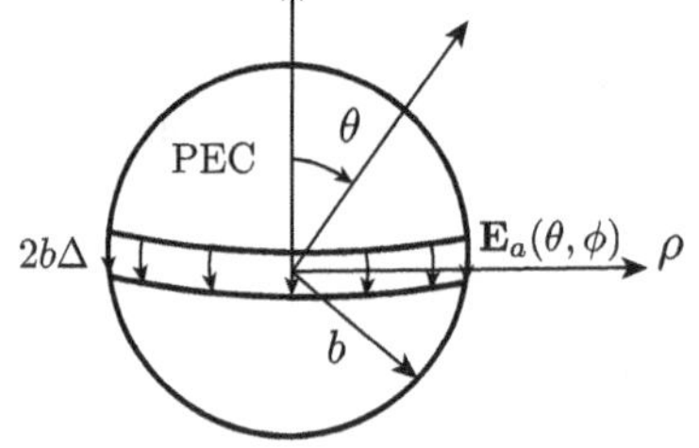


**Figure 8.2.** A spherical metallic antenna fed by an impulsive voltage via a coaxial cable to result in a frequency independent aperture field over the band of interest.

$$E_\theta = -\frac{1}{R}\sum_{n=1}^{\infty} A_n \hat{H}_n^{(2)'}(R) P_n^1(\cos\theta), \tag{8.30}$$

$$H_\phi = \frac{j}{R\eta_0}\sum_{n=1}^{\infty} A_n \hat{H}_n^{(2)}(R) P_n^1(\cos\theta), \tag{8.31}$$

where $R = kr$, $P_n(\cos\theta)$ is Legendre polynomial of degree $n$, $P_n^1(\cos\theta) = \mathrm{d}P_n(\cos\theta)/\mathrm{d}\theta$ and $\hat{H}_n^{(2)}(R) = Rh_n^{(2)}(R)$ is the spherical Ricatti–Hankel function of the second kind of order $n$. Multiplying both sides of equation (8.30) with $\sin\theta \mathrm{d}P_\mathrm{m}(\cos\theta)/\mathrm{d}\theta$, integrating over $\theta \in (0, \pi)$, evaluating at $r = b$ and using the result (D.121) yields

$$\begin{aligned} A_n &= \frac{kV_0(2n+1)}{2n(n+1)\hat{H}_n^{(2)'}(\beta)} \frac{1}{2\Delta} \int_{\pi/2-\Delta}^{\pi/2+\Delta} \sin\theta \frac{\mathrm{d}P_n(\cos\theta)}{\mathrm{d}\theta}\, \mathrm{d}\theta \\ &=: \frac{kV_0(2n+1)\alpha_n}{2n(n+1)\hat{H}_n^{(2)'}(\beta)}, \end{aligned} \tag{8.32}$$

where $\beta = kb$. The derivative w.r.t. $\theta$ of $P_n(\cos\theta)$ is even (odd) for $n$ odd (even) around $\theta = \pi/2$. Therefore all even harmonics drop out. For $n$ odd we have

$$\begin{aligned} \alpha_n &= \frac{1}{\Delta}\int_{\pi/2}^{\pi/2+\Delta} \sin\theta \frac{\mathrm{d}P_n(\cos\theta)}{\mathrm{d}\theta}\, \mathrm{d}\theta \\ &= \frac{1}{\Delta}\int_0^{\sin\Delta} \sqrt{1-x^2}\left[-\frac{\mathrm{d}}{\mathrm{d}x}P_n(x)\right] \mathrm{d}x, \\ &= -\frac{\cos\Delta P_n(\sin\Delta)}{\Delta} - \frac{1}{\Delta}\int_0^{\Delta} \sin y P_n(\sin y)\, \mathrm{d}y, \end{aligned} \tag{8.33}$$

$$=: -\frac{\cos\Delta P_n(\sin\Delta)}{\Delta} + R_n(\Delta) \approx -\frac{\cos\Delta P_n(\sin\Delta)}{\Delta}, \tag{8.34}$$

where the second line follows from integration by parts. To a very good approximation $\alpha_n$ is determined by the first term on the rhs of equation (8.33) as indicated in equation (8.34) because the remainder term $R_n$ satisfies

$$\begin{aligned} |R_n(\Delta)| &= \frac{1}{\Delta}\left|\int_0^{\Delta} \sin y\ P_n(\sin y)\, \mathrm{d}y\right| \\ &\approx \begin{cases} \dfrac{\sin\Delta}{\Delta}\left(\dfrac{1}{n}\sqrt{\dfrac{2}{n\pi}}\right)\sin\left[\left(n+\dfrac{1}{2}\right)\Delta\right], & n \gg 1 \\ < \dfrac{\sin\Delta}{\Delta}|P_n(\Delta)| = \mathcal{O}(\Delta^2), & \Delta \ll 1 \end{cases}. \end{aligned} \tag{8.35}$$

**Table 8.1.** Exact versus approximate values of $\alpha_n$ for $\Delta = 0.2$.

| $n$ | Exact $\alpha_n$ from (8.33) | Approximate $\alpha_n$ from (8.34) | Absolute % error |
|---|---|---|---|
| 1 | −0.9866 | −0.9732 | 1.36 |
| 3 | 1.3818 | 1.3625 | 1.40 |
| 5 | −1.5188 | −1.4964 | 1.48 |
| 7 | 1.4516 | 1.4282 | 1.61 |
| 9 | −1.2194 | −1.1968 | 1.85 |
| 11 | 0.8685 | 0.8482 | 2.34 |
| 13 | −0.4535 | −0.4365 | 3.75 |
| 15 | $3.185 \times 10^{-2}$ | $1.8826 \times 10^{-2}$ | — |
| 17 | 0.3421 | 0.3509 | 2.57 |
| 19 | −0.6238 | −0.6285 | 0.75 |
| 21 | 0.7837 | 0.7848 | 0.14 |

Making use of equation (D.91), it is straightforward to show that for $n \gg 1$

$$\alpha_n^2 \sim \frac{\cos\Delta}{n\pi}\left[\frac{1-\cos(2n+1)\Delta}{\Delta^2}\right] = \begin{cases} \mathcal{O}(n^{-1}), & \Delta \neq 0 \\ \mathcal{O}(n), & \Delta = 0 \end{cases}. \tag{8.36}$$

Indeed for $\Delta \to 0$, the aperture electric field approaches a delta function, which will render the energy density non-integrable and manifest in a diverging $\alpha_n = \mathrm{d}P_n(0)/\mathrm{d}\theta \sim\mathcal{O}(n^{1/2})$, $n \gg 1$. Table 8.1 shows a comparison between the exact $\alpha_n$ as computed from equation (8.33) with the approximate $\alpha_n$ given in equation (8.34). Results show that the approximate form has an error less than 4% even for a $\Delta$ as large as 0.2.[4] We shall assume henceforth that $\Delta > 0$ and also use $n = 2m + 1$, $m = 0, 1, \ldots$.

Using the asymptotic value of the Hankel function $\hat{H}^{(2)\prime}_{2m+1}(R) \sim j(-1)^m \mathrm{e}^{-jR}$, $R \gg 1$, the vector radiation pattern of the antenna is

$$\mathbf{F} = -j\hat{\theta}\frac{V_0}{4}\sum_{m=0}^{\infty}\frac{(-1)^m(4m+3)\alpha_{2m+1}}{(2m+1)(m+1)}\frac{P^1_{2m+1}(\cos\theta)}{\hat{H}^{(2)\prime}_{2m+1}(\beta)}. \tag{8.37}$$

The frequency dependence of $\mathbf{F}$ comes entirely from the variation of the antenna electrical dimension, which, in this example, is the radial dimension $\beta = kb$. Even though equation (8.37) has been derived for real frequencies with $\beta = b\omega/c$, it continues to be valid for the whole complex frequency plane since $1/\hat{H}^{(2)\prime}_{2m+1}(\beta)$ is an analytic function of its complex argument $\beta$ corresponding to the complex frequency via $\beta = b(\omega_r + j\omega_i)/c$. Figure 8.3 shows sample radiation patterns of the antenna for two different electrical sizes: $b = \lambda/10$ and $b = \lambda$. The current density induced on the

[4] The error is not shown near a zero crossing which occurs for $n = 15$ at the chosen $\Delta = 0.2$. The error is expected to be high when $\alpha_n$ is around zero. But those $\alpha_n$s are insignificant in the overall sum.

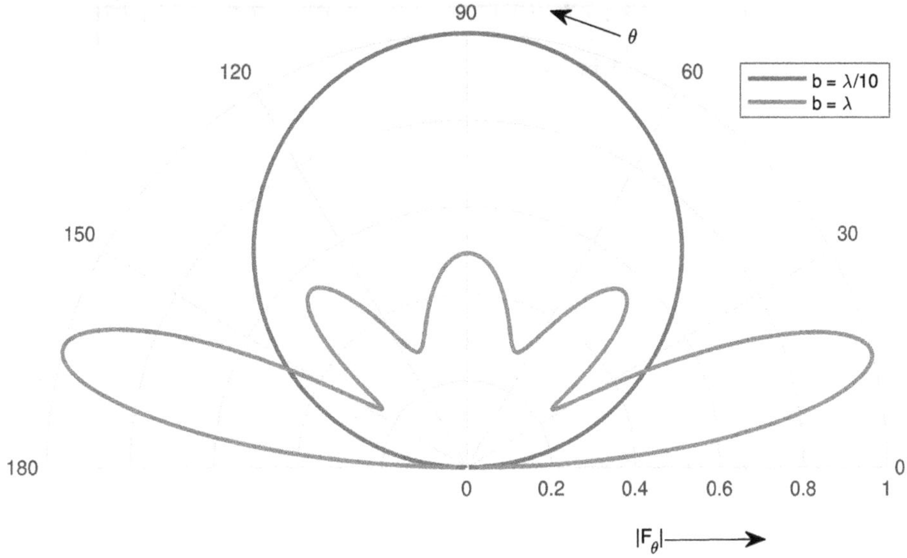


**Figure 8.3.** Frequency dependent radiation pattern of spherical antenna.

surface of the conductors is given by $\mathbf{J}_s = \hat{\mathbf{n}} \times \mathbf{H}\,|_{r=b} = -\hat{\theta} J_\theta(\theta)$ and flows along the $\hat{\theta}$-direction. Magnitude and phase of the current density for two electrical sizes are shown in figures 8.4 and 8.5. As expected the phase is nearly uniform for the electrically smaller antenna. It is interesting to note that the minima in the radiation pattern within the range $0 < \theta < \pi$ are not deep for the larger antenna. This is despite the fact that the phase of the current changes significantly in going from the south pole (or north pole) to the equator along a longitude of the spherical antenna.

Using the fields given in equations (8.29)–(8.31), it can be seen that

$$\begin{aligned}\frac{1}{c} P_{\text{rad}} &= \frac{V_0^2 \pi \varepsilon_0}{4} \sum_{m=0}^{\infty} \frac{(4m+3)\alpha_{2m+1}^2}{(2m+1)(m+1)\left|\hat{H}_{2m+1}^{(2)\prime}(\beta)\right|^2} \\ &= \frac{V_0^2 \eta_0 G_{\text{in}} \varepsilon_0}{2},\end{aligned} \tag{8.38}$$

$$r^2 \oiint_\Omega \bar{w}_{\text{m}}\, \mathrm{d}\Omega = \frac{V_0^2 \pi \varepsilon_0}{8} \sum_{m=0}^{\infty} \frac{(4m+3)\alpha_{2m+1}^2 \left|\hat{H}_{2m+1}^{(2)}(R)\right|^2}{(2m+1)(m+1)\left|\hat{H}_{2m+1}^{(2)\prime}(\beta)\right|^2}, \tag{8.39}$$

$$\sim \frac{1}{2c} P_{\text{rad}} + \mathcal{O}\left(\frac{1}{R^2}\right),\; R \to \infty, \tag{8.40}$$

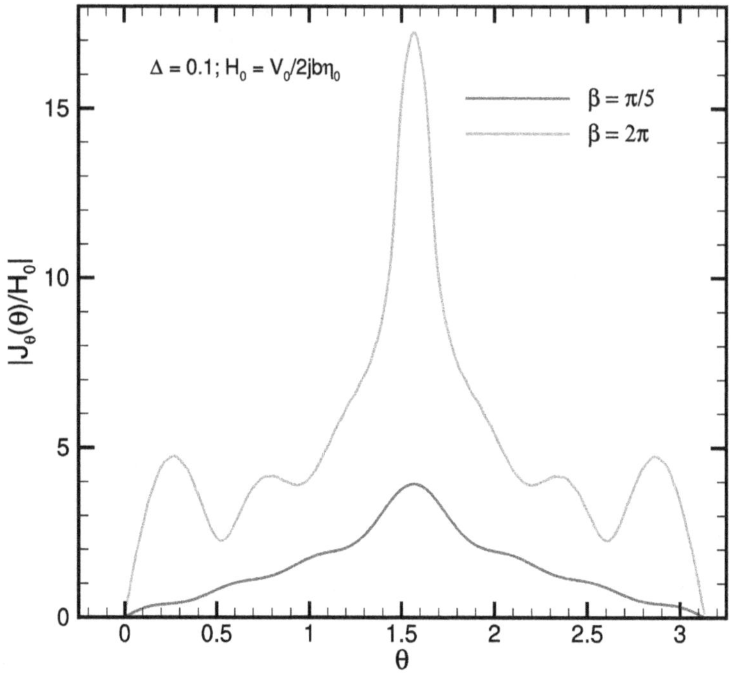


**Figure 8.4.** Magnitude $|J_\theta(\theta)|/H_0$ of the normalized current density for electrically small and electrically large spherical antennas, $H_0 = V_0/j2b\eta_0$.

$$r^2 \oiint_\Omega \frac{\varepsilon_r\,|E_r|^2}{4}\,\mathrm{d}\Omega = \frac{V_0^2 \pi\varepsilon_0}{4R^2}\sum_{m=0}^{\infty}\frac{(4m+3)\alpha_{2m+1}^2\left|\hat{H}_{2m+1}^{(2)}(R)\right|^2}{\left|\hat{H}_{2m+1}^{(2)'}(\beta)\right|^2}, \tag{8.41}$$

$$\sim \mathcal{O}\left(\frac{1}{R^2}\right),\ R \to \infty, \tag{8.42}$$

$$r^2 \oiint_\Omega \frac{\varepsilon_r\,|E_\theta|^2}{4}\,\mathrm{d}\Omega = \frac{V_0^2 \pi\varepsilon_0}{8}\sum_{m=0}^{\infty}\frac{(4m+3)\alpha_{2m+1}^2\left|\hat{H}_{2m+1}^{(2)'}(R)\right|^2}{(2m+1)(m+1)\left|\hat{H}_{2m+1}^{(2)'}(\beta)\right|^2}, \tag{8.43}$$

$$\sim \frac{1}{2c}P_{\text{rad}} + \mathcal{O}\left(\frac{1}{R^2}\right),\ R \to \infty, \tag{8.44}$$

where $G_{\text{in}} = 2P_{\text{rad}}/V_0^2$ is the input conductance. The above results follow from the identity (D.116). The asymptotic forms in equations (8.40), (8.42), (8.44) are true due to the large argument approximations of $\hat{H}_\nu^{(2)}(z)$ given in equations (C.151) and (C.153). Expressions for the frequency derivative of susceptance and the susceptance itself can be obtained directly from the definition in equation (8.24). Denoting the integrand in equation (8.43) as $\bar{w}_{e\theta}$ and that in equation (8.41) as $\bar{w}_{er}$ one sees that

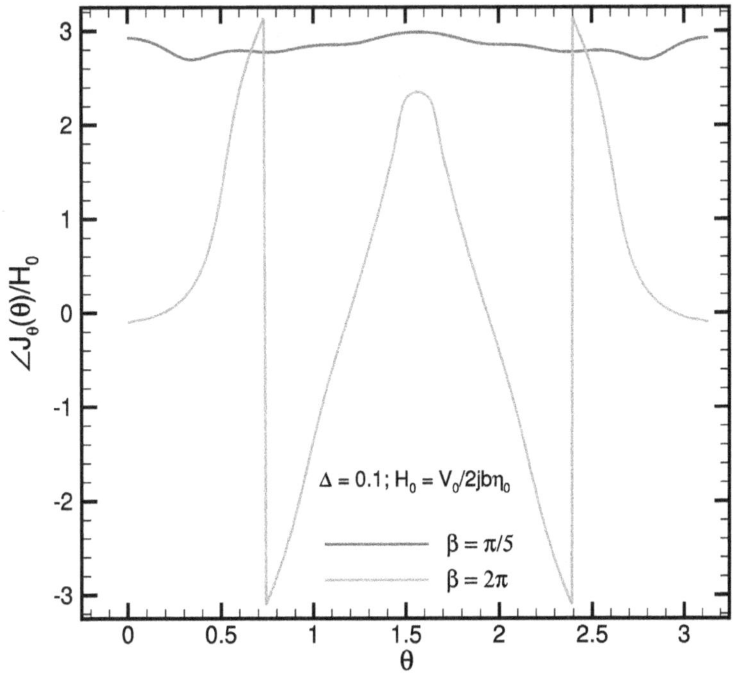


**Figure 8.5.** Phase $\angle J_\theta(\theta)$ of the current density for electrically small and electrically large spherical antennas.

$$r^2 \oiint_\Omega (\bar{w}_\mathrm{m} - \bar{w}_{\mathrm{e}\theta} - \bar{w}_{\mathrm{e}r})\,\mathrm{d}\Omega = \frac{V_0^2 \pi \varepsilon_0}{8} \sum_{m=0}^{\infty} \frac{(4m+3)\alpha_{2m+1}^2}{(m+1)(2m+1)} \times \left[\left(1 - \frac{2(2m+1)(m+1)}{R^2}\right)\left|\hat{H}_{2m+1}^{(2)}(R)\right|^2 - \left|\hat{H}_{2m+1}^{(2)\prime}(R)\right|^2\right] \frac{1}{\left|\hat{H}_{2m+1}^{(2)\prime}(\beta)\right|}. \tag{8.45}$$

The total energy density of the electric field is $\bar{w}_\mathrm{e} = \bar{w}_{\mathrm{e}\theta} + \bar{w}_{\mathrm{e}r}$. Using the defining equation (C.128) for the spherical Ricatti–Bessel function and the complex conjugate relation between $\hat{H}_\nu^{(2)}(x)$ and $\hat{H}_\nu^{(1)}(x)$, the quantity within the square brackets above can be written as

$$T(R) = \left(1 - \frac{2(2m+1)(m+1)}{R^2}\right)\left|\hat{H}_{2m+1}^{(2)}(R)\right|^2 - \left|\hat{H}_{2m+1}^{(2)\prime}(R)\right|^2 = -\left[\hat{H}_{2m+1}^{(2)}(R)\hat{H}_{2m+1}^{(1)\prime}(R)\right]'.$$

The asymptotic value of $T(R)$ for large argument can be deduced from equations (C.153) and (C.154) and is

$$T(R) \sim \frac{1}{4R^2} \to 0 \text{ as } R \to \infty, \tag{8.46}$$

indicating that $T(R)$ is integrable over $R \in (\beta, \infty)$. The integral is

$$\int_{\beta}^{\infty} T(R)\,\mathrm{d}R = \hat{H}_{2m+1}^{(2)}(\beta)\hat{H}_{2m+1}^{(1)'}(\beta). \tag{8.47}$$

Inserting this into equation (8.45), the difference in magnetic and electric energies exterior to the antenna is

$$\begin{aligned}\bar{W}_\mathrm{m} - \bar{W}_\mathrm{e} &= \int_{\beta}^{\infty} r^2 \oiint_{\Omega} (\bar{w}_\mathrm{m} - \bar{w}_\mathrm{e})\,\mathrm{d}\Omega\,\mathrm{d}r \\ &= \frac{V_0^2\pi\varepsilon_0}{8k}\sum_{m=0}^{\infty}\frac{(4m+3)\alpha_{2m+1}^2}{(m+1)(2m+1)}\frac{1}{\left|\hat{H}_{2m+1}^{(2)'}(\beta)\right|}\int_{\beta}^{\infty} T(R)\,\mathrm{d}R \\ &= \frac{V_0^2\pi\varepsilon_0}{8k}\sum_{m=0}^{\infty}\frac{(4m+3)\alpha_{2m+1}^2}{(m+1)(2m+1)}\Re\left(\frac{\hat{H}_{2m+1}^{(2)}(\beta)}{\hat{H}_{2m+1}^{(1)'}(\beta)}\right).\end{aligned} \tag{8.48}$$

The susceptance of the antenna is from the definition (8.24)

$$-B_\mathrm{in}(\omega) = -\frac{4\omega}{V_0^2}(\bar{W}_\mathrm{m} - \bar{W}_\mathrm{e}) = -\frac{\pi}{2\eta_0}\sum_{m=0}^{\infty}\frac{(4m+3)\alpha_{2m+1}^2}{(2m+1)(m+1)}\Re\left[\frac{\hat{H}_{2m+1}^{(2)}(\beta)}{\hat{H}_{2m+1}^{(2)'}(\beta)}\right]. \tag{8.49}$$

The conductance is obtainable from equation (8.38)

$$G_\mathrm{in}(\omega) = \frac{2P_\mathrm{rad}}{V_0^2} = \frac{\pi}{2\eta_0}\sum_{m=0}^{\infty}\frac{(4m+3)\alpha_{2m+1}^2}{(2m+1)(m+1)}\frac{1}{\left|\hat{H}_{2m+1}^{(2)'}(\beta)\right|^2} > 0, \tag{8.50}$$

and together with equation (8.49) completely determines the input admittance $Y_\mathrm{in}(\omega) = G_\mathrm{in}(\omega) - jB_\mathrm{in}(\omega)$ of the antenna. The frequency dependence of admittance is governed by the Bessel functions via their argument $\beta$. Note that $\hat{H}_{\nu}^{(2)}(\beta)/\hat{H}_{\nu}^{(2)'}(\beta) \sim -(\frac{\beta}{\nu})$, $\nu > 0$, $\beta \to 0$ so that $B_\mathrm{in}(\omega)$ has a simple zero at $\omega = 0$. In particular, the input admittance has no branch-point singularities even though the loss mechanism here is due to radiation[5]. Since the Ricatti–Hankel functions are entire functions of their complex-valued arguments, there are no singularities in the complex $\omega$-plane (or equivalently the complex $k$-plane) of this admittance function. Figure 8.6 shows the variation of antenna input admittance as a function of the electrical size of the antenna. It is interesting to note that the susceptance for this antenna is all positive over a wide range of frequencies; however, the fluctuations in the susceptance could be either positive or negative.

[5] Radiation is normally associated with branch-point singularities in the complex $k$-domain.

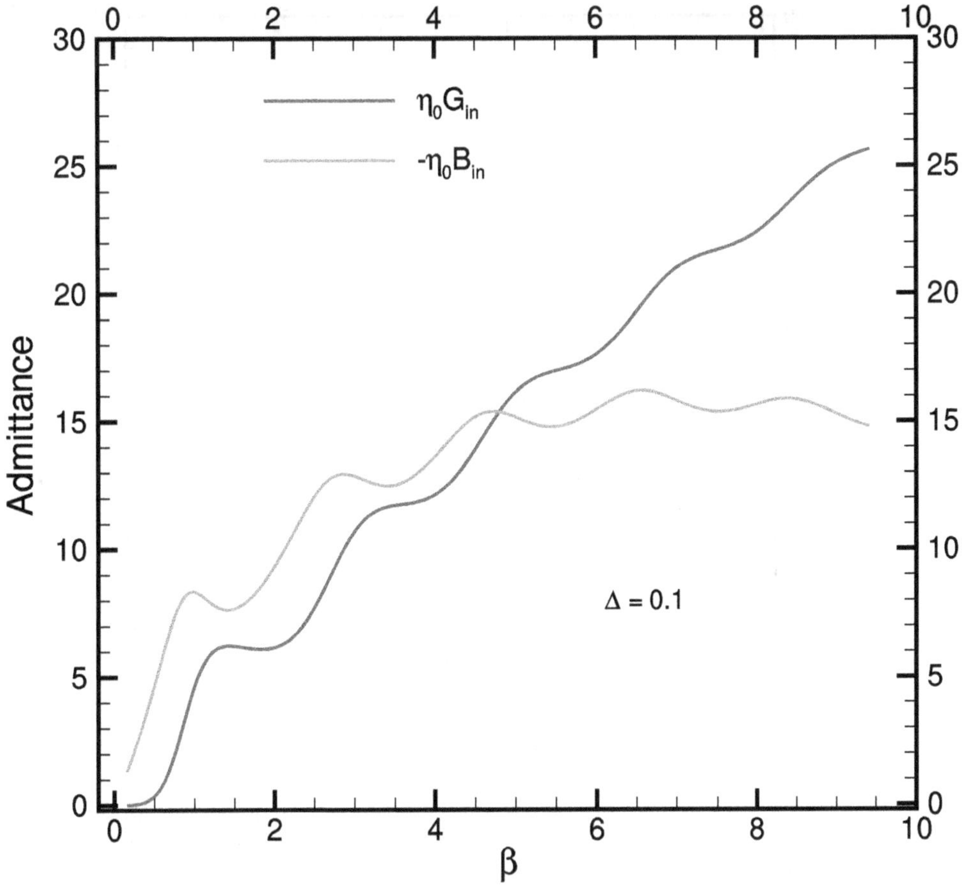


**Figure 8.6.** Frequency variation of input admittance as a function of electrical size $\beta = kb$ of the spherical antenna. Note that susceptance is equal to $-B_{\text{in}}$.

One may obtain frequency derivative $-\mathrm{d}B_{\text{in}}(\omega)/\mathrm{d}\omega$ directly from equation (8.49) for this antenna. The frequency derivative of equation (8.49) gives

$$-\frac{\mathrm{d}B_{\text{in}}(\omega)}{\mathrm{d}\omega} = \frac{C_0\pi}{2}\sum_{m=0}^{\infty}\frac{(4m+3)\alpha_{2m+1}^2}{(2m+1)(m+1)}\left[\left(\frac{2(2m+1)(m+1)}{\beta^2}-1\right)\right. \\ \left.\times\,\Re\left(\frac{\hat{H}_{2m+1}^{(2)}(\beta)}{\hat{H}_{2m+1}^{(2)\prime}(\beta)}\right)^2 - 1\right], \tag{8.51}$$

$$\sim \frac{C_0\pi}{2}\sum_{m=0}^{\infty}\frac{(4m+3)\alpha_{2m+1}^2}{(2m+1)^2(m+1)} > 0,\ \beta \to 0, \tag{8.52}$$

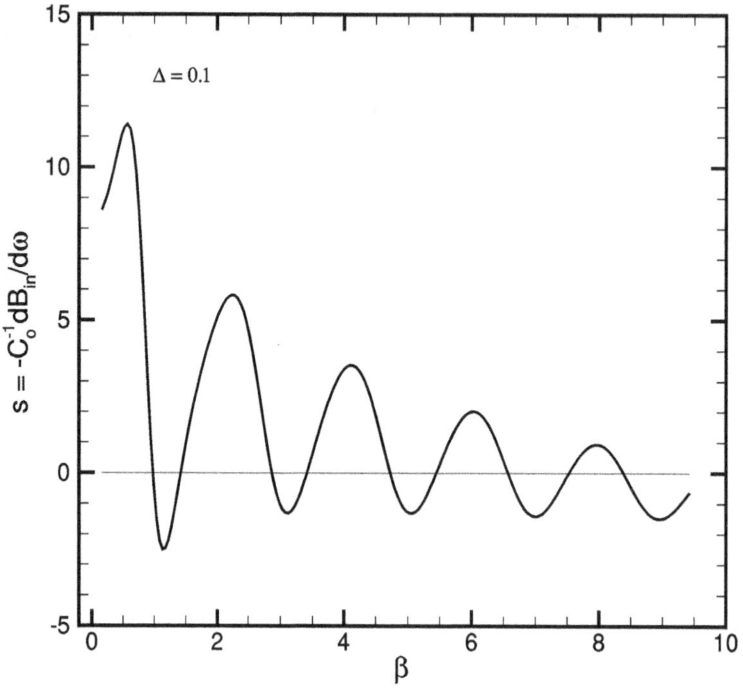


**Figure 8.7.** Frequency derivative of negative input susceptance as a function of electrical size $\beta = kb$ of the spherical antenna. Note that susceptance is equal to $-B_{\text{in}}$ and that $C_0 = b\varepsilon_0$.

where $C_0 = b\varepsilon_0$ is a frequency independent, size-related capacitance pertaining to the spherical shape. It is seen that $-\mathrm{d}B_{\text{in}}/\mathrm{d}\omega$ could be positive or negative depending on the value of $\beta = kb$. For an electrically small spherical antenna with $\beta$ small, the quantity in the square brackets within the summation sign reduces to $(2m + 1)^{-1}$. Consequently, as shown in equation (8.52), $-\mathrm{d}B_{\text{in}}/\mathrm{d}\omega > 0$ at very low frequencies. Figure 8.7 shows the variation of $s = -C_0^{-1}\mathrm{d}B_{\text{in}}/\mathrm{d}\omega$ as $\beta = kb$ is varied. It is seen that the frequency derivative of the antenna susceptance is neither positive definite nor negative definite in contrast to that of a lossless network satisfying the Foster's reactance theorem. Here, radiative loss produced by the antenna plays the same role as dissipation and makes the frequency variations in the susceptance function be non-monotonic. In addition, it is seen that $|\mathrm{d}B_{\text{in}}/\mathrm{d}\omega|$ has an envelope that generally decreases with $\beta$. Hence equation (8.5), when applied to the admittance function, implies that the antenna bandwidth increases as $\beta$ increases since $G_{\text{in}}$ also simultaneously increases with $\beta$.

Let us now explore the finiteness and definiteness of the effective energy defined in equation (8.20). It can be established in a manner similar to equation (8.46) that the quantity

$$\left[ r^2 \oiint_{\Omega} (\bar{w}_{\text{m}} + \bar{w}_{\text{e}})\, \mathrm{d}\Omega - \frac{P_{\text{rad}}}{c} \right] \to \mathcal{O}\left(\frac{1}{r^2}\right),\ r \to \infty, \tag{8.53}$$

clearly indicating that the infinities associated with individual terms within the square brackets above are canceled off exactly in the combination, thus rendering it integrable in $r \in (b, \infty)$. Consider the integrals

$$W_{\mathrm{eff}}^{(1)} = \int_b^{\infty}\left[r^2 \oiint_{\Omega} (\bar{w}_{\mathrm{m}} + \bar{w}_{\mathrm{e}\theta}) - \frac{1}{c}P_{\mathrm{rad}}\right]\mathrm{d}r$$
$$= \frac{V_0^2 \pi\varepsilon_0}{8k}\sum_{m=0}^{\infty}\frac{(4m+3)\alpha_{2m+1}^2}{(2m+1)(m+1)\left|\hat{H}_{2m+1}^{(2)\prime}(\beta)\right|^2} \tag{8.54}$$
$$\times \int_{\beta}^{\infty}\left[\left|\hat{H}_{2m+1}^{(2)\prime}(R)\right|^2 + \left|\hat{H}_{2m+1}^{(2)}(R)\right|^2 - 2\right]\mathrm{d}R,$$

and

$$W_{\mathrm{eff}}^{(2)} = \int_{r=b}^{\infty} r^2 \bar{w}_{\mathrm{e}r}\,\mathrm{d}r = \frac{V_0^2 \pi\varepsilon_0}{8k}\sum_{m=0}^{\infty}\frac{(4m+3)\alpha_{2m+1}^2}{(2m+1)(m+1)\left|\hat{H}_{2m+1}^{(2)\prime}(\beta)\right|^2}$$
$$\times \int_{\beta}^{\infty}\hat{H}_{2m+1}^{(2)}(R)\frac{(2m+1)2(m+1)}{R^2}\hat{H}_{2m+1}^{(1)}(R)\,\mathrm{d}R$$
$$= \frac{V_0^2 \pi\varepsilon_0}{8k}\sum_{m=0}^{\infty}\frac{(4m+3)\alpha_{2m+1}^2}{(2m+1)(m+1)\left|\hat{H}_{2m+1}^{(2)\prime}(\beta)\right|^2} \tag{8.55}$$
$$\times \int_{\beta}^{\infty}\hat{H}_{2m+1}^{(2)}(R)\left[\hat{H}_{2m+1}^{(1)}(R) + \hat{H}_{2m+1}^{(1)\prime\prime}(R)\right]\mathrm{d}R,$$

where the double-primed superscript denotes second derivative. Combining the integrands in equations (8.54) and (8.55) and using the result (C.132) as well as the Wronskian (C.131), we write with $\nu = 2m + 1$ that

$$\left|\hat{H}_{\nu}^{(2)\prime}(R)\right|^2 + 2\left|\hat{H}_{\nu}^{(2)}(R)\right|^2 + \hat{H}_{\nu}^{(2)}(R)\hat{H}_{\nu}^{(1)\prime\prime}(R) - 2$$
$$= \left|\hat{H}_{\nu}^{(2)\prime}(R) + \mathrm{i}\hat{H}_{\nu}^{(2)}(R)\right|^2 + \frac{\nu(\nu+1)}{R^2}\left|\hat{H}_{\nu}^{(2)}(R)\right|^2 > 0 \tag{8.56}$$

$$= \left(\hat{J}_{\nu}(R)\hat{J}_{\nu}'(R) + \hat{Y}_{\nu}(R)\hat{Y}_{\nu}'(R)\right)' + 2\left(\hat{J}_{\nu}^2(R) + \hat{Y}_{\nu}^2(R) - 1\right). \tag{8.57}$$

Note that the quantities within the parenthesis of both terms on the rhs of equation (8.57) are integrable over $R \in (\beta > 0, \infty)$ in view of the asymptotic forms given in equations (C.153) and (C.154). Now the indefinite integral

$$I_1 = \int \left( \hat{J}_\nu(x)\hat{J}'_\nu(x) + \hat{Y}_\nu(x)\hat{Y}'_\nu(x) \right)' \,\mathrm{d}x = \left( \hat{J}_\nu(x)\hat{J}'_\nu(x) + \hat{Y}_\nu(x)\hat{Y}'_\nu(x) \right), \tag{8.58}$$

and, using the identity (C.138), the indefinite integral

$$\begin{aligned} I_2 &= 2\int \left( \hat{J}_\nu^{\,2}(x) + \hat{Y}_\nu^{\,2}(x) - 1 \right) \mathrm{d}x \\ &= \left\{ \hat{J}_\nu^{\prime 2}(x) + \hat{Y}_\nu^{\prime 2}(x) + \left( 1 - \frac{\nu(\nu+1)}{x^2} \right) \left[ \hat{J}_\nu^{\,2}(x) + \hat{Y}_\nu^{\,2}(x) \right] - 2 \right\} x \\ &\quad - \left( \hat{J}_\nu(x)\hat{J}'_\nu(x) + \hat{Y}_\nu(x)\hat{Y}'_\nu(x) \right). \end{aligned} \tag{8.59}$$

Adding the two integrals yields

$$I_1 + I_2 = \left\{ \left| \hat{H}_\nu^{(2)\prime}(x) \right|^2 + \left( 1 - \frac{\nu(\nu+1)}{x^2} \right) \left| \hat{H}_\nu^{(2)}(x) \right|^2 - 2 \right\} x, \tag{8.60}$$

$$\sim \mathcal{O}\left( \frac{1}{x} \right), \quad |x| \gg \nu \text{ (fixed)}, \tag{8.61}$$

$$\sim \left( 1 - \frac{\nu}{x^2} \right) \frac{k_\nu^2}{x^{2\nu-1}}, \quad x \ll 1,\ \nu \text{ fixed}, \tag{8.62}$$

$$\begin{aligned} &\sim \left( \frac{x^3}{\nu^2} - \frac{x}{\nu} - \frac{2x}{\left| \hat{H}_\nu^{(2)\prime}(x) \right|^2} \right) \left| \hat{H}_\nu^{(2)\prime}(x) \right|^2 \\ &\sim -\frac{x}{\nu} \left| \hat{H}_\nu^{(2)\prime}(x) \right|^2, \quad \nu \gg \sqrt{x}. \end{aligned} \tag{8.63}$$

where $k_\nu$ is defined in equation (C.140). The latter three forms follow from the asymptotic expressions given in equations (C.140)–(C.145), (C.153), and (C.157). Hence the sum of the two integrals in equations (8.54) and (8.55) vanishes at the upper limit and converges at the lower end. Using equation (8.60) in equation (8.54) plus equation (8.55) finally yields

$$\begin{aligned} 2W_{\mathrm{eff}} &= 2\left( W_{\mathrm{eff}}^{(1)} + W_{\mathrm{eff}}^{(2)} \right) \\ &= W_0 \sum_{m=0}^{\infty} \frac{(4m+3)\alpha_{2m+1}^2}{(2m+1)(m+1)\left| \hat{H}_{2m+1}^{(2)\prime}(\beta) \right|^2} \left[ 2 - \left| \hat{H}_{2m+1}^{(2)\prime}(\beta) \right|^2 \right. \\ &\quad \left. - \left| \hat{H}_{2m+1}^{(2)}(\beta) \right|^2 + \frac{2(2m+1)(m+1)}{\beta^2} \left| \hat{H}_{2m+1}^{(2)}(\beta) \right|^2 \right], \end{aligned} \tag{8.64}$$

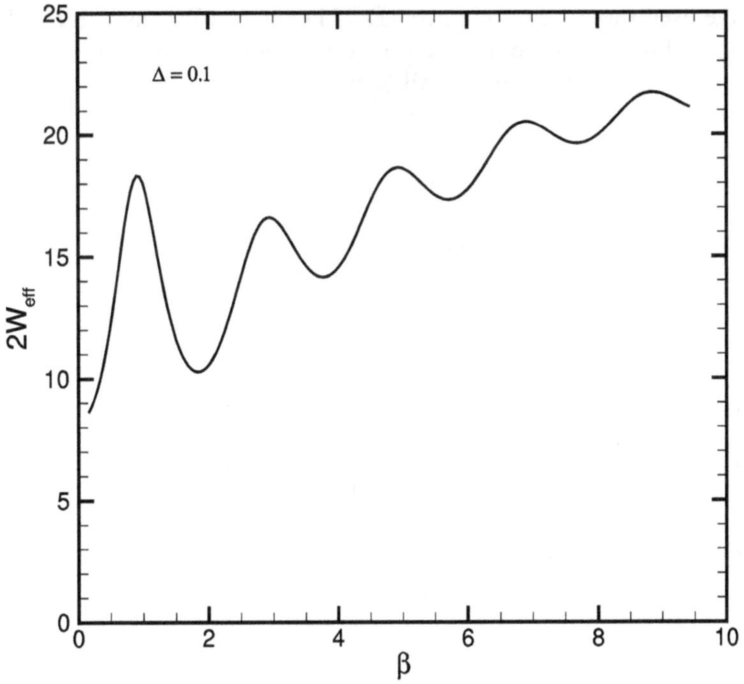


**Figure 8.8.** Frequency variation of the effective stored energy as a function of electrical size $\beta = kb$ of the spherical antenna with $\Delta = 0.1$ and $W_0 = 1$.

where $W_0 = V_0^2 \pi b \varepsilon_0 / 4$. It is easy to see that for small $\beta$

$$2W_{\text{eff}} \sim W_0 \sum_{m=0}^{\infty} \frac{(4m+3)\alpha_{2m+1}^2}{(2m+1)^2(m+1)} \left[ 1 + \frac{3\beta^2}{(4m+1)(2m+1)} + \mathcal{O}(\beta^4) \right] \tag{8.65}$$

$$= W_0 \sum_{m=0}^{\infty} \frac{(4m+3)\alpha_{2m+1}^2}{(2m+1)^2(m+1)}, \quad \beta = 0. \tag{8.66}$$

It is seen that the effective energy is finite even in the static case of $\beta = 0$ because $\alpha_{2m+1}^2 \sim \mathcal{O}[(2m+1)^{-1}]$, $m \gg 1$ provided $\Delta > 0$. In addition, $2W_{\text{eff}}$ is always positive definite because the quantity within the square brackets in equation (8.64) is the result of integration of an integrand which was shown to be positive in equation (8.56) for any $m$ and $\beta$. Indeed, positivity of this quantity results in the inequality (C.148). Figure 8.8 shows a plot of the effective energy versus frequency for a sample antenna with $\Delta = 0.1$ and $W_0 = 1$.

It is straightforward to see that

$$\begin{aligned} W_\delta &= \frac{1}{\eta_0} \Im \oint\!\!\!\oint_{\Omega} \mathbf{F} \cdot \frac{\partial \mathbf{F}^*}{\partial \omega} \, \mathrm{d}\Omega \\ &= 2W_0 \sum_{m=0}^{\infty} \frac{(4m+3)\alpha_{2m+1}^2}{\left| \hat{H}_{2m+1}^{(2)\prime}(\beta) \right|^4} \left[ \frac{2}{\beta^2} - \frac{1}{(2m+1)(m+1)} \right], \end{aligned} \tag{8.67}$$

and we have used the integral identity (D.121) and the Wronskian result (C.129). It is seen that this quantity could be positive or negative depending on the value of $\beta = kb$. For an electrically small spherical antenna such that $\beta^2 < 2$, the quantity in the square brackets within the summation sign is positive for all $m$ so that $W_\delta > 0$.

Other examples studied in the literature involving non-symmetric structures such as a dipole antenna and a Yagi array [10] with numerically generated data also show that the frequency derivative of reactance for an antenna is neither positive definite nor negative definite. Thus Foster's reactance theorem is not expected to work for passive antennas. ■■

## References


[1] Montgomery C G, Dicke R H and Purcell E M 1948 *Principles of Microwave Circuits, Radiation Laboratory Series* vol 8 (New York: McGraw-Hill)

[2] Guillemin E A 1935 *Communication Networks* vol 2 (New York: Wiley)

[3] Foster R M 1924 A reactance theorem *Bell Syst. Tech. J.* **3** 259–67

[4] Rhodes D R 1976 Observable stored energies of electromagnetic systems *J. Franklin Inst.* **302** 225–37

[5] Yaghjian A D and Best S R 2005 Impedance, bandwidth, and *Q* of antennas *IEEE Trans. Antennas Propag.* **53** 1298–324

[6] Levis C A 1957 A reactance theorem for antennas *Proc. IRE* **45** 1128–34

[7] Rhodes D R 1977 A reactance theorem *Proc. R. Soc. Lond.* A **353** 1–10

[8] Stratton J A 1941 *Electromagnetic Theory* (New York: McGraw-Hill)

[9] Geyi W 2016 Reply to Comments on 'Stored energies and radiation *Q*' *IEEE Trans. Antennas Propag.* **64** 4577–80

[10] Capek M and Jelinek L 2016 Comments on 'Stored energies and radiation *Q*' *IEEE Trans. Antennas Propag.* **64** 4575–6

**IOP** Publishing

Engineering Electrodynamics
A collection of theorems, principles and field representations
**Ramakrishna Janaswamy**

# Chapter 9

# Geometrical optics and Fermat's principle

Maxwell's equations can only be solved exactly in the presence of material objects for a few well-defined geometries. For complicated and irregularly shaped objects one often resorts to numerical approaches. When the principal radii of curvature on an object size are large compared to wavelength, reasonably approximate solutions to Maxwell's equations can be obtained by the geometrical optics technique [1], which has been in existence for a number of centuries. Geometrical optics also provides a basis and bridge to diffraction optics, wherein one describes waves reaching regions in the shadow portion of a lit object by means of diffraction by geometric singularities and curvature [2].

The rigorous derivation of a geometrical optics field comprised of wavefronts and rays and the various laws, such as the eikonal equation, Sommerfeld–Runge law, Snell's law, intensity law, are all discussed in the present chapter. A widely occurring variational principle known as Fermat's principle, which has its origin in geometrical optics, is derived. With the recent emergence of electromagnetic metasurfaces for the control of reflected and transmitted electromagnetic energies, there has been a renewed interest in geometrical optics that witnessed some generalizations to the classical laws. The related topics of gradient metasurfaces and generalized Snell's law are also treated in this chapter.

## 9.1 Geometrical optics and Fermat's principle

The most rigorous and useful definition of geometrical optics is provided by Kline and Kay [3, pp 14–16] and involves the time-domain Maxwell's equations. The geometrical optics field at any point $(x, y, z)$ of space is that field that corresponds to finite discontinuities[1] of the electric and magnetic fields, $\mathscr{E}$ and $\mathscr{H}$, on some hypersurface $\Phi(x, y, z; t) = \psi(x, y, z) - ct = 0$, where $c$ is the speed of light in

[1] By finite discontinuities we mean singularities which are finite discontinuities with respect to the time variable of $\mathscr{E}$ and $\mathscr{H}$ and their successive time derivatives.

 

free space. The family of surfaces $\psi(x, y, z) = ct$ are known as *wavefronts* and the normals erected on the wavefronts are termed *rays*, see figure 9.1. In the above representation, the hypersurface $\Phi(x, y, z; t) = 0$ is solved explicitly for $t$ and cast in the form $\Phi(x, y, z; t) = \psi(x, y, z) - ct$ and the speed of light in free space is merely used to scale the time variable. In particular, $c$ is *not* the speed of the wavefront.

### 9.1.1 Field discontinuities, wavefronts, and the eikonal equation

Consider a family of wavefronts $\psi = ct$ emanating from some source. At any specific time $t = t_0$, there is a point $(x_0, y_0, z_0)$ on the wavefront such that the electric field $\mathscr{E}$ (and $\mathscr{H}$) is zero for $t < t_0$; at $t = t_0$, $\mathscr{E}$ jumps from zero to some non-zero value, and for $t > t_0$, $\mathscr{E}$ is non-zero. The geometrical optics field $(\tilde{\mathscr{E}}, \tilde{\mathscr{H}})$ exists at $(x_0, y_0, z_0)$ *only* at the instant $t_0$. In particular, the values of $\mathscr{E}$ and $\mathscr{H}$ for $t > t_0$ at $(x_0, y_0, z_0)$ belong to the time-dependent field satisfying Maxwell's equations, but not to the geometrical optics field. If we follow the geometrical optics field along a ray, that is, let $x$, $y$, and $z$ take on values along a ray, then the corresponding value of $t$ at which the geometrical optics field exists is given by $t = \psi(x, y, z)/c$. Hence

$$\tilde{\mathscr{E}}(x, y, z) = \mathscr{E}(x, y, z; \psi/c); \quad \tilde{\mathscr{H}}(x, y, z) = \mathscr{H}(x, y, z; \psi/c). \tag{9.1}$$

In the following we show that such a definition of a geometrical optics field follows rigorously from Maxwell's equations [3]. We would like to know what the wavefronts are and how the values of $(\widetilde{\mathbf{E}}, \widetilde{\mathbf{H}})$ vary along a ray. In this regard the generalizations of divergence and the curl theorems to four-dimensional spacetime are useful. Let $G$ be a region in spacetime and let the hypersurface $\Gamma$ be its boundary,

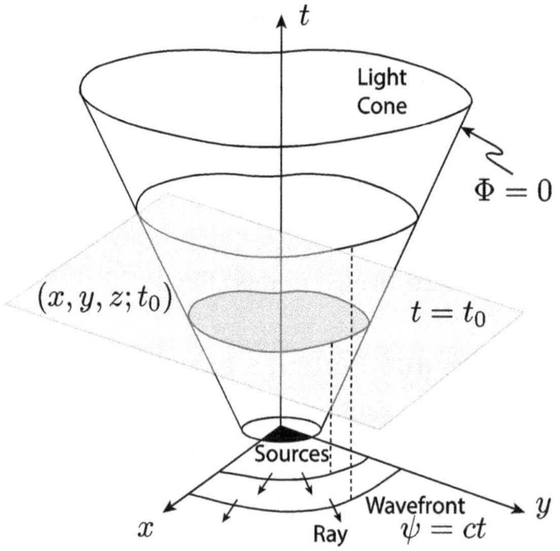


**Figure 9.1.** Wavefronts emanating from a finite-support source. Fields are zero outside the light cone and non-zero inside the light cone. The front across which the fields are discontinuous is described by $\Phi = 0$.

assumed to be continuous with piece-wise continuous derivatives, figure 9.2. The surface Γ is defined by $\chi(x, y, z; t) = 0$ and an outward unit normal $\hat{\mathbf{n}}$ has components

$$n_x = \lambda\hat{\mathbf{x}} \cdot \nabla\chi,\ n_y = \lambda\hat{\mathbf{y}} \cdot \nabla\chi,\ n_z = \lambda\hat{\mathbf{z}} \cdot \nabla\chi,\ n_t = \lambda\dot{\chi}, \tag{9.2}$$

where

$$\lambda = \pm[|\nabla\chi|^2 + \dot{\chi}^2]^{-1/2}, \tag{9.3}$$

and an over-dot represents a partial derivative w.r.t. $t$. The sign is $\lambda$ is chosen to fit the requirement that the normal be directed outward from $G$. Let $\mathbf{m} = n_x\hat{\mathbf{x}} + n_y\hat{\mathbf{y}} + n_z\hat{\mathbf{z}} = \lambda\nabla\chi$ be the projection of $\hat{\mathbf{n}}$ onto the $(x, y, z)$-space. Note that $\mathbf{m}$ is not a unit vector in space coordinates. Let d$w$ be an elemental volume of spacetime $G$ and let d$v$ be the elemental surface of the hypersurface Γ. Consider a vector field $\mathbf{A}(x, y, z; t) = \hat{\mathbf{x}}A_1(x, y, z; t) + \hat{\mathbf{y}}A_2(x, y, z; t) + \hat{\mathbf{z}}A_3(x, y, z; t)$ defined over $G$. Then [3, p 41]

$$\iiiint_G \nabla \times \mathbf{A}\, \mathrm{d}w = \iiint_\Gamma \mathbf{m} \times \mathbf{A}\, \mathrm{d}v, \tag{9.4}$$

$$\iiiint_G \nabla \cdot \mathbf{A}\, \mathrm{d}w = \iiint_\Gamma \mathbf{m} \cdot \mathbf{A}\, \mathrm{d}v, \tag{9.5}$$

where the curl and divergence operations are the usual three-dimensional curl and divergence.

Consider now the time-domain Maxwell's equations in an isotropic medium with electrical parameters $\varepsilon = \varepsilon_0\varepsilon_\mathrm{r}$, $\mu = \mu_0\mu_\mathrm{r}$, $\sigma$ (non-dispersive) and with no magnetic sources

$$\nabla \times \mathscr{E} = -\dot{\mathscr{B}};\quad \mathscr{B} = \mu\mathscr{H},\quad \mathscr{D} = \varepsilon\mathscr{E},\quad \mathscr{J}_\mathrm{c} = \sigma\mathscr{E}, \tag{9.6}$$

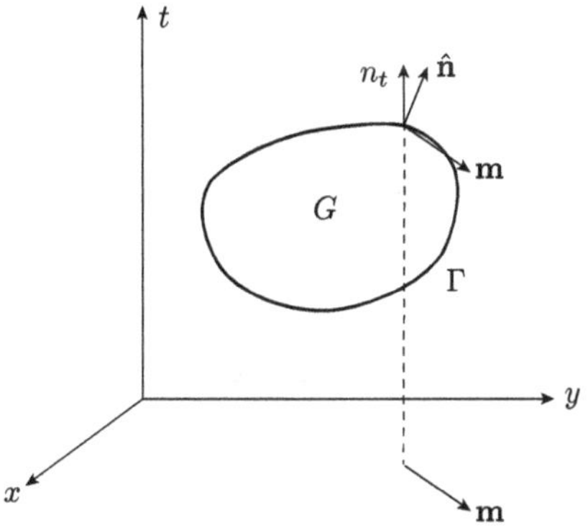


**Figure 9.2.** A four-dimensional domain $G$ bounded by a surface Γ. The normal unit normal $\hat{\mathbf{n}}$ has a component $n_t$ along the time-axis. The projection of $\hat{\mathbf{n}}$ onto space is the vector $\mathbf{m}$.

$$\nabla \times \mathscr{H} = \dot{\mathscr{D}} + \mathscr{J}_c + \mathscr{J}_i; \quad \mathscr{J}_i = \dot{\mathscr{Q}}_e, \quad \nabla \cdot \mathscr{Q}_e = -q_{ev}, \tag{9.7}$$

$$\nabla \cdot \mathscr{B} = 0; \quad \nabla \cdot (\mathscr{J}_i + \mathscr{J}_c + \dot{\mathscr{D}}) = 0, \tag{9.8}$$

where $\mathscr{Q}_e$ is the electric steam potential introduced in equation (3.1) and we have chosen the magnetic stream potential $\mathscr{Q}_m = 0$. The quantities $\mathscr{J}_i$ and $\mathscr{J}_c$ are, respectively, the impressed current density and the conduction current density. These equations can be cast in an integral form by making use of the above four-dimensional curl and divergence theorems. Applying equation (9.4) to the two Maxwell's curl equations and evaluating the integral with respect to time for time-derivative quantities yields

$$\iiint_{\Gamma} [\mathbf{m} \times \mathscr{H} - n_t \mathscr{D} - n_t \mathscr{Q}_e]\, \mathrm{d}v = \iiiint_{G} \mathscr{J}_c\, \mathrm{d}w, \tag{9.9}$$

$$\iiint_{\Gamma} [\mathbf{m} \times \mathscr{E} + n_t \mathscr{B}]\, \mathrm{d}v = 0. \tag{9.10}$$

Applying equation (9.5) to the two Maxwell's divergence equations (9.8) leads to

$$\iiint_{\Gamma} \mathbf{m} \cdot (\mathscr{J}_i + \mathscr{J}_c + \dot{\mathscr{D}})\, \mathrm{d}v = 0, \tag{9.11}$$

$$\iiint_{\Gamma} \mathbf{m} \cdot \mathscr{B}\, \mathrm{d}v = 0. \tag{9.12}$$

Consider now a hypersurface $\Phi = 0$ that divides the region $G$ into two subregions $G_1$ and $G_2$ as shown in figure 9.3. The boundary $\Gamma$ is also subdivided into two parts $\Gamma_1$

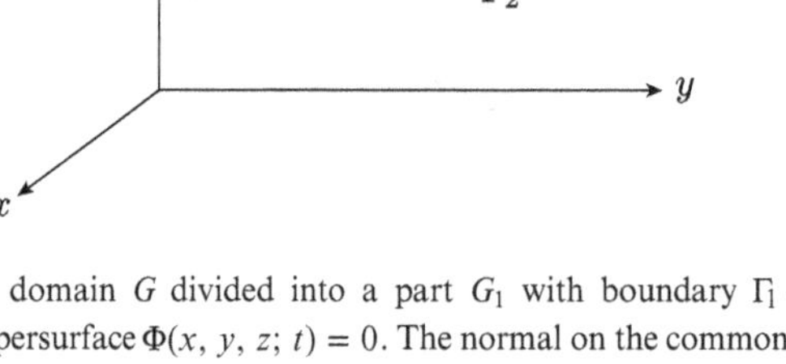

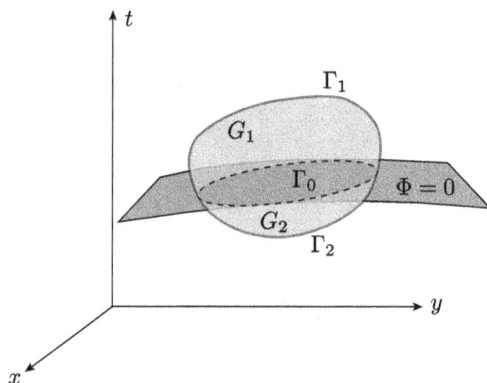

**Figure 9.3.** The spacetime domain $G$ divided into a part $G_1$ with boundary $\Gamma_1 + \Gamma_0$ and a part $G_2$ with boundary $\Gamma_2 + \Gamma_0$ by the hypersurface $\Phi(x, y, z; t) = 0$. The normal on the common boundary $\Gamma_0$ points out of region $G_i$, $i = 1, 2$.

and $\Gamma_2$ such that $\Gamma_1 \cup \Gamma_2 = \Gamma$. The boundary of $G_1$ is the portion $\Gamma_1$ of $\Gamma$ and the portion $\Gamma_0$ of $\Phi = 0$ which lies in $G$. Likewise the boundary of $G_2$ is $\Gamma_2$ and $\Gamma_0$. Note that if the vector $\mathbf{m} = \mathbf{m}_0$ associated with $\Gamma_0$ is taken as $+\lambda_0\nabla\Phi$ for $G_1$, it will be $-\lambda_0\nabla\Phi$ for $G_2$, where $\lambda_0 = [|\nabla\Phi|^2 + \dot{\Phi}^2]^{-1/2}$ and $n_t = \lambda_0\dot{\Phi}$. We assume that the field vectors are discontinuous on $\Gamma_0$. Let $(\mathscr{E}_i, \mathscr{H}_i, \mathscr{D}_i, \mathscr{B}_i, \mathscr{Q}_{ei})$ denote the values of the various field components when approaching $\Gamma_0$ through $G_i$, $i = 1, 2$.

We now apply the integral relations (9.9)–(9.12) to (i) the domain $G$ bounded by $\Gamma_1 \cup \Gamma_2$, both associated with $\mathbf{m} = \lambda\nabla\chi$, $\lambda = [|\nabla\chi|^2 + \dot{\chi}^2]^{-1/2}$, (ii) the domain $G_1$ bounded by $\Gamma_1 \cup \Gamma_0$, and (iii) the domain $G_2$ bounded by $\Gamma_2 \cup \Gamma_0$. Application to equation (9.9) yields

$$\iiint\limits_{\Gamma_1+\Gamma_2} [\mathbf{m} \times \mathscr{H} - n_t\mathscr{D} - n_t\mathscr{Q}_e]\,\mathrm{d}v = \iiiint\limits_{G} \mathscr{J}_c\,\mathrm{d}w, \tag{9.13}$$

$$\begin{aligned}&\iiint\limits_{\Gamma_1} [\mathbf{m} \times \mathscr{H} - n_t\mathscr{D} - n_t\mathscr{Q}_e]\,\mathrm{d}v \\ &\quad + \iiint\limits_{\Gamma_0} [\nabla\Phi \times \mathscr{H}_1 - \dot{\Phi}\mathscr{D}_1 - \dot{\Phi}\mathscr{Q}_{e1}]\lambda_0\,\mathrm{d}v = \iiiint\limits_{G_1} \mathscr{J}_c\,\mathrm{d}w,\end{aligned} \tag{9.14}$$

$$\begin{aligned}&\iiint\limits_{\Gamma_2} [\mathbf{m} \times \mathscr{H} - n_t\mathscr{D} - n_t\mathscr{Q}_e]\,\mathrm{d}v \\ &\quad - \iiint\limits_{\Gamma_0} [\nabla\Phi \times \mathscr{H}_2 - \dot{\Phi}\mathscr{D}_2 - \dot{\Phi}\mathscr{Q}_{e2}]\lambda_0\,\mathrm{d}v = \iiiint\limits_{G_2} \mathscr{J}_c\,\mathrm{d}w.\end{aligned} \tag{9.15}$$

Subtracting equation (9.13) from the sum of equations (9.14) and (9.15) results in

$$\iiint\limits_{\Gamma_0} \left\{\nabla\Phi \times \Delta[\mathscr{H}] - \dot{\Phi}\,\Delta[\mathscr{D}] - \dot{\Phi}\Delta[\mathscr{Q}_e]\right\}\lambda_0\,\mathrm{d}v = 0, \tag{9.16}$$

where $\Delta[\mathscr{H}] = \mathscr{H}_1 - \mathscr{H}_2$ is the discontinuity in $\mathscr{H}$ across the hypersurface $\Phi = 0$ and similarly the other quantities. Since $G$ can be taken arbitrarily small to enclose an arbitrary small section of $\Gamma_0$, the integrand of equation (9.16) must be zero. Thus

$$\nabla\Phi \times \Delta[\mathscr{H}] - \dot{\Phi}\Delta[\mathscr{D}] - \dot{\Phi}\Delta[\mathscr{Q}_e] = 0 \quad \text{on } \Phi = 0. \tag{9.17}$$

Repeating the process for equations (9.10)–(9.12) we obtain the additional discontinuity relations

$$\nabla\Phi \times \Delta[\mathscr{E}] + \dot{\Phi}\Delta[\mathscr{B}] = 0 \quad \text{on } \Phi = 0, \tag{9.18}$$

$$\nabla\Phi \cdot \left\{\Delta[\mathscr{J}_i] + \Delta[\mathscr{J}_c] + \Delta[\dot{\mathscr{D}}]\right\} = 0 \quad \text{on } \Phi = 0, \tag{9.19}$$

$$\nabla\Phi \cdot \Delta[\mathscr{B}] = 0 \quad \text{on } \Phi = 0. \tag{9.20}$$

When these discontinuity equations are applied to a static boundary separating two material media, one obtains the usual boundary conditions on the tangential components of $(\mathscr{E}, \mathscr{H})$ and the normal components of $(\mathscr{D}, \mathscr{B}, \mathscr{J})$. When they are applied to a special hypersurface $\Phi = t = 0$ $(\Longrightarrow \nabla\Phi = 0, \dot{\Phi} = 1)$ assuming that the source function $\mathscr{Q}_e = 0$ for $t < 0$ (which implies that the fields $\mathscr{E}$ and $\mathscr{H}$ are also zero for $t < 0$) and not equal to 0 for $t \geqslant 0$, one obtains from these discontinuity relations the initial conditions

$$\mathscr{E}(x, y, z; 0+) = -\frac{1}{\varepsilon}\mathscr{Q}_e(x, y, z, 0+), \tag{9.21}$$

$$\mathscr{H}(x, y, z; 0+) = 0. \tag{9.22}$$

Thus the electric field responds instantaneously, but the magnetic field continues to be zero after the electric source is switched on. The situation is similar to the voltage response across a capacitor and the current response through an inductor when the elements are excited by a step charge source.

The more interesting case of field discontinuities that exist even in the absence of material discontinuities arises when the fields spread out from the source into the $(x, y, z; t)$-spacetime at finite speed. As before the source is assumed to be located within a finite spatial domain and switched on at $t = 0$. Because of this finite speed the field at $(x, y, z; t)$ will be zero until $t$ is greater than some $t_0$ which is determined by its distance to the nearest source point. Hence there is a region of spacetime in which the field is non-zero and outside this a region in which the field is zero. In an isotropic medium (even in an inhomogeneous medium), the surface $\Phi(x, y, z; t) = 0$ separating the two regions will be a truncated light cone as shown in figure 9.1. The fields are zero outside the light cone and non-zero inside. The stream potential $\mathscr{Q}_e = 0$ for any $t > 0$ on this light cone. The discontinuity relations (9.17) and (9.18) then yield the two homogeneous, linear vector equations for the quantities $\Delta[\mathscr{E}]$ and $\Delta[\mathscr{H}]$

$$\nabla\Phi \times \Delta[\mathscr{H}] - \varepsilon\dot{\Phi}\Delta[\mathscr{E}] = 0, \tag{9.23}$$

$$\nabla\Phi \times \Delta[\mathscr{E}] + \mu\dot{\Phi}\Delta[\mathscr{H}] = 0. \tag{9.24}$$

Taking a scalar product of these equations with $\nabla\Phi$ gives[2]

$$\nabla\Phi \cdot \Delta[\mathscr{E}] = 0 = \nabla\Phi \cdot \Delta[\mathscr{H}], \tag{9.25}$$

implying that $\Delta[\mathscr{E}] \perp \nabla\Phi$ and $\Delta[\mathscr{H}] \perp \nabla\Phi$. In addition taking a scalar product of equation (9.23) with $\Delta[\mathscr{H}]$ gives $\Delta[\mathscr{E}] \cdot \Delta[\mathscr{H}] = 0$, which implies that $\Delta[\mathscr{E}] \perp \Delta[\mathscr{H}]$. Hence the vectors $\Delta[\mathscr{E}]$, $\Delta[\mathscr{H}]$ and $\nabla\Phi$ are mutually perpendicular at any point on the hypersurface $\Phi = 0$. Taking a vector product of equation (9.24) by $\nabla\Phi$ and utilizing equation (9.23) gives

[2] Note that $\dot{\Phi} \neq 0$ since we are dealing with a dynamic hypersurface here.

$$\begin{aligned} &\nabla\Phi \times (\nabla\Phi \times \Delta[\mathscr{E}]) + \mu\varepsilon\dot{\Phi}^2\Delta[\mathscr{E}] = 0 \\ &\Longrightarrow (-|\nabla\Phi|^2 + \mu\varepsilon\dot{\Phi}^2)\Delta[\mathscr{E}] = 0. \end{aligned} \tag{9.26}$$

Since $\Delta[\mathscr{E}] \neq 0$ we arrive at

$$|\nabla\Phi|^2 - \mu\varepsilon\dot{\Phi}^2 = 0 \quad \text{on } \Phi = 0. \tag{9.27}$$

This is a condition of the characteristic for the hyperbolic system of Maxwell's equations [4] and $\Phi = 0$ are known as the *characteristics*. Writing the light cone as $\Phi(x, y, z; t) = \psi(x, y, z) - ct = 0$ and noting that $\nabla\Phi = \nabla\psi$, $\dot{\Phi} = -c$, we finally arrive at the *eikonal differential equation of geometrical optics*

$$|\nabla\psi|^2 - c^2\mu\varepsilon = 0. \tag{9.28}$$

It is interesting to note that the eikonal equation is governed only by the permittivity and permeability of the medium. In particular, it is independent of the conductivity of the medium. Note that the derivation of this eikonal equation is exact and no approximations concerning field behavior have been made. The only assumption that has been made is that the media constants ($\varepsilon_r$, $\mu_r$, $\sigma$) are all frequency independent. Since the fields are zero outside the light cone and non-zero just inside the light cone, the field discontinuities may well be taken to be $\Delta[\mathscr{E}] = \tilde{\mathscr{E}}$ and $\Delta[\mathscr{H}] = \tilde{\mathscr{H}}$, where $\tilde{\mathscr{E}}$ and $\tilde{\mathscr{H}}$ are defined in equation (9.1). In this sense, geometrical optics is concerned with the field vectors on the wavefronts as the wavefront spreads out into the $(x, y, z)$-space as time progresses. These specific values of field are indeed termed as *signals* in the literature [5][3].

### 9.1.2 Ray equations

As remarked previously, the normals erected on the family of wavefronts $\psi = ct$ are the rays[4]. The rays will have a direction proportional to $\nabla\psi$. Let $\mathbf{r}(t) = (x = x(t), y = y(t), z = z(t))$ be the equation of one of the rays passing through a point $P$ on the family of wavefronts, figure 9.4. The velocity $\mathbf{v}$ of the wavefront is defined as the velocity with which the point $P$ moves along the ray as $t$ increases. Let us write

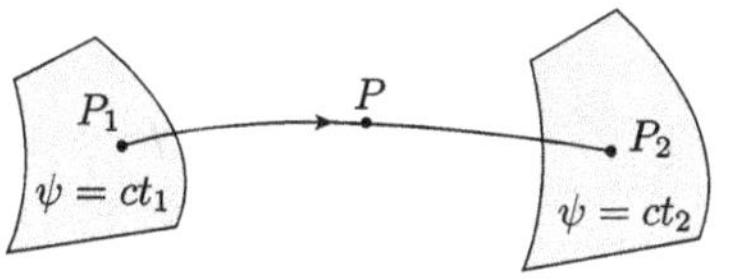


**Figure 9.4.** A ray emanating from a wavefront $\psi = ct_1$ at the point $P_1$, passing through an intermediate point $P$ and terminating on the wavefront $\psi = ct_2$ at the point $P_2$.

[3] By considering a sinusoidal waveform modulated by a step function (a quasi time-periodic signal), Sommerfeld and Brillouin have shown that the signals travel with the velocity $c$ even in a dispersive medium.
[4] Each value of $t$ will generate a surface in ordinary space defined by $\psi = ct$.

$$\mathbf{v} = \frac{\mathrm{d}\mathbf{r}(t)}{\mathrm{d}t} =: v\frac{\nabla\psi}{|\nabla\psi|}, \tag{9.29}$$

where $v$ is the speed to be determined. Along the ray we have

$$\psi(x(t), y(t), z(t)) = ct.$$

Differentiating this w.r.t. time we obtain upon using equations (9.28) and (9.29) that

$$\begin{aligned} c = \frac{\mathrm{d}\psi}{\mathrm{d}t} &= \frac{\partial\psi}{\partial x}\frac{\mathrm{d}x}{\mathrm{d}t} + \frac{\partial\psi}{\partial y}\frac{\mathrm{d}y}{\mathrm{d}t} + \frac{\partial\psi}{\partial z}\frac{\mathrm{d}z}{\mathrm{d}t} \\ &= \nabla\psi\cdot\mathbf{v} = v|\nabla\psi| = vc\sqrt{\mu\varepsilon} \Longrightarrow v = 1/\sqrt{\mu\varepsilon}. \end{aligned} \tag{9.30}$$

Thus $v = 1/\sqrt{\mu\varepsilon}$ is a speed associated with the wavefront $\Phi = 0$ and it depends only on the permittivity and permeability of the medium and is true even in the case of perfect or lossy conductors. The refractive index of the medium is $n = c/v = \sqrt{\mu\varepsilon/\mu_0\varepsilon_0} = \sqrt{\mu_r\varepsilon_r}$. In terms of the refractive index, the eikonal equation may be rewritten as

$$|\nabla\psi|^2 - n^2 = 0, \tag{9.31}$$

which is a non-linear first order partial differential equation. We introduce a unit vector along the ray as $\hat{\mathbf{p}} = \nabla\psi/|\nabla\psi| = \nabla\psi/n$ and define the *wave normal*

$$\mathbf{p} = \nabla\psi = \hat{\mathbf{p}}\,|\nabla\psi| = n\hat{\mathbf{p}}. \tag{9.32}$$

Note that the wave normal (a.k.a. *ray vector*) is not a unit vector.

The directional derivative of the wavefront $\psi = ct$ ($\Longrightarrow \mathrm{d}\psi = c\mathrm{d}t$) with respect to distance $\rho$ in any arbitrary direction with unit vector $\hat{\rho}$ is

$$\frac{\mathrm{d}\psi}{\mathrm{d}\rho} = \frac{\partial\psi}{\partial x}\frac{\mathrm{d}x}{\mathrm{d}\rho} + \frac{\partial\psi}{\partial y}\frac{\mathrm{d}y}{\mathrm{d}\rho} + \frac{\partial\psi}{\partial z}\frac{\mathrm{d}z}{\mathrm{d}\rho} = c\frac{\mathrm{d}t}{\mathrm{d}\rho} = \frac{c}{v}v\frac{\mathrm{d}t}{\mathrm{d}\rho} = n\frac{\mathrm{d}s}{\mathrm{d}\rho}, \tag{9.33}$$

where $\mathrm{d}s/\mathrm{d}t = v = c/n$ is the speed of the wavefront and $\mathrm{d}s$ is the arclength along the orthogonal trajectory between two neighboring wavefronts. In general $\mathrm{d}s \leqslant \mathrm{d}\rho$ as clear geometrically from figure 9.5. The *optical path length* between $P_1$ and $P_2$ along this new curve is defined as

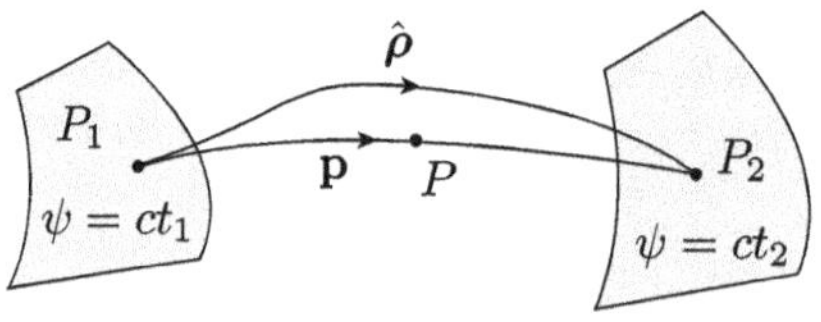


**Figure 9.5.** Any arbitrary path with a unit normal $\rho$ between $P_1$ and $P_2$. The actual ray between $P_1$ and $P_2$ has a wave normal $\mathbf{p}$.

$$\begin{aligned}\psi(P_2) - \psi(P_1) &= \int_{P_1}^{P_2} \frac{d\psi}{d\rho} d\rho = \int_{P_1}^{P_2} n(x, y, z)ds = c\int_{P_1}^{P_2} dt, \\ &= \int_{\rho_1}^{\rho_2} n(x(\rho), y(\rho), z(\rho))\sqrt{\dot{x}^2(\rho) + \dot{y}^2(\rho) + \dot{z}^2(\rho)}\, d\rho,\end{aligned} \tag{9.34}$$

where an overhead dot represents derivative with respect to argument, $ds$ is the elemental arclength along the arbitrary curve with parameter $\rho$ and $nds = cdt$. The geometrical optics energy flux density vector is defined from the Poynting vector on the wavefront as

$$\tilde{\mathscr{S}} = \tilde{\mathscr{E}} \times \tilde{\mathscr{H}}.$$

From the discussion above $\tilde{\mathscr{S}}$ points along $\nabla\psi$ so that flux density vector can be written as $\tilde{\mathscr{S}} = \hat{\mathbf{p}}\tilde{S}$. Now carrying out (9.23) $\times\, \tilde{\mathscr{E}}$ – (9.24) $\times\, \tilde{\mathscr{H}}$ and making use of the following facts that (i) $\tilde{\mathscr{E}}$, $\tilde{\mathscr{H}}$, and $\nabla\psi$ are mutually perpendicular, (ii) the field energy density is $\widetilde{W} = \frac{1}{2}(\tilde{\mathscr{E}} \cdot \tilde{\mathscr{D}} + \tilde{\mathscr{H}} \cdot \tilde{\mathscr{B}})$ from the definition in theorem 5.2.1, (iii) $\nabla\Phi = \nabla\psi$, $\dot{\Phi} = -c$, and (iv) the interchange of the dot and cross in a vector triple product is valid, we obtain

$$\nabla\psi \cdot \tilde{\mathscr{S}} = c\widetilde{W} \Longrightarrow \tilde{\mathscr{S}} = \hat{\mathbf{p}}\frac{c}{|\nabla\psi|}\widetilde{W} = \hat{\mathbf{p}}v\widetilde{W}; \quad \tilde{S} = v\widetilde{W}. \tag{9.35}$$

The geometrical optics energy density vector $\tilde{S}$ carries energy along the rays together with $\tilde{\mathscr{E}}$, $\tilde{\mathscr{H}}$.

### 9.1.3 Ray tracing

It is possible to obtain differential equations satisfied by the rays from the eikonal equation. The evolution parameter along a ray could be time $t$ or arclength $s$ or a particular spatial coordinate or any other quantity $\tau$. Different initial points $\mathbf{r}_0$ on a wavefront will generate different rays. Thus there will be a two-parameter (two initial coordinates contained in $\mathbf{r}_0$) family of rays belonging to the family of wavefronts $\psi = ct$. Let us represent the spatial coordinates of a ray by

$$x = x(\tau), \quad y = y(\tau), \quad z = z(\tau).$$

Since the path of the ray is orthogonal to the wavefront (i.e. along $\nabla\psi$) we have the dynamical equations

$$\frac{dx}{d\tau} = \lambda(\tau)\psi_x, \tag{9.36}$$

$$\frac{dy}{d\tau} = \lambda(\tau)\psi_y, \tag{9.37}$$

$$\frac{dz}{d\tau} = \lambda(\tau)\psi_z, \tag{9.38}$$

where $\psi_x = \partial\psi/\partial x$ and so on and $\lambda(\tau) \neq 0$ is a quantity that depends on the choice of $\tau$ (or vice versa). Indeed from the eikonal equation $\psi_x^2 + \psi_y^2 + \psi_z^2 = n^2$. Hence

$$\left(\frac{\mathrm{d}x}{\mathrm{d}\tau}\right)^2 + \left(\frac{\mathrm{d}y}{\mathrm{d}\tau}\right)^2 + \left(\frac{\mathrm{d}z}{\mathrm{d}\tau}\right)^2 = \left(\frac{\mathrm{d}s}{\mathrm{d}\tau}\right)^2 = \lambda^2 n^2 = c\frac{\mathrm{d}t}{\mathrm{d}\tau}, \tag{9.39}$$

where $\mathrm{d}s = \sqrt{\mathrm{d}x^2 + \mathrm{d}y^2 + \mathrm{d}z^2}$ is the elemental arclength and the last form follows from taking the derivative w.r.t. $\tau$ on the wavefront $\psi(x(\tau),\ y(\tau),\ z(\tau)) = ct$ and making use of the eikonal equation and equations (9.36)–(9.38). The last form also suggests that $\mathrm{d}^2\tau/\mathrm{d}t^2 = 0$ since $n$, $\lambda$ are assumed to be independent of time. In other words, $\tau$ has to be a linear function of the time variable. For instance, if one chooses $\lambda = 1$, then the last form in equation (9.39) implies that the corresponding $\tau$ is $\tau = ct/n^2$. On the other hand, if one chooses $\tau$ to be the arclength parameter $s$, then the lhs of equation (9.39) is unity and the corresponding $\lambda$ is $\lambda(s) = 1/n(s)$ that can vary arbitrarily with position.

Our goal now is to eliminate the wavefront $\psi$ from the ray equations. Dividing both sides of equation (9.36) with $\lambda$ and taking the derivative with respect to $\tau$ gives

$$\begin{aligned}\frac{\mathrm{d}}{\mathrm{d}\tau}\left(\frac{1}{\lambda}\frac{\mathrm{d}x}{\mathrm{d}\tau}\right) &= \psi_{xx}\frac{\mathrm{d}x}{\mathrm{d}\tau} + \psi_{xy}\frac{\mathrm{d}y}{\mathrm{d}\tau} + \psi_{xz}\frac{\mathrm{d}z}{\mathrm{d}\tau}\\ &= \frac{\lambda}{2}\frac{\partial}{\partial x}\left(\psi_x^2 + \psi_y^2 + \psi_z^2\right) = \frac{\lambda}{2}\frac{\partial}{\partial x}n^2,\end{aligned}$$

where equations (9.36)–(9.38) are used in the top line on the rhs to obtain the bottom line. Similar equations are obtained for the $y$- and $z$-coordinates. Therefore the differential equations for the ray coordinates are

$$\frac{2}{\lambda}\frac{\mathrm{d}}{\mathrm{d}\tau}\left(\frac{1}{\lambda}\frac{\mathrm{d}x}{\mathrm{d}\tau}\right) = \frac{\partial}{\partial x}n^2, \tag{9.40}$$

$$\frac{2}{\lambda}\frac{\mathrm{d}}{\mathrm{d}\tau}\left(\frac{1}{\lambda}\frac{\mathrm{d}y}{\mathrm{d}\tau}\right) = \frac{\partial}{\partial y}n^2, \tag{9.41}$$

$$\frac{2}{\lambda}\frac{\mathrm{d}}{\mathrm{d}\tau}\left(\frac{1}{\lambda}\frac{\mathrm{d}z}{\mathrm{d}\tau}\right) = \frac{\partial}{\partial z}n^2. \tag{9.42}$$

These are non-homogeneous, linear, second order ordinary differential equations with non-constant coefficients. The ray is determined by the refractive index of the medium and the initial conditions. Equation (9.39) constrains the three ray velocity components and will eliminate one of the six constants of integration arising from the solution of the ray equations. Furthermore, since the parameter $\tau$ can be chosen rather arbitrarily without changing the geometric trajectory of the ray, an additional constant of integration can be eliminated by an appropriate transformation of the parameter $\tau$.

**Example 9.1.3.1.** Ray equations in cylindrical coordinates.
It is possible to cast the ray equations given in equations (9.40)–(9.42) for Cartesian coordinates in any other coordinate system. In this example we demonstrate this for a cylindrical system with coordinates ($\rho$, $\phi$, $z$). To this end the total and partial derivatives need to be expressed in terms of the new coordinates. The transformation relations facilitating the conversion are $\partial\rho/\partial x = \cos\phi$, $\partial\rho/\partial y = \sin\phi$, $\partial\phi/\partial x = -\sin\phi/\rho$, $\partial\phi/\partial y = \cos\phi/\rho$ and the inverse relations are $\partial x/\partial\rho = \cos\phi$, $\partial x/\partial\phi = -\rho\sin\phi$, $\partial y/\partial\rho = \sin\phi$, $\partial y/\partial\phi = \rho\cos\phi$. Using these it is straightforward to establish that

$$\frac{\mathrm{d}x}{\mathrm{d}\tau} = \cos\phi\frac{\mathrm{d}\rho}{\mathrm{d}\tau} - \rho\sin\phi\frac{\mathrm{d}\phi}{\mathrm{d}\tau}, \tag{9.43}$$

$$\frac{\mathrm{d}y}{\mathrm{d}\tau} = \sin\phi\frac{\mathrm{d}\rho}{\mathrm{d}\tau} + \rho\cos\phi\frac{\mathrm{d}\phi}{\mathrm{d}\tau}, \tag{9.44}$$

$$\frac{\partial}{\partial x} = \cos\phi\frac{\partial}{\partial\rho} - \frac{\sin\phi}{\rho}\frac{\partial}{\partial\phi}, \tag{9.45}$$

$$\frac{\partial}{\partial y} = \sin\phi\frac{\partial}{\partial\rho} + \frac{\cos\phi}{\rho}\frac{\partial}{\partial\phi}. \tag{9.46}$$

Substituting these into equations (9.40) and (9.41) results in

$$\frac{2}{\lambda}\frac{\mathrm{d}}{\mathrm{d}\tau}\left[\frac{1}{\lambda}\left(\cos\phi\frac{\mathrm{d}\rho}{\mathrm{d}\tau} - \rho\sin\phi\frac{\mathrm{d}\phi}{\mathrm{d}\tau}\right)\right] = \cos\phi\frac{\partial n^2}{\partial\rho} - \frac{\sin\phi}{\rho}\frac{\partial n^2}{\partial\phi}, \tag{9.47}$$

$$\frac{2}{\lambda}\frac{\mathrm{d}}{\mathrm{d}\tau}\left[\frac{1}{\lambda}\left(\sin\phi\frac{\mathrm{d}\rho}{\mathrm{d}\tau} + \rho\cos\phi\frac{\mathrm{d}\phi}{\mathrm{d}\tau}\right)\right] = \sin\phi\frac{\partial n^2}{\partial\rho} + \frac{\cos\phi}{\rho}\frac{\partial n^2}{\partial\phi}, \tag{9.48}$$

Taking the sum of equation (9.47) multiplied by $\cos\phi$ and equation (9.48) multiplied by $\sin\phi$ and carrying out some simplification gives

$$\frac{2}{\lambda}\frac{\mathrm{d}}{\mathrm{d}\tau}\left(\frac{1}{\lambda}\frac{\mathrm{d}\rho}{\mathrm{d}\tau}\right) - \frac{2\rho}{\lambda^2}\left(\frac{\mathrm{d}\phi}{\mathrm{d}\tau}\right)^2 = \frac{\partial n^2}{\partial\rho}. \tag{9.49}$$

Likewise (9.47) $\times \sin\phi$ − (9.48) $\times \cos\phi$ and simplifications give

$$\frac{2}{\lambda}\frac{\mathrm{d}}{\mathrm{d}\tau}\left(\frac{\rho}{\lambda}\frac{\mathrm{d}\phi}{\mathrm{d}\tau}\right) + \frac{2}{\lambda^2}\frac{\mathrm{d}\phi}{\mathrm{d}\tau}\frac{\mathrm{d}\rho}{\mathrm{d}\tau} = \frac{1}{\rho}\frac{\partial n^2}{\partial\phi}. \tag{9.50}$$

Note that new ray equations (9.49) and (9.50) in the $\rho$–$\phi$-plane are now coupled, non-linear, second order ordinary differential equations. Simultaneously, equation (9.39) is modified to

$$\left(\frac{\mathrm{d}\rho}{\mathrm{d}\tau}\right)^2 + \left(\rho\frac{\mathrm{d}\phi}{\mathrm{d}\tau}\right)^2 + \left(\frac{\mathrm{d}z}{\mathrm{d}\tau}\right)^2 = \lambda^2 n^2, \tag{9.51}$$

at all points along the ray. ■■

**Example 9.1.3.2** Ray tracing in a standard atmosphere.
Consider the propagation of rays in a standard atmosphere where the refractive index in a certain geographical region changes only along the vertical axis, which is taken here as the $z$-axis. In a standard atmosphere, the refractive index varies linearly with height as [6, p 203], [7, p 112] $n(z) = n_0 + n_1 z$, where $n_0$ and $n_1 \neq 0$ are constants. For example, over the continental USA, a refractive index commonly used after correcting for the Earth's curvature is $n_0 = 1 + 289 \times 10^{-6}$, $n_1 = 118 \times 10^{-6}\ \text{km}^{-1}$.

Let the initial conditions in $\phi$ be $\phi(\tau = 0) = 0 = \mathrm{d}\phi(\tau = 0)/\mathrm{d}\tau$. A valid solution for $\phi$ satisfying equation (9.50) along the ray is $\phi(\tau) = 0$ so that a ray initially in the meridian plane $\phi = 0$ continues to remain in it along its entire trajectory. Inserting this solution into equation (9.49) determines the ratio $\mathrm{d}\rho/\mathrm{d}\tau$ as

$$\frac{1}{\lambda}\frac{\mathrm{d}\rho}{\mathrm{d}\tau} = \frac{1}{c_0}, \tag{9.52}$$

where $c_0 \neq 0$ is some constant. Let us choose $\tau = \rho$ along the ray. Then the function $\lambda$ reduces to the constant $c_0$ from equation (9.52). The $z$ coordinate as a function of $\rho$ along the ray is determined from equation (9.42):

$$\frac{\mathrm{d}^2 z}{\mathrm{d}\rho^2} - (c_0 n_1)^2 z = n_0 n_1 c_0^2. \tag{9.53}$$

The general solution of this second order differential equation is

$$z(\rho) = A\cosh(c_0 n_1 \rho) + B\sinh(c_0 n_1 \rho) - \frac{n_0}{n_1}.$$

If $z(\rho = 0) = z_0$ and $\mathrm{d}z/\mathrm{d}\rho\,|_{\rho=0} = \tan\theta_0$ are specified, then upon substituting into the above equation we obtain $A = z_0 + n_0/n_1$ and $B = \tan\theta_0/c_0 n_1$. Therefore the ray path is given by the closed-form expression

$$z(\rho) = \left(z_0 + \frac{n_0}{n_1}\right)\cosh(c_0 n_1 \rho) + \frac{\tan\theta_0}{c_0 n_1}\sinh(c_0 n_1 \rho) - \frac{n_0}{n_1}. \tag{9.54}$$

The constant $c_0$ can be determined from equation (9.51) which now reads

$$1 + \left(\frac{\mathrm{d}z}{\mathrm{d}\rho}\right)^2 = c_0^2(n_0 + n_1 z)^2. \tag{9.55}$$

Evaluating this at $\rho = 0$ gives $c_0 = \pm\sec\theta_0/(n_0 + n_1 z_0)$. Since $z(\rho)$ is an even function of $c_0$, the sign of $c_0$ is immaterial. The special case of $\theta_0 = 0$ gives $c_0 = (n_0 + n_1 z_0)^{-1}$. Figure 9.6 shows the path of the ray for two different values of $\theta_0$. Note that the scale of the range axis (that is, the $\rho$-axis) is in kilometers, while that of the altitude axis (that is, the $z$-axis) is in meters. For $\theta_0 = -1$ mrad, the ray is initially launched with a negative slope, but it eventually turns back up. Thus in both cases rays have a non-zero curvature and they eventually travel towards regions of higher refractive index values. ■■

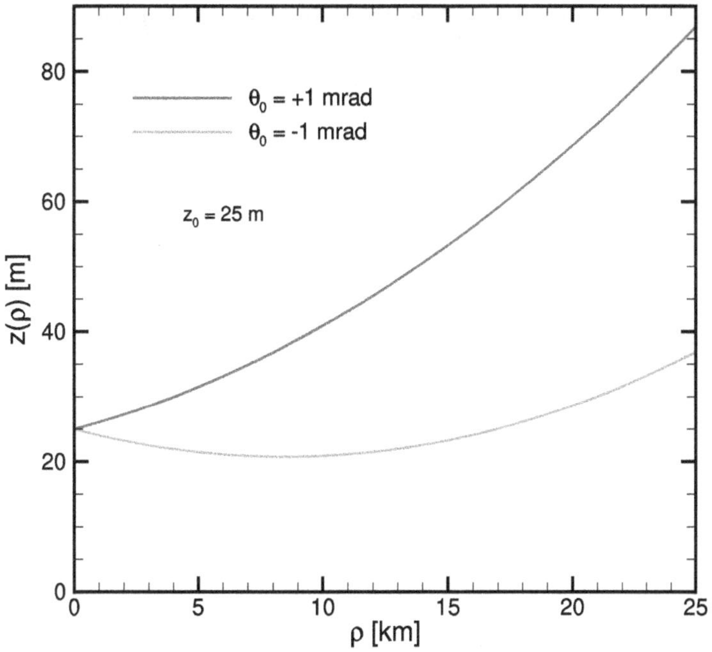


**Figure 9.6.** Ray tracing in a standard atmosphere with a refractive index $(n - 1) \times 10^6 = 289 + 118z$.

Additional examples involving a geometrical optics interpretation of the solution of scattering by smooth convex bodies can be found in the books [8] and [2].

### 9.1.4 Mechanical interpretation of the ray path

Consider the special choice $\tau = ct/n^2$ (units of m), $\lambda = 1$ in the ray equations (9.39)–(9.40). Denoting the derivative w.r.t. $\tau$ with an overhead dot, the ray equations are

$$\ddot{x}(\tau) = \frac{\partial}{\partial x}\left(\frac{n^2}{2}\right), \quad \ddot{y}(\tau) = \frac{\partial}{\partial y}\left(\frac{n^2}{2}\right), \quad \ddot{z}(\tau) = \frac{\partial}{\partial z}\left(\frac{n^2}{2}\right). \tag{9.56}$$

The associated constraining equation (9.39) can be rewritten in the form

$$V = \frac{1}{2}(\dot{x}^2 + \dot{y}^2 + \dot{z}^2) = \frac{1}{2}n^2. \tag{9.57}$$

Equations (9.56) can be thought as the equations of motion of a particle ($m\mathbf{a} = \mathbf{F} = -\nabla U$, where $m$ is the mass, $\mathbf{a}$ is the acceleration, $\mathbf{F}$ is the force, $U$ is the potential energy) of unit mass subject to a potential energy $U = -\frac{1}{2}n^2$ (unitless). Then the kinetic energy of motion, $V$ (unitless), is given by equation (9.57). The total energy of the particle $E = U + V = 0$. Thus electromagnetic rays in a medium of refractive index $n$ can be thought of as the paths taken by particles which move with a total energy $E$ in a potential field $U = -n^2/2$. The total energy of the particle is conserved at $E = 0$ as it moves along the path. Conversely, if a charged particle with a given energy $C$ enters an electrostatic field with potential energy $U$ then its kinetic energy $V = C - U$. We can then define a refractive index $n$ from equation (9.57) as $n = \sqrt{2(C - U)}$ and consider ray paths by equation (9.56) whose trajectories will then mimic the actual paths taken by particles while conserving the total energy at $C$. ■■

### 9.1.5 Optical path length and Fermat's principle

With reference to figure 9.5, Fermat's principle states that [3, p 72], [1, p 128], [9, p 607]:

*The rays between $P_1$ and $P_2$ are those curves along which the optical path length (9.34)*

$$I = \int_{\sigma_1}^{\sigma_2} n(x(\sigma), y(\sigma), z(\sigma))\sqrt{\dot{x}^2(\sigma) + \dot{y}^2(\sigma) + \dot{z}^2(\sigma)}\, \mathrm{d}\sigma$$

*is stationary with respect to infinitesimal variations in the path.*

*Proof.* The integral $I$ may be viewed as a cost function that depends not only on the 'particle' position $(x, y, z)$ but also on its velocity $(\dot{x}, \dot{y}, \dot{z})$. The goal is to find the trajectory about which the cost function remains constant up to a first order in changes. Let us choose three functions $\xi(\sigma)$, $\eta(\sigma)$, $\zeta(\sigma)$ with continuous first order derivatives and which vanish at the end points

$$\xi(\sigma_1) = 0 = \xi(\sigma_2) = \eta(\sigma_1) = \eta(\sigma_2) = \zeta(\sigma_1) = \zeta(\sigma_2). \tag{9.58}$$

Let the extremal values we seek be $x, y, z$ and the variations about the extremal values be $x + \varepsilon_1\xi,\ y + \varepsilon_2\eta,\ z + \varepsilon_3\zeta$, where $\varepsilon_1, \varepsilon_2, \varepsilon_3$ are small parameters. We can then view $I$ as a function of $\varepsilon_i,\ i = 1, 2, 3$ on these varied paths:

$$I(\varepsilon_1, \varepsilon_2, \varepsilon_3) = \int_{\sigma_1}^{\sigma_2} F(x + \varepsilon_1\xi, y + \varepsilon_2\eta, z + \varepsilon_3\zeta; \dot{x} + \varepsilon_1\dot{\xi}, \dot{y} + \varepsilon_2\dot{\eta}, \dot{z} + \varepsilon_3\dot{\zeta})\, \mathrm{d}\sigma, \tag{9.59}$$

where $F(x, y, z; \dot{x}, \dot{y}, \dot{z}) = n(x, y, z)[\dot{x}^2 + \dot{y}^2 + \dot{z}^2]^{1/2}$. Note that all varied paths, including the extremal path, start and terminate at the same point owing to the boundary conditions in equation (9.58). The *first variation* of $I$ w.r.t. $x$ is defined as $(\delta I)_x = \partial I/\partial\varepsilon_1|_{\varepsilon_i=0, i=1,2,3}$ and is

$$\begin{aligned}(\delta I)_x &= \int_{\sigma_1}^{\sigma_2} \left(\xi\frac{\partial F}{\partial x} + \dot{\xi}\frac{\partial F}{\partial \dot{x}}\right) \mathrm{d}\sigma \\ &= \int_{\sigma_1}^{\sigma_2} \left(F_x - \frac{\mathrm{d}F_{\dot{x}}}{\mathrm{d}\sigma}\right)\xi\, \mathrm{d}\sigma,\end{aligned} \tag{9.60}$$

where integration by parts followed by the boundary conditions (9.58) were used in the second term on the rhs in the bottom row and we use the notation that $F_x = \partial F/\partial x$, $F_{\dot{x}} = \partial F/\partial \dot{x}$. Setting the first variation to zero for obtaining the extremum values and noting that $\xi$ is arbitrary it follows that

$$F_x - \frac{\mathrm{d}F_{\dot{x}}}{\mathrm{d}\sigma} = 0 \Longrightarrow \frac{\partial n}{\partial x}\sqrt{\dot{x}^2 + \dot{y}^2 + \dot{z}^2} - \frac{\mathrm{d}}{\mathrm{d}\sigma}\left[\frac{n\dot{x}}{\sqrt{\dot{x}^2 + \dot{y}^2 + \dot{z}^2}}\right] = 0$$

$$\Longrightarrow \frac{\partial n}{\partial x}\frac{\mathrm{d}s}{\mathrm{d}\sigma} = \frac{\mathrm{d}}{\mathrm{d}\sigma}\left[n\frac{\mathrm{d}x}{\mathrm{d}\sigma}\frac{\mathrm{d}\sigma}{\mathrm{d}s}\right]$$

$$\Longrightarrow \frac{\partial n}{\partial x} = \frac{\mathrm{d}}{\mathrm{d}s}\left[n\frac{\mathrm{d}x}{\mathrm{d}s}\right].$$

The left-hand side equation above is known as the *Euler equation* for the variational problem being studied here. We also know from equation (9.39) that $\lambda(s) = [n(s)]^{-1}$. Therefore the last equation above may be also be written as

$$\frac{2}{\lambda(s)}\frac{\mathrm{d}}{\mathrm{d}s}\left(\frac{1}{\lambda(s)}\frac{\mathrm{d}x}{\mathrm{d}s}\right) = \frac{\partial n^2}{\partial x}, \tag{9.61}$$

which is precisely the ray equation (9.40) expressed in terms of the arclength ray coordinate $\tau = s$. The other two ray equations (9.41) and (9.42) follow similarly from setting the first variation with respect to $y$ and $z$ also to zero. ■

### 9.1.6 Sommerfeld–Runge and Snell's laws, the Lagrange integral invariant

On account of the vector identity curl grad ≡ 0, the wave normal $\mathbf{p} = \hat{\mathbf{p}}n$ defined in equation (9.32) satisfies

$$\nabla \times \mathbf{p} = \nabla \times (\hat{\mathbf{p}}n) = 0, \tag{9.62}$$

a fact first discovered by Sommerfeld and Runge [10] and later came to be known as the *Sommerfeld–Runge law*. One can construct a vector field $\hat{\mathbf{p}}(x, y, z)n(x, y, z)$ on knowing the refractive index and ray directions, which are in turn dictated by the initial conditions for launching rays (initial location and 'velocities of rays' as in example 9.1.3.1). The unit vector $\hat{\mathbf{p}}$ also has special properties. In a homogeneous medium $n$ is a constant and equation (9.62) implies $\nabla \times \hat{\mathbf{p}} = 0$. Otherwise, since $\nabla \times n\hat{\mathbf{p}} = n\nabla \times \hat{\mathbf{p}} + \nabla n \times \hat{\mathbf{p}}$, taking a scalar product with $\hat{\mathbf{p}}$ yields

$$\hat{\mathbf{p}} \cdot \nabla \times \hat{\mathbf{p}} = 0. \tag{9.63}$$

Assuming that $n$ is a continuous function of position, one may apply Stokes' theorem to an open surface $S$ bounded by a closed curve $C$ and deduce the *Lagrange's integral invariant* associated with this field $\hat{\mathbf{p}}n$

$$\oint_C n\hat{\mathbf{p}} \cdot \mathrm{d}\ell = \iint_S \nabla \times (\hat{\mathbf{p}}n)\, \mathrm{d}s = 0. \tag{9.64}$$

Lagrange's integral invariant continues to hold good for piece-wise continuous media as the contour $C$ can be deformed in that case to include material boundary as shown in figure 9.7, $C = (C_1 - K) \cup (C_2 + K)$. The contributions to the integral arising from the boundary $K$ will be zero since the vector $(n_2\hat{\mathbf{p}}_2 - n_1\hat{\mathbf{p}}_1) \perp \mathrm{d}\ell$ as is shown in example 9.1.6.1, where $\mathbf{p}_i$ is the wave normal at the interface in medium $i$, $i = 1, 2$.

The field $\mathbf{p} = \hat{\mathbf{p}}n$ thus acts like a conservative force field and implies that the integral

$$J(P_1, P_2) = \int_{P_1}^{P_2} n(x, y, z)\hat{\mathbf{p}}(x, y, z) \cdot \mathrm{d}\ell, \tag{9.65}$$

taken between any two points $P_1$ and $P_2$ is independent of the path of integration (see figure 9.8).

**Example 9.1.6.1.** Snell's law.
In this example we start from the Sommerfeld–Runge law and derive the well known Snell's law in a stratified medium. Indeed we have from the vector identity $\nabla(\mathbf{a} \cdot \mathbf{b}) = \mathbf{a} \times \nabla \times \mathbf{b} + \mathbf{b} \times \nabla \times \mathbf{a} + (\mathbf{a} \cdot \nabla)\mathbf{b} + (\mathbf{b} \cdot \nabla)\mathbf{a}$ and equation (9.62) that

$$\begin{aligned} \hat{\mathbf{p}} \cdot \nabla \times \mathbf{p} = 0 &\Longrightarrow 2(\mathbf{p} \cdot \nabla)\mathbf{p} = \nabla(\mathbf{p} \cdot \mathbf{p}) \\ &= 2n\nabla n \Longrightarrow (\hat{\mathbf{p}} \cdot \nabla)\mathbf{p} = \nabla n. \end{aligned} \tag{9.66}$$

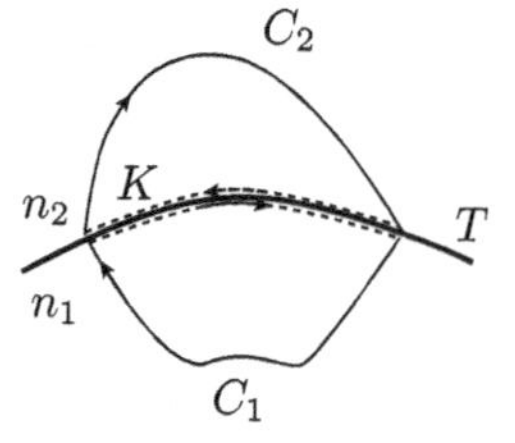


**Figure 9.7.** Lagrange's integral invariant in the presence of piece-wise continuous media.

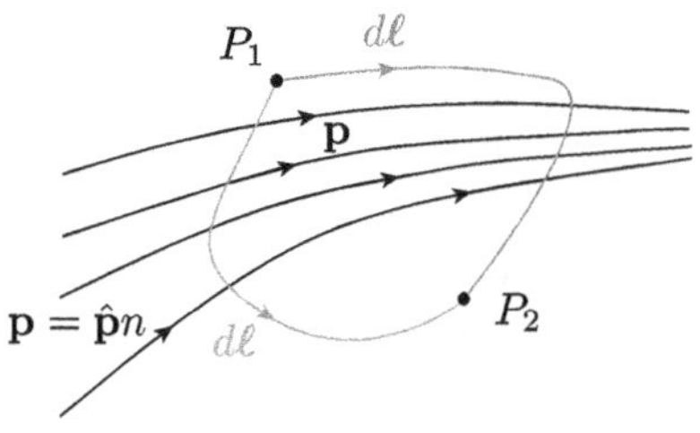


**Figure 9.8.** The integral $J(P_1, P_2)$ between two points is independent of the path taken.

Consider now a horizontally stratified medium with a refractive index changing only along one axis (say $z$). Then $\frac{\partial}{\partial x} = 0 = \frac{\partial}{\partial y}$ and $\nabla n = \hat{\mathbf{z}}\frac{\partial n}{\partial z}$. If the ray makes an angle $\theta(z)$ with the $z$-axis and has an azimuthal angle $\phi$ (independent of $z$), then in a local cylindrical coordinate system erected at any point $\hat{\mathbf{p}} = \hat{\rho}\sin\theta + \hat{\mathbf{z}}\cos\theta$ and $(\hat{\mathbf{p}}\cdot\nabla)\mathbf{p} = \hat{\rho}\cos\theta\frac{\partial}{\partial z}(n(z)\sin\theta(z)) + \hat{\mathbf{z}}\cos\theta\frac{\partial}{\partial z}(n(z)\cos\theta(z))$. Since $\nabla n$ has no radial component we conclude upon comparing both sides of equation (9.66) that

$$\frac{\partial}{\partial z}(n(z)\sin\theta(z)) = 0 \Longrightarrow n(z)\sin\theta(z) = \text{constant}, \tag{9.67}$$

which is a mathematical statement of Snell's law in a horizontally stratified medium. Note that the frequency of the wave does not explicitly enter into this equation. In the case of a boundary between two piece-wise continuous media with refractive indices $n_1$ and $n_2$ (see figure 9.7) and ray angles with respect to normal of $\theta_1$ and $\theta_2$, respectively, Snell's law reduces to the familiar result $n_1\sin\theta_1 = n_2\sin\theta_2$.

If on the two sides of the boundary $K$ between two different homogeneous media (see figure 9.7) $\mathbf{p} = \hat{\mathbf{p}}_1 n_1$ and $\mathbf{p} = \hat{\mathbf{p}}_2 n_2$, then $\nabla\times\mathbf{p} = 0 \Longrightarrow (\hat{\mathbf{p}}_2 n_2 - \hat{\mathbf{p}}_1 n_1)\cdot d\ell = 0$ from an application of Stokes' theorem to a closed rectangular contour consisting of the two dashed lines across $K$ in figure 9.7 connected by vanishingly small vertical segments. Therefore the vector $(\hat{\mathbf{p}}_2 n_2 - \hat{\mathbf{p}}_1 n_1)$ has no tangential component at the interface, or, equivalently, it is perpendicular to the interface.

### 9.1.7 Transport of the geometrical optics field

The propagation of electric and magnetic field vectors and that of the Poynting vector in geometrical optics is most easily carried out in the frequency domain. We will first derive the transport equation for the time-harmonic electric field $\mathbf{E}$. The equation for the magnetic field follows from duality. From Maxwell's equations in a source-free and non-conducting region we have $\nabla\times\nabla\times\mathbf{E} - k^2\mathbf{E} = 0$, $\nabla\cdot(\varepsilon\mathbf{E}) = 0$ $=\varepsilon\nabla\cdot\mathbf{E} + \mathbf{E}\cdot\nabla\varepsilon$, where $k = k_0 n$, $k_0$ is the free-space wavenumber and $n$ is the refractive index. Using the vector identity $\nabla\times\nabla\times\mathbf{E} = \nabla\nabla\cdot\mathbf{E} - \nabla^2\mathbf{E}$ and the above divergence equation we obtain the modified Helmholtz equation for the electric field

$$\nabla^2\mathbf{E} + k^2\mathbf{E} = -\nabla\left(\mathbf{E}\cdot\frac{\nabla\varepsilon}{\varepsilon}\right). \tag{9.68}$$

Let us consider the normalized space variables $\xi = k_0 x$, $\eta = k_0 y$, $\zeta = k_0 z$ and the corresponding gradient operator $\tilde{\nabla}$. In the normalized variables equation (9.68) reads

$$\tilde{\nabla}^2\mathbf{E} + n^2\mathbf{E} = -\tilde{\nabla}\left(\mathbf{E}\cdot\frac{\tilde{\nabla}\varepsilon}{\varepsilon}\right). \tag{9.69}$$

If the permittivity has a characteristic length scale $L$ (that is, the permittivity varies non-trivially over a distance $L$)[5] then $|\tilde{\nabla}\varepsilon|/\varepsilon \sim (k_0 L)^{-1}$. Consequently, for $k_0 L \gg 1$, the rhs of equation (9.69) and hence that of equation (9.68) may be ignored. Therefore for a slowly varying permittivity function the electric field satisfies the homogeneous vector Helmholtz equation

$$\nabla^2 \mathbf{E} + n^2 k_0^2 \mathbf{E} \approx 0, \quad k_0 L \gg 1. \tag{9.70}$$

In a Cartesian coordinate system, each component of the electric field satisfies the scalar Helmholtz equation $\nabla^2 E + n^2 k_0^2 E = 0$. We now express each field component as

$$E(\mathbf{r}) = \mathrm{e}^{-jk_0 \mathscr{S}(\mathbf{r})} E_0(\mathbf{r}), \tag{9.71}$$

where the phase function $\mathscr{S}(\mathbf{r})$ (units of m) and the amplitude function $E_0(\mathbf{r})$ (units of V m$^{-1}$) are assumed to be slowly varying functions of position and independent of $k_0$. Now

$$\nabla^2 E = \nabla \cdot \nabla E = \big[\nabla^2 E_0 - 2jk_0 \nabla\mathscr{S} \cdot \nabla E_0 - jk_0 E_0 \nabla^2 \mathscr{S} \\ - k_0^2 (\nabla\mathscr{S})^2 E_0\big] \mathrm{e}^{-jk_0 \mathscr{S}},$$

where $(\nabla\mathscr{S})^2 = \nabla\mathscr{S} \cdot \nabla\mathscr{S}$. Substituting into the scalar Helmholtz equation we obtain

$$[(\nabla\mathscr{S})^2 - n^2] E_0 + \frac{j}{k_0}\left(2\nabla\mathscr{S} \cdot \nabla E_0 + E_0 \nabla^2 \mathscr{S}\right) - \frac{1}{k_0^2} \nabla^2 E_0 = 0. \tag{9.72}$$

We choose the phase function $\mathscr{S}$ such that the term within the square bracket above vanishes. Then

$$(\nabla\mathscr{S})^2 = n^2, \tag{9.73}$$

which is seen to be the same equation as the eikonal equation satisfied by the function $\psi$ in equation (9.31). Hence the family of surfaces $\mathscr{S}=$ constant constitute the wavefronts. Indeed we set $\mathscr{S} = \psi$ in the ensuing analysis. With this choice of the phase function, the amplitude function now satisfies

$$(2\nabla\psi \cdot \nabla E_0 + E_0 \nabla^2 \psi) = \frac{1}{jk_0} \nabla^2 E_0. \tag{9.74}$$

If the amplitude function has a length scale of $\ell$, then $\nabla^2 E_0 \sim \mathcal{O}(\ell^{-2})$. At high enough frequencies such that $k_0 \ell \gg 1$, the term on the rhs of equation (9.74) may then be ignored. Therefore the amplitude function satisfies the simpler equation

$$\nabla\psi \cdot \nabla \ln(E_0) = -\frac{1}{2} \nabla^2 \psi, \tag{9.75}$$

[5] Note that if $\varepsilon$ varies as $\gamma_0\, \mathrm{e}^{-x/L_x}$ in the $x$-direction, then $\varepsilon^{-1} \partial\varepsilon/\partial\xi = (k_0 L_x)^{-1}$.

where we have utilized the fact that $\nabla \ln(E_0) = E_0^{-1}\nabla E_0$. Since $\nabla\psi = \mathbf{p} = \hat{\mathbf{p}}n$ (see equation (9.32)) the above equation for determining the amplitude function can also be written as

$$\hat{\mathbf{p}} \cdot \nabla \ln(E_0) = -\frac{1}{2n}\nabla \cdot \mathbf{p} = -\frac{1}{2n}\nabla \cdot (\hat{\mathbf{p}}n). \tag{9.76}$$

Similar equations are obtained for the components of a magnetic field vector. Once the gradient of the wavefront is calculated from the medium refractive index, the amplitude $E_0$ along the ray can be determined using equation (9.76). This procedure of obtaining the transport equations for field vectors is equivalent to extracting the leading term after expanding the time-harmonic fields ($\mathbf{E}(\mathbf{r})$, $\mathbf{H}(\mathbf{r})$) in inverse powers of the free-space wavenumber $k_0$ as done by Luneberg [11] and Kline [12].

The wave intensity $I$ is defined as the absolute value of the time average of the Poynting vector and is $I = \frac{1}{2}\Re(\mathbf{E} \times \mathbf{H}^*) \cdot \hat{\mathbf{p}}$. In a source-free and non-conducting region we have from equation (5.17) that

$$\nabla \cdot (I\hat{\mathbf{p}}) = 0. \tag{9.77}$$

Consider a narrow tube formed by rays exiting an elemental area $\mathrm{d}S_1$ of a wavefront $\psi = c_1$ and arriving at an elemental area $\mathrm{d}S_2$ of the wavefront $\psi = c_2$, where $c_1$ and $c_2$ are constants. The rays are all parallel to the sidewalls of the tube as indicated in figure 9.9. Integrating the divergence equation (9.77) over the volume comprised by the tube and applying Gauss' divergence theorem we obtain

$$I_2\mathrm{d}S_2 = I_1\mathrm{d}S_1, \tag{9.78}$$

after noting that $\hat{\mathbf{n}} \cdot \hat{\mathbf{p}} = 0$ on the portion of surface along the tube, $\hat{\mathbf{n}} \cdot \hat{\mathbf{p}} = -1$ on $\mathrm{d}S_1$, $\hat{\mathbf{n}} \cdot \hat{\mathbf{p}} = +1$ on $\mathrm{d}S_2$ and $I_1$ and $I_2$ are the intensities on $\mathrm{d}S_1$ and $\mathrm{d}S_2$, respectively. Hence $I\mathrm{d}S$ remains constant along a tube of rays. The intensity decreases along a ray if the rays diverge between two wavefronts and intensity increases if the rays converge. It is possible to obtain an explicit expression for the intensity of the ray in terms of the wavefront $\psi =$ constant. We know from equation (9.32) that $\hat{\mathbf{p}} = (\nabla\psi)/n$. Therefore from $\nabla \cdot (\hat{\mathbf{p}}I) = 0$ we have

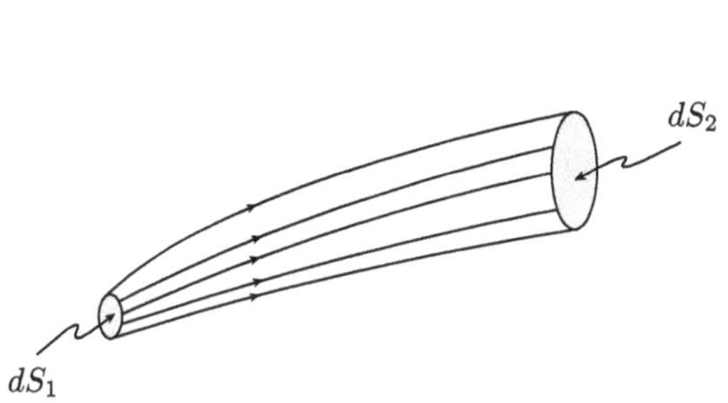


**Figure 9.9.** A tube of rays connecting two elemental areas $\mathrm{d}S_1$ and $\mathrm{d}S_2$ in an isotropic medium.

$$\left(\frac{I}{n}\right)\nabla^2\psi + \nabla\psi \cdot \nabla\left(\frac{I}{n}\right) = 0 \Longrightarrow \frac{1}{n(s)}\nabla\psi \cdot \nabla \ln\left(\frac{I}{n}\right) = -\frac{1}{n(s)}\nabla^2\psi, \tag{9.79}$$

where a factor $1/n(s)$, $s$ = arclength parameter along the ray, was included on both sides in the last equation for convenience. We also know from (9.36)–(9.38) that when the arbitrary parameter $\tau$ along the ray is chosen to be the arclength parameter the quantity $\lambda(\tau)$ there becomes equal to $n^{-1}(s)$. Therefore along the ray with a parametric representation $x = x(s),\ y = y(s),\ z = z(s)$, the operator

$$\frac{1}{n(s)}\nabla\psi \cdot \nabla = \frac{\mathrm{d}x(s)}{\mathrm{d}s}\frac{\partial}{\partial x} + \frac{\mathrm{d}y(s)}{\mathrm{d}s}\frac{\partial}{\partial y} + \frac{\mathrm{d}z(s)}{\mathrm{d}s}\frac{\partial}{\partial z} = \frac{\mathrm{d}}{\mathrm{d}s},$$

in view of equations (9.36)–(9.38). Using this in equation (9.79) leads to

$$\frac{\mathrm{d}}{\mathrm{d}s}\ln\left(\frac{I(s)}{n(s)}\right) = -\frac{\nabla^2\psi}{n(s)}. \tag{9.80}$$

Integrating this between two points gives the intensity law for geometrical optics

$$\frac{I(s_2)}{I(s_1)} = \frac{n(s_2)}{n(s_1)}\mathrm{e}^{-\int_{s_1}^{s_2}\frac{\nabla^2\psi}{n(s)}\mathrm{d}s}. \tag{9.81}$$

The integral inside the exponent of equation (9.81) forms the basis for defining Radon transform (see chapter 14), which has utility in x-ray computed tomography.

**Example 9.1.7.1.** Intensity law in a homogeneous medium.
We would like to show in this example the intensity law for geometrical optics in a homogeneous medium. In a homogeneous medium the refractive index is constant and the rays propagate along straight lines. It is sufficient to consider the case of a vacuum ($n = 1$). Let us assume that the principal radii of curvature of the wavefront at $s = 0$ are given to be $\rho_1(s = 0) = \rho_{10}$ and $\rho_2(s = 0) = \rho_{20}$ and that the intensity at $s = 0$ is unity. For convenience we take $s = z$ and express the wavefront in terms of a local Cartesian system erected around the central ray (in the so-called Dupin coordinate system). We can approximately construct the wavefront as a quadratic function in terms of the radii of curvature as

$$\psi(x, y, z):\ z + \frac{x^2}{2\rho_1(z)} + \frac{y^2}{2\rho_2(z)} = 0,$$

see figure 9.10. If $\rho_1$ and $\rho_2$ are positive the wavefront will be convex in the direction of propagation as shown in the figure. If they are negative, the wavefront will be concave in the direction of propagation. For this wavefront

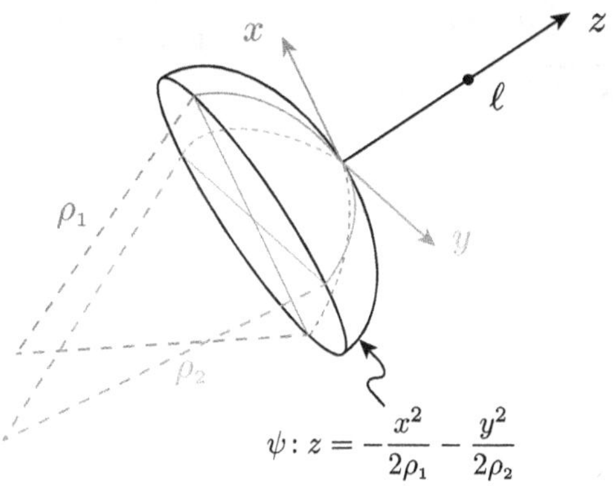


**Figure 9.10.** Wavefront having principal radii of curvature $\rho_1$ and $\rho_2$.

$$\nabla\psi = \hat{\mathbf{x}}\frac{x}{\rho_1(z)} + \hat{\mathbf{y}}\frac{y}{\rho_2(z)} + \hat{\mathbf{z}}\left(1 - \frac{x^2}{2\rho_1^2(z)}\rho_1'(z) - \frac{y^2}{2\rho_2^2(z)}\rho_2'(z)\right).$$

For the eikonal equation $|\nabla\psi|^2 = 1$ to be satisfied we require

$$\frac{x^2}{\rho_1^2}(1 - \rho_1') + \frac{y^2}{\rho_2^2}(1 - \rho_2') + \frac{1}{4}\left[\frac{x^2}{\rho_1^2}\rho_1' + \frac{y^2}{\rho_2^2}\rho_2'\right]^2 = 0.$$

Choosing $\rho_1' = 1 = \rho_2'$ yields $\rho_1(z) = \rho_{10} + z$; $\rho_2(z) = \rho_{20} + z$. However with this choice, the eikonal equation is satisfied exactly only for the central ray ($x = 0,\ y = 0$) and up to fourth order accuracy (that is the error will be proportional to $\mathcal{O}[(x/\rho_1)^4]$, $\mathcal{O}[(y/\rho_2)^4]$, $\mathcal{O}[(xy/\rho_1\rho_2)^2]$ for the other rays. Our main focus here is to derive the intensity law for the central ray and in that respect the above wavefront model is valid. The Laplacian of the wavefront in vacuum is

$$\begin{aligned}\nabla^2\psi &= \frac{1}{\rho_1(z)} + \frac{1}{\rho_2(z)} - \left(\frac{x^2}{\rho_1^3(z)} + \frac{y^2}{\rho_2^3(z)}\right)\\ &= \frac{1}{\rho_1(z)} + \frac{1}{\rho_2(z)} + \mathcal{O}\left(\frac{x^2}{\rho_1^3(z)}, \frac{y^2}{\rho_2^3(z)}\right)\\ &\sim \frac{1}{\rho_1(z)} + \frac{1}{\rho_2(z)},\end{aligned} \tag{9.82}$$

which is simply the mean curvature of the wavefront along the central ray[6]. Using this in equation (9.81) and integrating between $z = 0$ and $z = \ell$ gives

[6] It is shown in [13] that the above relation, i.e. $n^{-1}(s)\nabla^2\psi = \rho_1^{-1}(s) + \rho_2^{-1}(s)$ is valid also for an inhomogeneous isotropic medium. However, the ray trajectory will be non-rectilinear and the radii of curvature in an inhomogeneous medium will no longer be linear functions of arclength.

$$\int_0^\ell \left( \frac{1}{\rho_{10} + z} + \frac{1}{\rho_{20} + z} \right) \mathrm{d}z = \ln \left( \left[ \frac{\rho_{10} + \ell}{\rho_{10}} \right] \left[ \frac{\rho_{20} + \ell}{\rho_{20}} \right] \right)$$

and consequently

$$I(\ell) = \frac{\rho_{10}\rho_{20}}{(\rho_{10} + \ell)(\rho_{20} + \ell)}. \tag{9.83}$$

The square root of this quantity can be taken to be the factor affecting the amplitude of the field components along the central ray. On recognizing that $\psi = z = \ell$ along the ray, the electric field in a vacuum for the central ray under geometrical optics can then be written as

$$\mathbf{E}(\mathbf{r}) \sim \mathbf{E}(\mathbf{r}_o) \sqrt{\frac{\rho_{10}\rho_{20}}{(\rho_{10} + \ell)(\rho_{20} + \ell)}} \, \mathrm{e}^{-jk_0\ell}, \tag{9.84}$$

where $\mathbf{r}_o$ denotes the point on the initial wavefront and $\mathbf{E}(\mathbf{r}_o)$ is the initial electric field at that point. If the initial radii of curvature are positive (diverging rays), the electric field amplitude falls along the propagating ray. On the other hand, if the initial radii of curvature are negative (converging rays), the electric field amplitude increases along the propagating ray. If the radii of curvature are infinite (parallel rays; plane wavefront) the amplitude stays constant. ■■

### 9.1.8 Other theorems of geometrical optics

**Definition 9.1.1** *Congruence.* A system of curves which fill a portion of space in such a way that in general a single curve passes through each point of the region is called a congruence.

**Definition 9.1.2** *Normal congruence.* If there exists a family of surfaces which cut each of the curves of a congruence orthogonally, the congruence is said to be normal.

**Definition 9.1.3** *Rectilinear congruence.* If each curve of a congruence is a straight line, the congruence is said to be rectilinear.

**Theorem 9.1.1** Theorem of Malus and Dupin [1, p 131].
*A normal rectilinear congruence remains normal after any number of refractions or reflections.*

*Proof.* It is sufficient to establish the theorem for a single interface. Consider a normal rectilinear congruence in a homogeneous medium of refractive index $n_1$ and assume that the rays undergo a refraction at an interface $T$ which separates this medium from another homogeneous medium of refractive index $n_2$. Let $S_1$ be a

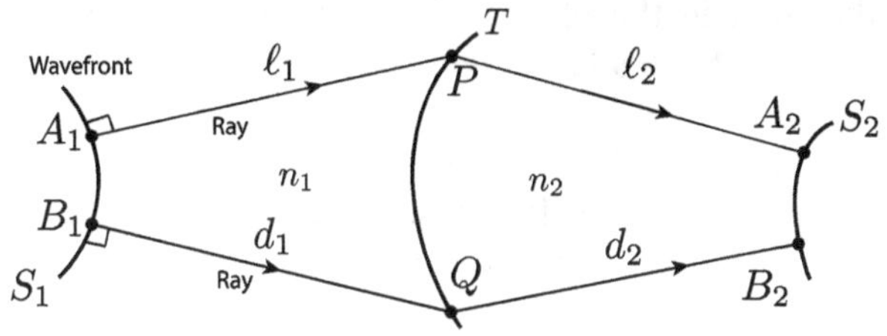


**Figure 9.11.** Geometry relating to the proof of the theorem of Malus and Dupin.

wavefront in medium 1, let $A_1$ and $P$ be points of intersection on surfaces $S_1$ and $T$, and let $A_2$ be any point on the refracted ray. The optical path length between $A_1$ and $A_2$ is designated as $[A_1PA_2]$ and equals $\ell_1 n_1 + \ell_2 n_2$. Consider a second point $B_1Q$ hitting the interface at $Q$ and let $B_2$ be a point such that $[B_1QB_2] = \mathrm{d}_1 n_1 + \mathrm{d}_2 n_2 = [A_1PA_2] = \ell_1 n_1 + \ell_2 n_2$. As the point $B_1$ is varied on $S_1$, the corresponding point $B_2$ traverses a surface $S_2$. It is required to show that the ray $QB_2 \perp S_2$. Applying Lagrange's integral invariant to the closed path $A_1PA_2B_2QB_1A_1$ (see figure 9.11) we obtain

$$[A_1PA_2] + \int_{A_2B_2\in S_2} n_2\hat{\mathbf{p}}_2 \cdot \mathbf{dr} - [B_2QB_1] + \int_{B_1A_1\in S_1} n_1\hat{\mathbf{p}}_1 \cdot \mathbf{dr} = 0.$$

Using $[A_1PA_2] = [B_1QB_2] = [B_2QB_1]$ and $\hat{\mathbf{p}}_1 \perp \mathbf{dr}$ on $B_1A_1 \in S_1$ (since $A_1$ and $B_1$ are rays originating from the wavefront $S_1$) gives

$$\int_{A_2B_2\in S_2} n_2\hat{\mathbf{p}}_2 \cdot \mathbf{dr} = 0,$$

which implies that $\hat{\mathbf{p}}_2 \perp \mathbf{dr}$ on $S_2$ since $A_2$, and consequently, $S_2$ is arbitrary. Therefore the refracted rays are orthogonal to $S_2$; in other words refracted rays form a *normal* congruence, and $S_2$ also constitutes a wavefront. It also follows from the proof presented that the optical path length between any two wavefronts (whether they are in the same medium or not) is the same for all *rays* connecting the two. Concave and convex lenses designed to project objects to image locations are examples of equality of optical path lengths between wavefronts located across various medium inhomogeneities. ■

## 9.2 Gradient metasurfaces and the generalized Snell's law

When a ray is reflected from an ordinary isotropic material half-space, the plane of incidence, the plane of reflection, and the plane of transmission remain the same and the angles of reflection and transmission are related to the angle of incidence via Snell's laws. Lately, researchers have exploited the additional degree of freedom provided by artificially modifying the properties of the surface of material discontinuity and shown that it is possible to have the plane of reflection and/or transmission different from the plane of incidence [14]. Below we discuss the mechanism that is responsible for the twist in the planes of reflection and refraction and derive the equations of generalized Snell's law.

Consider a smooth surface $S$ separating two piece-wise continuous media with refractive indices $n_i$, $i = 1, 2$. Let the surface be structured (by the inclusion of a two-dimensional array of sub-wavelength scatterers) to introduce a *constant* surface phase gradient $\mathbf{k}_s = \nabla\Psi$ rad m$^{-1}$ oriented in an arbitrary direction along the surface. If $\hat{\mathbf{n}}$ is the unit normal on the surface then $\hat{\mathbf{n}} \cdot \mathbf{k}_s = 0$. Consider the optical path length $[P_1AP_2]$ between a point $P_1$ located in medium 1 and a point $P_2$ located in medium 2 via a point $A$ on the surface. The line $P_1A$ has a direction given by the unit vector $\hat{\mathbf{k}}_1(\mathbf{r}_s) = -(\hat{\mathbf{x}}\sin\theta_1\cos\phi_1 + \hat{\mathbf{y}}\sin\theta_1\sin\phi_1 + \hat{\mathbf{z}}\cos\theta_1)$, while the line $AP_2$ has a direction given by the unit vector $\hat{\mathbf{k}}_2(\mathbf{r}_s) = (\hat{\mathbf{x}}\sin\theta_2\cos\phi_2 + \hat{\mathbf{y}}\sin\theta_2\sin\phi_2 + \hat{\mathbf{z}}\cos\theta_2)$. The point $A$ has a position vector $\mathbf{r}_s = \mathbf{s}_A + \hat{\mathbf{n}}r_n$ and the various angles are defined with respect to a *local origin* at $A$. The surface introduces an extra phase shift of $\mathbf{r}_s \cdot \mathbf{k}_s = \mathbf{s}_A \cdot \mathbf{k}_s$ on the propagating ray. Let the wavenumber in free space at the operating frequency be $k_0$. Then

$$k_0[P_1AP_2] = \int_{P_1}^{A} \mathbf{k}_1 \cdot d\mathbf{r} + \mathbf{k}_s \cdot \mathbf{s}_A + \int_{A}^{P_2} \mathbf{k}_2 \cdot d\mathbf{r},$$

where $\mathbf{k}_i = k_0 n_i \hat{\mathbf{k}}_i$, $i = 1, 2$. Our goal here is to determine the relation between $\mathbf{k}_1$, $\mathbf{k}_2$, $\hat{\mathbf{n}}$, and $\mathbf{k}_s$. Applying Fermat's principle, we take the first variation $\delta_A$ of the optical path length with respect to the coordinates of $A$ and set it to zero. In the local coordinate system at $A$ it is straightforward to see that $\delta_A\mathbf{r}_s = \delta_A\mathbf{s}_A + \mathcal{O}(\delta_A^2)$. With reference to figure 9.12 we also note

$$\delta_A \int_{P_1}^{A} \mathbf{k}_1 \cdot d\mathbf{r} = k_0 n_1 \delta_A \ell_1 = k_0 n_1 \hat{\mathbf{k}}_1 \cdot \delta_A \mathbf{s}_A = \mathbf{k}_1 \cdot \delta_A \mathbf{s}_A$$

$$\delta_A \int_{A}^{P_2} \mathbf{k}_2 \cdot d\mathbf{r} = k_0 n_2 \delta_A \ell_2 = k_0 n_2 \hat{\mathbf{k}}_2 \cdot \delta_A \mathbf{s}_A = \mathbf{k}_2 \cdot \delta_A \mathbf{s}_A.$$

Therefore we have from Fermat's principle that

$$(\mathbf{k}_1 + \mathbf{k}_s - \mathbf{k}_2) \cdot \delta_A \mathbf{s}_A = 0.$$

Since $\delta_A\mathbf{s}_A$ is arbitrary and lies on the surface, we have the condition that the quantity within the parenthesis above must be perpendicular to $\mathbf{s}_A$ or, equivalently, must be along the normal $\hat{\mathbf{n}}$ so that

$$\hat{\mathbf{n}} \times (\mathbf{k}_2 - \mathbf{k}_1) = \hat{\mathbf{n}} \times \mathbf{k}_s. \tag{9.85}$$

In a similar fashion it can be established by applying Fermat's principle to the optical path $P_1AQ$ pertaining to the reflection problem that

$$\hat{\mathbf{n}} \times (\mathbf{k}_r - \mathbf{k}_1) = \hat{\mathbf{n}} \times \mathbf{k}_s, \tag{9.86}$$

where $\mathbf{k}_r = k_0 n_1 \hat{\mathbf{k}}_r$ is the wave vector in the direction of the reflected ray and $\hat{\mathbf{k}}_r$ is a unit vector in the direction of the reflected ray.

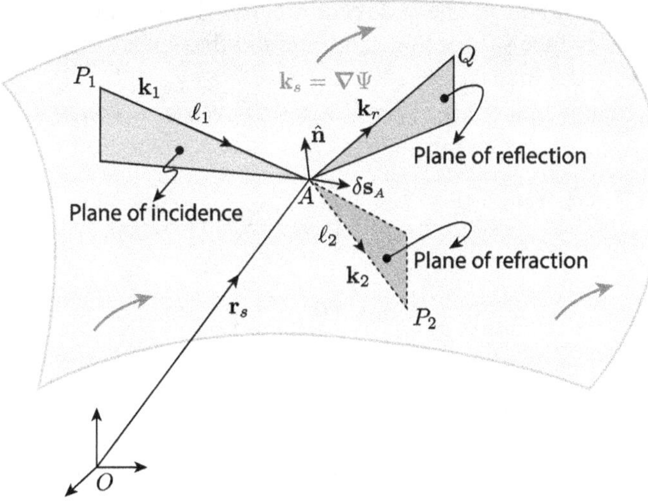


**Figure 9.12.** Reflection and refraction of light from a metasurface with phase discontinuity.

**Example 9.2.0.1.** Reflection and refraction from flat phase-gradient interface. When the interface is flat, we choose it to be the $(z = 0)$-plane on which $\hat{\mathbf{n}} = \hat{\mathbf{z}}$. Let the incident ray be in the $xz$-plane such that $\phi = \phi_1 = 0$, and $\theta = \theta_1$. In that case $\mathbf{k}_1 = -k_0 n_1(\hat{\mathbf{x}} \sin\theta_1 + \hat{\mathbf{z}} \cos\theta_1)$. Let the metasurface have a phase gradient given by $\mathbf{k}_s = k_s(\hat{\mathbf{x}} \cos\phi_s + \hat{\mathbf{y}} \sin\phi_s)$ so that the gradient makes an angle $\phi_s$ with the $xz$-plane and $k_s$ is its magnitude. Let the refracted ray have the wave vector $\mathbf{k}_2 = k_0 n_2(\hat{\mathbf{x}} \sin\theta_2 \cos\phi_2 + \hat{\mathbf{y}} \sin\theta_2 \sin\phi_2 + \hat{\mathbf{z}} \cos\theta_2)$. Then equation (9.85) suggests that

$$k_0 n_2(\hat{\mathbf{y}} \sin\theta_2 \cos\phi_2 - \hat{\mathbf{x}} \sin\theta_2 \sin\phi_2) = (k_0 n_1 \sin\theta_1 + k_s \cos\phi_s)\hat{\mathbf{y}} - k_s \hat{\mathbf{x}} \sin\phi_s$$

$$\Longrightarrow \sin\theta_2 \sin\phi_2 = \frac{k_s}{k_0 n_2} \sin\phi_s$$

$$\text{and } \sin\theta_2 \cos\phi_2 = \frac{n_1}{n_2} \sin\theta_1 + \frac{k_s}{k_0 n_2} \cos\phi_s.$$

Squaring and adding the last two equations gives

$$\sin\theta_2 = \sqrt{\left(\frac{n_1}{n_2}\right)^2 \sin^2\theta_1 + \left(\frac{k_s}{k_0 n_2}\right)^2 + \frac{2 n_1 k_s}{k_0 n_2^2} \sin\theta_1 \cos\phi_s}, \tag{9.87}$$

and dividing the second of the last two by the first gives

$$\cot\phi_2 = \frac{k_0 n_1}{k_s} \frac{\sin\theta_1}{\sin\phi_s} + \cot\phi_s, \tag{9.88}$$

For a normally incident wave $\theta_1 = 0$, but equation (9.87) clearly shows that $\theta_2 \neq 0$ unless the phase gradient $k_s = 0$. In general the angle $\theta_2$ also depends on the frequency of operation. Furthermore, equation (9.88) gives $\phi_2 = \phi_s \neq 0$. In other words, refraction takes place out of the $xz$-plane if $\phi_s$ is different from zero, that is, if a non-zero phase gradient in the $y$-direction exists. In a sense, the phase gradient surface acts like a surface supporting running fluid that twists the plane of incidence upon refraction by the interface.

In a similar manner, denoting $\mathbf{k}_r = k_0 n_1(\hat{\mathbf{x}} \sin\theta_r \cos\phi_r + \hat{\mathbf{y}} \sin\theta_r \sin\phi_r + \hat{\mathbf{z}} \cos\theta_r)$ gives

$$\sin\theta_r = \sqrt{\sin^2\theta_1 + \left(\frac{k_s}{k_0 n_1}\right)^2 + \frac{2k_s}{k_0 n_1}\sin\theta_1 \cos\phi_s}, \tag{9.89}$$

$$\cot\phi_r = \frac{k_0 n_1}{k_s}\frac{\sin\theta_1}{\sin\phi_s} + \cot\phi_s. \tag{9.90}$$

Equation (9.89) shows that the angle of reflection is not equal to the angle of incidence unless the phase gradient $k_s = 0$. Furthermore, reflection takes place out of the plane of incidence since $\phi_r \neq 0$ unless $k_s = 0$. It is also seen once again that in general the twist in the plane of reflection depends on the frequency of operation. ■■

## References


[1] Born M and Wolf E 1980 *Principles of Optics* 6th edn (Elmsford, NY: Pergamon)

[2] Fock V A 1965 *Electromagnetic Diffraction and Propagation Problems* (New York: Pergamon)

[3] Kline M and Kay I W 1979 *Electromagnetic Theory and Geometrical Optics* (Huntington, NY: Kreiger)

[4] Courant R and Hilbert D 1962 *Methods of Mathematical Physics* vol 2 (New York: McGraw-Hill)

[5] Brillouin L 1960 *Wave Propagation and Group Velocity* (New York: Academic)

[6] Craig R A, Katz I, Montgomery R B and Rubenstein P J 1990 Meteorology of the refraction problem *Propagation of Short Radio Waves* (London: Peter Peregrinus)

[7] Craig K H 2003 Clear-air characteristics of the troposphere *Propagation of Radiowaves* (London: Institution of Electrical Engineers)

[8] Felsen L B and Marcuwitz N 1994 *Radiation and Scattering of Waves* (Piscataway, NJ: IEEE)

[9] Jones D S 1979 *Methods in Electromagnetic Wave Propagation* (Oxford: Clarendon)

[10] Sommerfeld A and Runge J 1911 The application of vector calculus to the foundations of geometrical optics *Ann. Phys.* **35** 277–98

[11] Luneberg R K 1964 *Mathematical Theory of Optics* (Providence, RI: Brown University Publishers)

[12] Kline M 1951 An asymptotic solution of Maxwell's equations *Commun. Pure Appl. Math.* **4** 225–63

[13] Kline M 1961 A note on the expansion coefficient of geometrical optics *Commun. Pure Appl. Math.* **14** 473–9

[14] Aieta F *et al* 2012 Reflection and refraction of light from metasurfaces with phase discontinuities *J. Nanophoton.* **6** 063532-1–7

**IOP** Publishing

Engineering Electrodynamics
A collection of theorems, principles and field representations
**Ramakrishna Janaswamy**

# Chapter 10

# Integral field representations

Integral representations are obtained by directly integrating Maxwell's equations in homogeneous regions that are populated by islands of inhomogeneities (containing a scatterer, an antenna, etc). Two such useful representations are the Stratton–Chu integral representation and the Franz integral representation for relating the fields produced inside or outside the island in terms of incident fields and fields on the surface of the island. Such representations are very useful for formulating integral equations in the study of antennas, radar cross-section, inverse scattering, etc, or for obtaining the fields in the near- or far-zone due to surface sources. Several theorems result from such representations including the equivalence principle, the Huygen's principle, the Ewald–Oseen extinction theorem, and the Bojarski's identity to name a few. These are all discussed in the present chapter along with some practical examples detailing the use of integral representations.

## 10.1 Integral representation of fields

Among the several representations that exist, the Stratton–Chu and the Franz integral representations are most commonly used in the electromagnetics community. We derive these integral representations starting from Maxwell's equations in the following and our development for the former parallels that given in [1] and [2].

### 10.1.1 Stratton–Chu representation

Consider sources $(\mathbf{J}, \mathbf{M})$ radiating in a medium of volume $V$ with complex, constitutive parameters $(\varepsilon, \mu)$, figure 10.1. These parameters are assumed to be uniform with respect to spatial location in $V$. It is desirable to find the field $(\mathbf{E}, \mathbf{H})$ at the observation point $\mathbf{r}$ relative to an arbitrary coordinate system within $V$. The volume $V$ is bounded by a small sphere, $S_\delta$, of radius $\delta$ surrounding the observation point, by a surface, $S$, in the near zone, and by a large spherical surface $S_\infty$ at infinity. The unit normal pointing out of the volume $V$ at an arbitrary point on its boundary is denoted by $\hat{\mathbf{n}}_s$ and the corresponding inner normal by $\hat{\mathbf{n}}$. Obviously $\hat{\mathbf{n}}_s = -\hat{\mathbf{n}}$.

doi:10.1088/978-0-7503-1716-0ch10

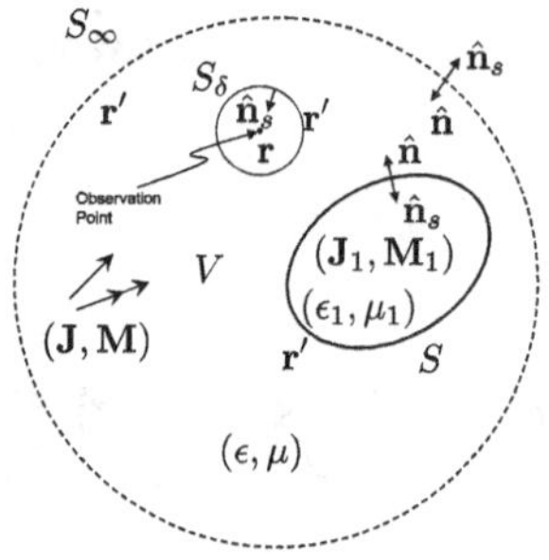


**Figure 10.1.** Integral representation of electromagnetic fields.

The closed surface $S$ may include either a second set of sources ($\mathbf{J}_1$, $\mathbf{M}_1$) or a physical object or both. As such the surface fields that reside on $S$ will depend on what is inside. We apply Green's second identity (B.15) with $\mathbf{P} = \mathbf{E}$ and $\mathbf{Q} = \hat{\mathbf{a}}e^{-jkR}/4\pi R \triangleq \hat{\mathbf{a}}\psi(\mathbf{r}, \mathbf{r}')$, where $R = |\mathbf{r} - \mathbf{r}'|$ and $\hat{\mathbf{a}}$ is an arbitrary constant unit vector. Note that $\mathbf{Q}$ is non-singular in $V$ because the point $R = 0$ is excluded from it through $S_\delta$. The following relations easily follow from standard vector identities:

$$\nabla \times \mathbf{Q} = \nabla \times (\hat{\mathbf{a}}\psi) = \nabla\psi \times \hat{\mathbf{a}}, \tag{10.1}$$

$$\nabla \times \nabla \times \mathbf{Q} = \nabla \times (\nabla\psi \times \hat{\mathbf{a}}) = -\hat{\mathbf{a}}\nabla \cdot \nabla\psi + (\hat{\mathbf{a}} \cdot \nabla)\nabla\psi, \tag{10.2}$$

$$\nabla(\hat{\mathbf{a}} \cdot \nabla\psi) = (\hat{\mathbf{a}} \cdot \nabla)\nabla\psi + \hat{\mathbf{a}} \times \nabla \times \nabla\psi) = (\hat{\mathbf{a}} \cdot \nabla)\nabla\psi, \tag{10.3}$$

$$\nabla \times \nabla \times \mathbf{Q} = \hat{\mathbf{a}}k^2\psi + \nabla(\hat{\mathbf{a}} \cdot \nabla\psi), \tag{10.4}$$

$$\text{also} \quad \nabla^2\psi + k^2\psi = 0, \ \ \mathbf{r} \neq \mathbf{r}', \tag{10.5}$$

where $k = \omega\sqrt{\mu\varepsilon}$ is the wavenumber in $V$ at the radian frequency $\omega$. Using equations (10.1)–(10.5), (1.36), (1.37), and (1.40) in equation (B.15), we arrive at

$$\begin{aligned} &\iiint_V [\psi\hat{\mathbf{a}} \cdot (k^2\mathbf{E} - j\omega\mu\mathbf{J} - \nabla' \times M) - \mathbf{E} \cdot (\hat{\mathbf{a}}k^2\psi + \nabla'(\hat{\mathbf{a}} \cdot \nabla'\psi))]\, dv' \\ &= \iint_{S_\delta + S + S_\infty} [\mathbf{E} \times (\nabla'\psi \times \hat{\mathbf{a}}) - \psi\hat{\mathbf{a}} \times (\nabla' \times \mathbf{E})] \cdot \hat{\mathbf{n}}_s'\, ds'. \end{aligned} \tag{10.6}$$

Using the relation $\mathbf{E} \cdot \nabla'(\hat{\mathbf{a}} \cdot \nabla'\psi) + (\hat{\mathbf{a}} \cdot \nabla'\psi)\nabla' \cdot \mathbf{E} = \nabla' \cdot [(\hat{\mathbf{a}} \cdot \nabla'\psi)\mathbf{E}]$, where $\nabla'$ is the gradient operator with respect to the coordinates $r'$, we can simplify the last term in the volume integral as

[1] $S$ is considered to be a *Lyapunov* surface, see B.1.1

$$
\begin{aligned}
-\iiint_V \mathbf{E} \cdot \nabla'(\hat{\mathbf{a}} \cdot \nabla'\psi)\, \mathrm{d}v' &= \iiint_V (\hat{\mathbf{a}} \cdot \nabla'\psi)\nabla' \cdot \mathbf{E}\, \mathrm{d}v' - \iiint_V \nabla' \cdot [(\hat{\mathbf{a}} \cdot \nabla'\psi)\mathbf{E}]\, \mathrm{d}v' \\
&= \hat{\mathbf{a}} \cdot \left[ \iiint_V \nabla'\psi\nabla' \cdot \mathbf{E}\, \mathrm{d}v' - \iint_{S_\delta+S+S_\infty} \nabla'\psi\mathbf{E} \cdot \hat{\mathbf{n}}_s'\, \mathrm{d}s' \right],
\end{aligned}
\tag{10.7}
$$

where the last equality follows from the application of the Gauss divergence theorem. The first term within the surface integral in equation (10.6) can also be simplified using the relation $\hat{\mathbf{n}}_s' \cdot [\mathbf{E} \times (\nabla'\psi \times \hat{\mathbf{a}})] = (\hat{\mathbf{n}}_s' \times \mathbf{E}) \cdot (\nabla'\psi \times \hat{\mathbf{a}}) = \hat{\mathbf{a}} \cdot [(\hat{\mathbf{n}}_s' \times \mathbf{E}) \times \nabla'\psi]$. Using these in equation (10.6), we arrive at

$$
\begin{aligned}
&\hat{\mathbf{a}} \cdot \iiint_V \{(-j\omega\mu\mathbf{J} - \nabla' \times M)\psi + (\nabla' \cdot \mathbf{E})\nabla'\psi\}\, \mathrm{d}v' \\
&= \hat{\mathbf{a}} \cdot \iint_{S_\delta+S+S_\infty} \left\{ \left[ \left( \hat{\mathbf{n}}_s' \times \mathbf{E} \right) \times +(\hat{\mathbf{n}}_s' \cdot \mathbf{E}) \right] \nabla'\psi + \left( j\omega\mu\mathbf{H} \times \hat{\mathbf{n}}_s' + \mathbf{M} \times \hat{\mathbf{n}}_s' \right)\psi \right\} \mathrm{d}s'.
\end{aligned}
\tag{10.8}
$$

Because $\hat{\mathbf{a}}\cdot$ is arbitrary, it can be safely removed from the above equality. Furthermore, applying the Curl theorem (B.5) to $\nabla' \times (\psi\mathbf{M}) = \psi\nabla' \times \mathbf{M} - \mathbf{M} \times \nabla'\psi$ yields

$$
\iiint_V \psi\nabla' \times \mathbf{M}\, \mathrm{d}v' = \iiint_V \mathbf{M} \times \nabla'\psi\, \mathrm{d}v' + \oiint_S \hat{\mathbf{n}}_s' \times \mathbf{M}\psi\, \mathrm{d}s'.
$$

Gathering all of these results in equation (10.8) and explicitly showing the spatial dependences of the various functions, we finally obtain

$$
\begin{aligned}
&\iiint_V \left\{ -j\omega\mu\mathbf{J}(\mathbf{r}')\psi(\mathbf{r}, \mathbf{r}') + \frac{q_{\mathrm{ev}}(\mathbf{r}')}{\varepsilon}\nabla'\psi(\mathbf{r}, \mathbf{r}') - \mathbf{M}(\mathbf{r}') \times \nabla'\psi(\mathbf{r}, \mathbf{r}') \right\} \mathrm{d}v' \\
&= \iint_{S_\delta+S+S_\infty} \{-j\omega\mu[\hat{\mathbf{n}}_s(\mathbf{r}') \times \mathbf{H}(\mathbf{r}')]\psi(\mathbf{r}, \mathbf{r}') + [\hat{\mathbf{n}}_s(\mathbf{r}') \times \mathbf{E}(\mathbf{r}')] \times \nabla'\psi(\mathbf{r}, \mathbf{r}') \\
&\quad + [\hat{\mathbf{n}}_s(\mathbf{r}') \cdot \mathbf{E}(\mathbf{r}')]\nabla'\psi(\mathbf{r}, \mathbf{r}')\}\, \mathrm{d}s',
\end{aligned}
\tag{10.9}
$$

where the redundant prime on $\hat{\mathbf{n}}_s$ has been dropped since the prime is contained in its argument. Note that Maxwell divergence equations have been used to replace $\nabla \cdot \mathbf{E}$ in terms of $q_{\mathrm{ev}}$. We will now evaluate the integrals on the rhs over the surfaces $S_\delta$, and $S_\infty$. On the large outer sphere $S_\infty$ where $r' \gg r$, the following are true:

$$
\begin{aligned}
&\hat{\mathbf{n}}_s(\mathbf{r}') = \hat{\mathbf{r}}', \quad \hat{\mathbf{R}} - \hat{\mathbf{r}}', \text{ where } \mathbf{R} = \mathbf{r} - \mathbf{r}' \\
&\nabla'\psi = \frac{1 + jkR}{4\pi R^2}\mathrm{e}^{-jkR}\hat{\mathbf{R}} - jk\frac{\mathrm{e}^{-jkR}}{4\pi R}\hat{\mathbf{r}}' = -jk\hat{\mathbf{r}}'\psi.
\end{aligned}
$$

Hence

$$\begin{aligned}-j\omega\mu[\hat{\mathbf{n}}_s \times \mathbf{H}]\psi + [\hat{\mathbf{n}}_s \times \mathbf{E})] \times \nabla'\psi + [\hat{\mathbf{n}}_s \cdot \mathbf{E}]\nabla'\psi &= -jk[\eta\hat{\mathbf{r}}' \times \mathbf{H} + (\hat{\mathbf{r}}' \times \mathbf{E}) \times \hat{\mathbf{r}}' \\ &\quad + (\hat{\mathbf{r}}' \cdot \mathbf{E})\hat{\mathbf{r}}']\psi \\ &= -jk[\eta\hat{\mathbf{r}}' \times \mathbf{H} + \mathbf{E}]\psi \\ &\to 0, \text{ as } r' \to \infty,\end{aligned}$$

where the last limit is true due to the radiation condition (1.69). Hence the contribution from $S_\infty$ vanishes as $r' \to \infty$ on $S_\infty$. Consider the contribution from the small sphere $S_\delta$ of radius $R = \delta$ with $kR \ll 1$ centered at $\mathbf{r}$. On the sphere $\mathrm{d}s' = R^2 \sin\theta' \mathrm{d}\theta' \mathrm{d}\phi'$, $\hat{\mathbf{n}}_s(\mathbf{r}') = \hat{\mathbf{R}}$, $\nabla'\psi \sim \hat{\mathbf{R}}/4\pi R^2$. Therefore using $[\hat{\mathbf{n}}_s \times \mathbf{E})] \times \nabla'\psi + [\hat{\mathbf{n}}_s \cdot \mathbf{E}]\nabla'\psi \sim \mathbf{E}/R^2$

$$\begin{aligned}\lim_{R=\delta\to 0} \oiint_{S_\delta} \left[(\hat{\mathbf{n}}_s \times \mathbf{E}) \times \nabla'\psi + (\hat{\mathbf{n}}_s \cdot \mathbf{E})\nabla'\psi\right] \mathrm{d}s' &= \frac{1}{4\pi} \lim_{R=\delta\to 0} \oiint_{S_\delta} \mathbf{E}(\mathbf{r}')\sin\theta' \,\mathrm{d}\theta'\mathrm{d}\phi' \\ &= \mathbf{E}(\mathbf{r}), \\ \lim_{R=\delta\to 0} \oiint_{S_\delta} j\omega\mu(\hat{\mathbf{n}}_s \times \mathbf{H})\psi \,\mathrm{d}s' &= \frac{1}{4\pi} \lim_{R=\delta\to 0} \oiint_{S_\delta} \hat{\mathbf{R}} \times \mathbf{H}(\mathbf{r}') R \sin\theta' \,\mathrm{d}\theta'\mathrm{d}\phi' \\ &= 0.\end{aligned}$$

If the point $\mathbf{r}$ lies *outside* the volume $V$, such as a point *within* $S$, then the Green's function $\psi$ will no longer be singular and the surface integral on $S_\delta$ will be zero. If the point $\mathbf{r}$ lies *on* the smooth surface $S$,[1] then it is excluded from $V$ by means of hemisphere $H_\delta$ instead of the sphere $S_\delta$. In that case the contribution from $H_\delta$ to the rhs will be $\mathbf{E}(\mathbf{r})/2$. The various cases can be summarized by the introduction of a location dependent parameter $T_V$, where

$$T_V = \begin{cases} 1, & \text{if } \mathbf{r} \in V \\ \dfrac{1}{2}, & \text{if } \mathbf{r} \in S, \\ 0, & \text{otherwise} \end{cases} \tag{10.10}$$

and writing

$$\lim_{R=\delta\to 0} \oiint_{S_\delta} [-j\omega\mu(\hat{\mathbf{n}}_s \times \mathbf{H})\psi + (\hat{\mathbf{n}}_s \times \mathbf{E}) \times \nabla'\psi + (\hat{\mathbf{n}}_s \cdot \mathbf{E})\nabla'\psi] \,\mathrm{d}s' = T_V\mathbf{E}(\mathbf{r}). \tag{10.11}$$

Note that the contribution from the surface integral over $S_\delta$ comes, if any, from the coefficients of the $\nabla'\psi$ term, but not of the $\psi$ term, as the former is clearly more singular in nature than the latter. The only surface integral that remains on the rhs of equation (10.9) is then that over $S$. Gathering all of these results and using the outward normal $\hat{\mathbf{n}} = -\hat{\mathbf{n}}_s'$ on $S$, we arrive at

[1] $S$ is considered to be a *Lyapunov* surface, see B.1.1

$$\left.\begin{aligned}&\iiint_V \left\{-j\omega\mu\mathbf{J}(\mathbf{r}')\psi(\mathbf{r},\mathbf{r}') + \frac{q_{\text{ev}}(\mathbf{r}')}{\varepsilon}\nabla'\psi(\mathbf{r},\mathbf{r}') - \mathbf{M}(\mathbf{r}')\times\nabla'\psi(\mathbf{r},\mathbf{r}')\right\}\mathrm{d}v' \\ &+\oiint_S \{-j\omega\mu[\hat{\mathbf{n}}(\mathbf{r}')\times\mathbf{H}(\mathbf{r}')]\psi(\mathbf{r},\mathbf{r}') + [\hat{\mathbf{n}}(\mathbf{r}')\times\mathbf{E}(\mathbf{r}')]\times\nabla'\psi(\mathbf{r},\mathbf{r}') \\ &+ [\hat{\mathbf{n}}(\mathbf{r}')\cdot\mathbf{E}(\mathbf{r}')]\nabla'\psi(\mathbf{r},\mathbf{r}')\}\,\mathrm{d}s'\end{aligned}\right\} = T_V\mathbf{E}(\mathbf{r}). \quad (10.12)$$

Similarly, by duality we write an expression for the magnetic field

$$\left.\begin{aligned}&\iiint_V \left\{-j\omega\varepsilon\mathbf{M}(\mathbf{r}')\psi(\mathbf{r},\mathbf{r}') + \frac{q_{\text{mv}}(\mathbf{r}')}{\mu}\nabla'\psi(\mathbf{r},\mathbf{r}') + \mathbf{J}(\mathbf{r}')\times\nabla'\psi(\mathbf{r},\mathbf{r}')\right\}\mathrm{d}v' \\ &+\oiint_S \{j\omega\varepsilon[\hat{\mathbf{n}}(\mathbf{r}')\times\mathbf{E}(\mathbf{r}')]\psi(\mathbf{r},\mathbf{r}') + [\hat{\mathbf{n}}(\mathbf{r}')\times\mathbf{H}(\mathbf{r}')]\times\nabla'\psi(\mathbf{r},\mathbf{r}') \\ &+ [\hat{\mathbf{n}}(\mathbf{r}')\cdot\mathbf{H}(\mathbf{r}')]\nabla'\psi(\mathbf{r},\mathbf{r}')\}\,\mathrm{d}s'\end{aligned}\right\} = T_V\mathbf{H}(\mathbf{r}). \quad (10.13)$$

Equations (10.12) and (10.13) are known as *Stratton–Chu* integral representation of the electric and magnetic fields and serve as starting points for the derivation of various integral equations. It is important to note that the surface integrals in equations (10.12) and (10.13) are of *Cauchy's principal value* type which excludes contribution from the singular point $\mathbf{r}' = \mathbf{r}$. Furthermore, the quantities in the integrands can be either the total or scattered fields. The volume integral terms on the lhs of equations (10.12) and (10.13) may be termed as the incident fields $\mathbf{E}^{\text{i}}(\mathbf{r})$ and $\mathbf{H}^{\text{i}}(\mathbf{r})$, respectively, as these are the fields generated in an unbounded medium of parameters $(\varepsilon, \mu)$ by the currents $\mathbf{J}$, $\mathbf{M}$ in the absence of $S$. The contribution from the surface integral part comes from additional sources or scatterers enclosed by $S$. By comparing the surface integral terms to the corresponding volume integral terms, it is clear that $-\hat{\mathbf{n}}\times\mathbf{E}$ is analogous to a magnetic current density, $\hat{\mathbf{n}}\times\mathbf{H}$ is analogous to an electric current density, $\varepsilon\hat{\mathbf{n}}\cdot\mathbf{E}$ is analogous to electric charge density, and $\mu\hat{\mathbf{n}}\cdot\mathbf{H}$ is analogous to the magnetic charge density. It is then appropriate to define the surface currents via $\hat{\mathbf{n}}\times\mathbf{E} = -\mathbf{M}_{\text{s}}$, $\hat{\mathbf{n}}\times\mathbf{H} = \mathbf{J}_{\text{s}}$. Note that the surface currents in the Stratton–Chu formulation are derived from the fields that lie just exterior to $S$ (i.e. limiting values of the tangential fields in $V$ as one approaches $S$). These tangential fields also define the normal field components of fields on the surface according to equations (B.21) and (B.23): $\nabla_{\text{s}}\cdot(\hat{\mathbf{n}}\times\mathbf{E}) = j\omega\mu\hat{\mathbf{n}}\cdot\mathbf{H}$, $\nabla_{\text{s}}\cdot(\hat{\mathbf{n}}\times\mathbf{H}) = -j\omega\varepsilon\hat{\mathbf{n}}\cdot\mathbf{E}$. Using these we rewrite the Stratton–Chu integral representations as

$$\begin{aligned}T_V\mathbf{E}(\mathbf{r}) = &-\iiint_V \left\{j\omega\mu\mathbf{J}(\mathbf{r}')\psi(\mathbf{r},\mathbf{r}') + \left[-\frac{q_{\text{ev}}(\mathbf{r}')}{\varepsilon} + \mathbf{M}(\mathbf{r}')\times\right]\nabla'\psi(\mathbf{r},\mathbf{r}')\right\}\mathrm{d}v' \\ &-\oiint_S \left\{j\omega\mu\mathbf{J}_{\text{s}}(\mathbf{r}')\psi(\mathbf{r},\mathbf{r}') + \left[\frac{\nabla_{\text{s}}'\cdot\mathbf{J}_{\text{s}}(\mathbf{r}')}{j\omega\varepsilon} + \mathbf{M}_{\text{s}}(\mathbf{r}')\times\right]\nabla'\psi(\mathbf{r},\mathbf{r}')\right\}\mathrm{d}s' \qquad (10.14) \\ &\triangleq \mathbf{E}^{\text{i}}(\mathbf{r}) - \oiint_S \{j\omega\mu\mathbf{J}_{\text{s}}(\mathbf{r}')\psi(\mathbf{r},\mathbf{r}') + \left[\frac{\nabla_{\text{s}}'\cdot\mathbf{J}_{\text{s}}(\mathbf{r}')}{j\omega\varepsilon} + \mathbf{M}_{\text{s}}(\mathbf{r}')\times\right]\nabla'\psi(\mathbf{r},\mathbf{r}')\right\}\mathrm{d}s',\end{aligned}$$

$$T_V\mathbf{H}(\mathbf{r}) = -\iiint_V \left\{ j\omega\varepsilon\mathbf{M}(\mathbf{r}')\psi(\mathbf{r}, \mathbf{r}') - \left[\frac{q_{\text{mv}}(\mathbf{r}')}{\mu} + \mathbf{J}(\mathbf{r}')\times\right]\nabla'\psi(\mathbf{r}, \mathbf{r}') \right\} \mathrm{d}v'$$

$$- \oiint_S \left\{ j\omega\varepsilon\mathbf{M}_\mathrm{s}(\mathbf{r}')\psi(\mathbf{r}, \mathbf{r}') - \left[\frac{\nabla_\mathrm{s}' \cdot \mathbf{M}_\mathrm{s}(\mathbf{r}')}{-j\omega\mu} + \mathbf{J}_\mathrm{s}(\mathbf{r}')\times\right]\nabla'\psi(\mathbf{r}, \mathbf{r}') \right\} \mathrm{d}s' \qquad (10.15)$$

$$\triangleq \mathbf{H}^i(\mathbf{r}) - \oiint_S \{j\omega\varepsilon\mathbf{M}_\mathrm{s}(\mathbf{r}')\psi(\mathbf{r}, \mathbf{r}') - \left[\frac{\nabla_\mathrm{s}' \cdot \mathbf{M}_\mathrm{s}(\mathbf{r}')}{-j\omega\mu} + \mathbf{J}_\mathrm{s}(\mathbf{r}')\times\right]\nabla'\psi(\mathbf{r}, \mathbf{r}')\}\mathrm{d}s'.$$

It is seen that the fields at an arbitrary point in space are completely determined if the surface electric and magnetic currents are known on an arbitrary closed surface. The statement that the surface integrals in equations (10.14) and (10.15) produce a null field for points inside $V$ is known as the *vector extinction theorem* (i.e. *vector Ewald–Oseen extinction theorem* or the *vector null-field theorem)* [3].

#### *10.1.1.1 No source but a scatterer present inside* S

In this case the surface integral terms in equations (10.14) and (10.15) produce the fields scattered by the material object present inside $S$. The presence of the obstacle inside $S$ is accounted for in the surface currents $\mathbf{J}_\mathrm{s} = \hat{\mathbf{n}} \times \mathbf{H}$ and $\mathbf{M}_\mathrm{s} = -\hat{\mathbf{n}} \times \mathbf{E}$ that appear on $S$. Note that these surface currents are defined in terms of the *true* fields that exist just exterior to $S$. From the expressions of the surface integrals it is clear that these currents radiate in a homogeneous medium whose electrical parameters are equal to $(\varepsilon, \mu)$ everywhere. In other words, the equivalent surface currents $\mathbf{J}_\mathrm{s}$ and $\mathbf{M}_\mathrm{s}$ radiate in a medium whose electrical parameters are the same as that of the medium *exterior* to $S$.

#### *10.1.1.2 No source and scatterer inside* S, *Huygen's principle*

If there are no sources and scatterers inside $S$ (i.e. if $(\mathbf{J}_1, \mathbf{M}_1) = (\mathbf{0}, \mathbf{0})$ and $(\varepsilon_1, \mu_1) = (\varepsilon, \mu)$ figure 10.2) the surface integrals in equations (10.12) and (10.13) must vanish since the total field everywhere must then be equal to the incident field. The incident fields are due to the sources $(\mathbf{J}, \mathbf{M})$ present in $V$ and radiating in a homogeneous medium with parameters $(\varepsilon_1, \mu_1) = (\varepsilon, \mu)$. For reasons that will become clear shortly we take the media constants to be $(\varepsilon_1, \mu_1)$ both inside and outside of $S$ and recast in terms of the free-space Green's function $\psi_1 = \mathrm{e}^{-jk_1R}/4\pi R$. We then obtain the identities

$$\oiint_S \{-j\omega\mu[\hat{\mathbf{n}}' \times \mathbf{H}^\mathrm{i}]\psi_1 + [\hat{\mathbf{n}}' \times \mathbf{E}^\mathrm{i}] \times \nabla'\psi_1 + [\hat{\mathbf{n}}' \cdot \mathbf{E}^\mathrm{i}]\nabla'\psi_1\}\, \mathrm{d}s' \equiv \begin{cases} 0, & \mathbf{r} \in V \\ -\mathbf{E}^\mathrm{i}(\mathbf{r}), & \mathbf{r} \notin V \end{cases}, \qquad (10.16)$$

and

$$\oiint_S \{j\omega\varepsilon[\hat{\mathbf{n}}' \times \mathbf{E}^\mathrm{i}]\psi_1 + [\hat{\mathbf{n}}' \times \mathbf{H}^\mathrm{i}] \times \nabla'\psi_1 + [\hat{\mathbf{n}}' \cdot \mathbf{H}^\mathrm{i}]\nabla'\psi_1\}\, \mathrm{d}s' \equiv \begin{cases} 0, & \mathbf{r} \in V \\ -\mathbf{H}^\mathrm{i}(\mathbf{r}), & \mathbf{r} \notin V \end{cases}, \qquad (10.17)$$

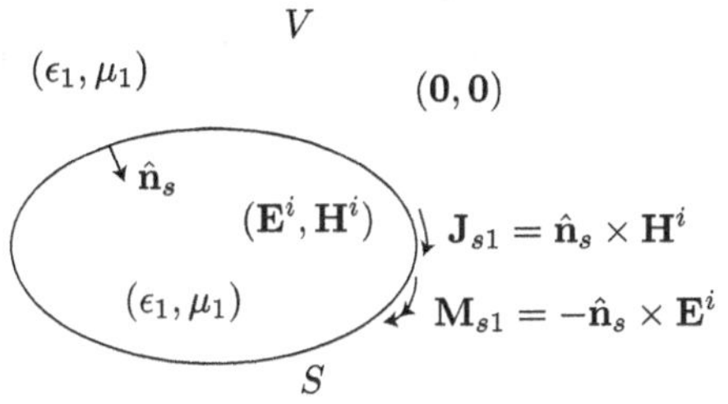


**Figure 10.2.** Equivalence principle for the interior of $S$.

where $\hat{\mathbf{n}}' = \hat{\mathbf{n}}(\mathbf{r}')$. The result for $\mathbf{r} \notin V$ is true because the totality of the fields generated by the sources $(\mathbf{J}, \mathbf{M})$ *and* the equivalent currents $\mathbf{J}_s = \hat{\mathbf{n}} \times \mathbf{H}^i$ and $\mathbf{M}_s = -\hat{\mathbf{n}} \times \mathbf{E}^i$ placed on $S$ due to the incident fields must produce a zero field inside $S$ as dictated by the vanishing of the parameter $T_V$ there.

Equations (10.16) and (10.17) may be interpreted in a different way on making use of $\hat{\mathbf{n}}_s = -\hat{\mathbf{n}}$ and defining the densities $\mathbf{M}_{s1} = -\hat{\mathbf{n}}_s \times \mathbf{E}^i$, $\mathbf{J}_{s1} = \hat{\mathbf{n}}_s \times \mathbf{H}^i$, $\hat{\mathbf{n}}_s \cdot \mathbf{E}^i = q_{es1}/\varepsilon_1$, and $\hat{\mathbf{n}}_s \cdot \mathbf{H}^i = q_{ms1}/\mu_1$ and rewriting them as

$$\mathbf{E}(\mathbf{r}) = \oiint_S \left\{ -j\omega\mu \mathbf{J}_{s1}\psi_1 - \mathbf{M}_{s1} \times \nabla'\psi_1 + \frac{q_{es1}}{\varepsilon}\nabla'\psi_1 \right\} \mathrm{d}s' = \mathbf{E}^i(\mathbf{r}), \quad \mathbf{r} \text{ inside } S, \tag{10.18}$$

$$\mathbf{H}(\mathbf{r}) = \oiint_S \left\{ -j\omega\varepsilon \mathbf{M}_{s1} + \mathbf{J}_{s1} \times \nabla'\psi_1 + \frac{q_{ms1}}{\mu}\nabla'\psi_1 \right\} \mathrm{d}s' = \mathbf{H}^i(\mathbf{r}), \quad \mathbf{r} \text{ inside } S. \tag{10.19}$$

Hence the equivalent densities $\mathbf{J}_{s1}$, $\mathbf{M}_{s1}$, $q_{es1}$, $q_{ms1}$ residing on $S$ and a radiating in a homogeneous medium with parameters $(\varepsilon_1, \mu_1)$ generate the fields $(\mathbf{E}^i, \mathbf{H}^i)$ in a source-free, scatter-free region interior to $S$. These would have been the true fields generated inside $S$ by the volumetric sources $(\mathbf{J}, \mathbf{M})$ contained in $V$ had the medium parameters inside $S$ matched those outside it. Equations (10.18) and (10.19) constitute a mathematical statement of the *equivalence principle* for the interior of $S$. They are sometimes also regarded as a mathematical statement of the vector *Huygens' principle* [3].

### *10.1.1.3 No sources in* V, *equivalence principle: form 1*

If there are no sources within $V$ (i.e. if $(\mathbf{J}, \mathbf{M}) \equiv (\mathbf{0}, \mathbf{0})$) the volume integral terms in equations (10.14) and (10.15) vanish and we obtain

$$T_V\mathbf{E}(\mathbf{r}) = \oiint_S \left\{ -j\omega\mu \mathbf{J}_s(\mathbf{r}')\psi(\mathbf{r}, \mathbf{r}') - \left[ \frac{\nabla_s' \cdot \mathbf{J}_s(\mathbf{r}')}{j\omega\varepsilon} + \mathbf{M}_s(\mathbf{r}')\times \right] \nabla'\psi(\mathbf{r}, \mathbf{r}') \right\} \mathrm{d}s', \tag{10.20}$$

$$T_V\mathbf{H}(\mathbf{r}) = \oiint_S \left\{ -j\omega\varepsilon \mathbf{M}_s(\mathbf{r}')\psi(\mathbf{r}, \mathbf{r}') + \left[ \frac{\nabla_s' \cdot \mathbf{M}_s(\mathbf{r}')}{-j\omega\mu} + \mathbf{J}_s(\mathbf{r}')\times \right] \nabla'\psi(\mathbf{r}, \mathbf{r}') \right\} \mathrm{d}s'. \tag{10.21}$$

The non-zero fields present in $V$ then arise from sources that reside *outside* $V$, i.e. from sources that reside *inside* $S$ or, equivalently, from surface currents $\mathbf{J}_s$ and $\mathbf{M}_s$ placed on $S$ that serve as proxy sources for the true sources inside $S$. If the point $\mathbf{r}$ is brought onto the surface from the volume $V$ of interest, the tangential electric field at a point $\mathbf{r}_0$ will be $\hat{\mathbf{n}} \times \mathbf{E}(\mathbf{r}_0) = -\mathbf{M}_s(\mathbf{r}_0)$ and the tangential magnetic field $\hat{\mathbf{n}} \times \mathbf{H}(\mathbf{r}_0) = \mathbf{J}_s(\mathbf{r}_0)$. So the surface currents are determined by the *true* fields that exist just exterior to $S$. The surface integrals in equations (10.20) and (10.21) also indicate a value of zero for points inside $S$ since $T_V$ is zero there. In other words, the equivalents currents ($\mathbf{J}_s$, $\mathbf{M}_s$) radiating in a homogeneous medium with parameters ($\varepsilon$, $\mu$) per equations (10.20) and (10.21) generate null fields in $S$. These equations may also be taken as the mathematical statements constituting the *equivalence principle*, this time for the exterior of $S$, which relate the fields outside $S$ to the equivalent surface currents $\mathbf{J}_s$ and $\mathbf{M}_s$ that flow on $S$ (figure 10.3). Summarizing the results in equations (10.18)–(10.21) we arrive at the following equivalence principle.

**Theorem 10.1.1** Equivalence principle [4], form 1.
*A distribution of electric and magnetic currents on a given surface S can be found such that outside S it produces the same field as that produced by given sources inside S; and also the field inside S is the same as that produced by given sources outside S. One of these system of sources can be identically zero.*

**Example 10.1.1.1.** Equivalence principle example.
We illustrate the equivalence principle by means of an elementary example[2]. Consider the problem of reflection of oblique plane waves by a flat material interface as shown in figure 10.4. Let the polarization of the plane wave be perpendicular to the plane of incidence. A field

$$
\begin{aligned}
\mathbf{E}^{\mathrm{i}} &= \hat{\mathbf{y}} E_0 e^{-jk_0(x \sin\theta_{\mathrm{i}} + z\cos\theta_{\mathrm{i}})}; \\
\mathbf{H}^{\mathrm{i}} &= \frac{E_0}{\eta_0}(-\hat{\mathbf{x}}\cos\theta_{\mathrm{i}} + \hat{\mathbf{z}}\sin\theta_{\mathrm{i}}) e^{-jk_0(x\sin\theta_{\mathrm{i}} + z\cos\theta_{\mathrm{i}})},
\end{aligned}
\tag{10.22}
$$

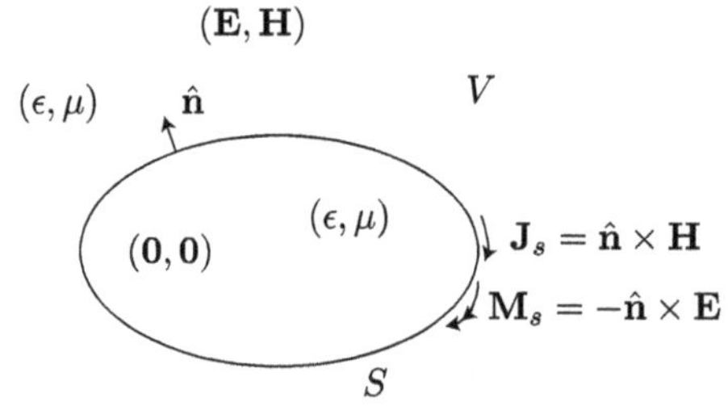


**Figure 10.3.** Equivalence principle for the exterior of $S$.

[2] This example is a generalization of the normally incident plane wave example considered in [5].

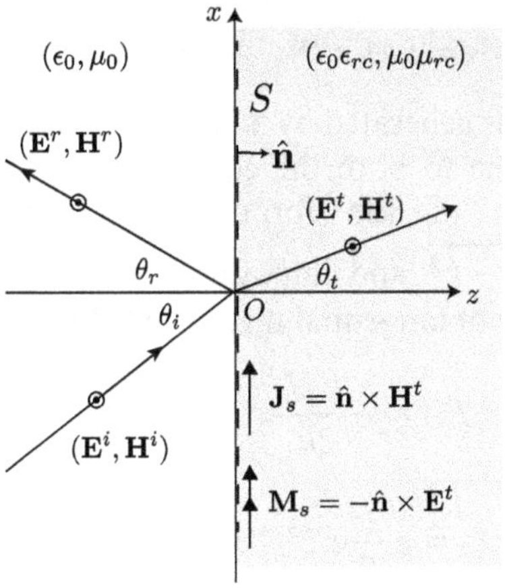


**Figure 10.4.** Plane wave incident on a material half-space.

is incident from free space ($z < 0$) onto the interface $z = 0$ formed by a dielectric medium ($\varepsilon_0\varepsilon_{\rm rc}$, $\mu_0\mu_{\rm rc}$) occupying the region $z > 0$. The medium parameters are its wavenumber $k = k_0 n_{\rm c}$, complex refractive index $n_{\rm c} = \sqrt{\varepsilon_{\rm rc}\mu_{\rm rc}}$ and its intrinsic impedance $\eta = \eta_0\sqrt{\mu_{\rm rc}/\varepsilon_{\rm rc}}$, where $k_0 = \omega\sqrt{\varepsilon_0\mu_0}$ and $\eta_0 = \sqrt{\mu_0/\varepsilon_0}$ are the wavenumber and intrinsic impedance of free space, and $\omega$ is the radian frequency of operation. The exact solution for the total field ($\mathbf{E}^{\rm i} + \mathbf{E}^{\rm r}$, $\mathbf{H}^{\rm i} + \mathbf{H}^{\rm r}$) for $z < 0$ and ($\mathbf{E}^{\rm t}$, $\mathbf{H}^{\rm t}$) for $z > 0$ is worked out in a number of books including [6, p 413] and is

$$\mathbf{E}^{\rm r} = \hat{\mathbf{y}} E_0 \Gamma_\perp {\rm e}^{-jk_0(x\sin\theta_{\rm r} - z\cos\theta_{\rm r})};\ \mathbf{H}^{\rm t} = \frac{\Gamma_\perp E_0}{\eta_0}(\hat{\mathbf{x}}\cos\theta_{\rm r} + \hat{\mathbf{z}}\sin\theta_{\rm r}){\rm e}^{-jk_0(x\sin\theta_{\rm r} - z\cos\theta_{\rm r})}, \quad (10.23)$$

$$\mathbf{E}^{\rm t} = \hat{\mathbf{y}} E_0 T_\perp {\rm e}^{-jk(x\sin\theta_{\rm t} + z\cos\theta_{\rm t})};\ \mathbf{H}^{\rm t} = \frac{T_\perp E_0}{\eta}(-\hat{\mathbf{x}}\cos\theta_{\rm t} + \hat{\mathbf{z}}\sin\theta_{\rm t}){\rm e}^{-jk(x\sin\theta_{\rm t} + z\cos\theta_{\rm t})}, \quad (10.24)$$

where the transmission coefficient $T_\perp$ and the complex transmission angle $\theta_{\rm t}$ are

$$T_\perp = \frac{2\eta\cos\theta_{\rm i}}{\eta\cos\theta_{\rm i} + \eta_0\cos\theta_{\rm t}}, \qquad n_{\rm c}\sin\theta_{\rm t} = \sin\theta_{\rm i} = \sin\theta_{\rm r}, \quad (10.25)$$

and the reflection coefficient $\Gamma_\perp = T_\perp - 1$.

Let us now explore the fields predicted by the equivalence principle in the regions $z \gtrless 0$. We take the infinite plane $z = 0+$ for the closed surface $S$ on which we erect the equivalent surface current densities $\mathbf{J}_{\rm s} = \hat{\mathbf{z}} \times \mathbf{H}^{\rm t}\,|_{z=0}$, $\mathbf{M}_{\rm s} = -\hat{\mathbf{z}} \times \mathbf{E}^{\rm t}\,|_{z=0}$. These surface currents radiate in a homogeneous medium with parameters ($\varepsilon_0\varepsilon_{\rm rc}$, $\mu_0\mu_{\rm rc}$) corresponding to the region $z > 0$. Using the total fields in equation (10.24) the equivalent current densities are

$$\mathbf{J}_{\rm s} = \hat{\mathbf{y}} J_0 {\rm e}^{-j\alpha x};\ J_0 = -\frac{M_0}{\eta}\cos\theta_{\rm t},\ \alpha = k\sin\theta_{\rm t} \quad (10.26)$$

$$\mathbf{M}_s = \hat{\mathbf{x}} M_0 e^{-j\alpha x}; \; M_0 = T_\perp E_0. \tag{10.27}$$

We now evaluate the fields generated by $\mathbf{J}_s$ and $\mathbf{M}_s$ separately and use superposition to determine the fields due to both the equivalent currents. Letting the magnetic vector potential to be of the form $A_y = A_0^\pm e^{-jk_x x} e^{-jk_z|z|}$ for $z \gtrless 0$ with $k_z = \sqrt{k^2 - k_x^2} = -j\sqrt{k_x^2 - k^2}$ and imposing the continuity of tangential electric field and the discontinuity of tangential magnetic field at $z = 0$ we obtain $k_x = \alpha$ and

$$E_{y1} = -\frac{k\eta J_0}{2k_z} e^{-j(\alpha x + k_z|z|)} \tag{10.28}$$

$$H_{x1} = \pm\frac{J_0}{2} e^{-j(\alpha x + k_z|z|)}, \; z \gtrless 0 \tag{10.29}$$

$$H_{z1} = -\frac{\alpha J_0}{2k_z} e^{-j(\alpha x + k_z|z|)}, \tag{10.30}$$

for the fields produced by $\mathbf{J}_s$ alone. These fields are $TM_y$. Likewise letting the electric vector potential be of the form $F_x = F_0^\pm e^{-jk_x x} e^{-jk_z|z|}$ for $z \gtrless 0$ with $k_z = \sqrt{k^2 - k_x^2} = -j\sqrt{k_x^2 - k^2}$ and imposing the discontinuity of the tangential electric field and the continuity of the tangential magnetic field at $z = 0$, we obtain $k_x = \alpha$ and

$$E_{y2} = \pm\frac{M_0}{2} e^{-j(\alpha x + k_z|z|)}, \; z \gtrless 0 \tag{10.31}$$

$$H_{x2} = -\frac{k_z M_0}{2k} e^{-j(\alpha x + k_z|z|)} \tag{10.32}$$

$$H_{z2} = \pm\frac{M_0 \alpha}{2k\eta} e^{-j(\alpha x + k_z|z|)}, \; z \gtrless 0, \tag{10.33}$$

for the fields produced by $\mathbf{M}_s$ alone. These fields are $TE_x$. Adding these two sets of fields and substituting for $M_0$, $J_0$, and $\alpha$ we obtain

$$E_y = E_{y1} + E_{y2} = \begin{cases} T_\perp E_0 e^{-jk(x \sin\theta_t + z\cos\theta_t)}, & z > 0 \\ 0, & z < 0 \end{cases} \tag{10.34}$$

$$H_x = H_{x1} + H_{x2} = \begin{cases} -\dfrac{T_\perp E_0}{\eta} \cos\theta_t e^{-jk(x \sin\theta_t + z\cos\theta_t)}, & z > 0 \\ 0, & z < 0 \end{cases} \tag{10.35}$$

$$H_z = H_{z1} + H_{z2} = \begin{cases} \dfrac{T_\perp E_0}{\eta} \sin\theta_t e^{-jk(x \sin\theta_t + z\cos\theta_t)}, & z > 0 \\ 0, & z < 0, \end{cases} \tag{10.36}$$

on recognizing that $k_z = \sqrt{k^2 - k^2 \sin^2\theta_t} = k \cos\theta_t$. Thus equivalent surface currents placed on the surface $z = 0$ and radiating in a homogeneous medium having electrical parameters the same as those of the half-space $z > 0$ generate the true fields in $z > 0$ and zero fields in $z < 0$. ■■

### 10.1.2 Franz representation

An alternative representation known as the Franz representation [7] which offers some computational advantages relative to the Stratton–Chu representation is now discussed.

With reference to figure 10.1 consider the scalar-vector Green's theorem B.12 with $\psi$ being the scalar Green's function satisfying $\nabla'^2\psi + k^2\psi = -\delta(\mathbf{r} - \mathbf{r}')$ and $\mathbf{A} = \mathbf{E}$, the electric field, satisfying $\nabla' \times \nabla' \times \mathbf{E} - k^2\mathbf{E} = -j\omega\mu\mathbf{J}(\mathbf{r}')$, where the prime on the del operator indicates that it operates on the primed coordinates. We assume that only impressed electric current exists. The presence of impressed magnetic current can be handled in a similar manner. Substituting these into equation (B.12) and using the two Maxwell's equations $\nabla' \cdot \mathbf{E}(\mathbf{r}') = \rho_v(\mathbf{r}')/\varepsilon$ and $\nabla' \times \mathbf{E}(\mathbf{r}') = -j\omega\mu\mathbf{H}(\mathbf{r}')$ we obtain

$$\iiint_V \left[\frac{\rho_v}{\epsilon}\nabla'\psi - j\omega\mu\mathbf{J}\psi\right] \mathrm{d}v' - \mathbf{E}(\mathbf{r}) = \oiint_S \left[(\hat{\mathbf{n}}_s' \cdot \mathbf{E})\nabla'\psi + (\hat{\mathbf{n}}_s' \times \mathbf{E}) \times \nabla'\psi \right. \tag{10.37}$$

$$\left. -j\omega\mu(\hat{\mathbf{n}}_s' \times \mathbf{H})\psi\right] \mathrm{d}s'. \tag{10.38}$$

Employing the identity (10.1), the fact that $\nabla'\psi = -\nabla\psi$ and $\hat{\mathbf{n}}_s' = -\hat{\mathbf{n}}'$ into the above equation and extracting the $\nabla$ operator outside the integrals yields

$$\begin{aligned} \mathbf{E}(\mathbf{r}) &= -\frac{1}{\varepsilon}\nabla \iiint_V \rho_v\psi \,\mathrm{d}v' - j\omega\mu \iiint_V \mathbf{J}\psi \,\mathrm{d}v' \\ &\quad - \nabla \oiint_S (\hat{\mathbf{n}}' \cdot \mathbf{E})\psi \,\mathrm{d}s' + \nabla \times \oiint_S (\hat{\mathbf{n}}' \times \mathbf{E})\psi \,\mathrm{d}s' - j\omega\mu \oiint_S (\hat{\mathbf{n}}' \times \mathbf{H})\psi \,\mathrm{d}s' \\ &=: \mathbf{E}^{\mathrm{i}}(\mathbf{r}) - \frac{1}{\varepsilon}\nabla \oiint_S q_{\mathrm{es}}\psi \,\mathrm{d}s' - \nabla \times \oiint_S \mathbf{M}_{\mathrm{s}}\psi \,\mathrm{d}s' - j\omega\mu \oiint_S \mathbf{J}_{\mathrm{s}}\psi \,\mathrm{d}s', \end{aligned} \tag{10.39}$$

where the volume integrals generate the incident field $\mathbf{E}^{\mathrm{i}}$, $q_{\mathrm{es}}(\mathbf{r}') = \varepsilon(\hat{\mathbf{n}}' \cdot \mathbf{E}(\mathbf{r}'))$, $\mathbf{M}_{\mathrm{s}}(\mathbf{r}') = -\hat{\mathbf{n}}' \times \mathbf{E}(\mathbf{r}')$, and $\mathbf{J}_{\mathrm{s}}(\mathbf{r}') = \hat{\mathbf{n}}' \times \mathbf{H}(\mathbf{r}')$ are the surface electric charge density, the surface magnetic current density, and the surface electric current density, respectively. Similarly, upon using $\mathbf{A} = \mathbf{H}$, $\nabla' \cdot \mathbf{H}(\mathbf{r}') = 0$, and $\nabla' \times \nabla' \times \mathbf{H} - k^2\mathbf{H} = \nabla' \times \mathbf{J}$ in the scalar-vector Green's theorem and carrying out simplifications as above we obtain

$$\mathbf{H}(\mathbf{r}) = \iiint_V (\nabla' \times \mathbf{J})\psi \, \mathrm{d}v' - \nabla \oiint_S (\hat{\mathbf{n}}' \cdot \mathbf{H})\psi \, \mathrm{d}s' + \nabla \times \oiint_S (\hat{\mathbf{n}}' \times \mathbf{H})\psi \, \mathrm{d}s' + j\omega\varepsilon \oiint_S (\hat{\mathbf{n}}' \times \mathbf{E})\psi \, \mathrm{d}s' \tag{10.40}$$
$$=: \mathbf{H}^{\mathrm{i}}(\mathbf{r}) - \frac{1}{\mu}\nabla \oiint_S q_{\mathrm{ms}}\psi \, \mathrm{d}s' + \nabla \times \oiint_S \mathbf{J}_{\mathrm{s}}\psi \, \mathrm{d}s' - j\omega\varepsilon \oiint_S \mathbf{M}_{\mathrm{s}}\psi \, \mathrm{d}s',$$

where $q_{\mathrm{ms}}(\mathbf{r}') = \mu(\hat{\mathbf{n}}' \cdot \mathbf{H}(\mathbf{r}'))$ is the magnetic surface charge induced on $S$ and the volume integral generates the incident magnetic field $\mathbf{H}^{\mathrm{i}}$. We would like to express the fields explicitly in terms of $\mathbf{J}_{\mathrm{s}}$ and $\mathbf{M}_{\mathrm{s}}$ only. To this end we take the curl of equations (10.39) and (10.40) and use Maxwell's equations to finally yield

$$\mathbf{H}(\mathbf{r}) = \mathbf{H}^{\mathrm{i}}(\mathbf{r}) + \frac{1}{j\omega\mu}\nabla \times \nabla \times \oiint_S \mathbf{M}_{\mathrm{s}}\psi \, \mathrm{d}s' + \nabla \times \oiint_S \mathbf{J}_{\mathrm{s}}\psi \, \mathrm{d}s', \tag{10.41}$$

$$\mathbf{E}(\mathbf{r}) = \mathbf{E}^{\mathrm{i}}(\mathbf{r}) + \frac{1}{j\omega\varepsilon}\nabla \times \nabla \times \oiint_S \mathbf{J}_{\mathrm{s}}\psi \, \mathrm{d}s' - \nabla \times \oiint_S \mathbf{M}_{\mathrm{s}}\psi \, \mathrm{d}s', \tag{10.42}$$

upon noting that $\nabla \times \nabla\xi \equiv 0$. Equations (10.42) and (10.41) together constitute the *Franz representation.* The advantage of this representation compared to the Stratton–Chu representation is that the fields are expressed entirely in terms of surface electric and magnetic currents and not their derivatives. The Franz formulation is more convenient in situations where the current density exhibits discontinuities arising from the object geometry (edges, corners, etc) or from application to approximate methodologies (such as the physical optics approximation).

### 10.1.3 Surface functions and their fields, equivalence principle: forms 2, 3

We now ask the interesting question as to how the fields represented by the surface integral terms in equations (10.20) and (10.21) behave if the surface currents $\mathbf{J}_{\mathrm{s}}$ and $\mathbf{M}_{\mathrm{s}}$ are replaced by *arbitrary* vector surface functions $\mathfrak{J}_{\mathrm{s}}$ and $\mathfrak{M}_{\mathrm{s}}$, respectively, and the surface charges $q_{\mathrm{es}}$ and $q_{\mathrm{ms}}$ by arbitrary surface functions $q_{\mathrm{e}}$ and $q_{\mathrm{m}}$, respectively. From the analysis of this subsection we will also present an alternative proof of the most general form of the surface equivalence principle. We are particularly interested in points close to the surface $S$. For a point close to the surface, let us write $\mathbf{r} = \mathbf{r}_0 + \hat{\mathbf{n}}_0 u_3$, where $\mathbf{r}_0$ is a point on the surface $S$ and $u_3$ is the distance along the normal $\hat{\mathbf{n}}_0$ at $\mathbf{r}_0$, figure 10.5. The surface $S$ is described by $u_3 = 0$. Let us denote by

$$\oiint_{S_-} \quad \text{and} \quad \oiint_{S_+} \tag{10.43}$$

the limits of surface integrals which we obtain if $\mathbf{r}$ approaches the point $\mathbf{r}_0$ of the surface $S$ from its interior or exterior respectively, figure 10.6.

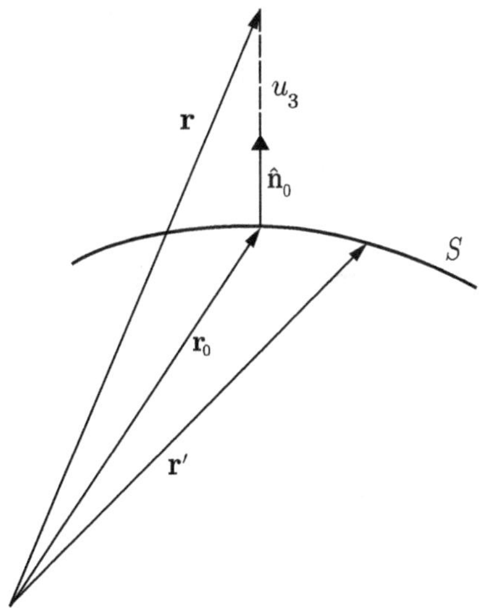


**Figure 10.5.** An exterior point $\mathbf{r}$ tending to $\mathbf{r}_0$ on a surface $S$.

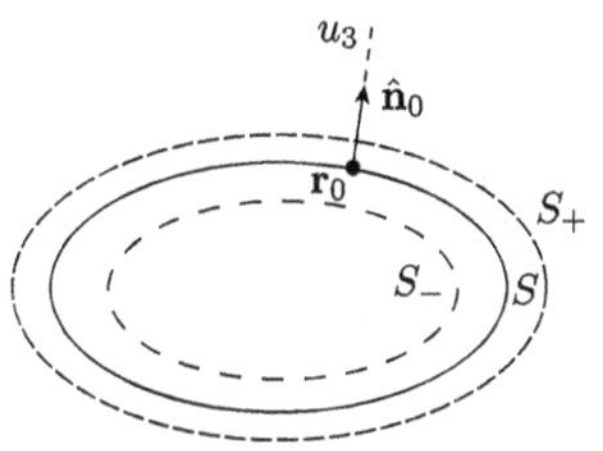


**Figure 10.6.** Point $\mathbf{r}$ tending to $\mathbf{r}_0$ on a surface $S$ from the exterior and interior.

Hence $u_3$ is positive for $S_+$ and negative for $S_-$. Stated differently, the 'plus' side refers to the side of the boundary into which the normal at $\mathbf{r}_0$ points and the 'minus' side refers to the other side. Consider a circular disk $D_\delta$ of radius $\delta$ with a center at $\mathbf{r}_0$ on the tangent plane drawn at $\mathbf{r}_0$, figure 10.7. Because the surface $S$ is considered smooth, the radius $\delta$ can be chosen such that a disk is a very good approximation to the surface at $\mathbf{r}_0$. For convenience, we designate as $(\rho, \phi)$ the local polar coordinates on the disk. From the previous section, $\nabla'\psi = \mathbf{R}(1 + jkR)\exp(-jkR)/4\pi R^3$, where $\mathbf{R} = \mathbf{r} - \mathbf{r}' = \hat{\mathbf{n}}_0 u_3 - \hat{\rho}\rho$, $R = |\mathbf{R}| = \sqrt{u_3^2 + \rho^2}$. It is also clear that $\mathrm{d}s' = \rho \mathrm{d}\rho \mathrm{d}\phi$, $\hat{\rho} = \hat{\mathbf{x}}\cos\phi + \hat{\mathbf{y}}\sin\phi$. For $R$ small, $\nabla'\psi \approx \mathbf{R}/4\pi R^3$. For $\oiint_{S_+}(\cdot)$, we need to evaluate the surface integral on $S$ for a fixed and small $\delta$ as the distance $u_3 \to 0^+$. A second limit is then taken as $\delta \to 0$. Therefore for any continuous function $f(\mathbf{r}')$

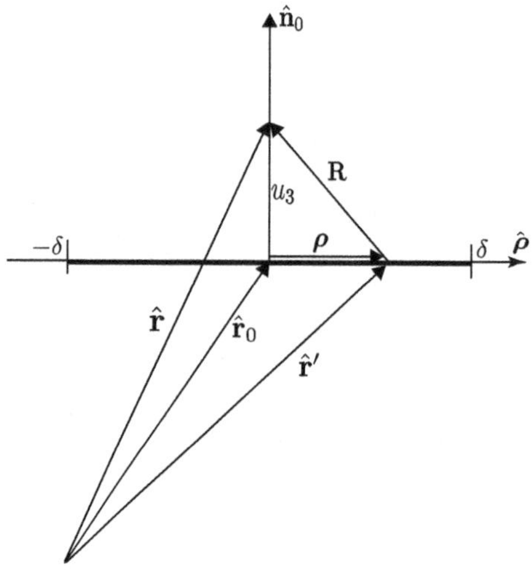


**Figure 10.7.** Integration over the disk $D_\delta$.

$$\begin{aligned}
\oiint_{S_+} f(\mathbf{r}')\nabla'\psi(\mathbf{r},\mathbf{r}')\,ds' &= \lim_{\substack{u_3\to 0^+\\ \delta\to 0}} \oiint_S f(\mathbf{r}')\nabla'\psi(\mathbf{r},\mathbf{r}')\,ds' \\
&= \lim_{\substack{u_3\to 0^+\\ \delta\to 0}}\left[\oiint_{D_\delta} f(\mathbf{r}')\nabla'\psi(\mathbf{r},\mathbf{r}')\,ds' + \iint_{S-D_\delta} f(\mathbf{r}')\nabla'\psi(\mathbf{r},\mathbf{r}')\,ds'\right] \\
&= \lim_{\substack{u_3\to 0^+\\ \delta\to 0}} \oiint_{D_\delta} f(\mathbf{r}')\nabla'\psi(\mathbf{r},\mathbf{r}')\,ds' + \oiint_S f(\mathbf{r}')\nabla'\psi(\mathbf{r}_0,\mathbf{r}')\,ds',
\end{aligned} \tag{10.44}$$

where the surface integral over $S$ in the last line is to be interpreted as the *Cauchy's principal value*. Now

$$\begin{aligned}
\lim_{\substack{u_3\to 0^+\\ \delta\to 0}} \oiint_{D_\delta} f(\mathbf{r}')\nabla'\psi(\mathbf{r},\mathbf{r}')\,\mathrm{d}s' &= \lim_{\substack{u_3\to 0^+\\ \delta\to 0}} f(\mathbf{r}_0)\int_0^\delta\int_0^{2\pi}\frac{u_3\hat{\mathbf{n}}_0 - \rho(\hat{\mathbf{x}}\cos\phi + \hat{\mathbf{y}}\sin\phi)}{4\pi(\rho^2+u_3^2)^{3/2}}\rho\,\mathrm{d}\rho\,\mathrm{d}\phi \\
&= \lim_{\substack{u_3\to 0^+\\ \delta\to 0}} \frac{1}{2}\hat{\mathbf{n}}_0 f(\mathbf{r}_0)\int_0^\delta \frac{u_3\rho\,\mathrm{d}\rho}{(\rho^2+u_3^2)^{3/2}} \\
&= \lim_{\substack{u_3\to 0^+\\ \delta\to 0}} \frac{1}{2}\hat{\mathbf{n}}_0 f(\mathbf{r}_0)\int_0^{\tan^{-1}(\delta/u_3)}\sin\alpha\,\mathrm{d}\alpha = \frac{1}{2}\hat{\mathbf{n}}_0 f(\mathbf{r}_0),
\end{aligned} \tag{10.45}$$

on employing $\rho = u_3\tan\alpha$. Using this we obtain

$$\oiint_{S_+} f(\mathbf{r}')\nabla'\psi(\mathbf{r},\mathbf{r}')\,\mathrm{d}s' = \frac{1}{2}f(\mathbf{r}_0)\hat{\mathbf{n}}_0 + \oiint_S f(\mathbf{r}')\nabla'\psi(\mathbf{r}_0,\mathbf{r}')\,\mathrm{d}s'. \tag{10.46}$$

Similarly,

$$\oiint_{S_-} f(\mathbf{r}')\nabla'\psi(\mathbf{r}, \mathbf{r}')\, ds' = -\frac{1}{2}f(\mathbf{r}_0)\hat{\mathbf{n}}_0 + \oiint_S f(\mathbf{r}')\nabla'\psi(\mathbf{r}, \mathbf{r}')\, ds', \tag{10.47}$$

and

$$\begin{aligned}\oiint_{S_+} f(\mathbf{r}')\psi(\mathbf{r}, \mathbf{r}')\, ds' &= \oiint_{S_-} f(\mathbf{r}')\psi(\mathbf{r}, \mathbf{r}')\, ds' \\ &= \oiint_S f(\mathbf{r}')\psi(\mathbf{r}_0, \mathbf{r}')\, ds'.\end{aligned} \tag{10.48}$$

Along the same lines,

$$\begin{aligned}\hat{\mathbf{n}}_0 \times \oiint_{S_+} \mathfrak{J}_s(\mathbf{r}') \times \nabla'\psi(\mathbf{r}, \mathbf{r}')\, ds' = \frac{1}{2}\mathfrak{J}_s(\mathbf{r}_0) + \hat{\mathbf{n}}_0 \times \oiint_S \mathfrak{J}_s(\mathbf{r}') \\ \times \nabla'\psi(\mathbf{r}_0, \mathbf{r}')\, ds'\end{aligned} \tag{10.49}$$

$$\begin{aligned}\hat{\mathbf{n}}_0 \times \oiint_{S_-} \mathfrak{J}_s(\mathbf{r}') \times \nabla'\psi(\mathbf{r}, \mathbf{r}')\, ds' = -\frac{1}{2}\mathfrak{J}_s(\mathbf{r}_0) + \hat{\mathbf{n}}_0 \times \oiint_S \mathfrak{J}_s(\mathbf{r}') \\ \times \nabla'\psi(\mathbf{r}_0, \mathbf{r}')\, ds'.\end{aligned} \tag{10.50}$$

Thus it is seen that manner in which one approaches the surface affects the value of the integrals. Equations (10.46) and (10.47) reveal that there is a discontinuity in the normal components. Equations (10.49) and (10.50) reveal that there is a discontinuity in the tangential components. Note that discontinuities only exist when the integrand involves the gradient of the Green's function (termed *double-layer potential*) rather that the Green's function itself (termed *single-layer potential*). The remaining integrals on $S$ in equations (10.46)–(10.50) are all of the principal value type.

Consider the surface integral part of equations (10.20) and (10.21) with densities $\mathfrak{J}_s$, $q_e$ and $\mathfrak{M}_s$, $q_m$

$$\mathbf{U}(\mathbf{r}) = -\oiint_S \left\{ j\omega\mu\mathfrak{J}_s(\mathbf{r}')\psi(\mathbf{r}, \mathbf{r}') - \frac{q_e(\mathbf{r}')}{\varepsilon}\nabla'\psi(\mathbf{r}, \mathbf{r}') + \mathfrak{M}_s(\mathbf{r}') \times \nabla'\psi(\mathbf{r}, \mathbf{r}') \right\} ds', \tag{10.51}$$

$$\mathbf{V}(\mathbf{r}) = -\oiint_S \left\{ j\omega\varepsilon\mathfrak{M}_s(\mathbf{r}')\psi(\mathbf{r}, \mathbf{r}') - \frac{q_m(\mathbf{r}')}{\mu}\nabla'\psi(\mathbf{r}, \mathbf{r}') - \mathfrak{J}_s(\mathbf{r}') \times \nabla'\psi(\mathbf{r}, \mathbf{r}') \right\} ds'. \tag{10.52}$$

In these equations, the point $\mathbf{r}$ can be either in the interior of $S$ or in the exterior of $S$ or on $S$. The Green's function and constitutive parameters ($\varepsilon$, $\mu$) pertain to a homogeneous medium. If $\mathbf{r}$ is not on $S$ (i.e. for $\mathbf{r} \neq \mathbf{r}_0$), no distinction needs to be made for the surface integral in terms of $S_+$ or $S_-$ and it is safe to use $\nabla'\psi = -\nabla\psi$. On using the identity $\nabla \times (f\mathbf{a}) = f\nabla \times \mathbf{a} + \nabla f \times \mathbf{a}$ we obtain, for $\mathbf{r} \neq \mathbf{r}_0$, that

$$-\mathbf{U}(\mathbf{r}) = j\omega\mu \oiint_S \mathfrak{J}_s(\mathbf{r}')\psi(\mathbf{r}, \mathbf{r}')\, ds' + \frac{1}{\varepsilon}\nabla \oiint_S q_e(\mathbf{r}')\psi(\mathbf{r}, \mathbf{r}')\, ds'$$
$$+\nabla \times \oiint_S \mathfrak{M}_s(\mathbf{r}')\psi(\mathbf{r}, \mathbf{r}')\, ds',$$
$$-\mathbf{V}(\mathbf{r}) = j\omega\varepsilon \oiint_S \mathfrak{M}_s(\mathbf{r}')\psi(\mathbf{r}, \mathbf{r}')\, ds' + \frac{1}{\mu}\nabla \oiint_S q_m(\mathbf{r}')\psi(\mathbf{r}, \mathbf{r}')\, ds'$$
$$-\nabla \times \oiint_S \mathfrak{J}_s(\mathbf{r}')\psi(\mathbf{r}, \mathbf{r}')\, ds'.$$

Consider $-\nabla \times \mathbf{U}$ for $\mathbf{r} \neq \mathbf{r}'$ together with $\nabla\psi = -\nabla'\psi$ and the relation $\nabla \times \nabla \times (\mathfrak{M}_s\psi) = -\mathfrak{M}_s\nabla^2\psi + \nabla[\nabla \cdot (\mathfrak{M}_s\psi)] = k^2\psi\, \mathfrak{M}_s - \nabla[\mathfrak{M}_s \cdot \nabla'\psi]$ since $\mathfrak{M}_s(\mathbf{r}')$ is regarded as a constant as far as $\nabla$ is considered. Now for $\mathbf{r} \neq \mathbf{r}'$

$$\begin{aligned}
\nabla \cdot \oiint_S \mathfrak{M}_s(\mathbf{r}')\psi(\mathbf{r}, \mathbf{r}')\, ds' &= -\oiint_S \mathfrak{M}_s(\mathbf{r}') \cdot \nabla'\psi(\mathbf{r}, \mathbf{r}')\, ds' \\
&= -\oiint_S \mathfrak{M}_s(\mathbf{r}') \cdot \left[\nabla_s'\psi + \hat{\mathbf{n}}_0'\frac{\partial\psi}{\partial n_0'}\right] ds' \\
&= -\oiint_S \mathfrak{M}_s(\mathbf{r}') \cdot \nabla_s'\psi\, ds' = \oiint_S \psi\nabla_s' \cdot \mathfrak{M}_s\, ds' \\
&= -j\omega \oiint_S q_m(\mathbf{r}')\psi(\mathbf{r}, \mathbf{r}')\, ds',
\end{aligned} \tag{10.53}$$

where we have used the result (B.19) [8, p 162] that

$$\oiint_S \nabla_s' \cdot (\psi\, \mathfrak{M}_s)\, ds' = \oiint_S \left[\psi\nabla_s' \cdot \mathfrak{M}_s + \mathfrak{M}_s \cdot \nabla_s'\psi\right] ds' = 0. \tag{10.54}$$

Using equation (10.53) we have for $\mathbf{r} \neq \mathbf{r}_0$ that

$$\begin{aligned}
-\nabla \times \mathbf{U} &= j\omega\mu\nabla \times \oiint_S \mathfrak{J}_s(\mathbf{r}')\psi(\mathbf{r}, \mathbf{r}')\, ds' + \nabla \times \nabla \times \oiint_S \mathfrak{M}_s(\mathbf{r}')\psi(\mathbf{r}, \mathbf{r}')\, ds' \\
&= j\omega\mu\nabla \times \oiint_S \mathfrak{J}_s(\mathbf{r}')\psi(\mathbf{r}, \mathbf{r}')\, ds' + k^2 \oiint_S \mathfrak{M}_s(\mathbf{r}')\psi(\mathbf{r}, \mathbf{r}')\, ds' \\
&\quad + \nabla\left(\nabla \cdot \oiint_S \mathfrak{M}_s(\mathbf{r}')\psi(\mathbf{r}, \mathbf{r}')\, ds'\right) \\
&= j\omega\mu\nabla \times \oiint_S \mathfrak{J}_s(\mathbf{r}')\psi(\mathbf{r}, \mathbf{r}')\, ds' + j\omega\mu(-j\omega\varepsilon)\oiint_S \mathfrak{M}_s(\mathbf{r}')\psi(\mathbf{r}, \mathbf{r}')\, ds' \\
&\quad - j\omega\mu\frac{1}{\mu}\nabla \oiint_S q_m(\mathbf{r}')\psi(\mathbf{r}, \mathbf{r}')\, ds' \\
&= j\omega\mu\mathbf{V}.
\end{aligned} \tag{10.55}$$

Similarly, $\nabla \times \mathbf{V} = j\omega\varepsilon\mathbf{U}$. It can also be shown that $(\mathbf{U}, \mathbf{V})$ satisfy the Silver–Müller radiation condition (1.69); see [8], for example. Thus the field generated by equations (10.51) and (10.52) with rather *arbitrary choice of densities* $\mathfrak{J}_s$, $q_e$, $q_m$, and $\mathfrak{M}_s$ satisfy

(i) *source-free* Maxwell's equations in the interior or exterior of $S$, with $\mathbf{U}$ representing the electric field and $\mathbf{V}$ representing the magnetic field,

(ii) all of the information needed of electromagnetic waves in homogeneous space is embedded in the free-space Green's function $\psi$ of the Helmholtz equation and the only requirement on $\mathfrak{J}_s$, $q_e$, $q_m$, and $\mathfrak{M}_s$ is that they satisfy the Hölder conditions uniformly on the Lyapunov surface $S$.

Let us denote by $(\mathbf{E}_-(\mathbf{r}_0), \mathbf{H}_-(\mathbf{r}_0))$ and $(\mathbf{E}_+(\mathbf{r}_0), \mathbf{H}_+(\mathbf{r}_0))$ the surface values of $(\mathbf{U}, \mathbf{V})$ generated by these densities by approaching $S$ from the interior and exterior side, respectively. Accordingly, the integral represented by $\mathbf{U}$ generates

$$\mathbf{E}_-(\mathbf{r}_0) = -\oiint_{S_-}\left\{j\omega\mu\mathfrak{J}_s(\mathbf{r}')\psi(\mathbf{r}, \mathbf{r}') - \frac{q_e(\mathbf{r}')}{\varepsilon}\nabla'\psi(\mathbf{r}, \mathbf{r}') + \mathfrak{M}_s(\mathbf{r}') \times \nabla'\psi(\mathbf{r}, \mathbf{r}')\right\}_{\mathbf{r}\to\mathbf{r}_0} ds', \tag{10.56}$$

$$\mathbf{E}_+(\mathbf{r}_0) = -\oiint_{S_+}\left\{j\omega\mu\mathfrak{J}_s(\mathbf{r}')\psi(\mathbf{r}, \mathbf{r}') - \frac{q_e(\mathbf{r}')}{\varepsilon}\nabla'\psi(\mathbf{r}, \mathbf{r}') + \mathfrak{M}_s(\mathbf{r}') \times \nabla'\psi(\mathbf{r}, \mathbf{r}')\right\}_{\mathbf{r}\to\mathbf{r}_0} ds', \tag{10.57}$$

and the integral represented by $\mathbf{V}$ on $S_-$ and $S_+$ generates the field $\mathbf{H}_-(\mathbf{r}_0)$ and $\mathbf{H}_+(\mathbf{r}_0)$, respectively. By equation (10.48), the first term in the surface integrals produces a field that is continuous across $S$. The second term produces a field that has discontinuous normal components across $S$, while the third term produces a field that has discontinuous tangential components across $S$. Applying equations (10.49), (10.50) to equations $\hat{\mathbf{n}}_0 \times$ (10.56) and $\hat{\mathbf{n}}_0 \times$ (10.57), we see that

$$\hat{\mathbf{n}}_0 \times [\mathbf{E}_+(\mathbf{r}_0) - \mathbf{E}_-(\mathbf{r}_0)] = -\mathfrak{M}_s(\mathbf{r}_0). \tag{10.58}$$

Similarly, by considering $\mathbf{H}_-$ and $\mathbf{H}_+$, it can be seen that

$$\hat{\mathbf{n}}_0 \times [\mathbf{H}_+(\mathbf{r}_0) - \mathbf{H}_-(\mathbf{r}_0)] = \mathfrak{J}_s(\mathbf{r}_0). \tag{10.59}$$

By applying equations (10.46), (10.47) to $\hat{\mathbf{n}}_0 \cdot$ (10.56) and $\hat{\mathbf{n}}_0 \cdot$ (10.57), we can see that

$$\hat{\mathbf{n}}_0 \cdot [\mathbf{E}_+(\mathbf{r}_0) - \mathbf{E}_-(\mathbf{r}_0)] = \frac{1}{\varepsilon}\, q_e(\mathbf{r}_0). \tag{10.60}$$

Likewise

$$\hat{\mathbf{n}}_0 \cdot [\mathbf{H}_+(\mathbf{r}_0) - \mathbf{H}_-(\mathbf{r}_0)] = \frac{1}{\mu} q_m(\mathbf{r}_0). \tag{10.61}$$

Thus the fields generated by the surface integrals in equations (10.51) and (10.52) produce discontinuous fields across the surface $S$. These discontinuities are due to the presence of the gradient term $\nabla'\psi$ in the integral representations and they depend solely on the densities $\mathfrak{J}_s$, $q_e$, $q_m$, and $\mathfrak{M}_s$ employed in the integrals and on the constitutive parameters $(\varepsilon, \mu)$ used for the homogeneous medium.

Let $(\mathbf{E}_+, \mathbf{H}_+)$ be the fields generated *exterior to S* (i.e. for $\mathbf{r} \in \mathbf{r}^+$) by the densities $\mathfrak{M}_s$, $q_m$, $\mathfrak{J}_s$, and $q_e$ *placed on* $S_+$ with $-j\omega q_m = \nabla_s \cdot \mathfrak{M}_s$, $-j\omega q_e = \nabla_s \cdot \mathfrak{J}_s$. Likewise let $(\mathbf{E}_-, \mathbf{H}_-)$ be the fields generated *interior to S* (i.e. for $\mathbf{r} \in \mathbf{r}^-$) by the densities $\mathfrak{M}_s$, $q_m$, $\mathfrak{J}_s$, and $q_e$ *placed on* $S_-$. Then we can write

$$\mathbf{E}_+(\mathbf{r}) = \oiint_{S_+} \left\{ -j\omega\mu\mathfrak{J}_s(\mathbf{r}')\psi(\mathbf{r}, \mathbf{r}') + \frac{q_e(\mathbf{r}')}{\varepsilon}\nabla'\psi(\mathbf{r}, \mathbf{r}') - \mathfrak{M}_s(\mathbf{r}') \times \nabla'\psi(\mathbf{r}, \mathbf{r}') \right\} \mathrm{d}s', \tag{10.62}$$

$$\mathbf{H}_+(\mathbf{r}) = \oiint_{S_+} \left\{ -j\omega\varepsilon\mathfrak{M}_s(\mathbf{r}')\psi(\mathbf{r}, \mathbf{r}') + \frac{q_m(\mathbf{r}')}{\mu}\nabla'\psi(\mathbf{r}, \mathbf{r}') + \mathfrak{J}_s(\mathbf{r}') \times \nabla'\psi(\mathbf{r}, \mathbf{r}') \right\} \mathrm{d}s', \tag{10.63}$$

$$\mathbf{E}_-(\mathbf{r}) = \oiint_{S_-} \left\{ -j\omega\mu\mathfrak{J}_s(\mathbf{r}')\psi(\mathbf{r}, \mathbf{r}') + \frac{q_e(\mathbf{r}')}{\varepsilon}\nabla'\psi(\mathbf{r}, \mathbf{r}') - \mathfrak{M}_s(\mathbf{r}') \times \nabla'\psi(\mathbf{r}, \mathbf{r}') \right\} \mathrm{d}s', \tag{10.64}$$

$$\mathbf{H}_-(\mathbf{r}) = \oiint_{S_-} \left\{ -j\omega\varepsilon\mathfrak{M}_s(\mathbf{r}')\psi(\mathbf{r}, \mathbf{r}') + \frac{q_m(\mathbf{r}')}{\mu}\nabla'\psi(\mathbf{r}, \mathbf{r}') + \mathfrak{J}_s(\mathbf{r}') \times \nabla'\psi(\mathbf{r}, \mathbf{r}') \right\} \mathrm{d}s'. \tag{10.65}$$

For a point $\mathbf{r}_0$ on $S$ we have the limit relations

$$\hat{\mathbf{n}}_0 \times \mathbf{E}_\pm(\mathbf{r}_0) \pm \frac{1}{2}\mathfrak{M}_s(\mathbf{r}_0) = \hat{\mathbf{n}}_0 \times \oiint_S \left[ -j\omega\mu\mathfrak{J}_s\psi + \frac{q_e}{\varepsilon}\nabla'\psi - \mathfrak{M}_s \times \nabla'\psi \right] \mathrm{d}s', \tag{10.66}$$

and

$$\hat{\mathbf{n}}_0 \times \mathbf{H}_\pm(\mathbf{r}_0) \mp \frac{1}{2}\mathfrak{J}_s(\mathbf{r}_0) = \hat{\mathbf{n}}_0 \times \oiint_S \left[ -j\omega\varepsilon\mathfrak{M}_s\psi + \frac{q_m}{\mu}\nabla'\psi + \mathfrak{J}_s \times \nabla'\psi \right] \mathrm{d}s'. \tag{10.67}$$

Two particular choices of these densities are important.

1. If the surface currents in equations (10.66) and (10.67) are chosen based on the fields exterior to $S$ such that

$$\mathfrak{M}_s(\mathbf{r}_0) = -\hat{\mathbf{n}}_0 \times \mathbf{E}_+(\mathbf{r}_0), \tag{10.68}$$

$$\mathfrak{J}_s(\mathbf{r}_0) = \hat{\mathbf{n}}_0 \times \mathbf{H}_+(\mathbf{r}_0), \tag{10.69}$$

$$\frac{1}{\varepsilon} q_e(\mathbf{r}_0) = \hat{\mathbf{n}}_0 \cdot \mathbf{E}_+(\mathbf{r}_0), \tag{10.70}$$

$$\frac{1}{\mu} q_m(\mathbf{r}_0) = \hat{\mathbf{n}}_0 \cdot \mathbf{H}_+(\mathbf{r}_0), \tag{10.71}$$

then it is easy to see from equations (10.66) and (10.67) that

$$\hat{\mathbf{n}}_0 \times \oiint_S \left[-j\omega\mu\mathfrak{J}_s\psi + \frac{q_e}{\varepsilon}\nabla'\psi - \mathfrak{M}_s \times \nabla'\psi\right]_{\mathbf{r}\to\mathbf{r}_0} ds' = -\frac{1}{2}\mathfrak{M}_s.$$

$$\hat{\mathbf{n}}_0 \times \oiint_S \left[-j\omega\varepsilon\mathfrak{M}_s\psi + \frac{q_m}{\mu}\nabla'\psi + \mathfrak{J}_s \times \nabla'\psi\right]_{\mathbf{r}\to\mathbf{r}_0} ds' = \frac{1}{2}\mathfrak{J}_s.$$

Consequently, $\hat{\mathbf{n}}_0 \times \mathbf{E}_-(\mathbf{r}_0) =: \mathfrak{M}_s^-(\mathbf{r}_0) = \mathbf{0}$ and $-\hat{\mathbf{n}}_0 \times \mathbf{H}_-(\mathbf{r}_0) =: \mathfrak{J}_s^-(\mathbf{r}_0) = \mathbf{0}$. The field $(\mathbf{E}_-, \mathbf{H}_-)$ in the interior to $S$ obtained with $\mathfrak{M}_s = \mathfrak{M}_s^-$, $\mathfrak{J}_s = \mathfrak{J}_s^-$ in equations (10.63) and (10.65) will then be identically zero. These results are consistent with what was obtained in equations (10.20) and (10.21) for the equivalence principle for the exterior to $S$.

2. If the densities are chosen based upon the fields interior to $S$ such that

$$\mathfrak{M}_s(\mathbf{r}_0) = \hat{\mathbf{n}}_0 \times \mathbf{E}_-(\mathbf{r}_0), \tag{10.72}$$

$$\mathfrak{J}_s(\mathbf{r}_0) = -\hat{\mathbf{n}}_0 \times \mathbf{H}_-(\mathbf{r}_0), \tag{10.73}$$

$$\frac{1}{\varepsilon} q_e(\mathbf{r}_0) = -\hat{\mathbf{n}}_0 \cdot \mathbf{E}_-(\mathbf{r}_0), \tag{10.74}$$

$$\frac{1}{\mu} q_m(\mathbf{r}_0) = -\hat{\mathbf{n}}_0 \cdot \mathbf{H}_-(\mathbf{r}_0), \tag{10.75}$$

then it is easy to see that from equations (10.66) and (10.67) that $-\hat{\mathbf{n}}_0 \times \mathbf{E}_+(\mathbf{r}_0) =: \mathfrak{M}_s^+(\mathbf{r}_0) = \mathbf{0}$ and $\hat{\mathbf{n}}_0 \times \mathbf{H}_+(\mathbf{r}_0) =: \mathfrak{J}_s^+(\mathbf{r}_0) = \mathbf{0}$. The field $(\mathbf{E}_+, \mathbf{H}_+)$ in the exterior to $S$ obtained with $\mathfrak{M}_s = \mathfrak{M}_s^+$, $\mathfrak{J}_s = \mathfrak{J}_s^+$ in equations (10.62) and (10.64) will then be identically zero. These results are consistent with equations (10.16) and (10.17) for the equivalence principle interior to $S$.

Note that the only requirement for these results to hold is that the medium parameters be constant throughout the region. Thus the choice of $(\varepsilon, \mu)$ used in equations (10.68)–(10.71) need not be the same as the ones used in equations (10.72)–(10.75) as long as they consistent with the choice $(\mathbf{E}_+, \mathbf{H}_+)$ and $(\mathbf{E}_-, \mathbf{H}_-)$, respectively. The true field is recovered from the surface integrals on the side of $S$ that contains the true limiting fields. On the other side of $S$ it generates zero fields. By juxtaposing the two choices and using the linear superposition of fields we are led to the following two versions of the most general surface equivalence principle:

**Theorem 10.1.2** Müller [8, p 217], equivalence principle, form 2.
*Let the currents $\mathfrak{J}_s$ and $\mathfrak{M}_s$ satisfy a Hölder condition on a smooth surface S. Let their surface divergences*

$$\nabla_s \cdot \mathfrak{J}_s = -j\omega q_e; \quad \nabla_s \cdot \mathfrak{M}_s = -j\omega q_m, \tag{10.76}$$

*exist and satisfy a Hölder condition as well. Then the fields generated by the surface integrals*

$$\mathbf{P} = \oiint_S \left[ -j\omega\mu\mathfrak{J}_s\psi - \mathfrak{M}_s \times \nabla'\psi + \frac{q_e}{\varepsilon}\nabla'\psi \right] ds', \tag{10.77}$$

$$\mathbf{Q} = \oiint_S \left[ -j\omega\varepsilon\mathfrak{M}_s\psi + \mathfrak{J}_s \times \nabla'\psi + \frac{q_m}{\mu}\nabla'\psi \right] ds', \tag{10.78}$$

*satisfy the source-free Maxwell's equations*

$$\nabla \times \mathbf{P} = -j\omega\mu\mathbf{Q}; \quad \nabla \times \mathbf{Q} = j\omega\varepsilon\mathbf{P}, \tag{10.79}$$

*for all* $\mathbf{r}$ *not lying on S, where* $\psi$ *is the free-space Green's function* $e^{-jk|\mathbf{r}-\mathbf{r}'|}/4\pi|\mathbf{r}-\mathbf{r}'|$ *and k is the free-space wavenumber pertaining to* $(\mu, \varepsilon)$ *at the frequency* $\omega$. *The boundary values of* $\mathbf{P}_i$, $\mathbf{P}_e$ *obtained on approaching the surface from, respectively, the interior and the exterior regions, and similarly of* $\mathbf{Q}_i$, $\mathbf{Q}_e$, *exist for all regular points* $\mathbf{r}_0$ *of S and satisfy the discontinuity relations*

$$\mathbf{P}_e(\mathbf{r}_0) - \mathbf{P}_i(\mathbf{r}_0) =: \Delta[\mathbf{P}(\mathbf{r}_0)] = \hat{\mathbf{n}}_0 \times \mathfrak{M}_s(\mathbf{r}_0) + \frac{q_e}{\varepsilon}\hat{\mathbf{n}}_0, \tag{10.80}$$

$$\mathbf{Q}_e(\mathbf{r}_0) - \mathbf{Q}_i(\mathbf{r}_0) =: \Delta[\mathbf{Q}(\mathbf{r}_0)] = -\hat{\mathbf{n}}_0 \times \mathfrak{J}_s(\mathbf{r}_0) + \frac{q_m}{\mu}\hat{\mathbf{n}}_0, \tag{10.81}$$

*where* $\hat{\mathbf{n}}_0$ *is the unit normal on S pointing from interior to exterior and* $\Delta[\cdot]$ *stands for surface discontinuity. Finally, the fields* $\mathbf{P}$ *and* $\mathbf{Q}$ *satisfy the Silver–Müller radiation conditions.*

Note that the fields $\mathbf{P}$ and $\mathbf{Q}$ are generated entirely by the postulated surface currents. In particular, there are no impressed sources inside or outside of $S$. We can also include sources inside (or outside with a suitable modification) the surface $S$ to result in the following modified form of the theorem.

**Theorem 10.1.3.** Equivalence principle, form 3. *Let sources* $(\mathbf{J}_1, \mathbf{M}_1)$ *placed in a region enclosed by a mathematical surface S in a homogeneous medium with parameters* $(\varepsilon, \mu)$ *produce a field* $(\mathbf{E}_1, \mathbf{H}_1)$. *Retain the original field exterior to S and postulate an arbitrary source-free field* $(\mathbf{E}_2, \mathbf{H}_2)$ *interior to S consistent with the parameters* $(\varepsilon, \mu)$. *To support such a field discontinuity, there must exist surface currents and charges*

$$\mathfrak{M}_s(\mathbf{r}_0) = -\hat{\mathbf{n}}_0 \times (\mathbf{E}_1(\mathbf{r}_0) - \mathbf{E}_2(\mathbf{r}_0)); \; \hat{\mathbf{n}}_0 \cdot [\mathbf{E}_1(\mathbf{r}_0) - \mathbf{E}_2(\mathbf{r}_0)] = \frac{1}{\varepsilon} q_e(\mathbf{r}_0), \tag{10.82}$$

$$\mathfrak{J}_s(\mathbf{r}_0) = \hat{\mathbf{n}}_0 \times (\mathbf{H}_1(\mathbf{r}_0) - \mathbf{H}_2(\mathbf{r}_0)); \quad \hat{\mathbf{n}}_0 \cdot [\mathbf{H}_1(\mathbf{r}_0) - \mathbf{H}_2(\mathbf{r}_0)] = \frac{1}{\mu} q_m(\mathbf{r}_0), \tag{10.83}$$

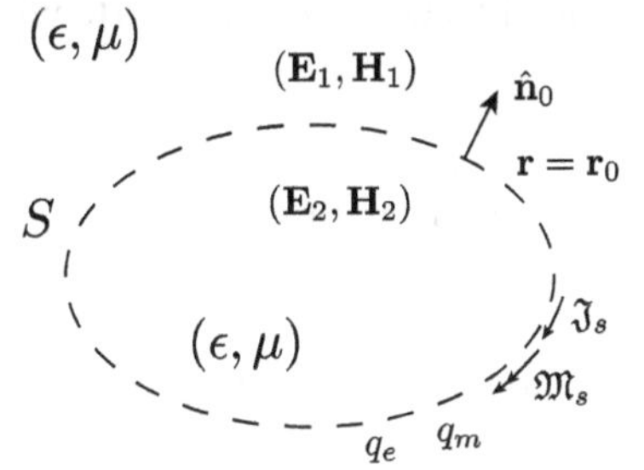


**Figure 10.8.** Equivalent sources on a surface arising due to field discontinuities.

*residing on S, where the unit normal* $\hat{\mathbf{n}}_0$ *at an arbitrary point* $\mathbf{r}_0$ *lying on S points from the interior to the exterior (figure 10.8). When these currents are allowed to radiate in a homogeneous medium with parameters* $(\varepsilon, \mu)$ *then they generate via the integrals (10.51) and (10.52) the fields* $(\mathbf{E}_1, \mathbf{H}_1)$ *exterior to S and the fields* $(\mathbf{E}_2, \mathbf{H}_2)$ *interior to S. The fields* $(\mathbf{E}_1, \mathbf{H}_1)$ *satisfy source-free Maxwell's equations exterior to S. Furthermore, they satisfy Silver–Müller radiation conditions at infinity.*

As in theorem 10.1.2, equations (10.82) and (10.83) can also be expressed in an inverse vectorial form as

$$\mathbf{E}_1(\mathbf{r}_0) - \mathbf{E}_2(\mathbf{r}_0) = \hat{\mathbf{n}}_0 \times \mathfrak{M}_s(\mathbf{r}_0) + \frac{q_e}{\varepsilon}\hat{\mathbf{n}}_0, \tag{10.84}$$

$$\mathbf{H}_1(\mathbf{r}_0) - \mathbf{H}_2(\mathbf{r}_0) = -\hat{\mathbf{n}}_0 \times \mathfrak{J}_s(\mathbf{r}_0) + \frac{q_m}{\mu}\hat{\mathbf{n}}_0. \tag{10.85}$$

## 10.2 Integral equations, physical optics, and Bojarski's identity

The surface equivalent representations discussed thus far provide convenient means for formulating integral equations for studying scattering from material objects or for developing analytical identities in the study of other electromagnetic problems. Three such examples are provided below. The first involves scattering by a homogeneous, penetrable object, wherein an integral equation suitable for very efficient numerical solution is derived. The derivation parallels that discussed in [8]. In the second example, we apply Franz's representation to the solution of a scattering problem by an approximate analytical technique known as *physical optics approximation*. In the third example we consider scattering by a homogeneous object under the physical optics approximation and derive the *Bojarski's identity* that finds use in inverse scattering.

### 10.2.1 Surface integral equations

Stratton–Chu, Franz, and surface field representations permit formulation of surface integral equations involving penetrable or impenetrable homogeneous objects. The formulation for homogeneous objects is more intricate which is what we consider in the following example.

**Example 10.2.1.1.** Müller-type integral equations for homogeneous objects.
Let sources (**J**, **M**) having local support in a region $G'$ with boundary $F'$ radiate in a homogeneous medium with parameters ($\varepsilon_e$, $\mu_e$) and generate the fields ($\mathbf{E}^i$, $\mathbf{H}^i$) (figure 10.9). A homogeneous scatterer having parameters ($\varepsilon_d$, $\mu_d$) occupies a region $G$ with boundary $F$. The presence of the scatterer yields additional fields ($\mathbf{E}^s$, $\mathbf{H}^s$). Let the region exterior to $G$ be labeled as $G_e$. Using the surface field formulations, derive integral equations for the unknown surface currents [3].

Let $k_e$ and $k_d$ be the wavenumbers in regions $G_e$ and $G$ and let the corresponding scalar Green's functions be

$$\psi_e = \frac{e^{-jk_e|\mathbf{r}-\mathbf{r}'|}}{4\pi|\mathbf{r}-\mathbf{r}'|}; \quad \psi_d = \frac{e^{-jk_d|\mathbf{r}-\mathbf{r}'|}}{4\pi|\mathbf{r}-\mathbf{r}'|}. \tag{10.86}$$

Let $\hat{\mathbf{n}}$ denote the unit normal to $F$ directed towards $G_e$. The scattered fields ($\mathbf{E}^s$, $\mathbf{H}^s$) satisfy source-free Maxwell's equations and the Silver–Muller radiation conditions exterior to $G$ with wavenumber $k_e$. Inside $F$ the true field is denoted by ($\mathbf{E}^d$, $\mathbf{H}^d$). We will now formulate integral equations that do not involve derivatives of the surface electric and magnetic currents.

Postulating zero fields interior to $G$ and scattered fields exterior to $G$ and constructing the surface sources $\mathfrak{J}_{se} = -\hat{\mathbf{n}} \times \mathbf{H}^s$, $\mathfrak{M}_{se} = \hat{\mathbf{n}} \times \mathbf{E}^s$, $q_{ee} = -\nabla_s \cdot \mathfrak{J}_{se}/j\omega$, $q_{me} = -\nabla_s \cdot \mathfrak{M}_{se}/j\omega$ we have from theorem 10.1.2 that exterior to $G$

$$\oiint_F \left[ -j\omega\mu_e \mathfrak{J}_{se}\psi_e - \mathfrak{M}_{se} \times \nabla'\psi_e + \frac{q_{ee}}{\varepsilon_e}\nabla'\psi_e \right] ds' = -\mathbf{E}^s, \tag{10.87}$$

$$\oiint_F \left[ -j\omega\varepsilon_e \mathfrak{M}_{se}\psi_e + \mathfrak{J}_{se} \times \nabla'\psi_e + \frac{q_{me}}{\mu_e}\nabla'\psi_e \right] ds' = -\mathbf{H}^s. \tag{10.88}$$

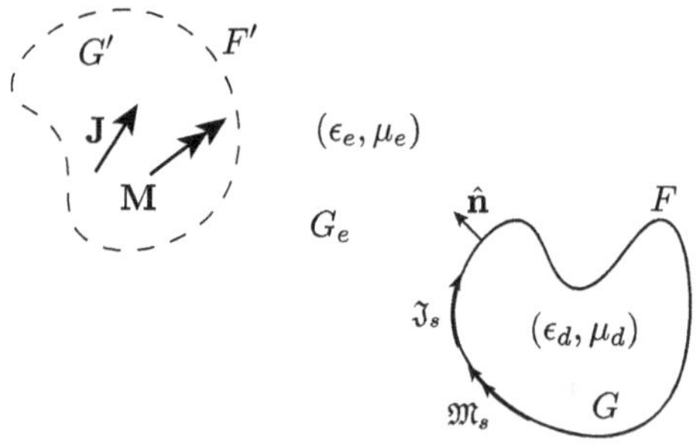


**Figure 10.9.** Finite-support currents radiating in the presence of a homogeneous scatterer.

In passing through $F$ from the interior to the exterior, the tangential components of fields computed by equations (10.87) and (10.88) have the following discontinuities in accordance with equations (10.80) and (10.81):

$$-\hat{\mathbf{n}} \times \mathfrak{J}_{se} = \hat{\mathbf{n}} \times (\hat{\mathbf{n}} \times \mathbf{H}^s) = \Delta[-\mathbf{H}^s]_{tang}; \ \hat{\mathbf{n}} \times \mathfrak{M}_{se} = \hat{\mathbf{n}} \times (\hat{\mathbf{n}} \times \mathbf{E}^s) = \Delta[-\mathbf{E}^s]_{tang}. \quad (10.89)$$

The above imply that the tangential field just inside $F$ is zero by construction of $\mathfrak{J}_{se}$ and $\mathfrak{M}_{se}$. Hence the field inside $F$ is identically zero as postulated since zero tangential fields on boundary implies zero fields everywhere interior to a boundary. The latter follows directly from equations (10.16) and (10.17).

With surface currents $\mathfrak{J}_{sd} = -\hat{\mathbf{n}} \times \mathbf{H}^d$, $\mathfrak{M}_{sd} = \hat{\mathbf{n}} \times \mathbf{E}^d$ and their divergences $q_{ed} = -\nabla_s \cdot \mathfrak{J}_{sd}/j\omega$ and $q_{md} = -\nabla_s \cdot \mathfrak{M}_{sd}/j\omega$, it follows from theorem 10.1.2 that exterior to $G$

$$\oiint_F \left[ -j\omega\mu_d \mathfrak{J}_{sd}\psi_d - \mathfrak{M}_{sd} \times \nabla'\psi_d + \frac{q_{ed}}{\varepsilon_d}\nabla'\psi_d \right] ds' = 0, \quad (10.90)$$

$$\oiint_F \left[ -j\omega\varepsilon_d \mathfrak{M}_{sd}\psi_d + \mathfrak{J}_{sd} \times \nabla'\psi_d + \frac{q_{md}}{\mu_d}\nabla'\psi_d \right] ds' = 0. \quad (10.91)$$

From the continuity of fields across $F$ we have the relation

$$\mathfrak{J}_{si} + \mathfrak{J}_{se} = \mathfrak{J}_{sd}; \quad \mathfrak{M}_{si} + \mathfrak{M}_{se} = \mathfrak{M}_{sd}, \quad (10.92)$$

where $\mathfrak{J}_{si} = -\hat{\mathbf{n}} \times \mathbf{H}^i$ and $\mathfrak{M}_{si} = \hat{\mathbf{n}} \times \mathbf{E}^i$. Furthermore, we have from equations (10.16) and (10.17)

$$\hat{\mathbf{n}} \times \oiint_{F_i} \left[ -j\omega\mu_e \mathfrak{J}_{si}\psi_e - \mathfrak{M}_{si} \times \nabla'\psi_e + \frac{q_{ei}}{\varepsilon_e}\nabla'\psi_e \right] ds' = \mathfrak{M}_{si}, \quad (10.93)$$

$$\hat{\mathbf{n}} \times \oiint_{F_i} \left[ -j\omega\varepsilon_e \mathfrak{M}_{si}\psi_e + \mathfrak{J}_{si} \times \nabla'\psi_e + \frac{q_{mi}}{\mu_e}\nabla'\psi_e \right] ds' = -\mathfrak{J}_{si}, \quad (10.94)$$

where $F_i$ is the integral obtained by approaching the surface $F$ from the inside. We also have from equations (10.87) and (10.88) that

$$\hat{\mathbf{n}} \times \oiint_{F_i} \left[ -j\omega\mu_e \mathfrak{J}_{se}\psi_e - \mathfrak{M}_{se} \times \nabla'\psi_e + \frac{q_{ee}}{\varepsilon_e}\nabla'\psi_e \right] ds' = 0, \quad (10.95)$$

$$\hat{\mathbf{n}} \times \oiint_{F_i} \left[ -j\omega\varepsilon_e \mathfrak{M}_{se}\psi_e + \mathfrak{J}_{se} \times \nabla'\psi_e + \frac{q_{me}}{\mu_e}\nabla'\psi_e \right] ds' = 0. \quad (10.96)$$

Adding equations (10.93) and (10.95) and equations (10.94) and (10.96) and results in

$$\hat{\mathbf{n}} \times \oiint_{F_i} \left[ -j\omega\mu_e \mathfrak{J}_{sd}\psi_e - \mathfrak{M}_{sd} \times \nabla'\psi_e + \frac{q_{ed}}{\varepsilon_e}\nabla'\psi_e \right] ds' = \mathfrak{M}_{si}, \tag{10.97}$$

$$\hat{\mathbf{n}} \times \oiint_{F_i} \left[ -j\omega\varepsilon_e \mathfrak{M}_{sd}\psi_e + \mathfrak{J}_{sd} \times \nabla'\psi_e + \frac{q_{md}}{\mu_e}\nabla'\psi_e \right] ds' = -\mathfrak{J}_{si}. \tag{10.98}$$

On approaching the surface $F$ from the exterior (denoted by $F_e$) we have from equations (10.90) and (10.91) that

$$\hat{\mathbf{n}} \times \oiint_{F_e} \left[ -j\omega\mu_d \mathfrak{J}_{sd}\psi_d - \mathfrak{M}_{sd} \times \nabla'\psi_d + \frac{q_{ed}}{\varepsilon_d}\nabla'\psi_d \right] ds' = 0, \tag{10.99}$$

$$\hat{\mathbf{n}} \times \oiint_{F_e} \left[ -j\omega\varepsilon_d \mathfrak{M}_{sd}\psi_d + \mathfrak{J}_{sd} \times \nabla'\psi_d + \frac{q_{md}}{\mu_d}\nabla'\psi_d \right] ds' = 0. \tag{10.100}$$

Now

$$\begin{aligned}
\hat{\mathbf{n}}(\mathbf{r}) \times \oiint_{F_i} q_{ed}(\mathbf{r}')\nabla'\frac{1}{|\mathbf{r}-\mathbf{r}'|}\, ds' &= -\,\hat{\mathbf{n}}(\mathbf{r}) \times \nabla \oiint_{F_i} q_{ed}(\mathbf{r}')\frac{1}{|\mathbf{r}-\mathbf{r}'|}\, ds' \\
&= -\,\hat{\mathbf{n}}(\mathbf{r}) \times \nabla \oiint_{F_e} q_{ed}(\mathbf{r}')\frac{1}{|\mathbf{r}-\mathbf{r}'|}\, ds' \\
&\quad \times \text{(from theorem B.2.4)} \\
&= \hat{\mathbf{n}}(\mathbf{r}) \times \oiint_{F_e} q_{ed}(\mathbf{r}')\nabla'\frac{1}{|\mathbf{r}-\mathbf{r}'|}\, ds'.
\end{aligned} \tag{10.101}$$

Furthermore, since $\nabla\left(\psi_e - \frac{1}{|\mathbf{r}-\mathbf{r}'|}\right)$ and $\nabla\left(\psi_d - \frac{1}{|\mathbf{r}-\mathbf{r}'|}\right)$ remain uniformly bounded in the limit of $\mathbf{r} \to \mathbf{r}'$, we have upon adding and subtracting terms of the form $1/|\mathbf{r}-\mathbf{r}'|$ in the integrands of the two integrals and using equation (10.101)

$$\begin{aligned}
\hat{\mathbf{n}}(\mathbf{r}) \times \oiint_{F_i} q_{ed}\nabla'\psi_e\, ds' - \hat{\mathbf{n}}(\mathbf{r}) \times \oiint_{F_e} q_{ed}\nabla'\psi_d\, ds' &= \hat{\mathbf{n}}(\mathbf{r}) \times \oiint_{F} q_{ed}\nabla'(\psi_e - \psi_d)\, ds' \\
&= -\,\hat{\mathbf{n}}(\mathbf{r}) \times \nabla \oiint_{F} q_{ed}(\psi_e - \psi_d)\, ds',
\end{aligned}$$

for a point $\mathbf{r}$ on $F$. Also, using $\nabla_s \cdot \mathfrak{J}_{sd} = -j\omega q_{ed}$ we have

$$\oiint_F q_{\rm ed}[\psi_{\rm e} - \psi_{\rm d}]\,{\rm d}s' = \frac{j}{\omega}\oiint_F \nabla_{\rm s}' \cdot \mathfrak{J}_{\rm sd}[\psi_{\rm e} - \psi_{\rm d}]\,{\rm d}s'$$
$$= \frac{-j}{\omega}\oiint_F \mathfrak{J}_{\rm sd} \cdot \nabla'_{\rm s}[\psi_{\rm e} - \psi_{\rm d}]\,{\rm d}s' \text{(from lemma B.2.2)} \qquad (10.102)$$
$$= \frac{-j}{\omega}\oiint_F \mathfrak{J}_{\rm sd} \cdot \nabla'[\psi_{\rm e} - \psi_{\rm d}]\,{\rm d}s',$$

since $\nabla = \nabla_{\rm s} + \hat{\mathbf{n}}\frac{\partial}{\partial n}$ and $\mathfrak{J}_{\rm sd} \cdot \hat{\mathbf{n}} = 0$. Therefore,

$$\hat{\mathbf{n}} \times \oiint_{F_{\rm i}} q_{\rm ed}\nabla'\psi_{\rm e}\,{\rm d}s' - \hat{\mathbf{n}} \times \oiint_{F_{\rm e}} q_{\rm ed}\nabla'\psi_{\rm d}\,{\rm d}s' = \frac{j}{\omega}\hat{\mathbf{n}} \times \nabla \oiint_F \mathfrak{J}_{\rm sd} \cdot \nabla'[\psi_{\rm e} - \psi_{\rm d}]\,{\rm d}s'$$
$$= \frac{j}{\omega}\hat{\mathbf{n}} \times \oiint_F \nabla[\mathfrak{J}_{\rm sd} \cdot \nabla'(\psi_{\rm e} - \psi_{\rm d})]\,{\rm d}s' \qquad (10.103)$$
$$= \frac{-j}{\omega}\hat{\mathbf{n}} \times \oiint_F [\mathfrak{J}_{\rm sd} \cdot \nabla']\nabla'(\psi_{\rm e} - \psi_{\rm d})\,{\rm d}s'.$$

Thus the difference in surface integrals obtained on approaching the surface from the interior and the exterior that involves electric charge (which is proportional to the surface derivative of electric current density) is converted to a surface integral involving only the current density. The derivate appearing in the electric charge density term has been effectively transferred to the difference in exterior and interior Green's functions, which is less singular than either Green's function. According to theorem B.2.5

$$\hat{\mathbf{n}} \times \oiint_{F_{\rm i}} \mathfrak{M}_{\rm sd} \times \nabla'\psi_{\rm e}\,{\rm d}s' = -\frac{1}{2}\mathfrak{M}_{\rm sd} + \hat{\mathbf{n}} \times \oiint_F \mathfrak{M}_{\rm sd} \times \nabla'\psi_{\rm e}\,{\rm d}s', \qquad (10.104)$$

$$\hat{\mathbf{n}} \times \oiint_{F_{\rm e}} \mathfrak{M}_{\rm sd} \times \nabla'\psi_{\rm e}\,{\rm d}s' = \frac{1}{2}\mathfrak{M}_{\rm sd} + \hat{\mathbf{n}} \times \oiint_F \mathfrak{M}_{\rm sd} \times \nabla'\psi_{\rm e}\,{\rm d}s'. \qquad (10.105)$$

Using

$$\oiint_{F_{\rm i}} \mathfrak{J}_{\rm sd}\psi_{\rm e}\,{\rm d}s' = \oiint_F \mathfrak{J}_{\rm sd}\psi_{\rm e}\,{\rm d}s'; \quad \oiint_{F_{\rm e}} \mathfrak{J}_{\rm s}\psi_{\rm d}\,{\rm d}s' = \oiint_F \mathfrak{J}_{\rm sd}\psi_{\rm d}\,{\rm d}s', \qquad (10.106)$$

considering $\varepsilon_{\rm e} \times (10.97) - \varepsilon_{\rm d} \times (10.99)$, we obtain upon utilizing equations (10.103)–(10.105)

$$\frac{\varepsilon_e+\varepsilon_d}{2}\mathfrak{M}_{sd} - \frac{j}{\omega}\hat{\mathbf{n}} \times \oint\!\!\oint_F [\mathfrak{J}_{sd} \cdot \nabla']\nabla'(\psi_e - \psi_d)\, ds'$$
$$+ \hat{\mathbf{n}} \times \oint\!\!\oint_F \left(\frac{-j}{\omega}\mathfrak{J}_{sd}[k_e^2\psi_e - k_d^2\psi_d] + \mathfrak{M}_{sd} \times \nabla'[\varepsilon_e\psi_e - \varepsilon_d\psi_d]\right) ds' = \varepsilon_e\mathfrak{M}_{si},$$

where $k_e^2 = \omega^2\mu_e\varepsilon_e$ and $k_d^2 = \omega^2\mu_d\varepsilon_d$. This leads to the integral equation

$$\begin{aligned}\mathfrak{M}_{sd} = {}& \frac{2\varepsilon_e}{\varepsilon_e + \varepsilon_d}\mathfrak{M}_{si} - \frac{2}{\varepsilon_e + \varepsilon_d}\hat{\mathbf{n}} \times \oint\!\!\oint_F \mathfrak{M}_{sd} \times \nabla'(\varepsilon_e\psi_e - \varepsilon_d\psi_d)\, ds' \\ & + \frac{2j}{\omega(\varepsilon_e + \varepsilon_d)}\hat{\mathbf{n}} \times \oint\!\!\oint_F [\mathfrak{J}_{sd}(k_e^2\psi_e - k_d^2\psi_d) + (\mathfrak{J}_{sd} \cdot \nabla')\nabla'(\psi_e - \psi_d)]\, ds'.\end{aligned} \tag{10.107}$$

In a similar fashion, considering the dual fields, we arrive at the dual integral equation

$$\begin{aligned}\mathfrak{J}_{sd} = {}& \frac{2\mu_e}{\mu_e + \mu_d}\mathfrak{J}_{si} - \frac{2}{\mu_e + \mu_d}\hat{\mathbf{n}} \times \oint\!\!\oint_F \mathfrak{J}_{sd} \times \nabla'(\mu_e\psi_e - \mu_d\psi_d)\, ds' \\ & \frac{2j}{\omega(\mu_e + \mu_d)}\hat{\mathbf{n}} \times \oint\!\!\oint_F [\mathfrak{M}_{sd}(k_e^2\psi_e - k_d^2\psi_d) + (\mathfrak{M}_{sd} \cdot \nabla')\nabla'(\psi_e - \psi_d)]\, ds'.\end{aligned} \tag{10.108}$$

Equations (10.107) and (10.108) constitute a system of vector integral equations for the unknown surface currents $\mathfrak{J}_{sd}$ and $\mathfrak{M}_{sd}$. In the literature, these are known as Müller-type integral equations. These integral equations have several advantages in numerical implementations:

(i) The equations do not involve derivatives of surface currents $\mathfrak{J}_s$ and $\mathfrak{M}_s$. Consequently, the currents can be expanded in more primitive basis functions.

(ii) The first derivatives $\psi_e$ and $\psi_d$ separately become singular as $1/|\mathbf{r} - \mathbf{r}'|^2$, but are, however, integrable over the surface.

(iii) The second derivatives of the difference term $\psi_e - \psi_d$ become singular at the most to the order $1/|\mathbf{r} - \mathbf{r}'|$ and surface integrals involving such terms are integrable.

In reference [9] integral equations of this type were used to study scattering by the shaped, absorbing walls of an electromagnetic anechoic chamber. ■■

### 10.2.2 Scattering by a homogeneous object, physical optics

In the physical optics approximation, one divides the object surface into two portions: a portion illuminated by the incident wave (known as the lit portion) and a portion that remains in the shadow of the source. Information about the interaction of the object with the incident wave is only retained for the lit portion. Furthermore, approximations that call for the principal radii of curvature of the object to be large compared to the wavelength of the incident wave are made to collect information on the lit portion. We consider below the example of the scattering of a plane wave by a homogeneous object to illustrate the technique.

**Example 10.2.2.1.** Scattering by a homogeneous object.
Consider an object with material parameters $(\varepsilon_2, \mu_2)$ and a bounding surface $S$ embedded in a homogeneous medium with parameters $(\varepsilon_1, \mu_1)$ (figure 10.10). Let sources $(\mathbf{J}, \mathbf{M})$ placed *exterior* to $S$ in the homogeneous medium generate the *incident* fields $(\mathbf{E}^i, \mathbf{H}^i)$ throughout. Let the presence of the object introduce the excess fields (i.e. the scattered fields) $(\mathbf{E}^s, \mathbf{H}^s)$ exterior to $S$ and the *total fields* $(\mathbf{E}, \mathbf{H})$ interior to $S$. From the boundary conditions we have $\hat{\mathbf{n}} \times (\mathbf{H}^s + \mathbf{H}^i) = \hat{\mathbf{n}} \times \mathbf{H}$ and $\hat{\mathbf{n}} \times (\mathbf{E}^s + \mathbf{E}^i) = \hat{\mathbf{n}} \times \mathbf{E}$ on $S$. We now wish to formulate an equivalent representation for the scattered field.

By choosing $(\mathbf{E}_2, \mathbf{H}_2) = (-\mathbf{E}^i, -\mathbf{H}^i)$ and $(\mathbf{E}_1, \mathbf{H}_1) = (\mathbf{E}^s, \mathbf{H}^s)$, both of which satisfy *source-free* Maxwell's equations in their respective regions and both of which are valid fields in a medium with parameters $(\varepsilon_1, \mu_1)$, we obtain for the surface currents $\mathfrak{J}_s = \hat{\mathbf{n}} \times (\mathbf{H}^s - (-\mathbf{H}^i)) = \hat{\mathbf{n}} \times \mathbf{H}$ and $\mathfrak{M}_s = -\hat{\mathbf{n}} \times (\mathbf{E}^s - (-\mathbf{E}^i)) = -\hat{\mathbf{n}} \times \mathbf{E}$ (figure 10.11). These currents are thus seen to depend on the total tangential electric and magnetic fields on the surface $S$. By the equivalence principle 10.1.3, these currents radiating in a homogeneous medium with parameters $(\varepsilon, \mu) = (\varepsilon_1, \mu_1)$ generate the fields $(\mathbf{E}^s, \mathbf{H}^s)$ exterior to $S$ when computed from equations (10.51) and (10.52). Thus

$$\mathbf{E}^s(\mathbf{r}) = -\oiint_S \left\{ j\omega\mu_1 \mathfrak{J}_s(\mathbf{r}')\psi_1(\mathbf{r}, \mathbf{r}') - \frac{q_e(\mathbf{r}')}{\varepsilon_1}\nabla'\psi_1(\mathbf{r}, \mathbf{r}') + \mathfrak{M}_s(\mathbf{r}') \times \nabla'\psi_1(\mathbf{r}, \mathbf{r}') \right\} \mathrm{d}s',$$

$$\mathbf{H}^s(\mathbf{r}) = -\oiint_S \left\{ j\omega\varepsilon_1 \mathfrak{M}_s(\mathbf{r}')\psi_1(\mathbf{r}, \mathbf{r}') - \frac{q_m(\mathbf{r}')}{\mu_1}\nabla'\psi_1(\mathbf{r}, \mathbf{r}') - \mathfrak{J}_s(\mathbf{r}') \times \nabla'\psi_1(\mathbf{r}, \mathbf{r}') \right\} \mathrm{d}s',$$

where the Green's function $\psi_1$ is the function $\psi$ evaluated at wavenumber $k = \omega\sqrt{\mu_1\varepsilon_1}$.

For the special case of a perfectly conducting object (for which $\mathfrak{M}_s \equiv 0,\ q_m \equiv 0$), the simpler expression

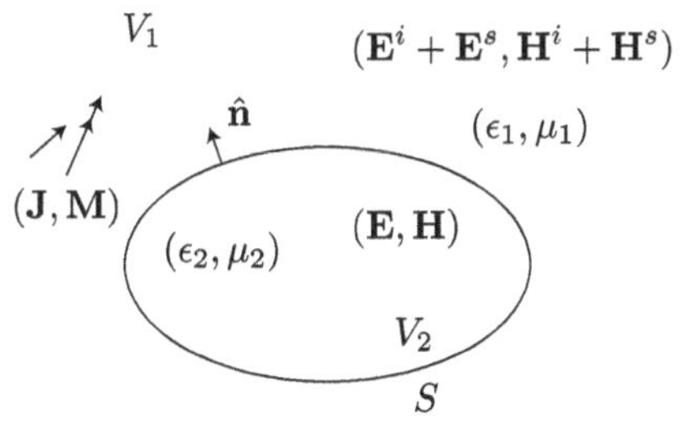


**Figure 10.10.** Sources radiating in the presence of an obstacle situated in an ambient medium.

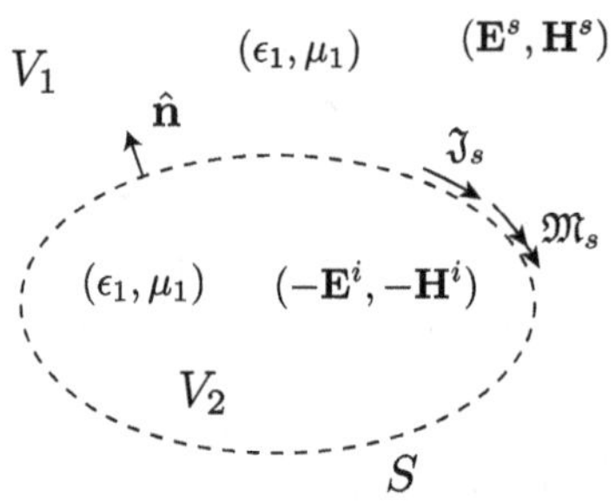


**Figure 10.11.** Physical equivalent for producing a scattered field outside $S$ and a negative incident field inside $S$.

$$\mathbf{H}^s(\mathbf{r}) = \oint\!\!\!\oint_S \mathfrak{J}_s(\mathbf{r}') \times \nabla'\psi_1(\mathbf{r}, \mathbf{r}')\, \mathrm{d}s'$$

holds. On dropping the current from the shadow region and assuming the current in the lit region to be equal to twice the tangential component of incident magnetic field

$$\mathfrak{J}_s \approx \begin{cases} 2(\hat{\mathbf{n}}' \times \mathbf{H}^i), & \mathbf{r}' \in S_{\text{lit}} \\ 0, & \mathbf{r}' \in S - S_{\text{lit}} \end{cases}, \tag{10.109}$$

we are led to the *physical optics approximation*

$$\mathbf{H}^s(\mathbf{r}) \approx \begin{cases} 2 \iint_{S_{\text{lit}}} [\hat{\mathbf{n}}' \times \mathbf{H}^i(\mathbf{r}')] \times \nabla'\psi_1(\mathbf{r}, \mathbf{r}')\, \mathrm{d}s', & \text{(Stratton–Chu formulation)} \\ 2\nabla \times \iint_{S_{\text{lit}}} [\hat{\mathbf{n}}' \times \mathbf{H}^i(\mathbf{r}')]\psi_1(\mathbf{r}, \mathbf{r}')\, \mathrm{d}s', & \text{(Franz formulation)} \end{cases}, \tag{10.110}$$

where $S_{\text{lit}}$ is the lit portion of $S$, i.e. the green portion in figure 10.12. The assumptions of the physical optics approximation hold quite accurately for electrically large and smooth (in the sense of Lyapunov) scatterers. It is interesting to compare this result with the result (11.13) obtained using an approximation of the induction theorem in example 11.1.1.2. In general, the two approximations will not coincide for all observation angles.

An expression for the scattered electric field under the physical optics approximation using Franz's formulation (10.42) is

$$\mathbf{E}^s(\mathbf{r}) = \frac{2}{j\omega\varepsilon_1}\nabla \times \nabla \times \iint_{S_{\text{lit}}} [\hat{\mathbf{n}}' \times \mathbf{H}^i(\mathbf{r}')]\psi_1(\mathbf{r}, \mathbf{r}')\, \mathrm{d}s', \tag{10.111}$$

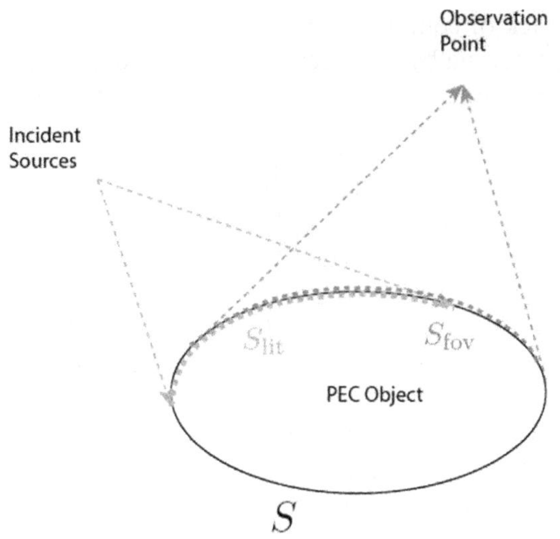


**Figure 10.12.** Portion $S_{\rm lit}$ (green color) of $S$ used in the physical optics approximation.

$$
\begin{aligned}
&\sim \frac{2}{j\omega\varepsilon_1}\nabla\times\nabla\times\left[\frac{\mathrm{e}^{-jk_1r}}{4\pi r}\iint_{S_{\rm lit}}[\hat{\mathbf{n}}'\times\mathbf{H}^{\rm i}(\mathbf{r}')]\mathrm{e}^{jk_1\hat{\mathbf{r}}\cdot\mathbf{r}'}\,\mathrm{d}s'\right],\ \text{(far-zone)}\\
&\sim jk_1\eta_1\frac{\mathrm{e}^{-jk_1r}}{2\pi r}\hat{\mathbf{r}}\times\hat{\mathbf{r}}\times\left[\iint_{S_{\rm lit}}[\hat{\mathbf{n}}'\times\mathbf{H}^{i}(\mathbf{r}')]\mathrm{e}^{jk_1\hat{\mathbf{r}}\cdot\mathbf{r}'}\,\mathrm{d}s'\right],\ \text{(far-zone)},
\end{aligned}
\tag{10.112}
$$

since $\nabla\times[\mathbf{a}\exp(-jk_1r)/r] = \nabla[\exp(-jk_1r)/r]\times\tilde{\mathbf{a}} - jk_1[\exp(-jk_1r)/r]\hat{\mathbf{r}}\times\mathbf{a}$, $k_1 = \omega\sqrt{\mu_1\varepsilon_1}$, $\eta_1 = \sqrt{\mu_1/\varepsilon_1}$, and $\omega\mu_1 = k_1\eta_1$. It is instructive to compare this expression (10.112) with the approximate solution (11.13) in example 11.1.1.2 obtained with an application of the induction theorem. The regions of integration on the surface are different in these two approaches and will only coincide in the back-scattered direction. In general, the two approximate solutions will not coincide in other directions. ■■

### 10.2.3 Bojarski's identity

In the forward scattering problem one is given the object and the incident wave and the goal is to determine the scattered fields everywhere. In an inverse scattering problem, the scattered field is measured over a number of frequencies and scattering angles by altering the properties of the incident wave and the goal is to infer the object shape and its constitution from the measurements. Inverse scattering is notoriously ill-conditioned due to imprecision of measurements, sensitivity of the inversion algorithms on initial conditions and small perturbations of object parameters, and a lack of complete data over the sample space. Below we consider the simple case of scattering by a perfectly conducting object and derive some useful results based on the approximate scattering technique of physical optics.

**Example 10.2.3.1** Inverse scattering and Bojarski's identity.
In this example we apply physical optics to a perfectly conducting object and show that it is possible to determine the object shape by measuring back-scattered fields over a range of angles and frequencies [10], [3, p 567].

Figure 10.13 shows a perfectly conducting object with volume $V$ and boundary $S$ illuminated by a plane wave with fields $(\mathbf{E}^{\mathrm{i}}, \mathbf{H}^{\mathrm{i}})$ propagating in the $\hat{\mathbf{k}}_{\mathrm{i}}$ direction. The radian frequency, wavenumber, intrinsic impedance, and media parameters exterior to the object are $\omega$, $k$, $\eta$, $(\varepsilon, \mu)$, respectively. Let the polarization of the electric field be described by the unit vector $\hat{\mathbf{e}}_{\mathrm{i}}$ and let the incident electric field be $\mathbf{E}^{\mathrm{i}} = \hat{\mathbf{e}}_{\mathrm{i}} E_0 \mathrm{e}^{-jk\hat{\mathbf{k}}_{\mathrm{i}}\cdot\mathbf{r}}$. Let the outward normal at an arbitrary point $\mathbf{r}'$ on the object be $\hat{\mathbf{n}}'$.

We consider the scattered wave in the back-scattered direction $\hat{\mathbf{r}} = -\hat{\mathbf{k}}_{\mathrm{i}}$. In view of equation (10.112), we note that

$$\begin{aligned}\eta\hat{\mathbf{k}}_{\mathrm{i}} \times \hat{\mathbf{k}}_{\mathrm{i}} \times (\hat{\mathbf{n}}' \times \mathbf{H}^{\mathrm{i}}(\hat{\mathbf{r}}')) &= (\hat{\mathbf{k}}_{\mathrm{i}} \cdot \hat{\mathbf{n}}')\hat{\mathbf{k}}_{\mathrm{i}} \times [-\eta\mathbf{H}^{\mathrm{i}}] \\ &= (\hat{\mathbf{k}}_{\mathrm{i}} \cdot \hat{\mathbf{n}}')\mathbf{E}^{\mathrm{i}}(\mathbf{r}') = \hat{\mathbf{e}}_{\mathrm{i}} E_0 (\hat{\mathbf{k}}_{\mathrm{i}} \cdot \hat{\mathbf{n}}')\mathrm{e}^{-jk\hat{\mathbf{k}}_{\mathrm{i}}\cdot\mathbf{r}'}.\end{aligned} \tag{10.113}$$

We therefore have from equation (10.112) (after dropping the extraneous subscript '1' from various terms) that in the back-scattered direction

$$\begin{aligned}\mathbf{E}^{\mathrm{s}}(\hat{\mathbf{r}} = -\hat{\mathbf{k}}_{\mathrm{i}}) &= jE_0\hat{\mathbf{e}}_{\mathrm{i}}\frac{\mathrm{e}^{-jkr}}{4\pi r}\iint_{S_{\mathrm{lit}}} (2k\hat{\mathbf{k}}_{\mathrm{i}} \cdot \hat{\mathbf{n}}')\mathrm{e}^{-j2k\hat{\mathbf{k}}_{\mathrm{i}}\cdot\mathbf{r}'}\,\mathrm{d}s' \\ &= jE_0\hat{\mathbf{e}}_{\mathrm{i}}\frac{\mathrm{e}^{-jkr}}{4\pi r}\iint_{S_{\mathrm{lit}}} (\boldsymbol{\kappa} \cdot \hat{\mathbf{n}}')\mathrm{e}^{-j\boldsymbol{\kappa}\cdot\mathbf{r}'}\,\mathrm{d}s',\end{aligned}$$

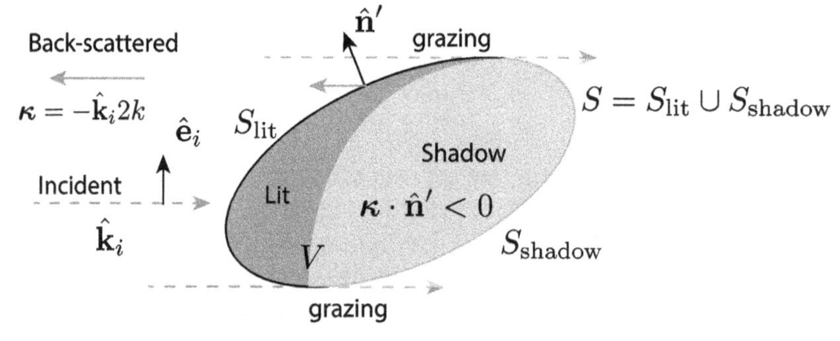


**Figure 10.13.** Lit portion $S_{\mathrm{lit}}$ $(\boldsymbol{\kappa} \cdot \hat{\mathbf{n}}' > 0)$ and shadow portion $S_{\mathrm{shadow}}$ $(\boldsymbol{\kappa} \cdot \hat{\mathbf{n}}' < 0)$ of $S$ due to an incident plane wave.

where $\boldsymbol{\kappa} = 2k\hat{\mathbf{k}}_i$ is *not* a unit vector. It is seen that *under the physical optics approximation* the scattered electric field is polarized in the same direction as the incident electric field in the back-scattered direction. This is despite the fact that the object is finite and could cause depolarization in general.

Let us now define a complex-valued scattering amplitude $F_1(\boldsymbol{\kappa})$ from the above such that $\mathbf{E}^s(\hat{\mathbf{r}} = -\hat{\mathbf{k}}_i) = F_1(\boldsymbol{\kappa})E_0\hat{\mathbf{e}}_i e^{-jkr}/r\sqrt{4\pi}$, or conversely, $F_1(\boldsymbol{\kappa}) = \sqrt{4\pi}re^{jkr}\hat{\mathbf{e}}_i \cdot \mathbf{E}^s(\hat{\mathbf{r}} = -\hat{\mathbf{k}}_i)/E_0$ (m). Then

$$F_1(\boldsymbol{\kappa}) = \frac{j}{\sqrt{4\pi}} \iint_{S_{\text{lit}}} [\boldsymbol{\kappa} e^{-j\boldsymbol{\kappa}\cdot\mathbf{r}'}] \cdot \hat{\mathbf{n}}' \mathrm{d}s'. \tag{10.114}$$

If the direction of the incident wave is reversed such that $\hat{\mathbf{k}}_i \to -\hat{\mathbf{k}}_i$, $\boldsymbol{\kappa} \to -\boldsymbol{\kappa}$, the roles of the shadow and lit portions of $S$ are reversed, see figure 10.13. Taking a complex conjugate of the resulting scattering amplitude gives

$$F_1^*(-\boldsymbol{\kappa}) = \frac{j}{\sqrt{4\pi}} \iint_{S_{\text{shadow}}} [\boldsymbol{\kappa} e^{-j\boldsymbol{\kappa}\cdot\mathbf{r}'}] \cdot \hat{\mathbf{n}}' \mathrm{d}s'. \tag{10.115}$$

It is seen that the integrand for $F_1^*(-\boldsymbol{\kappa})$ remains the same as that for $F_1(\boldsymbol{\kappa})$. Adding $F_1(\boldsymbol{\kappa})$ and $F_1^*(-\boldsymbol{\kappa})$ and noting that $S_{\text{lit}} \cup S_{\text{shadow}} = S$ (true for convex bodies, or bodies without reentrant cavities) one obtains

$$\begin{aligned} F_1(\boldsymbol{\kappa}) + F_1^*(-\boldsymbol{\kappa}) &= \frac{j}{\sqrt{4\pi}} \oiint_S [\boldsymbol{\kappa} e^{-j\boldsymbol{\kappa}\cdot\mathbf{r}'}] \cdot \hat{\mathbf{n}}' \mathrm{d}s' \\ &= \frac{j}{\sqrt{4\pi}} \iiint_V \nabla' \cdot [\boldsymbol{\kappa} e^{-j\boldsymbol{\kappa}\cdot\mathbf{r}'}] \,\mathrm{d}v', \end{aligned} \tag{10.116}$$

where the last step follows from an application of the divergence theorem. Now $\nabla' \cdot [\boldsymbol{\kappa} e^{-j\boldsymbol{\kappa}\cdot\mathbf{r}'}] = \boldsymbol{\kappa} \cdot \nabla' e^{-j\boldsymbol{\kappa}\cdot\mathbf{r}'} = -j\boldsymbol{\kappa} \cdot \boldsymbol{\kappa} e^{-j\boldsymbol{\kappa}\cdot\mathbf{r}'} = -4jk^2 e^{-j\boldsymbol{\kappa}\cdot\mathbf{r}'}$. Let $\gamma(\mathbf{r}')$ be the indicator function of the object (the probability of finding the object at $\mathbf{r}'$; must be real and non-negative), i.e.

$$\gamma(\mathbf{r}') = \begin{cases} 0, & \text{if } \mathbf{r}' \notin V \\ 1, & \text{if } \mathbf{r}' \in V \end{cases},$$

and let

$$F(\boldsymbol{\kappa}) = \frac{1}{4\pi\sqrt{2}k^2}[F_1(\boldsymbol{\kappa}) + F_1^*(-\boldsymbol{\kappa})], \quad (\mathrm{m}^3)$$

be the scaled Hermitian symmetric scattering amplitude of the object, that is, with the property $F(-\boldsymbol{\kappa}) = F^*(\boldsymbol{\kappa})$. Inserting all of these into equation (10.116) we arrive at the following relation between $F(\boldsymbol{\kappa})$ and $\gamma(\mathbf{r}')$:

$$F(\boldsymbol{\kappa}) = \frac{1}{(2\pi)^{3/2}} \iiint_{-\infty}^{\infty} \gamma(\mathbf{r}') e^{-j\boldsymbol{\kappa}\cdot\mathbf{r}'} \,\mathrm{d}v'. \tag{10.117}$$

The 3D Fourier transform relation between the indicator function $\gamma(\mathbf{r}')$ and the scattering amplitude $F(\boldsymbol{\kappa})$ is apparent. The dual variables are the space variable $\mathbf{r}'$ and the wavenumber variable $\boldsymbol{\kappa}$, both of which are three-dimensional in nature. Note that $F(\boldsymbol{\kappa} = 0) = V$, which is assumed finite here. The inverse of equation (10.117) gives

$$\gamma(\mathbf{r}') = \frac{1}{(2\pi)^{3/2}} \iiint_{-\infty}^{\infty} F(\boldsymbol{\kappa}) e^{j\boldsymbol{\kappa}\cdot\mathbf{r}'} \, d\boldsymbol{\kappa}. \tag{10.118}$$

Equation (10.118) is known as the *physical optics inverse scattering identity* or the *Bojarski's identity* and says that the object shape can be determined on knowing the back-scattered scattering amplitude over all $\boldsymbol{\kappa}$. The latter involves all frequencies and all observation angles. The totality of all scattering amplitudes when plugged into equation (10.118) will determine the function $\gamma(\mathbf{r}')$, which is non-negative. It should be borne in mind that equation (10.118) is based on the physical optics approximation, which assumes that the principal radii of curvature on the object are large compared to wavelength of operation. Consequently, it becomes invalid at low frequencies.

It is possible to convert the Fourier transform relations in equations (10.117) and (10.118) to expressions involving other transforms in non-Cartesian coordinates. In spherical coordinates we define $\mathbf{r}' = r'(\hat{\mathbf{x}} \sin\theta' \cos\phi' + \hat{\mathbf{y}} \sin\theta' \sin\phi' + \hat{\mathbf{z}} \cos\theta')$, $\boldsymbol{\kappa} = \kappa(\hat{\mathbf{x}} \sin\alpha \cos\beta + \hat{\mathbf{y}} \sin\alpha \sin\beta + \hat{\mathbf{z}} \cos\alpha)$ so that $\boldsymbol{\kappa} \cdot \mathbf{r}' = \kappa r'(\sin\alpha \sin\theta' \cos(\phi' - \beta) + \cos\alpha \cos\theta')$, $dv' = r'^2 \sin\theta' \, dr' d\theta' d\phi'$, and $d\boldsymbol{\kappa} = \kappa^2 \sin\alpha \, d\kappa d\alpha d\beta$. We present $\gamma(\mathbf{r}')$ and $F(\boldsymbol{\kappa})$ in spherical harmonics as

$$F(\boldsymbol{\kappa}) = \sum_{n,m} (-j)^n F_{nm}(\kappa) Y_n^m(\alpha, \beta), \tag{10.119}$$

$$\gamma(\mathbf{r}') = \sum_{n,m} \gamma_{nm}(r') Y_n^m(\theta', \phi'), \tag{10.120}$$

where $Y_n^m(\cdot, \cdot)$ is the surface spherical harmonic defined in equation (1.103). The coefficients of expansion $F_{nm}$ and $\gamma_{nm}$ can be determined from the orthogonality relationship in equation (1.105). Accordingly,

$$F_{nm}(\kappa) = j^n \int_{\alpha=0}^{\pi} \int_{\beta=0}^{2\pi} F(\kappa, \alpha, \beta) \bar{Y}_n^m(\alpha, \beta) \sin\alpha \, d\beta d\alpha, \tag{10.121}$$

$$\gamma_{nm}(r') = \int_{\theta'=0}^{\pi} \int_{\phi'=0}^{2\pi} \gamma(r', \theta', \phi') \bar{Y}_n^m(\theta', \phi') \sin\theta' \, d\phi' d\theta'. \tag{10.122}$$

Substituting equations (10.119) and (10.120) into equations (10.118) and (10.117), respectively, and applying the result (1.74) and the fact $j_n(-z) = (-1)^n j_n(z)$ for the spherical Bessel function $j_n(\cdot)$, we obtain the relations

$$\gamma_{nm}(r') = \sqrt{\frac{2}{\pi}} \int_{\kappa=0}^{\infty} \kappa^2 j_n(\kappa r') F_{nm}(\kappa) \, d\kappa, \tag{10.123}$$

$$F_{nm}(\kappa) = \sqrt{\frac{2}{\pi}} \int_{r'=0}^{\infty} r'^2 j_n(\kappa r') \gamma_{nm}(r')\, \mathrm{d}r'. \tag{10.124}$$

Thus $\gamma_{nm}(r')$ and $F_{nm}(\kappa)$ are spherical Bessel transforms of each other. Substituting equation (10.123) into equation (10.124) and changing the orders of integration gives

$$F_{nm}(\kappa) = \int_{\kappa_c=0}^{\infty} F_{nm}(\kappa_c) \left[ \frac{2}{\pi} \kappa_c^2 \int_{r'=0}^{\infty} r'^2 j_n(\kappa r') j_n(\kappa_c r')\, \mathrm{d}r' \right] \mathrm{d}\kappa_c$$

which implies that

$$\int_{r'=0}^{\infty} r'^2 j_n(\kappa r') j_n(\kappa_c r')\, \mathrm{d}r' = \frac{\pi}{2} \frac{\delta(\kappa_c - \kappa)}{\kappa_c^2}, \tag{10.125}$$

since the first equation is an identity for $F_{nm}(\kappa)$. Equation (10.125) is the completeness relation for the spherical Bessel functions and is simply a restatement of identity (C.54). On making measurements at all frequencies and over all observation angles for the far-zone scattering amplitude $F(\boldsymbol{\kappa})$, one obtains the coefficients $F_{nm}(\kappa)$ through equation (10.121). Then the coefficients $\gamma_{nm}(r')$ are determined using equation (10.123). Finally one obtains the object indicator function $\gamma(\mathbf{r}')$ from equation (10.120).

If the scattering amplitude is not known exactly, but only a filtered version is known, then the inversion algorithm will generate a distorted version of the object shape. If the filter with transfer function $H(\boldsymbol{\kappa})$ (unitless) and 'impulse response'

$$h(\mathbf{r}') = \frac{1}{(2\pi)^{3/2}} \iiint_{-\infty}^{\infty} H(\boldsymbol{\kappa}) \mathrm{e}^{j\boldsymbol{\kappa}\cdot\mathbf{r}'}\, \mathrm{d}\boldsymbol{\kappa},$$

is used in the inversion algorithm, the resulting object indicator function is

$$\begin{aligned} \tilde{\gamma}(\mathbf{r}') &= \frac{1}{(2\pi)^{3/2}} \iiint_{-\infty}^{\infty} H(\boldsymbol{\kappa}) F(\boldsymbol{\kappa}) \mathrm{e}^{j\boldsymbol{\kappa}\cdot\mathbf{r}'}\, \mathrm{d}\boldsymbol{\kappa} \\ &= \frac{1}{(2\pi)^{3/2}} \iiint_{-\infty}^{\infty} h(\mathbf{r}' - \mathbf{r}) \gamma(\mathbf{r})\, \mathrm{d}\mathbf{r}, \end{aligned} \tag{10.126}$$

where the last quantity is the convolutional integral for the symmetrically defined Fourier transform pair. Thus the object shape will be smeared by the impulse response of the filter. In addition, *with an incomplete scattering amplitude data, there is no a longer a guarantee that the distorted indicator function will be non-negative or even real-valued.* Of course if $H(\boldsymbol{\kappa}) \equiv 1$, then $h(\mathbf{r}') = \delta(\mathbf{r}')$ and one recovers the exact shape as expected. ■■

*10.2.3.1 Special case: single frequency measurement*

If measurements are performed only at a single radian frequency $\omega_c$ (corresponding $\kappa = \kappa_c = 2k_c = 2\omega_c\sqrt{\mu\varepsilon}$), but over all observation angles, then the filter transfer function is $H(\boldsymbol{\kappa}; \kappa_c) = H_0(\kappa_c)\delta(\kappa - \kappa_c)$. The constant $H_0(\kappa_c)$ is shown to depend on $\kappa_c$ and is not critical. It will be chosen suitably to facilitate comparison with the exact object shape. Substituting into equation (10.126)

$$\tilde{\gamma}(\mathbf{r}'; \kappa_c) = \frac{H_0(\kappa_c)\kappa_c^2}{(2\pi)^{3/2}} \oiint_{\alpha,\beta} F(\kappa_c, \alpha, \beta)\mathrm{e}^{j\kappa_c r' \cos\xi} \sin\alpha \, \mathrm{d}\beta \mathrm{d}\alpha, \tag{10.127}$$

where $\cos\xi = \sin\theta' \sin\alpha \cos(\phi' - \beta) + \cos\theta' \cos\alpha$. Making use of equation (10.119) for $F(\kappa_c, \alpha, \beta)$ and utilizing the identity (1.74) gives

$$\begin{aligned}
\tilde{\gamma}(\mathbf{r}'; \kappa_c) &= \sqrt{\frac{2}{\pi}} H_0(\kappa_c) \sum_{n,m} Y_n^m(\theta', \phi')\kappa_c^2 j_n(\kappa_c r') F_{nm}(\kappa_c) \\
&= \frac{2}{\pi} H_0(\kappa_c) \sum_{n,m} Y_n^m(\theta', \phi')\kappa_c^2 j_n(\kappa_c r') \int_{r_1=0}^{\infty} r_1^2 j_n(\kappa_c r_1) \\
&\times \left[ \int_{\theta_1=0}^{\pi} \int_{\phi_1=0}^{2\pi} \gamma(r_1, \theta_1, \phi_1) \bar{Y}_n^m(\theta_1, \phi_1) \sin\theta_1 \, \mathrm{d}\phi_1 \mathrm{d}\theta_1 \right] \mathrm{d}r_1,
\end{aligned} \tag{10.128}$$

where equations (10.124) and (10.122) have been used in the last step. Equation (10.128) relates the predicted object shape with the true object shape $\gamma(\mathbf{r}')$ for a single frequency measurement. The strength of the surface spherical harmonics is modified when frequencies are discarded from measurement.

If the true object is a sphere of radius $a$, its indicator function is

$$\gamma(\mathbf{r}') = \begin{cases} 1, & r' < a \\ 0, & r' > a \end{cases}. \tag{10.129}$$

Substituting this in equation (10.128) and noting that

$$\int_{\theta_1=0}^{\pi} \int_{\phi_1=0}^{2\pi} \bar{Y}_n^m(\theta_1, \phi_1) \sin\theta_1 \, \mathrm{d}\phi_1 \mathrm{d}\theta_1 = \delta_n^0 \delta_m^0,$$

finally yields the following expression for the predicted object shape with a single frequency measurement:

$$\begin{aligned}
\tilde{\gamma}(\mathbf{r}'; \kappa_c) &= \frac{2H_0(\kappa_c)\kappa_c^2}{\pi} j_0(\kappa_c r') \int_{r_1=0}^{a} r_1^2 j_0(\kappa_c r_1) \, \mathrm{d}r_1 \\
&= \frac{2H_0(\kappa_c)}{\pi} \frac{\sin\kappa_c r'}{r'} \int_{r_1=0}^{a} r_1 \sin\kappa_c r_1 \, \mathrm{d}r_1, \\
&= \frac{2H_0(\kappa_c)a}{\pi} \left[ \frac{\sin\kappa_c a}{\kappa_c a} - \cos\kappa_c a \right] \frac{\sin\kappa_c r'}{\kappa_c r'},
\end{aligned} \tag{10.130}$$

since $j_0(z) = \sin(z)/z$. As a check, integration of the top line over all $\kappa_c \in (0, \infty)$ with $H_0(\kappa_c) = 1$ gives

$$
\begin{aligned}
\int_{\kappa_c=0}^{\infty} \tilde{\gamma}(\mathbf{r}'; \kappa_c)\, d\kappa_c &= \int_{r_1=0}^{a} \frac{r_1}{r'} \int_{\kappa_c=0}^{\infty} \frac{2}{\pi} \sin \kappa_c r' \sin \kappa_c r_1 \, d\kappa_c dr_1 \\
&= \int_{r_1=0}^{a} \frac{r_1}{r'} \lim_{\kappa_c \to \infty} \left[ \frac{\sin \kappa_c (r' - r_1)}{\pi (r' - r_1)} + \frac{\sin \kappa_c (r' + r_1)}{\pi (r' + r_1)} \right] dr_1 \qquad (10.131) \\
&= \int_{r_1=0}^{a} \frac{r_1}{r'} [\delta(r' - r_1) + \delta(r' + r_1)]\, dr_1 = \begin{cases} 1, & r' < a \\ 0, & r' > a \end{cases},
\end{aligned}
$$

upon using the result [11, p 30] $\lim_{\Omega \to \infty} \frac{\sin \Omega(t' - t)}{\pi(t' - t)} = \delta(t' - t)$. As remarked earlier, one recovers the exact shape in the case when all frequencies with equal weight are used in the measurement.

It is obvious from equation (10.130) that with this single frequency measurement, the distorted indicator function $\tilde{\gamma}(\mathbf{r}'; \kappa_c)$ is not always non-negative or even non-zero. In fact if $\kappa_c a$ is such that the quantity within the square brackets of equation (10.130) is zero, then $\tilde{\gamma}(\mathbf{r}'; \kappa_c)$ will be zero for all $\mathbf{r}'$. Barring these pathological cases, we may choose $H_0(\kappa_c)$ such that $\tilde{\gamma}(\mathbf{r}' = 0; \kappa_c) = 1$, in which case

$$
\tilde{\gamma}(\mathbf{r}'; \kappa_c) = \frac{\sin \kappa_c r'}{\kappa_c r'}. \qquad (10.132)
$$

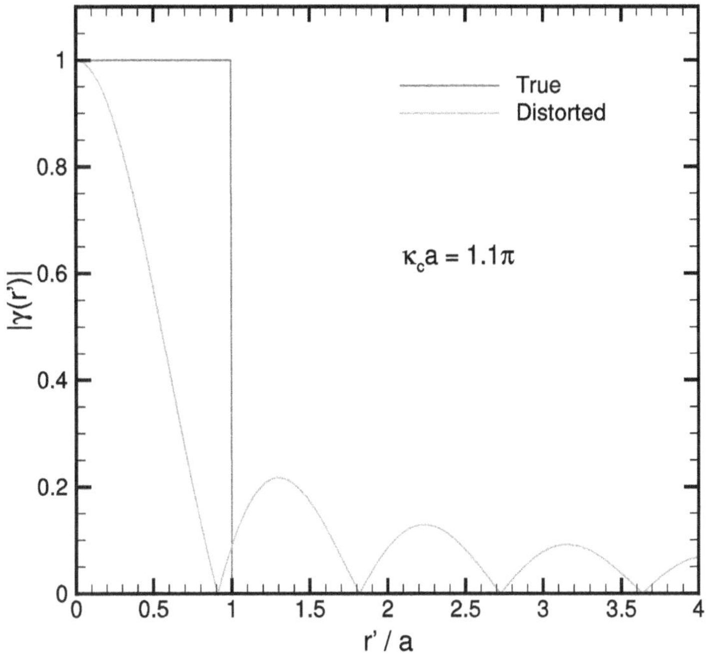


**Figure 10.14.** Absolute indicator function for a sphere and distorted object inverted from single frequency measurement.

The probability of finding the object in space is no longer of compact support. Furthermore, $\tilde{\gamma}(\mathbf{r}')$ is negative in certain regions. To physically interpret it in such regions, one may take the probability of finding the object at a location $\mathbf{r}'$ to be $|\tilde{\gamma}(\mathbf{r}')|$. Figure 10.14 shows the true indicator function for the sphere along the radial direction and the absolute value $|\tilde{\gamma}(\mathbf{r}')|$ obtained with a single frequency measurement at $\kappa_c a = 1.1\pi$ (equivalently, $2k_c a = 2.2\pi$). Smearing of the object due to limited frequency data is clearly seen.

## References


[1] Stratton J A 1941 *Electromagnetic Theory* (New York: McGraw-Hill)

[2] Poggio A J and Miller E 1973 Integral equation solution of three dimensional scattering problems *Computer Techniques in Electromagnetics* (New York: Elsevier)

[3] Ishimaru A 2017 *Electromagnetic Wave Propagation, Radiation, and Scattering* (Hoboken, NJ: Wiley)

[4] Schelkunoff S A 1936 Some equivalence theorems of electromagnetics and their application to radiation problems *Bell Syst. Tech. J.* **15** 92–112

[5] Rengarajan S R and Rahmat-Samii Y 2000 The field equivalence principle: illustration of the establishment of non-intuitive null fields *IEEE Antennas Propag. Mag.* **42** 122–8

[6] Cheng D K 1992 *Field and Wave Electromagnetics* 2nd edn (Reading, MA: Addison-Wesley)

[7] Tai C T 1972 Kirchhoff theory: scalar, vector, or dyadic *IEEE Trans. Antennas Propag.* **20** 114–5

[8] Müller C 1969 *Foundations of the Mathematical Theory of Electromagnetic Waves* (New York: Springer)

[9] Janaswamy R 1992 Oblique scattering from lossy periodic surfaces with application to anechoic chamber absorbers *IEEE Trans. Antennas Propag.* **40** 162–9

[10] Bojarski N N 1982 A survey of the physical optics inverse scattering identity *IEEE Trans. Antennas Propag.* **AP-30** 980–9

[11] Papoulis A D 1962 *The Fourier Integral and its Applications* (New York: McGraw-Hill)

# Chapter 11

# Induction theorem and optical theorem

## 11.1 Induction and forward scattering theorems

A theorem closely related to the equivalence principle is the induction theorem, which allows a scattering formulation in terms of known currents (proportional to the incident fields) radiating in the presence of the scattering object. It is widely used in radar cross section studies. The forward scattering theorem, also known as the optical theorem, relates the extinction cross section (the sum of absorption and scattering cross sections of a penetrable scatterer) to the scattering amplitude in the direction of the incident wave. It has many applications in the areas of microwave and optical remote sensing [1, 2]. Both of these theorems can be formulated in terms of the integral representations discussed in chapter 10 and are presented below.

### 11.1.1 Induction theorem

**Theorem 11.1.1** Induction theorem [3, 4, p 113].
*Let a closed surface* S *separate two homogeneous regions* $V_1$ *and* $V_2$ *having constitutive parameters* $(\varepsilon_1, \mu_1)$ *and* $(\varepsilon_2, \mu_2)$ *respectively. Let the sources* $(\mathbf{J}, \mathbf{M})$ *contained in region* $V_1$ *produce a field* $(\mathbf{E}^i, \mathbf{H}^i)$ *in an infinite homogeneous medium with parameters* $(\varepsilon_1, \mu_1)$*. Let the additional fields produced in region* $V_1$ *due to the presence of an object inside* S *be* $(\mathbf{E}^s, \mathbf{H}^s)$*. Let the total field inside* S *in the presence of the object be* $(\mathbf{E}, \mathbf{H})$*. See figure 11.1. Then the fields* $(\mathbf{E}^s, \mathbf{H}^s)$ *and* $(\mathbf{E}, \mathbf{H})$ *are source-free everywhere except on* S*. The distribution of electric and magnetic sources* $(\mathbf{J}_s, \mathbf{M}_s)$ *required on* S *to support a field* $(\mathbf{E}^s, \mathbf{H}^s)$ *outside* S *having constitutive parameters* $(\varepsilon_1, \mu_1)$ *and* $(\mathbf{E}, \mathbf{H})$ *inside* S *having constitutive parameters* $(\varepsilon_2, \mu_2)$ *are calculable from the boundary conditions on* S *and are* $\mathbf{J}_s = -\hat{\mathbf{n}} \times \mathbf{H}^i$, $\mathbf{M}_s = \hat{\mathbf{n}} \times \mathbf{E}^i$. See figure 11.2.

*Proof.* Note that boundary conditions dictate that $\hat{\mathbf{n}} \times [\mathbf{E}^i + \mathbf{E}^s] = \hat{\mathbf{n}} \times \mathbf{E}$, $\hat{\mathbf{n}} \times [\mathbf{H}^i + \mathbf{H}^s] = \hat{\mathbf{n}} \times \mathbf{H}$ on *S*. The surface currents in the equivalent problem are entirely in terms of the known fields $(\mathbf{E}^i, \mathbf{H}^i)$. However, these known currents

doi:10.1088/978-0-7503-1716-0ch11

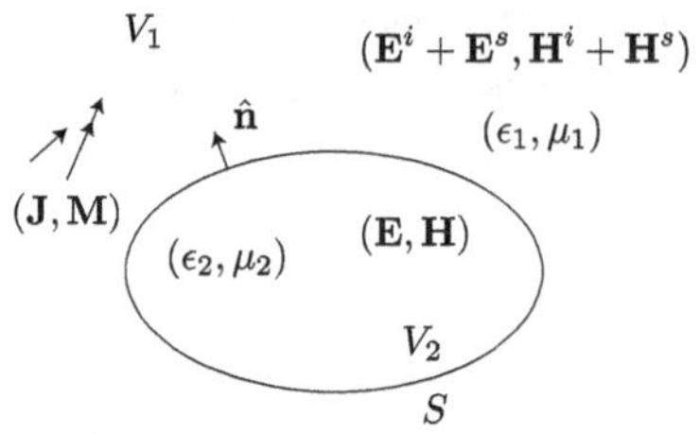


**Figure 11.1.** Sources radiating in the presence of an obstacle situated in an ambient medium.

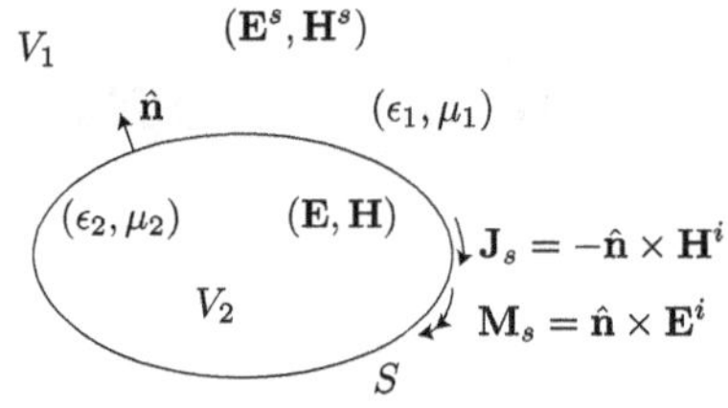


**Figure 11.2.** Induction equivalent for producing scattered field outside $S$ and total field inside $S$.

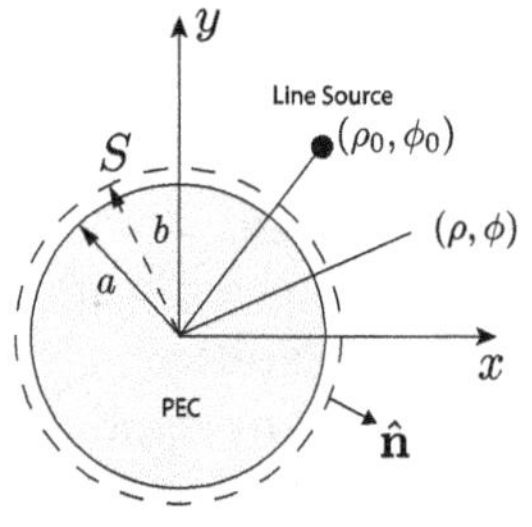


**Figure 11.3.** Cross section of line source radiating in the presence of a PEC circular cylinder.

radiate, not in a homogeneous medium, but in the presence of the object to produce the desired fields in the two regions. In other words the surface integrals in equations (10.14) and (10.15) cannot be used to produce the scattered fields here as they are applicable to a homogeneous medium. Because the original problem and the equivalent problem impose the same boundary conditions on the respective source-free fields (scattered fields outside $S$ and total fields inside $S$) they must have identical solutions due to the uniqueness theorem.

**Example 11.1.1.1.** Induction theorem.
Consider the problem of scattering of time-harmonic line source fields by a circular PEC cylinder of radius $a$. The line source is located at ($\rho_0 > a$, $\phi_0$) and carries a current $I$ in the $\hat{\mathbf{z}}$ direction (figure 11.3). The circular cylinder is placed in a homogeneous medium with electrical parameters ($\varepsilon$, $\mu$) and has its axis coinciding with the $z$-axis.

The problem can be worked out exactly using cylindrical harmonics and is available in many texts including [4, p 236]. The solution for the incident, $A_z^{\mathrm{i}}$, and scattered, $A_z^{\mathrm{s}}$, vector potential components is

$$A_z^{\rm i} = \frac{\mu I}{4j}\sum_{n=-\infty}^{\infty} J_n(kr_<)H_n^{(2)}(kr_>)e^{jn(\phi-\phi_0)}, \tag{11.1}$$

$$A_z^{\rm s} = -\frac{\mu I}{4j}\sum_{n=-\infty}^{\infty} \frac{J_n(ka)H_n^{(2)}(k\rho_0)}{H_n^{(2)}(ka)}H_n^{(2)}(k\rho)e^{jn(\phi-\phi_0)}, \tag{11.2}$$

where $r_< = \min(\rho, \rho_0)$, $r_> = \max(\rho, \rho_0)$. The sum of equations (11.1) and (11.2) constitutes the two-dimensional Green's function $\mathcal{G}_A(\rho, \phi; \rho_0, \phi_0)$ for the vector potential for an electric line source located at $(\rho_0, \phi_0)$. The fields generated will be $\text{TM}_z$ and are furthermore $z$-invariant.

We will now work out the same problem using the induction theorem, wherein external currents are defined and placed on the surface $S$ of radius $b$ just outside the PEC scatterer to generate a scattered field. The scattered field will be shown to be the same as in equation (11.2). On the surface of the cylinder $\hat{\mathbf{n}} = \hat{\rho}$ and the external surface currents determined by the incident fields are $\mathbf{J}_s(\phi) = -\hat{\rho} \times \mathbf{H}^{\rm i} = -\hat{\mathbf{z}}\frac{1}{\mu}\frac{\partial A_z^{\rm i}}{\partial \rho}|_{\rho=a}$ and $\mathbf{M}_s(\phi) = \hat{\rho} \times \mathbf{E}^{\rm i} = \hat{\boldsymbol{\phi}} j\omega A_z^{\rm i}|_{\rho=a}$. These currents radiate in the presence of the PEC cylinder to generate fields exterior to it. However, it is known that electric currents tangential to a PEC structure do not radiate any fields exterior to the PEC (see example 7.1.1.1)[1]. Hence it is sufficient to consider the fields generated by $\mathbf{M}_s$ alone, which, on using equation (11.1) is

$$\mathbf{M}_s(\phi) = \hat{\boldsymbol{\phi}}\frac{k\eta I}{4}\sum_{n=-\infty}^{\infty} J_n(ka)H_n^{(2)}(k\rho_0)e^{jn(\phi-\phi_0)} := \hat{\boldsymbol{\phi}}\sum_{n=-\infty}^{\infty} K_n\, e^{jn(\phi-\phi_0)}. \tag{11.3}$$

The current density is $z$-invariant and is in the form of Fourier series with respect to the azimuthal direction with known coefficients $K_n$. Our goal is to determine the total fields generated by this current density in the presence of the PEC cylinder. We will first need the solution for a cylindrical sheet of magnetic current of the form $\mathbf{M}_s(\phi; p) = \hat{\boldsymbol{\phi}}K_p e^{jp(\phi-\phi_0)} := \hat{\boldsymbol{\phi}}M_{s\phi}(\phi; p)$, with $p$ an integer and $K_p$ a constant, placed coaxially on a circular cylinder of radius $b > a$. This corresponds to the current density of one Fourier component in equation (11.3). Note that because of $z$-invariance we will still generate a $\text{TM}_z$ field with this magnetic current. In the end we take the limit as $b \to a$ to obtain the desired fields.

Let $A_1(\rho, \phi; p)$ and and $A_2(\rho, \phi; p)$ be the incident and scattered magnetic vector field components generated by $M_{s\phi}(\phi; p)$ in the presence of the PEC cylinder. Because of the orthogonality of the functions $e^{jn\phi}$ in $\phi \in (0, 2\pi)$ it is sufficient to set

$$A_1(\rho, \phi; p) = \frac{\mu I}{4j}B_p^{\lessgtr}J_p(kr_<)H_p^{(2)}(kr_>)e^{jp(\phi-\phi_0)}, \quad \rho \lessgtr b, \tag{11.4}$$

[1] This can also be shown here directly by integrating $J_s(\phi')$ with respect to the Green's function $\mathcal{G}_A(\rho, \phi; \phi', a)$.

$$A_2(\rho, \phi; p) = -\frac{\mu I}{4j} C_p \frac{J_p(ka) H_p^{(2)}(kb)}{H_p^{(2)}(ka)} H_p^{(2)}(k\rho) e^{jp(\phi - \phi_0)}, \tag{11.5}$$

where $r_< = \min(\rho, b)$, $r_> = \max(\rho, b)$ and determine the unknown coefficients $B_p^{\lessgtr}$ and $C_p$ by imposing boundary conditions (i) $H_\phi(\rho = b+) = H_\phi(\rho = b-)$, $E_z(b+) - E_z(b-) = M_{s\phi}(\phi; p)$ and (ii) $E_z = 0$ at $\rho = a$. The former boundary conditions give

$$-\frac{k\eta I}{4} J_p(kb) H_p^{(2)}(kb) \left[ B_p^> - B_p^< \right] = K_p, \tag{11.6}$$

$$B_p^> J_p(kb) H_p^{(2)'}(kb) - B_b^< J_p'(kb) H_p^{(2)}(kb) = 0, \tag{11.7}$$

where a prime in the Bessel functions denotes differentiation with respect to the entire argument. By setting

$$B_p^> J_p(kb) H_p^{(2)'}(kb) = B_p^< J_p'(kb) H_p^{(2)}(kb) = B_p J_p(kb) H_p^{(2)'}(kb) J_p'(kb) H_p^{(2)}(kb), \tag{11.8}$$

using the Wronskian for the Bessel functions and using equations (11.6) and (11.7), it is easy to see that $B_p^> - B_p^< = \frac{2j}{\pi kb} B_p$ with

$$B_p = \frac{2\pi jbK_p}{\eta I} \frac{1}{J_p(kb) H_p^{(2)}(kb)} \tag{11.9}$$

$$= \frac{j\pi ka}{2} \frac{H_p^{(2)}(k\rho_0)}{H_p^{(2)}(ka)}, \quad \text{for } b \to a. \tag{11.10}$$

Enforcement of boundary condition (ii) at $\rho = a$ gives $C_p = B_p^<$. Having determined the total field generated by one Fourier component $M_{sp}(\phi; p)$ of the surface current density in equation (11.3) we now sum up due to all components contained therein to obtain the total potential for $\rho > a$ and $b \to a$ as

$$\begin{aligned} A_z &= \sum_{p=-\infty}^{\infty} [A_1(\rho, \phi; p) + A_2(\rho, \phi; p)] \\ &= \frac{\mu I}{4j} \sum_{p=-\infty}^{\infty} (B_p^> - B_p^<) J_p(ka) H_p^{(2)}(k\rho) e^{jp(\phi - \phi_0)} \\ &= -\frac{\mu I}{4j} \sum_{p=-\infty}^{\infty} \frac{H_p^{(2)}(k\rho_0)}{H_2^{(2)}(ka)} J_p(ka) H_p^{(2)}(k\rho) e^{jp(\phi - \phi_0)}, \end{aligned} \tag{11.11}$$

which is identical to equation (11.2). Thus surface currents determined solely from the incident fields and radiating in the presence of the scattering object generate the exact scattered fields exterior to the object. ■■

**Example 11.1.1.2** Approximate formulation for PEC scatterer.
In this example we will arrive at an approximate formulation for determining the scattered electric field, $\mathbf{E}^s(\mathbf{r})$, using the induction theorem. Since an external electric current density placed tangentially on an arbitrary PEC object does not generate any net field exterior to the object by example 7.1.1.1, only the magnetic current density contributes to the scattered field. Denoting the dyadic Green's function [5] for the electric field of a magnetic current element located at $\mathbf{r}'$ and radiating in the presence of the object $S$ as $\mathcal{G}_E(\mathbf{r};\, \mathbf{r}')$,[2] the scattered electric field can be written as

$$\mathbf{E}^s(\mathbf{r}) = \oiint_S \mathbf{M}_s(\mathbf{r}') \cdot \mathcal{G}_E(\mathbf{r};\, \mathbf{r}')\, \mathrm{d}s'. \tag{11.12}$$

If the object is smooth and the mean curvature, $\kappa$, on it satisfies $\kappa\lambda \ll 1$, where $\lambda$ is the operating wavelength, then most of the contribution to the integral comes from the portion $S_{\text{fov}}$ of $S$ that is visible from the observation point (figure 11.4). Furthermore, image theory (see example 5.1.0.1) can be used to approximately account for the presence of the PEC obstacle on the magnetic current under the assumption $\kappa\lambda \ll 1$. Since the image of a magnetic dipole parallel to a PEC has the same orientation as the source dipole, image theory applied to the magnetic current density $\mathbf{M}_s$ will amplify it by a factor of two. Having accounted for the presence of the PEC obstacle, the amplified current density now radiates in the absence of the obstacle, i.e. in a homogeneous medium. Hence upon using equation (10.20) we obtain

$$\mathbf{E}^s(\mathbf{r}) \approx 2 \iint_{S_{\text{fov}}} \mathbf{M}_s(\mathbf{r}') \times \nabla'\psi(\mathbf{r}, \mathbf{r}')\, \mathrm{d}s', \tag{11.13}$$

where $\mathbf{M}_s = \hat{\mathbf{n}}' \times \mathbf{E}^i$. As the observation point is moved, $S_{\text{fov}}$ changes and different portions of the surface are used in the integral (11.13). It is instructive to compare

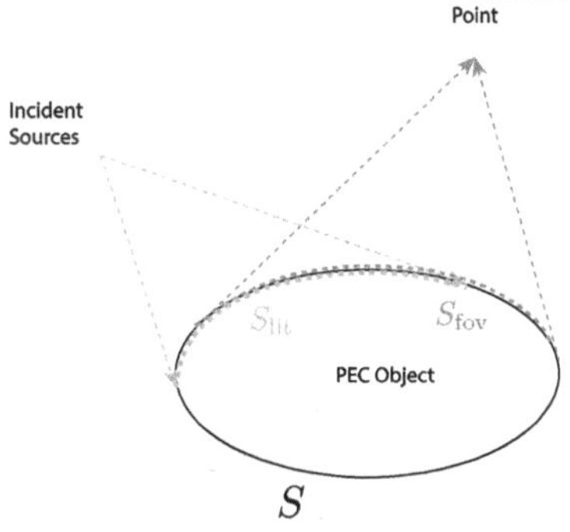


**Figure 11.4.** Portion $S_{\text{fov}}$ (red color) of $S$ used in the approximation.

[2] We do not need the explicit form of this Green's function in the ensuing analysis. Suffice it to know that $\mathcal{G}_E$ satisfies the boundary condition $\hat{\mathbf{n}} \times \mathcal{G}_E = 0$ on $S$.

this approximate result with the result (10.112) obtained using the physical optics approximation of example 10.2.2.1. The regions of integration on the surface are different in these two approaches and will only coincide in the back-scattered direction. In general, the two approximate solutions will not coincide in other directions. ■■

### 11.1.2 Forward scattering theorem

**Theorem 11.1.2.** Optical theorem [6, p 325].
*Let a unit amplitude ($E_0 = 1$ V $m^{-1}$) linearly polarized plane wave with unit polarization vector* $\hat{\mathbf{e}}_i$ *and propagating along a direction governed by the unit vector* $\hat{\mathbf{k}}_i$ *in an ambient lossless medium with constitutive parameters* $(\varepsilon_0, \mu_0)$ *strike an obstacle bounded by a closed surface* S *(figure 11.5). Let* $(\mathbf{E}^i, \mathbf{H}^i)$, $(\mathbf{E} = \mathbf{E}^i + \mathbf{E}^s,\ \mathbf{H} = \mathbf{H}^i + \mathbf{H}^s)$ *be, respectively, the incident field and the total field at the operating frequency* ω. *Define the incident Poynting vector,* $\mathbf{S}_i$, *total power absorbed by the obstacle,* $P_a$, *and the power scattered by the obstacle,* $P_s$, *as*

$$\mathbf{S}_i = \frac{1}{2}\mathbf{E}^i \times \mathbf{H}^{i*} = \hat{\mathbf{k}}_i S_i$$

$$P_a = -\frac{1}{2}\Re\left[\oiint_S \mathbf{E} \times \mathbf{H}^* \cdot \hat{\mathbf{n}}'\, ds'\right] := \sigma_a S_i, \tag{11.14}$$

$$P_s = \frac{1}{2}\Re\left[\oiint_S \mathbf{E}^s \times \mathbf{H}^{s*} \cdot \hat{\mathbf{n}}'\, ds'\right] := \sigma_s S_i, \tag{11.15}$$

*where* $\sigma_a$ *and* $\sigma_s$ *are known as the* absorption cross section *and the* scattering cross section, *respectively, and* $S_i$ *is the incident power density. Let* $\mathbf{F}^s(\hat{\mathbf{k}}_s, \hat{\mathbf{k}}_i)$ *(units of volts) be the vector scattering amplitude in a unit direction* $\hat{\mathbf{k}}_s$ *such that the far-zone scattered fields at a distance* r *from an arbitrary origin* O *inside* S *are*

$$\mathbf{E}^s \sim \frac{e^{-jkr}}{r}\mathbf{F}^s(\hat{\mathbf{k}}_s, \hat{\mathbf{k}}_i), \quad \mathbf{H}^s \sim \frac{1}{\eta}\left(\hat{\mathbf{k}}_s \times \mathbf{E}^s\right), \tag{11.16}$$

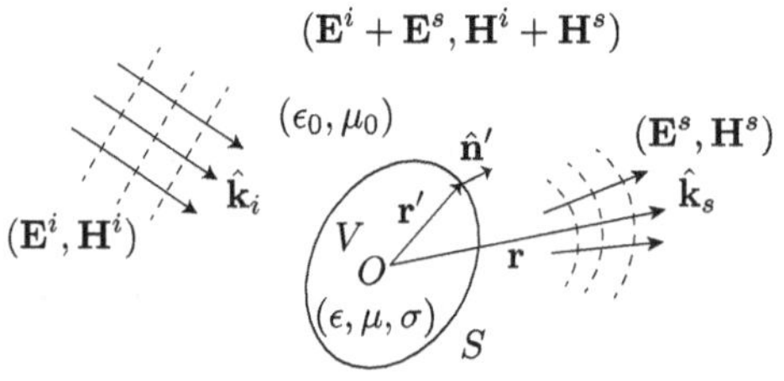


**Figure 11.5.** Plane wave striking a lossy scatterer bounded by surface $S$ and immersed in an ambient medium with constitutive parameters $(\varepsilon_0, \mu_0)$.

*where* $k = \omega\sqrt{\varepsilon_0\mu_0}$ *is the wavenumber,* $\eta = \sqrt{\mu_0/\varepsilon_0}$ *is the intrinsic impedance in the ambient medium. Then*

$$\sigma_s + \sigma_a = -\frac{4\pi}{k}\Im\left[\hat{\mathbf{e}}_i^* \cdot \frac{\mathbf{F}^s(\hat{\mathbf{k}}_i, \hat{\mathbf{k}}_i)}{E_0}\right], \quad (\mathrm{m}^2). \tag{11.17}$$

*Proof.* Note first by power conservation that

$$\oiint_S \mathbf{E}^i \times \mathbf{H}^{i*} \cdot \hat{\mathbf{n}}' \, \mathrm{d}s' = 0, \tag{11.18}$$

because the (a) the incident field has no sources inside $S$, (b) the incident field experiences no dissipation inside $S$, and (c) the incident electric and magnetic fields have identical energies in $S$. From Stratton–Chu integral representations (10.14), (10.15) and the results (10.16), (10.17) for incident fields we have for the scattered electric and magnetic fields

$$\mathbf{E}^s = \oiint_S \left\{-j\omega\mu_0(\hat{\mathbf{n}}' \times \mathbf{H}^s)\psi_0 + (\hat{\mathbf{n}}' \times \mathbf{E}^s) \times \nabla'\psi_0 + (\hat{\mathbf{n}}' \cdot \mathbf{E}^s)\nabla'\psi_0\right\} \mathrm{d}s', \tag{11.19}$$

$$\mathbf{H}^s = \oiint_S \left\{j\omega\varepsilon_0(\hat{\mathbf{n}}' \times \mathbf{E}^s)\psi_0 + (\hat{\mathbf{n}}' \times \mathbf{H}^s) \times \nabla'\psi_0 + (\hat{\mathbf{n}}' \cdot \mathbf{H}^s)\nabla'\psi_0\right\} \mathrm{d}s'. \tag{11.20}$$

Since we are interested in the region $S^+$ exterior to $S$ where $\psi_0$ and its derivatives are continuous, alternative forms are obtained by taking the curl of these equations and using Maxwell's equations, and freely moving the gradient operator from inside to outside of the integrals and vice versa. For instance, the scattered electric field can be written alternatively using Franz representation (10.42), [7]

$$\begin{aligned}\mathbf{E}^s = \frac{1}{j\omega\varepsilon_0}\nabla \times \mathbf{H}^s &= \nabla \times \oiint_S (\hat{\mathbf{n}}' \times \mathbf{E}^s)\psi_0 \, \mathrm{d}s' \\ &- \frac{j\eta}{k}\nabla \times \nabla \times \oiint_S (\hat{\mathbf{n}}' \times \mathbf{H}^s)\psi_0 \, \mathrm{d}s'.\end{aligned} \tag{11.21}$$

In the far-zone where $r \gg r'$ and

$$\psi_0 = \frac{e^{-jk|\mathbf{r}-\mathbf{r}'|}}{4\pi|\mathbf{r}-\mathbf{r}'|} \sim \frac{e^{-jkr}}{4\pi r}e^{jk\hat{\mathbf{k}}_s\cdot\mathbf{r}'}, \quad \nabla'\psi_0 = \nabla'\frac{e^{-jk|\mathbf{r}-\mathbf{r}'|}}{4\pi|\mathbf{r}-\mathbf{r}'|} \sim jk\hat{\mathbf{k}}_s\psi_0,$$

we obtain from equation (11.19) that

$$\mathbf{E}^s \sim \frac{e^{-jkr}}{r}\frac{jk}{4\pi}\oiint_S \left\{-\eta(\hat{\mathbf{n}}' \times \mathbf{H}^s) + (\hat{\mathbf{n}}' \times \mathbf{E}^s) \times \hat{\mathbf{k}}_s + (\hat{\mathbf{n}}' \cdot \mathbf{E}^s)\hat{\mathbf{k}}_s\right\} e^{jk\hat{\mathbf{k}}_s\cdot\mathbf{r}'} \mathrm{d}s', \tag{11.22}$$

$$:= \frac{e^{-jkr}}{r}\mathbf{F}^s(\hat{\mathbf{k}}_s, \hat{\mathbf{k}}_i). \tag{11.23}$$

Here $k$ is the wavenumber exterior to $S$ and is considered real. The dependence on $\hat{\mathbf{k}}_i$ in $\mathbf{F}^s$ is due to the dependence of the surface values of $\mathbf{E}^s$ and $\mathbf{H}^s$ on $\hat{\mathbf{k}}_i$ inside the integral in equation (11.22). The alternative form in equation (11.21) gives

$$\mathbf{F}^s(\hat{\mathbf{k}}_s, \hat{\mathbf{k}}_i) = \frac{-jk}{4\pi} \hat{\mathbf{k}}_s \times \left[ \oint\!\!\oint_S (\hat{\mathbf{n}}' \times \mathbf{E}^s) e^{jk\hat{\mathbf{k}}_s \cdot \mathbf{r}'} \, ds' - \eta \hat{\mathbf{k}}_s \times \oint\!\!\oint_S (\hat{\mathbf{n}}' \times \mathbf{H}^s) e^{jk\hat{\mathbf{k}}_s \cdot \mathbf{r}'} \, ds' \right] \tag{11.24}$$

upon using

$$\nabla \psi_0 \sim -jk\hat{\mathbf{r}}\psi_0, \qquad \nabla \times \nabla \times (\psi_0 \mathbf{a}) \sim k^2 \hat{\mathbf{r}} \times (\mathbf{a} \times \hat{\mathbf{r}})\psi_0,$$

where $\mathbf{a}$ is a vector independent of the observation coordinates. In this alternative form it is clear that the vector scattering amplitude is transverse to the direction of propagation. In the static limit of $k \to 0$, the two surface integrals vanish as is clear from the identity

$$\oint\!\!\oint_S (\hat{\mathbf{n}}' \times \mathbf{G}) \, ds' = \iiint_V \nabla \times \mathbf{G} \, dv' = 0 \text{ for a field satisfying } \nabla \times \mathbf{G} = 0 \text{ in } S^+.$$

The latter is true as both the static electric field and the static magnetic field outside the source region are conservative. This suggests that the electric and magnetic fields in the far-zone behave as $k^{1+\alpha}$, $\alpha > 0$. As a matter of fact, using the curl lemma, (B.6), and the form given in equation (11.24) it has been shown in [8] and later in [9] that the scattering amplitude can be rewritten as

$$\begin{aligned}\mathbf{F}^s(\hat{\mathbf{k}}_s, \hat{\mathbf{k}}_i) = \frac{k^2}{4\pi} \hat{\mathbf{k}}_s \times \Bigg[ \hat{\mathbf{k}}_s \times \oint\!\!\oint_S \mathbf{r}'(\eta \hat{\mathbf{k}}_s \cdot \hat{\mathbf{n}}' \times \mathbf{H}^s - \hat{\mathbf{n}}' \cdot \mathbf{E}^s) e^{jk\hat{\mathbf{k}}_s \cdot \mathbf{r}'} \, ds' \\ - \oint\!\!\oint_S \mathbf{r}'(\hat{\mathbf{k}}_s \cdot \hat{\mathbf{n}}' \times \mathbf{E}^s + \eta \hat{\mathbf{n}}' \cdot \mathbf{H}^s) e^{jk\hat{\mathbf{k}}_s \cdot \mathbf{r}'} \, ds' \Bigg].\end{aligned} \tag{11.25}$$

*Thus the scattering amplitude for any scatterer has a zero of order two at the origin of the complex k-plane.* Noting that

$$\mathbf{E}^i = E_0 \hat{\mathbf{e}}_i e^{-jk\hat{\mathbf{k}}_i \cdot \mathbf{r}}, \quad \mathbf{H}^i = \frac{E_0}{\eta}(\hat{\mathbf{k}}_i \times \hat{\mathbf{e}}_i) e^{-jk\hat{\mathbf{k}}_i \cdot \mathbf{r}}, \quad S_i = \frac{|E_0|^2}{2\eta} \tag{11.26}$$

$$\Re(\mathbf{E} \times \mathbf{H}^*) = \Re(\mathbf{E}^i \times \mathbf{H}^{i*}) + \Re(\mathbf{E}^s \times \mathbf{H}^{s*}) + \Re(\mathbf{E}^{i*} \times \mathbf{H}^s) + (\mathbf{E}^s \times \mathbf{H}^{i*}), \tag{11.27}$$

and using equations (11.18), (11.14) and (11.15) we obtain

$$
\begin{aligned}
(\sigma_a + \sigma_s)S_i &= -\frac{1}{2\eta}\Re\left[E_0^* \oiint_S \left\{\eta\hat{\mathbf{e}}_i^* \cdot \mathbf{H}^s \times \hat{\mathbf{n}}' + \hat{\mathbf{n}}' \times \mathbf{E}^s \cdot (\hat{\mathbf{k}}_i \times \hat{\mathbf{e}}_i)^*\right\} e^{jk\hat{\mathbf{k}}_i\cdot\mathbf{r}'}\,\mathrm{d}s'\right] \\
&= -\frac{1}{2\eta}\Re\left[E_0^*\hat{\mathbf{e}}_i^* \cdot \oiint_S \left\{-\eta(\hat{\mathbf{n}}' \times \mathbf{H}^s) + (\hat{\mathbf{n}}' \times \mathbf{E}^s) \times \hat{\mathbf{k}}_i\right\} e^{jk\hat{\mathbf{k}}_i\cdot\mathbf{r}'}\,\mathrm{d}s'\right] \\
&= -S_i\Re\left[\hat{\mathbf{e}}_i^* \cdot \left\{\frac{4\pi}{jkE_0}\mathbf{F}^s(\hat{\mathbf{k}}_i, \hat{\mathbf{k}}_i)\right\}\right] = -S_i\frac{4\pi}{k}\Im\left[\hat{\mathbf{e}}_i^* \cdot \frac{\mathbf{F}^s(\hat{\mathbf{k}}_i, \hat{\mathbf{k}}_i)}{E_0}\right],
\end{aligned}
\tag{11.28}
$$

where we have utilized equations (11.22), (11.23) and the fact that $\hat{\mathbf{e}}_i \cdot \hat{\mathbf{k}}_i = 0$. Therefore equation (11.17) follows. ■

The theorem applies under a wide variety of situations irrespective of the scatterer size and its composition and says that the *extinction cross section*, $\sigma_a + \sigma_s$, is only dependent on the properties of the scattering amplitude in the forward scatter direction. Note the special importance given to the scattering amplitude in the forward direction. Since the lhs of equation (11.17) is positive, the quantity within the square brackets on the rhs is negative. The rhs arises entirely from the cross terms between the incident and scattered fields.

## References


[1] Ishimaru A 1978 *Wave Propagation and Scattering in Random Media* vol 1 and 2 (New York: Academic)

[2] Tsang L, Kong J A and Ding K-H 2000 *Scattering of Electromagnetic Waves: Theories and ApplicationsWiley Series in Remote Sensing* (New York: Wiley)

[3] Schelkunoff S A 1936 Some equivalence theorems of electromagnetics and their application to radiation problems *Bell Syst. Tech. J.* **15** 92–112

[4] Harrington R F 1961 *Time-Harmonic Electromagnetic Fields* (New York: McGraw-Hill)

[5] Tai C T 1994 *Dyadic Green Functions in Electromagnetic Theory* (Piscataway, NJ: IEEE)

[6] Ishimaru A 2017 *Electromagnetic Wave Propagation, Radiation, and Scattering* (Hoboken, NJ: Wiley)

[7] Jones D S 1964 *The Theory of Electromagnetism* (New York: Pergamon)

[8] de Hoop A T 1959 On the plane-wave extinction cross-section of an obstacle *Appl. Sci. Res.* B **7** 463–9

[9] Kleinman R E 1967 Far field scattering at low frequencies *Appl. Sci. Res.* **18** 1–8

# Chapter 12

# Eigenfunctions, Green's functions, and completeness

The concept of completeness is invaluable in the understanding and devising of solution techniques in electrodynamics and in answering the following relevant questions: (i) What is the justification for the series expansion or integral representation of solutions that is commonly employed in various canonical coordinate systems? (ii) How does one ascertain the correctness or not of a series or integral representation of a solution? (iii) Under what conditions is a unique solution possible for a differential equation with a given input function? (iv) What checks can be performed for determining if a Green's function has a correct representation? (v) Under what conditions is an eigenfunction expansion for a Green's function possible? (vi) How does one transition from results available in 1D to the more practical case of multiple dimensions? (vii) What is the justification of the separation of variables technique that is so commonly used in multiple dimensions with regular geometries, etc? The answers to all of these questions are provided by examining the solvability conditions and examining the eigenfunctions of integral/differential operators, commutative operators associated with electromagnetic problems. In this regard, theorems 12.1.9, 12.1.10, 12.1.12, 12.2.3, 12.3.2, and 12.4.1 are crucial in providing appropriate answers. To better appreciate the properties of various operators and their eigenfunctions, we first present some introductory and background material on vector spaces and Hilbert spaces [1–5]. Some of this background material is also available within the electrodynamics community in the books by Van Bladel [6], Dudley [7], and Hanson [8], to name a few. The concepts and theorems are then applied to several electromagnetic examples to help answer the above questions.

doi:10.1088/978-0-7503-1716-0ch12

## 12.1 Hilbert space

**Definition 12.1.1.** Metric space.
A set of elements (or vectors). $x, y, z, \ldots$ is a metric space $\mathscr{S}$ if, for each pair $x, y \in \mathscr{S}$, there is associated a real number $d(x, y)$ called a metric such that

(i) $d(x, x) = 0$,
(ii) $d(x, y) = d(y, x) \geqslant 0$,
(iii) $d(x, y) = 0 \Longrightarrow x = y$,
(iv) $d(x, y) \leqslant d(x, z) + d(z, y)$ (triangle inequality).

An example of a metric space is the set of all complex numbers $z = x + \mathrm{i}y$ under the definition $d(z_1, z_2) = |z_1 - z_2| = \sqrt{(x_1 - x_2)^2 + (y_1 - y_2)^2}$. The metric is used to measure how close or far an element in a metric space is to another.

**Definition 12.1.2.** Convergence.
A sequence of elements $x_1, x_2, \ldots$ in a metric space $(\mathscr{S}, d)$ is said to converge to *an element x in* $\mathscr{S}$ if, given $\varepsilon > 0$, there exists an integer $N(\varepsilon)$, depending on $\varepsilon$ such that $d(x, x_n) < \varepsilon$ for all $n > N(\varepsilon)$. The element $x$ is called the limit of the sequence, and we write

$$x = \lim_{n \to \infty} x_n.$$

**Definition 12.1.3.** Closure of a set.
Let $\mathscr{S}$ be a metric space and let $S$ be a set of points in $\mathscr{S}$. The closure of $S$ is the set $\bar{\mathscr{S}}$ consisting of the limits of all sequences which can be constructed from points of $S$.

For example, if $S$ is the set of real numbers $x$ for which $0 < x < 1$, then $\bar{S}$ is the set $0 \leqslant x \leqslant 1$. A set is *closed* if $\bar{S} = S$. Among other uses, the concept of convergence is important when one devises a series solution and wants to investigate how close or not the sequence of partial sums behaves relative to the limit.

**Definition 12.1.4** Cauchy sequence.
A sequence of elements $\{x_n\}$, $n = 1, 2, \ldots$ in a metric space $(\mathscr{S}, d)$ is said to be a Cauchy sequence if for a given $\varepsilon > 0$, there exists an integer $N(\varepsilon)$, depending on $\varepsilon$, such that $d(x_n, x_m) < \varepsilon$ for all $m, n > N(\varepsilon)$.

In other words, in a Cauchy sequence, the distance between a pair elements for large indices, $m$, $n$, not necessarily consecutive ones, can be made arbitrarily small by choosing an appropriate threshold $N$ that depends on the definition of the neighborhood $\varepsilon$.

**Definition 12.1.5.** Completeness of metric space.

A metric space $(\mathscr{S}, d)$ is said to be complete if *every Cauchy sequence* in the space converges in some sense[1] to an element *that belongs to* $\mathscr{S}$. Stated otherwise, a metric space is complete if every Cauchy sequence is a convergent sequence.

As a counterexample, the space of rational numbers with the metric $d(x, y) = | x - y |$ is a metric space. But it is not complete relative to the chosen metric as the rational sequence of partial sums $x_n = \sum_{k=0}^{n-1} \frac{1}{k!}$, $n = 1, 2, \ldots$ converges to the irrational number 'e'. The space of rational and irrational numbers (i.e. the entire real axis) is, however, complete under this metric. The set of all real-valued continuous functions $u(x)$ defined on an interval $a \leqslant x \leqslant b$ with a distance function defined as

$$d_\infty(u, v) = \max_{a \leqslant x \leqslant b} | u(x) - v(x)|,$$

is another example of a complete metric space.

**Definition 12.1.6.** Dense sets.
Let $(\mathscr{S}, d)$ be a metric space and let $S$ and $T$ be two sets in $\mathscr{S}$ such that $S \subset T$. We say that $S$ is dense in $T$ if, to every element $f$ of $T$ and to every $\varepsilon > 0$, we can find an element $e$ in $S$ such that $d(e, f) < \varepsilon$. In other words, to every element in $T$ we can find an arbitrarily close element in its *subset* $S$.

As an example, the set $S$ of rational numbers is dense in the real line $T = R$. The set of real-valued differentiable functions in the interval $(a, b)$ is dense in the set of all square integrable, real-valued functions, $\mathscr{L}_2^{(r)}(a, b)$, defined in $(a, b)$.

**Definition 12.1.7** Linear space (a.k.a. Vector space).
A linear space $\mathscr{S}$ is a set of elements $0, 1, x, y, z, \ldots,$ for which the operations of addition and scalar multiplication by a constant $c$ (real or complex) are defined satisfying the following properties

(i) $x + y = y + x$ (commutativity in addition),
(ii) $(x + y) + z = x + (y + z)$ (associativity in addition),
(iii) $x + 0 = x$ (presence of a null element 0),
(iv) $c(x + y) = (x + y)c = cx + cy$ (scalar multiplication by $c$ and distributivity in multiplication),
(v) $1 \cdot x = x$ (presence of an identity element 1).

In order to be classified as a vector space, all of the above properties must hold.

[1] In the sense of mean or in the sense of mean-square, etc.

**Definition 12.1.8.** Inner product space.
An inner product space is a linear space $\mathscr{S}$ endowed with an inner product, $\langle x, y\rangle \in \mathbb{C}$, to each ordered pair $(x, y) \in \mathcal{S}$ with the properties

(i) $\langle x, x\rangle = \langle x, x\rangle^* \geqslant 0$ and $\langle x, x\rangle = 0$ iff $x = 0$,
(ii) $\langle x, y\rangle = \langle y, x\rangle^*$,
(iii) $\langle cx, y\rangle = c\langle x, y\rangle$, $c \in \mathbb{C}$,
(iv) $\langle x + z, y\rangle = \langle x, y\rangle + \langle z, y\rangle$, for $x, y, z \in \mathscr{S}$.

A commonly used inner product in a linear space comprised of square-integrable functions is

$$\langle x, y\rangle = \int_{\Omega} x^T y^* \mathrm{d}\Omega, \tag{12.1}$$

where the superscript $T$ denotes transpose and $\Omega$ is the domain of the vector functions $x$, $y$. The entries of the column vector $x$ could, for instance, denote the Cartesian components of a vector field.

**Definition 12.1.9** Norm.
The norm of an element $x$ in a linear space $\mathscr{S}$ is a real-valued scalar function, $\| x \|$, such that

(i) $\|x\| \geq 0$ and $\|x\| = 0$ iff $x = 0$,
(ii) $\|cx\| = |c| \, \|x\|$,
(iii) $\|x + y\| \leqslant \|x\| + \|y\|$ (triangle inequality).

In a metric space, it is possible to define the norm as $\|x\| = d(x, 0)$. A useful inequality (known as the Cauchy–Schwarz inequality) in an inner-product space is that $|\langle x, y\rangle| \leqslant \|x\|\|y\|$, where the norm is defined via $\|x\|^2 = \langle x, x\rangle$. A real angle $\alpha$ can be associated between two non-trivial elements $x$, $y$ in an inner product space such that $\cos\alpha = \langle x, y\rangle/\|x\|\|y\|$. The two elements are *orthogonal* if $\langle x, y\rangle = 0$. If, in addition, $\|x\| = 1 = \|y\|$, then $x$ and $y$ are said to be *orthonormal.*

In any inner product space we also have the *parallelogram law*

$$\|x + y\|^2 + \|x - y\|^2 = 2(\|x\|^2 + \|y\|^2), \qquad \text{for all } x, y.$$

**Definition 12.1.10.** Hilbert space.
An inner product space complete in the norm $\|x\| = \langle x, x\rangle^{1/2}$ is called a Hilbert space $\mathscr{H}$. In other words, for every Cauchy sequence $\{x_n\}$ in $\mathscr{H}$, there exists an $x \in \mathscr{H}$ such that $\lim_{n\to\infty} \|x - x_n\| \to 0$.

As an example, the space $\mathscr{L}_2^{(r)}(a, b)$ of square-integrable, real valued functions $x(t)$, defined on the finite interval $a \leq t \leq b$ with the inner product defined as

$$\langle x, y\rangle = \int_a^b x(t)y(t)\,\mathrm{d}t$$

is complete and constitutes a Hilbert space. The corresponding norm is $\|x\| = \sqrt{\langle x, x\rangle}$, which generates the metric

$$d(x, y) = \|x - y\| = \left[\int_a^b (x(t) - y(t))^2\,\mathrm{d}t\right]^{1/2}. \tag{12.2}$$

The space of $n$-tuples of complex numbers with the definition

$$\langle x, y\rangle = \sum_{k=1}^{n} \xi_k\eta_k^*$$

is a Hilbert space and constitutes the complex Euclidean $n$-space, $E_n^{(c)}$, where $\xi = [\xi_1, \xi_2, \dots, \xi_n]^T$, $y = [\eta_1, \eta_2, \dots, \eta_n]^T$.

As an additional example, the space $\mathscr{L}_2^{(c)}(\Omega_k)$ of all square-integrable, complex--valued functions $x$ defined over a $k$-dimensional space $\Omega_k$ with an inner product of the form (12.1) is complete and constitutes a Hilbert space. Note that the completeness or not of a class of functions is intimately tied to the metric chosen. For instance, the space of all real-valued continuous functions defined on the interval $a \le t \le b$ under the metric (12.2) is *not* complete and does not constitute a Hilbert space because one can construct a sequence of partial sum of continuous functions which will not necessarily converge to a continuous function. The completion of such a space by adding all the limit points constitutes the Hilbert space $\mathscr{L}_2^{(r)}(a, b)$. The set of elements $S$ in a Hilbert space $\mathscr{S}$ is said to be *bounded* if there exists a constant $c$ such that for all $x$ in $S$, $\|x\| \le c$. Note that a bounded set could have an infinite number of elements.

**Definition 12.1.11** Linear manifold. If a collection $\mathscr{M}$ of vectors in a vector space $\mathscr{S}$ is such that for all scalars $\alpha$ and $\beta$ it contains the vectors $\alpha x + \beta y$ whenever it contains the vectors $x$ and $y$, then $\mathscr{M}$ is a linear manifold.

As an example, a plane passing through the origin in 3D is a linear manifold. *A linear manifold must necessarily contain the zero vector* which corresponds to the element $x - x$.

**Definition 12.1.12** Span, basis.
A set of vectors $x_1, x_2, \dots, x_k, \dots$ in $\mathscr{M}$ is said to span or to determine $\mathscr{M}$ if every vector $y$ in $\mathscr{M}$ can be represented as a linear combination of finite or infinite number of $x_1, x_2, \dots, x_k, \dots$, i.e. there exist scalars $\alpha_k$, $k = (1, 2, \dots, )$ such that

$$y = \sum_{k=1}^{\infty} \alpha_k x_k.$$

The vectors $x_k$ form a basis for $\mathscr{M}$ if they span $\mathscr{M}$ and are also linearly independent. In other words, a basis is obtained by discarding all of the dependent vectors from a span. A basis consisting of orthonormal vectors is known as an *orthonormal* basis or as a *complete orthonormal set.*

As an example, a basis in $\mathscr{L}_2^{(c)}$ must have a countably infinite number of independent elements.

**Definition 12.1.13** Reciprocal basis.
Given an arbitrary basis $x_i$, $i = 1, 2, \ldots$, there exists a reciprocal basis $z_j$, $j = 1, 2, \ldots$ such that

$$\langle z_j, x_i \rangle = \delta_j^i,$$

where $\delta_j^i$ is the Kronecker's delta.

If $x_i$ is regarded as a column vector, then $z_j$ becomes a row vector. Alternatively, $z_j$ can be regarded as a covariant vector and $x_i$ can be regarded as a contravariant vector (see chapter 17). An arbitrary vector can be decomposed equivalently in terms of a basis or a reciprocal basis.

**Definition 12.1.14.** Compact set.
A set is *compact* if each sequence of points in $S$ contains a convergent subsequence[2].

A set containing a finite number of elements is compact, since any sequence constructed from the set must include some element an infinite number of times and therefore contains a convergent subsequence. Stated otherwise, a finite set is compact because it is both bounded and closed. An infinite orthonormal set $\{\varphi_1(t), \varphi_2(t), \ldots\}$ with $\|\varphi_n\| = 1$ in $\mathscr{L}_2^{(c)}(a, b)$ is an example of a non-compact set, even though it is bounded. That is because the set does not contain a *convergent subsequence*.

[2] Given the sequence $s_1, s_2, \ldots, s_n, \ldots$ and a sequence of increasing positive integers $k_1 < k_2, \ldots, k_{n-1} < k_n, \ldots$ chosen from indices 1, 2, 3, ... of the original sequence, the sequence $s_{k_1}, s_{k_2}, \ldots, s_{k_n}, \ldots$ is said to be a subsequence of $s_1, s_2, \ldots, s_n, \ldots$. Thus, 1, 5, 9, 13, ... is a subsequence of 1, 3, 5, 7, ... obtained from $k_1 = 1, k_2 = 3, k_3 = 5, \ldots$. However, 1, 1, 1, ... is *not* a subsequence of 1, 3, 5, 7, ... since in that case the integers $k_1 = k_2 = k_3, \ldots = 1$ are not increasing. Compactness requires not only the presence of a valid subsequence, but one that is also convergent.

**Definition 12.1.15** Separable Hilbert space.
A Hilbert space, $\mathcal{H}$, is separable if there exists a *countable* set of elements $(f_1, f_2, \ldots, f_k, \ldots)$ whose *finite* linear combination $f$ lies within an $\varepsilon$ distance from every element $x$ of $\mathcal{H}$, i.e. $d(x, f) < \varepsilon$ for every $\varepsilon > 0$. That is, the span of $(f_1, f_2, \ldots, k_k, \ldots)$ is dense in $\mathcal{H}$.

**Definition 12.1.16.** Dimension of manifold.
A linear manifold is said to be of dimension $n$ if the basis consists of $n$ vectors. If no finite set of vectors spans the manifold, the dimension of the manifold is said to be infinite.

**Definition 12.1.17** Linear subspace.
A linear manifold is said to be *closed* if, whenever a sequence of vectors $x_1$, $x_2$, ... in $\mathcal{M}$ converges to a limit, the limit of the sequence belongs to $\mathcal{M}$. A closed linear manifold is called a linear subspace.

**Theorem 12.1.1** Projection theorem [2, p 18].
*If $\mathcal{M}$ is a linear subspace and $y$ is not in $\mathcal{M}$, then there exists a vector $w$ in $\mathcal{M}$, called the projection of $y$ on $\mathcal{M}$, such that $y - w$ is orthogonal to $\mathcal{M}$, i.e.*

$$\langle y - w, x \rangle = 0$$

*whenever $x$ belongs to $\mathcal{M}$.*

As an example, consider a vector $y$ in 3D, which does not lie in a plane $\mathcal{M}_2$ through the origin. Suppose the vector $y$ is projected on the plane; then the difference between $y$ and its projection on the plane $\mathcal{M}_2$ will be a vector perpendicular to $\mathcal{M}_2$, i.e. a vector perpendicular to all vectors $x$ lying in $\mathcal{M}_2$.

**Definition 12.1.18.** Direct sum.
The space $\mathcal{N}$ is said to be the *sum* of two subspaces $\mathcal{M}_1$ and $\mathcal{M}_2$ written

$$\mathcal{N} = \mathcal{M}_1 + \mathcal{M}_2$$

if every vector in $\mathcal{N}$ can be written as the sum of a vector in $\mathcal{M}_1$ and a vector in $\mathcal{M}_2$. If every vector in $\mathcal{N}$ can be written *in only one way* as the sum of a vector in $\mathcal{M}_1$ and a vector in $\mathcal{M}_2$, then $\mathcal{N}$ is called the direct sum of $\mathcal{M}_1$ and $\mathcal{M}_2$ and is written as

$$\mathcal{N} = \mathcal{M}_1 \oplus \mathcal{M}_2.$$

If $\mathcal{N}$ is the direct sum of $\mathcal{M}_1$ and $\mathcal{M}_2$, then the zero vector is the only vector which is in both $\mathcal{M}_1$ and $\mathcal{M}_2$. Barring this zero vector, the vectors in $\mathcal{M}_1$ and $\mathcal{M}_2$ are orthogonal complements of each other.

**Theorem 12.1.2.** Riesz–Fischer theorem [5, p 122].
*Let* $\{\varphi_1, \varphi_2, \ldots\}$ *be an infinite orthonormal set (not necessarily a basis) in* $\mathscr{L}_2^{(c)}(a, b)$ *and let* $\{a_n\}$ *be a sequence of complex numbers. Then the series* $\sum_{n=1}^{\infty}|a_n|^2$ *and* $\sum_{n=1}^{\infty}a_n\varphi_n$ *converge or diverge together, that is*

(i) *If* $\sum_{n=1}^{\infty}|a_n|^2$ *diverges, then* $\sum_{n=1}^{\infty}a_n\varphi_n$ *diverges.*

(ii) *If* $\sum_{n=1}^{\infty}|a_n|^2$ *converges, then* $\sum_{n=1}^{\infty}a_n\varphi_n$ *converges to some element g in* $\mathscr{L}_2^{(c)}(a, b)$ *and* $a_n = \langle g, \varphi_n\rangle$.

**Lemma 12.1.3.** Riemann–Lebesgue lemma [5, p 126].
*Let* $\{\varphi_1, \varphi_2, \ldots\}$ *be an infinite orthonormal set (not necessarily a basis) in* $\mathscr{L}_2^{(c)}(a, b)$. *For any element x in* $\mathscr{L}_2^{(c)}(a, b)$, *consider the expansion*

$$\sum_{n=1}^{\infty} c_n\varphi_n = \sum_{n=1}^{\infty}\langle x, \varphi_n\rangle\varphi_n.$$

*Then this series converges and*

$$\lim_{n\to\infty} c_n = 0, \tag{12.3}$$

$$\left\| x - \sum_{n=1}^{\infty} c_n\varphi_n \right\|^2 = \|x\|^2 - \sum_{n=1}^{\infty} c_n^2 \geq 0. \tag{12.4}$$

**Theorem 12.1.4** Characterization of an orthonormal basis [5, p 127].
*Let* $\mathscr{H}$ *be a separable Hilbert space. If any of the following criteria is met, the orthonormal set* $\{\varphi_n\}$ *is a basis for* $\mathscr{H}$.

(i) *For every x in* $\mathscr{H}$

$$x = \sum_{n=1}^{\infty}\langle x, \varphi_n\rangle\varphi_n.$$

*That is, the best approximation to* x *in terms of* $\{\varphi_n\}$ *is perfect.*

(ii) *For every* x *in* $\mathscr{H}$,

$$\|x\|^2 = \sum_{n=1}^{\infty}|\langle x, \varphi_n\rangle|^2.$$

(iii) *The only* x *in* $\mathscr{H}$ *for which all coefficients* $c_n$ *vanish is the* x = *0 function.*

(iv) *There exists no function* $\varphi(t)$ *in* $\mathscr{H}$ *such that the appended set* $\{\varphi, \varphi_1, \varphi_2, \ldots\}$ *is an orthonormal set.*

As an application of this theorem, let $\{\varphi_n(t)\}$ be an orthonormal basis in $\mathscr{L}_2^{(c)}(a, b)$. By definition this is a separable Hilbert space and the above theorem applies. Then for any $x(t)$ in $\mathscr{L}_2^{(c)}(a, b)$, we have

$$x(t) = \sum_{n=1}^{\infty} c_n \varphi_n(t), \qquad \text{where } c_n = \int_a^b x(t)\varphi_n^*(t)\,\mathrm{d}t.$$

As a second application, let $w(t)$ be a continuous, real, positive function in $a < t < b$. Consider the inner product

$$\langle x, y\rangle_w = \int_a^b x(t)y^*(t)w(t)\,\mathrm{d}t$$

and the corresponding norm

$$\|x\|_w = \left[\int_a^b |x(t)|^2\, w(t)\,\mathrm{d}t\right]^{1/2}.$$

Let $\{\varphi_n\}$ be a set of complete orthogonal functions such that

$$\int_a^b \varphi_i(t)\varphi_j^*(t)w(t)\,\mathrm{d}t = \delta_i^{\,j}.$$

Then every $x(t)$ belonging to the Hilbert space $\mathscr{H}_w$ of functions for which $\|x\|_w < \infty$ can be expanded in the form

$$x(t) = \sum_{n=1}^{\infty} c_n \varphi_n(t), \qquad c_n = \langle x, \varphi_n\rangle_w.$$

In practical electromagnetic problems, the basis is constructed from the eigenfunctions of the governing boundary value problem. The eigenfunctions are linearly independent solutions of the homogeneous differential equation (that is, of the differential equation with the source function set to zero). For instance, a suitable basis for the boundary value problem $\frac{\mathrm{d}^2\varphi}{\mathrm{d}t^2} + k^2\varphi = f$, $0 \le t \le \pi$ with the boundary conditions $\varphi(0) = 0 = \varphi(\pi)$ is $\varphi_n(t) = \sqrt{\frac{2}{\pi}}\sin nt$, $n = 1, 2, \ldots$. For an interval $-L < t < L$, a complete basis for the boundary value problem $\frac{\mathrm{d}^2\varphi}{\mathrm{d}t^2} + k^2\varphi = f$ with periodic boundary conditions at the ends is $\left\{\sqrt{\frac{1}{2L}}\mathrm{e}^{in\pi t/L}\right\}$, $n = 0, \pm 1, \pm 2, \ldots$. Note that these basis functions are indeed independent of the source function $f$.

### 12.1.1 Functionals and operators on Hilbert space

**Definition 12.1.19** Linear functional.
Suppose that to every vector $x$ in a linear vector space $\mathscr{S}$, we associate a scalar, which we denote by $f(x)$ and call it a functional. If in addition, the functional satisfies the property

$$f(\alpha x + \beta y) = \alpha f(x) + \beta f(y),$$

then $f(x)$ is called a linear functional.

As an example, the value of any one component of $x$ in $R^n$ is a linear functional. However, the length of a vector in $R^n$ (or equivalently its norm) is a functional, but not linear.

**Definition 12.1.20.** Continuous functional.
A functional $f(x)$ defined on $\mathscr{S}$ is continuous if

$$\lim_{n\to\infty} x_n = x \Longrightarrow \lim_{n\to\infty} f(x_n) = f(x).$$

That is, if a sequence $\{x_n\}$ converges to $x$, then the sequence of functionals $\{f(x_n)\}$ converges to the functional $f(x)$ of the limit $x$.

**Definition 12.1.21** Bounded functional.
The functional $f(x)$ defined on $\mathscr{S}$ is bounded if there exists a positive real constant $\mu$ such that

$$|f(x)| < \mu \, ||x||$$

for all $x$ in $\mathscr{S}$.

**Theorem 12.1.5.** [2, p 20]. *If $f(x)$ is a continuous linear functional, there exists a vector $z \in \mathscr{S}$ such that*

$$f(x) = \langle z, x \rangle.$$

In other words, one can always associate an inner product with a linear functional if the latter is continuous. As an example, if $f(x) = \alpha_1 + \alpha_2 + \alpha_3$ for $x = (\alpha_1, \alpha_2, \alpha_3)$ in 3D, then $z = (1, 1, 1)$. If, for real-valued, square-integrable functions over (0, 1), a linear functional is defined as

$$f(u) = \int_0^1 t u(t)\, \mathrm{d}t,$$

then $z = t$.

**Definition 12.1.22** Operator.
An operator $L$ (i.e. a transformation) is a mapping that assigns to a vector $x$ in $\mathscr{S}$ another vector $y$ in $\mathscr{S}$, which we denote by $Lx$. The set of vectors $x$ for which the mapping is defined is called the *domain* of the operator $L$. It is denoted by $\mathscr{D}_L$. The set of vectors $y$ (i.e. image vectors) is called the *range* of the operator $L$. It is denoted by $\mathscr{R}_L$. An operator is one-to-one if the image vector unambiguously determines the original vector, i.e. iff $Lf = Lg \Longrightarrow f = g$. The spectral norm of an operator, $||L||$, is defined as

$$||L|| = \underset{x\neq 0}{\text{l.u.b}} \frac{||Lx||}{||x||} = \underset{||x||=1}{\text{l.u.b}}\, ||Lx||,$$

where l.u.b stands for the least upper bound. A sequence of functions $\{x_n\}$ in $\mathscr{D}_L$ such that $\lim_{n\to\infty} x_n = 0$ is known as the *null sequence* in $\mathscr{D}_L$.

**Definition 12.1.23.** Closed range.
An operator $L$ is said to have a closed range $\mathscr{R}_L$, whenever there exists a sequence of elements $y_n = Lx_n$ in $\mathscr{R}_L$ converging to a limit $y$ *that is in* $\mathscr{R}_L$, i.e. the limit $y$ can be written as $y = Lx$ for an $x \in \mathscr{D}_L$.

An example of an operator that does not have a closed range is the operator $Lx = (\xi_1, \frac{1}{2}\xi_2, \frac{1}{3}\xi_3, \ldots)$ for $x = (\xi_1, \xi_2, \xi_3, \ldots)$ in an infinite dimensional Euclidean space $E_\infty$. In $E_\infty$, the inner product is defined as $\langle x, y\rangle = \sum_{n=1}^{\infty} \xi_n \eta_n$ for $y = (\eta_1, \eta_2, \ldots)$ and the length of a vector $|x|$ is defined as $|x| = [\sum_{n=1}^{\infty} x_n^2]^{1/2}$. The length of the vector is assumed to be finite and the space $E_\infty$ consists only those vectors $x = (\xi_1, \xi_2, \ldots)$ such that the infinite series $\xi_1^2 + \xi_2^2 + \ldots$ converges. If we take $y_n = (1, \frac{1}{2}, \ldots, \frac{1}{n}, 0, 0, \ldots)$, $x_n = (\underbrace{1, 1, \ldots, 1}_{n \text{ places}}, 0, 0, \ldots)$, $y = (1, \frac{1}{2}, \frac{1}{3}, \ldots)$, we would obtain $x = (1, 1, 1, \ldots)$, which is not in $E_\infty$ since its length is not finite.

**Definition 12.1.24.** Adjoint operator.
A linear operator $L^\dagger$ is said to be the adjoint of $L$ if, for all $x$ and $y$ in a space $\mathcal{S}$,

$$\langle y, Lx\rangle = \langle L^\dagger y, x\rangle.$$

Note that the order of the entries in the inner product may be switched on both sides so that the adjoint may also be defined such that $\langle Lx, y\rangle = \langle x, L^\dagger y\rangle$. The operator is *self-adjoint* if $L^\dagger = L$. In such a case

$$\langle y, Lx\rangle = \langle Ly, x\rangle.$$

An operator $L$ is self-adjoint iff
(a) $\mathscr{D}_L = \mathscr{D}_{L^\dagger}$ (Domains of $L$ and $L^\dagger$ must coincide), and
(b) $Lx = L^\dagger x$ for every $x$ in $\mathscr{D}_L$.

**Definition 12.1.25** Null space.
The null space, $\mathscr{N}_L$, of an operator $L$ is the subspace of all $z$ such that $Lz = 0$.

**Definition 12.1.26.** Invariant manifold.
A manifold $\mathscr{M}$ belonging to an operator $L$ is called an invariant manifold if, whenever $m$ is a vector in $\mathscr{M}$, then $Lm$ is also in $\mathscr{M}$. The definition also applies to an *invariant subspace*.

As an example, the manifold of functions satisfying $u(1) = u(-1) = 0$ remains invariant under the reflecting operator $Ru(t) = u(-t)$. As a second example, the manifold of functions defined over $(-\infty, \infty)$ remains invariant under finite translation operator $Tu(t) = u(t + h)$. An important result is that *if $\mathcal{M}$ is a finite-dimensional invariant subspace of L, the effect of L on $\mathcal{M}$ may be represented by a matrix* [2, p 59].

The following theorem concerning the decomposition of the space of an operator is very useful.

**Theorem 12.1.6.** Decomposition of space of an operator [2, p 58].
*The whole space $\mathcal{S}$, the space containing only the zero vector, the null-space of L, and the range space of L are invariant manifolds of L. If $\mathcal{S}$ is finite dimensional, the dimension of the null-space of L plus the dimension of the range space equals the dimension of $\mathcal{S}$.*

*12.1.1.1 Commuting operators and Floquet's theorem*

**Definition 12.1.27.** Commuting operators.
Two operators $L$ and $K$ are said to commute if $LK = KL$.

As an example if the operators $L = \partial/\partial x$ and $K = \partial/\partial y$ act on functions continuous in two independent variables $x$ and $y$, then $LK = KL = \partial^2/\partial x \partial y = \partial^2/\partial y \partial x$. An example of operators that do not commute are the raising and lowering operators, $\mathcal{L}_{\mp}$, of equation (15.119). It is clear in that case that

$$\mathcal{L}_+\mathcal{L}_- = \frac{\partial^2}{\partial \rho^2} + \frac{1}{\rho^2}\frac{\partial^2}{\partial \phi^2} - \frac{j}{\rho^2}\frac{\partial}{\partial \phi}$$

$$\mathcal{L}_-\mathcal{L}_+ = \frac{\partial^2}{\partial \rho^2} + \frac{1}{\rho^2}\frac{\partial^2}{\partial \phi^2} + \frac{j}{\rho^2}\frac{\partial}{\partial \phi}.$$

Consequently,

$$\mathcal{L}_-\mathcal{L}_+ - \mathcal{L}_+\mathcal{L}_- = \frac{2j}{\rho^2}\frac{\partial}{\partial \phi} \neq 0. \tag{12.5}$$

The following theorem applying to commuting operators is very handy.

**Theorem 12.1.7** Commuting operators [2, p 63].
*If two operators L and K commute, then the range and null space of L are invariant manifolds for K. Similarly, the range and null space of K are invariant manifolds of L.*

**Corollary 12.1.7.1.** Commuting operators [2, p 63].
*If L and K commute and if one of the operators has an eigenvalue of finite algebraic multiplicity[3], both operators have a common eigenvector; that is, there exists a vector $x$ such that $Lx = \lambda x$, $Kx = \mu x$, where $\lambda$ and $\mu$ are scalars.*

[3] The algebraic multiplicity of an eigenvalue is the number of times it repeats. For example, if an eigenvalue occurs only once during the solution of the eigenvalue problem, then its algebraic multiplicity is one. If an eigenvalue appears twice, then its algebraic multiplicity is two and so on.

**Example 12.1.1.1** Differential equation with periodic coefficients, *Floquet's theorem.* An important consequence of the above corollary is Floquet's theorem that is often employed in electromagnetic problems with periodic boundaries or coefficients. Consider a second order differential operator

$$L = -\frac{d^2}{dt^2} + q(t)$$

acting on a linear manifold of twice-differentiable functions defined over $-\infty < t < \infty$. Let the coefficient $q$ be periodic with period $p$ so that $q(t + p) = q(t)$. The function $q(t)$ can be real- or complex-valued. One may think of this equation as arising in wave propagation in one dimension where the medium permittivity or permeability or both vary periodically along the space coordinate $t$. Our goal is to determine the properties satisfied by a general solution of the homogeneous equation

$$-u''(t) + q(t)u(t) = 0. \tag{12.6}$$

Let $K$ denote the translational operator $Tu(t) = u(t + p)$. Note that

$$\begin{aligned}(KL - LK)u &= -u''(t + p) + q(t + p)u(t + p) - [-u''(t + p) + q(t)u(t + p)] \\ &= [q(t + p) - q(t)]u(t + p) = 0.\end{aligned}$$

Therefore $K$ and $L$ are commuting operators. By theorem 12.1.7, the null space of $L$ (i.e. the space $\mathcal{S}$ of solutions $u(t)$ of equation (12.6); a two-dimensional space) is an invariant manifold of $K$ (i.e. $K\mathcal{S} = \mathcal{S}$; $K$ is two-dimensional). In other words, if $u(t)$ is a solution, so is $u(t + p)$. Of course the latter need not coincide with the former, i.e. $u(t)$ is not necessarily periodic. Let $u_1(t)$ and $u_2(t)$ be two *linearly independent* solutions of equation (12.6). The problem of determining $u_1(t)$ and $u_2(t)$ may itself be a major task, but we assume here that they can be determined by some means. Then any arbitrary solution of equation (12.6) can be written as

$$u(t) = \beta_1 u_1(t) + \beta_2 u_2(t) = \begin{bmatrix}\beta_1 & \beta_2\end{bmatrix}\begin{bmatrix}u_1(t) \\ u_2(t)\end{bmatrix} := \boldsymbol{\beta}^T \mathbf{u}(t), \tag{12.7}$$

where $\beta_1$ and $\beta_2$ are complex constants and superscript $T$ stands for transpose. In particular, since $u(t + p)$, $u_1(t + p)$, and $u_2(t + p)$ are also solutions of equation (12.6), we may write

$$\begin{aligned}u(t + p) &= \gamma_1 u_1(t) + \gamma_2 u_2(t) \\ u_1(t + p) &= \alpha_{11} u_1(t) + \alpha_{12} u_2(t) \\ u_2(t + p) &= \alpha_{21} u_1(t) + \alpha_{22} u_2(t),\end{aligned}$$

where $\alpha_{ij}$, $\gamma_i$, $i, j = 1, 2$ are complex constants. Therefore the coefficients of the translated function $u(t + p)$ are related to the coefficients of the original function $u(t)$ upon using $u(t + p) = \beta_1 u_1(t + p) + \beta_2 u_2(t + p)$ by

$$\boldsymbol{\gamma} := \begin{bmatrix}\gamma_1 \\ \gamma_2\end{bmatrix} = \begin{bmatrix}\alpha_{11} & \alpha_{21} \\ \alpha_{12} & \alpha_{22}\end{bmatrix}\begin{bmatrix}\beta_1 \\ \beta_2\end{bmatrix} := \boldsymbol{\alpha}^T\boldsymbol{\beta}. \tag{12.8}$$

The matrix $\boldsymbol{\alpha}^T$ may be thought of as representing the two-dimensional, continuous-time, translational operator $K$ w.r.t. the basis $u_1(t)$ and $u_2(t)$. A change of basis will change the component values of this matrix. Let $[\lambda_1, \mathbf{e}_1] = (c_1, c_2)^T$ and $[\lambda_2, \mathbf{e}_2]= (d_1, d_2)^T$ be the two eigenpairs of the matrix $\boldsymbol{\alpha}^T$. Each eigenvector $\mathbf{e}_i$ will define the corresponding eigenfunction $v_i(t)$ of $K$ according to $v_i(t) = \mathbf{e}_i^T\mathbf{u}(t)$. By definition $v_i(t + p) = Kv_i(t) = \lambda_i v_i(t)$. Let $\kappa_i = p^{-1}\ln\lambda_i$. Then $v_i(t + p) = e^{\kappa_i p}v_i(t) \Longrightarrow e^{-\kappa_i(t+p)} v_i(t + p) = e^{-\kappa_i t}v_i(t)$. In other words, $w_i(t) = e^{-\kappa_i t}v_i(t)$ is a periodic function of $t$ with period $p$. Since commuting operators share the same eigenvectors by theorem 12.1.7, the functions $v_i(t) = e^{\kappa_i t}w_i(t)$ are also the eigenfunctions of the operator $L$. In other words, equation (12.6) has two linearly independent solutions $v_1(t)$ and $v_2(t)$ such that

$$v_1(t) = e^{\kappa_1 t}w_1(t), \qquad v_2(t) = e^{\kappa_2 t}w_2(t), \tag{12.9}$$

where $w_i(t)$, $i = 1, 2$ are periodic functions of $t$ with a period $p$. This result (12.9) is known as *Floquet's theorem*, where $\kappa_i$ play the role of complex propagation constants along $t$. A few observations are noteworthy:

1. The solutions of equation (12.6) are *not* necessarily periodic since $v_i(t) = w_i(t)e^{\kappa_i t}$ is the product of a periodic function $w_i(t)$ and a non-periodic function $e^{\kappa_i t}$. A periodic solution $v_i(t)$ is possible only if $\lambda_i = 1$ so that $\kappa_i = 0$.
2. The theorem enables us to determine the solution for all time if the solution is known only in one period. If $v_i(t)$, $i = 1, 2$ are known over $(t_0, t_0 + p)$, then since $v_i(t_0 + p) = e^{\kappa_i p}v_i(t_0)$, we have $\kappa_i = p^{-1}\ln[v_i^{-1}(t_0)v_i(t_0 + p)]$. Then $w_i(t) = v_i(t)e^{-\kappa_i t}$ is determined over $t \in (t_0, t_0 + p)$. But since $w_i(t)$ is periodic with a period $p$, it is known for all times $t \in (-\infty, \infty)$. Having determined $\kappa_i$ and $w_i(t)$ for all $t$, $v_i(t)$ is now given over $(-\infty, \infty)$ by $v_i(t) = w_i(t)e^{\kappa_i t}$.
3. The constants $\kappa_i$ depend on the period $p$ as well as on the eigenvalues of the matrix representation $\boldsymbol{\alpha}^T$ of $K$. The matrix representation is $t$-invariant, but depends on the basis functions $u_i(t)$ chosen to represent an arbitrary solution of equation (12.6).
4. The second-order differential equation (12.6) can be converted to a system of first order equations by defining the state variables $x_1(t) = u(t)$, $x_2(t) = u'(t)$. Then the original differential equation becomes the following system of first-order equations

$$\mathbf{x}'(t) := \begin{bmatrix} x'_1(t) \\ x'_2(t) \end{bmatrix} = \begin{bmatrix} 0 & 1 \\ q(t) & 0 \end{bmatrix}\begin{bmatrix} x_1(t) \\ x_2(t) \end{bmatrix} := \mathbf{A}(t)\mathbf{x}(t), \tag{12.10}$$

where the system matrix $\mathbf{A}(t)$ is time-varying and satisfies the periodic condition $\mathbf{A}(t + p) = \mathbf{A}(t)$. Let $\mathbf{x}^1(t)$ and $\mathbf{x}^2(t)$, (each of size $2 \times 1$), be two linearly independent solutions of equation (12.10). We choose a normalization such that $\mathbf{x}^i(0) = \mathbf{e}^i$, $i = 1, 2$, where $\mathbf{e}^i$ is a column vector with all zero entries except at the $i$th row, where it takes a value unity. The $\mathbf{x}^i(t)$ constitute an invertible matrix, $\mathbf{X}(t)$, known as the *fundamental matrix*, having columns $\mathbf{x}^1(t)$ and $\mathbf{x}^2(t)$: $\mathbf{X}(t) = [\mathbf{x}^1(t), \mathbf{x}^2(t)]$. The fundamental matrix satisfies the governing equation $\mathbf{X}'(t) = \mathbf{A}(t)\mathbf{X}(t)$. It is obvious that $\mathbf{X}(t + p)$ is also a fundamental matrix of solutions since $\mathbf{X}'(t + p) = \mathbf{A}(t + p)\mathbf{X}(t + p) = \mathbf{A}(t)\mathbf{X}(t + p)$. The state transition matrix (of size $2 \times 2$ here) is defined as $\mathbf{\Phi}(t, \tau) = \mathbf{X}(t)\mathbf{X}^{-1}(\tau)$. The transition matrix of a linear system defines the general solution via $\mathbf{x}(t) = \mathbf{\Phi}(t, \tau)\mathbf{x}(\tau)$. In the system case, the Floquet's decomposition of the transition matrix (Floquet's theorem) can be used to investigate the solution properties [9, p 81]. As a matter of fact, under Floquet's decomposition, the transition matrix $\mathbf{\Phi}(t, \tau)$ can be written as

$$\mathbf{\Phi}(t, \tau) = \mathbf{P}(t)\mathrm{e}^{\mathbf{R}(t-\tau)}\mathbf{P}^{-1}(\tau), \tag{12.11}$$

where $\mathbf{P}(t)$ is a continuously differentiable, invertible, periodic matrix satisfying $\mathbf{P}(t + p) = \mathbf{P}(t)$, $\mathbf{P}(0) = \mathbf{I}$, the identity matrix, and $\mathbf{R}$ is a *constant* matrix (of size $2 \times 2$ here) defined by $\mathrm{e}^{\mathbf{R}p} = \mathbf{\Phi}(p, 0) = \mathbf{X}(p)\mathbf{X}^{-1}(0) = \mathbf{X}(p)$ since $\mathbf{X}(0) = \mathbf{I}$. Note that the matrix $\mathbf{R}$ is completely determined by the fundamental matrix evaluated at $t = p$. Equation (12.11) also implies that the fundamental matrix can be expressed as

$$\mathbf{X}(t) = \mathbf{P}(t)\mathrm{e}^{\mathbf{R}t}, \tag{12.12}$$

which is regarded as Floquet's theorem for a system of equations. As with the scalar case, a *periodic solution* is possible only if one of the eigenvalues of the matrix $\mathrm{e}^{\mathbf{R}p}$ is unity. Also, if $\mathbf{X}(t)$ in known only over the interval $[t_0, t_0 + T]$, then it is easy to see that $\mathbf{C} := \mathbf{X}^{-1}(t_0)\mathbf{X}(t_0 + p) = \exp(\mathbf{R}p)$ or the constant matrix $\mathbf{R} = p^{-1} \ln \mathbf{C}$. Thus the matrix $\mathbf{R}$ is completely determined from the knowledge of the fundamental matrix available at two points, $t_0$ and $t_0 + p$, separated by one period. Furthermore, on using equation (12.12), the matrix $\mathbf{P}(t) = \mathbf{X}(t)\mathrm{e}^{-t\mathbf{R}}$ is known over $t \in [t_0, t_0 + p]$. But since $\mathbf{P}(t)$ is periodic, it is known for all $t \in (-\infty, \infty)$. Equation (12.12) then gives $\mathbf{X}(t)$ for all $t$. Therefore, determination of fundamental matrix over one time period leads at once its determination over all $t \in (-\infty, \infty)$. There is complete parallel with the scalar case. However, in its system form, the Floquet's theorem (12.12) is also applicable to higher order differential equations as long as $\mathbf{A}(t)$ is any continuous periodic matrix.

In the propagation of time-harmonic waves in a periodic structure that may be characterized by either a periodically varying boundary shape or periodically varying material constants, the phasor wave amplitude $u(z)$ as a function of space $z$ is written following Floquet's theorem as

$$u(z) = \mathrm{e}^{-\gamma z}P(z),$$

where $P(z)$ is periodic in $z$ with period $L$ and $\gamma = \alpha + j\beta$, $\alpha > 0$ is a complex propagation constant. In general, the factor $\mathrm{e}^{-\gamma z}$ is thus not periodic. Owing to its periodic nature, the function $P(z)$ may be expanded in a Fourier series as

$$P(z) = \sum_{n=1}^{\infty} A_n e^{-j2n\pi z/L},$$

with complex coefficients $A_n$. Consequently the field $u$ may be represented as

$$u(z) = \sum_{n=-\infty}^{\infty} A_n e^{-\alpha z} e^{-j\beta_n z}, \quad \beta_n = \beta + 2n\pi/L.$$

The $n$th term above is called the $n$th *space or Hartree harmonic.* ■■

### 12.1.2 Uniqueness, existence, and Hilbert–Schmidt operators

We first state a uniqueness theorem and an existence theorem related to linear operators and then progress towards the important case of Hilbert–Schmidt operators that are widely encountered in physical problems. In particular, the special properties brought out in theorems 12.1.11–12.1.13 concerning the eigenfunctions and eigenvalues of Hilbert–Schmidt operators are relevant to understanding the completeness properties of Green's functions.

**Theorem 12.1.8.** Uniqueness of the solution of $Lx = y$ [2, p 45].
*Consider the non-homogeneous linear equation $Lx = y$ and the corresponding homogeneous linear equation $Lz = 0$. If the homogeneous equation has a non-trivial solution, then the solution of the non-homogeneous equation is not unique. Conversely, if the solution of the non-homogeneous equation is non-unique, there exists a non-trivial solution of the homogeneous equation.*

**Theorem 12.1.9.** Existence of the solution of $Lx = y$ [2, p 46, 48].
*If $L$ has a closed range, the whole space $\mathscr{S}$ is the direct sum of the range space, $\mathscr{R}_L$, of $L$ and the null space, $\mathscr{N}_{L^\dagger}$, of $L^\dagger$*

$$\mathscr{S} = \mathscr{R}_L \oplus \mathscr{N}_L^\dagger;$$

*in other words, the range space of $L$ is the orthogonal complement of the null space of $L^\dagger$. Stated otherwise, if the range of $L$ is closed, the non-homogeneous equation $Lx = y$ has a solution for a given $y$ iff $y$ is orthogonal to every solution of the adjoint homogeneous equation $L^\dagger z = 0$:*

$$0 = \langle x, L^\dagger z \rangle = \langle Lx, z \rangle = \langle y, z \rangle.$$

**Definition 12.1.28** Bounded operator.
An operator is bounded if its domain is the entire space $\mathscr{S}$ and if there exists a positive real constant $C$ such that

$$\|Lx\| < C\|x\|$$

for all $x$ in $\mathscr{S}$.

An example of a bounded linear operator on $\mathscr{L}_2^{(c)}(0, 1)$ is the transformation $y(t) = Lx = tx(t)$. Using the Cauchy–Schwarz inequality, it is easy to see that $\|Lx\| \le \|x\|$. A counterexample of a bounded linear operator is the differential operator[4] . A second counterexample is the operator $Lx = x(t)/t$ on $\mathscr{L}_2^{(c)}(0, 1)$. The image $x(t)/t$ is in $\mathscr{L}_2$ only if $x(t)$ behaves as $t^\alpha$, $\alpha > 1/2$. For instance, if

$$x(t) = \begin{cases} 0, & 0 \le t \le \varepsilon; \\ 1, & \varepsilon < t \le 1, \end{cases}$$

then $\|x\|^2 = 1 - \varepsilon$ and $\|Lx\|^2 = 1/\varepsilon - 1 = \|x\|^2/\varepsilon$. Clearly $\|Lx\|/\|x\| = 1/\sqrt{\varepsilon}$ that can be made as large as possible by choosing $\varepsilon$ appropriately. Consequently, $L$ is unbounded.

**Definition 12.1.29** Continuous operator.
A linear bounded operator is continuous at $x$ in $\mathscr{D}_L$ if, whenever a sequence of vectors $x_n$ converges to $x$, then $Lx_n$ converges to $Lx$. The operator is continuous in $\mathscr{D}_L$ if it is continuous at every point of $\mathscr{D}_L$.

An operator is continuous iff it is bounded. Furthermore, if an operator is continuous at a single point, it is continuous on all of $\mathscr{D}_L$.

**Definition 12.1.30** Closed operator.
A linear operator $L$ is closed if it has the following property: whenever a sequence of vectors $x_n$ is in $\mathscr{D}_L$ and converges to $x$ and $Lx_n$ converges to $f$, then $x$ is in $\mathscr{D}_L$ and $Lx = f$.

The distinction between closed and continuous operators is the following. If $L$ is continuous, $x_n \to x \Longrightarrow Lx_n \to Lx$, whereas if $L$ is merely closed all we know is that different sequences approaching $x$ cannot yield different limiting values for the transformation. An example of an operator that is closed but not continuous is $L = \mathrm{d}/\mathrm{d}t$ defined on the space $\mathscr{L}_2^{(c)}(0, 1)$. A bounded linear operator defined on a linear manifold $\mathscr{D}_L$, which is not closed, can always be extended to a larger domain $\bar{\mathscr{D}}_L$ by continuity. That is, if a sequence $\{x_n\}$ of functions in $\mathcal{D}_L$ converges to a limit $x$ that is not in $\mathscr{D}_L$, the limit can always be included in the extended domain $\bar{\mathscr{D}}_L$ and the limit of the image, i.e. $f = \lim_{n\to\infty} Lx_n$ is used as the definition of $Lx$. In this way, bounded linear operators can always be regarded as defined on the whole Hilbert space $\mathscr{S}$. Some properties of closed operators are

(i) A closed operator on a closed domain is continuous.
(ii) The null-space of a closed operator is a closed set.

[4] Consider the sequence of functions $x_n(t) = \sqrt{2}\sin n\pi t$ in the space $\mathscr{L}_2^{(r)}(0, 1)$ and the differential operator $Lx = \mathrm{d}x/\mathrm{d}t$. The $x_n$ all have a norm 1, yet $\| Lx \| = \| \mathrm{d}x_n/\mathrm{d}t \| = n\pi \to \infty$ as $n \to \infty$.

**Definition 12.1.31** Completely continuous operator.
An operator $L$ is called completely continuous if it transforms bounded sets into compact sets.

An example of a completely continuous operator is the *Hilbert–Schmidt* linear integral operator

$$y(t) = Kx = \int_a^b k(t, \tau)x(\tau)\, d\tau, \tag{12.13}$$

defined on $\mathscr{L}_2^{(c)}(a, b)$ with the property

$$\|K\|^2 \le \int_a^b \int_a^b |k(t, \tau)|^2\, dt d\tau < \infty. \tag{12.14}$$

The limits are allowed to be $a = -\infty$ or $b = \infty$. The function $k(t, \tau)$ is known as the *kernel* of the transformation. Whenever $x$ belongs to $\mathscr{L}_2^{(c)}(a, b)$, the transformed function $Kx$ belongs to $\mathscr{L}_2^{(c)}(a, b)$. Thus the Hilbert–Schmidt operator maps a space onto itself. The inner product of $Kx$ and $y$ is

$$\begin{aligned}\langle Kx, y\rangle &= \int_a^b \int_a^b y^*(t)k(t, \tau)x(\tau)\, d\tau dt \\ &= \int_a^b \int_a^b x(t)[k^*(\tau, t)y(\tau)]^*\, dt d\tau = \langle x, K^\dagger y\rangle,\end{aligned} \tag{12.15}$$

which also defines the adjoint operator

$$K^\dagger y = \int_a^b k^*(\tau, t)y(\tau)\, d\tau. \tag{12.16}$$

A kernel with the property $k(t, \tau) = k^*(\tau, t)$ is known as a *symmetric kernel.* For example $k(t, \tau) = 1/|t - \tau|^\alpha$, $0 \le \alpha < 1/2$ is a symmetric kernel, but $k(t, \tau) = e^{-j|t-\tau|}/|t - \tau|^\alpha$, $j = \sqrt{-1}$, $0 \le \alpha < 1/2$ is not. A symmetric kernel generates an self-adjoint integral operator with inner product $\langle Kx, y\rangle = \langle x, K^\dagger y\rangle = \langle x, Ky\rangle$. A kernel which is expressed as a finite sum of a product of a function of $t$ and a function of $\tau$ in the form $k(t, \tau) = \sum_{n=1}^N p_n(t)q_n(\tau)$ such that $\int_a^b |p_n(t)|^2\, dt < \infty$ and $\int_a^b |q_n(t)|^2\, dt < \infty$ is known as a *separable or degenerate kernel.*

A completely continuous operator is continuous, but the converse is not true. An example is the identity operator $Ix = x$ on an infinite dimensional space, which is not completely continuous. For instance, the bounded infinite orthonormal set

$\{\varphi_1, \varphi_2, \ldots\}$ is mapped onto itself by the identity operator, but the latter is not compact (see definition 12.1.14). The following properties of a completely continuous operator $L$ are worth noting:

(i) If $\{\varphi_n\}$ is an infinite orthonormal sequence in a Hilbert space $\mathscr{S}$, then for a completely continuous operator $\lim_{n\to\infty} L\varphi_n = 0$.

(ii) If $L$ is a completely continuous, invertible operator on an infinite-dimensional space, its inverse $L^{-1}$ is unbounded.

(iii) If an operator $L$ can be approximated in norm by a sequence of completely continuous operators $\{L_n\}$, i.e. $\|L - L_n\| \le 1/n$, then $L$ is completely continuous.

In addition we have the following alternative theorem for completely continuous operators.

**Theorem 12.1.10.** Alternative theorem for completely continuous operators, [5, p 187]. *Let $\lambda \neq 0$ be fixed and $L_c$ be a completely continuous operator in a Hilbert space $\mathscr{S}$. Let $L_c^\dagger$ be its adjoint such that $\langle L_c x, y\rangle = \langle x, L_c^\dagger y\rangle$. Consider the three related equations*

$$L_c x = \lambda x; \tag{12.17}$$

$$L_c^\dagger z = \lambda^* z; \tag{12.18}$$

$$L_c y - \lambda y = f. \tag{12.19}$$

*These three equations are known as the direct homogeneous equation, the adjoint homogeneous equation, and the direct non-homogeneous equation, respectively. Then:*

*Either equations (12.17) and (12.18) both have only the trivial solution, or both have different nontrivial solutions.*

*The necessary and sufficient condition for equation (12.19) to have a non-trivial solution is that* f *be orthogonal to every solution of equation (12.18). In particular, if λ is not an eigenvalue of equation (12.17), then $\lambda^*$ is not an eigenvalue of equation (12.18), and equation (12.19) has a unique solution for every $f$ in $\mathscr{S}$.*

Note that part (ii) of this theorem is consistent with theorem 12.1.9 if one replaces the operator $L$ there with $L_c - \lambda I$.

Because self-adjoint, Hilbert–Schmidt linear operator (the HS-operator for short) assumes a special place in electromagnetic applications, we gather below some important properties pertaining to it. The equation $Kx = \mu x$ is known as the eigenvalue problem for the HS-operator.

**Theorem 12.1.11.** Existence of eigenfunctions of a self-adjoint HS-operator [5, p 217].

*There exist a finite or countably infinite number of complex-valued eigenvalues* $\mu_n$ *and corresponding eigenfunctions* $\varphi_n$ *such that* $K\varphi_n = \mu_n\varphi_n$, $|\mu_{n+1}| \le |\mu_n|$, *and* $\langle\varphi_m, \varphi_n\rangle = \delta_m^n$.

**Theorem 12.1.12.** Hilbert–Schmidt theorem [5, p 218]. *Every function of the form* $y = Kx$ *of a self-adjoint HS-operator can be expanded in a Fourier series in the eigenfunctions of* $K$ *corresponding to nonzero eigenvalues of the form* $y = \sum_{n=1}^{\infty}\langle Kx, \varphi_n\rangle\varphi_n$. *Thus, the orthonormal set* $\{\varphi_n\}$ *forms a basis for the range of* $K$.

**Theorem 12.1.13.** Hilbert–Schmidt spectral theorem [5, p 219]. *The set of all eigenfunctions of a self-adjoint HS-operator* $K$ *(including the eigenfunctions corresponding to* $\mu = 0$*) forms a complete set. The set* $\{\varphi_n\}$ *of eigenfunctions corresponding to nonzero eigenvalues forms a complete set iff* $\mu = 0$ *is not an eigenvalue of* $K$.

## 12.2 Sturm–Liouville problem and Green's functions

### 12.2.1 Second order linear differential equations, Green's functions

Consider a second-order, linear, ordinary differential equation of the form

$$Lu = a_2(t)u'' + a_1(t)u' + a_0(t)u, \tag{12.20}$$

where $a_0$, $a_1$, $a_2$ are *real* continuous functions in $a \le t \le b$ and $u^n = \frac{d^n u}{dt^n}$, $n = 1, 2$ is a notation for the $n$th derivative. It is possible to cast this in the standard form of a Sturm–Liouville operator [1]

$$Lu = -\frac{1}{s(t)}\frac{d}{dt}[p(t)u'(t)] + q(t)u(t), \tag{12.21}$$

by making the transformations

$$q(t) = a_0(t); \quad s(t) = -\frac{p(t)}{a_2(t)}; \quad p(t) = \exp\left[\int_{-\infty}^{t}\frac{a_1(\tau)}{a_2(\tau)}\,d\tau\right].$$

We further require $u$ to satisfy two homogeneous boundary conditions of the form

$$B_1(u) = 0\colon \alpha_{11}u(a) + \alpha_{12}u'(a) + \beta_{11}u(b) + \beta_{12}u'(b) = 0, \tag{12.22}$$

$$B_2(u) = 0\colon \alpha_{21}u(a) + \alpha_{22}u'(a) + \beta_{21}u(b) + \beta_{22}u'(b) = 0, \tag{12.23}$$

where $\alpha_{ij}$, $\beta_{ij}$ are some given complex numbers such that the vectors $[\alpha_{11}, \alpha_{12}, \beta_{11}, \beta_{12}]$ and $[\alpha_{21}, \alpha_{22}, \beta_{21}, \beta_{22}]$ are linearly independent. We denote by $D_L$ the set of all twice-differentiable functions which satisfy the boundary conditions $B_1(u) = 0$, $B_2(u) = 0$. If $\beta_{11} = 0 = \beta_{12} = \alpha_{21} = \alpha_{22}$, the conditions are *unmixed* or *pure boundary conditions.* If $\alpha_{11} = \alpha_{22} = 1$ and all other coefficients are zero the resulting conditions are known as homogeneous *initial conditions.* If $\alpha_{11} = -\beta_{11} = \alpha_{22} = -\beta_{22}$ and all other coefficients are zero, the conditions are known as *periodic boundary conditions.*

**Definition 12.2.1** Green's function of a second order differential operator.
Suppose we can find a function $g(t, \tau)$ such that $Lg = \delta(t - \tau)$ with $B_1(g) = 0$, $B_2(g) = 0$. Here, the operators $L$, $B_1$, $B_2$ apply to $g$ as a function of $t$ with $\tau$ treated as a parameter. Then $g(t, \tau)$ is known as the Green's function of the operator $L$[5].

An important result is that a *Green's function exists and is unique if the completely homogeneous system $Lu = 0$, $a < t < b$; $B_1(u) = 0 = B_2(u)$ has only the trivial solution.*

### 12.2.2 Self-adjoint operators

**Definition 12.2.2** Formal adjoint of a second order differential operator.
The formal adjoint differential operator $L^\dagger$ is defined by considering the following inner product and employing integration by parts

$$\begin{aligned}\langle Lu, v\rangle &= \int_a^b [a_2u'' + a_1u' + a_0u]v^* \, \mathrm{d}t \\ &= \underbrace{\int_a^b u[(a_2v)'' - (a_1v)' + a_0v]^* \, \mathrm{d}t}_{\langle u, L^\dagger v\rangle} + \Bigg[\underbrace{a_2(v^*u' - uv^{*\prime}) + uv^*(a_1 - a_2')}_{J(u,v)}\Bigg]_a^b \\ &= \langle u, L^\dagger v\rangle + J(u, v)]_a^b,\end{aligned} \tag{12.24}$$

where the scalar function $J(u, v)$ is known as the *conjunct* or *bilinear concomitant.* The formal adjoint differential operator $L^\dagger$ is defined by

$$L^\dagger = a_2\frac{\mathrm{d}^2}{\mathrm{d}t^2} + (2a_2' - a_1)\frac{\mathrm{d}}{\mathrm{d}t} + (a_2'' - a_1' + a_0), \tag{12.25}$$

and we have the relation

$$\langle Lu, v\rangle - \langle u, L^\dagger v\rangle = J(u, v)]_a^b. \tag{12.26}$$

Equation (12.26) is known as *Green's formula.* If $a_2'(t) = a_1(t)$, then it is seen that $L^\dagger = L$. Such a differential operator is said to be *formally self-adjoint.* A formally self-adjoint operator can be written as

[5] An equivalent specification without recourse to a delta function is the following. The Green's function is a fundamental solution satisfying $Lg = 0$, $a \le t < \tau$ and $\tau < t \le b$; $B_1(g) = 0 = B_2(g)$, $g$ is continuous at $t = \tau$ i.e. $g(t = \tau+, \tau) = g(t = \tau-, \tau)$ and satisfies the discontinuity property $\frac{dg}{dt}|_{t=\tau+} - \frac{dg}{dt}|_{t=\tau-} = \frac{1}{a_2(\tau)}$. The latter follows directly upon integrating the defining equation of the Green's function.

$$L = \frac{\mathrm{d}}{\mathrm{d}t}\left(a_2(t)\frac{\mathrm{d}}{\mathrm{d}t}\right) + a_0(t), \tag{12.27}$$

in which case the conjunct becomes

$$\begin{aligned} J(u, v) &= a_2(t)[u'(t)v^*(t) - u(t)v^{*\prime}(t)] \\ &=: -a_2(t)\mathscr{W}(u, v^*;t) = -J^*(v, u), \end{aligned} \tag{12.28}$$

where $\mathscr{W}(u, v^*;t)$ is the Wronskian of $u$ and $v^*$ evaluated at $t$. Thus the conjunct of a formally self-adjoint operator exhibits skew-Hermitian symmetry with respect to its arguments. For a formally self-adjoint operator the Green's formula becomes

$$\int_a^b [v^*Lu - uLv^*]\,\mathrm{d}t = -\{a_2(t)\mathscr{W}(u, v^*;t)\}_a^b. \tag{12.29}$$

Let $D_L^\dagger$ denote the set of all twice-differentiable functions $v(t)$ that satisfy the boundary conditions $B_1^\dagger(v) = 0$ and $B_2^\dagger(v) = 0$ such that the conjunct in the general case vanishes whenever $u$ is $D_L$. These boundary conditions will also be of the type (12.22) and (12.23), but with possibly different coefficients, i.e. of the form

$$B_1^\dagger(v) = 0{:}\ \alpha_{11}^\dagger v(a) + \alpha_{12}^\dagger v'(a) + \beta_{11}^\dagger v(b) + \beta_{12}^\dagger v'(b) = 0, \tag{12.30}$$

$$B_2^\dagger(v) = 0{:}\ \alpha_{21}^\dagger v(a) + \alpha_{22}^\dagger v'(a) + \beta_{21}^\dagger v(b) + \beta_{22}^\dagger v'(b) = 0, \tag{12.31}$$

with a new set of complex coefficients $\alpha_{ij}^\dagger$, $\beta_{ij}^\dagger$. In general $D_L^\dagger \neq D_L$ even if the operator $L$ is formally self-adjoint. For instance, if $a_1(t) = a_2'(t)$ so that $L^\dagger = L$ and if $B_1(u) = u(a) + ju'(a) = 0$, $B_2(u) = u'(b) = 0$, $j = \sqrt{-1}$, then $B_1^\dagger(v) = v(a) - jv'(a) = 0$, $B_2^\dagger(v) = v'(b) = 0$ would make the conjunct zero. It is seen that $B_1$ and $B_1^\dagger$ have different coefficients. Another example of a formally self-adjoint operator with $D_L^\dagger \neq D_L$ is the initial value problem $B_1(u) = u(a) = 0$, $B_2(u) = u'(a) = 0$, which results in the final conditions $B_1^\dagger(v) = v(b) = 0$ and $B_2^\dagger(v) = v'(b) = 0$ for the adjoint problem.

**Definition 12.2.3** Self-adjoint differential operator.
The boundary value problem

$$Lu = f,\ a < t < b; \quad B_1(u) = 0,\ B_2(u) = 0, \tag{12.32}$$

is said to be self-adjoint if $L^\dagger = L$ *and* $D_L^\dagger = D_L$. This implies that $\langle Lu, v\rangle = \langle u, Lv\rangle$.

For a formally self-adjoint operator, the following boundary conditions result in a self-adjoint operator.

(i) Unmixed boundary conditions: $\alpha_{11}^\dagger = \alpha_{11}^*$, $\alpha_{12}^\dagger = \alpha_{12}^*$, $\beta_{21}^\dagger = \beta_{21}^*$, $\beta_{22}^\dagger = \beta_{22}^*$.

(ii) Periodic boundary conditions: $u(a) = u(b)$, $u'(a) = u'(b) \Longrightarrow v(a) = v(b)$, $v'(a) = v'(b)$ only if $a_2(a) = a_2(b)$, that is, the leading coefficient function of the differential operator must also be periodic.

In both of these situations, adjoint boundary conditions of the form $B_1^\dagger(v) = 0$, $B_2^\dagger(v) = 0$ reduce to $B_1(v^*) = 0$, $B_2(v^*) = 0$. Thus $u$ and $v^*$ satisfy the same boundary conditions and $\langle Lu, v\rangle = \langle u, Lv\rangle$. As already remarked at the end of definition 12.2.2, the initial value problem will not be self-adjoint even if the operator is formally self-adjoint. The following theorem relates the Green's function of a differential operator with the Green's function of the adjoint operator.

### 12.2.3 Relation between the Green's function of the original and adjoint equations

**Theorem 12.2.1** Green's function of the adjoint equation, [2, p 173].
*Let $g(t, \tau)$ be the Green's function of a second-order differential operator $L$ defined on $D_L$*

$$Lg = \delta(t - \tau),\ a < t, \tau < b; \quad B_1(g) = 0,\ B_2(g) = 0, \tag{12.33}$$

*and let $h(t, \tau)$ be the Green's function of the adjoint differential operator $L^\dagger$ defined on $D_L^\dagger$*

$$L^\dagger h = \delta(t - \tau),\ a < t, \tau < b; \quad B_1^\dagger(h) = 0,\ B_2^\dagger(h) = 0. \tag{12.34}$$

*The adjoint boundary conditions are such that the conjunct vanishes. Then*

$$g(t, \tau) = h^*(\tau, t); \quad a < t, \tau < b. \tag{12.35}$$

*In particular, if the differential system is self-adjoint, then $h(t, \tau) = g^*(t, \tau)$ and we have the symmetry (or the reciprocal) property*

$$g(t, \tau) = g(\tau, t); \quad a < t, \tau < b. \tag{12.36}$$

Consider now the solution of the non-homogeneous boundary value problem $Lu = f$, $a < t < b$; $B_1(u) = 0$, $B_2(u) = 0$ and the allied Green's function of the adjoint equation $L^\dagger h = \delta(t - \tau)$, $a < t, \tau < b$; $B_1^\dagger(h) = 0 = B_2^\dagger(h)$. Then

$$\begin{aligned} u(\tau) &= \int_a^b u(t)\delta(t - \tau)\,\mathrm{d}t = \langle u, L^\dagger h\rangle \\ &= \langle Lu, h\rangle = \langle f, h\rangle = \int_a^b f(t)h^*(t, \tau)\,\mathrm{d}t = \int_a^b f(t)g(\tau, t)\,\mathrm{d}t. \end{aligned} \tag{12.37}$$

Interchanging the dummy variables we write

$$u(t) = \int_a^b g(t, \tau) f(\tau)\, \mathrm{d}\tau := L^{-1}(f). \tag{12.38}$$

It is seen that a Green's function generates the inverse operator of $L$. Indeed for an eigenvalue problem where $f = \lambda u$, equation (12.38) implies

$$u(t) = \lambda \int_a^b g(t, \tau) u(\tau)\, \mathrm{d}\tau. \tag{12.39}$$

In many instances involving formally self-adjoint differential operators coupled with boundary conditions having all real coefficients, the Green's function works out to be purely real-valued. In that case, the symmetry of the Green's function also implies symmetry of the kernel $g(t, \tau) = g^*(\tau, t)$ required of self-adjoint integral equations. In other words, if $L$ is a self-adjoint differential operator and if $g$ is its real-valued Green's function subject to appropriate boundary conditions, then there is a corresponding Hilbert–Schmidt linear integral operator generated by the symmetric kernel $g(t, \tau)$. In view of the conditions on the existence of the Green's function (see the discussion under definition 12.2.1), we may restate theorem 12.1.10 directly for differential operators

**Theorem 12.2.2** Alternative theorem for self-adjoint differential operator, [1, p 355]. *Consider the non-homogeneous boundary value problem* $Lu = f$, $a < t < b$; $B_1(u) = 0$, $B_2(u) = 0$ *with* $\langle Lu, v\rangle = \langle u, Lv\rangle$. *Then the following alternative exists:*

(i) *Either the equation* $Lu = f$ *has a uniquely determined solution for every given* $f$, *or*

(ii) *The homogeneous equation* $Lu = 0$ *has a non-trivial solution. The non-homogeneous equation has then a solution iff the solutions* $u_0(t)$ *of the homogeneous equation* $Lu_0 = 0$ *are orthogonal to* $f$, *i.e.* $\langle f, u_0\rangle = 0$.

In the most general case of non-self-adjoint operators or if the adjoint boundary conditions are not chosen to make the conjunct vanish, one may express the solution of the problem $Lu = f$ by using the Green's formula (12.26) and considering the adjoint Green's function equation $L^{\dagger}h = \delta(t - \tau)$

$$u(\tau) = \int_a^b f(t) h^*(t, \tau)\, \mathrm{d}t + \left\{ a_2(t) \mathscr{W}(u, h^*; t, \tau) + u(t) h^*(t, \tau)[a_1(t) - a_2'(t)] \right\}_{t=a}^{t=b}, \tag{12.40}$$

where $\mathscr{W}$ is the Wronskian of $u(t)$ and $h^*(t; \tau)$. The result will coincide with equation (12.37) if the boundary conditions for the adjoint Green's function are chosen to make the conjunct vanish.

**Example 12.2.3.1.** Initial value problem; non self-adjoint case.

Consider the differential equation

$$Lu := -\frac{d^2u}{dx^2} - k^2u = f, \ 0 < x < b; \quad u(0) = \alpha, \ u'(0) = \beta,$$

where $\alpha, \beta$ are complex constants and $k$ is a real constant. Here $a_2(x) = -1$, $a_1(x) = 0$, and $a_0(x) = -k^2$. The domain $D_L$ consists of all functions $u(x)$ in $\mathscr{L}_2^{(c)}(0, b)$ that satisfy the above boundary conditions. The adjoint operator is $L^\dagger = -\frac{d^2}{dx^2} - k^2 = L$. The adjoint Green's function satisfies

$$L^\dagger h = \delta(x - \xi)$$

and if the boundary conditions are chosen to be

$$h(b, \xi) = 0, \ h'(b, \xi) = 0$$

then

$$u(\xi) = \int_0^b f(x)h^*(x, \xi)\,dx - \beta h^*(0, \xi) + \alpha h^{*\prime}(0, \xi).$$

The domain $D_{L^\dagger}$ is the set of all functions $v(x)$ belonging to $\mathscr{L}_2^{(c)}(0, b)$ that satisfy $v(b) = 0$, $v'(b) = 0$. Clearly, $D_L \neq D_{L^\dagger}$. The adjoint Green's function can be easily shown to be

$$h(x, \xi) = \Theta(\xi - x)\frac{\sin k(x - \xi)}{k},$$

where $\Theta(\cdot)$ is the unit step function. Using this we obtain

$$h(0, \xi) = -\frac{\sin k\xi}{k}, \ h'(0, \xi) = \cos k\xi.$$

Therefore

$$u(\xi) = \frac{1}{k}\int_0^\xi \sin k(x - \xi)f(x)\,dx + \alpha\cos k\xi + \frac{\beta}{k}\sin k\xi.$$

The total solution of the differential equation is seen to consist of a particular solution and a homogeneous solution. The integral term on the rhs gives the particular solution, while the last two terms arise from the two independent homogeneous solutions. Note that the differential system here is not self-adjoint as $D_L \neq D_{L^\dagger}$. This is despite the fact that the operator $L$ is formally self-adjoint. It is worth noting that the conjunct vanishes if $\alpha = \beta = 0$. In that case the Green's function $g(x, \xi) = h^*(\xi, x) \neq g(\xi, x)$. But even in this case, the problem is non self-adjoint because the boundary conditions for the original and adjoint problems are different. ■■

### 12.2.4 Completeness theorem for self-adjoint differential operators

Consider now the eigenvalue problem

$$Lu = \lambda u,\ B_1(u) = 0 = B_2(u), \tag{12.41}$$

where $L$ is a self-adjoint differential operator of second order. The system generates a set of eigenvalues $\{\lambda_k\}$ and corresponding normalized eigenfunctions $\{u_k\}$. We would like to relate these to the eigenpair of the equivalent Hilbert–Schmidt integral operator. To this end, we consider a slightly modified problem

$$Mu = \mu u;\ B_1(u) = 0 = B_2(u), \tag{12.42}$$

where $\mu = \lambda - \bar{\lambda}$, $M = L - \bar{\lambda}I$, $\bar{\lambda}$ is a fixed value that is *not* an eigenvalue of the system (12.41). The system (12.42) has the same set of eigenfunctions as equation (12.41), but its eigenvalues are $\mu_k = \lambda_k - \bar{\lambda}$. Any possible zero eigenvalue of equation (12.41) is converted to the non-zero eigenvalue $-\bar{\lambda}$ of equation (12.42) by this device. Let $g(t, \tau)$ be the Green's function of $M$ with the boundary conditions $B_1(g) = 0 = B_2(g)$. The Green's function exists and is unique because $\mu = 0$ is not an eigenvalue of equation (12.42), which in turn implies that the homogeneous equation $Mu = 0$ with the associated boundary conditions has only the trivial solution. By equation (12.39), the system (12.42) is equivalent to the integral equation

$$\theta u(t) := \mu^{-1}u(t) = \int_a^b g(t, \tau)u(\tau)\,\mathrm{d}\tau := Gu, \tag{12.43}$$

with the Hilbert–Schmidt operator $G$. By the HS-spectral theorem 12.1.13, the set of eigenfunctions $\{u_k\}$ of $Gu = \theta u$ (and hence those of systems (12.42), (12.41)) form a complete orthonormal set if $\theta = 0$ is not an eigenvalue of $G$. But $\theta = 0$ is not an eigenvalue because setting $\theta = 0$ results in $Gu = 0$, which in turn implies that the solution $z = Gu$ of the system $Mz = u$, $B_1(z) = 0 = B_2(z)$ is zero. Therefore $u = Mz = M0 = 0$, thereby yielding the important result.

**Theorem 12.2.3** Completeness of the eigenfunctions of a self-adjoint second order differential operator, [5, p 220].
*The eigenfunctions of any self-adjoint differential system of the second order form a complete orthonormal set.*

A simple example will illustrate the ubiquity of integral equations and Green's functions in electrodynamics. Consider the scalar wave equation in 1D space–time $(x, t)$ satisfied by a field component $u(x, t)$

$$\frac{\partial^2 u}{\partial x^2} - \frac{1}{c^2}\frac{\partial^2 u}{\partial t^2} = 0;\quad u(0, t) = 0 = u(\ell, t),$$

where $c$ is the speed of light. We seek an orthonormal basis for field expansion and consider solutions of the form $u(x, t) = \mathrm{e}^{j\omega t}X(x)$, $j = \sqrt{-1}$. The partial differential equation above is converted to an ordinary differential equation for $X$

$$-\frac{d^2X}{dx^2} = \kappa X, \quad X(0) = 0 = X(\ell), \tag{12.44}$$

$\kappa = \omega^2/c^2 > 0$. The function $X(x)$ is taken to belong to $\mathscr{L}_2^{(c)}(0, \ell)$. The inverse of a differential operator can be obtained through its Green's function $g(x, \xi)$, which in this case satisfies the allied differential equation

$$-\frac{d^2g}{dx^2} = \delta(x - \xi), \quad g(0, \xi) = 0 = g(\ell, \xi). \tag{12.45}$$

This equation has the continuous and symmetric solution

$$g(x, \xi) = \frac{1}{\ell}x_<(\ell - x_>) = g(\xi, x), \quad 0 \le x, \xi \le \ell,$$

where $x_< = \min(x, \xi)$, $x_> = \max(x, \xi)$. Note that the Green's function of this self-adjoint boundary value problem is purely real. The solution of equation (12.44) can be formally represented using linearity and the sifting property of the delta function or directly from equation (12.38) as

$$X(x) = \kappa \int_0^\ell g(x, \xi)X(\xi)\, d\xi. \tag{12.46}$$

This is an eigenvalue integral equation of the self-adjoint, Hilbert–Schmidt type and all of the properties listed above directly apply. In particular, a complete set of eigenfunctions exist for the eigenvalue problem (12.46) and hence for equation (12.44). The normalized eigenfunctions $\varphi_n(x) = \sqrt{\frac{2}{\ell}} \sin\left(\frac{n\pi x}{\ell}\right)$, $n = 1, 2, \ldots$ form an orthonormal basis in $\mathscr{L}_2^{(c)}(0, \ell)$. The eigenvalues are the constants $\kappa_n = (n\pi/\ell)^2$, $n=$ 1, 2, …. A few other differential systems and their eigenfunction system $\{\varphi_n, \lambda_n\}$ are shown in the following examples.

**Example 12.2.4.1.**

$$Lu = -\frac{d^2}{dt^2}, \ 0 \le t \le \ell, \ u'(0) = 0 = u'(\ell).$$

$$\varphi_n = \left(\frac{\varepsilon_n}{\ell}\right)^{1/2} \cos\left(\frac{n\pi t}{\ell}\right), \ \lambda_n, \ n = 0, 1, \ldots, \ \varepsilon_n = 1, n = 0, \ \varepsilon_n = 2, n > 0.$$

Eigenfunctions of this kind are encountered in a perfectly conducting parallel-plate waveguide where $u$ is a component of tangential magnetic field. ■■

**Example 12.2.4.2.**

$$L = -\frac{d^2}{dt^2}, \ 0 \le t \le \ell, \ u(0) = u(\ell), \ u'(0) = u'(\ell).$$

$$\varphi_0 = 1,\ \lambda_0 = 0$$

$$\varphi_{2n-1} = \sqrt{\frac{2}{\ell}}\cos\left(\frac{n\pi t}{\ell}\right),\ \lambda_{2n-1} = (2n\pi/\ell)^2,\ n = 1, 2, \ldots \tag{12.47}$$

$$\varphi_{2n} = \sqrt{\frac{2}{\ell}}\sin\left(\frac{n\pi t}{\ell}\right),\ \lambda_{2n} = (2n\pi/\ell)^2,\ n = 1, 2, \ldots .$$

■■

**Example 12.2.4.3.**

$$Lu = -\frac{1}{t}(tu')',\ 0 < a \le t \le b,\ u(a) = 0 = u(b).$$

$$\langle u, v\rangle = \int_a^b tu(t)v(t)\,\mathrm{d}t,$$

$$\varphi_n = \frac{\pi k_n}{\sqrt{2(R_n^2 - 1)}}[J_0(k_n t)Y_0(k_n a) - J_0(k_n a)Y_0(k_n t)],\ \lambda_n = k_n^2,\ n = 1, 2, \ldots, \tag{12.48}$$

$$J_0(k_n a)Y_0(k_n b) - J_0(k_n b)Y_0(k_n a) = 0,\ 0 < k_1 < k_2\cdots,$$

$$R_n = \frac{J_0(k_n a)}{J_0(k_n b)} = \frac{Y_0(k_n a)}{Y_0(k_n b)}.$$

Eigenfunctions of this kind are encountered inside a coaxial cable with inner and outer radii $a$, $b$, respectively, and having no azimuthal variation of fields. ■■

**Example 12.2.4.4.**

$$Lu = -\frac{1}{t}(tu')',\ 0 < t < a,\ u(a) = 0 = \lim_{t\to 0} tu'.$$

$$\langle u, v\rangle = \int_0^a tu(t)v(t)\,\mathrm{d}x,$$

$$\varphi_n = A_n J_0(k_n t),\ \lambda_n = k_n^2,\ J_0(k_n a) = 0, \tag{12.49}$$

$$A_n^2\int_0^a tJ_0^2(k_n t)\,\mathrm{d}t = 1,\ n = 1, 2, \ldots.$$

Eigenfunctions of this kind are encountered inside a circular waveguide of radius $a$ having no azimuthal variation of fields. ■■

## 12.3 Classification of operators and their properties

**Definition 12.3.1.** Closable operator.
A linear operator $L$ is closable if for every null sequence $\{x_n\}$ in $\mathscr{D}_L$, either $Lx_n = 0$ or else $Lx_n$ has no limit. In particular, the null sequence of a

closable operator will not produce a finite limit $Lx_n \to f$ that is different from zero.

The domain $\mathscr{D}_L$ of closable operators can also be extended to $\bar{\mathscr{D}}_L$ by including the limit points of sequences $\{x_n\}$ as was done for bounded linear operators provided that the limits $Lx_n$ exist.

**Definition 12.3.2.** A one-to-one operator $L$ has an inverse $L^{-1}$ whose domain is $\mathscr{R}_L$ and whose range is $\mathscr{D}_L$. The inverse is defined as

$$L^{-1}f = g \text{ iff } Lg = f.$$

A linear operator $L$ is one-to-one and has an inverse $L^{-1}$ iff $Lx = 0 \Longrightarrow x = 0$.

**Definition 12.3.3.** Regular operator.
An operator $L$ on a Hilbert space $\mathscr{S}$ is regular if all the following conditions are satisfied

- (a) $Lx = 0$ has only a trivial solution (i.e. $\lambda = 0$ is not an eigenvalue of the system $Lx = \lambda x$).
- (b) $L^{-1}$ is bounded.
- (c) $\mathscr{R}_L = \mathscr{S}$.

An example of a regular operator is the identity operator $Ix = x$.

**Definition 12.3.4.** Essentially regular operator.
An operator is essentially regular if

- (a) $Lx = 0$ has only a trivial solution (therefore $L^{-1}$ exists on $\mathscr{R}_L$).
- (b) $L^{-1}$ is bounded.
- (c) $\mathscr{R}_L \neq \mathscr{S}$, but $\bar{\mathscr{R}}_L = \mathscr{S}$ (i.e. $\mathscr{R}_L$ can be extended to include the entire Hilbert space by considering the limit points of convergent sequences $L^{-1}(y_n)$).

**Definition 12.3.5** Singular operator.
If the linear operator $L$ is neither regular nor essentially regular, it is said to be singular.

A closed operator must fall into one of the following mutually exclusive categories:

- (i) $L$ is regular. An example is the identity operator $Lx = x$.
- (ii) $Lx = 0$ has a nontrivial solution (therefore $L^{-1}$ does not exist). An example is the zero operator, which is bounded, but not one-to-one.
- (iii) $Lx = 0$ has only the trivial solution, $L^{-1}$ is unbounded, and $\mathscr{R}_L \neq \mathscr{S}$, but $\bar{\mathscr{R}}_L = \mathscr{S}$. An example is $Lx = tx(t)$; $\mathscr{S} = \mathscr{L}_2^{(c)}(0, 1)$. The inverse operator is $L^{-1}y = y(t)/t$; $\mathscr{R}_L \neq \mathscr{S}$ since the range only includes those elements which

vanish in the neighborhood of $t = 0$. But those elements are *dense* in $\mathscr{S}$. Hence $\bar{\mathscr{R}}_L = \mathscr{S}$. Furthermore $L^{-1}$ is unbounded.

(iv) $Lx = 0$ has only the trivial solution and $\bar{\mathscr{R}}_L \neq \mathscr{S}$ ($L^{-1}$ can be either bounded or unbounded).

An example is the *raising operator*. Let $\mathscr{S} = \mathscr{L}_2^{(c)}(0, 1)$, and consider an orthonormal basis $\varphi_1, \varphi_2, \ldots$ on $\mathscr{S}$. If $x$ is an arbitrary element of $\mathscr{S}$, then

$$x = \sum_{n=1}^{\infty} c_n\varphi_n; \quad c_n = \langle x, \varphi_n\rangle, \ \|x\|^2 = \sum_{n=1}^{\infty} |c_n|^2.$$

Define the linear transformation $L_\uparrow$ by $y = L_\uparrow x = \sum_{n=1}^{\infty} c_n\varphi_{n+1}$, i.e. $L_\uparrow\varphi_n = \varphi_{n+1}$, $n = 1, 2, \ldots$. By considering results of theorem 12.1.4 it is straightforward to see that $\|L_\uparrow x\| = \|x\|$. Hence the operator is bounded with $\|L_\uparrow\| = 1$. The range space $\mathscr{R}_{L_\uparrow}$ is the closed set consisting of all vectors in $\mathscr{S}$ having a 0 component along the first basis function $\varphi_1(t)$. Hence $\mathscr{R}_{L_\uparrow} = \bar{\mathscr{R}}_{L_\uparrow} \neq \mathscr{S}$. Let $f$ be in $\mathscr{R}_{L_\uparrow}$. Then $f = \sum_{n=2}^{\infty} a_n\varphi_n$ with $\sum_{n=2}^{\infty} |a_n|^2 = \|f\|^2$ and $L_\uparrow x = f$ has the only solution $x = L_\uparrow^{-1} f = \sum_{n=1}^{\infty} b_n\varphi_n$ such that $b_n = a_{n+1}$, $n = 1, 2, \ldots$. Also $\left\|L_\uparrow^{-1} f\right\|^2 = \sum_{n=1}^{\infty} |b_n|^2 = \sum_{n=1}^{\infty} |a_{n+1}|^2 = \sum_{n=2}^{\infty} |a_n|^2 = \|f\|^2 \Longrightarrow \left\|L_\uparrow^{-1}\right\| = 1$. Hence the inverse operator $L_\uparrow^{-1}$ is bounded.

As a second example consider the modified raising operator $L_e x = \sum_{n=1}^{\infty} c_n\varphi_{n+1}/n^2$ for $x = \sum_{n=1}^{\infty} c_n\varphi_n$, i.e. one with $L_e\varphi_n = \varphi_{n+1}/n^2$, $n = 1, 2, \ldots$. In addition to raising, this modified operator attenuates the higher-order mode content algebraically. In a sense the input–output system does the task of frequency up-converting, coupled with low-pass filtering. Note that

$$\begin{aligned} \| L_e x \|^2 &= \sum_{m=1}^{\infty}\sum_{n=1}^{\infty} \frac{c_m c_n^*}{m^2 n^2} \int_0^1 \varphi_{m+1}(t)\varphi_{n+1}^*(t)\,\mathrm{d}t \\ &= \sum_{n=1}^{\infty} \frac{|c_n|^2}{n^4} \end{aligned} \tag{12.50}$$

$$\leq \sum_{n=1}^{\infty} |c_n|^2 = \|x\|^2. \tag{12.51}$$

The operator $L_e$ is defined for all $x$ in $\mathscr{S} = \mathscr{L}_2^{(c)}(0, 1)$ and is bounded. Hence it is closed. In addition, it is completely continuous. The homogeneous equation $L_e x = 0 \Longrightarrow c_n = 0$, $n = 1, 2, \ldots$ and hence has only the trivial solution $x = 0$. However, the inverse is unbounded and the range is not closed. To see this, notice that every vector $f$ in $\mathscr{R}_{L_e}$ has the property $\langle f, \varphi_1\rangle = 0$. Let $\bar{\mathscr{R}}_{L_e}$ be the closure of $\mathscr{R}_{L_e}$ containing its limit points and let $\mathscr{A}$ be the set of all vectors such that $\langle f, \varphi_1\rangle = 0$. Obviously $\mathscr{A} \subset \mathscr{S}$ as it lacks $\varphi_1$. If $f$ is in $\bar{\mathscr{R}}_{L_e}$, there exists a sequence $\{f_n\}$ of vectors in $\mathscr{R}_{L_e}$ such that $f_n \to f$. Also since $\langle f_n, \varphi_1\rangle = 0$ and the inner product is continuous, we also have $\langle f, \varphi_1\rangle = 0$. Hence $f \in \mathscr{A}$ and $\bar{\mathscr{R}}_{L_e} \subset \mathscr{A}$. If $f$ is in $\mathscr{A}$, then it can be expressed as $f = \sum_{n=2}^{\infty} a_n \varphi_n$ and $f = \lim_{p\to\infty} f_p$, where $f_p = \sum_{n=2}^{p} a_n \varphi_n$ is the partial sum. The element $f_p$ is the image of $x_p = \sum_{n=1}^{p-1} n^2 a_{n+1} \varphi_n \in \mathscr{S}$ since $L_e x_p = \sum_{n=1}^{p-1} a_{n+1} \varphi_{n+1} = \sum_{n=2}^{p} a_n \varphi_n = f_p$. Hence $f_p \in \mathscr{R}_{L_e}$ and $f \in \bar{\mathscr{R}}_{L_e}$. Therefore, $\mathscr{A} \subset \bar{\mathscr{R}}_{L_e}$, which together with $\bar{\mathscr{R}}_{L_e} \subset \mathscr{A}$ implies that $\bar{\mathscr{R}}_{L_e} = \mathscr{A}$. The element $f = \sum_{n=2}^{\infty} \frac{1}{n} \varphi_n$ obtained with $a_n = 1/n$ is in $\mathscr{A}$, but not in $\mathscr{R}_{L_e}$ and we conclude $\mathscr{R}_{L_e} \neq \mathscr{A} = \bar{\mathscr{R}}_{L_e}$. Let $f = \sum_{n=2}^{\infty} a_n \varphi_n$ be in $\mathscr{R}_{L_e}$ with $\|f\|^2 = \sum_{n=2}^{\infty} |a_n|^2$. The equation $L_e x = f$ has the only solution $x = L_e^{-1} f = \sum_{n=1}^{\infty} c_n \varphi_n$ with $c_n = n^2 a_{n+1}$. But $\|L_e^{-1} f\|^2 = \sum_{n=1}^{\infty} n^4 |a_{n+1}|^2 = \sum_{n=2}^{\infty} (n-1)^4 |a_n|^2$. If $a_n = \delta_n^{N+1}$ for some integer $N \geq 1$, then $\|f\| = 1$ and $\|L_e^{-1} f\| / \|f\| = N^2$, which can be made as large as possible by choosing appropriate $N$. Hence the inverse operator $L_e^{-1}$ is unbounded.

### 12.3.1 Spectral representation of operators

By *spectral representation of an operator* $L$ is meant being able to expand *any vector* in its domain $D_L$ and in its range $R_L$ in terms of a complete set of vectors related to the operator or in terms of its eigenvectors if available. An equivalent interpretation is to be able to expand a delta function in terms of a complete set of vector functions related to the operator. If $L$ is a self-adjoint operator, a complete set of eigenvectors exists by theorem 12.2.3. If $L$ is not self-adjoint, the eigenvectors of $L$ may not span the entire space. A simple example is $L = A$, where $A$ is a matrix in $R^n$ that has repeated eigenvalues. The operator in that case may not have a complete set of eigenvectors. However, the set of orthonormal vectors available for expansion can be made complete by including the *generalized eigenvectors* of $L$. The eigenvectors together with the generalized eigenvectors will thus give a spectral representation of $L$.[6]

The spectral representation of an operator depends on the study of the inverse of the operator $L - \lambda I$ for all complex values of $\lambda$. The inverse is relevant for the solution of $(L - \lambda I)x = y$ in the space $\mathscr{S}$. Because the inverse of a linear operator exists only if the homogeneous equation has only a trivial solution; the operator $L - \lambda I$ will have an inverse iff $\lambda$ is not an eigenvalue of $L$. In that case three mutually exclusive possibilities arise [2, p 126]:

1. The closure of the range of $L - \lambda I$ is whole space $\mathscr{S}$ and the inverse is bounded. In other words, $L - \lambda I$ is an essentially regular operator (see definition 12.3.4). In that case $\lambda$ is said to belong to the *resolvent set* of the operator $L$ and is denoted by $\rho(L)$. For $\lambda \in \rho(L)$, the operator $R_\lambda = (\lambda I - L)^{-1}$ is known as the *resolvent of L*.

[6] The matrix

$$A = \begin{bmatrix} 2 & 0 & -1 \\ 0 & 2 & 1 \\ -1 & -1 & 2 \end{bmatrix}$$

has repeated eigenvalues $\lambda = 2, 2, 2$. It has only one normalized eigenvector corresponding to $\lambda = 2$, which is $\varphi_1 = \begin{bmatrix} 1/\sqrt{2} & -1/\sqrt{2} & 0 \end{bmatrix}'$, where the superscript $'$ stands for transpose. It has two generalized eigenvectors $\varphi_{2g} = [0 \;\; 0 \;\; -1]'$, $\varphi_{3g} = [1 \;\; 0 \;\; 0]'$. The three vectors define an invertible, $(3 \times 3)$ matrix $T = [\varphi_1 \;\; \varphi_{2g} \;\; \varphi_{3g}]$, which will cast $T^{-1}AT$ in its Jordan canonical form.

2. The range or closure of the range of $L - \lambda I$ is the whole space $\mathscr{S}$, but the inverse is an unbounded operator. In other words, $L - \lambda I$ is a singular operator (see definition 12.3.5). In this case $\lambda$ is said to belong to the *continuous spectrum* of $L$.
3. The range or closure of the range of $L - \lambda I$ is a *proper subset* of $\mathscr{S}$. In this case $\lambda$ is said to belong to the *residual spectrum* of $L$.

The spectrum of $L$ consists of all values of $\lambda$ which belong to either the resolvent set, the continuous spectrum (singular spectrum), or the residual spectrum[7].

### 12.3.2 Spectral theory of second order differential operators

Consider the self adjoint differential operator

$$Lu = \frac{1}{s}[-(pu')' + qu], \; a < x < b,$$

under the inner product

$$\langle u, v \rangle_s = \int_a^b s(x)u(x)v^*(x)\,\mathrm{d}x,$$

where the weight function $s(x)$ is real and $>0$ in $a < x < b$. The weight function has been added in the definition of the operator and in the corresponding definition of the inner product to accommodate problems encountered in various coordinates in electrodynamics. We are interested in the solution of the related problem $(L - \lambda I)u = 0$, where $\lambda$ is a given complex number. Let $H_s$ be the Hilbert space consisting of all functions $u(x)$ for which $\int_a^b s\,|u|^2\,\mathrm{d}x < \infty$. The coefficients $p > 0$ and $q$ are real-valued functions in $a < x < b$ and $p$, $p'$, $q$ are continuous in $a < x < b$. If the interval is finite and all assumptions on the coefficients hold for the closed interval $a \le x \le b$, the problem is said to be *regular*; otherwise it is *singular*. In the singular problem, the solutions of $(L - \lambda I)u = 0$ need not lie in $H_s$. The Green's function is taken to satisfy the equation $(L - \lambda I)g = \frac{\delta(x-\xi)}{s(x)}$ or equivalently, the equation $-(pg')' + qg - \lambda sg = \delta(x - \xi)$, $a < x$, $\xi < b$ subject to appropriate boundary conditions of the form $B_1(g) = 0 = B_2(g)$. For $\lambda \in \rho(L)$ it may also be written formally in terms of the resolvent operator $R_\lambda$ as

$$g(x, x'; \lambda) = -(\lambda I - L)^{-1}\frac{\delta(x - x')}{s(x)} = -R_\lambda \frac{\delta(x - x')}{s(x)}. \tag{12.52}$$

[7] Usage of the term 'spectrum of $L$' is not consistent within the mathematics community. For instance Lorch [3, p 89] defines the spectrum of $L$ as the set of all complex numbers $\lambda$ for which the inverse $(L - \lambda I)$ fails to exist. In this regard we shall use the qualifier 'singular' in front of the word 'spectrum' to distinguish regular spectra (resolvent set) from non-regular spectra.

*12.3.2.1 Regular differential operators*

For the regular problem we consider two associated boundary conditions

$$B_1(u) = \cos\alpha u(a) - \sin\alpha u'(a) = 0 \tag{12.53}$$

$$B_2(u) = \cos\beta u(b) + \sin\beta u'(b) = 0. \tag{12.54}$$

The boundary conditions $B_1$ and $B_2$ are of the unmixed type and the signs of the coefficients have been chosen so that the eigenvalues decrease as the angles $0 \le \alpha < \pi$, $0 \le \beta < \pi$ increase. In view of the direct connection of this problem with the self-adjoint, Hilbert–Schmidt integral operator, the following properties follow for the eigenvalues, eigenvectors of the regular problem $Lu = \lambda u$ with the above definitions of boundary conditions and inner product:

(i) The eigenvalues $\lambda_n$ are real, simple, and countable infinite at most with $\lambda_1 < \lambda_2 < \lambda_3 \ldots < \lambda_n \ldots$. This implies that $|\lambda_n| \to \infty$ and that there are only finitely many negative eigenvalues.

(ii) The eigenfunctions corresponding to different eigenvalues are orthogonal.

(iii) The eigenfunctions $\{\varphi_n\}$ form an orthonormal basis for $H_s$ as enunciated in theorem 12.2.3.

For instance, example 12.2.4.3 gives eigenvalues and eigenvectors for $s(x) = x = p(x)$, $q(x) = 0$, $\alpha = 0 = \beta$. By an application of the Hilbert–Schmidt theorem (12.1.45), it is straightforward to show that the Green's function is

$$g(x, \xi; \lambda) = -\sum_n \frac{\varphi_n(x)\varphi_n^*(\xi)}{\lambda - \lambda_n} = g^*(\xi, x; \lambda^*). \tag{12.55}$$

Using linearity and the sifting property of delta function or directly from equation (12.38), the non-homogeneous equation $s(L - \lambda I)w = f$, or equivalently, the equation $-(pw')' + qw - \lambda sw = f$, $a < x < b$; $B_1(w) = 0 = B_2(w)$ has the solution

$$\begin{aligned} w(x, \lambda) &= \int_a^b g(x, \xi; \lambda) f(\xi)\, d\xi = -\sum_n \varphi_n(x) \frac{\int_a^b f(\xi)\varphi_n^*(\xi)\, d\xi}{\lambda - \lambda_n} \\ &= -\sum_n \frac{\langle f, \varphi_n\rangle}{\lambda - \lambda_n}\varphi_n, \end{aligned} \tag{12.56}$$

where $\langle .,. \rangle$ is the usual inner-product with $s = 1$. The representation (12.55) reveals that as a function of the complex parameter $\lambda$, the Green's function $g(x, \xi; \lambda)$ has simple poles along the real axis at $\lambda = \lambda_n$ with residues $-\varphi_n(x)\varphi_n^*(\xi)$. Note that we also have from the expansion theorem 12.1.45 that for each $f$ in $L_2^{(c)}(a, b)$, $f/s$ will be in $H_s$ and

$$\frac{f}{s} = \sum_n \left\langle \frac{f}{s}, \varphi_n \right\rangle_s \varphi_n$$
$$= \sum_n \varphi_n(x) \int_a^b f(\xi)\varphi_n^*(\xi)\,\mathrm{d}\xi = \sum_n \langle f, \varphi_n\rangle \varphi_n := \sum_n f_n \varphi_n. \tag{12.57}$$

We also note from this equation that

$$\int_a^b \frac{1}{s}|f|^2\,\mathrm{d}x = \sum_n |f_n|^2. \tag{12.58}$$

This is known as *Parseval's identity* and can be thought of as demonstrating the equality of energy of a function in either of its dual representations. It some instances it is also known by the terms *sum rule* or *summation rule* (see sections 4.3.2 and 13.1). For instance, if $p = 1 = s$, $a = 0$, $b = 2\pi$, $\varphi_n(x) = \mathrm{e}^{-jnkx}/\sqrt{2\pi}$, one obtains the Fourier series expansion

$$f(x) = \frac{1}{\sqrt{2\pi}} \sum_{n=-\infty}^{\infty} f_n \mathrm{e}^{-jknx}; \quad f_n = \frac{1}{\sqrt{2\pi}} \int_0^{2\pi} f(x)\mathrm{e}^{jnkx}\,\mathrm{d}x$$

and the associated Parseval's relation

$$\int_{x=0}^{2\pi} |f(x)|^2\,\mathrm{d}x = \sum_{n=-\infty}^{\infty} |f_n|^2.$$

Apart from its various other utilities, one can conclude from the Parseval's relation that

$$|f_n| \leq \sqrt{\int_{x=a}^{b} \frac{1}{s(x)}|f(x)|^2\,\mathrm{d}x}, \quad \text{for every } n.$$

In fact the Parseval's relation remains valid in multiple dimensions (see section 13.1) and even in the case of eigenfunction expansion associated with singular operators such as those involved in Fourier transform, Fourier sine/cosine transforms, Hankel transform, Hilbert transform, etc [10].

Picking $f = \delta(x - \xi)$ in equation (12.57) leads to[8]

$$\frac{\delta(x-\xi)}{s(x)} = \sum_n \varphi_n(x)\varphi_n^*(\xi) = \sum_n \varphi_n^*(x)\varphi_n(\xi). \tag{12.59}$$

[8] Even though the delta function does not belong to $L_2^{(c)}(a, b)$, we can imagine a sequence of functions in $L_2(a, b)$ whose limit converges to the delta function and equation (12.59) applies to such a limit.

Integrating equation (12.55) over a large circle $P_R$ with center at the origin and having a radius $R$ in the complex $\lambda$-plane and using equation (12.59) brings us to the central relation for a *regular* eigenvalue problem

$$\lim_{R\to\infty}\frac{1}{2\pi j}\oint_{P_R} g(x,\xi;\lambda)\,\mathrm{d}\lambda = -\sum_n \varphi_n(x)\varphi_n^*(\xi) = -\frac{\delta(x-\xi)}{s(x)}. \tag{12.60}$$

Equation (12.60) is known as the completeness relation for the Green's function or for the basis of eigenfunctions of a regular, self-adjoint, second-order differential system. In this case the constant $\lambda$ belongs to the resolvent set of $L$ and the Green's function of $s(L - \lambda I)$ can be constructed directly from the eigenfunctions as in equation (12.55). The fidelity of a Green's function expression that may have been derived by some other means can be directly tested using equation (12.60).

Even if merely one term, $\varphi_k(x)$, is excluded from a countably infinite set $\{\varphi_n\}$ of eigenfunctions, a series constructed from the remaining infinite set of eigenfunctions $\{\varphi_n\}_{n\neq k}$ will fail to satisfy the completeness relation. In this case the rhs of a Parseval type identity such as equation (12.58) will fall short of the lhs and the set $\{\varphi_n\}_{n\neq k}$ becomes 'incomplete'. If a 'Green's function' is constructed from such an incomplete set, it will also be incomplete. This despite the fact that the incomplete series will still satisfy the boundary conditions at hand (see [11] for an example of such an incomplete representation). Thus the completeness requirement goes beyond the satisfaction of boundary conditions.

*12.3.2.2 Singular differential operators*

Consider the differential equation

$$L_\lambda u := Lu - \lambda su := -(pu')' + qu - \lambda su = 0, \quad a < x < b, \tag{12.61}$$

where one of the endpoints is regular, but the other endpoint is singular. Then either the interval is semi-infinite or, if finite, the coefficient $p(x)$ vanishes at one point (concurrently, $q$ or $s$ may be unbounded at that same point). Solutions of such differential equations are no longer necessarily in $H_s$. The following are some examples of singular points:

1. Simple harmonic equation $-u'' - \lambda u = 0$ on the semi-infinite interval $0 < a < x < \infty$. Here $p = s = 1$ and $H_s = \mathscr{L}_2^{(c)}(a, \infty)$. The point $a$ is regular, but the point $b = \infty$ is singular.
2. Bessel equation for arbitrary order $-(xu')' + \frac{\nu^2}{x}u - \lambda xu = 0$ on the finite interval $0 < x < b < \infty$ with $\nu \geq 0$. Here $p(x) = x = s(x)$. Then $x = 0$ is a singular point since $p(0) = 0$. Concurrently $q(0) \to \infty$ if $\nu \neq 0$.
3. Legendre equation $-[(1 - x^2)u']' + \lambda u = 0$ on the interval $-1 < x < 1$. Here $s(x) = 1$ and $p(x) = 1 - x^2$, which is zero at (a) $x = -1$ on the subinterval interval $-1 < x < \ell$ and (b) $x = 1$ on the subinterval $\ell < x < 1$ for any $\ell$, $-1 < \ell < 1$. Thus both endpoints $x = \pm 1$ are singular.

The theorem of Weyl allows one to classify the singular endpoint according to the number of solutions in $H_s$.

**Theorem 12.3.1.** Weyl's theorem [12, p 438]. *Consider equation (12.61) where the parameter* $\lambda$ *is allowed to be a variable complex number and one of the endpoints is singular. Then the following possibilities hold:*

1. *If for some particular value of* $\lambda$*, every solution of equation (12.61) is in* $H_s$*, then for every other value of* $\lambda$*, every solution is again in* $H_s$*.*
2. *For every* $\lambda$ *with* $\Im(\lambda) \neq 0$*, there exists at least one solution of equation (12.61) that is in* $\mathrm{H_s}$*.*

*Furthermore, the singular point falls into one of the two mutually exclusive categories:*

(i) *All solutions are in* $\mathrm{H_s}$ *for all* $\lambda$*. Then the singular point is said to belong to the limit-circle*[9]*.*

(ii) *For* $\lambda$ *with* $\Im(\lambda) \neq 0$ *there is exactly one independent solution in* $H_s$ *and then for* $\Im(\lambda) = 0$ *there may be one or no solution in* $H_s$*. Then the singular point is said to belong to the* limit-point *case.*

In the following examples the abbreviation SO stands for singular operators.

**Example 12.3.2.1.** SOs with one singular point; limit point case.
For the simple-harmonic equation of item 12.3.2 (1), the two independent solutions with $\lambda = 0$ are $u_1 = 1$, $u_2 = x$. Neither of these is in $H_s = L_2^{(c)}(a, \infty)$. For a general $\lambda$ with $\Im(\lambda) \neq 0$, assuming that square root function on the Riemannian sheet of interest (top-sheet) is defined as

$$\sqrt{\lambda} = |\lambda|^{1/2}\, \mathrm{e}^{j\theta/2} := k, \tag{12.62}$$

for $\lambda = |\lambda|\, \mathrm{e}^{j\theta}$, $-2\pi < \theta \leq 0$, the two independent solutions are $u_\pm = \mathrm{e}^{\pm jkx}$. Let $\Gamma_b$ denote the real-axis interval $0 \leq \Re(\lambda) < \infty$. It represents the branch cut $\Im(k) = 0$ in the complex $\lambda$-plane. For $\lambda \notin \Gamma_b$, $k$ has a negative imaginary part on the entire top-sheet and $u_-$ is in $H_s$, while $u_+$ is not. If $\Im(\lambda) = 0$ and $\lambda \in \Gamma_b$ (i.e. $\theta = 0$), neither of the solutions is in $H_s$. If $\Im(\lambda) = 0$ and $\lambda \notin \Gamma_b$ (i.e. $\theta = \pi$) there is one solution (i.e. $u_-$) that is in $H_s$. Thus the singular point $b = \infty$ belongs to the limit-point case. Furthermore, $\Gamma_b$ corresponds to the continuous spectrum of the operator $L = -\frac{\mathrm{d}^2}{\mathrm{d}x^2}$. ■■

---

[9] *The 'circle' here refers to a circle in the complex* $\gamma$*-plane, wherein a solution* $\psi_c \in H_s$ *of the equation* $L_\lambda u = 0$*, subject to a boundary condition of the type* $B(u; \beta, c) := \cos\beta u(c) + p(c)\sin\beta u'(c) = 0$ *at* $x = c$ *with real* $\beta$ *can be written as a linear combination* $\psi_c = \theta_1 + \gamma\theta_2$ *of the solutions* $\theta_1 \in H_s$ *and* $\theta_2 \in H_s$ *which satisfy* $L_\lambda u = 0$ *and the end conditions* $B(u; \alpha + \pi/2, e) = 0$ *and* $B(u; \alpha, e) = 0$*, respectively, enforced at a regular point* $x = e < c$*, for some real* $\alpha$*, all of this assuming that the singular point is at a right endpoint* $x = b > c$*. As the point* c *is made to approach* b*, the circle of convergence within which* $\gamma$ *can lie either tends to a circle of finite radius or to a circle of zero radius corresponding to the two cases of limit circle or limit point. All of these tests are conducted for a complex* $\lambda$ *with* $\Im(\lambda) \neq 0$*.*

**Example 12.3.2.2.** SOs with two singular points; limit circle/limit point and limit point cases.

For the Bessel equation of item 12.3.2 (2), the space $H_s$ consists of functions $u$ for which $\int_0^b x|u|^2 \, dx < \infty$. The two independent solutions with $\lambda = 0$, $\nu \neq 0$ are $u_1 = x^{\nu}$, $u_2 = x^{-\nu}$, of which only $u_1$ is in $H_s$ if $\nu > 1$, but both are in $H_s$ if $\nu < 1$. For $\nu = 0$, both $u_1 = 1$ and $u_2 = \ln(x)$ are in $H_s$. Thus if $\nu < 1$, we have a limit-circle case at $x = 0$, whereas for $\nu \geq 1$ we have the limit-point case at $x = 0$. For $\lambda \neq 0$ and for $\nu < 1$ the two independent solutions are Bessel functions $J_\nu(kx)$ and $J_{-\nu}(kx)$, where $k$ is as defined in equation (12.62). These are both in $H_s$ in accordance with Weyl's theorem. If $\nu = 0$, the two independent solutions are $J_0(kx)$ and $Y_0(kx)$, both of which once again are in $H_s$.

If the interval is $0 < a < x < \infty$, then the space $H_s$ consists of $u$ for which $\int_a^\infty x|u|^2 \, dx < \infty$. Now the left endpoint is regular, but the right endpoint is singular. For $\lambda = 0$, $0 < \nu \leq 1$, neither of the two independent solutions $u_1 = x^{\nu}$, $u_2 = x^{-\nu}$ is in $H_s$. Likewise, for $\lambda = 0$, $\nu = 0$, neither of the two solutions $u_1 = 1$, $u_2 = \ln(x)$ belong to $H_s$. For $\lambda = 0$ and $\nu > 1$, only $u_2$ is in $H_s$. For a general $\lambda$ with $\mathfrak{J}(\lambda) \neq 0$, the two independent solutions are the Hankel functions $u_1 = H_\nu^{(1)}(kx)$ and $u_2 = H_\nu^{(2)}(kx)$, where $k$ is defined as in equation (12.62). Out of these $u_2$ belongs to $H_s$, while $u_1$ does not. Thus we have the limit-point case at $x = \infty$ for all $\nu \geq 0$.

If the interval is $0 < x < \infty$, then combining the above two we conclude that $x = 0$ is either a limit circle case (if $\nu < 1$) or a limit point case (if $\nu \geq 1$) and a limit point case at $x = \infty$. ■■

(a) *Construction of Green's function in the limit point case*:

The following steps are adopted in finding the Green's function $g(x, \xi; \lambda)$ of the singular operator equation $L_\lambda g = \delta(x - \xi)$, $a < x < b$, when one endpoint, say, $x = a$, is regular and the other endpoint $x = b$ is singular. It is assumed that $\mathfrak{J}(\lambda) \neq 0$. Solution for a purely real $\lambda = \lambda_r$ may be considered as a limiting case of $\lambda = \lambda_r(1 \pm j\varepsilon)$, $\varepsilon > 0$ as appropriate.

(i) In the subinterval $a < x < \xi$ the Green's function satisfies the homogeneous equation subject to the boundary condition $B_1(u) = 0$ at the regular point $x = a$. There will be one multiplicative constant in the expression for Green's function in this subregion.

(ii) In the subinterval $\xi < x < b$, the Green's function satisfies the homogeneous equation and the requirement that $g$ lies in $H_s(\xi, b)$. (The latter is guaranteed to exist by Weyl's theorem.) There will be one multiplicative constant in the expression for Green's function in this subregion.

(iii) The two floating constants are evaluated by imposing the continuity condition $g(x = \xi -, \xi; \lambda) = g(x = \xi +, \xi; \lambda)$ and the discontinuity relation in the derivative $g'(x = \xi +, \xi; \lambda) - g'(x = \xi -, \xi; \lambda) = -\frac{1}{p(\xi)}$.

If both endpoints are singular, then no boundary conditions are used in either of the subintervals $(a, \xi)$ and $(\xi, b)$, but the requirement $g \in H_s(a, \xi)$ and $g \in H_s(\xi, b)$ is separately imposed. The determination of constants proceeds along the same lines as above.

**Example 12.3.2.3.** Green's function for singular operator-1.
Find the Green's function of

$$-\frac{d^2 g}{dx^2} - \lambda g = \delta(x - \xi),\ 0 < x, \xi < \infty; \quad g(x = 0, \xi; \lambda) = 0,\ g \in \mathscr{L}_2^{(c)}(0, \infty).$$

The left point $x = 0$ is regular, while the right endpoint $x = \infty$ is singular belonging to the limit point case. Assuming that $k = \sqrt{\lambda}$ is defined as in equation (12.62), we have the following solution satisfying the above two requirements

$$g(x, \xi; \lambda) = \begin{cases} A \sin kx, & x < \xi, \\ Be^{-jkx}, & x > \xi. \end{cases} \tag{12.63}$$

Continuity of $g$ at $x = \xi$ gives $A \sin k\xi = Be^{-jk\xi}$. The discontinuity relation in the derivative of $g$ at $x = \xi$ gives $k[-jBe^{-jk\xi} - A\cos k\xi] = -1$. Therefore

$$g(x, \xi; \lambda) = \frac{1}{k}\begin{cases} e^{-jk\xi} \sin kx, & x < \xi, \\ \sin k\xi e^{-jkx}, & x > \xi. \end{cases} \tag{12.64}$$

It is seen that Green's function is symmetric, i.e. $g(x, \xi, \lambda) = g(\xi, x; \lambda)$. The positive real axis $\Gamma_b = [0, \infty)$ defines the branch cut dividing the proper-sheet (top sheet) and the improper sheet of the two-sheeted complex $\lambda$ domain. Integrating the Green's function on the top-sheet of the complex $\lambda$-plane over a large circle $P_R$ of radius $R$ and center at the origin and punctured on the positive real axis (as in figure 12.1), we see that the Green's function satisfies the relation

$$\frac{1}{2\pi j} \lim_{R\to\infty} \oint_{P_R} g(x, \xi; \lambda)\, d\lambda = -\delta(x - \xi), \tag{12.65}$$

where Cauchy's theorem and the identity

$$\frac{2}{\pi} \int_0^\infty \sin kx \sin k\xi \, dk = \delta(x - \xi), \tag{12.66}$$

have been used. ■■

**Example 12.3.2.4** Green's function for singular operator-2.

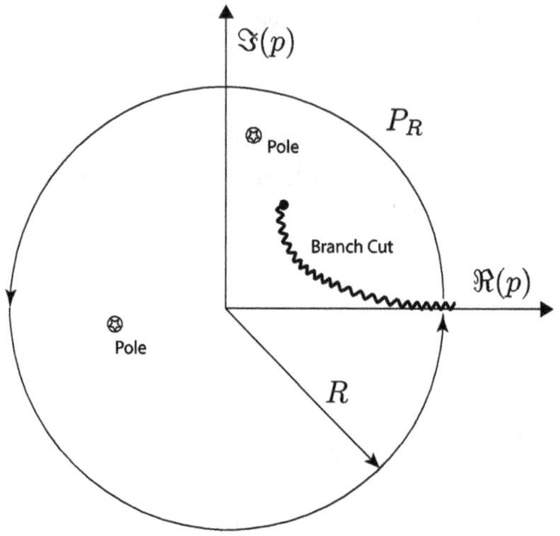


**Figure 12.1.** Contour of integration for the completeness theorem enclosing all singularities.

Find the Green's function of

$$-\frac{\mathrm{d}^2 g}{\mathrm{d}x^2} - \lambda g = \delta(x - \xi), \; -\infty < x, \xi < \infty; \quad g \in \mathscr{L}_2^{(c)}(-\infty, \infty).$$

Proceeding as in the previous example, but imposing no boundary condition at either endpoint, the solution that remains in $\mathscr{L}_2^{(c)}(-\infty, \mathfrak{l})$ and $\mathscr{L}_2^{(c)}(\mathfrak{l}, \infty)$, $\infty < \ell = \xi < \infty$ is

$$g(x, \xi; \lambda) = \begin{matrix} A\mathrm{e}^{jkx}, & -\infty < x < \xi, \\ B\mathrm{e}^{-jkx}, & \xi < x < \infty. \end{matrix} \tag{12.67}$$

Continuity of $g$ at $x = \xi$ gives $A\mathrm{e}^{jk\xi} = B\mathrm{e}^{-jk\xi}$. Discontinuity relation in the derivative of $g$ at $x = \xi$ gives $-jk[B\mathrm{e}^{-jk\xi} + A\mathrm{e}^{jk\xi}] = -1$. Therefore

$$g(x, \xi; \lambda) = \frac{1}{2jk}\mathrm{e}^{-jk|x-\xi|}. \tag{12.68}$$

Note once again that the Green's function is symmetric, $g(x, \xi; \lambda) = g(\xi, x; \lambda)$. As in the previous example, it is easy to verify that the Green's function satisfies the relation

$$\frac{1}{2\pi j}\lim_{R\to\infty}\oint_{P_R} g(x,\xi;\lambda)\,d\lambda = -\delta(x-\xi), \tag{12.69}$$

where the identity

$$\int_{-\infty}^{\infty} \mathrm{e}^{-jk(x-\xi)}\,\mathrm{d}k = 2\pi\delta(x - \xi), \tag{12.70}$$

has been used. ■■

(a) *Construction of Green's function in the limit circle case*:

The following steps are adopted [12, p 445] in finding the Green's function $g(x, \xi; \lambda)$ of the singular operator equation $L_\lambda g = \delta(x - \xi)$, $a < x < b$, when one endpoint, say, $x = a$, is regular and the other endpoint $x = b$ is singular.

(i) In the subinterval $a < x < \xi$ the Green's function satisfies the homogeneous equation subject to the boundary condition $B_1(u) = 0$ at the regular point $x = a$. Let the solution be denoted by $\psi(x, \lambda)$. There will be one multiplicative constant in the expression for Green's function in this subregion and we write $g(x, \xi; \lambda) = \psi(x, \lambda)$.

(ii) In the subinterval $\xi < x < b$, because all solutions of the homogeneous equation already lie in $H_s(\xi, b)$ for a limit circle poi nt $b$, we need to impose an additional condition at $x = b$ to determine $g$ uniquely. Define $[u, v]_x = p(x)\mathscr{W}(u, v; x)$. Let $\lambda_0$ be a fixed value of $\lambda$ such that $\mathfrak{J}(\lambda_0) \neq 0$ and $z(x, \lambda_0)$ be a solution of $Lz - \lambda_0 sz = 0$ satisfying $\lim_{x\to b}[z, z^*]_x = \lim_{x\to b}[p(x)\mathscr{W}(z, z^*;x) = 0$, where $x < b$. It is permissible to use a real value of $\lambda_0$ as long as the solution $z$ is chosen independent of $\psi$. The new boundary condition $B'_2(u) = \lim_{x\to b}[u, z^*]_x = 0$ is used at the singular point. Let $v(x, \lambda)$ be the solution of $Lv = \lambda v$, $B'_2(v) = 0$. The Green's function is then $g = \psi(x_<, \lambda)v(x_>, \lambda)$, where $x_< = \min(x, \xi)$, $x_> = \max(x, \xi)$ and the combined multiplicative constant is determined from the usual derivative discontinuity condition of the Green's function.

(iii) As an alternative to using the new boundary condition in the previous step, assuming that $\mathfrak{J}(\lambda_0) \neq 0$ has been used in the determination of $z$, we set $v = z(x, \lambda_0)$ as it qualifies as a solution. We choose the combined multiplicative constant from the derivative discontinuity of Green's function which directly translates to $[\psi, z]_x = -1$. Note that if the singular point is the left endpoint, then the derivative discontinuity of the Green's function that determines the multiplicative constant suggests $[z, \psi]_x = -1$ or $[\psi, z]_x = +1$ instead. The Green's function is given by $g(x, \xi; \lambda_0) = \psi(x_<, \lambda_0)z(x_>, \lambda_0)$.

**Example 12.3.2.5.** Green's function for singular operator-3. Find the Green's function of the zeroth-order Bessel equation $-(xu')' - \lambda xu = 0$ in the interval $0 < x < \ell$ subject to the boundary condition $u(x = \ell) = 0$ and belonging to $\mathscr{L}_2^{(c)}(0, \ell)$.

From example 12.3.2.2, the operator has regular point at $x = \ell$ and a limit circle point at $x = 0$. Here $p(x) = x$. At the singular point we could apply a boundary condition of the form $\lim_{x\to 0}[u, z^*]_x = 0$ in accordance with step (ii) above using any solution $z$ of the Bessel function for a $\lambda$ (including $\lambda = 0$) such that $z$ is independent of $\psi$. Alternatively, using step (iii) above, we can proceed on identifying an appropriate $z(x, \lambda_0)$ for a $\lambda = \lambda_0$ such that $\Im(\lambda_0) \neq 0$. Let $k = \sqrt{\lambda}$ be as defined in equation (12.62). Then a solution of the Bessel equation satisfying the boundary condition at the regular point is

$$\psi(x, \lambda) = A[J_0(kx)Y_0(k\ell) - Y_0(kx)J_0(k\ell)],\ A \text{ being a constant.}$$

Consider $\lambda = \lambda_0$ with $\Im(\lambda_0) \neq 0$. Then $k_0 = \sqrt{\lambda_0}$ will have a negative imaginary part. The two independent solutions of the Bessel equation are $J_0(k_0x)$ and $Y_0(k_0x)$, both of which belong to $\mathscr{L}_2^{(c)}(0, \ell)$. If $z = Y_0(k_0x)$, then $\lim_{x\to 0}[z, z^*]_x = -\lim_{x\to 0}x[k_0^* Y_1(k_0^*x)Y_0(k_0x) - k_0Y_1(k_0x)Y_0(k_0^*x)] = 0 \Longrightarrow k_0 = k_0^*$, which is invalid since $\Im(k_0) \neq 0$. Therefore $Y_0(k_0x)$ will not qualify as $z$. However, if $z = J_0(k_0x)$, then $\lim_{x\to 0}[z, z^*]_x = -\lim_{x\to 0}x[k_0^*J_1(k_0^*x)J_0(k_0x) - k_0J_1(k_0x)J_0(k_0^*x)] = 0$ remains valid even if $k_0$ is complex. Therefore $z(x, \lambda_0) = J_0(k_0x)$ is the admissible solution. Now $[\psi(x, \lambda_0), z(x, \lambda_0)]_x = Ax[z'\psi - \psi'z] = Axk_0J_0(k_0\ell)2/(\pi k_0x) = 2AJ_0(k_0\ell)/\pi$. Setting this to $+1$ (since the singular point is the left endpoint) gives $A = \pi/2J_0(k_0\ell)$. Therefore, for a general $\lambda$

$$g(x, \xi; \lambda) = z(x, \lambda)\psi(x_>, \lambda) = \frac{\pi}{2J_0(k\ell)}J_0(kx_<)\left[J_0(kx_>)Y_0(k\ell) - Y_0(kx_>)J_0(k\ell)\right].$$

By considering specific case of $\lambda_0 = 0$ the boundary condition $\lim_{x\to 0}[u, z^*]_x = 0$ can be shown to be equivalent to $\lim_{x\to 0}xu'(x) = 0$ in this example [12, p 448]. The solution $Y_0(kx) \sim C + \ln(x)$ violates this condition and once again one is led to use $v(x) = J_0(kx)$ in step (ii). It can be verified once again that

$$\frac{1}{2\pi j}\lim_{R\to\infty}\oint_{P_R} g(x,\xi;\lambda)\,d\lambda = -\delta(x-\xi), \tag{12.71}$$

on noting that the integrand has simple poles on the positive real axis at $\lambda_p = k_p^2 = (\chi_p/\ell)^2$ corresponding to the zeros, $\chi_p$, of the Bessel function: $J_0(\chi_p) = 0$, $p = 1, 2, \ldots$. The residue at $\lambda = \lambda_p$ is

$$-\frac{2}{[J_0'(k_p\ell)]^2}J_0(k_px)J_0(k_p\xi).$$

Note further that $\lambda = 0$ is not a branch point since the Green's function is an even function of $k$. The completeness relation defined for regular operators in equation (12.60) continues to be valid here with the orthonormal eigenfunctions $\varphi_p(x) = [\sqrt{2}/J_0'(k_p\ell)]J_0(k_px)$. ■■

**Example 12.3.2.6.** Green's function for singular operator 4.
Determine the Green's function that lies in $H_s = \mathscr{L}_2^{(c)}(0, \infty)$ of the Bessel operator $L_\lambda u = -(xu')' + \frac{\nu^2}{x}u - \lambda xu$ over the interval $0 < x < \infty$ for all $\nu \geq 0$.

From example 12.3.2.2 it is clear that $x = 0$ is a limit circle case if $0 < \nu < 1$, but a limit point case if $\nu > 1$. The point $x = \infty$ is a limit point case for all $\nu > 0$. Proceeding as in the previous example, the function $\psi$ for $x > \xi$ that lies in $H_s$ is proportional to the Hankel function $H_\nu^{(2)}(kx)$ for a $k$ defined as in equation (12.62). So $\psi(x, \lambda) = AH_\nu^{(2)}(kx)$, where $A$ is a constant. As in the previous example, of the two possibilities $J_\nu(kx)$, $Y_\nu(kx)$, the only admissible solution is $z(x, \lambda) = J_\nu(kx)$, satisfying either of the two criteria that (i) $z$ be in $H_s$ for a limit point criterion, (ii) $\lim_{x\to 0}[z, z^*] = 0$ for a limit circle criterion. The constant $A$ is picked according to $[z, \psi]_x = -1$ or from the equation $Akx[H_\nu^{(2)\prime}(kx)J_\nu(kx) - J_\nu'(kx)H_\nu^{(2)}(kx)] = -1$. Using the Wronskian (C.6) we obtain $2A/j\pi = -1$. Therefore

$$g(x, \xi; \lambda) = z(x_<, \lambda)\psi(x_>, \lambda) = \frac{\pi}{2j}J_\nu(kx_<)H_\nu^{(2)}(kx_>). \tag{12.72}$$

The $\lambda$ in this example belongs to the continuous spectrum of the singular operator $\frac{1}{s(x)}L$. The Green's function has a branch point singularity at $\lambda = 0$. The branch cut $\Re(\lambda) > 0$, $\Im(\lambda) = 0$ separates the top sheet (proper, $\Im(k) < 0$) from the bottom sheet (improper, $\Im(k) > 0$). It is easy to verify that Green's function satisfies the completeness relation

$$\frac{1}{2\pi j}\lim_{R\to\infty}\oint_{P_R} g(x, \xi; \lambda)\, d\lambda = -\delta(x - \xi), \tag{12.73}$$

upon deforming the integral along the branch-cut and carrying out the integration. As with the previous example, by considering the specific case of $\lambda_0 = 0$ the boundary condition $\lim_{x\to 0}[u, z^*]_x = 0$ can be shown to be equivalent to $\lim_{x\to 0} xu'(x) = 0$ here leading to the only possibility that $v(x, \lambda) = J_\nu(kx)$ in step (ii). ■■

**Example 12.3.2.7** Green's function for singular operator 5. Find the Green's function of the operator $L_\lambda = -\frac{d^2}{dx^2} - \lambda$ in the semi-infinite interval $0 < x < \infty$ with boundary condition $u'(0) + \alpha u(0) = 0$, where $\alpha \in \mathbb{C}$.

The operator is non self-adjoint for two reasons: (i) the boundary conditions are of initial value type and (ii) the boundary conditions involve a complex number, both of which render $D_{L^\dagger} \neq D_L$. The left endpoint $x = 0$ is a regular point, but the right endpoint belongs to the limit point type of singularity.

Let $k$ be as defined in equation (12.62). The solution $\psi(x, \lambda) = A[\cos kx - \frac{\alpha}{k}\sin kx]$, satisfying the boundary condition with $A$ constant, belongs to $H_s = \mathscr{L}_2^{(c)}(0, \xi < \infty)$. The solution $v(x, \lambda) = Be^{-jkx}$, $B$ constant, is the only one that belongs to $H_s = \mathscr{L}_2^{(c)}(\xi, \infty)$. Writing $g = \psi(x_<, \lambda)v(x_>, \lambda)$ and enforcing the continuity of $g$ at $x = \xi$ and the discontinuity in the derivative by the relation $[\psi, v]_\xi = -1$ yields

$$g(x, \xi; \lambda) = \frac{1}{jk - \alpha}\left[\cos kx_< - \frac{\alpha}{k}\sin kx_<\right]e^{-jkx_>}, \; 0 < x, \xi < \infty. \tag{12.74}$$

This type of problem is encountered by radiowaves propagating over an impedance ground plane situated at $x = 0$, where $u$ may be a component of magnetic field (say $H_z$) and $j\eta_0(k_0)^{-1}\mathrm{d}u/\mathrm{d}x$ may be the component of the electric field ($E_y$) perpendicular to both $x$ and the orientation of the magnetic field, where $\eta_0$ is the free-space impedance and $k_0$ is the wavenumber in the medium $x > 0$. Then $Z_{-x} = jk_0^{-1}\alpha$ is the normalized surface impedance of the ground plane [13]. For passive media, the real part of the impedance must be greater than zero requiring $\mathfrak{J}(\alpha) > 0$. Furthermore, if the surface impedance is inductive, then $\mathfrak{R}(\alpha) > 0$. Note that the symmetry relation $g(x, \xi; \lambda) = g(\xi, x, ;\lambda)$ still holds despite the operator being non-self-adjoint. Also $\lambda$ belongs both to the discrete spectrum $\lambda = -\alpha^2$, corresponding to the pole $k = -j\alpha$, as well to the continuous spectrum represented by the branch cut $\lambda \in \Gamma_b$: $\mathfrak{R}(\lambda)= [0, \infty)$, $\mathfrak{J}(\lambda) = 0$. For an inductive surface, the pole $k = -j\alpha$ lies on the top-sheet. For a capacitive surface, the pole does not lie on the top sheet.

It is straightforward to see that the Green's function once again satisfies the completeness relation

$$\frac{1}{2\pi j}\lim_{R\to\infty}\oint_{P_R} g(x,\xi;\lambda)\,d\lambda = -\delta(x-\xi). \tag{12.75}$$

■■

### 12.3.3 Completeness theorem in higher dimensions

The completeness relation for a Green's function is a recurring theme applying in all of these examples, whether the spectrum is discrete or continuous or a combination, whether the differential system is self-adjoint or not. In fact it continues to be valid in higher dimensions as the next theorem shows.

**Theorem 12.3.2.** Completeness theorem [2, p 214, 284], [12, p 416]. *Suppose that $\mathcal{G}(\mathbf{r}|\mathbf{r}'; p)$ is the Green's function in the complex* p*-plane of the scalar Helmholtz equation $\nabla^2\mathcal{G} + p\mathcal{G} = -\delta(\mathbf{r} - \mathbf{r}')$ subject to appropriate boundary conditions. Then the integral of the Green's function around a large circle*[10] *$P_R$ of radius $R$ in the p-domain (see figure 12.1) (that is, $P_R \in \rho(-\nabla^2)$ in the p-domain and encloses the singular spectra of the Laplacian operator) must equal*

$$\lim_{R\to\infty}\frac{1}{2\pi j}\oint_{P_R}\mathcal{G}(\mathbf{r}|\mathbf{r}';p)\,dp = -\delta(\mathbf{r}-\mathbf{r}'), \tag{12.76}$$

[10] *If the integrand has branch points in the complex* p*-plane and the branch cuts extend to infinity, then the isolated points of intersection of the large circle $P_R$ with the branch cuts are to be excluded. That is, the large circle $P_R$ is punctured at the branch cut intersections. For example, in figure* 12.2*; $P_R$ consists of the punctured circle ABC.*

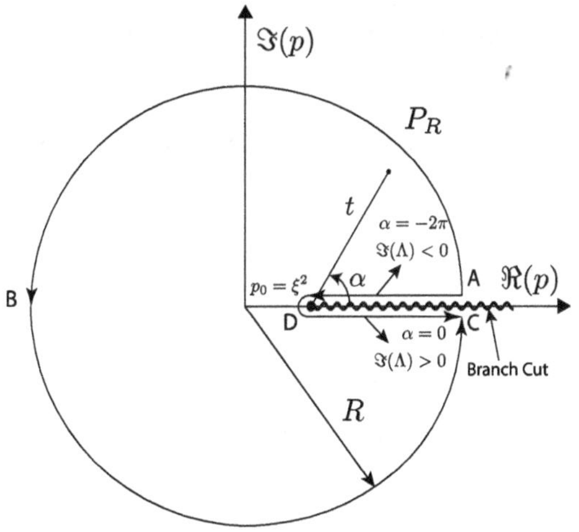


**Figure 12.2.** Contour of integration $P_R$ and the branch cut ($\Re(\Lambda) = 0$) on the top sheet with $\Re(\Lambda) > 0$. Copyright 2020 IEEE. Reprinted with permission from [18].

*Proof.* We will show the proof here for the case when boundary conditions render the boundary value problem self-adjoint and the singular spectrum is discrete[11]. In that case we have a complete set of eigenfunctions for the equation $\nabla^2 u_n + p_n u_n = 0$ with $u_n$ subject to the same boundary conditions as $\mathscr{G}$. The Green's function, being square integrable, can then be expressed as [14]

$$\mathscr{G} = -\sum \frac{u_n(\mathbf{r})u_n^*(\mathbf{r}')}{p - p_n}, \tag{12.77}$$

where the eigenfunctions satisfy the completeness relationship

$$\sum u_n(\mathbf{r})u_n^*(\mathbf{r}') = \delta(\mathbf{r} - \mathbf{r}'). \tag{12.78}$$

If the system of eigenfunctions is *separable* then the summation in equations (12.77) and (12.78) decomposes into three summations corresponding to the three coordinates. In a Cartesian system

$$\sum = \sum_l \sum_m \sum_n, \tag{12.79}$$

where $(l, m, n)$ are discrete indices denoting countably infinite modes along the $(x, y, z)$ coordinates. However, the summation signs in equations (12.77) and (12.78) are also used symbolically to indicate the situation of a continuous spectrum of eigenvalues, in which case the summation reduces to an integral in the space where the eigenvalue becomes continuous. Equation (12.78) guarantees that any square integrable function belonging to the space of admissible functions for the problem at hand can be represented in terms of the available eigenfunctions. If equation (12.77) is integrated in the complex $p$-plane about a contour $P_R$ enclosing

[11] In general the Green's function $\mathscr{G}$ will have pole and branch-point singularities in the complex $p$-plane.

all the singularities of the Green's function, then an application of Cauchy's theorem yields

$$\lim_{R\to\infty} \frac{1}{2\pi j} \oint_{P_R} \mathscr{G}(\mathbf{r}|\mathbf{r}';p)\,dp = -\sum u_n(\mathbf{r})u_n^*(\mathbf{r}') = -\delta(\mathbf{r}-\mathbf{r}'). \tag{12.80}$$

Note that the integration is carried out in the wavenumber squared domain over a counterclockwise circle and the size of the contour must be large enough to enclose all the singularities (poles and branch points in the spectral domain) of the Green's function representation. Also note the presence of the limiting operation on the left-hand side which indicates that the end result will be independent of $R$. In practice a principal asymptotic form of the Green's function in the variable $p$ of the form $\mathscr{G}O(R^{-1-\alpha})$, $\alpha > 0$ is all that is needed to pass to the limit as $R \to \infty$.

**Example 12.3.3.1.** Proof of completeness relation in the general case.
Consider the scalar Green's function of a 3D linear transformation $L$ satisfying

$$(L - \lambda I)\mathscr{G} = \delta(\mathbf{r} - \mathbf{r}'), \tag{12.81}$$

subject to appropriate boundary conditions. Taking the inverse of the transformation on the lhs results in the following formal definition of the Green's function

$$\mathscr{G}(\mathbf{r}|\mathbf{r}'; \lambda) = -(\lambda I - L)^{-1}\delta(\mathbf{r} - \mathbf{r}') = -R_\lambda\delta(\mathbf{r} - \mathbf{r}'), \tag{12.82}$$

where $R_\lambda$ is the resolvent operator of $L$ (see equation (12.52)). Let $C_\infty$ be a simple closed curve which lies on the resolvent set $\rho(L)$ of $L$ such that the singular spectrum of $L$ lies entirely in the interior, $G$, of $C_\infty$. If the singular spectra extend to infinity in the form of branch cuts, then $C_\infty$ is punctured to avoid the branch cuts. Let $f(\lambda)$ be a function analytic in $G$, that is, at every point $\lambda \in G$, the derivative $f'(\lambda)$ exists. Then the operator formula of Riesz, Gelfand, and Duncan states that [3, p 103]

$$f(L) = \frac{1}{2\pi j} \oint_{C_\infty} R_\lambda f(\lambda)\,d\lambda, \tag{12.83}$$

Note that the lhs is a function of the operator $L$. Specializing to $f(\lambda) = 1$ results in the identity

$$\frac{1}{2\pi j} \oint_{C_\infty} R_\lambda\,d\lambda = I. \tag{12.84}$$

Both sides of this equation are operators. Operating the lhs and the rhs separately on the distribution $-\delta(\mathbf{r} - \mathbf{r}')$, making use of equation (12.82) for the definition of the Green's function, and equating both sides gives the desired result

$$\frac{1}{2\pi j} \oint_{C_\infty} \mathscr{G}(\mathbf{r}|\mathbf{r}';\lambda)\,d\lambda = -\delta(\mathbf{r}-\mathbf{r}'). \tag{12.85}$$

An interesting observation can be made concerning the analytical properties of the Laplacian of the Green's function based on the completeness relation of the Green's function of the Helmholtz equation. Dividing the defining equation $\nabla^2\mathcal{G} + \lambda\mathcal{G} = -\delta(\mathbf{r} - \mathbf{r}')$ by $\lambda$ and performing a contour integration directly yields

$$-\delta(\mathbf{r}-\mathbf{r}')\frac{1}{2\pi j}\oint_{C_\infty}\frac{1}{\lambda}d\lambda = \frac{1}{2\pi j}\oint_{C_\infty}\left[\frac{\nabla^2\mathcal{G}}{\lambda} + \mathcal{G}(\mathbf{r}|\mathbf{r}';\lambda)\right]d\lambda$$

The integrand on the lhs has a pole at $\lambda = 0$ with residue 1. Upon using equation (12.85) we obtain

$$\oint_{C_\infty}\frac{\nabla^2\mathcal{G}(\mathbf{r}|\mathbf{r}';\lambda)}{\lambda}d\lambda = 0, \tag{12.86}$$

where $C_\infty$ encloses all singularities of $\mathcal{G}$ in the complex $\lambda$-plane. This result is true for points $\mathbf{r}$, $\mathbf{r}'$. ■■

**Example 12.3.3.2.** Completeness of spectral forms of free-space Green's function. In this example we would like to show that the popular integral representations of the free-space Green's function satisfy the completeness relation as well as Sommerfeld's radiation condition (Janaswamy [18]).

We consider the free-space Green's function $\mathcal{G}_0(\mathbf{r}|\mathbf{0}; p)$ for a point source placed at the origin and the response desired at the position vector represented by $\mathbf{r} = (\rho, \phi, z) = (r, \theta, \phi)$ in the usual cylindrical, spherical coordinates, respectively. Note that $p = k^2$. The Green's function has the following integral representations

$$\mathcal{G}_0(\mathbf{r}|\mathbf{0}; p) = \frac{e^{-jkr}}{4\pi r} = \frac{1}{8\pi j}\int_{-\infty}^{\infty} H_0^{(2)}(\lambda\rho)e^{-j\xi|z|}\frac{\lambda\, d\lambda}{\xi}, \tag{12.87}$$

$$= \frac{1}{2\pi^2}\int_0^{\infty} K_0(\Lambda\rho)\cos(\xi z)\, d\xi, \tag{12.88}$$

where $\xi^2 = p - \lambda^2$, $\Lambda^2 = \xi^2 - p$, $H_0^{(2)}(\cdot)$, and $K_0(\cdot)$ are, respectively, the Hankel function and modified Bessel function of the second kind of order zero. Representation (12.87) is due to Sommerfeld [14] and (12.88) is due to Schelkunoff [15]. The two integral representations are equivalent to each other and one form can be derived from the other by performing integration in the complex plane while properly deforming the contours. Our goal is to explore whether the integral representations given in equations (12.87) and (12.88) satisfy the completeness theorem as well as Sommerfeld's radiation condition

$$\lim_{r\to\infty} r\left(\frac{\partial \psi}{\partial r} + jk\psi\right) = 0, \tag{12.89}$$

At the outset it is very important to specify the range of values of the complex variables $\xi$ and $\Lambda$. In order that the integrals converge, we specify that the imaginary part $\mathfrak{J}(\xi) < 0$ and the real part $\mathfrak{R}(\Lambda) > 0$ along the axes $-\infty < \lambda < \infty$ and $0 \le \xi < \infty$, respectively. We treat lossless medium as the limiting case of a lossy medium with vanishingly small loss. Accordingly, we take $k = k_0(1 - j\varepsilon)$ where $\varepsilon > 0$ is a vanishingly small number and $k_0 > 0$ is purely real. The function $\mathcal{G}_0$ satisfies the Helmholtz equation $\nabla^2\mathcal{G}_0 + p\mathcal{G}_0 = -\delta(\mathbf{r} - \mathbf{0})$. It is seen that the Sommerfeld representation is an expansion in terms of outgoing and incoming (because $H_0^{(2)}(-z) = -H_0^{(1)}(z)$) waves along the radial direction $\rho$ and outward propagating and evanescent waves along the vertical direction $z$. In contrast the Schelkunoff representation is in terms of propagating (since $K_0(j\lambda\rho) = -j\pi H_0^{(2)}(\lambda\rho)/2$) plus evanescent waves along the radial direction and standing waves along the vertical direction. If one merely looks at the integral representations alone it is not clear if the function $\mathcal{G}_0$ satisfies the completeness relation (12.76) or the Sommerfeld radiation condition (12.89).

As an illustration we first show that the function defined by the Schelkunoff integral representation (12.88) satisfies the completeness relation in free-space. Completeness of the other form (12.87) can then be inferred as the two forms are mathematically equivalent (see [15] for a proof of the equivalence). Consider the function $K_0(\sqrt{\xi^2 - p}\,\rho)$. In the complex $p$-plane, it has branch points at $p = \xi^2 := p_0$ and at $p = \infty$. We let $p - p_0 = te^{j\alpha}$, $t$ real as indicated in figure 12.2. For a branch cut drawn from $p_0$ to $\infty$ as shown in figure 12.2 we restrict the argument $\alpha$ of $p - p_0$ to the range $-2\pi < \alpha < 0$ so that $\Lambda = \sqrt{\xi^2 - p} = te^{j(\pi+\alpha)/2}$ will have a real part greater than zero on this top sheet. Note that in this example the contour $P_R$ arising in the completeness theorem consists of the punctured circle ABC as shown in figure 12.2. The curve ADC wraps around the branch cut and the closed contour ABCDA is free of singularities.

Consider now the contour integral

$$\frac{1}{2\pi j}\oint_{P_R} \mathcal{G}_0(\mathbf{r}|\mathbf{0};p)\,dp = \frac{1}{2\pi^2}\int_0^\infty \cos(\xi z)\frac{1}{2\pi j}\oint_{P_R} K_0(\Lambda\rho)\,dp\,d\xi \tag{12.90}$$

where the right-hand side is obtained on interchanging the orders of integration. Applying Cauchy's theorem to the closed contour ABCDA, the integral over $P_R$ on the right-hand side can be converted to that along the branch cut ADC. On the branch cut $\Lambda = j\lambda$, $K_0(\Lambda\rho) = \pi H_0^{(2)}(\lambda\rho)/2j$, $dp = 2\lambda d\lambda$. Furthermore, on using $H_0^{(2)}(-\zeta) = -H_0^{(1)}(\zeta)$, $H_0^{(2)}(\zeta) + H_0^{(1)}(\zeta) = 2J_0(\zeta)$, where $J_\nu(\cdot)$ is the Bessel function of the first kind of order $\nu$, we obtain

$$\lim_{R\to\infty}\frac{1}{2\pi j}\oint_{P_R} K_0(\Lambda\rho)\,dp = -\int_0^\infty \lambda J_0(\lambda\rho)\,d\lambda = -\frac{\delta(\rho)}{\rho}, \tag{12.91}$$

where we have used the result [16, p 237, 238]:

$$\int_0^\infty \lambda J_\nu(\lambda\rho)J_\nu(\lambda\rho')\,\mathrm{d}\lambda = \frac{\delta(\rho-\rho')}{\rho}, \tag{12.92}$$

for $\nu = 0$ and $\rho' = 0$. If we additionally use [17, p 36]

$$\int_0^\infty \cos(\xi z)\,\mathrm{d}\xi = \pi\delta(z), \tag{12.93}$$

we finally obtain

$$\lim_{R\to\infty}\frac{1}{2\pi j}\oint_{P_R}\mathcal{G}_0(r,\phi,z|\mathbf{0};p)\,dp = -\frac{\delta(\rho)\delta(z)}{2\pi\rho} = -\delta(\mathbf{r}) \tag{12.94}$$

for this axisymmetric case. The integral representations shown in equations (12.87) and (12.88) are thereby complete in free space.

Even though it is clear from the closed form of $\mathcal{G}_0$ that it satisfies the Sommerfeld's radiation condition we will now establish the same using its integral representation. In many cases only an integral representation of the Green's function is available and it is imperative to examine the satisfaction or not of the Sommerfeld's radiation condition directly from it. To this end we will evaluate integral (12.87) for large values of $kr$ by the method of steepest descent discussed in detail in section 16.3 of appendix A and check the Sommerfeld's condition. We substitute $\rho = r\sin\theta_0$, $z = r\cos\theta_0$, $\lambda = k\sin\theta$, and $\xi = k\cos\theta$ in equation (12.87). The real axis in the $\lambda$-plane gets mapped to the contour $C$ in the complex $\theta$-plane as shown in figure 12.3. We rewrite the Hankel function using $\zeta = kr\sin\theta_0\sin\theta$ such that

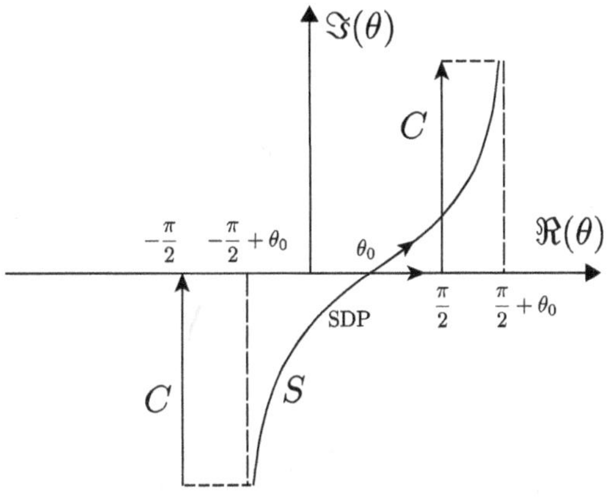


**Figure 12.3.** Mapping from the $\lambda$- to $\theta$-plane. Original contour $C$ and the steepest descent contour $S$. Copyright 2020 IEEE. Reprinted with permission from [18].

$$
\begin{aligned}
H_0^{(2)}(\zeta) &= e^{j\zeta}\left(\frac{\pi\zeta}{2j}\right)^{1/2} H_0^{(2)}(\zeta)\left(\frac{2j}{\pi\zeta}\right)^{1/2} e^{-j\zeta} \\
&:= g_0(\zeta)\left(\frac{2j}{\pi\zeta}\right)^{1/2} e^{-j\zeta}.
\end{aligned}
\tag{12.95}
$$

so that the auxiliary function $g_0(\zeta)$ is a slowly varying function of $\theta$ for large $kr \sin\theta_0$. Indeed using the asymptotic form of the Hankel function, it is easy to see that $g_0(\zeta = kr\sin\theta_0 \sin\theta) \to 1$ as $kr\sin\theta_0 \to \infty$, $\theta \neq m\pi$, $m = 0, \pm 1, \pm 2, \ldots$.

The integral in equation (12.87) for $z > 0$ can be written as

$$
\begin{aligned}
I(r, \theta_0) &= \frac{1}{8\pi j}\int_{-\infty}^{\infty} H_0^{(2)}(\lambda\rho) e^{-j\xi z}\frac{\lambda\, d\lambda}{\xi} \\
&= \frac{-jk e^{-jkr}}{8\pi}\sqrt{\frac{2j}{\pi kr\sin\theta_0}}\, I_1(kr, \theta_0),
\end{aligned}
\tag{12.96}
$$

where

$$
I_1(kr, \theta_0) = \int_C g_0(kr\sin\theta_0\sin\theta)\sqrt{\sin\theta}\, e^{kr\Phi_0(\theta)}\, d\theta, \tag{12.97}
$$

and $\Phi_0(\theta) = j[1 - \cos(\theta - \theta_0)]$. It is seen that $\Phi_0'(\theta_0) = 0$ and $\Phi_0''(\theta_0) = j \neq 0$, where a prime denotes a derivative. The contour integral $I_1$ over $C$ may be deformed into that over the steepest descent path (SDP) through the first-order saddle point $\theta := \theta_s = \theta_0$. Contributions from the two horizontal sections at infinity (shown as horizontal dashed lines in figure 12.3) between $C$ and the SDP may be ignored as the integrand becomes vanishingly small on those intervals. Making a second change of variable such that $-s^2/2 = \Phi_0$ with the corresponding $s = 2e^{-j\pi/4}\sin[(\theta - \theta_0)/2]$, $d\theta/ds = 2j/\sqrt{s^2 + 4j}$ and evaluating the integral on the SDP (on which $s$ will be purely real) we obtain

$$
I(r, \theta_0) \sim \frac{e^{-jkr}}{4\pi r} g_0(kr\sin^2\theta_0) \sim \frac{e^{-jkr}}{4\pi r} = \mathcal{G}_0, \text{ as } kr \to \infty. \tag{12.98}
$$

Detailed analysis reveals that the principal asymptotic value shown in equation (12.98) for $I(r, \theta_0)$ is accurate to within less than 5% as long as $kr \geq 2.5$ and $kr\sin^2\theta_0 \geq 2$.

We have thus shown that the integral representations given in equations (12.87) and (12.88) are not only complete but also satisfy the radiation condition at infinity. This is despite the fact that the Sommerfeld representation is comprised of standing waves along the radial direction for each spectral component and the Schelkunoff representation is comprised of standing waves along the vertical direction for each spectral component. In particular, the individual spectral components of Sommerfeld and Schelkunoff representations do *not* satisfy the radiation condition even though the aggregate sum over all spectral values does. Indeed for the spectral component $\widetilde{\psi} = H_0^{(2)}(\lambda\rho)e^{-j\xi z}$ of equation (12.87) that is valid for $z > 0$, we have with $\lambda = k\sin\theta$, $\xi = k\cos\theta$, $\rho = r\sin\theta_0$, $z = r\cos\theta_0$, and for $kr \gg 1$ and $\theta_0 \neq 0$ that

$$\left(\frac{\partial\widetilde{\psi}}{\partial r} + jk\widetilde{\psi}\right) = jk[1 - \cos(\theta - \theta_0)]\widetilde{\psi} + \mathcal{O}\left(\frac{1}{r^{3/2}}\right), \tag{12.99}$$

which clearly shows that Sommerfeld's radiation condition is not satisfied by this spectral component in general. The only spectral value for which the Sommerfeld's radiation condition is satisfied is for $\lambda = k\sin\theta_0$ that corresponds to $\theta = \theta_0$. *It is also important to observe that if the spectrum in equation (12.87) were truncated to some finite limits at the lower and upper ends in* $\xi$, *both the radiation condition and the completeness relation would be violated.* ■■

**Example 12.3.3.3.** Completeness of Green's function of conducting wedge.
A slightly more complicated example for the demonstration of the completeness relation is that of dipole radiation in the presence of a perfectly conducting wedge [18]. Figure 12.4 shows a vertical electric dipole, i.e. a dipole with its axis parallel to the vertical axis, $\hat{\mathbf{z}}$, and located at $(\rho', \phi', z')$ near a wedge. Figure 12.5 shows the cross section. For simplicity we take the dipole moment to be $I_0\ell = j\omega\varepsilon_0$.

The Green's function that corresponds to the $z$-component of the magnetic vector potential may be found by expanding the field in cylindrical harmonics subject to the boundary conditions on the wedge and at infinity. The result is derived in section 15.2.2 and given in equation (15.59)

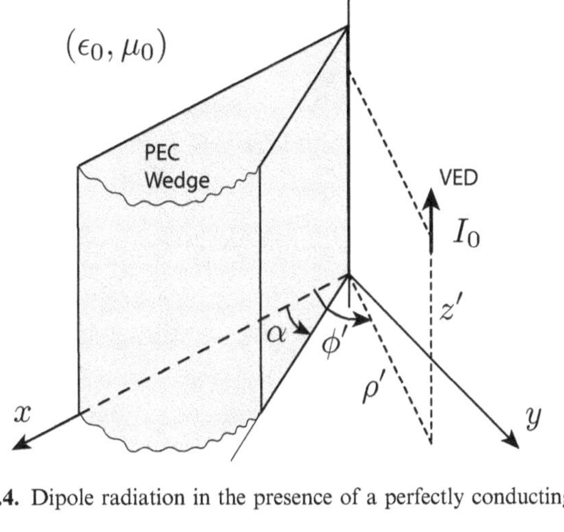


**Figure 12.4.** Dipole radiation in the presence of a perfectly conducting wedge.

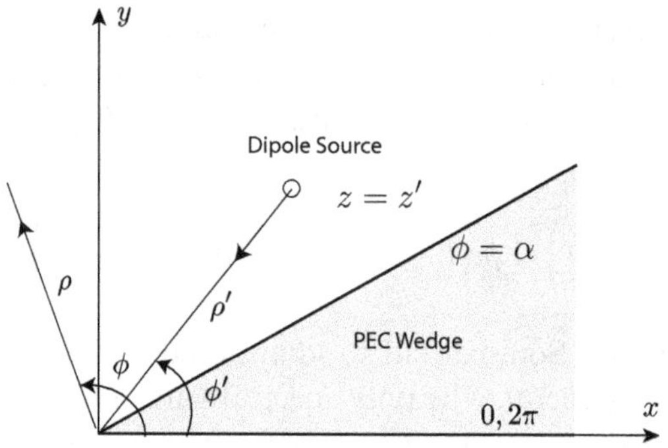


**Figure 12.5.** Cross-section of dipole radiation in the presence of a conducting wedge. Copyright 2020 IEEE. Reprinted with permission from [18].

$$\mathcal{G}_w(\mathbf{r}|\mathbf{r}'; p) = \frac{1}{2j(2\pi - \alpha)} \int_{\xi=-\infty}^{\infty} \mathrm{e}^{-j\xi(z-z')} \sum_{m=1}^{\infty} \sin \nu_m(\phi - \alpha) \sin \nu_m(\phi' - \alpha) \\ J_{\nu_m}(\lambda\rho_<) H^{(2)}_{\nu_m}(\lambda\rho_>) \,\mathrm{d}\xi, \tag{12.100}$$

where $\nu_m = m\pi/(2\pi - \alpha) > 0$, $m = 1, 2, \ldots$, $\lambda^2 + \xi^2 = k^2 = p$, $\rho_< = \min(\rho, \rho')$, $\rho_>$ $=\max(\rho, \rho')$, and $\alpha$ is the interior angle of the wedge. Along the contour of integration $\xi \in (-\infty, \infty)$, we set $\mathfrak{J}(\lambda) < 0$ for the radial wavenumber. The integral over $\xi \in (-\infty, \infty)$ can be folded over to that over $\xi \in (0, \infty)$ relying on the evenness in $\xi$ of the Fourier transform. But we shall retain the full form here and show that the representation (12.100) is complete. Here, only the radial functions are dependent on the wavenumber $k = \sqrt{p}$. The following orthogonality identity can be easily proven for the angular eigenfunctions $\Phi_{\nu_m}(\phi) = \sqrt{2/(2\pi - \alpha)} \sin \nu_m(\phi - \alpha)$:

$$\int_{\alpha}^{2\pi} \Phi_{\nu_m}(\phi)\Phi_{\nu_n}(\phi)\, \mathrm{d}\phi = \delta_m^n, \tag{12.101}$$

where $\delta_m^n$ is the Kronecker's delta. We expand a delta function in terms of the eigenfunction $\Phi_{\nu_m}(\phi)$ and use equation (12.101) to arrive at

$$\sum_{m=1}^{\infty} \Phi_{\nu_m}(\phi)\Phi_{\nu_m}(\phi') = \delta(\phi - \phi'). \tag{12.102}$$

For a fixed $\xi$, there is a branch point in the complex $p$-plane at $p = \xi^2$ due to the presence of the function $J_{\nu_m}(\lambda\rho_<)H^{(2)}_{\nu_m}(\lambda\rho_>)$. We erect a branch cut as in figure 12.2 and on it write $p = \xi^2 + t^2\mathrm{e}^{j\zeta}$, $-2\pi < \zeta < 0$, $0 < t < \infty$ and deform the contour of integration from a large punctured circle $P_R$ to that along the branch cut to result in

$$\lim_{R\to\infty}\frac{1}{2}\oint_{P_R} J_{\nu_m}(\lambda\rho_<)H^{(2)}_{\nu_m}(\lambda\rho_>)\,dp = \int_{t=0}^{\infty} tJ_{\nu_m}(t\rho_<)H^{(2)}_{\nu_m}(t\rho_>)\,dt$$
$$+\int_{t=\infty}^{0} tJ_{\nu_m}(\mathrm{e}^{-j\pi}t\rho_<)H^{(2)}_{\nu_m}(\mathrm{e}^{-j\pi}t\rho_>)\,dt$$
$$= 2\int_{0}^{\infty} tJ_{\nu_m}(t\rho_<)J_{\nu_m}(t\rho_>)\,dt$$
$$= 2\frac{\delta(\rho-\rho')}{\rho} \tag{12.103}$$

by equation (12.92). Note that the result (12.103) is independent of $m$. Therefore on using equations (12.93), (12.103) and (12.102) we finally obtain

$$\lim_{R\to\infty}\frac{1}{2\pi j}\oint_{P_R}\mathscr{G}_w(\mathbf{r}|\mathbf{r}';p)\,dp = -\frac{\delta(\rho-\rho')\delta(z-z')\delta(\phi-\phi')}{\rho}$$
$$= -\delta(\mathbf{r}-\mathbf{r}') \tag{12.104}$$

which proves that the representation (12.100) exterior to the wedge is complete. As with the free-space Green's function, it is also important to note that if the spectrum in equation (12.100) were truncated to some finite limits at the lower and upper ends in $\xi$, both the radiation condition and the completeness relation will be violated even though the boundary conditions on the faces of the wedge remain intact. Thus satisfaction of boundary conditions alone will not give an indication whether an integral representation is correct or not. A more complicated example of a dipole radiating in the presence of a conducting half-space is also considered in [18]. Reference [11] gives an example of a Green's function integral representation that is incomplete. ■■

**Example 12.3.3.4.** Satisfaction of radiation condition of Green's function of conducting wedge.

Because the integral representation (12.100) contains waves going both in the positive and negative $z$-directions, it raises a question on the possibility that the overall solution may violate Sommerfeld's radiation condition. Hence it is imperative to explore the far-field behavior of any integral representation. Show by considering the far-zone fields that the integral representation satisfies the Sommerfeld's radiation condition.

We perform an asymptotic analysis valid for large $\Omega = kr$ and show that the potential given in equation (12.100) indeed satisfies the radiation condition. We focus on the integral

$$I_w = \int_{\xi=-\infty}^{\infty} J_{\nu_m}(\lambda\rho_<)H^{(2)}_{\nu_m}(\lambda\rho_>)\mathrm{e}^{-j\xi(z-z')}\,\mathrm{d}\xi, \tag{12.105}$$

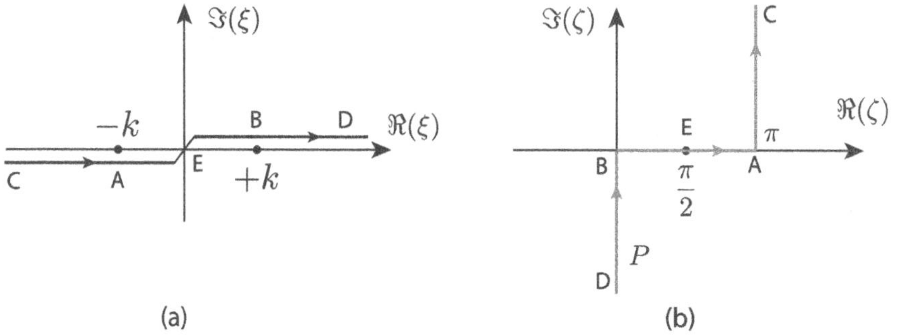


**Figure 12.6.** (a) Original contour of integration in the complex $\xi$-plane, (b) Transformed contour in the complex $\xi$-plane. The points A, E, and B which have coordinates $\xi = -k,\ 0,\ k$, respectively, in the $\xi$-plane are transformed to the respective points $\zeta = \pi,\ \pi/2,\ 0$ in the $\zeta$-plane.

for $\rho > \rho'$. The case $\rho < \rho'$ can be handled in a similar form. A change of variable $\xi = k\cos\zeta$, $\lambda = k\sin\zeta$, $\mathrm{d}\xi = -k\sin\zeta\ \mathrm{d}\zeta$ converts the contour ABCDE shown in figure 12.6(a) to that shown in figure 12.6(b). It will become clear shortly that the steepest descent path becomes easier to visualize in the transformed domain than in the original $\xi$ domain.

Let

$$g_{\nu_m}(\lambda\rho,\lambda\rho';\xi) = j^{-\nu_m}\mathrm{e}^{j\lambda\rho}\sqrt{\frac{\pi k\rho}{2j}}H^{(2)}_{\nu_m}(\lambda\rho)J_{\nu_m}(\lambda\rho'), \tag{12.106}$$

so that

$$g_{\nu_m}(\lambda\rho,\lambda\rho';\xi) \sim \frac{J_{\nu_m}(\lambda\rho')}{\sqrt{\lambda/k}},\ \text{as}\ \lambda\rho\to\infty, \tag{12.107}$$

and

$$J_{\nu_m}(\lambda\rho')H^{(2)}_{\nu_m}(\lambda\rho) = \sqrt{\frac{2j}{\pi k\rho}}j^{\nu_m}\mathrm{e}^{-j\lambda\rho}g_{\nu_m}(\lambda\rho,\lambda\rho';\xi). \tag{12.108}$$

Note that despite the presence of the Hankel function as a factor, the function $g_{\nu_m}(\lambda\rho,\lambda\rho';\xi)$ is well behaved at the origin in the $\xi$-plane[12] since, by equation (C.12), we have for $\nu_m > 0$ that

[12] The integrand of $I_w$ is free of any branch points due to the result (12.109). However, branch points are introduced when the exponential phase factor $\mathrm{e}^{j\lambda\rho}$ is included by the definition in equation (12.106). These branch points at $\xi = \pm k$ corresponding to $\lambda = 0$. It is possible to choose branch cuts emanating from $\pm k$ as in example 16.4.0.1 such that they will lie outside the region enclosed by the original contour and the SDP. Therefore, branch points will not play any material role in the ensuing analysis. These branch cuts will appear in the second and fourth quadrants of the $\xi$-plane for the $\mathrm{e}^{j\omega t}$ time convention as opposed to the branch cuts shown in figure 16.16 of appendix A for the $\mathrm{e}^{-i\omega t}$ time convention. For this reason the original contour in figure 12.6(a) is shown slightly offset from the real-axis for clarity.

$$J_{\nu_m}(\lambda\rho')H^{(2)}_{\nu_m}(\lambda\rho) \sim \frac{j}{\pi\nu_m}\left(\frac{\rho'}{\rho}\right)^{\nu_m}, \lambda \to 0. \tag{12.109}$$

Expressing the observation coordinates in spherical coordinates $\rho = r\sin\theta$, $z = r\cos\theta$ converts the integral (12.105) to

$$I_w = k\sqrt{\frac{2j}{\pi\Omega\sin\theta}}\, j^{\nu_m}\int_P \left[g_{\nu_m}(\zeta)e^{jkz'\cos\zeta}\sin\zeta\right]e^{\Omega[-j\cos(\zeta-\theta)]}\,d\zeta, \tag{12.110}$$

where we simply write $=g_{\nu_m}(\zeta)$ for $g_{\nu_m}(\Omega\sin\theta\sin\zeta,\, k\rho'\sin\zeta;\, k\cos\zeta)$ and the contour $P$ is shown in figure 12.6(b). The integral in equation (12.110) is still exact and is merely in terms of a new independent variable $\zeta$. Comparing equation (12.110) with equation (16.9) lets us identify $q(\zeta) = -j\cos(\zeta - \theta) = -j\cos[(\zeta_r - \theta) + j\zeta_i]$. We make the following observations:

(a) The function $q(\zeta) = -\sin(\zeta_r - \theta)\sinh\zeta_i - j\cos(\zeta_r - \theta)\cosh\zeta_i$ with its derivative $q'(\zeta) = j\sin(\zeta - \theta)$. The saddle point is at $\zeta_s = \theta$, which is purely real and lies in the range $0 \leq \zeta_s \leq \pi$. The second derivative of $q(\zeta)$ at the saddle point is $q''(\zeta_s) = j = e^{j\pi/2} \neq 0$. Hence the saddle point is first order and $\arg(-q''(\zeta_s)) = -\pi/2$.

(b) The steepest descent path (SDP) passing through the saddle point is defined by $\Im[q(\zeta)]=$ constant $= \Im[q(\zeta_s)] = -1$, or in other words by $\cos(\zeta_r - \theta)\cosh\zeta_i = 1$. Equivalently, it is defined by $\cos(\zeta_r - \theta)\sinh\zeta_i = \sin(\zeta_r - \theta)$. The asymptotes to the SDP are at $\zeta_r = -\pi/2 + \theta$ and $\zeta_r = \pi/2 + \theta$. The complete SDP in the $\zeta$-plane along with the asymptotes are shown in figure 12.7. The function $q(\zeta)$ on the SDP is $q(\zeta) = -j - \sin(\zeta_r - \theta)\sinh\zeta_i = -j - \sin^2(\zeta_r - \theta)\sec(\zeta_r - \theta) = -j - \cos(\zeta_r - \theta)\sinh^2\zeta_i$. It is seen that the real part of $q(\zeta)$ remains negative for all $\zeta_r \neq \theta$ on the SDP and, furthermore, it takes large negative values quickly away from $\zeta_r = \theta$.

(c) For $\zeta$ close to the saddle point it is clear from the Taylor series expansion of the cosine function that $q(\zeta) \approx -j + j(\zeta - \zeta_s)^2/2 = -j + q''(\zeta_s)(\zeta - \zeta_s)^2/2$. It

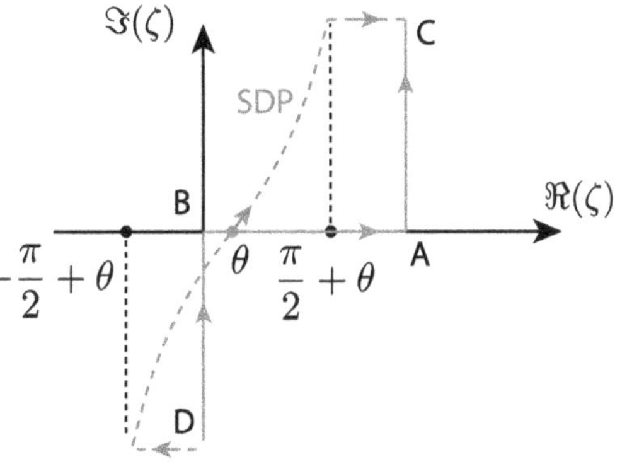


**Figure 12.7.** Steepest descent path (shown in red) in the $\zeta$-plane passing through the first-order saddle point at $\zeta = \theta$. The horizontal sections are added to the SDP at infinity to connect to the original contour (shown in green) but do not contribute to the integral as the integrand there vanishes exponentially.

is then prudent to define a new integration variable $s = (\zeta - \zeta_s)\sqrt{-q''(\zeta_s)/2}$ so that $q(\zeta) = -j - s^2$. For $s$ to be real on the SDP, we need $\arg(d\zeta) = -\frac{1}{2}\arg(-q''(\zeta_s)) = \frac{\pi}{4}$.

Substituting all of these results and the asymptotic value of equation (12.107) with $\zeta = k\cos\theta$, $\lambda = k\sin\theta$ into equation (16.23) we arrive at

$$I_w = \frac{2je^{-jkr}}{r} j^{\nu_m} J_{\nu_m}(k\rho' \sin\theta) e^{jkz'\cos\theta}, \tag{12.111}$$

and finally at the following asymptotic value of the Green's function for $kr \gg 1$

$$\mathcal{G}_w \sim \frac{e^{-jkr}}{r} \frac{e^{jkz'\cos\theta}}{(2\pi - \alpha)} \sum_{m=1}^{\infty} j^{\nu_m} \sin\nu_m(\phi' - \alpha) J_{\nu_m}(k\rho' \sin\theta) \sin\nu_m(\phi - \alpha). \tag{12.112}$$

The quantity defined by the summation is purely a function of the surface spherical coordinates $(\theta, \phi)$ and all of the radial dependence of $\mathcal{G}_w$ is contained in the spherical wave factor $e^{-jkr}/r$. Clearly, $\lim_{r\to\infty} r\left(\frac{\partial \mathcal{G}_w}{\partial r} + jk\mathcal{G}_w\right) \to 0$ and, consequently, Sommerfeld's condition is satisfied by the integral representation (12.100). ■■

**Example 12.3.3.5** Impedance, Fourier sine, Fourier cosine transforms.
By explicitly evaluating the contour integral (12.75) using the Green's function of example 12.3.2.7, define an impedance transform for a solution $f \in \mathcal{L}_2^{(c)}(0, \infty)$ of $L_\lambda f = 0$ and satisfying the boundary condition $f'(0) + \alpha f(0) = 0$.

Recall from example 12.3.2.7 that the surface is inductive (capacitive) if $\Re(\alpha) > 0$ ($\Re(\alpha) < 0$). For passive surfaces, we impose the additional condition that $\Im(\alpha) > 0$. The Green's function (12.74) has the following properties in the complex $\lambda$-plane:

(i) For an inductive surface, it has a simple pole on the top-sheet at $\lambda = -\alpha^2$, $k = -j\alpha$ with residue $\lim_{\lambda\to-\alpha^2}(\lambda + \alpha^2)g(x, \xi; \lambda) = -\lim_{k\to-j\alpha}(jk + \alpha)(jk - \alpha)g(x, \xi; -\alpha^2) = -2\alpha e^{-\alpha(x+\xi)}$. For a capacitive surface, there is no pole on the top sheet.

(ii) The wavenumber takes the value $k = -|\lambda|^{1/2} := -q$ above the branch-cut $\Gamma_b$: $\Re(\lambda) > 0$, $\Im(\lambda) = 0$ and $k = |\lambda|^{1/2} = +q$ below the branch-cut. The factor inside the square brackets in the expression for $g$, i.e.

$$K(x_<; k) = \cos kx_< - \frac{\alpha}{k}\sin kx_<$$

is an even function of $k$. Consequently,

$$\begin{aligned} g(x, \xi; k = q) - g(x, \xi, k = -q) &= K(x_<; q)\left(\frac{e^{-jqx_>}}{jq - \alpha} - \frac{e^{jqx_>}}{-jq - \alpha}\right) \\ &= -\frac{2jq}{q^2 + \alpha^2} K(x_<; q) K(x_>; q). \end{aligned} \tag{12.113}$$

The product $K(x_<; q)K(x_>; q)$ is symmetric in $x$ and $\xi$ and may be written as $K(x; q)K(\xi; q)$.

Deforming the contour $P_R$ around the pole and the branch cut (see figure 12.8) and applying Cauchy's theorem coupled with the substitution $d\lambda = 2q dq$ leads to the following spectral representation

$$\delta(x - \xi) = 2\alpha e^{-\alpha(x+\xi)} + \frac{2}{\pi}\int_0^\infty K(x; q)K(\xi; q)\frac{q^2}{q^2 + \alpha^2}\, dq. \tag{12.114}$$

For a solution $f \in \mathscr{L}_2^{(c)}(0, \infty)$, we may use the sifting property of the delta function and write $f(x) = \int_0^\infty f(\xi)\delta(x - \xi)\, d\xi$. Substituting equation (12.114) for the delta function we arrive at

$$f(x) = 2\alpha F_s\, e^{-\alpha x} + \frac{2}{\pi}\int_0^\infty F(q)K(x; q)\frac{q}{\sqrt{q^2 + \alpha^2}}\, dq, \tag{12.115}$$

where

$$F_s = \int_0^\infty f(\xi)e^{-\alpha\xi}\, d\xi, \tag{12.116}$$

and

$$F(q) = \frac{q}{\sqrt{q^2 + \alpha^2}}\int_0^\infty K(\xi; q)f(\xi)\, d\xi. \tag{12.117}$$

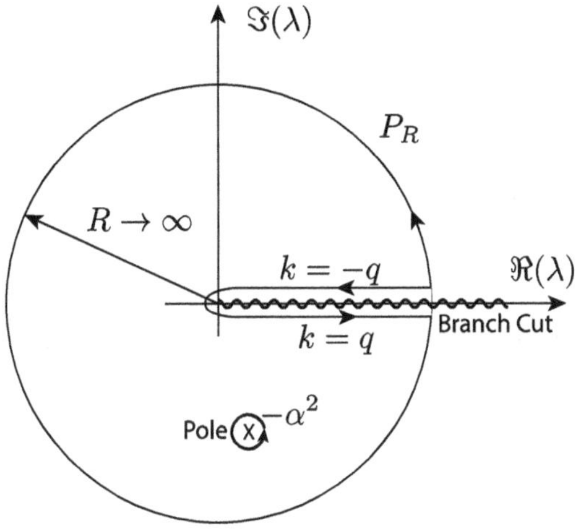


**Figure 12.8.** Contour of integration $P_R$ (punctured circle of radius $R$) along with the deformed contour consisting of branch cut $\Im(k) = 0$ and pole $\lambda = -\alpha^2$ on the top-sheet.

The quantities $F_s$ and $F(q)$ together constitute the *impedance transform* of a function belonging to $\mathscr{L}_2^{(c)}(0, \infty)$. The part $F_s$ relates to a surface wave and the part $F(q)$ to a space wave. It can be shown that $\langle e^{-\alpha x}, K(x; q)\rangle = 0$. Thus the surface wave and the space wave parts are orthogonal. Equations (12.116) and (12.117) define the forward impedance transform, while equation (12.115) defines the inverse impedance transform. In a higher dimensional space, an expansion of this kind corresponds to expanding the function in terms of standing waves in the $x$-direction.

*Special cases:*

(i) For $\alpha = 0$, the surface wave part vanishes and the impedance transform reduces to a *Fourier cosine transform* for the tangential magnetic field component $f$. The function satisfies the boundary condition $f'(0) = 0$ and the impedance surface corresponds to a PEC.

(ii) For $|\alpha| \to \infty$, the surface wave part vanishes and the impedance transform reduces to a *Fourier sine transform* for the tangential magnetic field component $f$ after replacing $qK/\sqrt{\alpha^2 + q^2}$ with $Q(x; q) = -\lim_{|\alpha|\to\infty} qK(x; q)/\sqrt{\alpha^2 + q^2} = \sin qx$. The boundary condition in this case is $f(0) = 0$ and the impedance surface corresponds to a PMC.

Consider now the solution of the equation $-u'' - \kappa^2 u = f$, $0 < x < \infty$, $\kappa \in \mathbb{C}$, $\Im(\kappa) < 0$ subject to the boundary condition $u(0) = 0$. Assuming that $f, u \in \mathscr{L}_2^{(c)}(0, \infty)$ and $f(0) = 0$, we expand both $f$ and $u$ in terms of their sine transforms

$$f(x) = \frac{2}{\pi}\int_0^\infty F(q)\sin qx\, \mathrm{d}q; \quad F(q) = \int_0^\infty f(x)\sin qx\, \mathrm{d}x \tag{12.118}$$

$$u(x) = \frac{2}{\pi}\int_0^\infty U(q)\sin qx\, \mathrm{d}q; \quad U(q) = \int_0^\infty u(x)\sin qx\, \mathrm{d}x. \tag{12.119}$$

By substituting these into the governing differential equation, we see that $U(q) = F(q)/(q^2 - \kappa^2)$. Thus

$$u(x) = \frac{2}{\pi}\int_0^\infty \frac{F(q)}{q^2 - \kappa^2}\sin qx\, \mathrm{d}q, \tag{12.120}$$

$$\begin{aligned} &= \frac{1}{\pi}\int_{-\infty}^\infty \frac{F(q)}{q^2 - \kappa^2}\sin qx\, \mathrm{d}q, \ (\text{upon defining } F(-q) = -F(q),\ q > 0) \\ &= \frac{1}{\pi j}\int_{-\infty}^\infty \frac{F(q)}{q^2 - \kappa^2}e^{jqx}\, \mathrm{d}q. \end{aligned} \tag{12.121}$$

Interestingly, the solution $u(x)$ satisfies Sommerfeld's radiation condition $\lim_{x\to\infty}\left(\frac{du}{dx} + j\kappa u\right) \to 0$ even though the form (12.120) indicates that the solution is comprised of pure standing waves $\sin qx$ at wavenumber $q$ along $x$. This is clear from considering an impulsive source $f = \delta(x - \xi)$, $\xi > 0$ in which case $u(x)$ reduces to the Green's function $g(x, \xi; \lambda = \kappa^2)$ listed in equation (12.64). The case for general $f$ follows from linear superposition. But let us demonstrate satisfaction of Sommerfeld's radiation condition directly from the integral representations (12.120)–(12.121). For $f = \delta(x - \xi)$ we have $F(q) = \sin q\xi = [e^{jq\xi} - e^{-jq\xi}]/2j$. Substituting this in equation (12.121) we obtain

$$u(x) = \frac{1}{2\pi j}\int_{-\infty}^{\infty} \frac{1}{j(q-\kappa)(q+\kappa)}[e^{jq(x+\xi)} - e^{jq(x-\xi)}]\,dq. \tag{12.122}$$

The integrand has simple poles at $q = \pm\kappa$ in the complex $q$-plane. The residue at the pole $q = -\kappa$ in the upper half-plane (recall that $\mathfrak{J}(\kappa < 0)$ is $e^{-j\kappa x}\sin\kappa\xi/\kappa$). The contour may be closed off in the upper half of the complex $q$-plane by a large semicircle. On the semicircle the integrand vanishes exponentially provided $x > \xi$. Hence applying Cauchy's residue theorem we obtain

$$u(x) = \frac{1}{\kappa}\sin\kappa\xi e^{-j\kappa x}, \quad x > \xi, \tag{12.123}$$

which coincides with equation (12.64) and satisfies the condition $\lim_{x\to\infty}\left(\frac{du}{dx} + j\kappa u\right) = 0$. Hence just because an integral representation such as equation (12.120) or equation (12.121) contains waves going back and forth (in this case pure standing waves) does not preclude the integrated quantity to satisfy the outgoing traveling wave condition. The incoming and outgoing waves are both needed in the region between the source and the impedance boundary to satisfy the boundary conditions, but only the outgoing waves prevail at large distances away from the source. Similar conclusions hold for the cosine transform as well as the impedance transform. ■■

## 12.4 Sum of two commutative operators

We end this chapter with theorems related to finding spectral representations of a sum of two commuting operators (see definition 12.1.27). The results have direct bearing on finding the inverse, changing one representation to another, or finding the spectral representation of the sum of two commuting operators in higher dimensions. The separation of variables technique that is commonly used in multiple dimensions for regular geometries owes its justification to the theorem on commutative operators.

Let $N_1$ be a self-adjoint differential operator acting on the independent variable $x$ alone and $N_2$ be a self-adjoint differential operator acting on the independent variable $y$ alone. Let $\lambda_1, \lambda_2, \ldots$ be the eigenvalues and $v_1(y), v_2(y), \ldots$ be the corresponding normalized eigenfunctions of $N_2$. Let $\mu_1, \mu_2, \ldots$ be the eigenvalues and $u_1(x), u_2(x), \ldots$ be the corresponding normalized eigenfunctions of $N_1$. Assume that the sets $\{u_j(x)\}$ and $\{v_j(x)\}$ are complete in their respective Hilbert-spaces (taken as $\mathscr{L}_2^{(c)}(a, b)$ here along each independent variable). Also, the inverse of $N_1 + \lambda_k$ (respectively, inverse of $N_2 + \mu_k$) is a bounded operator for every value of $\lambda_k$ (respectively, $\mu_k$), real or complex, with a bound independent of $\lambda_k$ (respectively, $\mu_k$). Let $\mathscr{S}$ be the space of complex-valued functions $f(x, y)$ which are square integrable over a rectangle $a_x < x < b_x$, $a_y < y < b_y$ whose sides are parallel to the $x$- and $y$-axes, respectively. For every function $f(x, y) \in \mathcal{S}$ the expansions $\sum \alpha_k(x) v_k(y)$ and $\sum \beta_k(y) u_k(x)$ converge to $f(x, y)$ in the sense of the norm $\iint |f(x, y)|^2 \,dx dy$ in $\mathscr{S}$, where $\alpha_k(x) = \int f(x, y) v_k^*(y)\, dy$, $\beta_k(y) = \int f(x, y) u_k^*(x)\, dx$. The following three theorems hold.

**Theorem 12.4.1.** Inverse of the sum of two commutative operators, [2, p 264].
*If $N_1$ and $N_2$ are operators described above and if a function $f(x, y) \in \mathscr{S}$, then*

$$(N_1 + N_2)^{-1} f = (N_1 + N_2)^{-1} \sum_{k=1}^{\infty} \alpha_k(x) v_k(y) = \sum_{k=1}^{\infty} (N_1 + \lambda_k)^{-1} \alpha_k(x) v_k(y), \tag{12.124}$$

$$= (N_1 + N_2)^{-1} \sum_{k=1}^{\infty} \beta_k(y) u_k(x) = \sum_{k=1}^{\infty} (\mu_k + N_2)^{-1} \beta_k(y) u_k(x). \tag{12.125}$$

*In other words, in finding the inverse of $N_1 + N_2$, $N_2$ (respectively, $N_1$) may be considered as a constant. The result will be a function of the operator* N$_2$ *(respectively, $N_1$) and should be interpreted by using the spectral representation of* N$_2$ *(respectively, $N_1$).*

**Theorem 12.4.2** Change of representation [2, p 273].
*If $N_1$ and $N_2$ are operators as described above and if $f(x, y) \in \mathscr{S}$, then*

$$\begin{aligned}(N_1 + N_2)^{-1} f &= \sum_{k=1}^{\infty} (N_1 + \lambda_k)^{-1} \alpha_k(x) v_k(y) \\ &= \lim_{n \to \infty} \frac{1}{2\pi j} \oint_{\Gamma_n} (N_1 + \lambda)^{-1} (\lambda - N_2)^{-1} f \, d\lambda \end{aligned} \tag{12.126}$$

*where $\Gamma_n$ is a closed contour in the $\lambda$-plane which contains the points $\lambda = \lambda_1, \ldots, \lambda_n$, and does not contain any point of the spectrum of $-N_1$. Equivalently,*

$$\begin{aligned}(N_1 + N_2)^{-1} f &= \sum_{k=1}^{\infty} (\mu_k + N_2)^{-1} \beta_k(y) u_k(x) \\ &= \lim_{n \to \infty} \frac{1}{2\pi j} \oint_{\Gamma_n} (\mu + N_2)^{-1} (\mu - N_1)^{-1} f \, d\mu \end{aligned} \tag{12.127}$$

*where $\Gamma_n$ is a closed contour in the $\mu$-plane which contains the points $\mu = \mu_1, \ldots, \mu_n$, and does not contain any point of the spectrum of $-N_2$.*

**Theorem 12.4.3** .Spectral representation for the sum of two commutative operators [2, p 278].
*The eigenvalues of* $N_1 + N_2$ *are* $\mu_j + \lambda_k (j, k = 1, 2, \ldots)$. *The corresponding eigenfunctions are* $u_j(x)v_k(y)$. *These eigenfunctions are complete in the linear vector space of all real-valued functions* $f(x, y)$ *such that*

$$\int_{a_y}^{b_y}\int_{a_x}^{b_x} |f(x, y)|^2 \, dxdy < \infty.$$

We illustrate these theorems by finding the Green's function of a vertical electric dipole (VED) over an impedance plane in the following example.

**Example 12.4.0.1.** Green's function for a VED over an impedance plane.
Consider a vertical dipole over an impedance plane as shown in figure 12.9. The upper medium is vacuum with permittivity $\varepsilon_0$ and permeability $\mu_0$ and wavenumber $k_0 = \omega\sqrt{\mu_0\varepsilon_0}$ at the radian frequency $\omega$.

The vertical component, $A_z(\rho, z)$, of the magnetic vector potential satisfies the equation

$$(N_1 + N_2)A_z = \mu_0 I_0 \ell \frac{\delta(\rho)\delta(z - z')}{2\pi\rho}, \; 0 < z, \rho < \infty, \tag{12.128}$$

where $N_2 = -\frac{1}{\rho}\frac{\partial}{\partial\rho}\left(\rho\frac{\partial}{\partial\rho}\right)$ is an operator involving only the radial coordinate $\rho$ and $N_1 = -\frac{\partial^2}{\partial z^2} - k_0^2$ is an operator involving only the vertical coordinate $z$. The potential satisfies an impedance boundary condition of the form $\frac{\partial A_z}{\partial z}(\rho, 0) + \alpha A_z(\rho, 0) = 0$ at $z = 0$ and the radiation condition at $z = \infty$. Note that these boundary conditions apply at all $\rho$ and are such that the self-adjointness and commutation of $N_1$ and $N_2$ are not breached. The two operators $N_1$ and $N_2$ commute and hence theorem 12.4.1 applies. We treat $N_2$ as a constant and find the inverse of $N_1$. Let $\kappa^2 = k_0^2 - N_2$ and $\psi = \mu_0 I_0 \ell \delta(\rho)/2\pi\rho = \psi_0\delta(\rho)/\rho$. Then the potential satisfies the equation

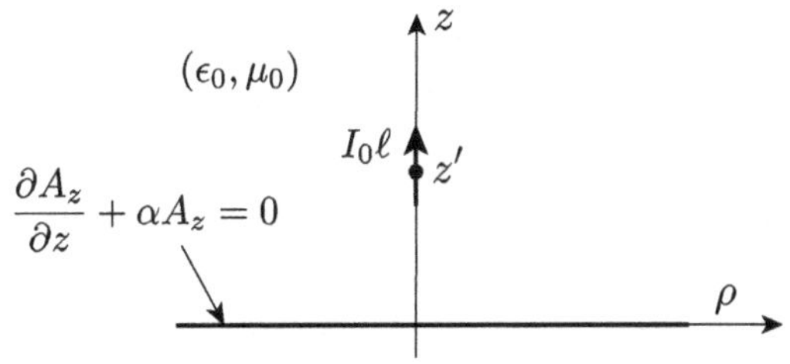


**Figure 12.9.** A vertical electric dipole placed over an impedance plane.

$$-\frac{\partial^2 A_z}{\partial z^2} - \kappa^2 A_z = \psi\delta(z - z'),\ 0 < z,\ z' < \infty. \tag{12.129}$$

This problem is very similar to that encountered in example 12.3.2.7 with $\kappa^2$ in place of $\lambda$ and $z$ in place of $x$. We assume that $\kappa$ has a vanishingly small, negative imaginary part to conform to the conditions of that example. Adapting from the solution there, the vector potential obeys

$$A_z(z, z'; \kappa) = \frac{\psi_0}{j\kappa - \alpha}\left[\cos\kappa z_< - \frac{\alpha}{\kappa}\sin\kappa z_<\right]\mathrm{e}^{-j\kappa z_>}\frac{\delta(\rho)}{\rho},\ 0 < z,\ z' < \infty. \tag{12.130}$$

The expression is a function of the operator $N_2$ through the variable $\kappa(N_2) = \sqrt{k_0^2 - N_2}$. The result must be interpreted by using the spectral representation of the operator $N_2$ (that is, finding the Green's function along $\rho$ or equivalently finding an appropriate representation of a delta function along $\rho$). Consider now the equation for the Green's function along the $\rho$-coordinate

$$(N_2 - \lambda)g_\rho = \frac{\delta(\rho)}{\rho} \Longrightarrow -(\rho g_\rho)' - \lambda\rho g_\rho = \delta(\rho). \tag{12.131}$$

This problem was addressed in example 12.3.2.6. The solution is obtained by substituting $\nu = 0$, $x = \rho$, $\xi = 0$, $k = k_\rho = \sqrt{\lambda}$ in equation (12.72) to result in

$$g_\rho(\rho; \lambda) = \frac{\pi}{2j}H_0^{(2)}(k_\rho\rho). \tag{12.132}$$

This Green's function satisfies the usual completeness relation (12.73). The contour integral is deformed around the branch cut along the positive real axis as done in example 12.3.3.2 to result in equation (12.91), which is reproduced below under the current notation

$$\lim_{R\to\infty}\frac{1}{2\pi j}\oint_{P_R} g_\rho(\rho;\lambda)\,d\lambda = -\frac{\delta(\rho)}{\rho} = -\int_0^\infty k_\rho J_0(k_\rho\rho)\,dk_\rho. \tag{12.133}$$

Using this in equation (12.130) finally gives

$$A_z(\rho, z; k_0) = \psi_0\int_0^\infty \frac{\mathrm{e}^{-jk_z z_>}}{jk_z - \alpha}\left[\cos k_z z_< - \frac{\alpha}{k_z}\sin k_z z_<\right]k_\rho J_0(k_\rho\rho)\,\mathrm{d}k_\rho, \tag{12.134}$$

where we have used the fact that $N_2\left[J_0(k_\rho\rho)\right] = k_\rho^2 J_0(k_\rho\rho)$. In addition, the effect of a function, $h(N_2)$, of the operator $N_2$ acting on the eigenfunction $J_0(k_\rho\rho)$ is equivalent to $h(k_\rho^2)J_0(k_\rho\rho)$. Consequently, $k_z = \kappa(N_2) = \sqrt{k_0^2 - k_\rho^2}$. It is noteworthy that equation (12.134) has been derived without formally using the separation of variable technique. The requirements for separation of variables technique to apply translate to the requirement here that the original operator appearing in the definition of multi-dimensional Green's function can be written as the sum of two operators $N_1$

and $N_2$ that commute with each other and the multi-dimensional boundary conditions are such that they can be written in a product form separately involving each independent variable. If the boundary condition in this example cannot be factorized as $B_1(z)\ B_2(\rho)$, then complete eigenfunctions cannot be found in each variable $z$ and $\rho$ and separation of variables technique does not apply. ■■

## References


[1] Courant R and Hilbert D 1953 *Methods of Mathematical Physics-I* vol 1 (New York: Wiley)

[2] Friedman B 1956 *Principles and Techniques of Applied Mathematics* (New York: Wiley)

[3] Lorch E R 1962 *Spectral Theory* (New York: Oxford University Press)

[4] Sobolev S L 1964 *Partial Differential equations of Mathematical Physics* (New York: Pergamon)

[5] Stakgold I 1967 *Boundary Valuen Problems of Mathematical Physics* vol I (New York: Macmillan)

[6] van Bladel J 1985 *Electromagnetic Fields* (New York: Hemisphere)

[7] Dudley D G 1994 *Mathematical Foundations for Electromagnetic Theory* (IEEE Press Series on Electromagnetic Wave Theory) (Piscataway, NJ: IEEE)

[8] Hanson G W and Yakovlev A B 2002 *Operator Theory for Electromagnetics: An Introduction* (New York: Springer)

[9] Rugh W J 1996 *Linear System Theory* (New York: Prentice-Hall)

[10] Debnath L 1995 *Integral Transfroms and Their Applications* (Boca Raton, FL: CRC Press)

[11] Janaswamy R 2018 Consistency requirements for integral representations of Green's functions. Part II: an erroneous representation *IEEE Trans. Antennas Propag.* **66** 4069–76

[12] Stakgold I 1979 *Green's Functions and Boundary Value Problems* (New York: Wiley)

[13] Janaswamy R 2001 Radiowave propagation over a non-constant immittance plane *Radio Sci.* **36** 387–405

[14] Sommerfeld A 1949 *Partial Differential equations in Physics* (New York: Academic)

[15] Schelkunoff S A 1936 Modified Sommerfeld's integral and its applications *Proc. IRE* **24** 1388–98

[16] Davies B 2002 *Integral Transforms and Their Applications* 3rd edn (New York: Springer)

[17] Papoulis A D 1962 *The Fourier Integral and its Applications* (New York: McGraw-Hill)

[18] Janaswamy R 2018 Consistency requirements for integral representations of Green's functions–part I *IEEE Trans. Antennas Propag.* **66** 4060–8

# Chapter 13

# Electromagnetic degrees of freedom

By electromagnetic degrees of freedom is meant the minimum amount of information needed to completely characterize the electromagnetic field in a certain finite region of space at a given frequency. This is specified in terms of an integer number that describes the minimum number of known field modes or field samples that can be used to completely characterize the field. The notion of the electromagnetic degree of freedom (DoF) is important in a number of areas including multiple-input multiple-output (MIMO) communication systems [1], antenna performance limitations [2], optical image processing and reconstruction [3], inverse scattering [4, 5], design of numerical methods in computational electromagnetics [6], etc, to name a few. We look at two central problems in this chapter (i) the DoF existing between two volumes communicating over a wireless link and (ii) the performance limitations of finite-sized antennas due to the limited DoF available, to illustrate the importance of the notion of degrees of freedom in the study of electromagnetic systems.

## 13.1 DoF between communicating volumes in free space

The limited number of DoFs available between two communicating volumes is reminiscent of the product between the *bandwidth* $W$ and the *time duration* $T$ that one associates with band-limited signals [7]. Indeed the dimension theorem states that [7] the degrees of a freedom of a signal is equal to $2WT$. There is a complete parallel of such a result with fields defined with respect to analogous conjugate variables of *wavenumber* and *space*. One would expect the electromagnetic degrees of freedom to be associated with the product of relevant quantities (such as the product between angular extent $\Theta$ and spatial extent $L$) in these two domains. In addition to the frequency of operation, the number of degrees of freedom depends on the size and nature of the primary and secondary sources of electromagnetic waves (that is, the size of the impressed sources and the size and nature of scatterers present in the environment that act like secondary sources), the distance between the sources and observation, etc.

doi:10.1088/978-0-7503-1716-0ch13  

In this section we discuss the dimension theorem for fields by considering the scalar Helmholtz equation. The treatment here is similar to that contained in [8]. A vector version can be carried out in a similar fashion [9]. The formulation itself is valid for any environment, but we shall show an example only in free space and derive results for the dimension theorem. Studies for DoFs in multipath environments have been carried out in [10] and [11].

Consider the scalar Helmholtz equation $\nabla^2\psi + k_0^2\psi = -h$, where $k_0 = \omega\sqrt{\mu_0\varepsilon_0}$ is the wavenumber at the operating frequency $\omega$ in a homogeneous medium with electrical parameters $(\varepsilon_0, \mu_0)$. The sources described by the function $h$ are all assumed to lie within the transmitting volume $V_T$ and transmitted field is sampled in another volume $V_R$ that is disjoint from $V_T$, see figure 13.1.

Using the homogeneous Green's function

$$\mathscr{G} = \frac{e^{-jk_0R}}{4\pi R}$$

where $R = |\mathbf{r} - \mathbf{r}'|$ is the distance between a point $\mathbf{r}'$ in $V_T$ and a point $\mathbf{r}$ in $V_R$, the fields generated by $h$ can be expressed as

$$\psi(\mathbf{r}) = \int_{V_T} \mathscr{G}(\mathbf{r}; \mathbf{r}')h(\mathbf{r}')\, d\mathbf{r}'. \tag{13.1}$$

Depending on whether the source $h(\mathbf{r}')$ is a line source or a surface source or a volumetric source, the integral over $V_T$ is a 1D integral or a 2D integral or a 3D integral, respectively. However, we use the same symbol $\int_{V_T}$ in all the three cases for the sake of brevity. Let $u_m(\mathbf{r}')$, $m = 1, 2, \ldots$ be a complex-valued orthonormal basis for the Hilbert space $\mathscr{L}_2^{(c)}(V_T)$ of functions defined over the volume $V_T$ and $v_m(\mathbf{r})$, $m = 1, 2, \ldots$ be a complex-valued orthonormal basis for the Hilbert space $\mathscr{L}_2^{(c)}(V_R)$ of functions defined over the volume $V_R$:

$$\int_{V_T} u_m(\mathbf{r}')u_n^*(\mathbf{r}')\, d\mathbf{r}' = \delta_m^n; \qquad \int_{V_R} v_m(\mathbf{r})v_n^*(\mathbf{r})\, d\mathbf{r} = \delta_m^n.$$

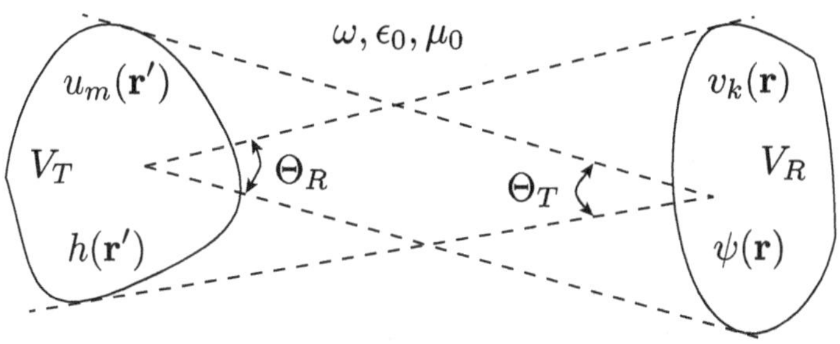


**Figure 13.1.** Sources $h(\mathbf{r}')$ distributed in a volume $V_T$ generating the field $\psi(\mathbf{r})$ in $V_R$. The surfaces bounding $V_T$ and $V_R$ need not be smooth.

We expand $h$ and $\psi$ in terms of these orthonormal bases in their respective volumes with coefficients $\mathbf{a} = [a_1, a_2, \ldots]'$ and $b = [b_1, b_2, \ldots]'$

$$h(\mathbf{r}') = \sum_m a_m u_m(\mathbf{r}'); \quad a_m = \int_{V_T} h(\mathbf{r}') u_m^*(\mathbf{r}')\, \mathrm{d}\mathbf{r}', \tag{13.2}$$

$$\psi(\mathbf{r}) = \sum_n b_n v_n(\mathbf{r}); \quad b_n = \int_{V_T} \psi(\mathbf{r}) v_n^*(\mathbf{r})\, \mathrm{d}\mathbf{r}. \tag{13.3}$$

As in the case of notation for the integrals we use only a single summation sign and a single index in the interest of brevity even if the summation extends over a multidimensional space. Substituting equations (13.2) and (13.3) into equation (13.1) and using the orthogonality condition on $u_m(\mathbf{r})$ gives the relation between the coefficients $\mathbf{b}$ and $\mathbf{a}$

$$b_k = \sum_m g_{km} a_m; \quad g_{km} = \int_{V_R} \int_{V_T} v_k^*(\mathbf{r}) \mathscr{G}(\mathbf{r}; \mathbf{r}') u_m(\mathbf{r}')\, \mathrm{d}\mathbf{r}' \mathrm{d}\mathbf{r}. \tag{13.4}$$

The coefficient $g_{km}$ can be thought of as a coupling strength between the transmission mode $m$ and receive mode $k$. We can write the above relation between the vectors $\mathbf{b}$ and $\mathbf{a}$ in matrix terms as

$$\mathbf{b} := \begin{bmatrix} b_1 \\ b_2 \\ \vdots \\ b_k \\ \vdots \end{bmatrix} = \begin{bmatrix} g_{11} & g_{12} \cdots g_{1m} & \cdots \\ g_{21} & g_{22} \cdots g_{2m} & \cdots \\ \vdots & & \\ g_{k1} & g_{k2} \cdots g_{km} & \cdots \\ \vdots & & \end{bmatrix} \begin{bmatrix} a_1 \\ a_2 \\ \vdots \\ a_m \\ \vdots \end{bmatrix} =: \mathbf{G}\, \mathbf{a}. \tag{13.5}$$

Since the volumes $V_T$ and $V_R$ are assumed disjoint, the Green's function is non-singular and may also be expanded in the bilinear basis $(u_i(\mathbf{r}'), v_j(\mathbf{r}))$

$$\mathscr{G}(\mathbf{r}; \mathbf{r}') = \sum_i \sum_j \gamma_{ji} u_i^*(\mathbf{r}') v_j(\mathbf{r}). \tag{13.6}$$

Substituting this into equation (13.4) gives

$$\begin{aligned} g_{km} &= \sum_i \sum_j \gamma_{ji} \int_{V_T} u_m(\mathbf{r}') u_i^*(\mathbf{r}')\, \mathrm{d}\mathbf{r}' \int_{V_R} v_k^*(\mathbf{r}) v_j(\mathbf{r})\, \mathrm{d}\mathbf{r} \\ &= \sum_i \sum_j \gamma_{ji} \delta_i^m \delta_j^k = \gamma_{km}. \end{aligned} \tag{13.7}$$

Thus $g_{km}$ may also be thought of as the coefficients of expansion of the Green's function in terms of the bilinear set $(u_i(\mathbf{r}'), v_j(\mathbf{r}))$. Based on the expansion in equation (13.6) and the relation (13.7), the norm square of the Green's function can be calculated as

$$
\begin{aligned}
\| \mathscr{G} \|^2 &= \int_{V_R}\int_{V_T} \mathscr{G}(\mathbf{r};\, \mathbf{r}')\mathscr{G}^*(\mathbf{r};\, \mathbf{r}')\, \mathrm{d}\mathbf{r}'\mathrm{d}\mathbf{r} = \int_{V_R}\int_{V_T} \frac{1}{|\mathbf{r}-\mathbf{r}'|^2}\, \mathrm{d}\mathbf{r}'\mathrm{d}\mathbf{r} \\
&= \sum_{i,j}\sum_{m,n} g_{ji} g_{nm}^* \int_{V_T} u_i^*(\mathbf{r}')u_m(\mathbf{r}')\, \mathrm{d}\mathbf{r}' \int_{V_R} v_j(\mathbf{r})v_n^*(\mathbf{r}')\, \mathrm{d}\mathbf{r} \\
&= \sum_{i,j} \left| g_{ji} \right|^2 =: \| \mathbf{G} \|_F^2,
\end{aligned}
\tag{13.8}
$$

where $\|\mathbf{G}\|_F$ is the Frobenius norm of the coupling matrix $\mathbf{G}$. This relation is reminiscent of the well-known Parseval's identity (see section 12.3.2.1). Thus the norm $\|\mathbf{G}\|_F$ of the coupling matrix (i) depends only on the norm of the Green's function (which in turn depends on the distance function between transmitting and receiving volumes) and (ii) *independent* of the bases chosen to represent $h(\mathbf{r}')$ in $V_T$ and $\psi(\mathbf{r}')$ in $V_R$. This is despite the fact that the individual coefficients $g_{km}$ *depend* on the choice of the basis functions. Note also that the off-diagonal terms $g_{km}$ are non-zero because the transformation $\mathscr{G}u_m = \int_{V_T} \mathscr{G}(\mathbf{r};\, \mathbf{r}')u_m(\mathbf{r}')\, \mathrm{d}\mathbf{r}'$ of a basis function $u_m(\mathbf{r}')$ via the Green's function does not necessarily render it orthogonal to $v_k(\mathbf{r})$. The sum rule (13.8) also implies that

$$
\left| g_{ji} \right| \leqslant \| \mathscr{G} \|,\ \forall\, i,\, j,
\tag{13.9}
$$

irrespective of the bases chosen. For the purpose of determining the degrees of freedom we are interested in choosing the bases such that the matrix $\mathbf{G}$ will become diagonal. In view of the upper bound (13.9) this also corresponds to choosing the basis functions to maximize the coupling strength $|g_{jj}|$ between a transmit basis function $u_j(\mathbf{r}')$ and the receive basis function $v_j(\mathbf{r})$.

Consider the transformation of a unit-norm transmit basis function $u_j(\mathbf{r}')$ via the Green's function to yield a function $\varphi_j(\mathbf{r})$ in the receiving volume $V_R$:

$$
\varphi_j(\mathbf{r}) = \mathscr{G}u_j = \int_{V_T} \mathscr{G}(\mathbf{r};\, \mathbf{r}')u_j(\mathbf{r}')\, \mathrm{d}\mathbf{r}'.
\tag{13.10}
$$

This function has a finite norm, which we denote as $|g|$, such that

$$
\begin{aligned}
|g_j|^2 &= \int_{V_R} \varphi_j^*(\mathbf{r})\varphi_j(\mathbf{r})\, \mathrm{d}\mathbf{r} = \int_{V_T} \mathrm{d}\mathbf{r}_2{}' u_j^*(\mathbf{r}_2{}') \int_{V_T} \underbrace{\int_{V_R} \mathscr{G}^*\left(\mathbf{r};\, \mathbf{r}_2'\right)\mathscr{G}\left(\mathbf{r};\, \mathbf{r}_1'\right)\mathrm{d}\mathbf{r}}_{k_T(\mathbf{r}_2';\mathbf{r}_1')}\ u_j(\mathbf{r}_1')\mathrm{d}\mathbf{r}_1' \\
&= \int_{V_T}\int_{V_T} u_j^*(\mathbf{r}_2')k_T(\mathbf{r}_2',\, \mathbf{r}_1')\ u_j(\mathbf{r}_1')\mathrm{d}\mathbf{r}_1'\mathrm{d}\mathbf{r}_2'.
\end{aligned}
\tag{13.11}
$$

The function $\varphi_j$ is not normalized and hence its norm is not necessarily unity. It is clear from the above definition that function $k_T$ satisfies the symmetry property

$k_T^*(\mathbf{r}_2', \mathbf{r}_1') = k_T(\mathbf{r}_1', \mathbf{r}_2')$. Furthermore $k_T$ is continuous in both of its arguments and is bounded. Indeed it forms the kernel of a self-adjoint Hilbert–Schmidt linear integral operator (see definition 12.1.31)

$$w_j(\mathbf{r}_2') = K_T u_j = \int_{V_T} k_T(\mathbf{r}_2', \mathbf{r}_1') u_j(\mathbf{r}_1')\, d\mathbf{r}_1', \quad w_j \in V_T, \tag{13.12}$$

and $\|g_j\|$ in equation (13.11) is simply the norm generated by the inner product $\langle w_j, u_j\rangle = \langle K_T u_j, u_j\rangle = \langle u_j, K_T u_j\rangle = \int_{V_T} w_j(\mathbf{r}')u_j^*(\mathbf{r}')\, d\mathbf{r}'$. Hence theorems 12.1.11–12.1.13 apply in this situation. In particular, countably infinite number of eigenvalues $\mu_j$ and corresponding eigenfunctions $\tilde{u}_j$ exist for the eigenvalue problem

$$K_T \tilde{u}_j = \int_{V_T} k_T(\mathbf{r}, \mathbf{r}')\tilde{u}_j(\mathbf{r}')\, d\mathbf{r}' = \mu_j \tilde{u}_j(\mathbf{r}), \tag{13.13}$$

with the property $|\mu_{j+1}| \le |\mu_j|$ and the orthonormality property $\langle \tilde{u}_j, \tilde{u}_k\rangle = \delta_j^k$. Furthermore, the eigenfunctions form a complete set in $V_T$ and hence constitute a basis in $V_T$. With this choice of basis $u_j(\mathbf{r}') = \tilde{u}_j(\mathbf{r}')$ the quadratic form on the rhs of equation (13.11) becomes

$$\int_{V_T}\int_{V_T} \tilde{u}_j^*(\mathbf{r}_2') k_T(\mathbf{r}_2', \mathbf{r}_1')\, u_j(\mathbf{r}_1') d\mathbf{r}_1' d\mathbf{r}_2 = \mu_j \int_{V_T} \tilde{u}_j^*(\mathbf{r}_2')\tilde{u}_j(\mathbf{r}_2')\, d\mathbf{r}_2' = \mu_j. \tag{13.14}$$

Comparing with the lhs of equation (13.11) we see that $\mu_j = |g_j|^2$ implying that $\mu_j$ must be real and positive and that $|g_j| \ge |g_{j+1}| \ge \cdots$. The eigenfunctions $\tilde{u}_j(\mathbf{r}')$ of the operator $K_T$ give rise to the dual functions $\tilde{\varphi}(\mathbf{r})$ in the receiving volume

$$\tilde{\varphi}_j(\mathbf{r}) = \int_{V_T} \mathcal{G}(\mathbf{r}; \mathbf{r}')\tilde{u}_j(\mathbf{r}')\, d\mathbf{r}', \tag{13.15}$$

such that

$$\begin{aligned}
\frac{1}{\sqrt{\mu_j\mu_k}} \int_{V_R} \tilde{\varphi}_j^*(\mathbf{r})\tilde{\varphi}_k(\mathbf{r})\, d\mathbf{r} &= \frac{1}{\sqrt{\mu_j\mu_k}} \int_{V_T} d\mathbf{r}_2'\, \tilde{u}_j^*(\mathbf{r}_2') \int_{V_T} \left[\int_{V_R} \mathcal{G}^*(\mathbf{r}; \mathbf{r}_2')\mathcal{G}(\mathbf{r}; \mathbf{r}_1')\, d\mathbf{r}\right] \tilde{u}_k(\mathbf{r}_1') d\mathbf{r}_1' \\
&= \frac{1}{\sqrt{\mu_j\mu_k}} \int_{V_T} d\mathbf{r}_2'\, \tilde{u}_j^*(\mathbf{r}_2') \int_{V_T} k_T(\mathbf{r}_2', \mathbf{r}_1')\tilde{u}_k(\mathbf{r}_1') d\mathbf{r}_1' \\
&= \frac{\mu_k}{\sqrt{\mu_j\mu_k}} \int_{V_T} d\mathbf{r}_2'\, \tilde{u}_j^*(\mathbf{r}_2')\tilde{u}_k(\mathbf{r}_2') = \frac{\mu_k}{\sqrt{\mu_j\mu_k}} \delta_k^j = \delta_k^j.
\end{aligned} \tag{13.16}$$

Thus the dual functions $\frac{1}{\sqrt{\mu_j}}\tilde{\varphi}_j(\mathbf{r})$ are orthonormal in the receiving volume. Furthermore, they constitute a complete set in $V_R$ since $\mathcal{G}$ is bounded and continuous. For the optimum pair $(\tilde{u}_m(\mathbf{r}'), \tilde{\varphi}_k(\mathbf{r})/\sqrt{\mu_k})$ the coupling strengths as defined in equation (13.4) are

$$\begin{aligned}\tilde{g}_{km} &= \frac{1}{\sqrt{\mu_k}} \int_{V_R} \int_{V_T} \tilde{\varphi}_k^*(\mathbf{r}) \mathcal{G}(\mathbf{r}; \mathbf{r}') \tilde{u}_m(\mathbf{r}') \, d\mathbf{r}' d\mathbf{r} \\ &= \frac{1}{\sqrt{\mu_k}} \int_{V_R} \tilde{\varphi}_k^*(\mathbf{r}) \tilde{\varphi}_m(\mathbf{r}) \, d\mathbf{r} = \sqrt{\mu_k} \delta_k^m,\end{aligned} \tag{13.17}$$

where a tilde over $g_{km}$ is now used to denote the optimum case (of maximizing $|g_{ii}|$) and relation (13.16) has been used in the last step. It is seen that the coupling matrix $\tilde{\mathbf{G}} = \{\tilde{g}_{km}\}$ is now diagonal. From the sum rule (13.8) we have for the optimum pair that

$$\sum_{k,m} |\tilde{g}_{km}|^2 = \sum_k \mu_k = \| \mathcal{G} \|^2 < \infty. \tag{13.18}$$

Since the eigenvalues $\mu_k$ form a descending sequence of $\mu_1 \geq \mu_2 \geq \mu_3, \cdots 0$, the number of degrees of freedom $N_{\text{DoF}}(\varepsilon)$ for a given precision $\varepsilon$ (equivalently, for a tolerance $\varepsilon$) is defined such that

$$\sum_{k=1}^{N_{\text{DoF}}(\varepsilon)} \mu_k \geq \| \mathcal{G} \|^2 - \varepsilon. \tag{13.19}$$

For infinite precision ($\varepsilon = 0$) the number of degrees of freedom is infinite. However, in the practical case of receivers corrupted by noise $\varepsilon > 0$ and $N_{\text{DoF}}$ is finite. The key here is to determine the eigenfunctions and eigenvalues for a given Green's function and transmit, receive volumes $V_T$ and $V_R$. Note that the eigenvalues and eigenfunctions are determined by the nature of both $V_T$ and $V_R$ as the kernel $k_T$ depends on $V_R$ per equation (13.11) and the definition (13.13) involves $V_T$.

If one makes the assumption that all $\mu_k$ below certain index $k = N_{\text{DoF}}$ have the same order of magnitude as the largest eigenvalue $\mu_1$ and beyond which have diminishing magnitude, a useful lower bound for the degrees of freedom is obtained from

$$\| \mathcal{G} \|^2 = \sum_{k=1}^{N_{\text{DoF}}} \mu_k \leq N_{\text{DoF}} \mu_1 \Longrightarrow N_{\text{DoF}} \geq \frac{\| \mathcal{G} \|^2}{\mu_1}. \tag{13.20}$$

### 13.1.1 DoF between planar rectangular apertures, prolate spheroidal functions

In this section we will determine the degrees of freedom between two broadside planar, rectangular apertures $S_T$ and $S_R$. The dimensions of the transmit and receive apertures are $2L_x \times 2L_y$ and $2R_x \times 2R_y$, respectively, and the horizontal distance between them is $D$, figure 13.2. We will consider the case of $D \gg L_x, L_y, R_x, R_y$ that will permit an analytical approximation. The norm of the Green's function is

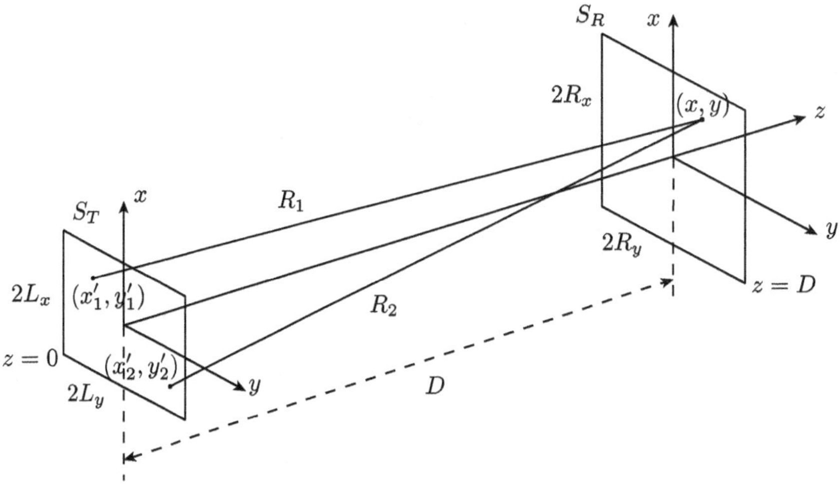


**Figure 13.2.** Two planar apertures communicating with each other.

$$\| \mathscr{G} \|^2 = \frac{1}{(4\pi)^2} \int_{S_T} \int_{S_R} \frac{\mathrm{d}x\mathrm{d}x'\mathrm{d}y\mathrm{d}y'}{[(x - x')^2 + (y - y')^2 + D^2]} \approx \frac{A_T A_R}{16\pi^2 D^2}, \tag{13.21}$$

where $A_T = 4L_xL_y$ is the area of the transmitting aperture and $A_R = 4R_xR_y$ is the area of the receiving aperture. The approximation holds under the stated conditions of $L_x$, $L_y$, $R_x$, $R_y \ll D$. The kernel $k_T$ for these planar regions is

$$k_T(\mathbf{r}_2', \mathbf{r}_1') = \left(\frac{1}{4\pi}\right)^2 \int_{x=-R_x}^{R_x} \int_{y=-R_y}^{R_y} \frac{\mathrm{e}^{jk_0(R_2 - R_1)}}{R_2 R_1} \mathrm{d}x\mathrm{d}y, \tag{13.22}$$

where $R_2^2 = (x - x_2')^2 + (y - y_2')^2 + D^2$, $R_1 = (x - x_1')^2 + (y - y_1')^2 + D^2$. Note that under the conditions stated $R_2 \approx D$ and $R_1 \approx D$ as far as the denominator of the integrand goes. In the phase term of the integrand

$$\begin{aligned} R_2 - R_1 = \frac{R_2^2 - R_1^2}{R_1 + R_1} &= \frac{(x_1' - x_2')(2x - x_1' - x_2') + (y_1' - y_2')(2y - y_1'y_2')}{R_2 + R_1} \\ &\approx \frac{x_1' - x_2'}{D}x + \frac{y_2' - y_1'}{D}y + \frac{x_2'^2 - x_1'^2}{2D} + \frac{y_2'^2 - y_1'^2}{2D}, \end{aligned} \tag{13.23}$$

where $R_2 \approx D$ and $R_1 \approx D$ have been used in the denominator. Substituting into equation (13.22) and evaluating the elementary integrals yields

$$k_T(\mathbf{r}_2', \mathbf{r}_1') \approx \frac{\mathrm{e}^{jk_0\left[(x_2'^2 - x_1'^2) + (y_2'^2 - y_1'^2)\right]/2D}}{4k_0^2} \frac{\sin \frac{k_0 R_x}{D}(x_2' - x_1')}{\pi(x_2' - x_1')} \frac{\sin \frac{k_0 R_y}{D}(y_2' - y_1')}{\pi(y_2' - y_1')}. \tag{13.24}$$

It is seen that $k_T$ depends on the parameters involving both the transmit and receive apertures $S_T$ and $S_R$ as previously remarked. The spatial dependence in the $x$- and $y$-directions is governed by a product of two 'sinc' functions that appear on the rhs.

The spatial frequency of the *sinc* function (and hence of $k_T$) in the $x$-direction over the source of dimension $2L_x$ is directly proportional to the receive aperture dimension $R_x$ and the distance $D$ in addition to the wavenumber $k$. Likewise, the spatial frequency of the sinc function (and hence of $k_T$) in the $y$-direction over the source of dimension $2L_y$ is directly proportional to the receive aperture dimension $R_y$ and the distance $D$ in addition to the wavenumber $k$. The overall expression for $k_T$ is separable w.r.t. the $x$- and $y$-coordinates. We can then rewrite the expression for $k_T$ in a separable form as

$$k_T(\mathbf{r}_2', \mathbf{r}_1') = \mathrm{e}^{j\phi(x_2')}k_{Tx}(x_2', x_1')\mathrm{e}^{-j\phi(x_1')} \cdot \mathrm{e}^{j\phi(y_2')}k_{Ty}(y_2', y_1')\mathrm{e}^{-j\phi(y_1')}, \tag{13.25}$$

where

$$k_{Tx}(x_2', x_1') = \frac{1}{2k_0L_x}\frac{\sin\left[\frac{k_0R_xL_x}{D}\left(\frac{x_2' - x_1'}{L_x}\right)\right]}{\pi\frac{x_2' - x_1'}{L_x}}, \tag{13.26}$$

$$k_{Ty}(y_2', y_1') = \frac{1}{2k_0L_y}\frac{\sin\left[\frac{k_0R_yL_y}{D}\left(\frac{y_2' - y_1'}{L_y}\right)\right]}{\pi\frac{y_2' - y_1'}{L_y}}, \tag{13.27}$$

and $\phi(z) = k_0z^2/2D$ is a quadratic phase term. For the eigenvalue problem let us express the eigenfunction to be determined as $\tilde{u}(\mathbf{r}') = \mathrm{e}^{j\phi(x')}\widetilde{X}(x') \cdot \mathrm{e}^{j\phi(y')}\widetilde{Y}(y')$ and the corresponding eigenvalue as $\mu = \mu_x\mu_y/(2k_0)^2$. Substituting these into equation (13.13) and making the change of variable $x = \xi L_x$, $y = \eta L_y$ results in

$$\begin{aligned}&\int_{-1}^{1}\widetilde{X}(\xi_1')\frac{\sin c_x(\xi_2' - \xi_1')}{\pi(\xi_2' - \xi_1')}\,\mathrm{d}\xi_1' \cdot \int_{-1}^{1}\widetilde{Y}(\eta_1')\frac{\sin c_y(\eta_2' - \eta_1')}{\pi(\eta_2' - \eta_1')}\,\mathrm{d}\eta_1' \\ &= \mu_x\widetilde{X}(\xi_2') \cdot \mu_y\widetilde{Y}(\eta_2'),\end{aligned} \tag{13.28}$$

where $c_x = k_0R_xL_x/D$ and $c_y = k_0R_yL_y/D$. It is seen that the solution of the eigenvalue problem is now related to the solution of the integral equation

$$\int_{-1}^{1}\frac{\sin\Omega(z - \zeta)}{\pi(z - \zeta)}\psi(\zeta)\,\mathrm{d}\zeta = \alpha\psi(z). \tag{13.29}$$

The solutions of this equation are the *prolate spheroidal wave functions* [12]. Let $\mathcal{B}_\Omega$ denote the space of functions, $f(z)$, in $\mathcal{L}_2^{(r)}(0, \infty)$, whose Fourier transforms, $F(\kappa)$, vanish in the interval $|\kappa| > \Omega$, where the function and its Fourier transform are related by

$$F(\kappa) = \int_{-\infty}^{\infty}f(z)\mathrm{e}^{-j\kappa z}\,\mathrm{d}z; \quad f(z) = \frac{1}{2\pi}\int_{\kappa=-\Omega}^{\Omega}F(\kappa)\mathrm{e}^{j\kappa z}\,\mathrm{d}\kappa.$$

Functions belonging to $\mathcal{B}_\Omega$ are called *band-limited* and the space $\mathcal{B}_\Omega$ itself is called the class of band-limited functions. Given an $\Omega$, we can find a countably infinite set of real functions $\psi_0(z)$, $\psi_1(z)$, $\psi_2(z)$, ... (the prolate spheroidal functions) and a set of real positive numbers

$$\alpha_0 > \alpha_1 > \alpha_2 \cdots$$

with the following properties:

(a) The $\psi_m(z)$ are band-limited, orthonormal on the real line and complete in $\mathcal{B}_\Omega$, i.e.

$$\int_{-\infty}^{\infty} \psi_m(z)\psi_n(z)\,\mathrm{d}z = \begin{cases} 0, & m \neq n \\ 1, & m = n \end{cases} \quad m, n = 0, 1, 2, \ldots$$

(b) In the interval $-1 \leq z \leq 1$, the $\psi_m$ are orthogonal and complete in $\mathcal{L}_2^{(r)}(-1, 1)$:

$$\int_{-1}^{1} \psi_m(z)\psi_n(z)\,\mathrm{d}z = \begin{cases} 0, & m \neq n \\ \alpha_m, & m = n \end{cases} \quad m, n = 0, 1, 2, \ldots$$

(c) For all values of real $z$: $-\infty < z < \infty$, the $\psi_m(z)$ satisfy equation (13.29) with $\alpha = \alpha_m$.

(d) The Fourier transforms satisfy

$$\int_{-\infty}^{\infty} \psi_m(z)\mathrm{e}^{-j\kappa z}\,\mathrm{d}z = \frac{j^{-m}\sqrt{2\pi}}{\sqrt{\Omega\alpha_m}}\psi_m\left(\frac{\kappa}{\Omega}\right)\chi\left(\frac{\kappa}{\Omega}\right), \tag{13.30}$$

$$\int_{-\infty}^{\infty} \chi(z)\psi_m(z)\mathrm{e}^{-j\kappa z}\,\mathrm{d}z = \int_{-1}^{1} \psi_m(z)\mathrm{e}^{-j\kappa z}\,\mathrm{d}z = \frac{j^{-m}\sqrt{2\pi}}{\sqrt{\Omega\alpha_m}}\psi_m\left(\frac{\kappa}{\Omega}\right), \tag{13.31}$$

where

$$\chi(\omega) = \begin{cases} 1, & |\omega| \leq 1 \\ 0, & |\omega| > 1 \end{cases}.$$

(e) The constants $\lambda_m$ depend on $\Omega$ and satisfy $1 > \alpha_m > \alpha_{m+1} > 0$, $m = 0, 1, \ldots$, $\lim_{m\to\infty} \alpha_m = 0$. Furthermore,

$$\lim_{\Omega\to\infty} \alpha_m = \begin{cases} 0, & m = \left\lfloor (1+\eta)\frac{2\Omega}{\pi} \right\rfloor \\ [1 + \mathrm{e}^{\pi b}]^{-1}, & m = \left\lfloor \frac{2\Omega}{\pi} + \frac{b}{\pi}\ln\frac{\Omega}{\pi} \right\rfloor \\ 1, & m = \left\lfloor (1-\eta)\frac{2\Omega}{\pi} \right\rfloor, \end{cases}$$

where $b$ and $\eta$ are numbers independent of $\Omega$.

(f) The $\alpha_m$ fall off to zero rapidly with increasing $m$ for $m > M_0 = \lfloor 2\Omega/\pi \rfloor$, where $\lfloor\cdot\rfloor$ denotes the floor function.

Table 13.1 shows the eigenvalues $\alpha_m$ for $\Omega$ in the range (0.5, 8) [12].

In view of the property in item (f), the constants $\mu_x = \mu_{xm}$ and $\mu_y = \mu_{yn}$, $m, n = 0, 1, 2, \ldots$ in equation (13.28) will diminish beyond $m = M_0 = \lfloor 2c_x/\pi \rfloor$ and $n = N_0 = \lfloor 2c_y/\pi \rfloor$. We may then take the number of degrees of freedom in our problem to be proportional to $M_0N_0$. Now

$$M_0 \sim \frac{2}{\pi}c_x = \frac{2}{\pi}\frac{kR_xL_x}{D} = \frac{2L_x}{D}\frac{2R_x}{\lambda_0}; \quad N_0 \sim \frac{2L_y}{D}\frac{2R_y}{\lambda_0},$$

where $\lambda_0$ is the wavelength in the medium. Writing $\Theta_x = 2L_x/D$ for the total angle subtended in the $x$-direction by the transmitting aperture as seen from the receiver, $\Theta_y = 2L_y/D$ as the total angle subtended in the $y$-direction by the transmitting aperture as seen from the receiver, $N_{sx} = 2R_x/\lambda_0$ as the number of wavelengths present in the receiving aperture in the $x$-direction, and $N_{sy} = 2R_y/\lambda_0$ as the number of wavelengths present in the receiving aperture in the $y$-direction, we may write the degrees of freedom between the communicating apertures as

$$N_{\mathrm{DoF}} \sim \lceil \Theta_x\Theta_y N_{sx}N_{sy} \rceil = \left\lceil \frac{A_T A_R}{\lambda_0^2 D^2} \right\rceil,$$

where $\lceil \cdot \rceil$ denotes the ceiling function. The degrees of freedom are seen to vary symmetrically with respect to the areas of the transmitting and receiving apertures. Furthermore, they decrease in inverse square with increasing range and wavelength. Note that this result is consistent with the lower bound given in equation (13.20) when one recognizes that (i) $\mu_k = \mu_{xm}\mu_{yn}/(2k_0)^2 = \lambda_0^2\mu_{xm}\mu_{yn}/16\pi^2$, (ii) $\mu_{xm}, \mu_{yn} < 1$, from item (e) and (iii) from equation (13.21) $\|\mathscr{G}\|^2 = A_T A_R/(16\pi^2 D^2)$. The lower bound upon using $\mu_{x0} = 1 = \mu_{y0}$ in $\mu_1 = \mu_{x0}\mu_{y0}/(2k_0)^2$ is

$$N_{\mathrm{DoF}} \geq \frac{A_T A_R}{16\pi^2 D^2}\frac{1}{\mu_1} = \frac{A_T A_R}{16\pi^2 D^2}\frac{16\pi^2}{\lambda_0^2} = \frac{A_T A_R}{\lambda_0^2 D^2}.$$

subject to the conditions that $L_x$, $L_y$, $R_x$, $R_y \ll D$. In some applications involving field sampling, one is interested in the number of degrees of freedom $n_{\mathrm{DoF}}$ per unit

**Table 13.1.** Values of $\alpha_m(\Omega) = L_m(\Omega) \times 10^{p_m(\Omega)}$.

| | $\Omega = 0.5$ | | $\Omega = 1$ | | $\Omega = 2$ | | $\Omega = 4$ | | $\Omega = 8$ | |
|---|---|---|---|---|---|---|---|---|---|---|
| $m$ | $L_m$ | $p_m$ | $L_m$ | $p_m$ | $L_m$ | $p_m$ | $L_m$ | $p_m$ | $L_m$ | $p_m$ |
| 0 | 3.10 | −1 | 5.73 | −1 | 8.81 | −1 | 9.96 | −1 | 1.0 | 0 |
| 1 | 8.58 | −3 | 6.28 | −2 | 3.56 | −1 | 9.12 | −1 | 1.0 | 0 |
| 2 | 3.92 | −5 | 1.24 | −3 | 3.59 | −2 | 5.19 | −1 | 9.97 | −1 |
| 3 | 7.21 | −8 | 9.20 | −6 | 1.15 | −3 | 1.10 | −1 | 9.61 | −1 |
| 4 | 7.27 | −11 | 3.72 | −8 | 1.89 | −5 | 8.83 | −3 | 7.48 | −1 |

solid angle subtended by the transmitting aperture. In terms of the communicating degrees of freedom it is

$$n_{\mathrm{DoF}} = N_{\mathrm{DoF}} = \frac{N_{sx}N_{sy}}{\frac{1}{2}\Theta_x\frac{1}{2}\Theta_y} = \frac{2R_x}{\frac{\lambda_0}{2}}\frac{2R_y}{\frac{\lambda_0}{2}},$$

which corresponds to the number of Nyquist samples (spaced at $\lambda_0/2$) contained in the receiving aperture. This quantity is only dependent on the electrical size of the receiving aperture.

As an example, if $L_x = L_y = R_x = R_y = 0.5$ m, $D = 10$ m, the angles subtended by the transmitter are $\Theta_x = \Theta_y = 0.1$ radians, the aperture areas are $A_T = A_R = 1$ m$^2$, the constants $c_x = c_y = 3\pi/20$, and the number of wavelengths in the receiving aperture are $N_{sx} = 30 = N_{sy}$ at a frequency of 9 GHz. Consequently, the number of degrees of freedom between the planar apertures is $N_{\mathrm{DoF}} = 9$. The values under $\Omega = 0.5$ of table 13.1 approximately pertain to this case. The table shows that $\alpha_3(\Omega)/\alpha_0(\Omega) \sim 2.3 \times 10^{-7}$ suggesting that only the $m = 0, 1, 2$ eigenmodes are substantial in this case between the communicating apertures. The number of degrees of freedom corresponding to Nyquist sampling at the receiving aperture is $n_{\mathrm{DoF}} = 60 \times 60 = 3600$. If the frequency is reduced to 3 GHz, while keeping all other parameters the same, the number of degrees of freedom will be reduced further to $N_{\mathrm{DoF}} = 1$.

## 13.2 Antenna gain limitations due to finite DoF

When a source radiating in free space at a frequency $\omega$ is confined to a sphere of radius $a$, the equivalents currents on the spherical surface $r = a$ can be represented accurately by a limited number of spherical harmonics for a given tolerance. The number of excited spherical harmonics will be limited to the order $ka$, where $k = \omega\sqrt{\mu_0\varepsilon_0}$ is the wavenumber in free space. This places a restriction on the amount of gain one can obtain from the radiating source. We show in the following the fundamental result that the gain is limited to $ka(ka + 2)$ [13] when the electromagnetic field constitutes an equipartition of energy between TE and TM modes.

Figure 13.3 shows an arbitrary antenna inside a circumscribing sphere of radius $a$. The wavenumber and intrinsic impedance of the medium are $k$ and $\eta$, respectively. The coordinate system is chosen such that $\theta = 0$ points in the direction of antenna maximum radiation.

The starting point in the analysis are the field expressions (1.134)–(1.146) valid outside $r = a$. With reference to those expressions we adopt the following notation $R = kr$, $' = \mathrm{d}/\mathrm{d}R$, $Rh_n^{(2)}(R) = \hat{H}_n^{(2)}(R)$, $C_{mn} = \eta A_{mn} I_n^m$ (V), $D_{mn} = B_{mn} L_n^m$ (V), where $m$, $n$ are the order and degree, respectively, of the surface spherical harmonic (1.103)

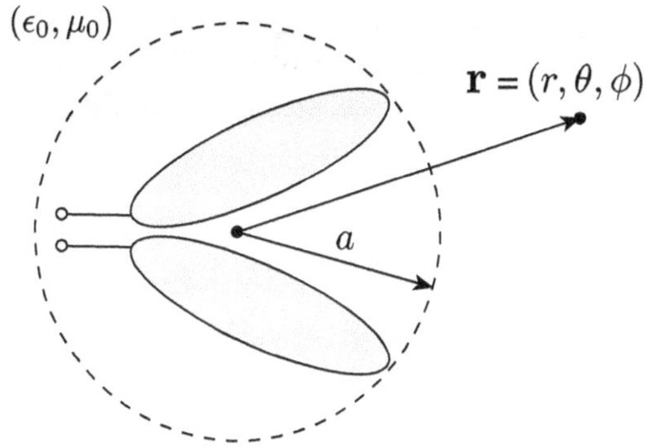


**Figure 13.3.** An antenna confined to a volume of radius $a$ and radiating in free space.

$$Y_n^m(\theta, \phi) = \left[\frac{2n+1}{4\pi}\frac{(n-m)!}{(n+m)!}\right]^{\frac{1}{2}} P_n^m(\cos\theta)e^{jm\phi}, \quad n = 0, 1, 2, \ldots, m = (-n, n), \tag{13.32}$$

with the conjugate harmonic

$$Y_n^{-m}(\theta, \phi) = (-1)^m \bar{Y}_n^m(\theta, \phi), \tag{13.33}$$

where the overhead bar denotes complex conjugation, and the orthogonality relationship (1.105)

$$\int_{\theta=0}^{\pi}\int_{\phi=0}^{2\pi} Y_n^m(\theta, \phi)\bar{Y}_{n'}^{m'}(\theta, \phi)\sin\theta \, d\phi d\theta = \delta_n^{n'}\delta_m^{m'}. \tag{13.34}$$

The antenna is assumed to to contain arbitrarily directed currents so that the field exterior to $r = a$ will be a combination of of $TM_r$ and $TE_r$ modes. The total surface field expressions are

$$rE_\theta = \sum_{n,m}\left[-C_{mn}\hat{H}_n^{(2)'}(R)\frac{\partial Y_n^m}{\partial\theta}(\theta, \phi) + jD_{mn}\hat{H}_n^{(2)}(R)\frac{1}{\sin\theta}\frac{\partial Y_n^m}{\partial\phi}(\theta, \phi)\right], \tag{13.35}$$

$$r\eta H_\phi = \sum_{n,m}\left[jC_{mn}\hat{H}_n^{(2)}(R)\frac{\partial Y_n^m}{\partial\theta}(\theta, \phi) - D_{mn}\hat{H}_n^{(2)'}(R)\frac{1}{\sin\theta}\frac{\partial Y_n^m}{\partial\phi}(\theta, \phi)\right], \tag{13.36}$$

$$rE_\phi = \sum_{n,m}\left[-C_{mn}\hat{H}_n^{(2)'}(R)\frac{1}{\sin\theta}\frac{\partial Y_n^m}{\partial\phi}(\theta, \phi) - jD_{mn}\hat{H}_n^{(2)}(R)\frac{\partial Y_n^m}{\partial\theta}(\theta, \phi)\right], \tag{13.37}$$

$$r\eta H_\theta = \sum_{n,m}\left[-jC_{mn}\hat{H}_n^{(2)}(R)\frac{1}{\sin\theta}\frac{\partial Y_n^m}{\partial\phi}(\theta, \phi) - D_{mn}\hat{H}_n^{(2)'}(R)\frac{\partial Y_n^m}{\partial\theta}(\theta, \phi)\right], \tag{13.38}$$

where the summation is over the indices $n \in (1, \infty)$, $m \in (-n, n)$. The strength of the TM$_r$ field is governed by $C_{mn}$ and that of the TE$_r$ field is governed by $D_{mn}$.

For simplicity of notation we drop the arguments of the various functions where it is not required. The gain of the antenna is defined as

$$G(R) = \frac{4\pi \Re[S_r(R, \theta \to 0, \phi)]}{\oiint\limits_{\theta,\phi} \Re[S_r(R, \theta, \phi)] \sin\theta \, \mathrm{d}\theta \mathrm{d}\phi} =: 4\pi \frac{U(R)}{V(R)}, \tag{13.39}$$

where $S_r(R, \theta, \phi) = (rE_\theta \, r\eta H_\phi^* - rE_\phi \, r\eta H_\theta^*)$ (V$^2$) is the complex-valued radiation intensity per unit solid angle. We note that $\partial Y_n^m / \partial \phi = jmY_n^m$ and recall the following results that will be used in the ensuing analysis:

(i) The Wronskian (C.131) for the spherical Ricatti–Hankel functions

$$\hat{H}_n^{(2)\prime}(R)\hat{H}_n^{(1)}(R) - \hat{H}_n^{(1)\prime}(R)\hat{H}_n^{(2)}(R) = -2j. \tag{13.40}$$

(ii) The limits (D.112) for the Legendre function

$$\lim_{\theta\to 0} \frac{mP_n^m(\cos\theta)}{\sin\theta} = \lim_{\theta\to 0} \frac{\mathrm{d}}{\mathrm{d}\theta} P_n^m(\cos\theta) = \begin{cases} 0, & m \neq 1 \\ -\dfrac{n(n+1)}{2} =: a_n, & m = 1 \end{cases}, \tag{13.41}$$

which imply that

$$\lim_{\theta\to 0} \left( \frac{m}{\sin\theta} Y_n^m(\theta, \phi) \right) = (\delta_m^1 \mathrm{e}^{j\phi} + \delta_m^{-1} \mathrm{e}^{-j\phi}) b_n, \tag{13.42}$$

$$\lim_{\theta\to 0} \frac{\partial Y_n^m}{\partial \theta}(\theta, \phi) = (\delta_m^1 \mathrm{e}^{j\phi} - \delta_m^{-1} \mathrm{e}^{-j\phi}) b_n, \tag{13.43}$$

where $b_n = a_n \sqrt{(2n+1)/4\pi n(n+1)} = -\sqrt{n(2n+1)(n+1)/16\pi}$.

(iii) The integrals (D.125) and (D.126)

$$\int_0^\pi \frac{\mathrm{d}}{\mathrm{d}\theta} \left[ P_n^m(\cos\theta) P_q^m(\cos\theta) \right] \mathrm{d}\theta = 0, \tag{13.44}$$

$$\int_0^\pi \left[ \frac{\mathrm{d}P_n^m(\cos\theta)}{\mathrm{d}\theta} \frac{\mathrm{d}P_q^m(\cos\theta)}{\mathrm{d}\theta} + \frac{m^2}{\sin^2\theta} P_n^m(\cos\theta) P_q^m(\cos\theta) \right] \sin\theta \, \mathrm{d}\theta = \delta_n^q \frac{2n(n+1)}{2n+1} \frac{(n+m)!}{(n-m)!}, \tag{13.45}$$

and

$$\int_0^{2\pi} \mathrm{e}^{j(m-p)\phi} \, \mathrm{d}\phi = 2\pi \delta_m^p. \tag{13.46}$$

(iv) The Cauchy–Schwarz inequality

$$\left| \sum_{n=1}^{N} u_n v_n^* \right|^2 \leq \sum_{n=1}^{N} |u_n|^2 \sum_{n=1}^{N} |v_n|^2, \tag{13.47}$$

with equality for $u_n = v_n$.

Using equations (13.35)–(13.38) the radiation intensity can be expressed as

$$\begin{aligned} S_\mathrm{r} = {} & j \sum_{n,m} \sum_{q,p} \left( C_{mn} C_{pq}^* \hat{H}_n^{(2)'} \hat{H}_q^{(1)} - D_{mn} D_{pq}^* \hat{H}_n^{(2)} \hat{H}_q^{(1)'} \right) \left( \frac{\partial Y_n^m}{\partial \theta} \frac{\partial \bar{Y}_q^p}{\mathrm{d}\theta} + \frac{mp}{\sin^2\theta} Y_n^m \bar{Y}_q^p \right) \\ & - \frac{1}{\sin\theta} \left( C_{mn} D_{pq}^* \hat{H}_n^{(2)'} \hat{H}_q^{(1)'} - D_{mn} C_{pq}^* \hat{H}_n^{(2)} \hat{H}_q^{(1)} \right) \left( p \frac{\partial Y_n^m}{\partial \theta} \bar{Y}_q^p + m Y_n^m \frac{\partial \bar{Y}_q^p}{\partial \theta} \right). \end{aligned} \tag{13.48}$$

Taking the limit $S_\mathrm{r}(R, \theta \to 0, \phi)$ and taking the real part gives for the numerator function in equation (13.39) as

$$\begin{aligned} U(R) = {} & \sum_{n,m} \sum_{q,p} b_n b_q \Big[ 2\left(\delta_m^1 \delta_p^1 + \delta_m^{-1} \delta_p^{-1}\right)\left(C_{mn} C_{pq}^* + D_{mn} D_{pq}^*\right) \\ & + j\left(\delta_m^1 \delta_p^1 - \delta_m^{-1} \delta_p^{-1}\right)(D_{mn} C_{pq}^* - C_{mn} D_{pq}^*)\left(\hat{H}_n^{(2)'} \hat{H}_q^{(1)'} + \hat{H}_n^{(2)} \hat{H}_q^{(1)}\right) \Big], \end{aligned} \tag{13.49}$$

upon using the results (13.42), (13.43) and the Wronskian (13.40). The real part is evaluated according to

$$\Re(S_\mathrm{r}) = \frac{S_\mathrm{r} + S_\mathrm{r}^*|_{mn \to pq}}{2},$$

where, in the complex conjugate term, the indices $mn$ are switched with $pq$ to result in the desired simplified expression as shown above. Notice that the rhs of equation (13.49) is independent of $\phi$ as assumed in the definition (13.39). Note also that the radial dependence of $U(R)$ arises from the term containing the coupling between the $\mathrm{TM}_r$ and $\mathrm{TE}_r$ modes, which is antisymmetric in $C_{mn}$ and $D_{mn}$ (that is, switching $C_{mn}$ with $D_{mn}$ produces a term that is negative of the original term). The term of equation (13.49) that is independent of $R$ is symmetric in $C_{mn}$ and $D_{mn}$ (that is, switching $C_{mn}$ with $D_{mn}$ maintains that term). The overall quantity $U(R)$ is independent of $R$ in three situations: (a) a pure $\mathrm{TM}_r$ field is excited ($D_{mn} = 0$), (b) a pure $\mathrm{TE}_r$ field is excited ($C_{mn} = 0$), and (c) equal strengths of $\mathrm{TE}_r$ and $\mathrm{TM}_r$ fields are excited ($C_{mn} = D_{mn}$). In all of these three situations the summations over $n$ and $q$ in $U(R)$ separate and the numerator can be written as a complex quantity times its complex conjugate.

The denominator term $V(R)$ in equation (13.39) is obtained by integrating the expression in equation (13.48), taking its real part and using the results (13.34), (13.44)–(13.46), and the Wronskian (13.40) to yield

$$V(R) = \sum_{n,m} n(n+1)\left(|C_{mn}|^2 + |D_{mn}|^2\right), \tag{13.50}$$

which is seen to be independent of $R$ and symmetric in $C_{mn}$ and $D_{mn}$. Since $U(R)$ is dependent only on the terms $m,\, p = \pm 1$ for an arbitrary antenna, the gain will be maximized of the remaining terms are also discarded in the denominator function $V(R)$. Accordingly, the excitations that maximize the gain will have the property $C_{mn} = 0 = D_{mn},\, m \neq \pm 1$. The gain will be further maximized if the antisymmetric part of $U(R)$ is set to zero, while increasing its symmetric part. Thus setting $D_{mn} = C_{mn}$ will maximize the gain. Inserting all of these features into the gain expression we obtain

$$G(R) = \frac{\sum_{n,m}\sum_{q,p} \sqrt{(2n+1)(2q+1)n(n+1)q(q+1)}\, C_{mn}C_{pq}^{*}(\delta_m^1\delta_p^1 + \delta_m^{-1}\delta_p^{-1})}{\sum_{n,m} n(n+1)|C_{mn}|^2(\delta_m^1 + \delta_m^{-1})}.$$

Let us define new excitation coefficients $F_{mn} = \sqrt{n(n+1)}\, C_{mn}, \|F_{\pm 1}\|^2 = \sum_n |F_{\pm 1n}|^2 < \infty$, $T_{\pm 1} = \sum_n \sqrt{2n+1}\, F_{\pm 1n}$. Then

$$G(R) = \frac{|T_1|^2 + |T_{-1}|^2}{\| F_1 \|^2 + \| F_2 \|^2}.$$

Since this is symmetric in $F_{1n}$ and $F_{-1n}$, it will be extremized if $F_{-1n} = F_{1n}$ in which case

$$G(R) = \frac{|T_1|^2}{\| F_1 \|^2} = \left| \sum_{n=1}^{\infty} \sqrt{2n+1}\, \frac{F_{1n}}{\| F_1 \|} \right|^2. \tag{13.51}$$

This gain can potentially go to infinity if all orders of spherical harmonics are permitted. However, a definite limit exists if the harmonics are restricted to orders $n \leq N$. In practice the spherical modes decay rapidly when the order $n$ exceeds $ka$. As an example, figure 13.4 shows the normalized excitation coefficients $C_{mn} = A_n$ of an equatorial slot antenna (even though this example antenna does not have a main beam at $\theta = 0$) of electrical radius $kb = 10$ and slot width $\Delta = 0.1$ rad. The excitation coefficients for the antenna are given in equation (8.32). It is seen that the normalized magnitude drop below 0.1 for $n \gtrsim ka$. This kind of behavior is typical of all antennas and one may take the highest harmonic excited as $N = \lceil ka \rceil$ and regard it as representing the effective number of degrees of freedom of the surface harmonics. Under this constraint

$$\begin{aligned} G(R) &= \left| \sum_{n=1}^{N} \sqrt{2n+1}\, \frac{F_{1n}}{\| F_1 \|} \right|^2 \\ &\leq \sum_{n=1}^{N}(2n+1) \sum_{n=1}^{N} \frac{|F_{1n}|^2}{\| F_1 \|^2} \\ &= N(N+2), \end{aligned}$$

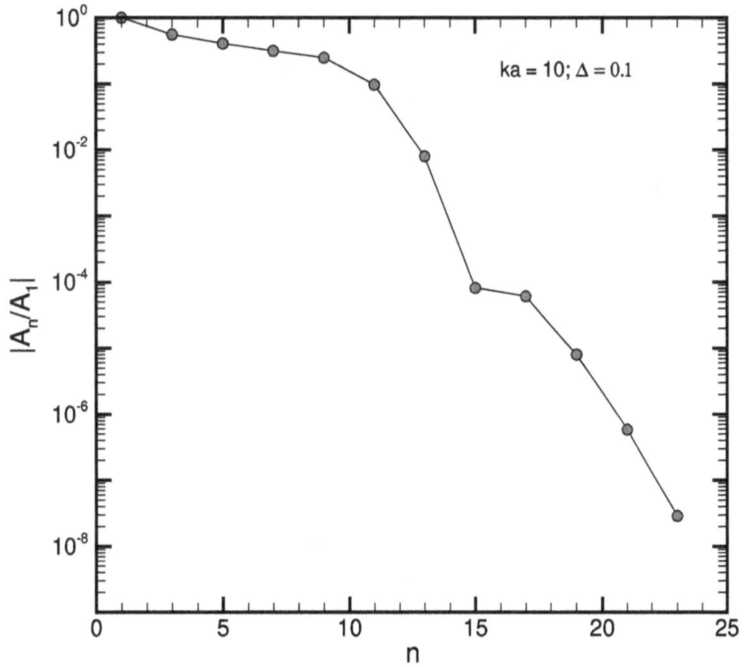


**Figure 13.4.** Normalized magnitude of excitation coefficients of an equatorial slot antenna of figure 8.2, $ka = kb = 10$.

where the second line above follows from Cauchy–Schwarz inequality (13.47). The equality happens when $F_{1n}/\|F_1\| = \sqrt{2n+1}$ or when

$$\frac{F_{1n}}{F_{11}} = \sqrt{\frac{2n+1}{3}}, \qquad n = 1, 2, \ldots.$$

In terms of the size of the circumscribing sphere the maximum possible gain is

$$G(R)|_{\max} = ka(ka+2), \tag{13.52}$$

and the maximum possible gain per effective degree of freedom, $g(R)|_{\max} = \frac{1}{N}G(R)|_{\max}$ is

$$g(R)|_{\max} = ka + 2.$$

Using the conversion from $F_{mn}$ to $C_{mn}$ we see that the maximum gain happens when the original excitation coefficients satisfy

$$C_{1n} = \sqrt{\frac{2(2n+1)}{3n(n+1)}}\, C_{11}, \qquad n = 1, 2\ldots N. \tag{13.53}$$

Thus the power in the excitation coefficients ($\propto |C_{1n}|^2$) decreases inversely with $n$ for large $n$ for the optimum antenna.

## References

[1] Poon A S Y, Brodersen R W and Tse D N C 2005 Degrees of freedom in multiple-antenna channels: a signal space approach *IEEE Trans. Inf. Theory* **51** 523–36
[2] Harrington R F 1961 *Time-Harmonic Electromagnetic Fields* (New York: McGraw-Hill)
[3] di Francia G T 1969 Degrees of freedom of an image *J. Opt. Soc. Am.* **59** 799–804
[4] Bucci O M and Franceschetti G 1989 On the degrees of freedom of scattered fields *IEEE Trans. Antennas Propag.* **37** 918–26
[5] Bucci O M, Crocco L and Isernia T 1999 Improving the reconstruction capabilities in inverse scattering problems by exploitation of close-proximity setups *J. Opt. Soc. Am.* A **16** 1788–98
[6] Cangellaris A C, Celik M, Pasha S and Zhao L 1999 Electromagnetic model order reduction for systems-level modeling *IEEE Trans. Microwave Theory Tech.* **47** 840–50
[7] Slepian D 1975 On bandwidth *Proc. IEEE* **64** 292–300
[8] Miller D A B 2000 Communicating with waves between volumes: evaluating orthogonal spatial channels and limits on coupling strengths *Appl. Opt.* **39** 1681–99
[9] Piestun R and Miller D A B 2000 Electromagnetic degrees of freedom of an optical system *J. Opt. Soc. Am.* **17** 892–902
[10] Xu J and Janaswamy R 2006 Electromagnetic degrees of freedom in 2-D scattering environments *IEEE Trans. Antennas Propag.* **54** 3882–92
[11] Janaswamy R 2011 On the EM degrees of freedom in scattering environments *IEEE Trans. Antennas Propag.* **59** 3872–81
[12] Slepian D and Pollak H O 1961 Prolate spheroidal wave functions, Fourier analysis and uncertainty—I *Bell Syst. Tech. J.* **40** 43–64
[13] Harrington R F 1960 Effect of antenna size on gain, bandwidth, and efficiency *J. Res. Natl Bur. Stand.* D **64D** 1–12

# Chapter 14

# Projection slice theorem and computed tomography

As remarked previously in chapter 10, the goal of inverse scattering is to determine the object shape and its constitution from measurements of transmitted field quantities (or field derived quantities) when the object is irradiated by an electromagnetic wave. The Radon transform and the projection slice theorem find extensive use in inverse scattering by x-ray irradiation [1]. Object features imaged with x-rays are often very large compared to the wavelength and as such the phase of the wave becomes less important, particularly when the detectors used are non-coherent. In such cases one relies on the measurement of the intensity of the transmitted radiation. Recall from equation (9.81) that the intensity ratio under geometrical optics considerations is proportional to the exponent of the integral of the optical ray path through the medium. The Radon transform and the projection slice theorem apply in such circumstances. Practical applications of the Radon transform and the projection slice theorem are in computed tomography (CT scans) and all of these topics will be considered in the present chapter. An example imaging problem is considered at the end of the chapter to illustrate the various steps.

## 14.1 Radon transform and projection slice theorem

The Radon transform is an integral transform which maps a continuous function $f(x, y)$ defined on a plane to another function $\mathscr{R}[f]$ defined on the two-dimensional space of parallel straight lines in the plane, whose value on a particular line is equal to the line integral of $f(x, y)$ over that line. With the reference to figure 14.1, consider a tilted straight line $L$ whose normal distance from the origin is $\xi$ and on which the arc length is denoted by $\eta$. The tilt of the straight line is governed by $\alpha$ ($\alpha = 0$ for a vertical line and $\alpha = \pi/2$ for a horizontal line). The normal to the straight line is $\hat{\boldsymbol{\xi}} = \hat{\mathbf{x}} \cos\alpha + \hat{\mathbf{y}} \sin\alpha$. For a fixed $\xi$, the Cartesian coordinates on the straight line are $\mathbf{r} = (x(\eta; \xi), y(\eta; \xi)) = (\xi \cos\alpha - \eta \sin\alpha, \xi \sin\alpha + \eta \cos\alpha)$. This

doi:10.1088/978-0-7503-1716-0ch14  

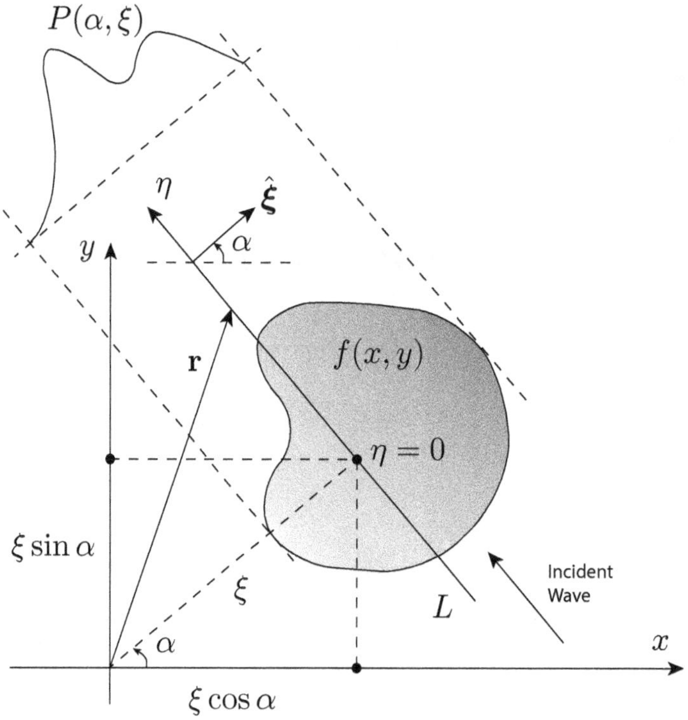


**Figure 14.1.** Transformation of an object with density $f(x, y)$ to a projection $P(\alpha, \xi)$ through the Radon transform. The color and shade denote the value of density.

simply corresponds to rotation of the $xy$-coordinate system through an angle $\alpha$ with respect to the $x$-axis. The inverse transformation yields $\xi = x\cos\alpha + y\sin\alpha = \mathbf{r}\cdot\hat{\boldsymbol{\xi}}$, $y = -x\sin\alpha + y\cos\alpha$. The Jacobian of the transformation is equal to one and $\mathrm{d}x\mathrm{d}y = \mathrm{d}\xi\mathrm{d}\eta$. Then the Radon transform $\mathscr{R}[f](\alpha, \xi)$ is defined in any of the three equivalent mathematical forms

$$\mathscr{R}[f](\alpha, \xi) =: P(\alpha, \xi) = \int_{\eta=-\infty}^{\infty} f(x(\eta;\,\xi),\, y(\eta;\,\xi))\,\mathrm{d}\eta, \tag{14.1}$$

$$\begin{aligned} &= \iint_{\xi,\eta} f(x(\eta;\,\xi),\, y(\eta;\,\xi))\delta(\xi - \mathbf{r}\cdot\hat{\boldsymbol{\xi}})\,\mathrm{d}\xi\mathrm{d}\eta \\ &= \iint_{x,y} f(x,\, y)\delta(\xi - \mathbf{r}\cdot\hat{\boldsymbol{\xi}})\,\mathrm{d}x\mathrm{d}y. \end{aligned} \tag{14.2}$$

For a fixed $\alpha$, the straight line is translated back and forth, which is represented by the value of the coordinate $\xi$, and each line gives rise to one value of $\mathscr{R}[f](\alpha, \xi)$ over $\infty < \xi < \infty$. To exhaust all points in the $xy$-plane it is clear that the angle $\alpha \in (0, \pi)$.

In applications, the function $f(x, y) > 0$ may represent attenuation (scattering plus absorption) of a high-frequency wave as it passes through a physical object and $\alpha$ may represent the direction of the incident wave. The Radon transform will then correspond to the normalized intensity at scan angle $\alpha$ after transmission through the object. If object is of finite support, then $P(\alpha, \xi)$ will also be of finite extent in $\xi$ and will be zero beyond the confines of the object.

The 1D Fourier transform, $\widetilde{P}(\alpha, \kappa)$, of the projection $P(\alpha, \xi)$ w.r.t. $\xi$ is[1]

$$\begin{aligned}\widetilde{P}(\alpha, \kappa) &= \frac{1}{2\pi}\int_{\xi=-\infty}^{\infty} P(\alpha, \kappa)\mathrm{e}^{-j\kappa\xi}\,\mathrm{d}\xi, \quad -\infty < \kappa < \infty \\ &= \frac{1}{2\pi}\iint_{x,y} f(x, y)\int_{\xi=-\infty}^{\infty} \delta(\xi - \mathbf{r}\cdot\hat{\boldsymbol{\xi}})\mathrm{e}^{-j\kappa\xi}\,\mathrm{d}\xi\,\mathrm{d}x\mathrm{d}y \\ &= \frac{1}{2\pi}\iint_{x,y} f(x, y)\mathrm{e}^{-j\kappa\hat{\boldsymbol{\xi}}\cdot\mathbf{r}}\,\mathrm{d}x\mathrm{d}y = F(\boldsymbol{\kappa} = \hat{\boldsymbol{\xi}}\kappa),\end{aligned} \tag{14.3}$$

where

$$F(\boldsymbol{\kappa}) = \frac{1}{2\pi}\iint_{x,y} f(x, y)\mathrm{e}^{-j\boldsymbol{\kappa}\cdot\mathbf{r}}\,\mathrm{d}x\mathrm{d}y \tag{14.4}$$

is the 2D Fourier transform of $f(x, y)$ in the transform variables $\boldsymbol{\kappa} = \hat{\mathbf{x}}\kappa_x + \hat{\mathbf{y}}\kappa_y = \kappa_r(\hat{\mathbf{x}}\cos\beta + \hat{\mathbf{y}}\sin\beta)$, $-\infty < \kappa_x, \kappa_y, \kappa < \infty$, $0 \leqslant \kappa_r < \infty$, $0 \leqslant \beta \leqslant 2\pi$. It may be noted that since $\kappa$ ranges over $(-\infty, \infty)$, it is not the same as the radial wavenumber, $\kappa_r$, that one associates with a 2D Fourier transform. In particular, $\kappa_r = \sqrt{\kappa_x^2 + \kappa_y^2} \neq \kappa$, but $\kappa_r = |\kappa|$. However, with $\kappa_x = \kappa\cos\alpha$ and $\kappa_y = \kappa\sin\alpha$, $0 \leqslant \alpha \leqslant \pi$, $-\infty < \kappa < \infty$, all points in the $\boldsymbol{\kappa}$-space are exhausted, but $\alpha \neq \beta$. Equation (14.3) states that *the 1D Fourier transform, $\widetilde{P}(\alpha, \kappa)$, of the projection, $P(\alpha, \xi)$, of a function, $f(x, y)$, at an angle $\alpha$ obtained through the Radon transform is the slice at angle $\alpha$ of the 2D Fourier transform $F(\boldsymbol{\kappa})$ of $f(x, y)$*. This is known as the *projection slice theorem*. This theorem allows us to retrieve the function $f(x, y)$ through the inverse Fourier transform

$$f(x, y) = \frac{1}{2\pi}\iint_{\kappa_x, \kappa_y=-\infty}^{\infty} F(\boldsymbol{\kappa})\mathrm{e}^{j\boldsymbol{\kappa}\cdot\boldsymbol{r}}\,\mathrm{d}\boldsymbol{\kappa} = \frac{1}{2\pi}\int_{\kappa_r=0}^{\infty}\int_{\beta=0}^{2\pi} \kappa_r F(\boldsymbol{\kappa})\mathrm{e}^{j\kappa_r(x\cos\beta + y\sin\beta)}\,\mathrm{d}\beta\mathrm{d}\kappa_r. \tag{14.5}$$

Note that $\boldsymbol{\kappa} = \kappa_r(\hat{\mathbf{x}}\cos\beta + \hat{\mathbf{y}}\sin\beta) = (-\kappa_r)(\hat{\mathbf{x}}\cos(\beta - \pi) + \hat{\mathbf{y}}\sin(\beta - \pi))$. Therefore the integral in equation (14.5) may be rewritten as

[1] We put the factor $1/2\pi$ in the 1D forward transform here so that the eventual result involving 2D forward and inverse transforms will be symmetric in the factor $1/2\pi$.

$$f(x, y) = \frac{1}{2\pi} \int_{\kappa_r=0}^{\infty} \left[ \int_{\beta=0}^{\pi} + \int_{\beta=\pi}^{2\pi} \right] \kappa_r F(\boldsymbol{\kappa}) e^{j\kappa_r(x\cos\beta + y\sin\beta)} \, d\beta d\kappa_r$$

$$= \frac{1}{2\pi} \int_{\kappa_r=0}^{\infty} \int_{\beta=0}^{\pi} \kappa_r F(\boldsymbol{\kappa}) e^{j\kappa_r(x\cos\beta + y\sin\beta)} \, d\beta d\kappa_r +$$

$$+ \frac{1}{2\pi} \int_{\kappa_r=0}^{\infty} \int_{\beta=\pi}^{2\pi} (-\kappa_r) F(\boldsymbol{\kappa}) e^{j(-\kappa_r)[x\cos(\beta-\pi) + y\sin(\beta-\pi)]} \, d\beta(-d\kappa_r)$$

$$= \frac{1}{2\pi} \int_{\kappa=-\infty}^{\infty} \int_{\alpha=0}^{\pi} |\kappa| \, F(\boldsymbol{\kappa}) e^{j\kappa(x\cos\alpha + y\sin\alpha)} d\alpha d\kappa$$

$$= \frac{1}{2\pi} \int_{-\infty}^{\infty} \int_{\alpha=0}^{\pi} |\kappa| \, F(\boldsymbol{\kappa}) e^{j\kappa\hat{\boldsymbol{\xi}}\cdot\mathbf{r}} \, d\alpha d\kappa,$$

where $\kappa_r = \kappa$, $\beta = \alpha$ was used in the first integral and $-\kappa_r = \kappa$, $\beta - \pi = \alpha$ was used in the second integral in the middle step to arrive at the end result. Combining this with the fact that $\widetilde{P}(\alpha, \kappa) = F(\boldsymbol{\kappa} = \hat{\boldsymbol{\xi}}\kappa)$ we finally obtain the inversion formula

$$f(x, y) = \frac{1}{2\pi} \int_{\alpha=0}^{\pi} \int_{\kappa=-\infty}^{\infty} |\kappa| \, \widetilde{P}(\alpha, \kappa) e^{j\kappa\hat{\boldsymbol{\xi}}\cdot\mathbf{r}} \, d\kappa d\alpha. \tag{14.6}$$

## 14.2 Computed tomography

In x-ray tomography, $f(x, y)$ represents the attenuation function presented by the object being scanned such that the intensity at the exit plane, $I_t(\alpha, \xi)$, for illumination at the angle $\alpha$ is related to the incident intensity $I_0$ via

$$\frac{I_t(\alpha, \xi)}{I_0} = e^{-\left[\int f(x,\, y)\, d\eta\right]} = e^{-P(\alpha,\, \xi)} \Longrightarrow P(\alpha, \xi) = -\ln \frac{I_t(\alpha, \xi)}{I_0}. \tag{14.7}$$

If measurements of intensity are only available over some finite bandwidth $-W \leqslant \kappa \leqslant W$ through a filter with transfer function $H(\kappa)$, then the reconstructed attenuation function will be smeared by the filter response. The reconstructed attenuation function, $\hat{f}(x, y)$, is

$$\hat{f}(x, y) = \frac{1}{2\pi} \int_{0}^{\pi} \int_{-W}^{W} |\kappa| \, H(\kappa) \widetilde{P}(\alpha, \kappa) e^{j\kappa\hat{\boldsymbol{\xi}}\cdot\mathbf{r}} \, d\kappa d\alpha. \tag{14.8}$$

Equation (14.8) is suitable for practical implementations once measurements of intensity at the exit plane are recorded. The Fourier transform of $P(\alpha, \xi)$ as well as the inverse transform in equation (14.8) can be conveniently evaluated using FFTs and inverse FFTs.

Substituting for $\widetilde{P}(\alpha, \xi)$ from equation (14.3), defining $R = \sqrt{(x - x')^2 + (y - y')^2}$, $(x - x') = R\cos\zeta$, $(y - y') = R\sin\zeta$, $\hat{\boldsymbol{\xi}} \cdot (\mathbf{r} - \mathbf{r}') = R\cos(\alpha - \zeta)$, utilizing the series expansion (C.71) and evaluating the integral w.r.t. $\alpha$, the reconstructed attenuation function can be recast as

$$\begin{aligned}\hat{f}(x, y) &= \frac{1}{(2\pi)^2} \iint_{x',y'} dx'dy' f(x', y') \int_{-W}^{W} |\kappa| H(\kappa) \int_{0}^{\pi} e^{j\kappa\hat{\boldsymbol{\xi}}\cdot(\mathbf{r}-\mathbf{r}')}\, d\alpha d\kappa \\ &= \frac{1}{(2\pi)^2} \iint_{x',y'} dx'dy' f(x', y') \int_{-W}^{W} \left[\pi J_0(\kappa R) + 4 \sum_{m=0}^{\infty} \frac{J_{2m+1}(\kappa R)}{2m + 1} \sin(2m + 1)\zeta\right] \\ &\quad \times |\kappa| H(\kappa)\, d\kappa.\end{aligned} \tag{14.9}$$

If the filter function is symmetric in $\kappa$ such that $H(-\kappa) = H(\kappa)$, then the series terms drops out on recalling that Bessel functions with odd order are odd functions of the argument. In that case

$$\hat{f}(x, y) = \frac{1}{4\pi} \iint_{x',y'} f(x', y') \left[\int_{-W}^{W} |\kappa| H(\kappa) J_0(\kappa R)\, \mathrm{d}\kappa\right] \mathrm{d}x'\mathrm{d}y'. \tag{14.10}$$

The quantity within the square brackets in equation (14.10) is a function of $R = |\mathbf{r} - \mathbf{r}'|$. Thus the spatial integral in equation (14.10) is of convolutional type. For an ideal low-pass filter $H(\kappa) = 1, |\kappa| < W$. The integral involving Bessel function can then be evaluated using identity (C.61) with the result

$$\hat{f}(x, y) = \frac{W}{2\pi} \iint_{x',y'} \frac{J_1(WR)}{R} f(x', y')\, \mathrm{d}x'\mathrm{d}y'. \tag{14.11}$$

Equation (14.11) gives $\hat{f}(x, y) = WJ_1(Wr)/2\pi r$ for a point attenuation function $f(x, y) = \delta(\mathbf{r})$. Note that $\lim_{W\to\infty} WJ_1(Wr) = \delta(r)$, so that one recovers the exact point for infinite bandwidth.

**Example 14.2.0.1.** Circular object with uniform attenuation.
Determine the reconstructed object if the attenuation within a circular object of radius $a$ is constant and an ideal low-pass filter is used in measurements.

Let us assume the constant attenuation to be unity. The most convenient form to proceed here is that given in equation (14.10). Using cylindrical coordinates, we have $f(x', y') = 1$, $\mathrm{d}x'\mathrm{d}y' = \rho'\mathrm{d}\rho'\mathrm{d}\phi'$, $R = [\rho^2 + \rho'^2 - 2\rho\rho'\cos(\phi' - \phi)]^{1/2}$. We next employ the addition theorem for Bessel functions (C.39), interchange the orders of integration in equation (14.10), evaluate the integral w.r.t. $\phi'$ and use the identity (C.61) to obtain the following expression for the reconstructed object

$$
\begin{aligned}
\hat{f}(\rho, \phi) &= \frac{1}{2\pi}\int_0^W \kappa d\kappa \int_0^a \rho' d\rho' \int_0^{2\pi} \Big[J_0(\kappa\rho)J_0(\kappa\rho') \\
&\qquad + 2\sum_{n=1}^{\infty} J_n(\kappa\rho)J_n(\kappa\rho')\cos n(\phi - \phi')\Big] d\phi' \\
&= \int_0^W \kappa J_0(\kappa\rho) d\kappa \int_0^a \rho' J_0(\kappa\rho')\, d\rho' \\
&= a\int_0^W J_1(\kappa a)J_0(\kappa\rho)\, d\kappa = \int_0^{Wa} J_1(x)J_0(xr)\, dx \\
&= \Theta(1 - r) - \int_{Wa}^{\infty} J_1(x)J_0(xr)\, dx,
\end{aligned}
\tag{14.12}
$$

where the final expression in the last step follows from the identity [2, 6.512–3], $r = \rho/a$ and $\Theta(\cdot)$ is the unit step function. Clearly, for $Wa \to \infty$ the asymptotic integral on the rhs of equation (14.12) vanishes and one recovers the exact shape. Figure 14.2 shows the reconstructed object as a function of $r = \rho/a$ for two different filter bandwidths of $Wa = 2\pi$ and $Wa = 4\pi$ after carrying out the integral in equation (14.12) numerically. The smearing of the object shape due to band-limited

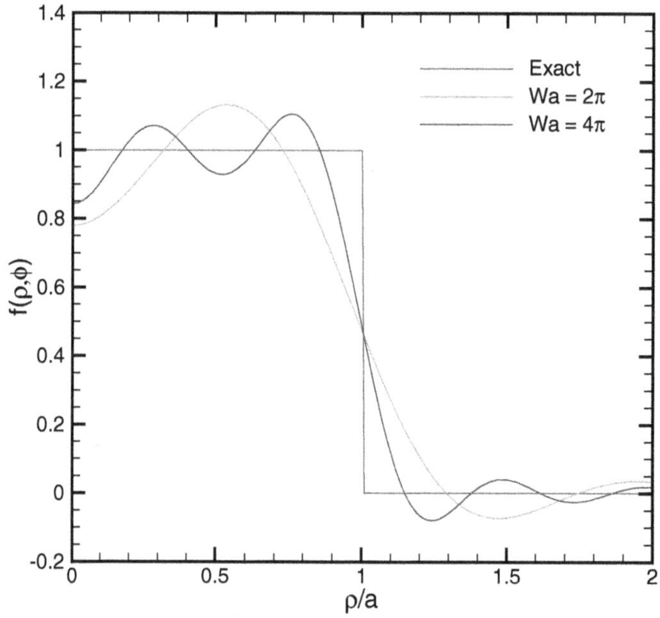


**Figure 14.2.** Exact object shape and reconstructed ones with $Wa = 2\pi$ and $Wa = 4\pi$ for a circular cylinder of radius $a$.

data is clearly evident. There is ringing as well as softening of edges in the reconstructed image with better results appearing for larger bandwidth of $Wa = 4\pi$ than for $Wa = 2\pi$. ■■

## References

[1] Kak A C 1979 Computerized tomography with x-ray emission and ultrasound sources *Proc. IEEE* **67** 1245–72

[2] Gradshteyn I S and Ryzhik I M 2015 Tables of Integrals *Series and Products* 8th edn (New York: Academic)

**IOP** Publishing

Engineering Electrodynamics
A collection of theorems, principles and field representations
**Ramakrishna Janaswamy**

# Chapter 15

# Free-space Green's function and its application in various coordinates

In this example-laden chapter we show various forms of the scalar Green's function in a homogeneous, unbounded medium and demonstrate its applications in solving boundary value problems in Cartesian, cylindrical, and spherical coordinates. Depending on the geometry, the expansion of fields is carried out in a continuous/ discrete spectrum in the longitudinal, lateral, or radial wavenumbers. The specific examples considered here are: (i) radiation by planar currents; and an electric/ magnetic dipole radiating in the presence of (ii) an infinite circular cylinder, (iii) an infinite conducting wedge, (iv) a conductor-backed stratified layer, and (v) a material sphere (both ordinary and plasmonic).

## 15.1 Various forms of the free-space Green's function

The following theorem summarizes various forms of the scalar Green's function in the frequency domain.

**Theorem 15.1.1** Scalar Green's function in a homogeneous medium.
*Consider a homogeneous medium characterized by a complex wavenumber* $k = k_r - jk_i = |k|e^{-j\phi_k}$, *where* $k_r$ *and* $k_i$ *are real positive numbers and* $\tan\phi_k = k_i/k_r$. *Let the field point be denoted by* $\mathbf{r} = (x, y, z)$, $(\rho, \phi, z)$, *or* $(r, \theta, \phi)$ *in Cartesian, cylindrical, or spherical coordinate systems, respectively. Likewise let* $\mathbf{r}' = (x', y', z')$, $(\rho', \phi', z')$, or $(r', \theta', \phi')$ *be the coordinates of the source point. Let* $R = |\mathbf{r} - \mathbf{r}'| = \sqrt{R_t^2 + (z - z')^2}$ *be the distance between the source and field points, where* $(x - x') = R_t \cos\alpha_t$, $(y - y_t) = R_t \sin\alpha_t$, $R_t^2 = (x - x')^2 + (y - y')^2 = \rho^2 + \rho'^2 - 2\rho\rho' \cos(\phi - \phi')$, $\tan\alpha_t = (y - y')/(x - x')$. *Then the following spectral forms of the Green's function (scalar radiation function)*

doi:10.1088/978-0-7503-1716-0ch15

$$\psi = \frac{e^{-jkR}}{4\pi R}, \tag{15.1}$$

*of the Helmholtz equation* $\nabla^2\psi + k^2\psi = -\delta(\mathbf{r} - \mathbf{r}')$ *are all equivalent:*

$$\psi = \frac{1}{8j\pi^2}\iint_{-\infty}^{\infty}\frac{1}{k_z}e^{\pm j[k_x(x-x')+k_y(y-y')]}e^{-jk_z|z-z'|}\,dk_x dk_y,\ \text{(Weyl)}, \tag{15.2}$$

$$= \frac{k}{8j\pi^2}\int_{\Gamma}\int_{\alpha=0}^{2\pi} e^{\pm jkR_t\sin\beta\cos(\alpha-\alpha_t)}e^{-jk|z-z'|\cos\beta}\sin\beta\,d\alpha d\beta,\ \text{(APWS)}, \tag{15.3}$$

$$= \frac{1}{8j\pi}\int_{-\infty}^{\infty}\frac{k_\rho}{k_z}H_0^{(2)}(k_\rho R_t)e^{-jk_z|z-z'|}\,dk_\rho,\ \text{(Sommerfeld 1)}, \tag{15.4}$$

$$= \frac{1}{4j\pi}\int_{0}^{\infty}\frac{k_\rho}{k_z}J_0(k_\rho R_t)e^{-jk_z|z-z'|}\,dk_\rho,\ \text{(Sommerfeld 2)}, \tag{15.5}$$

$$= \frac{1}{8j\pi}\int_{-\infty}^{\infty}H_0^{(2)}(k_\rho R_t)e^{\pm jk_z(z-z')}\,dk_z,\ \text{(longitudinal wave)}, \tag{15.6}$$

$$= \frac{1}{8j\pi}\int_{-\infty}^{\infty}\sum_{n=-\infty}^{\infty}\frac{k_\rho}{k_z}J_n(k_\rho\rho_<)H_n^{(2)}(k_\rho\rho_>)e^{jn(\phi-\phi')}e^{-jk_z|z-z'|}\,dk_\rho,\ \text{(CH-1)}, \tag{15.7}$$

$$= \frac{1}{4j\pi}\int_{0}^{\infty}\sum_{n=-\infty}^{\infty}\frac{k_\rho}{k_z}J_n(k_\rho\rho')J_n(k_\rho\rho)e^{jn(\phi-\phi')}e^{-jk_z|z-z'|}\,dk_\rho,\ \text{(CH-2)}, \tag{15.8}$$

$$= \frac{1}{8j\pi}\int_{-\infty}^{\infty}\sum_{n=-\infty}^{\infty}J_n(k_\rho\rho_<)H_n^{(2)}(k_\rho\rho_>)e^{jn(\phi-\phi')}e^{-jk_z(z-z')}\,dk_z,\ \text{(CH-3)}, \tag{15.9}$$

$$= \frac{1}{2\pi^2}\int_{0}^{\infty}K_0(\Lambda R_t)\cos(k_z(z-z'))\,dk_z,\ \text{(Schelkunoff)} \tag{15.10}$$

$$= \frac{1}{2\pi^{3/2}}\int_{\gamma}e^{-R^2s^2+\frac{k^2}{4s^2}}\,ds,\ \text{(Ewald)}, \tag{15.11}$$

$$= \frac{k}{4j\pi}\sum_{n=0}^{\infty}(2n+1)h_n^{(2)}(kr_>)j_n(kr_<)P_n(\cos\zeta),\ \text{(SH)}, \tag{15.12}$$

*where* $k_z = \sqrt{k^2 - (k_x^2 + k_y^2)} = -j\sqrt{(k_x^2 + k_y^2) - k^2}$ *in equation (15.2),* $k_z = \sqrt{k^2 - k_\rho^2} = -j\sqrt{k_\rho^2 - k^2}$ *in equation (15.4),* $k_\rho = \sqrt{k^2 - k_z^2} = -j\sqrt{k_z^2 - k^2}$ *in equations (15.6) and (15.9),* $\Lambda = \sqrt{k_z^2 - k^2} = j\sqrt{k^2 - k_z^2}$ *in equation (15.10),* $r_> = \max(r, r')$, $r_< = \min(r, r')$, $\rho_< = \min(\rho, \rho')$, $\rho_> = \max(\rho, \rho')$, $\cos\zeta = \cos\theta\cos\theta' + \sin\theta\sin\theta'$

$\cos(\phi - \phi')$, $J_n(\cdot)$ *is the Bessel function of the first kind of order n,* $H_n^{(2)}(\cdot)$ *is the Hankel function of the second kind of order n,* $K_0(\cdot)$ *is the modified Bessel function of the second kind of order zero,* $h_n^{(2)}(\cdot)$ *is the spherical Hankel function of the second kind of order n, and* $P_n(\cdot)$ *is the Legendre polynomial of the first kind of degree n. The abbreviation APWS stands for angular plane wave spectrum, CH stands for cylindrical harmonic, and SH stands for spherical harmonic. The contours* $\Gamma$ *and* $\gamma$ *are shown in figure 15.1(a) and (b).*

*Proof.* For proof see the references listed in table 15.1[1].
As an example we show the proof for the representation (15.9) first. The starting point is the non-homogeneous Helmholtz equation $\nabla^2\psi + k^2\psi = -\delta(x - x')$

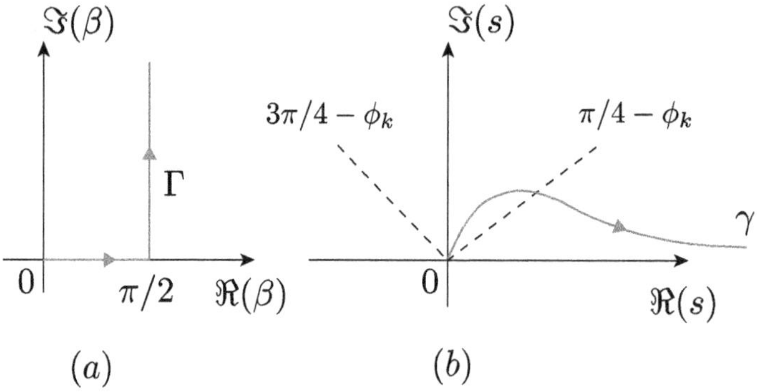


**Figure 15.1.** Contour of integration in the integral representations. For the contour shown in (b) the requirements are $-\pi/4 \leqslant \arg(s) \leqslant \pi/4$ for $s \to \infty$ and $\pi/4 - \phi_k \leqslant \arg(s) \leqslant 3\pi/4 - \phi_k$.

**Table 15.1.** Representations of scalar Green's functions.

| Equation | Proof in reference |
|---|---|
| Equation (15.2) | [1, p 229] |
| Equation (15.3) | [1, p 231] |
| Equation (15.4) | [2, p 242] |
| Equation (15.5) | [2, p 242] |
| Equation (15.6) | [3, p 244] |
| Equation (15.9) | [3, pp 232, 244] |
| Equation (15.10) | [4] |
| Equation (15.11) | [5] |
| Equation (15.12) | Example 15.1.0.1 |

[1] The form of representations contained in some of these references is for an $e^{-i\omega t}$ time convention, which we translated to conform to the $e^{j\omega t}$ time convention mostly adopted here.

$\delta(y - y')\delta(z - z') = \delta(\rho - \rho')\delta(\phi - \phi')\delta(z - z')/\rho$ satisfied by $\psi$. Taking the Fourier transform of this equation with respect to $z$ results in

$$\nabla_t^2\tilde{\psi}+k_\rho^2\tilde{\psi} = -e^{jk_z z'}\frac{\delta(\rho - \rho')\delta(\phi - \phi')}{\rho}, \tag{15.13}$$

where $\nabla_t^2 = \frac{\partial^2}{\partial x^2} + \frac{\partial^2}{\partial y^2} = \frac{1}{\rho}\frac{\partial}{\partial \rho}(\rho\frac{\partial}{\partial \rho}) + \frac{1}{\rho^2}\frac{\partial^2}{\partial \phi^2}$ is the transverse Laplacian,

$$\begin{aligned}\tilde{\psi}(x, y, k_z) &= \int_{z=-\infty}^{\infty} \psi(x, y, z)e^{jk_z z}\, dz;\\ \psi(x, y, z) &= \frac{1}{2\pi}\int_{k_z=-\infty}^{\infty} \tilde{\psi}(x, y, k_z)e^{-jk_z z}\, dk_z,\end{aligned} \tag{15.14}$$

and $k_\rho^2 = k^2 - k_z^2$. On physical grounds, one may identify (15.13) with the equation satisfied by the $z$-component of magnetic vector potential due to a $\hat{\mathbf{z}}$-directed electric line source of permeability times strength $\mu I = e^{jk_z z'}$, located at $\rho = \rho'$ and $\phi = \phi'$. The function $\tilde{\psi}$ could be expressed in terms of the complete set of cylindrical harmonics in the form

$$\tilde{\psi}(\rho, \phi, k_z) = \begin{cases}\sum\limits_{n=-\infty}^{\infty} A_n J_n(k_\rho\rho)H_n^{(2)}(k_\rho\rho')e^{jn(\phi-\phi')}, & \rho < \rho'\\ \sum\limits_{n=-\infty}^{\infty} B_n J_n(k_\rho\rho')H_n^{(2)}(k_\rho\rho)e^{jn(\phi-\phi')}, & \rho > \rho'\end{cases}, \tag{15.15}$$

which is intrinsically symmetric in $\rho$ and $\rho'$ but for the constants $A_n$, $B_n$. Continuity of $\tilde{\psi}$ (i.e. absence of a jump discontinuity) at $\rho = \rho'$ implies $A_n = B_n$. Continuing with the physical analogy, one may identify the continuity of $\tilde{\psi}$ as representing a continuity of the $z$-directed electric field due to the line source. Integrating $\rho\times$ (15.13) about $\rho \in (\rho' - \varepsilon, \rho' + \varepsilon)$ with $\varepsilon \to 0$ yields

$$\lim_{\varepsilon\to 0}\left[\rho\frac{\partial\tilde{\psi}}{\partial\rho}\right]_{\rho'-\varepsilon}^{\rho'+\varepsilon} = -e^{jk_z z'}\delta(\phi - \phi'). \tag{15.16}$$

On physical grounds, the above equation represents the discontinuity of the $\phi$-directed magnetic field due to the electric surface current on the circle $\rho = \rho'$. Substituting equation (15.15) into equation (15.16) and using the Wronskian $J_n(z)\frac{d}{dz}H_n^{(2)}(z) - H_n^{(2)}(z)\frac{d}{dz}J_n(z) = \frac{2}{j\pi z}$, we obtain

$$\sum_{n=-\infty}^{\infty} A_n\frac{2}{j\pi}e^{jn(\phi-\phi')} = -e^{jk_z z'}\delta(\phi - \phi'). \tag{15.17}$$

Multiplying both sides of this equation with $e^{-jm(\phi-\phi')}$ and integrating over $(\phi - \phi') \in (0, 2\pi)$ gives $4jA_m = e^{jk_z z'}$. Therefore

$$\psi = \frac{1}{8\pi j}\int_{k_z=-\infty}^{\infty}\sum_{n-=\infty}^{\infty} J_n(k_\rho\rho_<)H_n^{(2)}(k_\rho\rho_>)e^{jm(\phi-\phi')}e^{-jk_z(z-z')}\,dk_z. \tag{15.18}$$

Requiring finiteness of fields at $\rho \to \infty$ implies that the branch of $k_\rho$ representing proper waves must be such that $k_\rho = -j\sqrt{k_z^2 - k^2}$ for large values of $k_z$.

See also example 16.1.0.1 as to how to convert an expansion available in $k_\rho$ to an expansion in $k_z$ and vice versa. ■

*It is observed that the Green's function (15.1) is an analytic function in the lower half of the complex k-plane, which also follows from causality. Even in the presence of more complicated boundary conditions, the Green's function will always be analytic in the lower half of the complex k-plane owing to causality. Any singularities such as poles and branch points of a Green's function must all lie in the upper half of the complex k-plane.*

In the representation (15.12), the expansion theorem 1.3.2 for spherical harmonics may be applied to separate the source coordinates $(\theta', \phi')$ from the field coordinates $(\theta, \phi)$ in the function $P_n(\cos\zeta)$ to result in

$$\begin{aligned} P_n(\cos\zeta) &= \sum_{m=-n}^{n}(-1)^m P_n^m(\cos\theta)P_n^{-m}(\cos\theta')e^{jm(\phi-\phi')} \\ &= \sum_{m=-n}^{n}\frac{(n-m)!}{(n+m)!}P_n^m(\cos\theta)P_n^m(\cos\theta')e^{jm(\phi-\phi')} \\ &= P_n(\cos\theta)P_n(\cos\theta') + 2\sum_{m=1}^{n}\frac{(n-m)!}{(n+m)!}P_n^m(\cos\theta)P_n^m(\cos\theta') \\ &\quad \times \cos m(\phi-\phi'). \end{aligned} \tag{15.19}$$

Thus a more elementary expansion of the free-space Green's function in terms of the spherical harmonics is

$$\begin{aligned} \psi(r,\theta,\phi; r',\theta',\phi') = \frac{k}{4j\pi}\sum_{n=0}^{\infty}\sum_{m=-n}^{n}(2n+1)\frac{(n-m)!}{(n+m)!}h_n^{(2)}(kr_>)j_n(kr_<) \\ \times P_n^m(\cos\theta)P_n^m(\cos\theta')e^{jm(\phi-\phi')}, \end{aligned} \tag{15.20}$$

$$\begin{aligned} = \frac{k}{4j\pi}\sum_{n=0}^{\infty}\sum_{m=0}^{n}\varepsilon_m(2n+1)\frac{(n-m)!}{(n+m)!}j_n(kr_<)h_n^{(2)}(kr_>) \\ \times P_n^m(\cos\theta)P_n^m(\cos\theta')\cos m(\phi-\phi'), \end{aligned} \tag{15.21}$$

where $\varepsilon_m$ is the Neumann's number equal to 1 for $m = 0$ and equal to 2 for $m > 0$. For the special case of $k = 0$, see the generating function (D.115) of spherical harmonics.

**Example 15.1.0.1.** Spherical harmonics form of the free-space Green's function. In this example we derive the spherical harmonic form (15.21) directly from the solution of the Helmholtz equation in spherical coordinates.

The starting point in the derivation of the spherical harmonic form in equation (15.20) is to express the solution of the Helmholtz equation in terms of the spherical harmonics in the form

$$\psi = \begin{cases} \sum_{n=0}^{\infty} \sum_{m=-n}^{n} C^{<} j_n(kr) P_n^m(\cos\theta) e^{jm\phi}, & r < r' \\ \sum_{n=0}^{\infty} \sum_{m=-n}^{n} C^{>} h_n^{(2)}(kr) P_n^m(\cos\theta) e^{jm\phi}, & r > r'. \end{cases} \tag{15.22}$$

Spherical Bessel functions are used for $r < r'$ to satisfy the finiteness of fields at the origin, while spherical Hankel functions are used for $r > r'$ to satisfy the radiation condition at infinity. Furthermore, $m$ is an integer owing to periodicity in $\phi$ and $n$ is an integer owing to finiteness of fields along the $z$-axis (i.e. for $\theta = 0, \pi$). The index $m$ need not exceed $n$ because $P_n^m(\cos\theta) = 0,\ m > n$ (see identity (D.76)). Using the orthogonality of $e^{jm\phi}$, $\phi \in (0, 2\pi)$ and $P_n^m(\cos\theta)$, $\theta \in (0, \pi)$, imposition of continuity of $\psi$ at $r = r'$ implies

$$C^{<} j_n(kr') = C^{>} h_n^{(2)}(kr') =: C_n^m j_n(kr') h_n^{(2)}(kr'), \tag{15.23}$$

where $C_n^m$ is an undetermined constant. Multiplying the governing equation $\nabla^2\psi + k^2\psi = -\delta(r - r')\delta(\phi - \phi')\delta(\theta - \theta')/r^2 \sin\theta$ by $[r^2 \sin\theta e^{-j\mu\phi} P_\nu^\mu(\cos\theta)]$, $\mu, \nu$ integers and integrating over $r \in (r' - \Delta, r' + \Delta)$, $\theta \in (0, \pi)$, $\phi \in (0, 2\pi)$ gives the condition

$$\lim_{\Delta\to 0} (r')^2 \frac{d\tilde{\psi}_\nu^\mu}{dr}\Bigg|_{r'-\Delta}^{r'+\Delta} = -P_\nu^\mu(\cos\theta') e^{-j\mu\phi'}, \tag{15.24}$$

upon using the auxiliary condition $\lim_{\Delta\to 0} \int_{r'-\Delta}^{r'+\Delta} \psi \, dr = 0$ arising from the continuity of $\psi$, where

$$\tilde{\psi}_\nu^\mu(r) = \int_0^{2\pi} \int_0^{\pi} \psi(r, \theta, \phi) e^{-j\mu\phi} P_\nu^\mu(\cos\theta) \sin\theta \, d\theta d\phi. \tag{15.25}$$

In view of (i) the orthogonality of $e^{jm\phi}$, (ii) identity (D.116) and (iii) the relation (15.23), we obtain

$$\tilde{\psi}_\nu^\mu = \frac{4\pi C_\nu^\mu}{2\nu + 1} \frac{(\nu + \mu)!}{(\nu - \mu)!} j_\nu(kr_<) h_\nu^{(2)}(kr_>) \tag{15.26}$$

upon substituting equation (15.22) into equation (15.25). Using this relation in equation (15.24) and employing the Wronskian, (C.81), $\mathscr{W}\left(j_n(z), h_n^{(2)}(z)\right) = 1/jz^2$, the equation for determining the constant $C_\nu^\mu$ is obtained as

$$\frac{-j4\pi C_\nu^\mu}{2\nu+1}\frac{(\nu+\mu)!}{(\nu-\mu)!}\frac{1}{k} = -P_\nu^\mu(\cos\theta')\mathrm{e}^{-j\mu\phi'}. \tag{15.27}$$

Substituting this into equation (15.22) together with equation (15.23) finally yields

$$\begin{aligned}\psi(r,\theta,\phi;r',\theta',\phi') = \frac{k}{4j\pi}\sum_{n=0}^{\infty}\sum_{m=-n}^{n}(2n+1)\frac{(n-m)!}{(n+m)!}\\ h_n^{(2)}(kr_>)j_n(kr_<)\\ \times P_n^m(\cos\theta)P_n^m(\cos\theta')\mathrm{e}^{jm(\phi-\phi')},\end{aligned} \tag{15.28}$$

which is identical to equation (15.20). ■■

In the next section we make use of the various forms available for the free-space Green's function and demonstrate their application in solving various boundary value problems.

## 15.2 Canonical problems in various coordinate systems

### 15.2.1 Planar currents radiating in half-space

Consider a planar, time-harmonic electric surface current $\mathbf{J}_s = \hat{\mathbf{p}}J_s(x, y)$ radiating in a homogeneous medium with parameters $(\varepsilon, \mu)$. For the sake of illustration we choose $\hat{\mathbf{p}} = \hat{\mathbf{x}}$ as shown in figure 15.2. Other polarizations could be handled in a similar manner. Our goal is to determine the fields and power using an appropriate representation of the free-space Green's function. We first express the volume current density of the source as $\mathbf{J}(x, y, z) = \mathbf{J}_s(x, y)\delta(z - 0)$. The magnetic vector

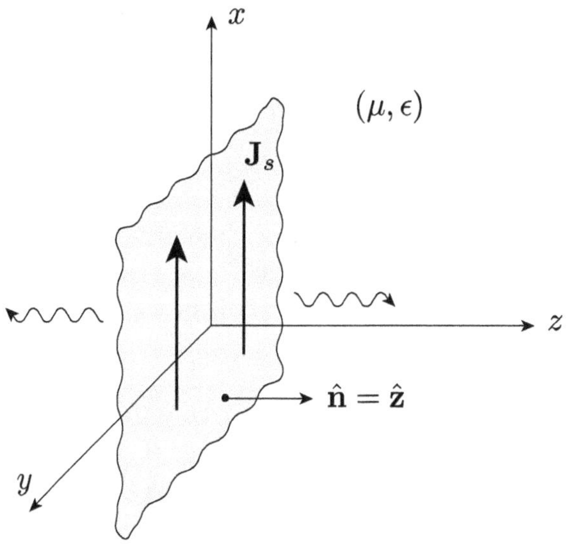


**Figure 15.2.** Planar current radiating in an unbounded medium.

potential under the Lorenz gauge satisfies $\nabla^2\mathbf{A} + k^2\mathbf{A} = -\mu\mathbf{J}$, $k^2 = \omega^2\mu\varepsilon$. Since $\mathbf{J}$ has only an $x$-component it is reasonable to assume that $\mathbf{A}$ will have only an $x$-component. The adequacy or not of this assumption is verified by the satisfaction or not of various field boundary conditions at $z = 0$. Accordingly, we let $\mathbf{A} = \hat{\mathbf{x}}A_x$. Then $A_x$ satisfies the scalar Helmholtz equation $\nabla^2 A_x + k^2 A_x = 0$, separately for $z \gtrless 0$. The complete field is given in terms of $\mathbf{A}$ via equations (1.44) and (1.45). In particular

$$E_x = \frac{1}{j\omega\mu\varepsilon}\left(\frac{\partial^2}{\partial x^2} + k^2\right)A_x, \quad E_y = \frac{1}{j\omega\mu\varepsilon}\frac{\partial^2 A_x}{\partial x \partial y}, \quad H_x = 0, \quad H_y = \frac{1}{\mu}\frac{\partial A_x}{\partial z}. \tag{15.29}$$

A field that has no $x$-component of magnetic field in all space is classified as *transverse magnetic to* $x$ or simply $\mathrm{TM}_x$. Assuming absolute integrability[2], the potential is expanded in terms of plane waves, or equivalently, in terms of a Fourier transform as

$$A_x(\mathbf{r}) = \frac{1}{4\pi^2}\iint_{-\infty}^{\infty} \widetilde{A}_x^{\pm}(k_x, k_y)\mathrm{e}^{j(k_x x + k_y y)}\mathrm{e}^{\mp jk_z z}\,\mathrm{d}k_x\mathrm{d}k_y, \quad z \gtrless 0, \tag{15.30}$$

where $k_z = \sqrt{k^2 - k_x^2 - k_y^2} = -j\sqrt{k_x^2 + k_y^2 - k^2}$ is chosen to have a positive real value inside the circle $k_x^2 + k_y^2 = k^2$ and a negative imaginary value outside it to satisfy the radiation condition. The potential automatically satisfies the Helmholtz equation in this form. Continuity of $E_y$ at $z = 0$ gives

$$\iint_{-\infty}^{\infty} k_x k_y\left(\widetilde{A}_x^{+}(k_x, k_y) - \widetilde{A}_x^{-}(k_x, k_y)\right)\mathrm{e}^{j(k_x x + k_y y)}\,\mathrm{d}k_x\mathrm{d}k_y = 0. \tag{15.31}$$

But from the completeness of Fourier transform in free-space we have

$$\frac{1}{2\pi}\int_{-\infty}^{\infty} \mathrm{e}^{j(k_x - k_x')x}\,\mathrm{d}x = \delta(k_x - k_x'); \quad \frac{1}{2\pi}\int_{-\infty}^{\infty} \mathrm{e}^{j(k_y - k_y')y}\,\mathrm{d}y = \delta\left(k_y - k_y'\right), \tag{15.32}$$

and from the sifting property of delta functions we have

$$\iint_{-\infty}^{\infty} \widetilde{\psi}(k_x, k_y)\delta(k_x - k_x')\delta\left(k_y - k_y'\right)\mathrm{d}k_x\mathrm{d}k_y = \widetilde{\psi}\left(k_x', k_y'\right). \tag{15.33}$$

Multiplying both sides of equation (15.31) with $\mathrm{e}^{-j\left(k_x' x + k_y' y\right)}/4\pi^2$ and integrating over $x \in (-\infty, \infty)$, $y \in (-\infty, \infty)$ and utilizing equations (15.32) and (15.33) results in

$$\widetilde{A}_x^{+}\left(k_x', k_y'\right) = \widetilde{A}_x^{-}\left(k_x', k_y'\right) =: \widetilde{A}_x\left(k_x', k_y'\right). \tag{15.34}$$

Continuity of $E_x$ at $z = 0$ leads to the same conclusion. Imposing the remaining boundary condition on the discontinuity of magnetic field, $\hat{\mathbf{n}} \times [\mathbf{H}(x, y, z = 0^+) - \mathbf{H}(x, y, z = 0^-)] = \mathbf{J}_s$ gives

$$\frac{j}{2\pi^2}\iint_{-\infty}^{\infty} k_z\widetilde{A}_x(k_x, k_y)\mathrm{e}^{j(k_x x + k_y y)}\,\mathrm{d}k_x\mathrm{d}k_y = \mu J_s(x, y). \tag{15.35}$$

[2] The Fourier transform is complete for absolutely integrable functions.

Multiplying both sides of equation (15.35) with $e^{-j(k_x'x+k_y'y)}$ and integrating over $x \in (-\infty, \infty)$, $y \in (-\infty, \infty)$ and utilizing equations (15.32) and (15.33) gives

$$k_z\widetilde{A}_x(k_x', k_y') = \frac{\mu}{2j}\iint_{-\infty}^{\infty} J_s(x, y)e^{-j(k_x'x+k_y'y)}\,dxdy \;=:\; \frac{\mu}{2j}\widetilde{J}_s(k_x', k_y'), \tag{15.36}$$

where $\widetilde{J}_s(k_x, k_y)$ is the 2D Fourier transform of the surface current distribution. Substituting into equation (15.30), the required component of the magnetic vector potential is determined for all $z$ in terms of the current distribution as

$$A_x(x, y, z) = \frac{\mu}{8\pi^2 j}\iint_{-\infty}^{\infty} \widetilde{J}_s(k_x, k_y)e^{j(k_x x+k_y y)}e^{-jk_z|z|}\,dk_x dk_y, \tag{15.37}$$

This single component of $A_x$ is adequate to satisfy all boundary conditions at $z = 0$ and at infinity. The complex power flowing through any $z$ = constant plane is

$$\begin{aligned}
P_f &= \frac{1}{2}\iint_{x,y=-\infty}^{\infty} \mathbf{E}\times\mathbf{H}^*\cdot\hat{\mathbf{z}}\,dxdy = \frac{1}{2}\iint_{-\infty}^{\infty} E_x H_y^*\,dxdy \\
&= \frac{1}{32\omega\mu^2\epsilon\pi^4}\iint_{-\infty}^{\infty}\iint_{-\infty}^{\infty}(k^2-k_x^2)(k_z')^*\widetilde{A}_x(k_x, k_y)\widetilde{A}_x^*(k_x', k_y')e^{-j\left(k_z-(k_z')^*\right)z} \\
&\quad \iint_{x,y=-\infty}^{\infty} e^{j\left[(k_x-k_x')x+(k_y-k_y')y\right]}\,dxdy\,dk_x'dk_y'\,dk_x dk_y \\
&= \frac{1}{8\omega\mu^2\epsilon\pi^2}\iint_{-\infty}^{\infty}(k^2-k_x^2)k_z^*\left|\widetilde{A}_x(k_x, k_y)\right|^2 e^{2\Im(k_z)z}\,dk_x dk_y \\
&= \frac{\eta}{32k\pi^2}\iint_{-\infty}^{\infty}\frac{(k^2-k_x^2)}{k_z}\left|\widetilde{J}_s(k_x, k_y)\right|^2 e^{2\Im(k_z)z}\,dk_x dk_y \\
&= \frac{\eta}{32k\pi^2}\left[\iint_{S_r}\frac{(k^2-k_x^2)}{k_z}\left|\widetilde{J}_s(k_x, k_y)\right|^2\,dk_x dk_y\right. \\
&\quad \left.+j\iint_{S_i}\frac{(k^2-k_x^2)}{|k_z|}\left|\widetilde{J}_s(k_x, k_y)\right|^2 e^{-2|k_z|z}\,dk_x dk_y\right] \\
&=: P_r + jP_i(z),
\end{aligned} \tag{15.38}$$

where $S_{\substack{r\\i}}: k_x^2 + k_y^2 \lessgtr k^2$ are regions inside and outside the circle $k_x^2 + k_y^2 = k^2$ (figure 15.3). It is seen that the real part, $P_r$, of the complex power flow is independent of distance, whereas the imaginary part, $P_i$, depends on $z$, having its largest value near the current source.

As an elementary example, for a vertical dipole located at $\mathbf{r}' = (x_0, y_0, 0)$ in the $xy$-plane, $J_s = I_0\ell\delta(x - x_0)\delta(y - y_0)$ (figure 15.4). The vector potential for this is already known to be (see equation (1.47))

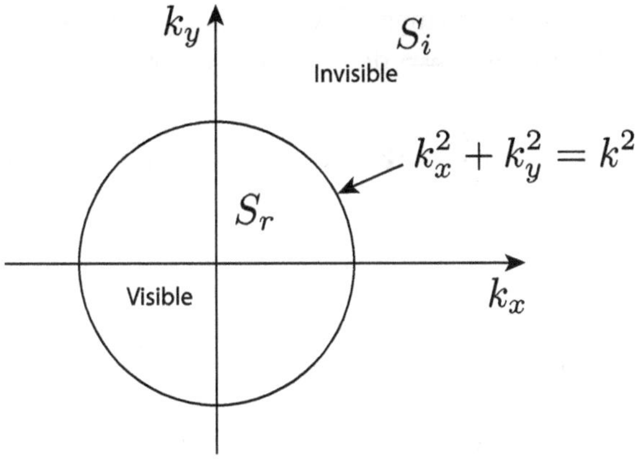


**Figure 15.3.** Integration domain showing visible ($S_r$: $k_x^2 + k_y^2 < k^2$) and invisible ($S_i$: $k_x^2 + k_y^2 > k^2$) regions in the $k_x$–$k_y$-plane.

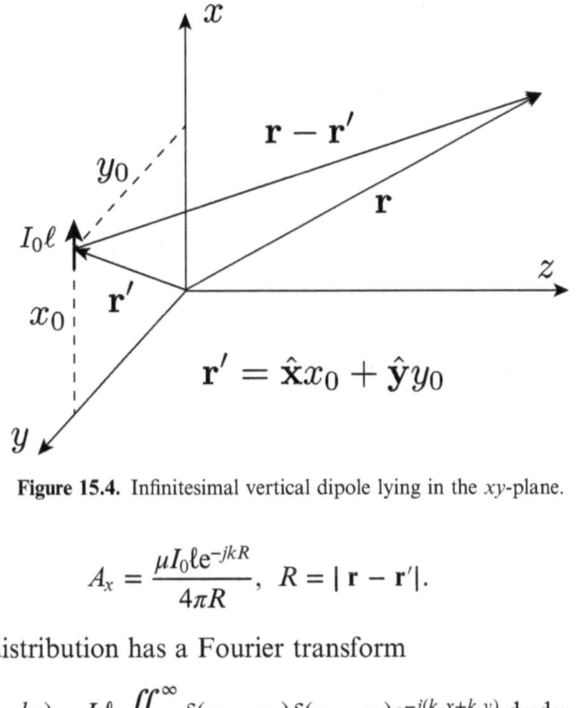


**Figure 15.4.** Infinitesimal vertical dipole lying in the $xy$-plane.

$$A_x = \frac{\mu I_0 \ell \mathrm{e}^{-jkR}}{4\pi R}, \quad R = |\mathbf{r} - \mathbf{r}'|. \tag{15.39}$$

This source distribution has a Fourier transform

$$\begin{aligned} \tilde{J}_s(k_x, k_y) &= I_0\ell \iint_{-\infty}^{\infty} \delta(x - x_0)\delta(y - y_0)\mathrm{e}^{-j(k_x x + k_y y)}\,\mathrm{d}x\mathrm{d}y \\ &= I_0\ell \mathrm{e}^{-j(k_x x_0 + k_y y_0)}. \end{aligned} \tag{15.40}$$

Substituting this into equation (15.37) and equating the two forms of $A_x$ we arrive at the spectral representation given in equation (15.3), i.e.

$$\frac{\mathrm{e}^{-jk|\mathbf{r}-\mathbf{r}'|}}{4\pi|\mathbf{r} - \mathbf{r}'|} = \frac{1}{8j\pi^2} \iint_{-\infty}^{\infty} \mathrm{e}^{j[k_x(x-x_0)+k_y(y-y_0)]} \frac{\mathrm{e}^{-jk_z|z|}}{k_z}\,\mathrm{d}k_x\mathrm{d}k_y. \tag{15.41}$$

Further, making use of the change of variables $k_x = k_\rho \cos\alpha,\ k_y = k_\rho \sin\alpha \Longrightarrow \mathrm{d}k_x\mathrm{d}k_y = k_\rho \mathrm{d}k_\rho \mathrm{d}\alpha,\ k_\rho = k\sin\beta \Longrightarrow \mathrm{d}k_\rho = k\cos\beta\,\mathrm{d}\beta$, we obtain

$$\begin{aligned} P_\mathrm{r} &= \frac{\eta \mid I_0\ell \mid^2}{32k\pi^2} \iint_{k_x^2+k_y^2<k^2} \frac{k^2 - k_x^2}{k_z} \mathrm{d}k_x\mathrm{d}k_y = \frac{\eta \mid I_0\ell \mid^2}{32k\pi} \int_0^k \frac{k_\rho\left(2k^2 - k_\rho^2\right)}{\sqrt{k^2 - k_\rho^2}} \mathrm{d}k_\rho \\ &= \frac{\eta \mid I_0 k\ell \mid^2}{32\pi} \int_0^{\pi/2} (1 + \cos^2\beta)\sin\beta \,\mathrm{d}\beta = \frac{1}{2}\eta\frac{\pi}{3} \left| \frac{I_0\ell}{\lambda} \right|^2, \end{aligned} \tag{15.42}$$

where $\lambda = 2\pi/k$ is the wavelength in the medium. This is identical to the expression provided in [6, p 436] for the power radiated by an elementary dipole in one half-space. Similarly, making use of successive change of variables, $k_\rho = k\cosh\alpha$, $\sinh\alpha = t$, we obtain for the imaginary part

$$\begin{aligned} P_\mathrm{i}(z) &= \frac{\eta \mid I_0\ell \mid^2}{32k\pi} \int_k^\infty \frac{k_\rho\left(2k^2 - k_\rho^2\right)}{\sqrt{k_\rho^2 - k^2}} \mathrm{e}^{-2\sqrt{k_\rho^2 - k^2}\,z} \,\mathrm{d}k_\rho \\ &= \frac{-\eta \mid kI_0\ell \mid^2}{32\pi} \int_0^\infty (t^2 - 1)\mathrm{e}^{-2kzt} \,\mathrm{d}t = -\frac{\eta\pi}{16} \left| \frac{I_0\ell}{\lambda} \right|^2 \frac{1}{kz}\left[1 + \frac{1}{2(kz)^2}\right]. \end{aligned} \tag{15.43}$$

It is seen that the imaginary part of the power flow is negative, indicating that the near fields of an electric dipole are capacitative in nature. Moreover, for large distances the imaginary part decays as $(kz)^{-1}$ so that only the real part prevails in the far-zone.

### 15.2.2 Dipole radiating in the presence of a conducting wedge

Consider a time-harmonic vertical electric dipole radiating in the presence of an infinitely long conducting wedge as shown in figure 15.5. The dipole axis is parallel to the edge of the wedge. The surrounding medium is assumed to be vacuum with electrical parameters $(\varepsilon_0, \mu_0)$. The internal angle of the wedge is $\alpha$ with $0 \leqslant \alpha < \pi$. It is most convenient to work out this problem in the cylindrical coordinate system. The dipole has current moment $I_0\ell$ and is located at a point $(\rho', \phi', z')$. The current density of the dipole is expressed as

$$\begin{aligned} \mathbf{J} &= \hat{\mathbf{z}} I_0\ell \frac{\delta(\rho - \rho')\delta(\phi - \phi')\delta(z - z')}{\rho'} \\ &=: \delta(\rho - \rho')\mathbf{J}_\mathrm{s} =: \hat{\mathbf{z}}\delta(\rho - \rho')\delta(z - z')J_\mathrm{s}(\phi). \end{aligned} \tag{15.44}$$

Due to the invariance of the geometry along the $z$-axis, it is reasonable to assume that the magnetic vector potential has only a $z$-component so that $\mathbf{A} = \hat{\mathbf{z}}A_z$. The correctness or not of this assumption is verified by the satisfaction or not of all the required boundary conditions on the wedge and at infinity. The vector potential satisfies $\nabla^2 A_z + k^2 A_z = -\mu_0\delta(\rho - \rho')\delta(z - z')J_\mathrm{s}$, where $k = \omega/c$, $c = 1/\sqrt{\mu_0\varepsilon_0}$. The complete fields are given by equations (1.44) and (1.45). In particular,

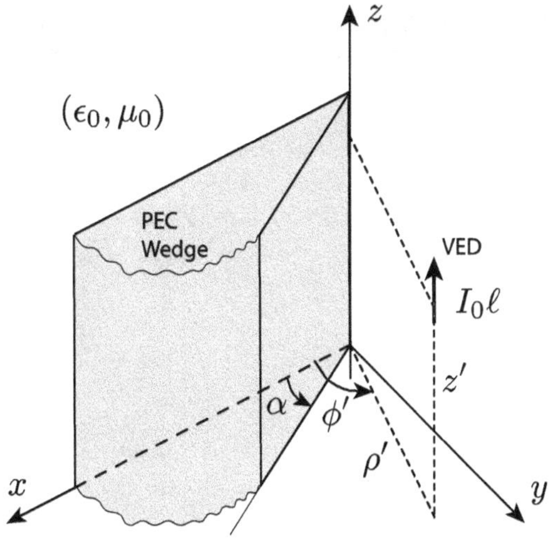


**Figure 15.5.** Dipole radiation in the presence of an infinitely long perfectly conducting wedge.

$$E_z = \frac{c}{jk}\left(\frac{\partial^2}{\partial z^2} + k^2\right)A_z, \quad H_\phi = -\frac{1}{\mu_0}\frac{\partial A_z}{\partial \rho}, \quad H_\rho = \frac{1}{\mu_0 \rho}\frac{\partial A_z}{\partial \phi}, \quad H_z = 0. \tag{15.45}$$

The absence of $H_z$ in all space means that the field can be classified as $\mathrm{TM}_z$. The invariance of geometry along the $z$-axis permits an expression of $A_z$ as a Fourier transform with respect to the variable $(z - z')$

$$A_z(\rho, \phi, z) = \frac{1}{2\pi}\int_{k_z=-\infty}^{\infty} \widetilde{A}_z(\rho, \phi, k_z)\mathrm{e}^{-k_z(z-z')}\,\mathrm{d}k_z, \tag{15.46}$$

where $\widetilde{A}_z$ is the transformed potential. Due to the orthogonality of the kernel $\mathrm{e}^{-jk_z z}$ over $z \in (-\infty, \infty)$ and completeness of the Fourier transform as expressed in equation (15.32), all derivatives and boundary conditions involving the $(\rho, \phi)$ coordinates to be satisfied by $A_z$ may be transferred over to the transformed potential $\widetilde{A}_z$. Introducing the transverse Laplacian $\nabla_t^2 = \nabla^2 - \frac{\partial^2}{\partial z^2}$, it is easy to see that $\widetilde{A}_z$ satisfies

$$\nabla_t^2 \widetilde{A}_z + k_\rho^2 \widetilde{A}_z = -\mu_0 \delta(\rho - \rho')J_s(\phi), \tag{15.47}$$

where $k_\rho^2 = k^2 - k_z^2$. We now expand $\widetilde{A}_z$ in cylindrical harmonics as

$$\widetilde{A}_z = \mu_0 I_0 \ell \sum_{\nu > 0} J_\nu(k_\rho \rho_<)H_\nu^{(2)}(k_\rho \rho_>)[B_\nu^\pm \cos \nu(\phi - \phi') + C_\nu^\pm \sin \nu(\phi - \phi')], \tag{15.48}$$

where $\rho_< = \min(\rho, \rho')$, $\rho_> = \max(\rho, \rho')$ and $B_\nu^\pm$, $C_\nu^\pm$ are unknown constants, defined for $\rho \gtrless \rho'$, to be determined by the boundary conditions. Note that the index $\nu$ is not necessarily an integer. Also the trigonometric form in $\phi$ (as opposed to an exponential form) suggests that negative values of $\nu$ are not required. To satisfy radiation conditions at infinity we require $k_\rho$ to have a positive real part and a negative imaginary part as $k_z$ traverses $(-\infty, \infty)$ along the real axis. The continuity of $E_z$ at $\rho = \rho'$ implies

$$B_\nu^+ = B_\nu^- =: B_\nu; \qquad C_\nu^+ = C_\nu^- =: C_\nu. \tag{15.49}$$

Discontinuity of the magnetic field $\hat{\mathbf{n}} \times (\mathbf{H}(\rho' +) - \mathbf{H}(\rho' -)) = \mathbf{J}_s$ implies $H_\phi(\rho' +) - H_\phi(\rho' -) = \delta(z - z')J_s(\phi)$, which when translated to $\widetilde{A}_z$ via (15.45) further implies

$$\sum_\nu [B_\nu \cos \nu(\phi - \phi') + C_\nu \sin \nu(\phi - \phi')] = \frac{\pi}{2j}\delta(\phi - \phi'), \tag{15.50}$$

where the Wronskian $\mathscr{W}\{J_\nu(z), H_\nu^{(2)}(z)\} = \frac{2}{j\pi z}$ from equation (C.6) was utilized in arriving at the final result. Finally the boundary condition on the PEC wedge $E_z(\phi = \alpha) = 0 = E_z(\phi = 2\pi)$ yields

$$B_\nu \cos \nu(\alpha - \phi') + C_\nu \sin \nu(\alpha - \phi') = 0, \tag{15.51}$$

$$B_\nu \cos \nu(2\pi - \phi') + C_\nu \sin \nu(2\pi - \phi') = 0. \tag{15.52}$$

Therefore
$\cos \nu(\alpha - \phi')\sin \nu(2\pi - \phi') - \sin \nu(\alpha - \phi')\cos \nu(2\pi - \phi') = \sin \nu(2\pi - \alpha) = 0$ or

$$\nu = \frac{m\pi}{2\pi - \alpha} =: \nu_m, \qquad m = 0, 1, \ldots. \tag{15.53}$$

From equation (15.51) we express $B_\nu$ and $C_\nu$ in terms of a single constant $D_\nu$ as

$$B_\nu \cos \nu(\alpha - \phi') = -C_\nu \sin \nu(\alpha - \phi') =: D_\nu \cos \nu(\alpha - \phi')\sin \nu(\alpha - \phi'), \tag{15.54}$$

so that $B_\nu \cos \nu(\phi - \phi') + C_\nu \sin \nu(\phi - \phi') = D_\nu \sin \nu(\alpha - \phi)$. Hence equation (15.50) may be expressed with $\nu = \nu_m$ as

$$\sum_{m=1}^{\infty} D_{\nu_m} \sin \nu_m(\alpha - \phi) = \frac{\pi}{2j}\delta(\phi' - \phi). \tag{15.55}$$

Note that

$$\begin{aligned}\int_{\phi=\alpha}^{2\pi} \sin \nu_m(\alpha - \phi)\sin \nu_n(\alpha - \phi)\, \mathrm{d}\phi &= \frac{1}{2}\left[\frac{\sin(\nu_m - \nu_n)(2\pi - \alpha)}{\nu_m - \nu_n} - \frac{\sin(\nu_m + \nu_n)(2\pi - \alpha)}{\nu_m + \nu_n}\right] \\ &= \frac{2\pi - \alpha}{2}\left[\frac{\sin(m - n)\pi}{(m - n)\pi} - \frac{\sin(m + n)\pi}{(m + n)\pi}\right] \\ &= \delta_m^n \frac{2\pi - \alpha}{2}, \qquad m, n \geqslant 1.\end{aligned} \tag{15.56}$$

Hence $\sin \nu_m(\alpha - \phi)$, $\nu_m = m\pi/(2\pi - \alpha)$, $m = 1, 2, \ldots$ are orthogonal functions in the interval $\phi \in (\alpha, 2\pi)$. Multiplying equation (15.55) with $\sin \nu_n(\alpha - \phi)$, integrating over $\phi \in (\alpha, 2\pi)$ and utilizing equation (15.56) gives

$$D_{\nu_n} = \frac{\pi}{j(2\pi - \alpha)} \sin \nu_n(\alpha - \phi'). \tag{15.57}$$

Inserting this into equation (15.46) finally yields the expression for the magnetic vector potential

$$A_z = A_0 \int_{-\infty}^{\infty} \sum_{m=1}^{\infty} J_{\nu_m}(k_\rho \rho_<) H_{\nu_m}^{(2)}(k_\rho \rho_>) \sin \nu_m(\phi' - \alpha) \sin \nu_m(\phi - \alpha) \mathrm{e}^{-jk_z(z-z')}\, \mathrm{d}k_z, \tag{15.58}$$

where $A_0 = \mu_0 I_0 \ell / 2j(2\pi - \alpha)$, $k_\rho = \sqrt{k^2 - k_z^2} = -j\sqrt{k_z^2 - k^2}$. The fields everywhere can be determined from equation (15.45). In particular, the various fields for $\rho < \rho'$ are

$$\begin{aligned} E_z &= \frac{A_0 c}{jk} \int_{-\infty}^{\infty} \sum_{m=1}^{\infty} J_{\nu_m}(k_\rho \rho) H_{\nu_m}^{(2)}(k_\rho \rho') \\ &\quad \times \sin \nu_m(\phi' - \alpha) \sin \nu_m(\phi - \alpha) \mathrm{e}^{-jk_z(z-z')}\, \mathrm{d}k_z \\ &=: \frac{1}{2\pi} \int_{k_z=-\infty}^{\infty} \widetilde{E}_z \mathrm{e}^{-jk_z z}\, \mathrm{d}k_z, \end{aligned} \tag{15.59}$$

$$\begin{aligned} H_\phi &= -\frac{A_0}{\mu_0} \int_{-\infty}^{\infty} \sum_{m=1}^{\infty} J_{\nu_m}'(k_\rho \rho) H_{\nu_m}^{(2)}(k_\rho \rho') \\ &\quad \times \sin \nu_m(\phi' - \alpha) \sin \nu_m(\phi - \alpha) \mathrm{e}^{-jk_z(z-z')}\, \mathrm{d}k_z \\ &=: \frac{1}{2\pi} \int_{k_z=-\infty}^{\infty} \widetilde{H}_\phi \mathrm{e}^{-jk_z z}\, \mathrm{d}k_z, \end{aligned} \tag{15.60}$$

$$\begin{aligned} H_\rho &= \frac{A_0}{\mu_0 \rho} \int_{-\infty}^{\infty} \sum_{m=1}^{\infty} \nu_m J_{\nu_m}(k_\rho \rho) H_{\nu_m}^{(2)}(k_\rho \rho') \\ &\quad \times \sin \nu_m(\phi' - \alpha) \cos \nu_m(\phi - \alpha) \mathrm{e}^{-jk_z(z-z')}\, \mathrm{d}k_z. \\ &=: \frac{1}{2\pi} \int_{k_z=-\infty}^{\infty} \widetilde{H}_\rho \mathrm{e}^{-jk_z z}\, \mathrm{d}k_z. \end{aligned} \tag{15.61}$$

Using the small argument approximations of the Bessel function from equation (C.9), it is seen that the transformed fields behave as

$$\widetilde{E}_z \sim O(\rho^{\nu_m}) \tag{15.62}$$

$$\widetilde{H}_\phi \sim O(\rho^{\nu_m - 1}) \tag{15.63}$$

$$\widetilde{H}_\rho \sim O(\rho^{\nu_m - 1}), \tag{15.64}$$

as $\rho \to 0$. Since the smallest value of $\nu_m$ is $0 < \nu_1 = \pi/(2\pi - \alpha) < 1$ and $-1 < \nu_1 - 1 = -(\pi - \alpha)/(2\pi - \alpha) < 0$, the components of magnetic field normal to the edge, viz., $\widetilde{H}_\phi$ and $\widetilde{H}_\rho$ are singular at $\rho = 0$. The component of electric field parallel to the edge, i.e. $\widetilde{E}_z$ is finite near the edge. These conditions are known as *edge conditions* for fields near a conducting edge. The magnetic energy density is also singular near the edge. However, the total energy contained in a unit-height pillbox centered around the edge $\alpha \leqslant \phi \leqslant 2\pi$, $0 \leqslant \rho \leqslant a$, $0 \leqslant z \leqslant 1$ is still finite as $\rho \widetilde{H}_\tau \widetilde{H}_\tau^* \sim O(\rho^{2\nu_1 - 1})$ for $\tau = \rho, \phi$ and

$$\int_{\rho=0}^{a} \int_{\phi=\alpha}^{2\pi} \left( \left| \widetilde{H}_\phi \right|^2 + \left| \widetilde{H}_\rho \right|^2 \right) \rho \, \mathrm{d}\rho \mathrm{d}\phi \sim O(a^{2\nu_1}) \to 0 \text{ as } a \to 0. \tag{15.65}$$

### 15.2.3 Circumferential magnetic dipole near a conducting circular cylinder

Figure 15.6 shows a time-harmonic, circumferential magnetic dipole of moment $V_0\ell$ and located at cylindrical coordinates $(\rho', \phi', z')$. The dipole radiates in the presence of an infinitely long PEC circular cylinder with its axis along the $z$-axis. The surrounding medium has constants $(\varepsilon, \mu)$. From the given geometry it clear that the most convenient coordinate system to address the problem is the cylindrical coordinate system. The current density of the source is

$$\mathbf{M} = \hat{\boldsymbol{\phi}}' V_0 \ell \frac{\delta(\phi - \phi')}{\rho'} \delta(z - z') \delta(\rho - \rho') =: \mathbf{m}_s(\rho', \phi', z') \delta(\rho - \rho'). \tag{15.66}$$

We would like to determine the fields at all points exterior to the cylinder by expanding the fields separately for $\rho < \rho'$ and $\rho > \rho'$ and determining the constants by imposing the appropriate boundary conditions at the source and on the cylinder surface. This problem is of interest in the understanding and study of slot antennas deployed on cylindrical structures [3, 7].

Because the source is of finite extent and its orientation does not coincide with the axis of the cylinder as in the previous case, both magnetic and electric vector potentials will be retained from the outset to accommodate all boundary conditions.

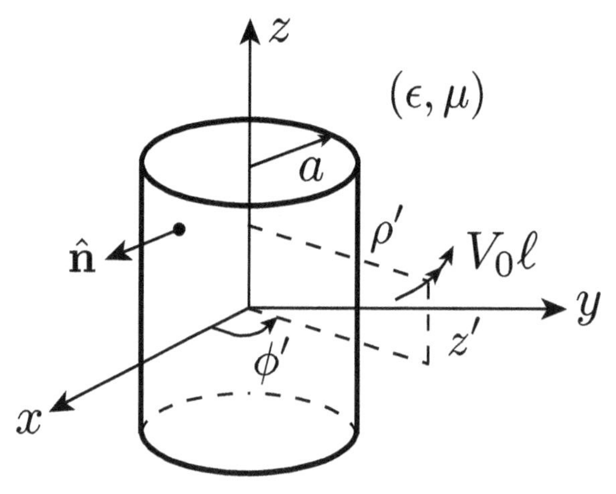


**Figure 15.6.** Circumferential magnetic dipole radiating in the presence of an infinitely long conducting circular cylinder.

In the event that the presence of both parts is not required for a specific orientation, the redundant component will drop out automatically. Accordingly, we let $\mathbf{A} = \hat{\mathbf{z}}A_z$ *and* $\mathbf{F} = \hat{\mathbf{z}}F_z$ and note that the field will be neither $TE_z$ nor $TM_z$. In each of the regions $\rho < \rho'$ and $\rho > \rho'$, the potentials $A_z$ and $F_z$ satisfy the homogeneous Helmholtz equation $\nabla^2\psi + k^2\psi = 0$, where $k = \omega/v, v = 1/\sqrt{\varepsilon\mu}$. The complete fields in terms of $\mathbf{A}$ and $\mathbf{F}$ are given by equations (1.44) and (1.45). In particular,

$$E_\rho = \frac{v}{jk}\frac{\partial^2 A_z}{\partial\rho\partial z} - \frac{1}{\varepsilon\rho}\frac{\partial F_z}{\partial\phi}; \quad H_\rho = \frac{1}{\mu\rho}\frac{\partial A_z}{\partial\phi} + \frac{v}{jk}\frac{\partial^2 F_z}{\partial\rho\partial z} \tag{15.67}$$

$$E_\phi = \frac{v}{jk\rho}\frac{\partial^2 A_z}{\partial\phi\partial z} + \frac{1}{\varepsilon}\frac{\partial F_z}{\partial\rho}; \quad H_\phi = -\frac{1}{\mu}\frac{\partial A_z}{\partial\rho} + \frac{v}{jk\rho}\frac{\partial^2 F_z}{\partial\phi\partial z} \tag{15.68}$$

$$E_z = \frac{v}{jk}\left(\frac{\partial^2}{\partial z^2} + k^2\right)A_z; \quad H_z = \frac{v}{jk}\left(\frac{\partial^2}{\partial z^2} + k^2\right)F_z. \tag{15.69}$$

Due to the invariance of the geometry along the $z$-axis, it is convenient to express the potentials in terms of Fourier transformation with respect to the $z$-axis. Furthermore, the periodicity of fields in the azimuthal direction permits a Fourier series expansion in the $\phi$-direction. Accordingly, we represent

$$A_z^- = \sum_{n=-\infty}^{\infty} \int_{k_z=-\infty}^{\infty} \left[C_n^- J_n(k_\rho\rho) + A_n^- H_n^{(2)}(k_\rho\rho)\right] \times \mathrm{e}^{jn(\phi-\phi')}\mathrm{e}^{-jk_z(z-z')}\,\mathrm{d}k_z, \quad a < \rho < \rho'$$

$$A_z^+ = \sum_{n=-\infty}^{\infty} \int_{k_z=-\infty}^{\infty} A_n^+ H_n^{(2)}(k_\rho\rho)\mathrm{e}^{jn(\phi-\phi')}\mathrm{e}^{-jk_z(z-z')}\,\mathrm{d}k_z, \quad \rho > \rho',$$

and

$$F_z^- = \sum_{n=-\infty}^{\infty} \int_{k_z=-\infty}^{\infty} \left[D_n^- J_n(k_\rho\rho) + B_n^- H_n^{(2)}(k_\rho\rho)\right] \times \mathrm{e}^{jn(\phi-\phi')}\mathrm{e}^{-jk_z(z-z')}\,\mathrm{d}k_z, \quad a < \rho < \rho'$$

$$F_z^+ = \sum_{n=-\infty}^{\infty} \int_{k_z=-\infty}^{\infty} B_n^+ H_n^{(2)}(k_\rho\rho)\mathrm{e}^{jn(\phi-\phi')}\mathrm{e}^{-jk_z(z-z')}\,\mathrm{d}k_z, \quad \rho > \rho',$$

where $k_\rho = \sqrt{k^2 - k_z^2} = -j\sqrt{k_z^2 - k^2}$ in order to satisfy radiation conditions at infinity. The potentials in this form automatically satisfy the homogeneous Helmholtz equation. Vanishing of $E_z$ at $\rho = a$ implies

$$C_n^- J_n(k_\rho a) = -A_n^- H_n^{(2)}(k_\rho a) =: K_n J_n(k_\rho a) H_n^{(2)}(k_\rho a), \tag{15.70}$$

where $K_n$ is a constant to be determined. This together with vanishing of $E_\phi$ at $\rho = 0$ implies

$$D_n^- J_n'(k_\rho a) = -B_n^- H_n^{(2)'}(k_\rho a) =: L_n J_n'(k_\rho a) H_n^{(2)'}(k_\rho a), \tag{15.71}$$

where $L_n$ is a constant to be determined. Substituting these, the potentials in the region $a < \rho < \rho'$ are simplified to

$$\begin{aligned} A_z^- = \sum_{n=-\infty}^{\infty} \int_{k_z=-\infty}^{\infty} &\left[J_n(k_\rho \rho) H_n^{(2)}(k_\rho a) - J_n(k_\rho a) H_n^{(2)}(k_\rho \rho)\right] \\ &\times K_n e^{jn(\phi-\phi')} e^{-jk_z(z-z')}\, dk_z, \\ F_z^- = \sum_{n=-\infty}^{\infty} \int_{k_z=-\infty}^{\infty} &\left[J_n(k_\rho \rho) H_n^{(2)'}(k_\rho a) - J_n'(k_\rho a) H_n^{(2)}(k_\rho \rho)\right] \\ &\times L_n e^{jn(\phi-\phi')} e^{-jk_z(z-z')}\, dk_z. \end{aligned}$$

Continuity of $H_z$ at $\rho = \rho'$ implies continuity of $F_z$ at $\rho = \rho'$, which, in turn, implies

$$B_n^+ = \frac{\left[J_n(k_\rho \rho') H_n^{(2)'}(k_\rho a) - J_n'(k_\rho a) H_n^{(2)}(k_\rho \rho')\right]}{H_n^{(2)}(k_\rho \rho')} L_n =: f_n L_n. \tag{15.72}$$

This together with continuity of $H_\phi$ at $\rho = \rho'$ implies

$$A_n^+ = \frac{\left[J_n'(k_\rho \rho') H_n^{(2)}(k_\rho a) - J_n(k_\rho a) H_n^{(2)'}(k_\rho \rho')\right]}{H_n^{(2)'}(k_\rho \rho')} K_n =: g_n K_n. \tag{15.73}$$

If the source is situated on the cylinder surface, then $\rho' = a$ and the quantities $f_n$ and $g_n$ simplify to $f_n = 2/j\pi k_\rho a H_n^{(2)}(k_\rho a)$ and $g_n = 2j/\pi k_\rho a H_n^{(2)'}(k_\rho a)$ owing to the Wronskian (C.6).

Continuity of $E_\phi$ at $\rho = \rho'$, together with substitution from equations (15.72) and (15.73), and the Wronskian (C.6) gives

$$k_\rho \frac{H_n^{(2)}(k_\rho a) J_n'(k_\rho \rho') - J_n(k_\rho a) H_n^{(2)'}(k_\rho \rho')}{H_n^{(2)'}(k_\rho a) J_n(k_\rho \rho') - J_n'(k_\rho a) H_n^{(2)}(k_\rho \rho')} H_n^{(2)'}(k_\rho a) B_n^+ = \frac{n k_z}{jk\rho' \eta} H_n^{(2)}(k_\rho a) A_n^+, \tag{15.74}$$

where $\eta = \sqrt{\mu/\varepsilon}$. Note that for $k_z = 0$, the coefficients $B_n^+ = 0$ and, hence, $L_n = 0$. Therefore, if the source is $z$-invariant, only $A_n^+$ and $K_n$ survive, i.e. only a TM$_z$ mode is excited. Finally, $E_z(\rho' +) - E_z(\rho' -) = \hat{\boldsymbol{\phi}} \cdot \mathbf{m}_s$ implies

$$\begin{aligned} \frac{V_0 \ell \delta(\phi - \phi')\delta(z - z')}{\rho'} &= \frac{v}{jk} \sum_{-\infty}^{\infty} \int_{-\infty}^{\infty} k_\rho^2 A_n^+ e^{jn(\phi-\phi')} e^{-jk_z(z-z')} \left[H_n^{(2)}(k_\rho \rho') - \frac{1}{g_n}\right] \\ &\quad \left[H_n^{(2)}(k_\rho a) J_n(k_\rho \rho') - J_n(k_\rho a) H_n^{(2)}(k_\rho \rho')\right] dk_z \\ &= \frac{2v}{\pi k \rho'} \sum_{-\infty}^{\infty} \int_{-\infty}^{\infty} \frac{k_\rho A_n^+ H_n^{(2)}(k_\rho a) e^{jn(\phi-\phi')} e^{-jk_z(z-z')}}{g_n H_n^{(2)'}(k_\rho \rho')} dk_z. \end{aligned}$$

Multiplying both sides of this equation with $e^{-jm(\phi-\phi')} e^{jk'z(z-z')}$, with $m$ an integer, and integrating over $\phi \in (0, 2\pi)$, $z \in (-\infty, \infty)$, and using the identities

$$\frac{1}{2\pi}\int_0^{2\pi} e^{j(n-m)(\phi-\phi')}\,d\phi = \delta_m^n; \quad \frac{1}{2\pi}\int_{-\infty}^{\infty} e^{j(k_z-k_z')(z-z')}\,dz = \delta(k_z - k_z'), \tag{15.75}$$

finally yields

$$A_n^+ = \frac{V_0\ell k}{8\pi v k_\rho}\frac{H_n^{(2)'}(k_\rho\rho')}{H_n^{(2)}(k_\rho a)}g_n. \tag{15.76}$$

The other coefficients are given from equations (15.72)–(15.74) and are

$$K_n = \frac{V_0\ell k}{8\pi v k_\rho}\frac{H_n^{(2)'}(k_\rho\rho')}{H_n^{(2)}(k_\rho a)}, \tag{15.77}$$

$$B_n^+ = \frac{jnk_z\varepsilon V_0\ell}{8\pi k_\rho^2\rho'}\frac{H_n^{(2)}(k_\rho\rho')}{H_n^{(2)'}(k_\rho a)}f_n, \tag{15.78}$$

$$L_n = \frac{jnk_z\varepsilon V_0\ell}{8\pi k_\rho^2\rho'}\frac{H_n^{(2)}(k_\rho\rho')}{H_n^{(2)'}(k_\rho a)}. \tag{15.79}$$

For the special case of $\rho' = a$, the magnetic dipole is mounted directly on the PEC cylinder and the coefficients become

$$A_n^+ = \frac{jV_0\ell k}{4\pi^2 k_\rho^2 a v H_n^{(2)}(k_\rho a)}, \tag{15.80}$$

$$K_n = \frac{V_0\ell k}{8\pi v k_\rho}\frac{H_n^{(2)'}(k_\rho a)}{H_n^{(2)}(k_\rho a)}, \tag{15.81}$$

$$B_n^+ = n\frac{k_z}{k_\rho}\frac{\varepsilon V_0\ell}{(2\pi k_\rho a)^2}\frac{1}{H_n^{(2)'}(k_\rho a)}, \tag{15.82}$$

$$L_n = jn\frac{k_z}{k_\rho}\frac{\varepsilon V_0\ell}{8\pi k_\rho a}\frac{H_n^{(2)}(k_\rho a)}{H_n^{(2)'}(k_\rho a)}. \tag{15.83}$$

Note that coefficients $A_n^+$ and $K_n$ have units of V s, whereas the coefficients $B_n^+$ and $L_n$ have units of A s.

As an application of the theory consider a uniformly excited circumferential slot that runs completely around the conducting cylinder. The slot has a height $2w \ll a, \lambda$ and has a specified tangential electric field $\mathbf{E}_a = \hat{\mathbf{z}}V/2w$, $0 \leqslant \phi \leqslant 2\pi$, $|z| < w$. The corresponding magnetic current in the slot is $\mathbf{M}_s = \hat{\rho} \times \mathbf{E}_a = \hat{\boldsymbol{\phi}}V/2w$, $0 \leqslant \phi, \leqslant 2\pi$, $|z| < w$. The quantity $V$ may be interpreted as the voltage across the slot. In terms of elementary circumferential dipoles the current $\mathbf{M}_s$ may be expressed as

$$\mathbf{M}_s = \frac{V}{2w}\int_{\phi'=0}^{2\pi}\int_{z'=-\infty}^{\infty} p_w(z')p_\pi(\phi' - \pi)\frac{\mathbf{m}_s(a, \phi', z')}{V_0\ell}dz'ad\phi',$$
$$= \frac{Va}{2wV_0\ell}\int_{\phi'=0}^{2\pi}\int_{z'=-w}^{w}\mathbf{m}_s(a, \phi', z')dz'd\phi' \tag{15.84}$$

where $p_\Delta(x)$ is a unit rectangular pulse of width $2\Delta$ centered at $x = 0$. By linear superposition the solution for the specified $\mathbf{M}_s$ is the integral of the solution for $\mathbf{m}_s$ with the weight factor $Va/2wV_0\ell$. As a result, integration of the factor $\frac{1}{2w}e^{jk_z z'}$ in the potentials w.r.t. $z'$ gives the factor $\sin(k_z w)/k_z w$. Integration of the factor $ae^{-jn\phi'}$ in the potentials w.r.t. $\phi'$ gives the factor $2\pi a\delta_n^0$. Hence only the $n = 0$ term is retained in the coefficients. In particular, all $B_n^+$ and $L_n$ vanish and the fields become purely $TM_z$ with

$$A_z = \frac{jVk}{2\pi v}\int_{-\infty}^{\infty}\frac{1}{k_\rho^2}\frac{H_0^{(2)}(k_\rho\rho)}{H_0^{(2)}(k_\rho a)}\frac{\sin k_z w}{k_z w}e^{-jk_z z}\,dk_z, \quad \rho \geqslant a \tag{15.85}$$

$$E_z = \frac{V}{2\pi}\int_{-\infty}^{\infty}\frac{H_0^{(2)}(k_\rho\rho)}{H_0^{(2)}(k_\rho a)}\frac{\sin k_z w}{k_z w}e^{-jk_z z}\,dk_z, \quad \rho \geqslant a. \tag{15.86}$$

For $\rho = a$ one recovers

$$E_z = V\frac{1}{2\pi}\int_{-\infty}^{\infty}\frac{\sin k_z w}{k_z w}e^{-jk_z z}\,dk_z = \frac{V}{2w}p_w(z). \tag{15.87}$$

### 15.2.4 Dipole radiating over a PEC-backed conducting slab

#### *15.2.4.1 Vertical electric dipole*

Figure 15.7 shows a *vertical electric dipole* (VED) placed over a PEC-backed lossy slab of thickness $d$. The source dipole is placed at $z'$ along the $z$-axis and is vertically

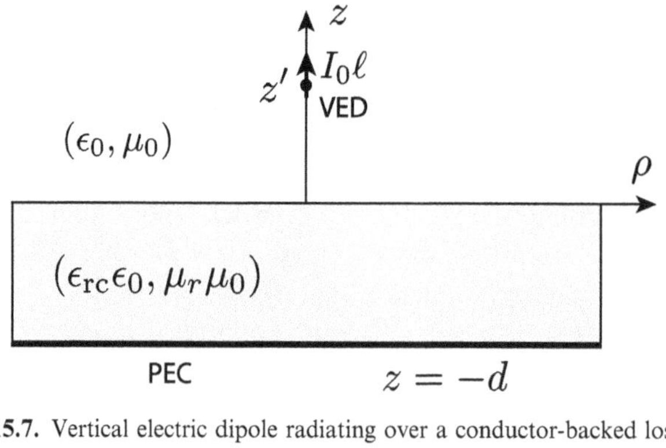


**Figure 15.7.** Vertical electric dipole radiating over a conductor-backed lossy slab.

polarized. The material is characterized by a complex relative permittivity $\varepsilon_{rc} = \varepsilon_r - j\sigma_r$, $\sigma_r > 0$ and a relative permeability $\mu_r$. For $d \to \infty$, the influence of the PEC diminishes in view of $\sigma_r > 0$, and the problem reduced to that of a dipole radiating over an imperfectly conducting half-space. The field is purely TM$_z$ with a $z$-directed magnetic vector potential $\mathbf{A} = \hat{\mathbf{z}}A_z$. Recognizing that the horizontal wavenumber must match in all regions and that the potential is $\phi$-invariant, we express the incident, $A_z^{\mathrm{i}}$, scattered, $A_z^{\mathrm{s}}$, and transmitted, $A_z^{\mathrm{t}}$, potentials in terms of the lateral wavenumber using the representation (15.8) as

$$A_z^{\mathrm{i}} = \frac{\mu_0 I_0 \ell}{4\pi j}\int_0^\infty \frac{k_\rho}{k_{z0}} J_0(k_\rho \rho)\mathrm{e}^{-jk_{z0}|z-z'|}\,\mathrm{d}k_\rho, \qquad z > 0, \tag{15.88}$$

$$A_z^{\mathrm{s}} = \frac{\mu_0 I_0 \ell}{4\pi j}\int_0^\infty \Gamma_{\mathrm{v}}(k_\rho)\frac{k_\rho}{k_{z0}} J_0(k_\rho \rho)\mathrm{e}^{-jk_{z0}(z+z')}\,\mathrm{d}k_\rho, \qquad z > 0, \tag{15.89}$$

where $\Gamma_{\mathrm{v}}(k_\rho)$ is the reflection coefficient for vertical polarization and

$$\begin{aligned} A_z^{\mathrm{t}} = {} & \frac{\mu_0 I_0 \ell}{4\pi j}\int_0^\infty \frac{k_\rho}{k_{z0}} J_0(k_\rho \rho)\mathrm{e}^{-jk_{z0}z'} \\ & \times [a\mathrm{e}^{-jk_z(z+d)} + b\mathrm{e}^{jk_z(z+d)}]\,\mathrm{d}k_\rho, \qquad -d \leqslant z \leqslant 0. \end{aligned} \tag{15.90}$$

Note that the expansions in the lateral wavenumber here involve Bessel functions of the first kind (as opposed to Hankel functions), whose presence, however, does not violate the radiation conditions (see example 16.5.0.1). The wavenumber in the upper medium is equal to $k = \omega\sqrt{\mu_0\varepsilon_0} =: k_0$ and that in the material medium is equal to $k = \omega\sqrt{\mu\varepsilon}$. Likewise the wave-speed in the upper medium is $v = 1/\sqrt{\mu_0\varepsilon_0} =: c$, while that in the slab medium is $v = c/\sqrt{\varepsilon_{rc}\mu_r}$. The total potential in the upper region is $A_z^{\mathrm{i}} + A_z^{\mathrm{s}}$. The various fields in terms of the potentials are

$$\begin{aligned} E_z &= \frac{v}{jk}\left(\frac{\partial^2}{\partial z^2} + k^2\right)A_z; \qquad H_z = 0, \\ E_\rho &= \frac{v}{jk}\frac{\partial^2 A_z}{\partial\rho\partial z}; \qquad H_\rho = \frac{1}{\mu\rho}\frac{\partial A_z}{\partial\phi} = 0, \\ E_\phi &= \frac{v}{jk\rho}\frac{\partial^2 A_z}{\partial\phi\partial z} = 0; \qquad H_\phi = -\frac{1}{\mu}\frac{\partial A_z}{\partial\rho}, \end{aligned} \tag{15.91}$$

where $k_{z0} = \sqrt{k_0^2 - k_\rho^2} = -j\sqrt{k_\rho^2 - k_0^2}$ and $k_z = \sqrt{k^2 - k_\rho^2} = -j\sqrt{k_\rho^2 - k^2}$. Due to the azimuthal symmetry, the only non-zero components of fields for a VED are $E_\rho$, $E_z$, and $H_\phi$. Note that since the dipole in free-space radiates power flowing both in the positive and negative $z$-directions about the plane $z = z'$ and the normal component of incident power flow $P^{\mathrm{i}} \propto E_\rho^{\mathrm{i}}H_\phi^{\mathrm{i}*}$, the components of the incident fields $E_\rho^{\mathrm{i}}$, $H_\phi^{\mathrm{i}}$ cannot both be continuous about the plane $z = z'$. It is seen from equation

(15.91) that $H_\phi^{\mathrm{i}}$ is continuous at $z = z'$ if $A_z^{\mathrm{i}}$ is continuous there, but $E_\rho^{\mathrm{i}}$ can have a discontinuity[3]. Imposing the boundary condition $E_\rho = 0$ at $z = -d$ implies, in view of the orthogonality condition (C.54), that

$$b = a =: T_{\mathrm{v}}(k_\rho)\mathrm{e}^{-jk_z d}, \tag{15.92}$$

where $T_{\mathrm{v}}$ is a quantity proportion to the transmission coefficient for vertical polarization. With this substitution

$$A_z^{\mathrm{t}} = \frac{\mu_0 I_0 \ell}{2\pi j}\int_0^\infty \frac{k_\rho}{k_{z0}} J_0(k_\rho \rho)\mathrm{e}^{-j(k_z d + k_{z0}z')} T_{\mathrm{v}} \cos[k_z(z + d)]\,\mathrm{d}k_\rho. \tag{15.93}$$

Note that the continuity of $E_\rho$ and $H_\phi$ at $z = 0$ implies

$$jk_{z0}(1 - \Gamma_{\mathrm{v}}) = \frac{-2T_{\mathrm{v}}k_z\mathrm{e}^{-jk_z d}}{\mu_{\mathrm{r}}\varepsilon_{\mathrm{rc}}}\sin k_z d, \tag{15.94}$$

$$(1 + \Gamma_{\mathrm{v}}) = \frac{2T_{\mathrm{v}}\mathrm{e}^{-jk_z d}}{\mu_{\mathrm{r}}}\cos k_z d. \tag{15.95}$$

Solving for $\Gamma_{\mathrm{v}}$ and $T_{\mathrm{v}}$ using equation (15.94) by equation (15.95) yields

$$\Gamma_{\mathrm{v}} = \frac{\varepsilon_{\mathrm{rc}}k_{z0} - jk_z \tan k_z d}{\varepsilon_{\mathrm{rc}}k_{z0} + jk_z \tan k_z d}, \tag{15.96}$$

$$T_{\mathrm{v}} = \frac{\mu_{\mathrm{r}}\varepsilon_{\mathrm{rc}}k_{z0}\mathrm{e}^{jk_z d}}{\varepsilon_{\mathrm{rc}}k_{z0}\cos k_z d + jk_z \sin k_z d}. \tag{15.97}$$

Writing $k_z = \beta - j\alpha$, $\alpha > 0$, it is easy to see that $\cos k_z d \to \mathrm{e}^{jk_z d}/2$, $\sin k_z d \to -j\mathrm{e}^{jk_z d}/2$, and $\tan k_z d \to -j$, as $d \to \infty$. Hence for a material half-space

$$\Gamma_{\mathrm{v}} = \frac{\varepsilon_{\mathrm{rc}}k_{z0} - k_z}{\varepsilon_{\mathrm{rc}}k_{z0} + k_z}, \tag{15.98}$$

$$T_{\mathrm{v}} = \frac{2\mu_{\mathrm{r}}\varepsilon_{\mathrm{rc}}k_{z0}}{\varepsilon_{\mathrm{rc}}k_{z0} + k_z}. \tag{15.99}$$

---

[3] Indeed by assuming continuity of $A_z^{\mathrm{i}}$ about $z = z'$, writing $\nabla^2 = \nabla_{\mathrm{t}}^2 + \partial^2/\partial z^2$, where $\nabla_{\mathrm{t}}^2$ is the transverse Laplacian involving derivatives only w.r.t. $(\rho, \phi)$, and integrating $(\nabla^2 + k^2)A_z^{\mathrm{i}} = \delta(z - z')q(\rho, \phi)$ with respect to $z$ about the interval $z \in (z' - , z' + )$, one obtains for the jump in $\partial A_z^{\mathrm{i}}/\partial z$ as

$$\Delta\left[\frac{A_z^{\mathrm{i}}}{\partial z}\right] = \left.\frac{\partial A_z^{\mathrm{i}}}{\partial z}\right|_{z'-}^{z'+} = q(\rho, \phi).$$

This jump discontinuity is already incorporated in the expansion (15.88). Using equation (15.91), this implies that $E_\rho^{\mathrm{i}}$ has a jump discontinuity at $z = z'$.

Therefore, the $z$-components of the potentials are

$$\begin{aligned} A_z^{\mathrm{s}} &= \frac{\mu_0 I_0 \ell}{4\pi j} \int_0^{\infty} \frac{k_\rho}{k_{z0}} \frac{\varepsilon_{\mathrm{rc}} k_{z0} - jk_z \tan k_z d}{\varepsilon_{\mathrm{rc}} k_{z0} + jk_z \tan k_z d} J_0(k_\rho \rho) \mathrm{e}^{-jk_{z0}(z+z')} \, \mathrm{d}k_\rho \\ &= \frac{\mu_0 I_0 \ell}{8\pi j} \int_{-\infty}^{\infty} \frac{k_\rho}{k_{z0}} \frac{\varepsilon_{\mathrm{rc}} k_{z0} - jk_z \tan k_z d}{\varepsilon_{\mathrm{rc}} k_{z0} + jk_z \tan k_z d} H_0^{(2)}(k_\rho \rho) \mathrm{e}^{-jk_{z0}(z+z')} \, \mathrm{d}k_\rho, \end{aligned} \tag{15.100}$$

$$= \frac{\mu_0 I_0 \ell}{8\pi j} \int_{-\infty}^{\infty} \frac{k_\rho}{k_{z0}} \frac{\varepsilon_{\mathrm{rc}} k_{z0} - k_z}{\varepsilon_{\mathrm{rc}} k_{z0} + k_z} H_0^{(2)}(k_\rho \rho) \mathrm{e}^{-jk_{z0}(z+z')} \, \mathrm{d}k_\rho; \quad d \to \infty, \tag{15.101}$$

$$\begin{aligned} A_z^{\mathrm{t}} &= \frac{\mu_0 \mu_{\mathrm{r}} \varepsilon_{\mathrm{rc}} I_0 \ell}{2\pi j} \int_0^{\infty} \frac{k_\rho J_0(k_\rho \rho) \mathrm{e}^{-jk_{z0} z'}}{\varepsilon_{\mathrm{rc}} k_{z0} \cos k_z d + jk_z \sin k_z d} \cos[k_z(z+d)] \, \mathrm{d}k_\rho, \\ &= \frac{\mu_0 \mu_{\mathrm{r}} \varepsilon_{\mathrm{rc}} I_0 \ell}{4\pi j} \int_{-\infty}^{\infty} \frac{k_\rho H_0^{(2)}(k_\rho \rho) \mathrm{e}^{-jk_{z0} z'}}{\varepsilon_{\mathrm{rc}} k_{z0} \cos k_z d + jk_z \sin k_z d} \cos[k_z(z+d)] \, \mathrm{d}k_\rho, \end{aligned} \tag{15.102}$$

$$= \frac{\mu_0 \mu_{\mathrm{r}} \varepsilon_{\mathrm{rc}} I_0 \ell}{4\pi j} \int_{-\infty}^{\infty} \frac{k_\rho H_0^{(2)}(k_\rho \rho) \mathrm{e}^{-jk_{z0} z'}}{\varepsilon_{\mathrm{rc}} k_{z0} + k_z} \mathrm{e}^{jk_z z} \, \mathrm{d}k_\rho, \quad d \to \infty, \tag{15.103}$$

where the second forms involving the Hankel function are obtained by expressing $J_0(k_\rho \rho)$ in terms of $H_0^{(2)}(k_\rho \rho)$ and $H_0^{(1)}(k_\rho \rho)$ and reflecting the latter over the interval $-\infty < k_\rho < 0$. It is seen that the integrands of $A_z^{\mathrm{i}}$, $A_z^{\mathrm{s}}$, and $A_z^{\mathrm{t}}$ are all multi-valued functions of the square root function $k_{z0}$, i.e. their values are altered as the complex square root, $k_{z0}$, changes from one branch to the other in the complex $k_\rho$-plane about the branch point $k_{z0} = 0$. However, they are single-valued w.r.t. $k_z$ because they depend on $k_z^2$ for small $k_z$ and, thus, remain the same even when the square root function, $k_z$, goes from one branch to the other in the complex $k_\rho$-plane about the point $k_z = 0$. For the integrals to converge we stipulate that $\Im(k_{z0}) < 0$ on the contour of integration $0 \leqslant k_\rho < \infty$. To maintain single-valued nature of the integrands, the complex $k_\rho$-plane is divided into two Riemann sheets and the original contour is taken to lie on the proper sheet (taken to be the top sheet), $\Im(k_{z0}) < 0$. In the case of a half-space, the integrands are multiple valued functions of $k_z$ as well. The various fields can be obtained by substituting equations (15.100)–(15.102) into equation (15.91).

It is also seen that the integrands of $A_z^{\mathrm{s}}$ and $A_z^{\mathrm{t}}$ have a singularity when the denominator vanishes. This happens at a $k_\rho = k_p$ with corresponding $k_{z0p}$, $k_{zp}$ such that $\varepsilon_{\mathrm{rc}} k_{z0p} = -jk_{zp} \tan k_{zp} d$ or equivalently at

$$k_p^{\pm} = \pm k_0 \sqrt{\frac{\varepsilon_{\mathrm{rc}}(\varepsilon_{\mathrm{rc}} - \mu_{\mathrm{r}})}{\varepsilon_{\mathrm{rc}}^2 - 1} + \frac{1}{\varepsilon_{\mathrm{rc}}^2 - 1} \left( \frac{k_{zp}}{k_0 \cos k_{zp} d} \right)^2}, \tag{15.104}$$

$$=\pm k_0\sqrt{\frac{\varepsilon_{rc}(\varepsilon_{rc}-\mu_r)}{\varepsilon_{rc}^2-1}}, \quad d\to\infty. \tag{15.105}$$

It can be shown that these are simple poles, with the $k_p^+$ lying on the top sheet.

*15.2.4.2 Horizontal electric dipole*

Figure 15.8 shows a *horizontal electric dipole* (HED) located at $(\rho', \phi', z')$ over a conductor-backed lossy material of thickness $d$. The dipole is oriented in a general direction $\hat{\mathbf{p}}$ in the horizontal plane so that we regard it as a *surface current* with density $\mathbf{J}_s = \hat{\rho}J_{s\rho} + \hat{\phi}J_{s\phi} = I_0\ell\hat{\mathbf{p}}\delta(\rho-\rho')\delta(\phi-\phi')/\rho'$ A m$^{-1}$. Here, the overall fields will have azimuthal dependence and will contain both $E_z$ and $H_z$. In view of this both types of vector potentials $\mathbf{A} = \hat{\mathbf{z}}A_z$ and $\mathbf{F} = \hat{\mathbf{z}}F_z$ are needed to expand the fields. The component $A_z$ alone will generate a TM$_z$ field and $F_z$ alone will generate a TE$_z$ field. The overall fields are given by

$$\begin{aligned}
E_\rho &= \frac{v}{jk}\frac{\partial^2 A_z}{\partial\rho\partial z} - \frac{1}{\varepsilon\rho}\frac{\partial F_z}{\partial\phi}; \quad & H_\rho &= \frac{v}{jk}\frac{\partial^2 F_z}{\partial\rho\partial z} + \frac{1}{\mu\rho}\frac{\partial A_z}{\partial\phi},\\
E_\phi &= \frac{v}{jk\rho}\frac{\partial^2 A_z}{\partial\phi\partial z} + \frac{1}{\varepsilon}\frac{\partial F_z}{\partial\rho}; \quad & H_\phi &= \frac{v}{jk\rho}\frac{\partial^2 F_z}{\partial\phi\partial z} - \frac{1}{\mu}\frac{\partial A_z}{\partial\rho},\\
E_z &= \frac{v}{jk}\left(\frac{\partial^2}{\partial z^2}+k^2\right)A_z; \quad & H_z &= \frac{v}{jk}\left(\frac{\partial^2}{\partial z^2}+k^2\right)F_z.
\end{aligned} \tag{15.106}$$

The following boundary conditions are to be met by the fields:

(i) $E_\rho(z'+) = E_\rho(z'-)$; $E_\phi(z'+) = E_\phi(z'-)$,

(ii) $E_\rho(0+) = E_\rho(0-)$, $E_\phi(0+) = E_\phi(0-)$;
$H_\rho(0+) = H_\rho(0-)$, $H_\phi(0+) = H_\phi(0-)$,

(iii) $E_\rho(-d) = 0 = E_\phi(-d)$,

(iv) $H_\rho(z'+) - H_\rho(z'-) = J_{s\phi}$; $H_\phi(z'+) - H_\phi(z'-) = -J_{s\rho}$.

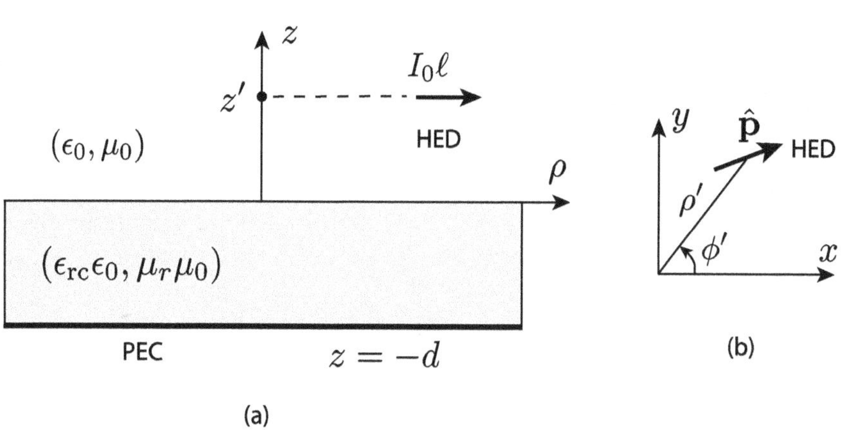


**Figure 15.8.** (a) Horizontal electric dipole radiating over a PEC-backed lossy slab. (b) Location and orientation of the dipole in the horizontal plane.

We expand the potentials in terms of cylindrical harmonics as

$$A_z = \frac{1}{2\pi}\int_0^\infty dk_\rho \sum_{n=-\infty}^{\infty} J_n(k_\rho \rho)e^{jn\phi}\begin{cases} A_1^+ e^{-jk_{z0}(z-z')}, & z > z' \\ A_1^- e^{-jk_{z0}z'}[e^{jk_{z0}z} + \Gamma_1 e^{-jk_{z0}z}], & 0 < z < z' \\ A_2\, e^{-jk_{z0}z'} \cos k_z(z+d), & -d \leqslant z \leqslant 0 \end{cases}, \quad (15.107)$$

$$F_z = \frac{1}{2\pi}\int_0^\infty \mathrm{d}k_\rho \sum_{n=-\infty}^{\infty} J_n(k_\rho \rho)\mathrm{e}^{jn\phi}\begin{cases} F_1^+ \mathrm{e}^{-jk_{z0}(z-z')}, & z > z' \\ F_1^- \mathrm{e}^{-jk_{z0}z'}[\mathrm{e}^{jk_{z0}z} + \Gamma_2 \mathrm{e}^{-jk_{z0}z}], & 0 < z < z' \\ F_2\, \mathrm{e}^{-jk_{z0}z'} \sin k_z(z+d), & -d \leqslant z \leqslant 0 \end{cases}, \quad (15.108)$$

Boundary conditions (i)–(iii) can be met separately by $\mathrm{TE}_z$ and $\mathrm{TM}_z$ modes and it is the last boundary conditions (iv) that couples the $\mathrm{TM}_z$ and $\mathrm{TE}_z$ sets. Note that the boundary conditions (iii) are automatically met in the form assumed in equations (15.107) and (15.108). Boundary conditions (i) imply $A_1^+ = -A_1^-(1 - \Gamma_1 \mathrm{e}^{-2jk_{z0}z'})$ and $F_1^+ = F_1^-(1 + \Gamma_2 \mathrm{e}^{-2jk_{z0}z'})$. Boundary conditions (ii) imply

$$jk_{z0}A_1^-(1 - \Gamma_1) = -\frac{A_2}{\varepsilon_{\mathrm{rc}}\mu_r}k_z \sin k_z d; \quad A_1^-(1 + \Gamma_1) = \frac{A_2}{\mu_r}\cos k_z d,$$

from which

$$\Gamma_1(k_{z0}, k_z) = \frac{\varepsilon_{\mathrm{rc}}k_{z0} - jk_z \tan k_z d}{\varepsilon_{\mathrm{rc}}k_{z0} + jk_z \tan k_z d}, \quad (15.109)$$

$$\rightarrow \frac{\varepsilon_{\mathrm{rc}}k_{z0} - k_z}{\varepsilon_{\mathrm{rc}}k_{z0} + k_z}, \quad d \rightarrow \infty, \quad (15.110)$$

and

$$F_1^-(1 + \Gamma_2) = \frac{F_2}{\varepsilon_{\mathrm{rc}}}\sin k_z d; \quad jk_{z0}F_1^-(1 - \Gamma_2) = \frac{F_2}{\varepsilon_{\mathrm{rc}}\mu_r}k_z \cos k_z d,$$

from which

$$\Gamma_2(k_{z0}, k_z) = \frac{\mu_{\mathrm{r}}k_{z0} + jk_z \cot k_z d}{\mu_{\mathrm{r}}k_{z0} - jk_z \cot k_z d}, \quad (15.111)$$

$$\rightarrow \frac{\mu_{\mathrm{r}}k_{z0} - k_z}{\mu_{\mathrm{r}}k_{z0} + k_z}, \quad d \rightarrow \infty. \quad (15.112)$$

Thus the coefficients $\Gamma_1$ and $\Gamma_2$ are completely determined by the, respectively, uncoupled $\mathrm{TM}_z$ and $\mathrm{TE}_z$ sets. As a result, the expression for $\Gamma_1$ here coincides with that for the reflection coefficient $\Gamma_{\mathrm{v}}$ of the VED. Note that the spectral reflection coefficients satisfy the conditions

$$\Gamma_1(k_{z0}, k_z)\Gamma_1(-k_{z0}, k_z) = 1; \quad \Gamma_2(k_{z0}, k_z)\Gamma_2(-k_{z0}, k_z) = 1. \quad (15.113)$$

The denominator of $\Gamma_2$ vanishes at a location $k_\rho = k_p$ and a corresponding $k_{z0p}$ and $k_{zp}$ such that $\mu_r k_{z0p} \sin k_{zp}d = jk_{zp} \cos k_{zp}d$, or equivalently when

$$k_p = \pm k_0 \left[ \frac{\mu_r(\mu_r - \varepsilon_{rc})}{\mu_r^2 - 1} + \frac{1}{\mu_r^2 - 1} \left( \frac{k_{zp}}{k_0 \sin k_{zp}d} \right)^2 \right]^{1/2}, \mu_r \neq 1, \tag{15.114}$$

$$\to \pm k_0 \sqrt{\frac{\mu_r(\mu_r - \varepsilon_{rc})}{\mu_r^2 - 1}}, \quad d \to \infty. \tag{15.115}$$

For non-magnetic materials with $\mu_r = 1$, a simplified expression for determining the pole location is

$$k_p = \pm k_0 \sqrt{\sin^2 k_{zp}d + \varepsilon_{rc} \cos^2 k_{zp}d} \to \pm\infty \text{ as } d \to \infty. \tag{15.116}$$

In the case of a half-space, it is clear from equation (15.112) that for a non-magnetic material, the pole does not lie on the top sheet as both $k_z$ and $k_{z0}$ will have negative imaginary parts and their sum, thereby, cannot vanish. At this point the coefficients $A_1^+$, $A_2$, $F_1^+$, and $F_2$ can be determined using

$$\begin{aligned} A_1^+ &= -(1 - \Gamma_1 e^{-j2k_{z0}z'})A_1^-, & A_2 &= \frac{\mu_r(1 + \Gamma_1)}{\cos k_z d} A_1^-, \\ F_1^+ &= (1 + \Gamma_2 e^{-j2k_{z0}z'})F_1^-, & F_2 &= \frac{\varepsilon_{rc}(1 + \Gamma_2)}{\sin k_z d} F_1^-, \end{aligned} \tag{15.117}$$

once the two remaining coefficients $A_1^-$ and $F_1^-$ are determined. Enforcement of the remaining boundary conditions (iv) enables the latter. To avoid dealing with the laborious manipulations involving both a Bessel function and its derivative in the same expression, it is convenient if one introduces the transverse fields [8]

$$H_{\substack{1\\2}} = H_\rho \pm jH_\phi = \mathscr{L}_\mp \left( \frac{1}{\mu} A_z \pm \frac{v}{k} \frac{\partial F_z}{\partial z} \right), \quad \mathscr{L}_\mp = \mp j \left( \frac{\partial}{\partial \rho} \pm j \frac{1}{\rho} \frac{\partial}{\partial \phi} \right), \tag{15.118}$$

and the corresponding transverse currents $J_{\substack{s1\\s2}} = J_{s\rho} \pm jJ_{s\phi}$ and recognizing that

$$\mathscr{L}_\mp \left[ J_n(k_\rho \rho) e^{jn\phi} \right] = jk_\rho J_{n\pm1}(k_\rho \rho) e^{jn\phi}, \tag{15.119}$$

which follows from the recursion relations (C.7) and (C.8) of Bessel functions. The operators $\mathscr{L}_\mp$ may be identified with a *raising/lowering operator* for a cylindrical harmonic of order $n$, with the operator $\mathscr{L}_-$ raising the order of the Bessel function (while maintaining the azimuthal harmonic) by one and the operator $\mathscr{L}_+$ lowering the order of the Bessel function by one. As an observation, because the azimuthal dependence has a different index than the order of the Bessel functions, the function resulting from raising or lowering operations will not be a cylindrical harmonic either of order $n + 1$ or $n - 1$. Consequently, subsequent operations of raising followed by lowering or lowering followed by raising will not yield the same end result. In other words, $\mathscr{L}_-$ and $\mathscr{L}_+$ do not commute.

In terms of these newly defined fields and currents, the boundary conditions (iv) become

$$H_{\substack{1\\2}}(z'+) - H_{\substack{1\\2}}(z'-) = \mp jJ_{\substack{s1\\s2}}. \tag{15.120}$$

Enforcing this using equations (15.107), (15.108), and (15.117) leads to

$$\frac{1}{\pi}\int_0^{\infty}\sum_{n=-\infty}^{\infty} k_\rho J_{n+1}(k_\rho\rho)\mathrm{e}^{jn\phi}\left(\frac{A_1^-}{\mu_0} - j\frac{k_{z0}v}{k_0}F_1^-\right)\mathrm{d}k_\rho = J_{s1}, \tag{15.121}$$

$$\frac{1}{\pi}\int_0^{\infty}\sum_{n=-\infty}^{\infty} k_\rho J_{n-1}(k_\rho\rho)\mathrm{e}^{jn\phi}\left(\frac{A_1^-}{\mu_0} + j\frac{k_{z0}v}{k_0}F_1^-\right)\mathrm{d}k_\rho = -J_{s2}. \tag{15.122}$$

Multiplying equation (15.121) with $\rho J_{m+1}(k_\rho{'}\rho)\mathrm{e}^{-jm\phi}$ and integrating over $\rho \in (0, \infty)$ and $\phi \in (0, 2\pi)$ and employing the identities (15.75) and (C.54) gives (on replacing $m$ with $n$ and $k_\rho'$ with $k_\rho$ in the final result)

$$\begin{aligned}\frac{A_1^-(n, k_\rho)}{\mu_0} - j\frac{k_{z0}v}{k_0}F_1^-(n, k_\rho) &= \frac{1}{2}\int_0^{\infty}\int_0^{2\pi} \rho J_{n+1}(k_\rho\rho)\mathrm{e}^{-jn\phi}J_{s1}(\rho, \phi)\,\mathrm{d}\phi\mathrm{d}\rho \\ &=: \frac{1}{2}\tilde{J}_{s1}(n, k_\rho),\end{aligned} \tag{15.123}$$

where we are explicitly showing the dependence of the various quantities on $n$, $k_\rho$. Similarly, multiplying equation (15.122) with $\rho J_{m-1}\left(k_\rho'\rho\right)\mathrm{e}^{-jm\phi}$ and integrating over $\rho \in (0, \infty)$ and $\phi \in (0, 2\pi)$ yields

$$\begin{aligned}\frac{A_1^-(n, k_\rho)}{\mu_0} + j\frac{k_{z0}v}{k_0}F_1^-(n, k_\rho) &= -\frac{1}{2}\int_0^{\infty}\int_0^{2\pi} \rho J_{n-1}(k_\rho\rho)\mathrm{e}^{-jn\phi}J_{s2}(\rho, \phi)\,\mathrm{d}\phi\mathrm{d}\rho \\ &=: -\frac{1}{2}\tilde{J}_{s2}(n, k_\rho).\end{aligned} \tag{15.124}$$

Adding and subtracting equations (15.123) and (15.124) finally gives

$$\begin{aligned}A_1^-(n, k_\rho) &= \tfrac{\mu_0}{4}\left[\tilde{J}_{s1}(n, k_\rho) - \tilde{J}_{s2}(n, k_\rho)\right] \\ F_1^-(n, k_\rho) &= \tfrac{jk_0}{4vk_{z0}}\left[\tilde{J}_{s1}(n, k_\rho) + \tilde{J}_{s2}(n, k_\rho)\right].\end{aligned}$$

It is seen that a horizontal current distribution will, in general, excite both a $\mathrm{TM}_z$ and a $\mathrm{TE}_z$ mode over a horizontally stratified medium. The expressions provided thus far are valid for any surface current distribution parallel to the conductor-backed slab. In particular, for a HED located at $(\rho', \phi')$ and having an orientation $\hat{\mathbf{p}}$, the surface current densities $J_{s\rho} = \hat{\mathbf{p}}\cdot\hat{\rho}I_0\ell\delta(\rho-\rho')\delta(\phi-\phi')/\rho'$ and $J_{s\phi} = \hat{\mathbf{p}}\cdot\hat{\phi}I_0\ell\delta(\rho-\rho')\delta(\phi-\phi')/\rho'$ and

$$
\begin{aligned}
A_1^-(n, k_\rho) &= \frac{\mu_0 I_0 \ell}{4}\hat{\mathbf{p}} \cdot \left[(\hat{\rho}' + j\hat{\phi}')J_{n+1}(k_\rho\rho') - (\hat{\rho}' - j\hat{\phi}')J_{n-1}(k_\rho\rho')\right]e^{-jn\phi'} \\
&= \frac{\mu_0 I_0 \ell}{2}\hat{\mathbf{p}} \cdot \left[-\hat{\rho}' J_n'(k_\rho\rho') + j\hat{\phi}'\frac{n}{k_\rho\rho'}J_n(k_\rho\rho')\right]e^{-jn\phi'},
\end{aligned}
\tag{15.125}
$$

$$
\begin{aligned}
F_1^-(n, k_\rho) &= \frac{jI_0 \ell k_0}{4vk_{z0}}\hat{\mathbf{p}} \cdot \left[(\hat{\rho}' + j\hat{\phi}')J_{n+1}(k_\rho\rho') + (\hat{\rho}' - j\hat{\phi}')J_{n-1}(k_\rho\rho')\right]e^{-jn\phi'} \\
&= \frac{jI_0 \ell k_0}{2vk_{z0}}\hat{\mathbf{p}} \cdot \left[\hat{\rho}'\frac{n}{k_\rho\rho'}J_n(k_\rho\rho') - j\hat{\phi}' J_n'(k_\rho\rho')\right]e^{-jn\phi'},
\end{aligned}
\tag{15.126}
$$

where recursion relations of Bessel functions, (C.7) and (C.8), were used to arrive at the latter forms. As an example, for a radial dipole located at $(\rho', \phi' = 0)$ (i.e. for an $\hat{\mathbf{x}}$-directed HED located on the $x$-axis) $\hat{\mathbf{p}} = \hat{\rho}'$ and

$$
A_1^-(n, k_\rho) = -\frac{\mu_0 I_0 \ell}{2}J_n'(k_\rho\rho')
\tag{15.127}
$$

$$
F_1^-(n, k_\rho) = \frac{jnI_0 \ell k_0}{2vk_\rho k_{z0}\rho'}J_n(k_\rho\rho').
\tag{15.128}
$$

If, in addition, the dipole is at the origin, then $\rho' = 0$ and only the $n = \pm 1$ modes are excited.

For an azimuthal dipole $\hat{\mathbf{p}} = \hat{\phi}'$ and

$$
A_1^-(n, k_\rho) = \frac{jn\mu_0 I_0 \ell}{2k_\rho\rho'}J_n(k_\rho\rho')e^{-jn\phi'}
\tag{15.129}
$$

$$
F_1^-(n, k_\rho) = \frac{I_0 \ell k_0}{2vk_{z0}}J_n'(k_\rho\rho')e^{-jn\phi'}.
\tag{15.130}
$$

If the current flows along a closed ring of radius $a$, then the current moment $I_0\ell = I_0 a d\phi'$ at any point on the ring. Integrating the expressions for $A_1^-$ and $F_1^-$ over $\phi' \in (0, 2\pi)$ retains only the $n = 0$ term. Consequently, $A_1^- = 0 = A_1^+ = A_2$ and the $\mathrm{TM}_z$ component vanishes. The field then becomes purely $\mathrm{TE}_z$ with

$$
F_1^-(n, k_\rho) = -\frac{I_0 \pi k_0 a}{vk_{z0}}J_1(k_\rho a)\delta_n^0,
\tag{15.131}
$$

since $J_0'(z) = -J_1(z)$. If the radius $a$ shrinks to zero such that $\pi a^2 I_0$ remains constant at $C_0$, then $F_1^-(n, k_\rho) \to -C_0 k_0 k_\rho \delta_n^0 / 2vk_{z0}$. The $\mathrm{TE}_z$ field is $\phi$-invariant and the infinitesimal ring of current may be interpreted as a *vertical magnetic dipole* (VMD) with

$$
F_z = -\frac{C_0 k_0}{4\pi v}\int_0^\infty \frac{k_\rho}{k_{z0}}J_0(k_\rho\rho)[e^{jk_{z0}(z-z')} + \Gamma_2 e^{-jk_{z0}(z+z')}]\,dk_\rho,\ 0 \leqslant z \leqslant z',
\tag{15.132}
$$

and similar expressions in the other regions.

### 15.2.5 Dipole radiating over a material sphere (ordinary and plasmonic)

Consider a material, non-magnetic sphere of radius $a$ having constitutive parameters $[\varepsilon_0\varepsilon_s, \mu_0]$ with $\varepsilon_s = (\varepsilon_{rs} - j\sigma_s)$ embedded in a background medium with constitutive parameters $(\varepsilon_0\varepsilon_b, \mu_0)$. It is assumed that the relative permittivity of the background medium $\varepsilon_b$ and that the relative conductivity of the sphere $\sigma_s$ are both >0. However, the relative permittivity of the sphere is permitted to be $\varepsilon_{rs} \gtrless 0$ corresponding respectively to a traditional- or a plasmonic medium. Let the sphere be driven by an electric dipole with current density $\mathbf{J} = \hat{\mathbf{p}} I_0\ell\delta(x-0)\delta(y-0)\delta(z-r')$. The orientation of the dipole is specified by the constant unit vector $\hat{\mathbf{p}}$ and its location by $\mathbf{r}' = (r', 0, 0)$ with $r' > a$. The orientation is chosen to be along the $z$-axis ($\hat{\mathbf{p}} = \hat{\mathbf{z}}$) and the case will be referred to as a vertical electric dipole (VED). Figure 15.9 shows the geometry for a radial dipole. For ease of notation we denote the spherical Hankel function of the second kind as $h_\nu(\cdot) =: h_\nu^{(2)}(\cdot)$. We will first determine the field of the dipole in the background medium and then add the scattered fields to satisfy the required boundary conditions on the surface of the sphere. The source is placed along the positive $z$-axis at $r' = b$. Symmetry of the source and the geometry suggests that the fields must be independent of $\phi$. Furthermore, symmetry demands that the source as well as the induced currents within the sphere do not have azimuthal components. Hence we expect the only non-zero field components of the overall fields to be $E_r$, $E_\theta$, and $H_\phi$ (the TM$_r$ mode). The total field exterior to the sphere is expressed as the sum of an incident part and a scattered part, while the field interior to the sphere is expressed as the total part. In a spherical coordinate system the electric and magnetic fields *in a source free region* can be expressed in terms of the magnetic vector potential $\mathbf{A} = \mathbf{r}\Pi$ as

$$\mathbf{H} = \frac{1}{\mu}\nabla \times \mathbf{A} =: \frac{1}{\mu}\mathbf{M},\ \mathbf{E} = \frac{-jv}{k}\nabla \times \nabla \times \mathbf{A} =: -jv\mathbf{N}, \tag{15.133}$$

where $v = 1/\sqrt{\mu\varepsilon}$ and $\mathbf{M} = \nabla \times \mathbf{A}$, $\mathbf{N} = \nabla \times \nabla \times \mathbf{A}/k$ are solenoidal vectors. The scalar function $\Pi$, satisfying the homogeneous Helmholtz equation $\nabla^2\Pi + k^2\Pi = 0$,

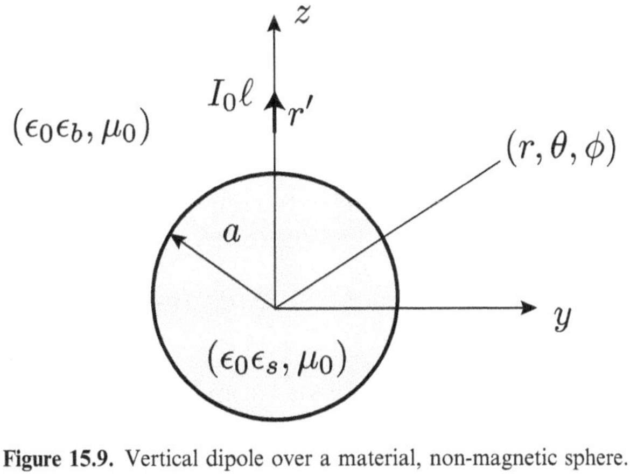


**Figure 15.9.** Vertical dipole over a material, non-magnetic sphere.

is the Debye potential with units of V s m$^{-2}$ (see example 1.3.1.2). It is thus a scalar wave function. The components of **M** and **N** in spherical coordinates are

$$M_r = 0, \quad M_\theta = \frac{1}{\sin\theta}\frac{\partial\Pi}{\partial\phi}, \quad M_\phi = -\frac{\partial\Pi}{\partial\theta}, \tag{15.134}$$

and

$$kN_r = \frac{\partial^2(r\Pi)}{\partial r^2} + k^2 r\Pi, \quad N_\theta = \frac{1}{kr}\frac{\partial^2(r\Pi)}{\partial r\partial\theta}, \quad N_\phi = \frac{1}{kr\sin\theta}\frac{\partial^2(r\Pi)}{\partial r\partial\phi}. \tag{15.135}$$

The starting point in our development in the spherical harmonic representation of the free-space Green's function (15.20)

$$\begin{aligned}\Pi_g(\mathbf{r};\mathbf{r}') = \frac{e^{-jk|\mathbf{r}-\mathbf{r}'|}}{4\pi|\mathbf{r}-\mathbf{r}'|} &= \frac{-jk}{4\pi}\sum_{n=0}^{\infty}(2n+1)j_n(kr_<)h_n(kr_>)\\ &\times \sum_{m=0}^{n}\frac{\varepsilon_m(n-m)!}{(n+m)!}P_n^m(\cos\theta)P_n^m(\cos\theta')\cos[m(\phi-\phi')],\end{aligned} \tag{15.136}$$

where $r_< = \min(r, r')$, $r_> = \max(r, r')$, $\varepsilon_m = 1$ for $m = 0$ and $\varepsilon_m = 2$ for $m > 0$. For $r \gtrless r'$, the scalar function $\Pi_g$ is a solution of the homogeneous Helmholtz equation $\nabla^2\Pi_g + k^2\Pi_g = 0$. A quantity proportional to $\Pi_g$ thus qualifies as a Debye potential provided it satisfies the required discontinuity conditions at the source. Note that the series representation in terms of spherical harmonics in equation (15.136) converges absolutely only for $|\mathbf{r}| \neq |\mathbf{r}'|$.

For any given spherical harmonic $\Pi_n^m$ of degree $n$ and order $m$ the radial component of $N_e$ reduces to the simpler form $krN_r = n(n+1)\Pi_n^m$. A few points are worth noting concerning the behavior of **M** and **N** around the source point. Denoting the discontinuity of a function $f$ about the point $r = r'$ as $\Delta'[f] = f(r'^+) - f(r'^-)$ we make the following observations.

1. Using equation (15.136) it is easy to see that $\Delta'[M_\theta] = 0)$ and $\Delta'[M_\phi] = 0$. Hence the tangential components of a magnetic field of a TM$_r$ mode are continuous across the sphere $r = r'$ passing through the current source. Likewise the tangential components of the electric field of a TE$_r$ mode are continuous across the sphere $r = r'$.
2. Using the Wronskian of spherical Bessel functions, it is straightforward to see that the tangential components of $\mathbf{N}_e$ satisfy the discontinuity relations

$$\Delta'[N_\theta] = -\sum_{n=0}^{\infty}\frac{(2n+1)}{4\pi kr'^2}\sum_{m=0}^{n}\frac{\varepsilon_m(n-m)!}{(n+m)!}\frac{\mathrm{d}P_n^m(\cos\theta)}{\mathrm{d}\theta}P_n^m(\cos\theta')\cos m(\phi-\phi'),$$

$$\Delta'[N_\phi] = \sum_{n=1}^{\infty}\frac{(2n+1)}{2\pi k\sin\theta r'^2}\sum_{m=1}^{n}\frac{m(n-m)!}{(n+m)!}\frac{\mathrm{d}P_n^m(\cos\theta)}{\mathrm{d}\theta}P_n^m(\cos\theta')\sin m(\phi-\phi').$$

Note that the summation for $N_\phi$ starts from $m$, $n = 1$ onwards. It is seen that the tangential components of electric field (magnetic field) of a TM$_r$ (TE$_r$)

> mode are *not* continuous across the spherical surface $r = r'$ irrespective of the source orientation. This makes sense on physical grounds on realizing that the situation for *radial* power flow out of a sphere of radius $r$ changes from enclosing no sources to enclosing the dipole source as one crosses from $r'^{-}$ to $r'^{+}$. Hence $E_\theta$ and $H_\phi$ cannot both be continuous on the sphere that touches the source[4].

We will find the appropriate Debye potential here by first assuming a form similar to equation (15.136) with an undetermined constant factor. The constant will then be found by matching the radial electric field generated by the assumed form with that already found in example 1.3.1.1 and given in equation (1.117) exterior to the sphere enclosing the source. That field is reproduced below considering $m = 0$ and $r' = b$:

$$E_r = -\frac{\eta I_o\ell}{4\pi br}\sum_{n=1}^{\infty} n(n+1)(2n+1)P_n(\cos\theta)j_n(kb)h_n(kr), \quad r > b. \tag{15.139}$$

Setting $k = k_b = \sqrt{\varepsilon_b}k_0$, $k_0 = \omega\sqrt{\mu_0\varepsilon_0}$, $\eta_b = \sqrt{\mu_0/\varepsilon_0\varepsilon_b} = \eta_0/\sqrt{\varepsilon_b}$, $\eta_0 = \sqrt{\mu_0/\varepsilon_0}$ and using $\psi$ from equation (15.136), an azimuthally symmetric Debye potential in the background medium in the absence of the sphere is assumed to be (choosing $m = 0$, $r' = b$ in equation (15.136))

$$\Pi^{\mathrm{i}} = \frac{-jk_b I_o \ell C}{4\pi b}\sum_{n=0}^{\infty}(2n+1)j_n(k_b r_<)h_n(k_b r_>)P_n(\cos\theta),$$

where $C$ is a constant (units of Ohm s m$^{-1}$) to be determined. The superscript 'i' present in $\Pi^{\mathrm{i}}$ denotes the incident field. The radial component of the incident electric field for $r > b$ is obtained from equation (15.135) (actually using $krN_r = n(n+1)\Pi_n^{\mathrm{i}}$ for each harmonic term of equation (15.136))

$$E_{\mathrm{r}}^{\mathrm{i}} = \frac{-jv_b}{k_b r}\frac{-jk_b I_o \ell C}{4\pi b}\sum_{n=0}^{\infty} n(n+1)(2n+1)j_n(\zeta_b)h_n(k_b r)P_n(\cos\theta), \tag{15.140}$$

where $\zeta_b = k_b b$, $v_b = c/\sqrt{\varepsilon_b}$, $c$ being the speed of light in free-space. Comparing equations (15.140) and (15.139) with $k = k_b$ and $\eta = \eta_b$ gives $v_b C = \eta_b$ or $C = \mu_0$. Hence

---

[4] Mathematically, it is straightforward to see for the solution of a Helmholtz equation with a source $q(\theta, \phi)$ distributed on the surface of a sphere of radius $r'$ that

$$\nabla^2\psi + k^2\psi = -\delta(r - r')q(\theta, \phi), \tag{15.137}$$

which, in turn, implies that

$$\Delta'[\psi] = 0; \quad \Delta'\left[\frac{\partial\psi}{\partial r}\right] = -q(\theta, \phi). \tag{15.138}$$

Hence the field $\psi$ is continuous across the spherical surface $r = r'$, but its normal derivative is discontinuous by the amount of source distribution $q(\theta, \phi)$. Equation (15.138) may be used to relate any unknown factors contained in $\psi$ to the source function $q(\theta, \phi)$.

$$\Pi^{\mathrm{i}} = \frac{-jk_b\mu_0 I_0\ell}{4\pi b}\sum_{n=0}^{\infty}(2n+1)P_n(\cos\theta)j_n(k_b r_<)h_n(k_b r_>)$$
$$=: -\mu_0 H_0 \sum_{n=0}^{\infty}(2n+1)P_n(\cos\theta)\begin{cases} j_n(k_b r)h_n(\zeta_b), & r<b \\ j_n(\zeta_b)h_n(k_b r), & r>b\end{cases}, \tag{15.141}$$

where $H_0 = jk_b I_0\ell/4\pi b$. It is interesting to consider the two special cases of $\theta = 0$, $\theta = \pi$ for $r \leqslant b$. Using $P_n(1) = 1$ and $P_n(-1) = (-1)^n$ we see that

$$\Pi^{\mathrm{i}}(\theta = 0) = \frac{\mu_0 I_0 \ell e^{-jk_b(b-r)}}{4\pi b(b-r)} = -\mu_0 H_0 \sum_{n=0}^{\infty}(2n+1)j_n(k_b r)h_n(\zeta_b), \tag{15.142}$$

$$\Pi^{\mathrm{i}}(\theta = \pi) = \frac{\mu_0 I_0 \ell e^{-jk_b(b+r)}}{4\pi b(b+r)} = -\mu_0 H_0 \sum_{n=0}^{\infty}(-1)^n(2n+1)j_n(k_b r)h_n(\zeta_b). \tag{15.143}$$

On a sphere containing the dipole the radial distance $r = b$. Noting that

$$(2n+1)j_n(z_1)h_n(z_2) \sim \frac{j}{z_2}\left(\frac{z_1}{z_2}\right)^n \text{ for } n \gg |z_1|, |z_2|,$$

one sees that the series in equation (15.142) diverges as it should when $r = b$. However, the alternating series in equation (15.143) should sum up to the finite left-hand side even though it does not converge absolutely. In this case the non-absolute convergence is due to the spherical harmonic representation and not due to Green's function singularity. Indeed one may use the following approximation for field computation

$$\Pi^{\mathrm{i}}(\theta = \pi, r = b) = \frac{\mu_0 I_0 \ell e^{-j2\zeta_b}}{8\pi b^2} \approx \frac{-j\zeta_b \mu_0 I_0 \ell}{4\pi b^2}\sum_{n=0}^{N}(-1)^n(2n+1)j_n(\zeta_b)h_n(\zeta_b), \tag{15.144}$$

where $N < \lceil \zeta_b \rceil$ and $\lceil \cdot \rceil$ denotes the ceiling operator.

The remaining components of the incident field are

$$E_\theta^{\mathrm{i}} = \frac{j\eta_b H_0}{k_b r}\sum_{n=0}^{\infty}(2n+1)\frac{\mathrm{d}P_n(\cos\theta)}{\mathrm{d}\theta}\begin{cases}[k_b r j_n(k_b r)]' h_n(\zeta_b), & r<b \\ j_n(\zeta_b)[k_b r h_n(k_b r)]', & r>b\end{cases}, \tag{15.145}$$

$$H_\phi^{i} = H_0 \sum_{n=0}^{\infty}(2n+1)\frac{dP_n(\cos\theta)}{d\theta}\begin{cases} j_n(k_b r)h_n(\zeta_b), & r<b \\ j_n(\zeta_b)h_n(k_b r), & r>b\end{cases}, \tag{15.146}$$

where the prime denotes the derivative with respect to the entire argument of the Bessel functions. Note that $E_\theta(r, 0, \phi) = 0 = E_\theta(r, \pi, \phi)$ because $\mathrm{d}P_n(\cos\theta)/\mathrm{d}\theta = 0$ for $\theta = 0, \pi$ for any $n$.

In the presence of the sphere we express the potential outside the sphere as the sum of incident part, $\Pi^{\mathrm{i}}$, and scattered part, $\Pi^{\mathrm{s}}$, and the potential inside the sphere as a total part $\Pi^{\mathrm{t}}$. In their respective regions $\Pi^{\mathrm{s}}$ and $\Pi^{\mathrm{t}}$ are the Debye potentials that satisfy the *homogeneous* Helmholtz equations. Denoting $\varepsilon_{rs} = -\varepsilon_p$, $k_s = \sqrt{\varepsilon_s} k_0 = j\sqrt{\kappa_s} k_0$, $\kappa_s = -\varepsilon_s = \varepsilon_p + j\sigma_s$, $\zeta_a = k_b a$, $\zeta_s = k_s a$, we write

$$\Pi^{\mathrm{s}} = -\mu_0 H_0 \sum_{n=0}^{\infty} a_n (2n+1) P_n(\cos\theta) \frac{h_n(k_b r)}{h_n(\zeta_a)} j_n(\zeta_a) h_n(\zeta_b), \qquad r > a, \tag{15.147}$$

$$\Pi^{\mathrm{t}} = -\mu_0 H_0 \sum_{n=0}^{\infty} b_n (2n+1) P_n(\cos\theta) \frac{j_n(k_s r)}{j_n(\zeta_s)} j_n(\zeta_a) h_n(\zeta_b), \qquad r < a. \tag{15.148}$$

where $a_n$ and $b_n$ are constants to be determined. The amplitude of the scattered part is governed by the coefficients $a_n$, while that of the total field inside the sphere is governed by the coefficients $b_n$. The coefficients are solved by imposing the continuity of $E_\theta$ and $H_\phi$ at the boundary $r = a$. Using equations (15.133)–(15.135) to evaluate the fields $\left(E_\theta^{\mathrm{s}}, H_\phi^{\mathrm{s}}\right)$, $\left(E_\theta^{\mathrm{t}}, H_\phi^{\mathrm{t}}\right)$ and imposing the continuity conditions $E_\theta^{\mathrm{i}}(r=a) + E_\theta^{\mathrm{s}}(r=a) = E_\theta^{\mathrm{t}}(r=a)$ and $H_\phi^{\mathrm{i}}(r=a) + H_\phi^{\mathrm{s}}(r=a) = H_\phi^{\mathrm{t}}(r=a)$ gives

$$\sum_{n=0}^{\infty} (2n+1) j_n(\zeta_a) h_n(\zeta_b) \frac{\mathrm{d}P_n(\cos\theta)}{\mathrm{d}\theta} [1 + a_n - b_n] = 0, \tag{15.149}$$

$$\sum_{n=0}^{\infty} (2n+1) j_n(\zeta_a) h_n(\zeta_b) \frac{\mathrm{d}P_n(\cos\theta)}{\mathrm{d}\theta} \left[ \frac{1}{\varepsilon_b} \left[ R_{nj}(\zeta_a) + a_n R_{nh}(\zeta_a) \right] - \frac{b_n}{\varepsilon_s} R_{nj}(\zeta_s) \right] = 0, \tag{15.150}$$

where

$$R_{nj}(x) := \frac{[x j_n(x)]'}{j_n(x)}, \qquad \text{and } R_{nh}(x) := \frac{[x h_n(x)]'}{h_n(x)}.$$

Multiplying both sides of equations (15.149) and (15.150) with $\sin\theta \partial P_m(\cos\theta)/\partial\theta$ and integrating over $\theta \in (0, \pi)$ and using the identity (D.121), the boundary conditions are seen to apply on a mode-by-mode basis resulting in

$$1 + a_n = b_n,$$
$$\varepsilon_s\left(R_{nj}(\zeta_a) + a_n R_{nh}(\zeta_a)\right) = \varepsilon_b R_{nj}(\zeta_s) b_n.$$

Solving we obtain

$$a_n = -\frac{\varepsilon_s R_{nj}(\zeta_a) - \varepsilon_b R_{nj}(\zeta_s)}{\varepsilon_s R_{nh}(\zeta_a) - \varepsilon_b R_{nj}(\zeta_s)}, \qquad b_n = \frac{\varepsilon_s\left(R_{nh}(\zeta_a) - R_{nj}(\zeta_a)\right)}{\varepsilon_s R_{nh}(\zeta_a) - \varepsilon_b R_{nj}(\zeta_s)}. \tag{15.151}$$

It is seen that the coefficients $a_n$ and $b_n$ are independent of the height of the dipole over the sphere. Note also that $a_n = 0$, $b_n = 1$ if $\varepsilon_{\mathrm{s}} = \varepsilon_{\mathrm{b}}$ or when $a \to 0$ (since $j_n(z) \to 0$ as $z \to 0$ as long as $n \geqslant 1$). For large arguments $|j_n(z)| \sim |z^{-1/2}\cos(z - n\pi/2 - \pi/4)|$, $|h_n(z)||e^{-iz} z^{-1/2}|$. Hence for a perfectly conducting sphere with $\sigma_s \to \infty$

$$a_n \to -\frac{R_{nj}(\zeta_a)}{R_{nh}(\zeta_a)}, \qquad b_n \to \frac{R_{nh}(\zeta_a) - R_{nj}(\zeta_a)}{R_{nh}(\zeta_a)}, \qquad \text{as } \sigma_s \to \infty. \tag{15.152}$$

However, the fields inside the PEC sphere vanish because of the presence of the factor $j_n(k_s r)/j_n(\zeta_s)$ in $\Pi^t$, which behaves as $|j_n(k_s r)/j_n(\zeta_s)|(a/r)^{1/2}e^{\Im(k_s)(r-a)} \to 0$ keeping in mind that the imaginary part of $k_s$, $\Im(k_s)$, is >0.

To determine the power dissipated in the conducting sphere we first look at the radial part of the complex power $P_e(a)$ flowing *out* of the spherical surface $r = a$. Using

$$H_\phi(r = a, \theta, \phi) = H_0 \sum_{n=0}^{\infty} (2n + 1) b_n j_n(\zeta_a) h_n(\zeta_b) \frac{dP_n(\cos\theta)}{d\theta}$$

$$E_\theta(r = a, \theta, \phi) = \frac{jH_0\eta_0}{\varepsilon_s k_0 a} \sum_{n=0}^{\infty} (2n + 1) b_n j_n(\zeta_a) h_n(\zeta_b) \frac{dP_n(\cos\theta)}{d\theta} R_{nj}(\zeta_s),$$

and equation (D.121), the complex power is

$$\begin{aligned} P_e(a) &= \frac{1}{2}\int_0^{2\pi}\int_0^{\pi} E_\theta(a, \theta, \phi) H_\phi^*(a, \theta, \phi)\, a^2 \sin\theta\, d\theta\, d\phi \\ &= I_0^2 \eta_0 \frac{j\varepsilon_b}{\varepsilon_s}\left(\frac{a\lambda_0}{b^2}\right)\left(\frac{\ell}{2\lambda_0}\right)^2 \sum_{n=1}^{\infty} n(n + 1)(2n + 1)|b_n|^2 j_n^2(\zeta_a)|h_n(\zeta_b)|^2 R_{nj}(\zeta_s). \end{aligned}$$

Now

$$R_{nh}(\zeta_a) - R_{nj}(\zeta_a) = \frac{[\zeta_a h_n(\zeta_a)]'}{h_n(\zeta_a)} - \frac{[\zeta_a j_n(\zeta_a)]'}{j_n(\zeta_a)} = \frac{-j}{\zeta_a h_n(\zeta_a) j_n(\zeta_a)}, \tag{15.153}$$

(by Wronskian (C.6)). Hence

$$P_e(a) = \frac{j\varepsilon_s^* \ell^2 I_0^2 \eta_0}{8\pi k_0 a b^2} \sum_{n=1}^{\infty} n(n + 1)(2n + 1)\left|\frac{h_n(\zeta_b)}{h_n(\zeta_a)}\right|^2 \frac{R_{nj}(\zeta_s)}{|\varepsilon_s R_{nh}(\zeta_a) - \varepsilon_b R_{nj}(\zeta_s)|^2}. \tag{15.154}$$

From power conservation, the power dissipated, $P_d$, within the conducting sphere is the negative real part of $P_e(a)$, $P_d = -\Re[P_e(a)]$. Note that if the sphere is lossless then $\sigma_s = 0$, $\varepsilon_s$ is purely real, $R_{nj}(\zeta_s)$ is purely real and hence $P_e(a)$ is purely imaginary, implying that $P_d = 0$. Also note that

$$R_{nj}(\zeta_s) = \frac{[\zeta_s j_n(\zeta_s)]'}{j_n(\zeta_s)} = \frac{1}{2} + \zeta_s \frac{J'_{n+\frac{1}{2}}(\zeta_s)}{J_{n+\frac{1}{2}}(\zeta_s)} \sim n + 1, \quad n \gg |\zeta_s|$$

$$R_{nh}(\zeta_a) = \frac{[\zeta_a h_n(\zeta_a)]'}{h_n(\zeta_a)} = \frac{1}{2} + \zeta_a \frac{H'_{n+\frac{1}{2}}(\zeta_a)}{H_{n+\frac{1}{2}}(\zeta_a)} \sim -n, \quad n \gg |\zeta_s|$$

$$\frac{h_n(\zeta_b)}{h_n(\zeta_a)} \sim \left(\frac{\zeta_a}{\zeta_b}\right)^{n+1} = \left(\frac{a}{b}\right)^{n+1}, \quad n \gg |\zeta_b|.$$

Hence the summand behaves as $\mathcal{O}\left[n^2\left(\frac{a}{b}\right)^{2n}\right]$ for $n \gg \max(|\zeta_b|, |\zeta_s|)$ clearly indicating that the $n$th term in the summation will diminish exponentially to zero as $n \to \infty$ as long as $b > a$.

For $r > b$, the scattered electric and magnetic fields are

$$E_\theta^s = \frac{jH_0\eta_b}{k_b r}\sum_{n=0}^{\infty}(2n+1)a_n[\zeta h_n(\zeta)]'\frac{j_n(\zeta_a)h_n(\zeta_b)}{h_n(\zeta_a)}\frac{\mathrm{d}P_n(\cos\theta)}{\mathrm{d}\theta}, \tag{15.155}$$

$$H_\phi^s = H_0\sum_{m=0}^{\infty}(2m+1)a_m h_m(\zeta)\frac{j_m(\zeta_a)h_m(\zeta_b)}{h_m(\zeta_a)}\frac{\mathrm{d}P_m(\cos\theta)}{\mathrm{d}\theta}, \tag{15.156}$$

where $\zeta = k_b r$. Once again, $E_\theta^s(r, 0, \phi) = 0 = E_\theta^s(r, \pi, \phi)$. The scattered power is

$$\begin{aligned} P_s(r) &= \frac{1}{2}\int_0^{2\pi}\int_0^{\pi} E_\theta^s(r, \theta, \phi)H_\phi^{s*}(r, \theta, \phi)r^2\sin\theta\,\mathrm{d}\theta\mathrm{d}\phi \\ &= \frac{j(I_0\ell)^2\eta_b\zeta}{8\pi b^2}\sum_{n=1}^{\infty}n(n+1)(2n+1)|a_n|^2 \\ &\quad\times\left|\frac{j_n(\zeta_a)h_n(\zeta_b)}{h_n(\zeta_a)}\right|^2[\zeta h_n(\zeta)]'h_n^*(\zeta). \end{aligned}$$

By making use of

$$[\zeta h_n(\zeta)]'h_n^*(\zeta) \sim \frac{-j}{\zeta}, \quad n \text{ fixed}, |\zeta| \gg 1,$$

the power flowing out of a sphere of radius $r > r'$, $k_0 r \gg 1$ is equal to

$$P_s(r) = \frac{(I_0\ell)^2\eta_b}{8\pi b^2}\sum_{n=1}^{\infty}n(n+1)(2n+1)|a_n|^2\left|\frac{j_n(\zeta_a)h_n(\zeta_b)}{h_n(\zeta_a)}\right|^2, \quad \zeta \gg \zeta_b, \tag{15.157}$$

which is seen to be purely real and *independent* of $r$. It represents the total power scattered by the sphere into the far-zone. In a similar fashion, the total power radiated by the dipole in the background medium in the absence of the sphere is

$$P_i(r) = \frac{(I_0\ell)^2\eta_b}{8\pi b^2}\sum_{n=1}^{\infty}n(n+1)(2n+1)j_n^2(\zeta_b), \quad \zeta \gg \zeta_b, \tag{15.158}$$

$$= \frac{(I_0\ell)^2\eta_b}{8\pi}\frac{2k_b^2}{3}, \quad \zeta_b \to 0. \tag{15.159}$$

Because the power radiated by the dipole in the background medium must be independent of its location, the quantities on the right-hand sides of equations (15.158) and (15.159) must be equal to each other *for any* $\zeta_b$. It is also clear from equation (15.158) that $P_i(r)$ is actually independent of $r$ as it should be. The ratio of the total power extinguished by the sphere, $P_d + P_s(r)$, relative to the power supplied by the dipole $P_i(r)$ is

$$W_e = \frac{P_s(r) + P_d}{P_i(r)} = U_s + \left(\frac{\sqrt{\varepsilon_b}}{k_0 a}\right)\Re[-j\varepsilon_s^* U_d], \tag{15.160}$$

where

$$U_s = \frac{3}{2\zeta_b^2}\sum_{n=1}^{\infty} n(n+1)(2n+1)|a_n|^2 \left|\frac{j_n(\zeta_a)h_n(\zeta_b)}{h_n(\zeta_a)}\right|^2, \tag{15.161}$$

and

$$U_d = \frac{3}{2\zeta_b^2}\sum_{n=1}^{\infty} n(n+1)(2n+1)\left|\frac{h_n(\zeta_b)}{h_n(\zeta_a)}\right|^2 \frac{R_{nj}(\zeta_s)}{|\varepsilon_s R_{nh}(\zeta_a) - \varepsilon_b R_{nj}(\zeta_s)|^2}. \tag{15.162}$$

Note that

$$\begin{aligned}\left|\frac{j_n(\zeta_a)h_n(\zeta_b)}{h_n(\zeta_a)}\right|^2 &= \frac{\pi}{2\zeta_b}\left|\frac{J_{n+\frac{1}{2}}(\zeta_a)H_{n+\frac{1}{2}}(\zeta_b)}{H_{n+\frac{1}{2}}(\zeta_a)}\right|^2 \\ &\sim \frac{1}{2\zeta_b(2n+1)}\left|\frac{e\zeta_a^2}{\zeta_b(2n+1)}\right|^{2n+1} \to 0 \text{ as } n \to \infty.\end{aligned} \tag{15.163}$$

In view of this super exponential decay and the fact that $|a_n| = |\varepsilon_s - \varepsilon_b|/|\varepsilon_s + \varepsilon_b| = \mathcal{O}(1)$ as $n \to \infty$, the convergence of the series in equation (15.161) is guaranteed. A good rule of thumb to truncate the series is to use

$$N \geqslant \left\lceil \frac{e\zeta_a^2}{2\zeta_b} - \frac{1}{2}\right\rceil \sim \left\lceil \frac{e\zeta_a}{2}\right\rceil \sim \lceil \zeta_b \rceil \tag{15.164}$$

for the upper limit.

Another quantity of interest is the radial impedance $Z_r$ on the conducting surface:

$$Z_r(\theta) = \frac{E_\theta(r=a)}{H_\phi(r=a)} = j\eta_s \left.\frac{\frac{\partial(\zeta H_\phi)}{\partial \zeta}}{\zeta H_\phi}\right|_{\zeta=\zeta_s} = j\eta_s \left.\frac{\frac{\partial^2(\zeta \Pi^t)}{\partial\zeta\partial\theta}}{\frac{\partial(\zeta\Pi^t)}{\partial\theta}}\right|_{\zeta=\zeta_s}, \tag{15.165}$$

where $\eta_s = \eta_0/\sqrt{\varepsilon_s}$. Note that for each given spherical harmonic $Z_{rn} = j\eta_s R_{nj}(\zeta_s)/\zeta_s$. However, $Z_r \neq \sum_{n=1}^{\infty} Z_{rn}$. Indeed using $\mathrm{d}P_n(\cos\theta)/\mathrm{d}\theta = P_n^1(\cos\theta)$ we have

$$Z_r(\theta) = \frac{\sum_{n=1}^{\infty}(2n+1)b_n j_n(\zeta_a)h_n(\zeta_b)P_n^1(\cos\theta)Z_{rn}}{\sum_{n=1}^{\infty}(2n+1)b_n j_n(\zeta_a)h_n(\zeta_b)P_n^1(\cos\theta)} =: R_r(\theta) + jX_r(\theta), \tag{15.166}$$

where $R_r$ and $X_r$ are the real and imaginary parts of $Z_r$. Note that

$$(2n+1)b_{nj_n}(\zeta_a)h_n(\zeta_b) \sim \frac{j}{\zeta_a}\frac{2\varepsilon_s}{\varepsilon_s+\varepsilon_b}\left(\frac{a}{b}\right)^{n+1}, \quad n \gg |\zeta_s| \tag{15.167}$$

$$\sin^{\frac{3}{2}}\theta \mid P_n^1(\cos\theta)| < \sqrt{\frac{2}{n\pi}}\frac{\Gamma(n+2)}{\Gamma(n+1)} \sim \sqrt{\frac{2n}{\pi}}, \quad n \gg 1. \tag{15.168}$$

Hence $(2n+1)b_{nj_n}(\zeta_a)h_n(\zeta_b)R_{nj}(\zeta_s)P_n^1(\cos\theta) = \mathcal{O}[n^{\frac{3}{2}}(a/b)^n] \to 0$ as $n \to \infty$ and the summations in equation (15.166) converge as long as $b > a$.

Figure 15.10 shows normalized power dissipated within the plasmonic nano-sphere at optical frequencies for $\varepsilon_b = 1$, $a = 400$ nm and for same material parameters as in figure 1.1. Also shown is the extinction ratio $W_e$. It is seen that the dissipated power has a major peak at around $\lambda_{max} = 325$ nm corresponding to a resonance wherein maximum power is coupled to the nano-sphere. This is also confirmed by the peak of $W_e$ which occurs at the same wavelength. At this resonant wavelength the sphere has a radius of $a/\lambda_0 \approx 1.22$. At this wavelength maximum power from the dipole is extinguished by the sphere. The peaked response found in figure 15.10 is characteristic of a plasmonic material that has negative real part for its permittivity. If $\varepsilon_{rs} > 0$ the response will no longer exhibit a resonance behavior. For comparison, we show the extinction ratio of an ordinary dielectric medium with

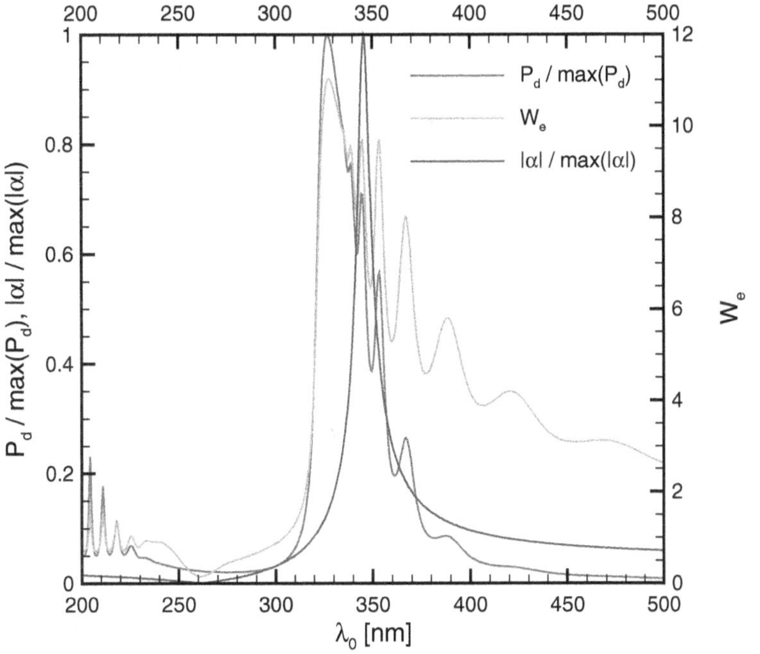


**Figure 15.10.** Dissipative loss, extinction ratio, and normalized electric polarizability of typical metallic media at optical wavelengths.

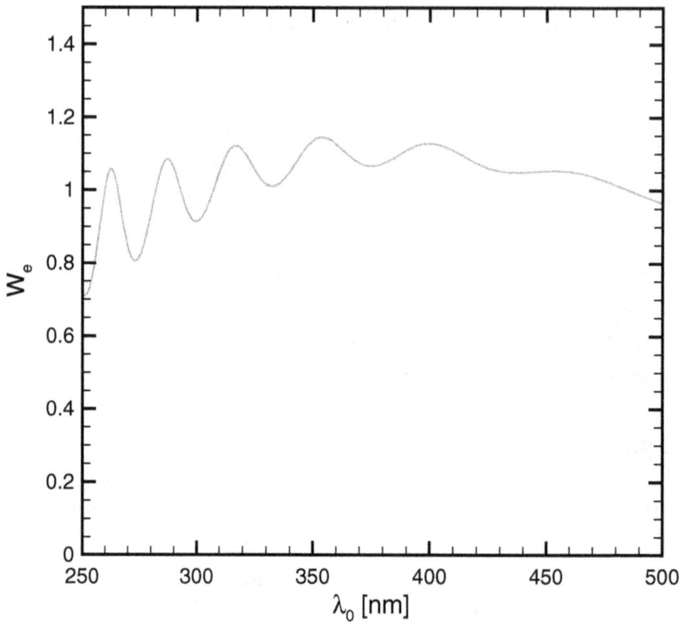


**Figure 15.11.** Extinction ratio of a lossless dielectric medium at optical wavelengths.

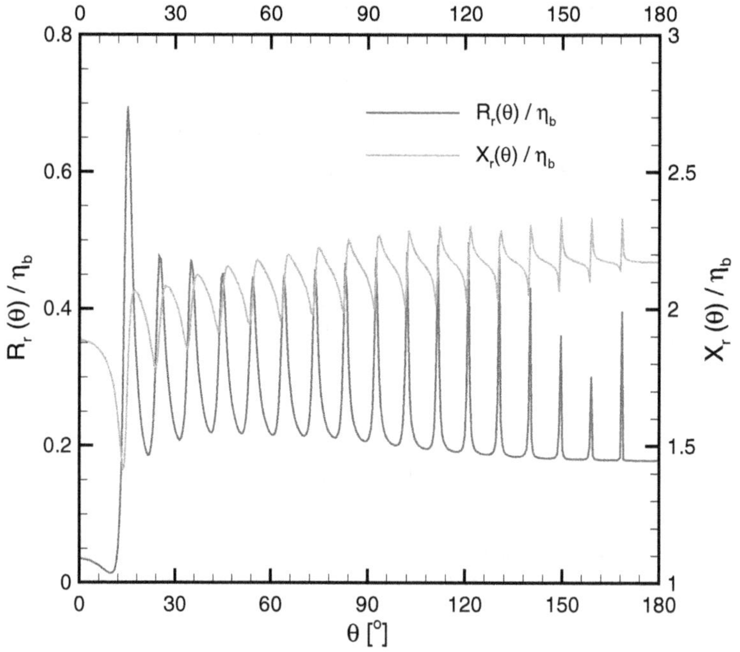


**Figure 15.12.** Normalized impedance on the surface of a plasmonic sphere.

$\varepsilon_s = 2$ in figure 15.11. The oscillations seen here are the normal oscillations associated with Rayleigh scattering.

If the sphere size is small compared to wavelength, i.e. $a/\lambda_0 \ll 1$, the plasmonic resonance phenomenon can approximately be predicted by examining the frequency response of the electric polarizability of the particle. The polarizability is proportional to the factor [9, chapter 10, p 148], [10]

$$\alpha = \frac{\varepsilon_s - \varepsilon_b}{\varepsilon_s + 2\varepsilon_b}. \tag{15.169}$$

For comparison we also show the normalized magnitude of the polarization factor $|\alpha|/\max(|\alpha|)$ in figure 15.10. The polarizability factor based on quasi-static approximation peaks at around $\lambda_0 = 345$ nm, which is not too from the true resonance of $\lambda_0 = 325$ nm.

Figure 15.12 shows the normalized impedance $Z_r(\theta)/\eta_b$ at $\lambda_0 = \lambda_{\max}$ and for the same remaining parameters as in figure 15.10.

## References


[1] Brekhovskikh L M 1980 *Waves in Layered Media* 2nd edn (Orlando, FL: Academic)

[2] Sommerfeld A 1949 *Partial Differential Equations in Physics* (New York: Academic)

[3] Harrington R F 1961 *Time-Harmonic Electromagnetic Fields* (New York: McGraw-Hill)

[4] Schelkunoff S A 1936 Modified Sommerfeld's integral and its application *Proc. IRE* **24** 1388–98

[5] Jordan K E, Richter G R and Sheng P 1986 An efficient numerical evaluation of the Green's function for the Helmholtz operator on periodic structures *J. Comput. Phys.* **63** 222–35

[6] Stratton J A 1941 *Electromagnetic Theory* (New York: McGraw-Hill)

[7] Silver S and Saunders W K 1950 The external field produced by a slot in an infinite cylinder *J. Appl. Phys.* **21** 153–8

[8] Araki K and Itoh T 1981 Hankel transform domain analysis of open circular microstrip radiating structures *IEEE Trans. Antennas Propag.* **AP-29** 84–9

[9] Bongersma M L and Kik P G 2007 *Surface Plasmon Nanophotonics* (Dordrecht: Springer)

[10] Ruppin R 1982 Decay of an excited molecule near a small metallic sphere *J. Chem. Phys.* **76** 1681–4

# Chapter 16

# Asymptotic analysis

We saw in previous chapters that in open problems, one expresses the total radiated field or the scattered field as a spectral integral. The spectral variable could be the longitudinal wavenumber or the lateral wavenumber. Often one is interested in the far-zone fields and it becomes imperative to evaluate those integrals asymptotically when one of the parameters in the integrand (in many cases it is the radial distance) becomes large. The complexity in the evaluation of integrals in wave propagation is compounded by the fact that the integrand may contain singular points such as poles and/or branch points in the complex spectral domain and one should pay close attention to these during the asymptotic evaluation process. In this chapter we will learn about the saddle point techniques (which encompasses the steepest descent (SDP) technique as well as the stationary phase method (SPM)) and the modified saddle point technique of asymptotically evaluating certain integrals for a large parameter $\Omega$. We will first look in detail at the intricacies introduced by branch-point singularities and the nature of waves associated with singularities. The saddle point technique is then applied to several practical problems in the following sections.

## 16.1 Branch cuts for wave propagation

Consider a plane wave of the form $\psi = \exp[\mathrm{i}(k_x x + k_z z)]$ propagating in the $xz$-plane in a homogeneous medium under the $\mathrm{e}^{-\mathrm{i}\omega t}$ time convention. We assume the medium to be slightly lossy with permittivity $\varepsilon_0$ and conductivity $\sigma_c$ so that the wavenumber $k = k_0(1 + \mathrm{i}\sigma_c/\omega\varepsilon_0)$, where $k_0$ is the wavenumber in free space at the radian frequency $\omega$. In this chapter we sometimes depart from the $\mathrm{e}^{j\omega t}$ time convention that we have been mostly following in the book and, instead, adopt the $\mathrm{e}^{-\mathrm{i}\omega t}$ dependence for time-harmonic waves. However, it is straightforward to convert the procedure to the $\mathrm{e}^{j\omega t}$ convention by simply replacing i with $-j$ and reflecting the various branch-cuts about the real axis. For instance the branch cuts shown in the first and third quadrants in figure 16.2 for the $\mathrm{e}^{-\mathrm{i}\omega t}$ time convention will appear, respectively, in the fourth and second quadrants under the $\mathrm{e}^{j\omega t}$ time convention.

doi:10.1088/978-0-7503-1716-0ch16

We are interested in the region $z > 0$ and like to study the plane wave function $\psi$ in the complex $k_x$ plane.

From the separation equation we have $k_z = \sqrt{k^2 - k_x^2}$. The plane wave function then has branch points at $\pm k$ in the complex $k_x$-plane (see section A.1.2.1). In order that $\psi$ represent a physical, non-growing wave at infinity, we impose the condition that the imaginary part $\mathfrak{J}(k_z) \geqslant 0$. The complex $k_x$-plane is divided into two Riemann sheets and we label the sheet on which $\mathfrak{J}(k_z) > 0$ as the proper sheet or the top-sheet and the other sheet as the improper or bottom sheet. The two sheets will be connected along the curve $\mathfrak{J}(k_z) = 0$. Writing $k_x + k = r_1 \exp(\mathrm{i}\phi_1)$ and $k - k_x = r_2 \exp(\mathrm{i}\phi_2)$, we may express $k_z$ as $k_z = \sqrt{(r_1 r_2)} \exp[\mathrm{i}(\phi_1 + \phi_2)/2]$. On the top sheet we restrict $0 \leqslant \phi_1 + \phi_2 \leqslant 2\pi$ so that the imaginary part of $k_z$ is non-negative. Writing $k_x = k_{x\mathrm{r}} + \mathrm{i}k_{x\mathrm{i}}$, $k_z = k_{z\mathrm{r}} + \mathrm{i}k_{z\mathrm{i}}$, and $k = k_\mathrm{r} + \mathrm{i}k_\mathrm{i}$ we obtain $k_{z\mathrm{r}}^2 - k_{z\mathrm{i}}^2 + 2\mathrm{i}k_{z\mathrm{r}}k_{z\mathrm{i}} = k_\mathrm{r}^2 - k_\mathrm{i}^2 - k_{x\mathrm{r}}^2 + k_{x\mathrm{i}}^2 + 2\mathrm{i}(k_\mathrm{r}k_\mathrm{i} - k_{x\mathrm{r}}k_{x\mathrm{i}})$. The branch-cuts will be defined by setting $\mathfrak{J}(k_z) = 0$. Using $k_{x\mathrm{r}}^2 - k_{x\mathrm{i}}^2 = k_\mathrm{r}^2 - k_\mathrm{i}^2 - k_{z\mathrm{r}}^2 + k_{z\mathrm{i}}^2$ and $k_{x\mathrm{r}}k_{x\mathrm{i}} + k_{z\mathrm{r}}k_{z\mathrm{i}} = k_\mathrm{r}k_\mathrm{i}$, this will result in

$$\begin{aligned} k_{x\mathrm{r}}^2 - k_{x\mathrm{i}}^2 = k_\mathrm{r}^2 - k_\mathrm{i}^2 - k_{z\mathrm{r}}^2 &< k_\mathrm{r}^2 - k_\mathrm{i}^2, && \text{region I} \\ k_{x\mathrm{r}}k_{x\mathrm{i}} &= k_\mathrm{r}k_\mathrm{i}, && \text{curve II.} \end{aligned}$$

Region I is shown in figure 16.1 as the hatched area and bounded by the hyperbola H1: $k_{x\mathrm{r}}^2 - k_{x\mathrm{i}}^2 = k_\mathrm{r}^2 - k_\mathrm{i}^2$. Curve II corresponds to the hyperbola H2: $k_{x\mathrm{r}}k_{x\mathrm{i}} = k_\mathrm{i}k_\mathrm{r}$. The branch cuts correspond to the solid line part of II. If we define the branch cuts based on setting $\mathfrak{R}(k_z) = 0$ instead, they correspond to the dashed line part of curve II that falls in region III: $k_{x\mathrm{r}}^2 - k_{x\mathrm{i}}^2 > k_\mathrm{r}^2 - k_\mathrm{i}^2$. In many problems it is more convenient to use branch cuts that separate a proper wave from an improper wave and a natural choice is the former one. Figure 16.2 shows the branch cuts as

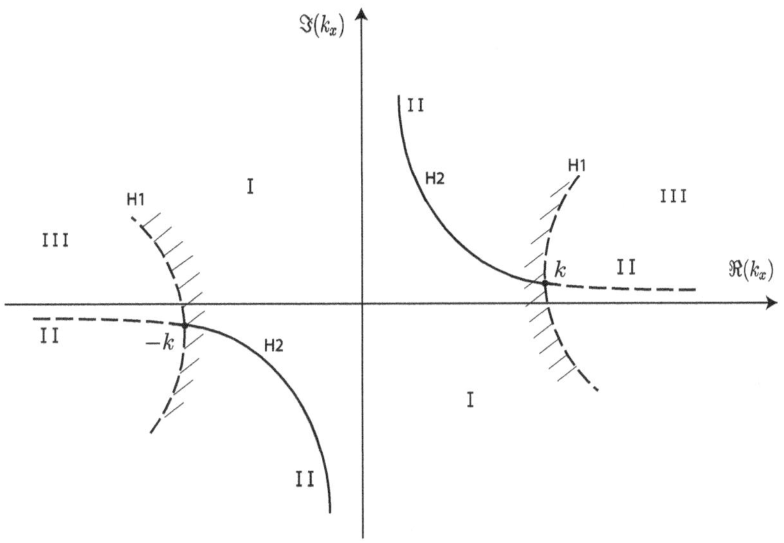


**Figure 16.1.** Curves defined by $\mathfrak{J}(k_z) = 0$ and $\mathfrak{R}(k_z) = 0$ in the complex $k_x$-plane.

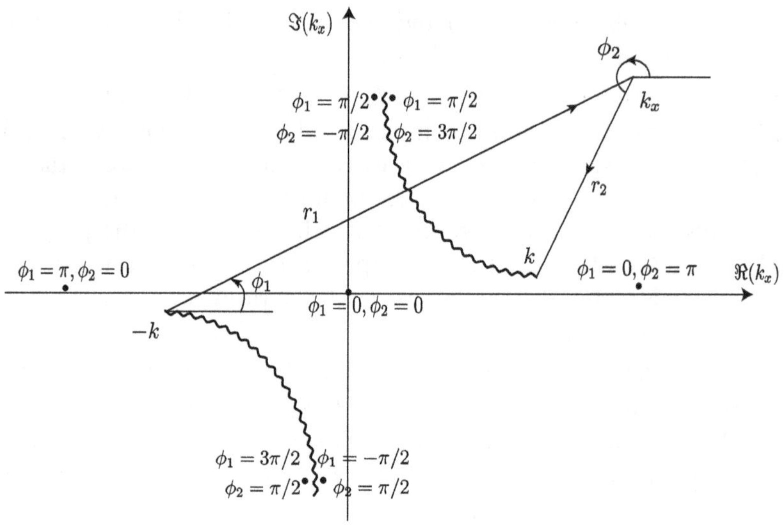


**Figure 16.2.** Branch cuts separating a proper sheet from an improper one.

wavy lines and they are defined by $\Im(k_z) = 0$. If the angles $\phi_1$ and $\phi_2$ are chosen for one point so that $k_x$ lies on the top sheet at that point, and if the angles are restricted as mentioned above, then $k_z$ will be defined uniquely as long as the branch cuts are not crossed. We choose $\phi_1 = 0$, $\phi_2 = 0$ at $k_x = 0$ so that $k_z = k$ at the origin on the top sheet. On the bottom sheet the angles are taken as $\phi_1 = 0$, $\phi_2 = 2\pi$, at $k_x = 0$, so that $k_z = -k$. Hence the points on the top sheet are defined so that

$$0 \leqslant \phi_1 + \phi_2 \leqslant 2\pi, \quad \phi_1 = 0, \phi_2 = 0 \text{ at } k_x = 0. \tag{16.1}$$

The angles $\phi_1$ and $\phi_2$ at various points on the top sheet are indicated in figure 16.2. Note that the values of $\phi_2$ are discontinuous across the upper branch cut, while the values of $\phi_1$ are discontinuous across the lower cut. Everywhere on the top sheet $\Im(k_z)$ will be greater than 0, except on the cuts, where it will be zero. If the branch cut is crossed once from the top sheet, then one enters the bottom sheet. If the medium becomes lossless so that imaginary part of $k$ vanishes, the branch cuts shown in figure 16.2 reduce to those shown in figure 16.3. We now work out a typical example from wave propagation where branch cuts are routinely encountered.

**Example 16.1.0.1.** Spectral representation of free-space Green's function, $e^{-i\omega t}$ time convention.

In this example, we wish to obtain various integral forms for the free-space scalar Green's function of the Helmholtz equation. For a point source located on the $z$-axis at $\mathbf{r}_0 = (x_0, y_0, z_0) = (0, 0, z_t)$, the scalar Green's function $G_0(x, y, x; x_0, y_0, z_0)= \exp(ik|\mathbf{r} - \mathbf{r}_0|)/4\pi|\mathbf{r} - \mathbf{r}_0|$ satisfies

$$\nabla^2 G_0 + k^2 G_0 = -\delta(\mathbf{r} - \mathbf{r}_0).$$

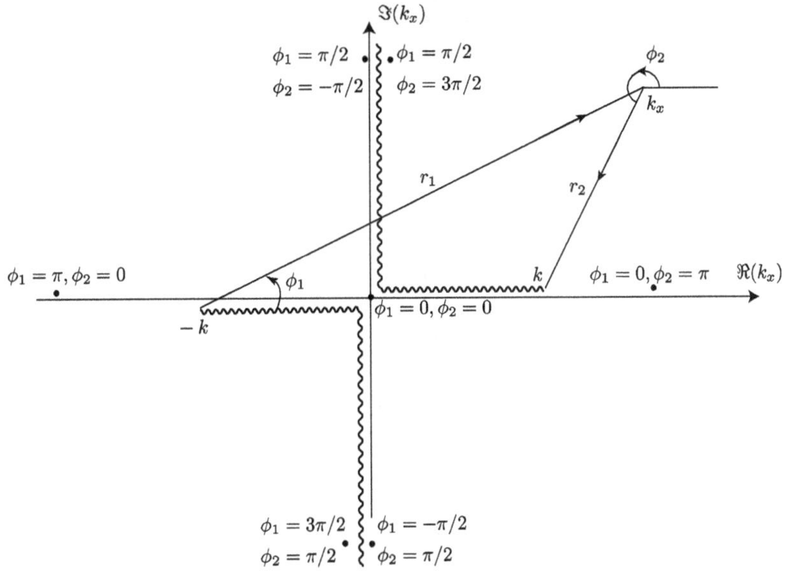


**Figure 16.3.** Branch cuts defined by $\Im(k_z) = 0$ when $k_i = 0$.

The Green's function can be expanded in terms of radial cylindrical waves and can be written as (see theorem 15.1.1)

$$G_0 = \frac{i}{8\pi} \int_{-\infty}^{\infty} \frac{k_\rho}{k_z} H_0^{(1)}(k_\rho \rho) e^{ik_z|z-z_t|} \, dk_\rho, \tag{16.2}$$

where $k_z = \sqrt{k^2 - k_\rho^2}$ and $\mathbf{r} = \hat{\boldsymbol{\rho}}\rho + \hat{\mathbf{z}}z$. For the integral to converge, we take $k_z$ to have a non-negative imaginary part. The form given in equation (16.2) is not suitable for numerical evaluation for small $|z - z_t|$ and large $\rho$ because of the slow decay and oscillatory nature of the integrand. Using contour integration techniques we will convert it to a form more suitable for numerical computations for large $\rho$ and small $|z - z_t|$. To this end, we draw branch cuts corresponding to $\Im(k_z) = 0$ and assume the contour along the real $k_\rho$ axis to lie on the top sheet, as shown in figure 16.4. It is to be noted that the function $H_0^{(1)}(k_\rho \rho)$ also has a branch point at the origin. We may choose its branch cut to run along the negative real axis and imagine the contour $k_\rho = -\infty$ to $\infty$ to run above it. This branch cut will not affect our discussion in the present example and will be ignored. We deform the contour $C_1$ to $C_\infty + C_2^- + C_2^+$ and note that no singularities are enclosed between the two. The semi-circular portion $C_\infty$ in the upper half-space avoids the branch cut by going around it via $C_2^-$ and $C_2^+$. Applying Cauchy's theorem (A.8) to the closed contour $C_1 - C_\infty - C_2^- - C_2^+$, we arrive at

$$\lim_{R\to\infty} \frac{i}{8\pi} \int_{-R}^{R} \frac{k_\rho}{k_z} H_0^{(1)}(k_\rho \rho) e^{ik_z|z-z_t|} \, dk_\rho = \lim_{R\to\infty} \frac{i}{8\pi} \int_{C_\infty + C_2^- + C_2^+} \frac{k_\rho}{k_z} H_0^{(1)}(k_\rho \rho) e^{ik_z|z-z_t|} \, dk_\rho. \tag{16.3}$$

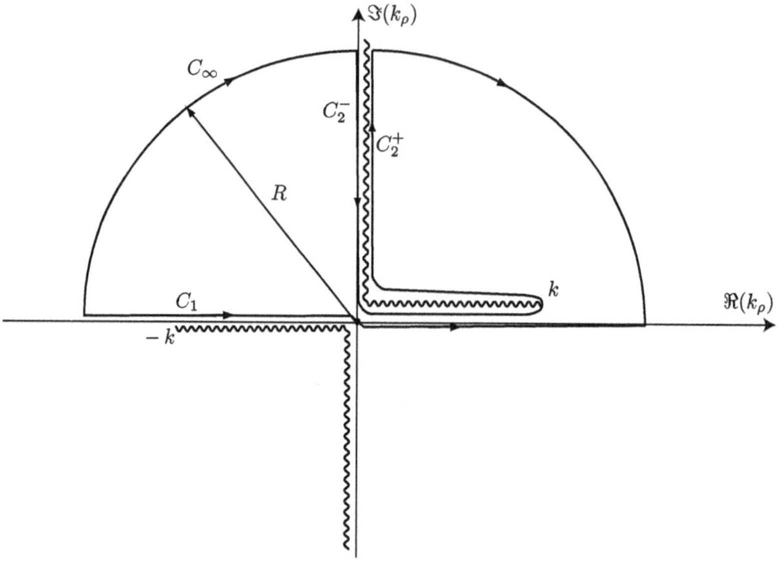


**Figure 16.4.** Branch cuts defined by $\Im(k_z) = 0$ in the top sheet of the complex $k_\rho$ plane.

We first consider the contribution to the rhs arising from $C_\infty$ on which $k_\rho = k_{\rho r} + \mathrm{i}k_{\rho i}$, $k_{\rho i} > 0$. Using the large argument approximation of the Hankel function and noting that $k_z = +\mathrm{i}k_\rho$ on the top sheet for large $k_\rho$, it is seen that

$$\left| \frac{k_\rho}{k_z} H_0^{(1)}(k_\rho \rho) \mathrm{e}^{\mathrm{i}k_z|z-z_t|} \right| \sim \sqrt{\frac{2}{\pi k_\rho \rho}} \left| \frac{k_\rho}{k_z} \right| \mathrm{e}^{-k_{\rho i}\rho} \mathrm{e}^{-k_{zi}|z-z_t|} \to 0 \text{ as } k_{\rho i} \to \infty. \tag{16.4}$$

Hence the contribution from $C_\infty$ vanishes in the limit of $R \to \infty$. On the branch cut, the imaginary part of $k_z$ is zero, making $k_z$ purely real. Referring to the angles specified in figure 16.3, we write

$$\begin{aligned} k_z &= -\sqrt{k_{\rho i}^2 + k^2} \text{ on } C_2^+, \text{ where } k_\rho = \mathrm{i}k_{\rho i},\ k_{\rho i} > 0 \\ &= +\sqrt{k_{\rho i}^2 + k^2} \text{ on } C_2^-, \text{ where } k_\rho = ik_{\rho i},\ k_{\rho i} > 0 \\ &= -\sqrt{k^2 - k_{\rho r}^2} \text{ on } C_2^+, \text{ where } k_\rho = k_{\rho r},\ |k_{\rho r}| < k \\ &= +\sqrt{k^2 - k_{\rho r}^2} \text{ on } C_2^-, \text{ where } k_\rho = |k_{\rho r}| < k. \end{aligned}$$

Also, $k_\rho\, \mathrm{d}k_\rho = -k_z\, \mathrm{d}k_z$ on $C_2^+$ and $C_2^-$. Therefore with $k_\rho = \sqrt{k^2 - k_z^2}$, we obtain

$$\begin{aligned} G_0 &= \frac{\mathrm{i}}{8\pi} \int_{-\infty}^{\infty} H_0^{(1)}(k_\rho \rho) \mathrm{e}^{\mathrm{i}k_z|z-z_t|}\, \mathrm{d}k_z \\ &= \frac{\mathrm{i}}{8\pi} \int_{-\infty}^{\infty} H_0^{(1)}(k_\rho \rho) \mathrm{e}^{\mathrm{i}k_z(z-z_t)}\, \mathrm{d}k_z, \end{aligned} \tag{16.5}$$

where $\Im(k_\rho) > 0$ and $\Re(k_\rho) > 0$. Because $k_z$ is all real and the limits extend over $(-\infty, \infty)$, there was no need to retain the absolute sign in equation (16.5). Equation (16.5) is preferable over equation (16.2) for large $\rho$ and small $|z - z_t|$ as the Hankel function provides exponential attenuation for large $|k_z|$. ■■

## 16.2 Complex waves

Consider a plane wave of the form $E = E_0 \exp[\mathrm{i}(k_x x + k_z z)]$ propagating in the $xz$-plane in a lossless medium, as shown in figure 16.5. The wave has an amplitude $E_0$ and wavenumber $k$. As in the previous section $\exp(-\mathrm{i}\omega t)$ time convention is assumed, where $\omega$ is the radian frequency and $t$ is the time variable. From the separation equation $k_x^2 + k_z^2 = k^2$. In order that this representation be valid in the upper half-space $z > 0$, we impose the condition that $\Im(k_z) > 0$ for the proper wave. For the improper wave $\Im(k_z) < 0$. The branch-cuts will be drawn as in the previous section and figure 16.6 shows the branch cut in the first quadrant of the $k_x$ plane. Let $k_x = k_{x\mathrm{r}} + \mathrm{i}k_{x\mathrm{i}}$, $k_z = k_{z\mathrm{r}} + \mathrm{i}k_{z\mathrm{i}}$, $\mathbf{r} = \hat{\mathbf{x}}x + \hat{\mathbf{z}}z$, $\mathbf{k}_\mathrm{r} = \hat{\mathbf{x}}k_{x\mathrm{r}} + \hat{\mathbf{z}}k_{z\mathrm{r}}$, and $\mathbf{k}_\mathrm{i} = \hat{\mathbf{x}}k_{x\mathrm{i}} + \hat{\mathbf{z}}k_{z\mathrm{i}}$. Then the plane wave could be written as

$$E = E_0 \mathrm{e}^{\mathrm{i}\mathbf{k}_\mathrm{r}\cdot\mathbf{r}} \mathrm{e}^{-\mathbf{k}_\mathrm{i}\cdot\mathbf{r}}. \tag{16.6}$$

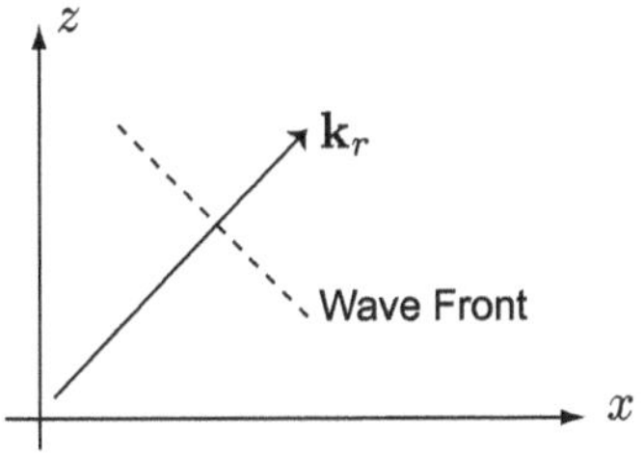


**Figure 16.5.** A plane wave propagating in the $xz$-plane.

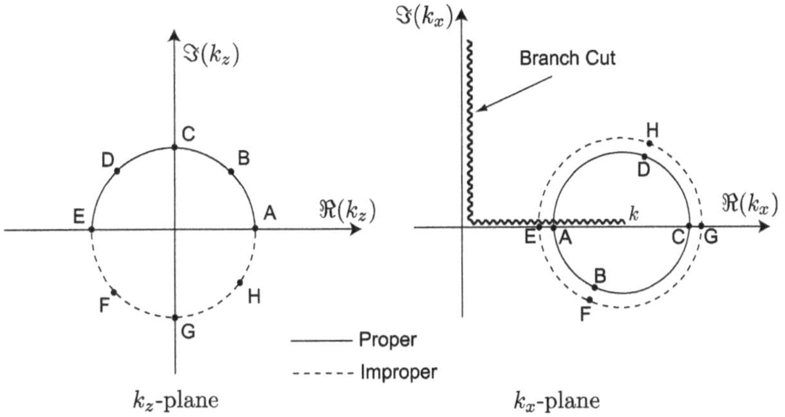


**Figure 16.6.** Complex waves in the $k_x$- and $k_z$-planes.

The phase of the plane wave is governed by the real part, $\mathbf{k}_r$, of the propagation vector and its attenuation is governed by the imaginary part $\mathbf{k}_i$ of the propagation vector. Note that the attenuation here takes place because of wave evanescence and not by real heat loss. From the separation equation we have

$$k_{xr}^2 + k_{zr}^2 - (k_{xi}^2 + k_{zi}^2) = k^2, \tag{16.7}$$

$$k_{xr}k_{xi} + k_{zr}k_{zi} = 0. \tag{16.8}$$

The constant phase surface, $\mathbf{r}_p$, is given by $\mathbf{k}_r \cdot \mathbf{r}_p = \psi_1$, where $\psi_1$ is some constant, and the constant magnitude surface, $\mathbf{r}_a$, is given by $\mathbf{k}_i \cdot \mathbf{r}_a = A_1$, where $A_1$ is some other constant. Since $\mathbf{k}_r \cdot \mathbf{k}_i = 0$ by equation (16.8), the constant phase surface and the constant magnitude surface are perpendicular to each other. Let us now consider $k_x$ values on a circle of radius $\kappa$ and centered at $k$ so that $k_x = k + \kappa \exp(\mathrm{i}\theta)$, with $\kappa < k$. We choose various values of $\theta$ so that the corresponding $k_x$ points (A through H) lie on the top and bottom sheets of the complex $k_x$-plane as shown in figure 16.6. The solid circle corresponds to the top sheet and the dashed line corresponds to the bottom sheet. The circle on the bottom sheet is shown larger only for clarity. Depending the angle $\theta$, the complex wave manifests in different forms. Table 16.1 shows the various $k_x$ values and the corresponding terminology for the complex waves. The real and imaginary parts of the propagation vector in space are shown in figure 16.7.

Points B–D lie on the top sheet and points F–G lie on the bottom sheet, while points A and E lie on the branch cut.

## 16.3 Asymptotic evaluation of integrals

In this section we describe both the steepest descent technique as well as the stationary phase method to asymptotically evaluate integrals of the type

$$I(\Omega) = \int_P f(z)\mathrm{e}^{\Omega q(z)}\,\mathrm{d}z, \tag{16.9}$$

**Table 16.1.** $k_x$, $k_z$ for various complex waves.

| Point | $k_{xr}$ | $k_{xi}$ | $k_{zr}$ | $k_{zi}$ | Complex wave |
|---|---|---|---|---|---|
| A | $<k$ | 0 | $>0$ | 0 | Outward plane wave |
| B | $>0$ | $<0$ | $>0$ | $>0$ | Backward leaky wave |
| C | $>k$ | 0 | 0 | $>0$ | Trapped surface wave |
| D | $>0$ | $>0$ | $<0$ | $>0$ | Zenneck wave |
| E | $<k$ | 0 | $<0$ | 0 | Incoming plane wave |
| F | $>0$ | $<0$ | $<0$ | $>0$ | Improper plane wave |
| G | $>k$ | 0 | 0 | $<0$ | Untrapped surface wave |
| H | $>0$ | $>0$ | $>0$ | $<0$ | Forward leaky wave |

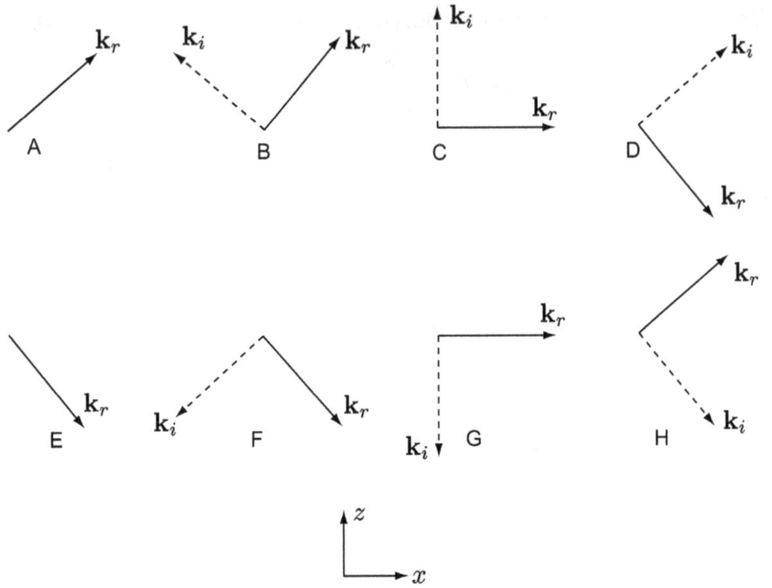


**Figure 16.7.** Complex waves in the $k_x$- and $k_z$-planes.

for a large positive real parameter $\Omega$, where $f(z)$ and $q(z)$ are analytic functions of the complex variable $z$ along the entire path $P$ of integration. These techniques are discussed in a number of books including [1–6].

### 16.3.1 Steepest descent technique

Suppose that a point $z_s$ can be found in the complex plane such that $\Re[q(z)]$ has a maximum value there with the property that $\Re[q(z)] < \Re[q(z_s)]$ at every point on another path $P_1$. Then the path $P$ may be deformed into $P_1$ with the inclusion of appropriate singularities of $f(z)$ and $q(z)$ between $P$ and $P_1$. Hence

$$I(\Omega) = \int_P f(z)\mathrm{e}^{\Omega q(z)}\,\mathrm{d}z = \int_{P_1} f(z)\mathrm{e}^{\Omega q(z)}\,\mathrm{d}z + \int_{P_s} f(z)\mathrm{e}^{\Omega q(z)}\,\mathrm{d}z, \tag{16.10}$$

where $P_s$ is the contribution coming from the singularities of $f(z)$ and $q(z)$ between $P$ and $P_1$. Since $\Omega$ is large,

$$|\mathrm{e}^{\Omega q(z)}| < |\mathrm{e}^{\Omega q(z_s)}|,$$

and the lhs decreases rapidly away from $z_s$ on the path $P_1$. It is then possible to approximate the integral by the contribution arising from the segment in the vicinity of $z_s$, since the other contribution will be exponentially small. We make a change of variable $s = (z - z_s)$ and write the Taylor series expansion

$$q(z) \approx \sum_{n=0}^{N} \frac{(z - z_s)^n}{n!}\frac{\mathrm{d}^n q(z_s)}{\mathrm{d}z^n} \;:=\; \tau(s), \tag{16.11}$$

$G(s) = f(z)\mathrm{d}z/\mathrm{d}s$. Here $\tau(s)$ is some polynomial in $s$ valid around $s = 0$. Then

$$\int_{P_1} f(z)\mathrm{e}^{\Omega q(z)}\,\mathrm{d}z \sim G(0)\int_{P_\tau} \mathrm{e}^{\Omega\tau(s)}\,\mathrm{d}s \text{ as } \Omega \to \infty, \tag{16.12}$$

where $P_\tau$ is the contour in the $s$-plane corresponding to $P_1$ in the $z$-plane. Appropriate choices of the contour $P_\tau$ are

(i) $\Re[\tau(s)]$ should decrease rapidly on $P_\tau$ away from $s = 0$.

(ii) $\frac{\mathrm{d}z}{\mathrm{d}s}$ must be finite near $s = 0$.

In either case, we hope to evaluate the rhs integral in equation (16.12) in a closed form. We next discuss the identification of the steepest descent path $P_1$.

*16.3.1.1 Steepest descent path*

For the analytic function $q(z) = u(x, y) + \mathrm{i}v(x, y)$, $z = x + \mathrm{i}y$, we recall from section A.1 that

$$q'(z) = \frac{\partial u}{\partial x} + \mathrm{i}\frac{\partial v}{\partial x} = \frac{\partial v}{\partial y} - \mathrm{i}\frac{\partial u}{\partial y}, \tag{16.13}$$

$$q''(z) = \frac{\partial^2 u}{\partial x^2} + \mathrm{i}\frac{\partial^2 v}{\partial x^2} = -\frac{\partial^2 u}{\partial y^2} - \mathrm{i}\frac{\partial^2 v}{\partial y^2}, \quad \nabla^2 u = 0, \quad \nabla^2 v = 0. \tag{16.14}$$

A point $z_s$ in the complex plane is a *stationary point* if $q'(z_s) = 0$. From equation (16.13), this also implies that the functions $u$ and $v$ are stationary at $z_s = (x_s, y_s)$. Assuming $q''(z_s) \neq 0$, we observe from the Laplacian equations in (16.14) that if the curvature ($\propto \partial^2/\partial x^2$ or $\partial^2/\partial y^2$) of the surface $u(x, y) =$ constant is $> 0$ (concave) along the $x$-direction, then it is $< 0$ (convex) along the (perpendicular) $y$-direction, as shown in figure 16.8. Hence stationary points of an analytic function are also *saddle points*. The order of the highest derivative that vanishes at the saddle point is known as the order of the saddle point. Depending on the choice of the path through $z_s$, $q(z)$ my remain constant, increase, or decrease along the path. Along the mountain (the

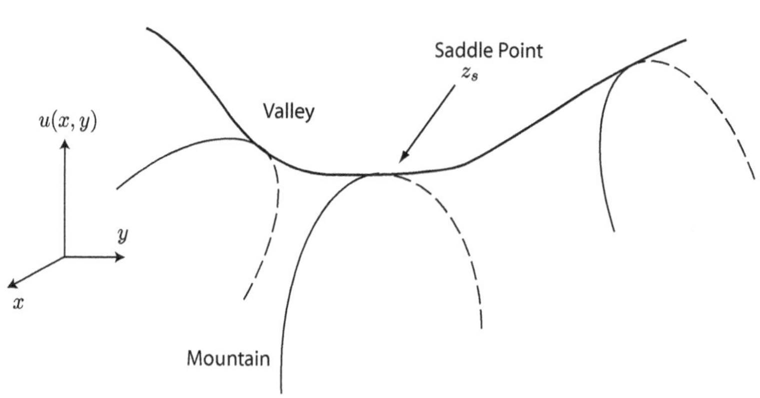


**Figure 16.8.** Behavior of an analytic function near a stationary point.

$x$-axis in the figure), the function decays most rapidly away from $z_s$; along the valley (the $y$-axis in the figure), the function rises most rapidly at $z_s$.

*16.3.1.2 Path of constant phase*

For $e^{\Omega q(z)} = e^{\Omega u(z)} e^{i\Omega v(z)}$, the paths along which the magnitude $e^{\Omega u(z)}$ changes most rapidly is the one along which $v(x, y)$ remains constant. To show this consider a path $\bar{P}$ with incremental arc length $d\gamma$, figure 16.9. The path makes an angle $\alpha$ with the $x$-axis at a general point $z$ on it. The increments along the $x$- and $y$-directions are $dx = \cos\alpha\, d\gamma$, $dy = \sin\alpha\, d\gamma$. Now

$$\frac{du}{d\gamma} = \frac{\partial u}{\partial x}\frac{dx}{d\gamma} + \frac{\partial u}{\partial y}\frac{dy}{d\gamma} = \cos\alpha\frac{\partial u}{\partial x} + \sin\alpha\frac{\partial u}{\partial y}, \tag{16.15}$$

$$\frac{dv}{d\gamma} = \frac{\partial v}{\partial x}\frac{dx}{d\gamma} + \frac{\partial v}{\partial y}\frac{dy}{d\gamma} = \cos\alpha\frac{\partial v}{\partial x} + \sin\alpha\frac{\partial v}{\partial y}. \tag{16.16}$$

The value of $\alpha$ for which $\frac{du}{d\gamma}$ is maximum is obtained by setting $\frac{\partial}{\partial\alpha}\left(\frac{du}{d\gamma}\right) = 0$. This implies that $-\sin\alpha\partial u/\partial x + \cos\alpha\partial u/\partial y = 0$. By C–R conditions (A.2) and (A.3), this implies that $-\sin\alpha\partial v/\partial y - \cos\alpha\partial v/\partial x = 0$, which in turn implies from equation (16.16) that $dv/d\gamma = 0$, or $v(x, y)$ = constant. Hence of all the paths that pass through the saddle point $z_s$, the one with $v(x, y)$ = constant will exhibit the most rapid change in $u(x, y)$, i.e. the path through $z_s$ along which the magnitude of $\exp[\Omega q(z)]$ changes most rapidly is the one that has a constant phase $v(x, y) = \text{constant} = v(x_s, y_s)$. Conversely, the paths through $z_s$ that have the most rapid phase variation are the ones that have constant $u(x, y) = u(x_s, y_s)$. These paths are used in the stationary phase method of evaluating integrals. *Near a first order saddle point*, we have the approximation $q(z) \approx q(z_s) + (z - z_s)^2 q''(z_s)/2$ and

$$e^{\Omega q(z)} \approx e^{\Omega q(z_s)} e^{\Omega q''(z_s)\frac{(z-z_s)^2}{2}}. \tag{16.17}$$

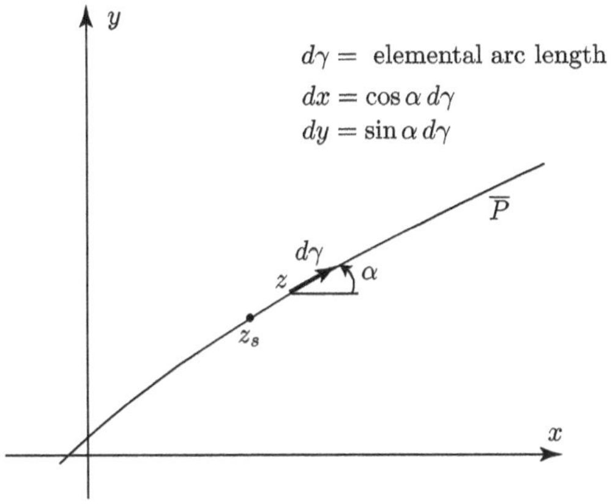


**Figure 16.9.** An arbitrary path that passes through the saddle point.

Using $q''(z_s)(z-z_s)^2 = |q''(z_s)(z-z_s)^2|\exp(\mathrm{i}2\psi)$, where $\psi = \arg(z-z_s) + \frac{1}{2}\arg[q''(z_s)]$, we obtain

$$\mathrm{e}^{\Omega q(z)} = \mathrm{e}^{\Omega q(z_s)}\mathrm{e}^{\frac{\Omega}{2}|q''(z_s)(z-z_s)^2|(\cos 2\psi + \mathrm{i}\sin 2\psi)}. \tag{16.18}$$

The paths $\psi = 0, \pi$ correspond to the steepest ascent path (SAP) along which the magnitude of $\exp[\Omega q(z)]$ increases most rapidly and the phase remains constant. The path $\psi = \pm\pi/2$ corresponds to the steepest descent path (SDP) along which the magnitude of $\exp[\Omega q(z)]$ decreases most rapidly and the phase remains constant. The paths $\psi = \pm\pi/4, \pm 3\pi/4$ correspond to those paths along which the magnitude of $\exp[\Omega q(z)]$ remains constant, but along which the phase varies most rapidly. These are known as the constant level paths. For the first order saddle point, the SAP and SDP are perpendicular to each other as illustrated in figure 16.10. For a saddle point of order $M$,

$$\mathrm{e}^{\Omega q(z)} = \mathrm{e}^{\Omega q(z_s)}\mathrm{e}^{\frac{\Omega}{(M+1)!}|q^{(M+1)}(z_s)(z-z_s)^{M+1}|[\cos(M+1)\psi + \mathrm{i}\sin(M+1)\psi]}, \tag{16.19}$$

and $(M+1)$ branches of SDPs emanate from the point $z_s$ with angles $\psi = (2m+1)\pi/M$, $m = 0, 1, \dots, M$.

*16.3.1.3 Asymptotic evaluation of integrals with first order saddle point*

We now consider the asymptotic evaluation of equation (16.10) when $q(z)$ has a first order saddle point. Near the saddle point we set

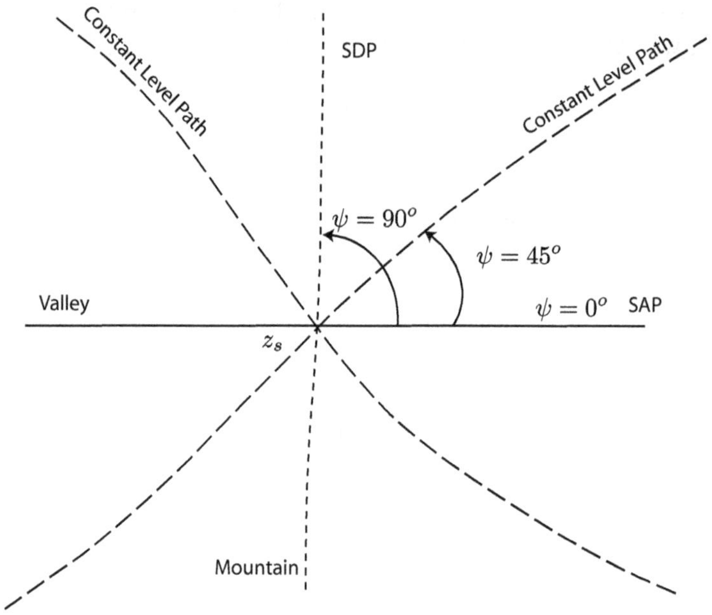


**Figure 16.10.** Orientation of SDP and SAP for a first order saddle point.

$$q(z) \approx q(z_s) + \frac{(z - z_s)^2}{2} q''(z_s) \;:=\; q(z_s) - s^2 \triangleq \tau(s), \tag{16.20}$$

where $s = (z - z_s)\sqrt{-q''(z_s)/2}$. In terms of the new variable $s$, $\mathrm{d}z = \sqrt{-2/q''(z_s)}\ \mathrm{d}s$. Along the SDP in the $s$-plane, $\mathfrak{J}[\tau(s)] = \mathfrak{J}[\tau(0)] =$ constant. Now $\pm\pi/2 + \arg(s) = \arg(z - z_s) + \arg[q''(z_s)]/2 = \pm\pi/2$ on the SDP. Hence $\arg(s) = 0$ and the SDP is along the real axis in the $s$-plane. Since the integrand decays very rapidly away from the saddle point $s = 0$, most of the contribution to the integral will come from the vicinity of $s = 0$ and it is reasonable to extend the limits on the SDP to $s = \pm\infty$, particularly for $\Omega \to \infty$. Furthermore, the rest of the integrand is assumed to be analytic about $s = 0$ so that a Taylor series expansion can be considered

$$\begin{aligned} G(s) = f(z)\frac{\mathrm{d}z}{\mathrm{d}s} = \sqrt{\frac{-2}{q''(z_s)}} f(z) = G(0) \\ + sG'(0) + \frac{s^2}{2}G''(0) + \cdots + \frac{s^n}{n!}G^{(n)}(0) + \cdots. \end{aligned} \tag{16.21}$$

Most of the time, a single term approximation is adequate so that $G(s) \approx G(0) = f(z_s)\sqrt{\frac{-2}{q''(z_s)}}$. Therefore

$$\begin{aligned} I(\Omega) &\sim \mathrm{e}^{\Omega q(z_s)} f(z_s)\sqrt{\frac{-2}{q''(z_s)}} \int_{-\infty}^{\infty} \mathrm{e}^{-\Omega s^2}\,\mathrm{d}s + \int_{P_s} f(z)\mathrm{e}^{\Omega q(z)}\,\mathrm{d}z \\ &= \mathrm{e}^{\Omega q(z_s)} f(z_s)\sqrt{\frac{-2\pi}{\Omega q''(z_s)}} + \int_{P_s} f(z)\mathrm{e}^{\Omega q(z)}\,\mathrm{d}z, \quad \Omega \to \infty. \end{aligned} \tag{16.22}$$

The complex square root in equation (16.22) must be determined so that $\arg(\mathrm{d}z) - \arg(\mathrm{d}s) = \arg\left[\frac{1}{\sqrt{-q''(z_s)}}\right]$. Since $\arg(\mathrm{d}s) = 0$ along the SDP in the s-plane, $\arg[1/\sqrt{-q''(z_s)}] = \arg(\mathrm{d}z)|_{z_s,\mathrm{SDP}}$. Therefore,

$$\begin{aligned} I(\Omega) \sim \mathrm{e}^{\Omega q(z_s)} f(z_s)\sqrt{\frac{2\pi}{\Omega |q''(z_s)|}}\ \mathrm{e}^{\mathrm{i}\arg(\mathrm{d}z)|_{z_s,\mathrm{SDP}}} \\ + \int_{P_s} f(z)\mathrm{e}^{\Omega q(z)}\,\mathrm{d}z, \quad \Omega \to \infty. \end{aligned} \tag{16.23}$$

**Example 16.3.1.1.** In this example we wish to obtain the asymptotic expression of the gamma function $\Gamma(x)$ for large argument $x$. We consider the integral representation of the gamma function

$$\Gamma(\Omega + 1) = \int_0^{\infty} \mathrm{e}^{-x} x^{\Omega}\,\mathrm{d}x, \tag{16.24}$$

The first step is to convert equation (16.24) into a form similar to equation (16.9). We rewrite equation (16.24) using the identity $x^\Omega = e^{\Omega \ln x}$ with a change of variable $x = \Omega z$ as

$$\Gamma(\Omega + 1) = \Omega^{\Omega+1} \int_0^\infty e^{\Omega(\ln z - z)}\, dz. \tag{16.25}$$

We consider the integral

$$I(\Omega) = \int_0^\infty e^{[\Omega(\ln z - z)]}\, dz, \tag{16.26}$$

as an integral in the complex plane and compare with the general expression (16.9). Clearly $f(z) = 1$, $q(z) = \ln z - z$. The function $q(z)$ has a branch point at $z = 0$ and we take the branch cut to be along the negative real axis. At the branch point $q(z) \to -\infty$, but the integrand vanishes. On the principal branch in which the contour of integration lies, we take $\pi \leqslant \arg(z) \leqslant \pi$. If $z$ is expressed in the polar form $z = r\exp(i\phi)$, then $\ln(z) = \ln r + i\phi$ and $q(z) = \ln r - r\cos\phi + i(\phi - r\sin\phi)$. The saddle point of $q(z)$ corresponds to $q'(z_s) = 0$ or $z_s = 1$. Also $q(z_s) = -1$, $q''(z_s) = -1/z_s^2 = -1 \neq 0$ and $\arg[q''(z_s)] = \pm\pi$. The SDP is obtained by imposing $\psi = \arg(z - z_s) + \arg[q''(z_s)]/2 = \pm\pi/2$ to result in $\arg(z - z_s) = 0$. Hence the SDP corresponds to the real $z$-axis on which $\arg(dz) = 0$. Alternatively, the SDP may be obtained by setting $\mathfrak{J}[q(z)] = \phi - r\sin\phi = \mathfrak{J}[q(z_s)] = 0$ to once again lead to $\phi = 0$ and $\arg(dz) = 0$. Because the SDP is already along the real axis, the $s$-plane is the same as the $z$-plane itself and no additional transformation is necessary. Using these values in equation (16.23), we arrive at

$$I(\Omega) = e^{-\Omega}\sqrt{\frac{2\pi}{\Omega|1|}}\, e^{i0} = \sqrt{\frac{2\pi}{\Omega}}\, e^{-\Omega}.$$

Therefore

$$\Gamma(\Omega + 1) \sim \sqrt{2\pi}\, \Omega^{\Omega+1/2}\, e^{-\Omega}. \tag{16.27}$$

This approximation is known as the *Stirling's* approximation of the gamma function. If $\Omega = 10$, $\Gamma(11) = 10! = 3\,628\,800$. The approximate value suggested by equation (16.27) is 3 598 695, which is true to within 0.83%. ■■

### 16.3.2 Stationary phase method

Consider the integral in equation (16.9) on the constant level contour $C$ which passes through the saddle point $z_s$ of order $(p - 1)$. Let the argument of $(z - z_s)$ on $C$ be denoted by $\beta_C$. The function $q(z) = u(x_s, y_s) + iv(x, y)$ has a varying imaginary part and the phase of $\exp[\Omega q(z)]$ varies most rapidly on $C$. Near the saddle point $q(z) \approx q(z_s) + (z - z_s)^p q^{(p)}(z_s)/p!$. The constant level contour $C$ is such that $\arg(z - z_s)^p + \arg[q^{(p)}(z_s)] = \pm\pi/2$ which implies that $p\beta_C = -\arg[q^{(p)}(z_s)] \pm \pi/2$. Let us set $q(z) = q(z_s) \pm it^p/p!$ with $t$ real so that

$$t = (z - z_s)[\mp \mathrm{i}q^{(p)}(z_s)]^{1/p}, \quad \mathrm{d}z = \frac{\mathrm{d}t}{[\mp \mathrm{i}q^{(p)}(z_s)]^{1/p}}. \tag{16.28}$$

Now $\arg(\mathrm{d}t) = 0 = \arg(\mathrm{d}z) + \arg[\mp \mathrm{i}q^{(p)}(z_s)]^{1/p}$. Hence the $p$th root of $[\mp \mathrm{i}q^{(p)}(z_s)]$ must be such that $\arg[\mp \mathrm{i}q^{(p)}(z_s)]^{1/p} = -\beta_C$. Using these

$$I(\Omega) = \int_C f(z)\mathrm{e}^{\Omega q(z)}\,\mathrm{d}z \sim \frac{\mathrm{e}^{\Omega q(z_s)}\mathrm{e}^{\mathrm{i}\beta_C}}{|q^{(p)}(z_s)|^{1/p}} \int_{t_1}^{t_2} f[z(t)]\mathrm{e}^{\pm \mathrm{i}\Omega t^p/p!}\,\mathrm{d}t, \tag{16.29}$$

where $(t_1, t_2)$ correspond to the end points of $C$. The phase of the exponential function within the integrand varies very rapidly, particularly for large $\Omega$, and its rate of change is equal to $\Omega t^{p-1}/(p-1)!$. Because of this rapid phase variation, the majority of the contribution to the integral comes from the vicinity of the stationary point $t = 0$ around which the phase variation is the least. Using

$$\int_0^\infty \mathrm{e}^{\pm \mathrm{i}at^p}\,\mathrm{d}t = \frac{\Gamma(1/p)}{pa^{1/p}}\mathrm{e}^{\pm \mathrm{i}\pi/2p}, \quad a > 0, \tag{16.30}$$

$$\int_{-\infty}^0 \mathrm{e}^{\pm \mathrm{i}at^p}\,\mathrm{d}t = \frac{\Gamma(1/p)}{pa^{1/p}}\mathrm{e}^{\pm \mathrm{i}(-1)^p\pi/2p}, \quad a > 0, \tag{16.31}$$

we obtain

$$\begin{aligned} I(\Omega) &\sim \frac{\mathrm{e}^{\Omega q(z_s)}\mathrm{e}^{\mathrm{i}\beta_C}}{|q^{(p)}(z_s)|^{1/p}} \int_{t_1}^{t_2} f[z(t)]\mathrm{e}^{\pm \mathrm{i}\Omega t^p/p!}\,\mathrm{d}t \\ &\approx \mathrm{e}^{\Omega q(z_s)} f(z_s)\mathrm{e}^{\mathrm{i}\beta_C}\frac{\Gamma(1/p)}{p}\left[\frac{p!}{\Omega|q^{(p)}(z_s)|}\right]^{1/p} \\ &\quad \times \begin{cases} \mathrm{e}^{\pm \mathrm{i}\pi/2p} & \text{if } t_1 = 0 \\ \mathrm{e}^{\pm \mathrm{i}(-1)^p\pi/2p} & \text{if } t_2 = 0 \end{cases}, \end{aligned} \tag{16.32}$$

where we have used the results in equations (16.30) and (16.31). The approximation in equation (16.32) arises because we have replaced $(t_1, t_2)$ with $(0, \infty)$ when $t_1 = 0$, $t_2 > 0$ or with $(-\infty, 0)$ when $t_1 < 0$, $t_2 = 0$ as the case may be. Note that the rate of decay with $\Omega$ is dictated by the order of the stationary point. The higher the order, the slower is the decay. Equation (16.32) is identical to equation (16.23) that is obtained using SDP. The difference between the steepest descent method and the stationary phase method is that, in the former, the contribution from the tail of the contour vanishes on account of magnitude decay along the SDP path, whereas, in the latter, the contribution from the tail diminishes because of the rapid phase variations.

**Example 16.3.2.1.** Find the leading behavior of $\int_0^{\pi/2} \mathrm{e}^{\mathrm{i}\Omega \cos z}\,\mathrm{d}z$ as $\Omega \to \infty$.
The function $q(z) = \mathrm{i}\cos z$ has a stationary point at $z = 0$. The contour $C$ has $z_1 = 0$, $z_2 = \pi/2$, $q(0) = \mathrm{i}$. Since $q''(0) = -\mathrm{i}$, we have a first order saddle point, $p = 2$. Also,

$\beta_C = 0$. Hence from equation (16.32) on using the bottom sign we obtain $I(\Omega) \sim e^{i\Omega}\Gamma(1/2)\sqrt{2/\Omega}\, e^{-i\pi/4}/2 = e^{i(\Omega-\pi/4)}\sqrt{\pi/2\Omega}$. ■■

**Example 16.3.2.2.** Obtain the leading behavior of $J_n(n)$ as $n \to \infty$.
When $n$ is an integer, the Bessel function $J_n(x)$ as the exact integral representation

$$J_n(x) = \frac{1}{\pi}\int_0^{\pi} \cos[x \sin\tau - n\tau]\, d\tau.$$

$$\text{Hence,} \quad J_n(n) = \Re\left[\frac{1}{\pi}\int_0^{\pi} e^{in(\sin\tau-\tau)}\, d\tau\right].$$

Comparing with the general form in equation (16.29), $f(\tau) = 1/\pi$, $q(\tau) = \sin\tau - \tau$, with the stationary point at $\tau = 0$. Furthermore, $\arg(\tau)|_C = 0$, $q(0) = q''(0) = 0$, $q^{(3)}(0) = -1$ implying that $p = 3$. Using these in equation (16.32), we obtain

$$J_n(n) \sim \Re\left[\frac{\Gamma(1/3)}{3\pi}\left(\frac{3!}{n}\right)^{1/3} e^{-i\pi/6}\right] = \frac{1}{\pi}2^{-2/3}3^{-1/6}n^{-1/3}\Gamma\left(\frac{1}{3}\right), \qquad n \to \infty.$$

■■

## 16.4 Examples in wave propagation

In this section we work out two examples pertaining to wave propagation illustrating the steepest descent technique. The problems considered are radiation by a planar aperture antenna and that of scattering of line source fields by a conducting wedge. In both cases the goal is to find the far-zone field starting from an exact integral representation.

**Example 16.4.0.1** Planar aperture antenna, $e^{-i\omega t}$ time convention. Consider a planar aperture antenna radiating in a homogeneous medium at wavenumber $k = \omega\sqrt{\mu\varepsilon}$. To simplify analysis, all fields and sources are assumed to be invariant of the $y$-axis and the aperture is located in the $x = 0$ plane. The $y$-component of the aperture electric field $E_y(x = 0, z)$ is specified in the $x = 0$ plane. Find an expression for the far-zone electric field in $x > 0$ using saddle point integration.

The equivalent magnetic current for this source flows in the $z$-direction and field produced will be $TE_z$. Owing to the invariance of the source w.r.t. the $y$-axis, the aperture field can be represented in terms of its single Fourier transform $\widetilde{E}_y(k_z)$ where

$$E_y(x = 0, z) = \frac{1}{2\pi}\int_{-\infty}^{\infty} \widetilde{E}_y(k_z)e^{ik_z z}\, dk_z, \quad \widetilde{E}_y(k_z) = \int_{-\infty}^{\infty} E_y(0, z)e^{-ik_z z}\, dz. \tag{16.33}$$

The field in the half-space $x > 0$ can be expressed in terms of the aperture field as

$$E_y(x, z) = \frac{1}{2\pi}\int_{-\infty}^{\infty} \widetilde{E}_y(k_z)e^{ik_z z}e^{-\gamma x}\, dk_z, \tag{16.34}$$

where

$$\gamma(k_z) = (k_z^2 - k^2)^{1/2} = \begin{cases} -\mathrm{i}\sqrt{k^2 - k_z^2}, & |k_z| < k \\ \sqrt{k_z^2 - k_2}, & |k_z| > k \end{cases}. \tag{16.35}$$

We require the real part $\Re(\gamma) > 0$ on the top sheet. The imaginary part $\Im(\gamma)$ can, however, take positive or negative values on the top sheet. Writing $k_z - k = r_2 \exp(\mathrm{i}\phi_2)$, $k_z + k = r_1 \exp(\mathrm{i}\phi_1)$, $\gamma = \sqrt{r_1 r_2} \exp[\mathrm{i}(\phi_1 + \phi_2)/2] = \gamma_{\text{top}}$, we require $-\pi < (\phi_1 + \phi_2) < \pi$ on the top sheet. The angles $\phi_1$ and $\phi_2$ are counted differently on the bottom sheet so that $\gamma_{\text{bot}} = -\gamma_{\text{top}}$. On the branch cuts $\Re(\gamma) = 0$. The value of $\gamma$ is uniquely defined on a sheet if the angles $\phi_1$ and $\phi_2$ are fixed at a point on it. We take $\phi_1 = 0$, $\phi_2 = -\pi$ to correspond to $\gamma = -ik$ on the top sheet when $k_z = 0$. On the bottom sheet $\phi_1 = 0$, $\phi_1 = +\pi$ when $k_z = 0$. It is seen that the range of $\phi_1$ and $\phi_2$ on the top sheet are $-\pi/2 < \phi_1 < 2\pi$ and $-3\pi/2 < \phi_2 < \pi$. Regions ① – ④ belong to the top sheet, while ⑤ – ⑧ belong to the bottom sheet, as shown in figures 16.11 and 16.12. Angle $\phi_1$ is identical in the pair of regions ① and ⑤ as well as in the pair ② and ⑥. In each pair, the angle $\phi_2$ differs by $\pm 2\pi$ between its members. Likewise angle $\phi_2$ is identical in the pair ③ and ⑦ as well as ④ and ⑧, where $\phi_1$ differs by $\pm 2\pi$.

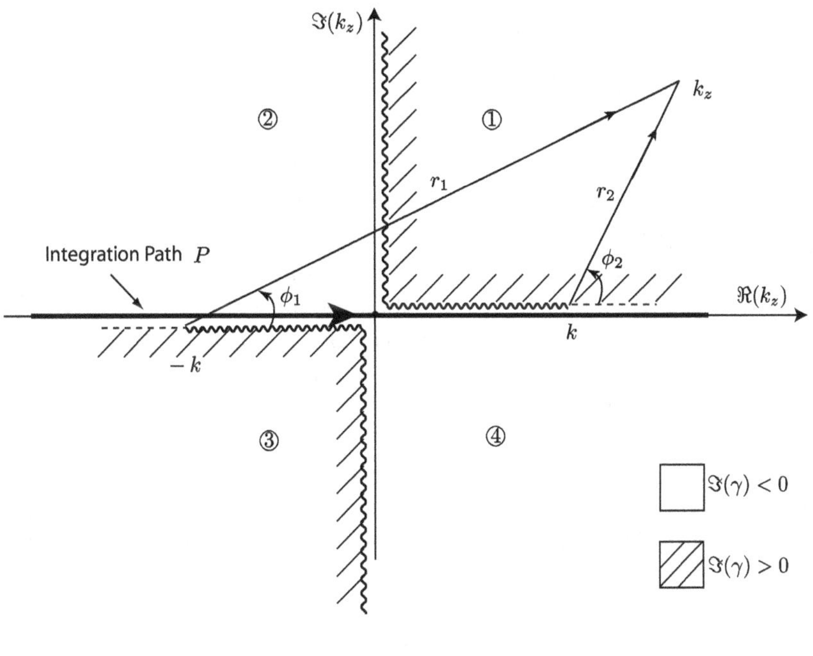


Top Sheet ($\Re(\gamma) > 0$)

**Figure 16.11.** Various regions of the top sheet.

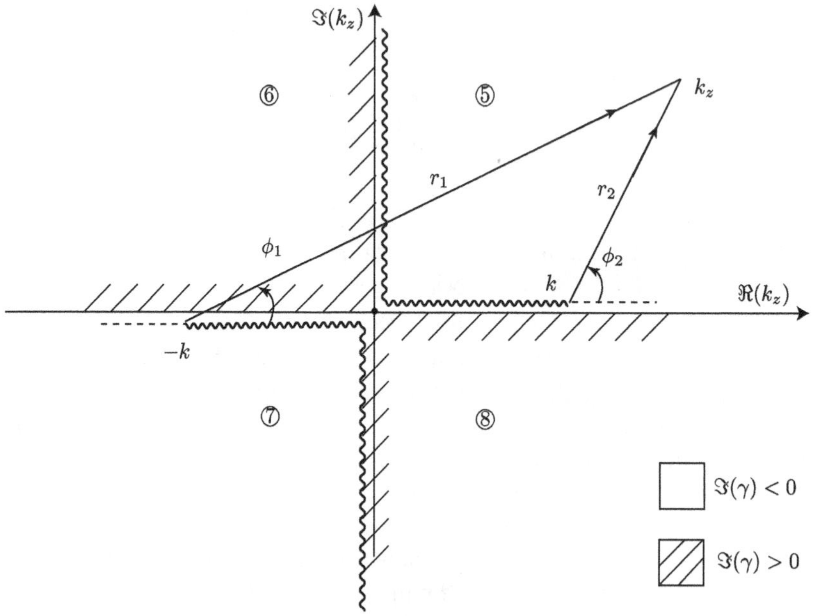


Bottom Sheet ($\Re(\gamma) < 0$)

**Figure 16.12.** Various regions of the bottom sheet.

Letting $x = \rho \cos\phi$, $z = \rho \sin\phi$ in equation (16.34) results in

$$E_y(x, z) = \frac{1}{2\pi}\int_{-\infty}^{\infty} \widetilde{E}_y(k_z)\mathrm{e}^{\rho(-\gamma\cos\phi + \mathrm{i}k_z\sin\phi)}\,\mathrm{d}k_z. \tag{16.36}$$

Hence, on comparing with the general form in equation (16.9), $f(k_z) = \widetilde{E}_y(k_z)/2\pi$, $q(k_z) = -\gamma\cos\phi + \mathrm{i}k_z\sin\phi$, which is analytic on the top sheet. We would like to evaluate $E_y(x, z)$ for large $\rho$ and $|\phi| < \pi/2$. Now $q'(k_z) = \mathrm{i}\sin\phi - k_z\cos\phi/\gamma$, $q''(k_z) = k^2\cos\phi/\gamma^3$, where $k^2 = k_z^2 - \gamma^2$ has been used in the latter. At the saddle point $q'(k_z = k_{zs}) = 0$. Hence saddle points are defined by $k_{zs} = \mathrm{i}\gamma_s\tan\phi$. Possible solutions are $k_{zs} = k\sin\phi$, $\gamma_s = -\mathrm{i}k\cos\phi$ and $k_{zs} = -k\sin\phi$, $\gamma_s = +\mathrm{i}k\cos\phi$. Referring to table 16.2, it is clear that the former belongs to region ④, whereas the latter belongs to region ⑥. Since our original integration contour is on the top sheet, we pick $k_{zs} = k\sin\phi$ and $\gamma_s = -\mathrm{i}k\cos\phi$. At this saddle point $q(k_{zs}) = \mathrm{i}k$, $q''(k_{zs}) = -\mathrm{i}\sec^2\phi/k$, $\arg[q''(k_{zs})] = -\pi/2$. Hence $k_{zs}$ is a saddle point of first order. The steepest descent path through $k_{zs}$ is defined as $\Im[q(k_z)] = \Im[q(k_{zs})] = k$. Therefore the SDP is given by $k_{z\mathrm{r}}\sin\phi - \gamma_{\mathrm{i}}\cos\phi = k$, where $\gamma = \gamma_{\mathrm{r}} + \mathrm{i}\gamma_{\mathrm{i}}$, $k_z = k_{z\mathrm{r}} + \mathrm{i}k_{z\mathrm{i}}$. Near the saddle point, the SDP is given by the line $\arg(k_z - k_{zs}) + \arg[q''(k_{zs})]/2 = \pm\pi/2 \Longrightarrow \arg(k_z - k_{zs}) = 3\pi/4, -\pi/4$. Choosing the latter (the original path traverses the real axis in the direction of increasing values of $\Re(k_z)$, hence the deformed SDP path must also traverse in the direction of increasing $\Re(k_z)$, which implies that $\arg(z - z_s) = -\pi/4$ and not $3\pi/4$) we obtain

**Table 16.2.** $k_z$ and $\gamma$ for various regions in the complex $k_z$-plane.

| Region | $\phi_1$ | $\phi_2$ | $k_z = k_{zr} + ik_{zi}$ | $\gamma = \gamma_r + i\gamma_i$ |
|---|---|---|---|---|
| ① | $0 < \phi_1 < \pi/2$ | $0 < \phi_2 < \pi$ | $k_{zr} > 0,\ k_{zi} > 0$ | $\gamma_r > 0, \gamma_i > 0$ |
| ② | $0 < \phi_1 < \pi$ | $-3\pi/2 < \phi_2 < -\pi$ | $k_{zr} < 0, k_{zi} > 0$ | $\gamma_r > 0, \gamma_i < 0$ |
| ③ | $\pi < \phi_1 < 2\pi$ | $-\pi < \phi_2 < -\pi/2$ | $k_{zr} < 0, k_{zi} < 0$ | $\gamma_r > 0, \gamma_i > 0$ |
| ④ | $-\pi/2 < \phi_1 < 0$ | $-\pi/2 < \phi_2 < 0$ | $k_{zr} > 0, k_{zi} < 0$ | $\gamma_r > 0, \gamma_i < 0$ |
| ⑤ | $0 < \phi_1 < \pi/2$ | $2\pi < \phi_2 < 3\pi$ | $k_{zr} > 0, k_{zi} > 0$ | $\gamma_r < 0, \gamma_i < 0$ |
| ⑥ | $0 < \phi_1 < \pi$ | $\pi/2 < \phi_2 < \pi$ | $k_{zr} < 0, k_{zi} > 0$ | $\gamma_r < 0, \gamma_i > 0$ |
| ⑦ | $-\pi < \phi_1 < 0$ | $-\pi < \phi_2 < -\pi/2$ | $k_{zr} < 0, k_{zi} > 0$ | $\gamma_r < 0, \gamma_i < 0$ |
| ⑧ | $3\pi/2 < \phi_1 < 2\pi$ | $-\pi/2 < \phi_2 < 0$ | $k_{zr} > 0, k_{zi} < 0$ | $\gamma_r < 0, \gamma_i > 0$ |

$\arg(dk_z)|_{z_s}$, SDP $= -\pi/4$. Using these results in equation (16.23) and assuming that there are no singularities between $P$ and the SDP, we obtain

$$\begin{aligned} E_y(x, z) &\sim \frac{1}{2\pi} \cos\phi \widetilde{E}_y(k\sin\phi)\sqrt{\frac{2\pi k}{\rho}}\, e^{ik\rho} e^{-i\pi/4} \\ &= \sqrt{\frac{k}{2\pi i\rho}}\, e^{ik\rho} \cos\phi \widetilde{E}_y(k\sin\phi). \end{aligned} \tag{16.37}$$

Thus the far-zone field behaves like a cylindrical wave whose amplitude is dictated by the Fourier transform of the aperture field. The factor $\cos\phi$ may be thought of as the element factor and the factor $\widetilde{E}_y(k\sin\phi)$ as the array factor.

A closer look at the SDP reveals that a portion of it resides on the bottom sheet and this was implicitly taken into account in formula (16.23) by means of working in the $s$-plane. The path taken by the entire SDP in the complex $k_z$-plane is made clear by studying its defining equation:

$$\gamma_i = k_{zr}\tan\phi - k\sec\phi \quad \text{(SDP in the complex } k_z\text{-plane)}. \tag{16.38}$$

For large negative values of $k_{zr}$, the SDP starts in region ② where $\gamma_i < 0$. The SDP intersects the imaginary $k_z$-axis at $k_{zi} = k\tan\phi$ where $\gamma_i = -k\sec\phi$. It intersects the real axis at $k_{zr} = k\sin\phi$ where $\gamma_i = -k\cos\phi$. For $k_{zr}$ in the range $0 < k_{zr} < k\sin\phi$, i.e. in the first quadrant of $k_z$ plane, $\gamma_i < 0$ on the SDP. But $\gamma_i < 0$ only in region ⑤ of the first quadrant. Hence that portion of the SDP lies on the bottom sheet. The SDP crosses the branch cut on the imaginary $k_z$-axis in the upper half space and enters the bottom sheet. At $k_{zr} = k\sin\phi$ it again crosses the branch cut on the real $k_z$-axis and re-enters the top sheet. The complete trajectory of the SDP is shown in figure 16.13.

A cleaner evaluation of the integral in equation (16.36) is facilitated by the transformation $k_z \sin\alpha$, which implies that $\gamma = \mp ik\cos\alpha$. When $\alpha = 0$, $k_z = 0$, and $\gamma = \mp ik$. Hence the upper sign corresponds to the top sheet and the lower one to the

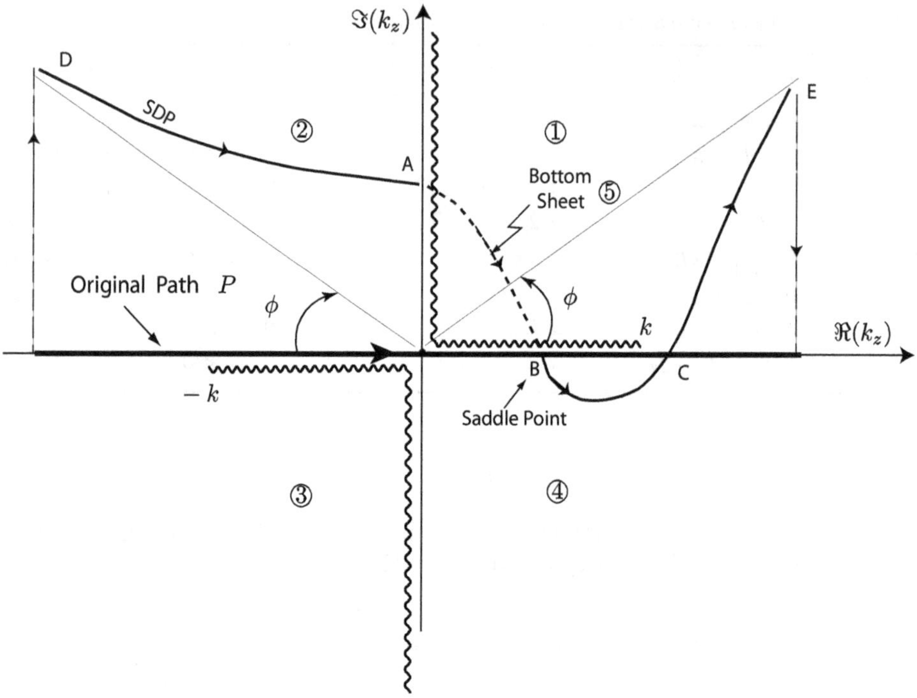


**Figure 16.13.** Original path and the complete SDP in the complex $k_z$-plane.

bottom sheet. Writing $\alpha = \alpha_r + i\alpha_i$, we explicitly write the transformation relations as

$$k_z = k_{zr} + ik_{zi} = k \sin\alpha = k \sin\alpha_r \cosh\alpha_i + ik \cos\alpha_r \sinh\alpha_i, \tag{16.39}$$

$$\begin{aligned}\gamma = \gamma_r + i\gamma_i &= -ik \cos\alpha \\ &= k \sin\alpha_r \sinh\alpha_i - ik \cos\alpha_r \cosh\alpha_i, \quad \text{(top sheet)}\end{aligned} \tag{16.40}$$

$$\begin{aligned}\gamma = \gamma_r + i\gamma_i &= ik \cos\alpha \\ &= -k \sin\alpha_r \sinh\alpha_i + ik \cos\alpha_r \cosh\alpha_i, \quad \text{(bottom sheet).}\end{aligned} \tag{16.41}$$

The inverse mapping for the top sheet is obtained by recognizing that $k_z + \gamma = -ik(\cos\alpha + i\sin\alpha) = -ik\exp(i\alpha) \Longrightarrow \alpha = -i\ln[i(k_z + \gamma)/k]$, where $\ln[r\exp(i\theta)] = \ln r + i\theta$, $-\pi < \theta < \pi$. To see how the various regions in the $k_z$ plane are mapped into the $\alpha$-plane, we consider $k_z = R\exp(i\theta) \approx R\exp[i(\phi_1 + \phi_2)/2]$, $R \gg k$. Table 16.3 shows the values of $k_z$, $\gamma$, and $\alpha$ where $\varepsilon$, $\varepsilon_1$, $\varepsilon_2$, and $\delta$ are small positive numbers. In terms of $\alpha$, the electric field is expressed as

**Table 16.3.** Mapping between the $k_z$-plane and the $\alpha$-plane.

| Region | $k_{zr}$ | $k_{zi}$ | $\gamma_r$ | $\gamma_i$ | $\alpha$ |
|---|---|---|---|---|---|
| ① | $R \gg k$ | $\delta > 0$ | $R$ | $\delta$ | $(\pi/2 + \varepsilon) - i\infty$ |
| ② | $-R$ | $\delta$ | $R - \varepsilon_1$ | $-(\delta + \varepsilon_2)$ | $\pi/4 + i\infty$ |
| ③ | $-R$ | $-\delta$ | $R - \varepsilon_1$ | $(\delta + \varepsilon_2)$ | $-3\pi/4 + i\infty$ |
| ④ | $R$ | $-\delta$ | $R - \varepsilon_1$ | $-(\delta + \varepsilon_2)$ | $(\pi/2 - \varepsilon) - i\infty$ |
| ⑤ | $R$ | $\delta$ | $-R(1 - \varepsilon_1)$ | $-i\delta/(1 - \varepsilon_1)$ | $(\pi/2 - \varepsilon) + i\infty$ |

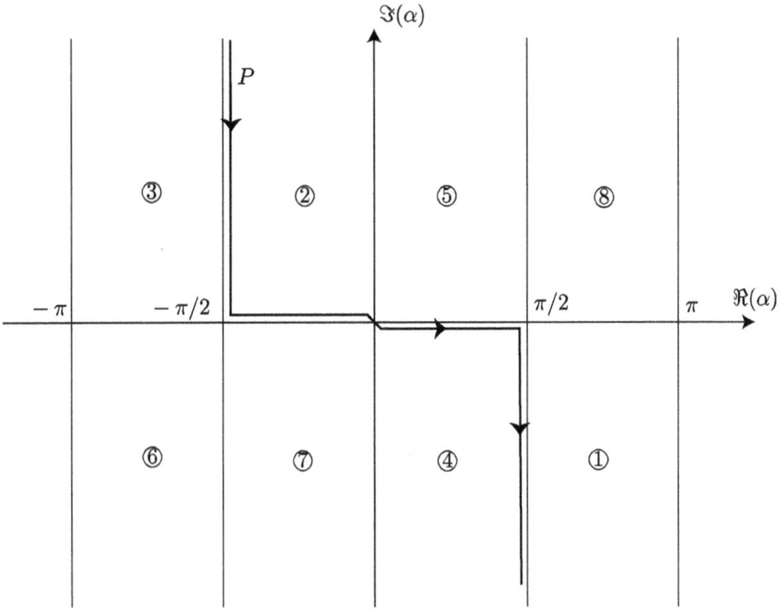


**Figure 16.14.** Mapping from the $k_z$-plane into the $\alpha$-plane and the original contour of integration.

$$
\begin{aligned}
E_y(x, z) &= \frac{1}{2\pi}\int_P \widetilde{E}_y(k_z = k\sin\alpha)\mathrm{e}^{\rho q_1(\alpha)}\frac{\mathrm{d}k_z}{\mathrm{d}\alpha}\,\mathrm{d}\alpha \\
&= \frac{k}{2\pi}\int_P \cos\alpha\widetilde{E}_y(\alpha)\mathrm{e}^{\rho q_1(\alpha)}\,\mathrm{d}\alpha,
\end{aligned}
\qquad (16.42)
$$

where $q_1(\alpha) = q(k_z)|_{k_z = k\sin\alpha} = -\gamma\cos\phi + ik_z\sin\phi = ik\cos(\alpha - \phi)$. The various regions of the complex $k_z$-plane as well as the original contour are mapped into the complex $\alpha$-plane as shown in figure 16.14. Notice that there are no branch points in the $\alpha$-plane and the two Riemann sheets of the $k_z$-plane are mapped distinctly in the $\alpha$-plane. The saddle point $\alpha_s$ in the $\alpha$-plane is the solution of $q_1'(\alpha_s) = 0 \Longrightarrow \alpha_s = \phi$. Furthermore, $q_1(\alpha_s) = ik$, $q_1''(\alpha_s) = -ik \neq 0$. The SDP is governed by $\Im[ik\cos(\alpha - \phi)] = k$ or $\cos(\alpha_r - \phi)\cosh\alpha_i = 1$ and is shown in figure 16.15. It passes through regions ①, ②, ④, and ⑤ as already remarked. The various points A,

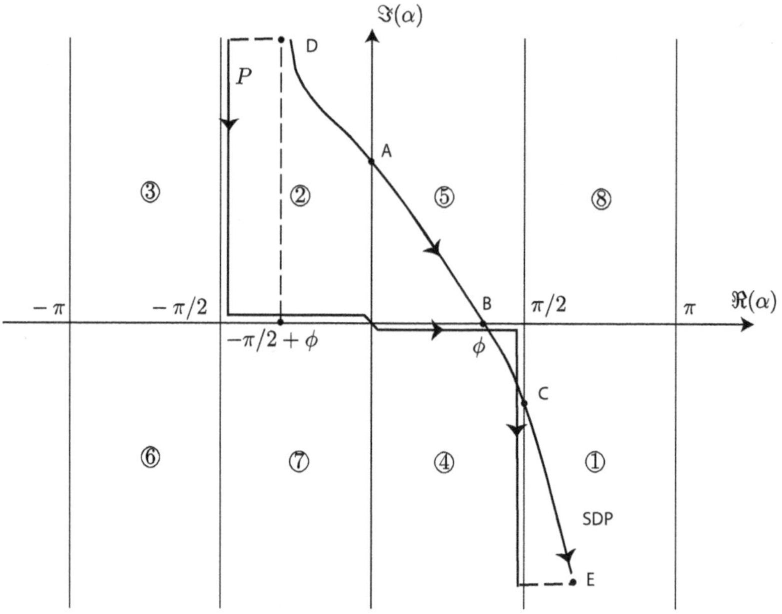


**Figure 16.15.** SDP in the $\alpha$-plane.

**Table 16.4.** Various points on the SDP.

| Point on SDP | $\alpha_r$ | $\alpha_i$ | $k_z/k$ | $\gamma/k$ |
|---|---|---|---|---|
| A | 0 | $\cosh^{-1}(\sec\phi)$ | $i\tan\phi$ | $-i\sec\phi$ |
| B | $\phi$ | 0 | $\sin\phi$ | $-i\cos\phi$ |
| C | $\pi/2$ | $-\cosh^{-1}(\csc\phi)$ | $\csc\phi$ | $\cot\phi$ |
| D | $-\pi/2+\phi$ | $\infty$ | $\infty e^{i(\pi-\phi)}$ | $\infty e^{-i\phi}$ |
| E | $\pi/2+\phi$ | $-\infty$ | $\infty e^{i\phi}$ | $\infty e^{i\phi}$ |

B, C, D, E, and their counterparts in the $k_z$-plane are calculated in table 16.4 and indicated in figures 16.15 and 16.13. Near the saddle point, the SDP behaves as $\arg(\alpha-\alpha_s)+\arg[q_1''(\alpha_s)]/2=\pm\pi/2 \Longrightarrow \arg(\alpha-\alpha_s)=3\pi/4,\ -\pi/4$. The SDP obviously has $\arg(d\alpha)=-\pi/4$ at the saddle point. Deforming the original contour $P$ into the SDP and assuming that there are no singularities of $\tilde{E}_y(\alpha)$ between them, we have

$$E_y(x, z) \sim \cos\phi\tilde{E}_y(k\sin\phi)e^{ik\rho}\sqrt{\frac{k}{2\pi\rho}}e^{-i\pi/4}, \quad \rho\to\infty, \tag{16.43}$$

which is the same as equation (16.37) that is obtained using $k_z$ integration.

It may be remarked that the branch cuts shown in figure 16.3 are not the only ones possible for solving this problem. Identical results will be obtained if the branch cuts are chosen according to $\Re(k_z)=\pm k$, although some of the regions will be redistributed as shown in figures 16.16 and 16.17. Note that some parts of the top sheet with these branch cuts will contain improper waves, but these will cause no difficulty in the saddle point integration. With this choice of the branch cuts, the SDP will lie entirely on the top sheet as shown in figure 16.18 and has been preferred by some workers while carrying out asymptotic evaluation [7]. ■■

**Example 16.4.0.2** Scattering of line-source fields by a conducting wedge, $e^{-i\omega t}$ time convention. Consider a $z$-polarized, time harmonic line source exciting a PEC wedge with a wedge angle $(2-\nu)\pi$ as shown in figure 16.19. The source is either an electric line source carrying a current $I_o$ or a magnetic line source carrying a voltage $V_o$ and is located at $(\rho', \phi')$. In the former case the field is $TM_z$, whereas in the latter it is $TE_z$. The medium surrounding the wedge is lossless with electrical parameters $(\varepsilon, \mu)$ and the radian frequency of operation is $\omega$. The wedge parameter $\nu$ need not be an integer. Denote the wavenumber as $k = \omega\sqrt{\varepsilon\mu}$ and the intrinsic impedance as $\eta = \sqrt{\mu/\varepsilon}$. It is desired to find the field at the point $(\rho, \phi)$ for large $k\rho'$ and $k\rho$.

Let $\rho_> = \max(\rho, \rho')$, $\rho_< = \min(\rho, \rho')$, $\xi^{\mp} = (\phi \mp \phi')$, $\varepsilon_m = 2$, $m \geqslant 2$, $\varepsilon_1 = 1$. The boundary conditions that need to be satisfied on the wedge surface $\phi = 0$, $\phi = \nu\pi$ are $E_z = 0$ for the $TM_z$ case and $\partial H_z/\partial\phi = 0$ for the $TE_z$ case. The solution can be obtained in a series form using cylindrical harmonics and is found in many books including Harrington [8]. For a time convention $\exp(-i\omega t)$, it is given by

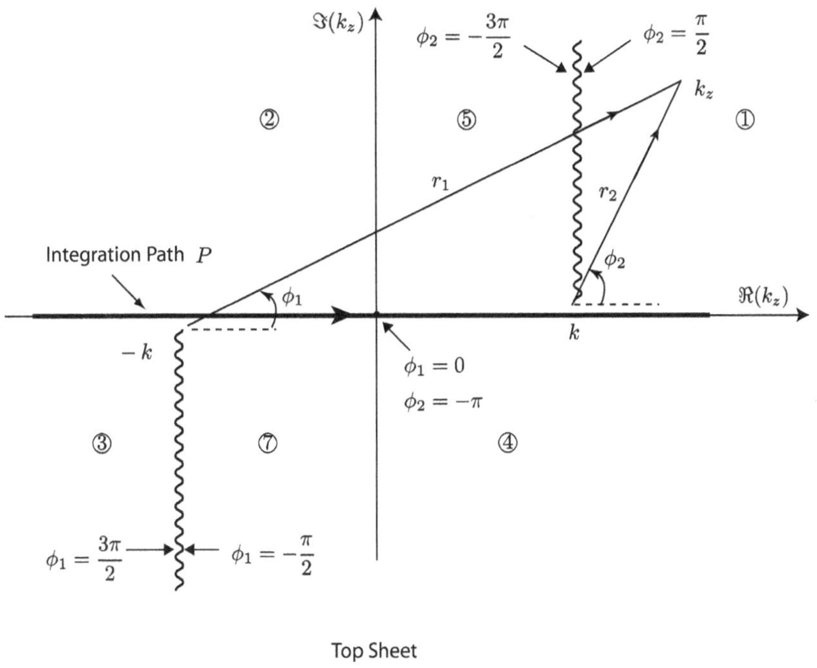


**Figure 16.16.** Top sheet of complex $k_z$-plane with branch cuts defined by $\Re(k_z) = \pm k$.

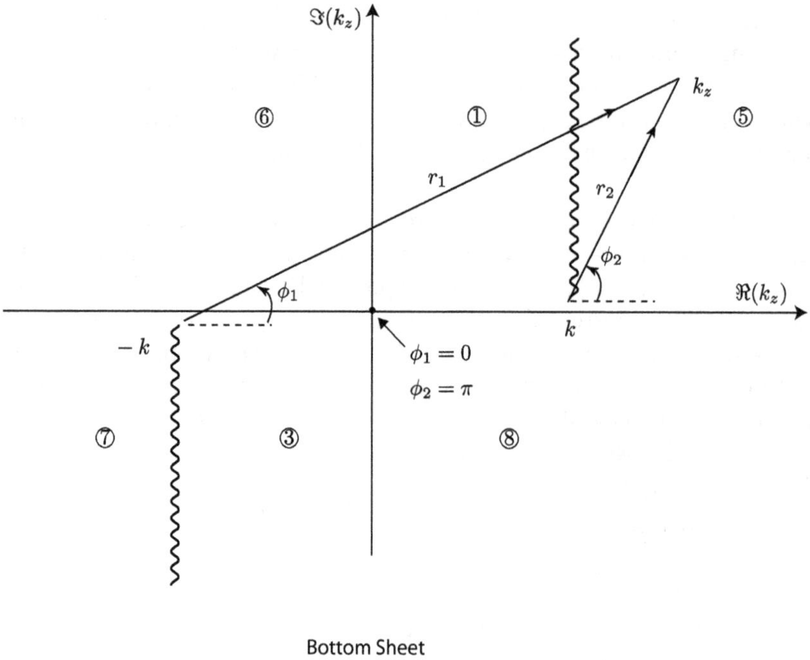


**Figure 16.17.** Bottom sheet of complex $k_z$-plane with branch cuts defined by $\Re(k_z) = \pm k$.

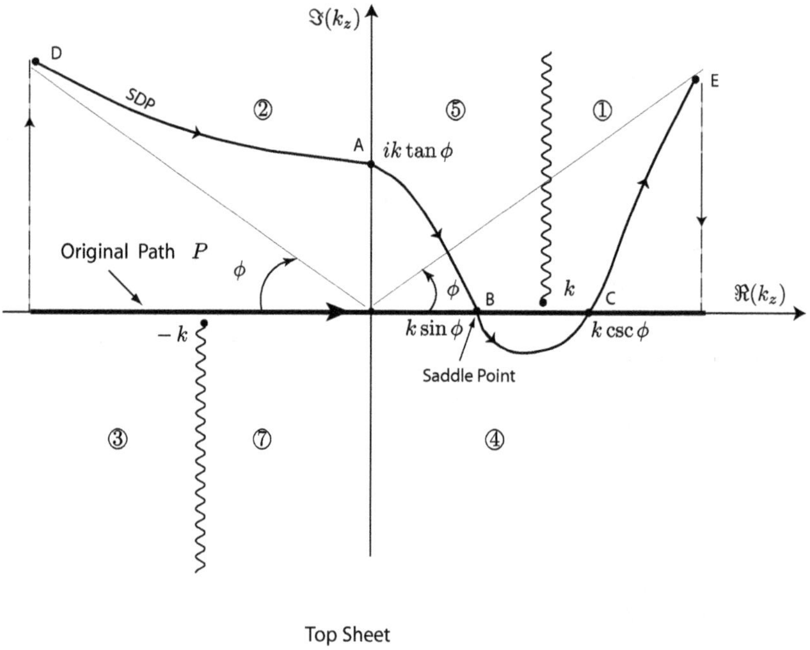


**Figure 16.18.** Original contour and the SDP for vertical branch cuts.

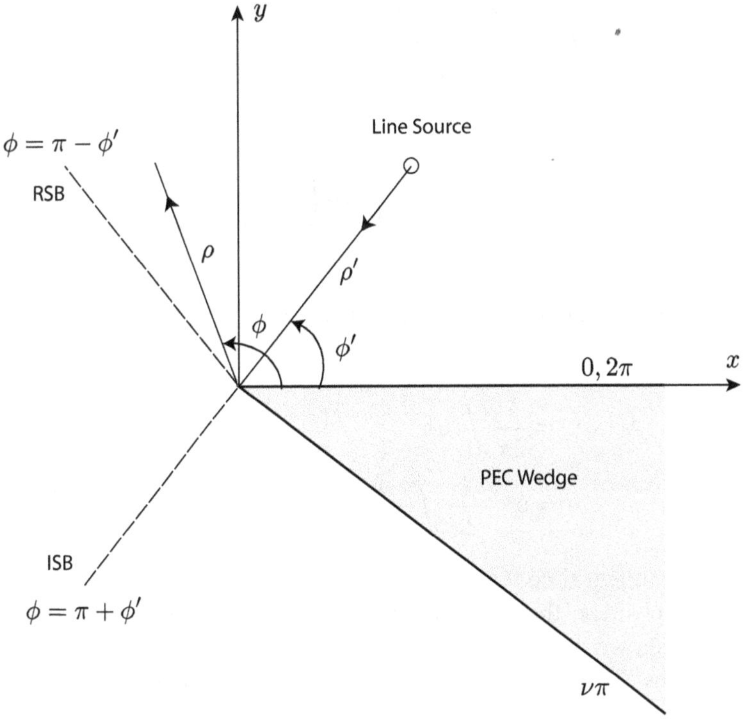


**Figure 16.19.** A perfectly conducting wedge of angle $(2 - \nu)\pi$ excited by a line source at $(\rho', \phi')$.

$$E_z = -\frac{I_o k\eta}{4} G_1, \quad H_z = \frac{V_o k}{4\eta} G_2,$$

where

$$G_{\frac{1}{2}} = \frac{1}{\nu} \sum_{m=0}^{\infty} \varepsilon_m J_{\frac{m}{\nu}}(k\rho_<) H^{(1)}_{\frac{m}{\nu}}(k\rho_>) \left[ \cos\left(\frac{m}{\nu}\xi^-\right) \mp \cos\left(\frac{m}{\nu}\xi^+\right) \right]. \tag{16.44}$$

It may be noted that the solution is well behaved at all points in the region exterior to the wedge; in particular, the field is finite along the incident shadow boundary (ISB) $\phi = \phi' + \pi$ and the reflection shadow boundary (RSB) $\phi = \pi - \phi'$. If $k\rho_> \gg 1$, then it is reasonable to use the asymptotic expression of the Hankel function

$$H^{(1)}_{\frac{m}{\nu}}(k\rho_>) \sim \sqrt{\frac{2}{\pi k\rho_>}}\, e^{i\left(k\rho_> - \frac{\pi}{4} - \frac{m}{\nu}\frac{\pi}{2}\right)}. \tag{16.45}$$

Even though equation (16.45) is only valid for $\frac{m}{\nu} < k\rho_>$, we will still continue with the approximation because it will lead to the ray-optics solution. The result of this approximation will be seen later to be the cause of the field becoming discontinuous at the ISB and RSB. With this approximation,

$$G_{\frac{1}{2}} \sim \sqrt{\frac{2}{\pi k\rho_>}}\,e^{i(k\rho_> - \pi/4)} K_{\frac{1}{2}}(k\rho_<), \tag{16.46}$$

where

$$K_{\frac{1}{2}}(k\rho_<) = \frac{1}{\nu}\sum_{m=0}^{\infty} \varepsilon_m J_{\frac{m}{\nu}}(k\rho_<) e^{-i\frac{m}{\nu}\frac{\pi}{2}}\left[\cos\left(\frac{m}{\nu}\xi^-\right) \mp \cos\left(\frac{m}{\nu}\xi^+\right)\right]. \tag{16.47}$$

Consider now the asymptotic evaluation if $K_{\frac{1}{2}}(k\rho_<)$ for large $k\rho_<$. To carry this out we need an appropriate integral representation of $K_{\frac{1}{2}}$. We use the following integral representations for $J_{\frac{m}{\nu}}(\cdot)$

$$\begin{aligned} J_{\frac{m}{\nu}}(k\rho_<) &= \frac{1}{2\pi}\int_{C_1} e^{i\left[k\rho_<\cos\alpha + \frac{m}{\nu}\left(\alpha - \frac{\pi}{2}\right)\right]}\,d\alpha \\ &= e^{i\nu 2\pi}\frac{1}{2\pi}\int_{C_2} e^{i\left[k\rho_<\cos\alpha - \frac{m}{\nu}\left(\alpha + \frac{\pi}{2}\right)\right]}\,d\alpha. \end{aligned} \tag{16.48}$$

In order that the integral converges for positive $k\rho_<$, we require that the imaginary part of $\cos\alpha$ be positive, that is $\sin\alpha_r \sinh\alpha_i < 0$, where $\alpha = \alpha_r + i\alpha_i$. The contours $C_1$ and $C_2$ are shown in figure 16.20 along with the regions (shaded) where the integral converges. The contour $C_1$ must be confined to the upper half-space, while $C_2$ must be confined to the lower half-space. Both $C_1$ and $C_2$ start and terminate at infinity in the shaded regions, but their exact shape in other regions is immaterial because the integrand is analytic in any finite region. We can use either $C_1$ or $C_2$ in equation (16.47) after expressing $\cos\left(\frac{m}{\nu}\xi^-\right)$ and $\cos\left(\frac{m}{\nu}\xi^+\right)$ as the sum of exponentials so that the resulting exponentials will all have arguments of either $\frac{m}{\nu}(\alpha + \xi^- - \pi)$ or $\frac{m}{\nu}(\alpha + \xi^+ - \pi)$. Also, the $m = 0$ term will be identical in either exponential term. Using these we obtain

$$\begin{aligned} K_{\frac{1}{2}} &= \frac{1}{\nu}\sum_{m=0}^{\infty} \varepsilon_m J_{\frac{m}{\nu}}(k\rho_<) e^{-i\frac{m\pi}{2\nu}}\frac{1}{2}\left[e^{i\frac{m}{\nu}\xi^-} + e^{-i\frac{m}{\nu}\xi^-} \mp e^{i\frac{m}{\nu}\xi^+} \mp e^{-i\frac{m}{\nu}\xi^+}\right] \\ &= \frac{1}{2\pi\nu}\left\{\int_{C_1} e^{ik\rho_<\cos\alpha}\sum_{m=0}^{\infty}\frac{1}{2}\varepsilon_m\left[e^{i\frac{m}{\nu}(\alpha+\xi^-)} \mp e^{i\frac{m}{\nu}(\alpha+\xi^+)}\right]e^{-i\frac{m}{\nu}\pi}\,d\alpha\right. \\ &\quad\left. + \int_{C_2} e^{ik\rho_<\cos\alpha}\sum_{m=0}^{\infty}\frac{1}{2}\varepsilon_m\left[e^{-i\frac{m}{\nu}(\alpha+\xi^-)} \mp e^{-i\frac{m}{\nu}(\alpha+\xi^+)}\right]e^{+i\frac{m}{\nu}\pi}\,d\alpha\right\} \\ &= \frac{1}{2\pi\nu}\left\{\int_{C_1} e^{ik\rho_<\cos\alpha}\sum_{m=1}^{\infty}\left[e^{i\frac{m}{\nu}(\alpha+\xi^- - \pi)} \mp e^{i\frac{m}{\nu}(\alpha+\xi^+ - \pi)}\right]d\alpha\right. \\ &\quad\left. + \int_{C_2} e^{ik\rho_<\cos\alpha}\sum_{m=0}^{\infty}\left[e^{-i\frac{m}{\nu}(\alpha+\xi^- - \pi)} \mp e^{-i\frac{m}{\nu}(\alpha+\xi^+ - \pi)}\right]d\alpha\right\}. \end{aligned}$$

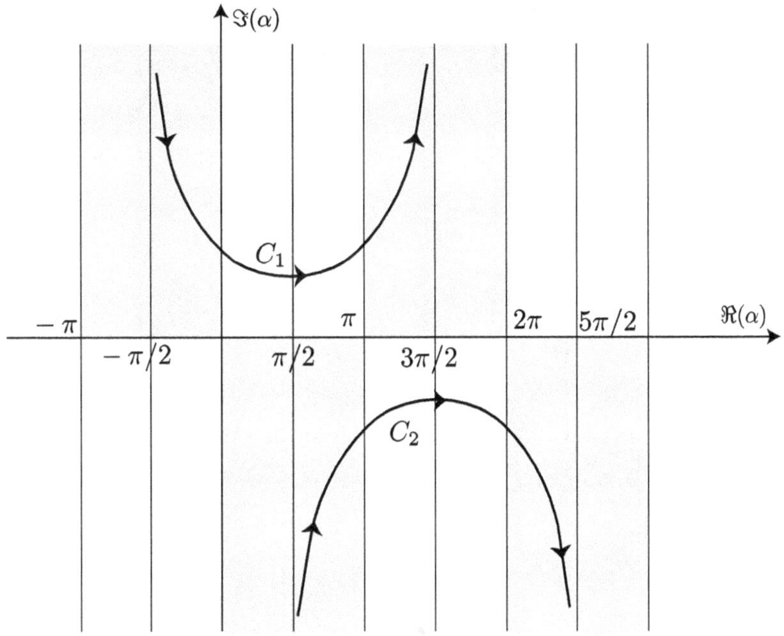


**Figure 16.20.** Contours for representing the Bessel function $J_{\frac{m}{\nu}}(x)$.

The following results follow easily from summation of geometric series:

$$\sum_{m=1}^{\infty} e^{i\frac{m}{\nu}(\alpha+\xi^- -\pi)} = \frac{1}{\left[e^{-i\left(\frac{\alpha+\xi^- -\pi}{\nu}\right)} - 1\right]} = -\frac{1}{2}\left[1 - i\cot\left(\frac{\alpha+\xi^- - \pi}{2\nu}\right)\right]$$

$$\sum_{m=0}^{\infty} e^{-i\frac{m}{\nu}(\alpha+\xi^- -\pi)} = \frac{1}{\left[1 - e^{-i\left(\frac{\alpha+\xi^- -\pi}{\nu}\right)}\right]} = \frac{1}{2}\left[1 - i\cot\left(\frac{\alpha+\xi^- - \pi}{2\nu}\right)\right].$$

Note that the above series diverge for $\alpha + \xi^- - \pi = 0, \pm\pi, \pm 2\pi\cdots$. On using these expressions and noting further that

$$\frac{1}{2}\int_{C_2-C_1} e^{ik\rho_< \cos\alpha}\, d\alpha = \pi\left[J_0(k\rho_<) - J_0(k\rho_<)\right] = 0,$$

we can write

$$K_{\frac{1}{2}} = I(k\rho_<, \nu, \xi^-) \mp I(k\rho_<, \nu, \xi^+), \tag{16.49}$$

where

$$I(k\rho_<, \nu, \xi) = \frac{1}{4\pi i\nu}\int_{C_2-C_1} e^{ik\rho_< \cos\alpha} \cot\left(\frac{\alpha + \xi - \pi}{2\nu}\right) d\alpha. \tag{16.50}$$

We wish to evaluate the integral $I$ for large values of the argument $k\rho_<$. It is to be borne in mind that the integrand has poles in the complex $\alpha$-plane and care must be

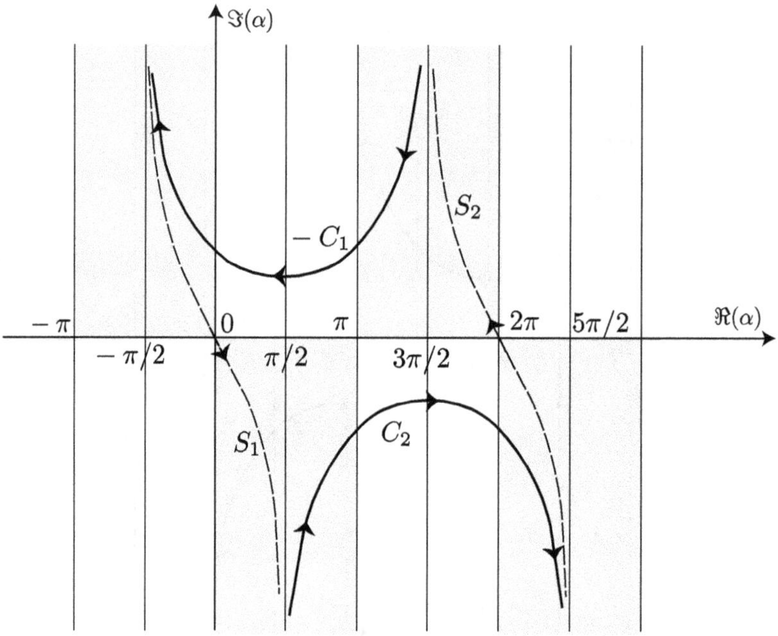


**Figure 16.21.** Adding the SDP contours $S_1$ and $S_2$ to make $C_2 - C_1$ a closed contour $C_T$.

exercised to include their contribution. For the function $q(\alpha) = \mathrm{i}\cos\alpha$ the stationary points are given by $q'(\alpha_s) = -\mathrm{i}\sin\alpha_s = 0 \Longrightarrow \alpha_s = 0, \pm\pi, \pm 2\pi, \cdots$, at which points $q''(\alpha_s) = -\mathrm{i}\cos\alpha_s = -\mathrm{i}, \mathrm{i}, -\mathrm{i}, \ldots$, and $q(\alpha_s) = \mathrm{i}, -\mathrm{i}, \mathrm{i}, \ldots$ respectively. We are particularly interested in the saddle points $\alpha_s = 0, 2\pi$. The SDPs through $\alpha_s = 0, 2\pi$ are given by $\Im[q(\alpha)] = \cos\alpha_r \cosh\alpha_i = \Im[q(\alpha_s)] = 1$. These SDPs $S_1$ and $S_2$ through $\alpha_s = 0$ and $\alpha_s = 2\pi$, respectively, are shown in figure 16.21. Consider the closed contour $C_T = C_2 - C_1 + S_1 + S_2$ with the orientations of $S_1$ and $S_2$ as shown in the figure. To determine the angle at which the SDP contours cross the real axis, we look at the small argument expansion of the SDP contours near $\alpha = \alpha_s$. Let us write $\alpha_r = \alpha_s + x$, $\alpha_i = y$. Then $\cos\alpha_r \cosh\alpha_i = 1 \Longrightarrow \cos x \cosh y = 1$. For $x$, $y$ small, $\cos x \approx 1 - x^2/2$, $\cosh y \approx 1 + y^2/2$. Therefore $y^2 = x^2/(1 - x^2/2) \approx x^2 \Longrightarrow y = \pm x$. Hence $\arg(\mathrm{d}\alpha)|_{\alpha_s, S_{1,2}} = -\pi/4, 3\pi/4$. We express the integral as

$$\begin{aligned} I(k\rho_<, \nu, \xi) &= \frac{1}{4\pi\mathrm{i}\nu}\int_{C_2 - C_1} \mathrm{e}^{\mathrm{i}k\rho_< \cos\alpha} \cot\left(\frac{\alpha + \xi - \pi}{2\nu}\right) \mathrm{d}\alpha, \\ &= \oint_{C_T} \frac{\mathrm{e}^{k\rho_< q(\alpha)}\,\mathrm{d}\alpha}{4\pi\mathrm{i}\nu \tan\left(\frac{\alpha+\xi-\pi}{2\nu}\right)} - \int_{S_1} \frac{\mathrm{e}^{k\rho_< q(\alpha)}\,\mathrm{d}\alpha}{4\pi\mathrm{i}\nu \tan\left(\frac{\alpha+\xi-\pi}{2\nu}\right)} \\ &\quad - \int_{S_2} \frac{\mathrm{e}^{k\rho_< q(\alpha)}\,\mathrm{d}\alpha}{4\pi\mathrm{i}\nu \tan\left(\frac{\alpha+\xi-\pi}{2\nu}\right)}. \end{aligned} \tag{16.51}$$

The integrand of equation (16.51) has simple poles on the real axis at $\alpha = \alpha_p$ given by $(\alpha_p + \xi - \pi)/2\nu = N\pi$, $N = 0, \pm 1, \pm 2, \cdots$. However, we are only interested in those poles that lie within $C_T$ or that lie in the range $0 \leqslant \alpha_p \leqslant 2\pi \Longrightarrow -\pi \leqslant -\xi + 2\nu N\pi \leqslant \pi$. Applying Cauchy's residue theorem

$$\oint_{C_T} \frac{e^{k\rho_< q(\alpha)}\, d\alpha}{4\pi i\nu \tan\left(\frac{\alpha+\xi-\pi}{2\nu}\right)} = \frac{1}{2\nu}\mathscr{R}\mathrm{es}\left[\frac{e^{k\rho_< q(\alpha)}\, d\alpha}{\tan\left(\frac{\alpha+\xi-\pi}{2\nu}\right)}\right]_{\alpha=\alpha_p}.$$

For the poles $\alpha_p = -\xi + 2\nu N\pi + \pi$, the residues are easily evaluated as

$$\mathscr{R}\mathrm{es} = 2\nu e^{ik\rho_< \cos(-\xi+2\pi\nu N+\pi)} = 2\nu e^{-ik\rho_< \cos(\xi-2\pi\nu N)}. \tag{16.52}$$

Therefore

$$\oint_{C_T} \frac{e^{k\rho_< q(\alpha)}\, d\alpha}{4\pi i\nu \tan\left(\frac{\alpha+\xi}{2\nu}\right)} = \sum_N e^{-ik\rho_< \cos(\xi-2\pi\nu N)} U[\pi - |\xi - 2\pi\nu N|], \tag{16.53}$$

where $U(\cdot)$ is a unit step function and is included to ensure that we only include the pole contribution if the poles are enclosed by $C_T$. For evaluating the contribution due the SDP paths $S_{\frac{1}{2}}$, we apply the formula (16.23) for large $k\rho_<$ to arrive at

$$\begin{aligned}\int_{S_1} \frac{e^{k\rho_< q(\alpha)}\, d\alpha}{4\pi i\nu \tan\left(\frac{\alpha+\xi-\pi}{2\nu}\right)} &\sim \frac{e^{k\rho_< q(\alpha_s)}}{4\pi i\nu \tan\left(\frac{\alpha_s+\xi-\pi}{2\nu}\right)} \times \sqrt{\frac{2\pi}{k\rho_< |q''(\alpha_s)|}}\, e^{i\arg(d\alpha)|_{\alpha_s, S_1}} \\ &= \frac{e^{ik\rho_<}}{4\pi i\nu \tan\left(\frac{\xi-\pi}{2\nu}\right)}\sqrt{\frac{2\pi}{k\rho_<}}\, e^{-i\pi/4}, \quad k\rho_< \gg 1.\end{aligned} \tag{16.54}$$

Similarly,

$$\int_{S_2} \frac{e^{k\rho_< q(\alpha)}\, d\alpha}{4\pi i\nu \tan\left(\frac{\alpha+\xi-\pi}{2\nu}\right)} \sim \frac{e^{ik\rho_<}}{4\pi i\nu \tan\left(\frac{\xi+\pi}{2\nu}\right)}\sqrt{\frac{2\pi}{k\rho_<}}\, e^{i3\pi/4}, \quad k\rho_< \gg 1. \tag{16.55}$$

Inserting these results into equation (16.51) we finally obtain for the integral

$$\begin{aligned} I(k\rho_<, \nu, \xi) &= \sum_N e^{-ik\rho_< \cos(\xi-2\pi\nu N)} U[\pi - |\xi - 2\pi\nu N|] + \frac{e^{i(k\rho_< - \frac{\pi}{4})}}{4\pi i\nu}\sqrt{\frac{2\pi}{k\rho_<}} \\ &\quad \left\{\frac{1}{\tan\left(\frac{\pi+\xi}{2\nu}\right)} + \frac{1}{\tan\left(\frac{\pi-\xi}{2\nu}\right)}\right\} \\ &= \sum_N e^{-ik\rho_< \cos(\xi-2\pi\nu N)} U[\pi - |\xi - 2\pi\nu N|] \\ &\quad + \frac{e^{i(k\rho_< + \frac{\pi}{4})}}{\sqrt{2\pi k\rho_<}} \frac{\frac{1}{\nu}\sin\left(\frac{\pi}{\nu}\right)}{\cos\left(\frac{\pi}{\nu}\right) - \cos\left(\frac{\xi}{\nu}\right)}. \end{aligned} \tag{16.56}$$

Note that this result is only valid when the pole $\alpha_p$ is not near the saddle point $\alpha_s$, i.e. when $\xi + 2\nu N\pi - \pi \neq 0, 2\pi$. Since there are two values of $\xi$, i.e. $\xi^-$ and $\xi^+$, this means that $\phi \neq \phi' - 2\pi\nu N \pm \pi$, or $\phi \neq -\phi' - 2\pi\nu N \pm \pi$. At the reflection shadow boundary, $\phi = \pi - \phi'$ or $\xi^+=\pi$. At the incident shadow boundary, $\phi = \phi' + \pi$ or $\xi^- = \pi$. Thus, at both of these shadow boundaries, the poles coincide with the saddle point at $\alpha_s = 0$ for $N = 0$ and the expressions for $I$ become infinity. The expression for $K_{\frac{1}{2}}$ becomes

$$\begin{aligned} K_{\frac{1}{2}} = \sum_{N=0,\pm1,\ldots} & \left\{ e^{-ik\rho_< \cos(\xi^- - 2\pi\nu N)} U[\pi - |\xi^- - 2\pi\nu N|] \right. \\ & \left. \mp e^{-ik\rho_< \cos(\xi^+ - 2\pi\nu N)} U[\pi - |\xi^+ - 2\pi\nu N|] \right\} \\ & + \frac{e^{i(k\rho_< + \frac{\pi}{4})}}{\sqrt{2\pi k\rho_<}} \frac{1}{\nu} \sin\left(\frac{\pi}{\nu}\right) \times \left\{ \frac{1}{\cos\left(\frac{\pi}{\nu}\right) - \cos\left(\frac{\xi^-}{\nu}\right)} \mp \frac{1}{\cos\left(\frac{\pi}{\nu}\right) - \cos\left(\frac{\xi^+}{\nu}\right)} \right\}. \end{aligned} \tag{16.57}$$

The above expression yields infinite values at the shadow boundaries $\xi^- = \pi$ or $\xi^+=\pi$. To obtain better approximations at the shadow boundaries, one needs to use the modified steepest descent method that addresses the presence of the poles near saddle points. The consequence of this is that certain transition functions are introduced that smoothly connect the field values across the shadow boundaries without making it discontinuous at the shadow boundaries.

The first term in the summation on the rhs of equation (16.57) is associated with the direct wave between the source and observation points. The second term is associated with the wave reflected off the face $\phi = 0$ of the wedge. The remaining terms are associated with the wave diffracted by the wedge. The first terms exists for $\xi^- < \pi$, the second for $\xi^+ < \pi$, but the diffracted term will be present at all angles. However, it will be significant only when the first two terms disappear, namely in the deep shadow region where $\phi > \pi + \phi'$. It is then clear that the pole terms contribute to the geometric optics field and the asymptotic contribution from the SDP contours gives rise to the diffracted field. It is convenient to express the final field in terms of the incident field. The incident fields $E_z^i$ and $H_z^i$ are given by

$$\begin{aligned} E_z^i(\rho, \phi) = \frac{-I_0 k\eta}{4} H_0^{(1)}(k|\boldsymbol{\rho} - \boldsymbol{\rho}'|) \sim \frac{-I_0 k\eta}{4} \sqrt{\frac{2}{i\pi k\rho'}} e^{ik\rho'} e^{-ik\rho\cos(\phi-\phi')} \\ := E_0\, e^{-ik\rho\cos\xi^-}, \; k\rho' \gg 1, \end{aligned} \tag{16.58}$$

$$H_z^i(\rho, \phi) = \frac{-V_0 k}{4\eta} H_0^{(1)}(k|\boldsymbol{\rho} - \boldsymbol{\rho}'|) \sim -\, H_0\, e^{-ik\rho\cos\xi^-}, \tag{16.59}$$

where

$$E_0 = E_z^{\mathrm{i}}(\rho = 0, \phi) = -\frac{I_0 k\eta}{4}\sqrt{\frac{2}{\mathrm{i}\pi k\rho'}}\,\mathrm{e}^{\mathrm{i}k\rho'},$$
$$H_0 = -H_z^{\mathrm{i}}(\rho = 0, \phi) = \frac{V_0 k}{4\eta}\sqrt{\frac{2}{\mathrm{i}\pi k\rho'}}\,\mathrm{e}^{\mathrm{i}k\rho'}. \tag{16.60}$$

For the case $\rho' > \rho$ so that $\rho_> = \rho'$, $\rho_< = \rho$, we then obtain

$$\begin{aligned} E_z = E_0 \sum_{N=0,\pm1,\ldots} &\left\{\mathrm{e}^{-\mathrm{i}k\rho\cos(\xi^- - 2\pi\nu N)}U[\pi - |\xi^- - 2\pi\nu N|]\right. \\ &\left. - \,\mathrm{e}^{-\mathrm{i}k\rho\cos(\xi^+ - 2\pi\nu N)}U[\pi - |\xi^+ - 2\pi\nu N|]\right\} \\ &+ \frac{E_0\mathrm{e}^{\mathrm{i}k\rho}}{\sqrt{\rho}}V_{\mathrm{s}}(\phi, \phi'), \end{aligned} \tag{16.61}$$

where

$$V_{\mathrm{s}}(\phi, \phi') = \frac{\sqrt{\mathrm{i}}}{\nu\sqrt{2\pi k}}\sin\left(\frac{\pi}{\nu}\right)\left\{\frac{1}{\cos\left(\frac{\pi}{\nu}\right) - \cos\left(\frac{\xi^-}{\nu}\right)} - \frac{1}{\cos\left(\frac{\pi}{\nu}\right) - \cos\left(\frac{\xi^+}{\nu}\right)}\right\}, \tag{16.62}$$

is the diffraction coefficient for the Dirichlet boundary condition. Similarly,

$$\begin{aligned} H_z = H_0 \sum_{N=0,\pm1,\ldots} &\left\{\mathrm{e}^{-\mathrm{i}k\rho\cos(\xi^- - 2\pi\nu N)}U[\pi - |\xi^- - 2\pi\nu N|]\right. \\ &\left. + \,\mathrm{e}^{-\mathrm{i}k\rho\cos(\xi^+ - 2\pi\nu N)}U[\pi - |\xi^+ - 2\pi\nu N|]\right\} \\ &+ \frac{H_0\mathrm{e}^{\mathrm{i}k\rho}}{\sqrt{\rho}}V_{\mathrm{h}}(\phi, \phi'), \end{aligned} \tag{16.63}$$

where

$$V_{\mathrm{h}}(\phi, \phi') = \frac{\sqrt{\mathrm{i}}}{\nu\sqrt{2\pi k}}\sin\left(\frac{\pi}{\nu}\right)\left\{\frac{1}{\cos\left(\frac{\pi}{\nu}\right) - \cos\left(\frac{\xi^-}{\nu}\right)} + \frac{1}{\cos\left(\frac{\pi}{\nu}\right) - \cos\left(\frac{\xi^+}{\nu}\right)}\right\}, \tag{16.64}$$

is the diffraction coefficient for the Neumann boundary condition. This same problem is worked in [9] for the $\mathrm{e}^{j\omega t}$ time convention. ■■

## 16.5 Modified saddle point technique

When the integrand of equation (16.9) has a pole in the vicinity of the saddle point, extra care must be taken to include its influence and a new asymptotic expression will result. It may be recalled from the standard SDP technique that the integrand was assumed to be smoothly varying in a region about the saddle point that permitted its expansion in the form of a Taylor series. This assumption breaks down if the integrand has a pole singularity in the vicinity of the saddle point and the procedure must be modified. Essentially, we subtract the pole term from the integrand and evaluate this term separately. The remaining part of the integrand will be free of poles and can be evaluated by the standard method of steepest descent. We will illustrate the procedure here by considering the example of the classical problem of dipole radiating in the presence of a conducting half-space.

**Example 16.5.0.1** Sommerfeld half-space problem, $e^{-i\omega t}$ time convention, hyperbolic branch cuts.
Consider a vertical Hertzian dipole of current moment $I_0\ell$ operating at a radian frequency $\omega$ and located at a height $h > 0$ above a lossy flat medium as shown in figure 16.22. Unless otherwise specified, the lower medium will be taken to be Earth. Both regions are assumed to be non-magnetic with permeability $\mu = \mu_0$, its value in free-space. The lower medium has permittivity $\varepsilon_r$, and a conductivity $\sigma$. A complex dielectric constant $\kappa = \varepsilon_r + i\frac{\sigma}{\omega\varepsilon_0}$ may be defined by combining both of these electrical parameters into one. Determine the far-zone fields in the upper half-space using modified saddle point technique.

Let the wavenumbers in the air region and in the lossy region be denoted by $k_0$ and $k_1 = \sqrt{\kappa}k_0$, respectively. The field will be $TM_z$ in both the upper and lower regions and, owing to the azimuthal symmetry, will depend only on the coordinates $(\rho, z)$ in the cylindrical coordinate system. Assuming an $e^{-i\omega t}$ time convention, the expression for the $z$-directed magnetic vector potential $A_z$ for $z > 0$ can be easily derived by enforcing the continuity of the tangential electric and magnetic fields at the interface $z = 0$ as

$$A_z = \frac{I_0\ell\mu_0}{4\pi R_1}e^{ik_0R_1} - \frac{iI_0\ell\mu_0}{4\pi}\int_0^\infty \frac{k_\rho\Gamma(k_\rho)}{k_{z0}}J_0(k_\rho\rho)e^{ik_{z0}(z+h)}\,dk_\rho, \tag{16.65}$$

where the reflection coefficient, $\Gamma(k_\rho)$, is given by

$$\Gamma(k_\rho) = \frac{k_{z1} - \kappa k_{z0}}{k_{z1} + \kappa k_{z0}}, \tag{16.66}$$

$k_{z0} = \sqrt{k_0^2 - k_\rho^2}$, $k_{z1} = \sqrt{\kappa k_0^2 - k_\rho^2}$, $k_0 = \omega\sqrt{\mu_0\varepsilon_0}$, $\rho$ is the radial distance from dipole, $R_1 = \sqrt{(z-h)^2 + \rho^2}$ is the distance between the source and observation points, and $J_0(\cdot)$ is the Bessel function of the first kind of order zero. The first term is the incident field produced by the dipole and the second term represents the field scattered from the interface. We may rewrite equation (16.65) by adding and subtracting the field scattered by a perfect magnetic conductor as

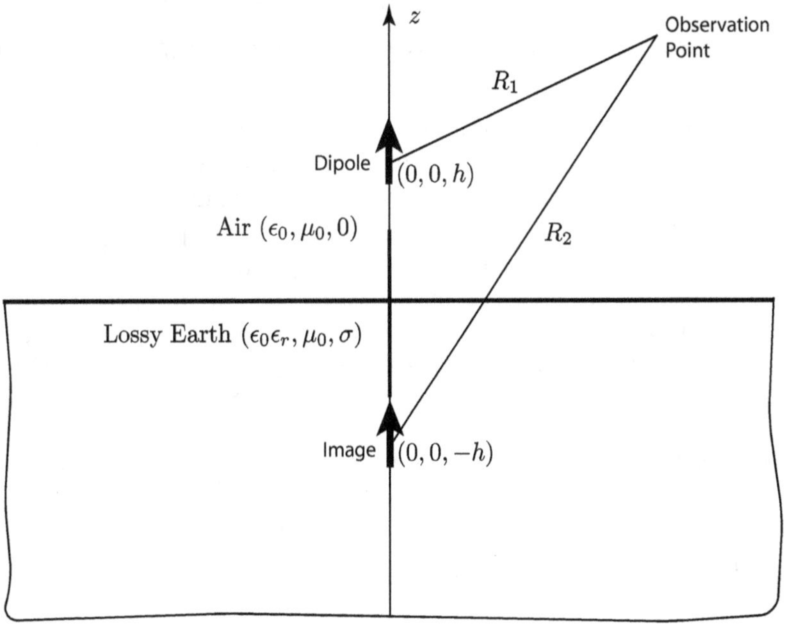


**Figure 16.22.** A Hertzian dipole located above a lossy Earth.

$$\begin{aligned} A_z = &\frac{I_0\ell\mu_0}{4\pi R_1}\mathrm{e}^{\mathrm{i}k_0R_1} - \frac{I_0\ell\mu_0}{4\pi R_2}\mathrm{e}^{\mathrm{i}k_0R_2} \\ &+ \frac{\mathrm{i}2I_0\ell\mu_0\kappa}{4\pi}\int_0^\infty \frac{k_\rho}{\kappa k_{z0} + k_{z1}} J_0(k_\rho\rho)\mathrm{e}^{\mathrm{i}k_{z0}(z+h)}\,\mathrm{d}k_\rho, \end{aligned} \tag{16.67}$$

where $R_2 = \sqrt{(z+h)^2 + \rho^2}$ is the distance from the image location of the dipole to the observation point. The integral in equation (16.67) will be labeled as $I$. By writing $2J_0(\cdot) = H_0^{(1)}(\cdot) + H_0^{(2)}(\cdot)$ and using $H_0^{(2)}(\cdot) = -H_0^{(1)}(\cdot)$, the integral $I$ can be rewritten as

$$I = \frac{\mathrm{i}I_0\ell\mu_0\kappa}{4\pi}\int_{-\infty}^\infty \frac{k_\rho}{\kappa k_{z0} + k_{z1}} H_0^{(1)}(k_\rho\rho)\mathrm{e}^{\mathrm{i}k_{z0}(z+h)}\,\mathrm{d}k_\rho. \tag{16.68}$$

We are interested in evaluating this integral in the far-zone. Multiplying and dividing the integrand with the large value approximation of the Hankel function will enable us to express equation (16.68) as

$$\begin{aligned} I = &\frac{\mathrm{i}I_0\ell\mu_0\kappa\sqrt{2}\,\mathrm{e}^{-\mathrm{i}\pi/4}}{4\pi\sqrt{\pi\rho}}\int_{-\infty}^\infty \frac{\sqrt{k_\rho}\,\mathrm{e}^{\mathrm{i}k_{z0}(z+h)+\mathrm{i}k_\rho\rho}}{\sqrt{k_0}\,(\kappa k_{z0} + k_{z1})} \\ &\times\left[H_0^{(1)}(k_\rho\rho)\mathrm{e}^{-\mathrm{i}k_\rho\rho+\mathrm{i}\pi/4}\sqrt{\frac{\pi k_\rho\rho}{2}}\right]\mathrm{d}k_\rho, \end{aligned} \tag{16.69}$$

$$\triangleq f(\rho)\int_{-\infty}^{\infty}\frac{\sqrt{k_\rho}}{\sqrt{k_0}\,(\kappa k_{z0}+k_{z1})}g_1(k_\rho,\,\rho)e^{ik_{z0}(z+h)+ik_\rho\rho}\,dk_\rho, \tag{16.70}$$

where $f(\rho)$ represents the factor on the right-hand side of equation (16.69) outside the integral and $g_1(k_\rho, \rho)$ represents the function within the square brackets. In the complex $k_\rho$ plane, $g_1(k_\rho, \rho)$ will be a smoothly varying function, free of any singularities, and approach a value 1 for large $|k_\rho|\rho$.

The integrand has branch points at $k_\rho = \pm k_0$ and $k_\rho = \pm k_1$ and as such the complex $k_\rho$-plane will be comprised of four sheets. On the top sheet (proper sheet), $k_{z0}$ and $k_{z1}$ will have positive imaginary parts and the branch cuts will be chosen to correspond to $\Im(k_{z0}) = 0$ and $\Im(k_{z1}) = 0$ and will appear as shown in figure 16.23. The field contribution arising from wrapping around the branch cut at $k_1$ will be less important than that arising from the branch cut at $k_0$ because $|k_1| \gg |k_0|$ and the field attenuates exponentially at least at the rate dictated by $\exp(k_1\,\rho)$. This can be easily demonstrated by considering the contour integral around the branch cut at $+k_1$ and using the following estimates: $|\sqrt{k_\rho/k_0}/(\kappa k_{z0} + k_{z1})| \leqslant r_1$, $|g_1(k_\rho, \rho| \approx 1$, $|\exp[ik_{z0}(z + h)]| \leqslant 1$, $|\exp(ik_\rho\rho)| = |\exp(ik_1\rho)|\exp[-(\Im(k_\rho - k_1)\rho]$ to result in

$$\begin{aligned}|I_1| &= \left|\int_{C_1}\frac{\sqrt{k_\rho}}{\sqrt{k_0}\,(\kappa k_{z0}+k_{z1})}g_1(k_\rho,\,\rho)e^{ik_{z0}(z+h)+ik_\rho\rho}\,dk_\rho\right| \\ &\leqslant r_1|e^{ik_1\rho}|\int_{C_1}e^{-\Im(k_\rho-k_1)\rho}\,dk_\rho = r_2|e^{ik_1\rho}|,\end{aligned} \tag{16.71}$$

where $r_1$ and $r_2$ are some positive real constants. For most cases of practical interest the wavenumber $k_1$ will have a positive imaginary part and the integral will decay exponentially with range $\rho$. For this reason we shall ignore the branch points $\pm k_1$ altogether in the subsequent analysis. Also, owing to the presence of the factor $\sqrt{k_\rho}$, the integrand in equation (16.70) will also have a branch point singularity at $k_\rho = 0$ with a branch cut running along the negative real-axis. But all of the contours encountered in the following will traverse the complex $k_\rho$-plane above this branch cut and consequently its presence will not affect the discussion. For this reason we will also ignore it altogether. The other singularities present are the poles of the integrand in equation (16.70), which are obtained by setting the denominator to zero: $\kappa k_{z0} = -k_{z1}$. In the $k_\rho$-plane, they occur at

$$k_\rho = \pm k_\text{p} = \pm\sqrt{\frac{\kappa}{\kappa+1}}k_0. \tag{16.72}$$

Because $\kappa$ has positive real and imaginary parts, these poles will lie in the first and third quadrants of the complex $k_\rho$-plane. For example for typical soil, $\varepsilon_\text{r} = 12$, $\sigma = 10$ mS m$^{-1}$. At a frequency of 10 MHz, $\kappa = 21.63\angle 0.983$ and $k_\text{p} = (0.973\,6 + i0.036\,5)k_0$. For sea water, $\varepsilon_\text{r} = 81$, $\sigma = 2000$ mS m$^{-}$. At a frequency of 10 MHz, $\kappa \approx 3600\angle 1.55$ and $k_\text{p} \approx 0.999\,994k_0$. So, the poles are very close to $\pm k_0$ for most cases encountered in radiowave propagation over the Earth.

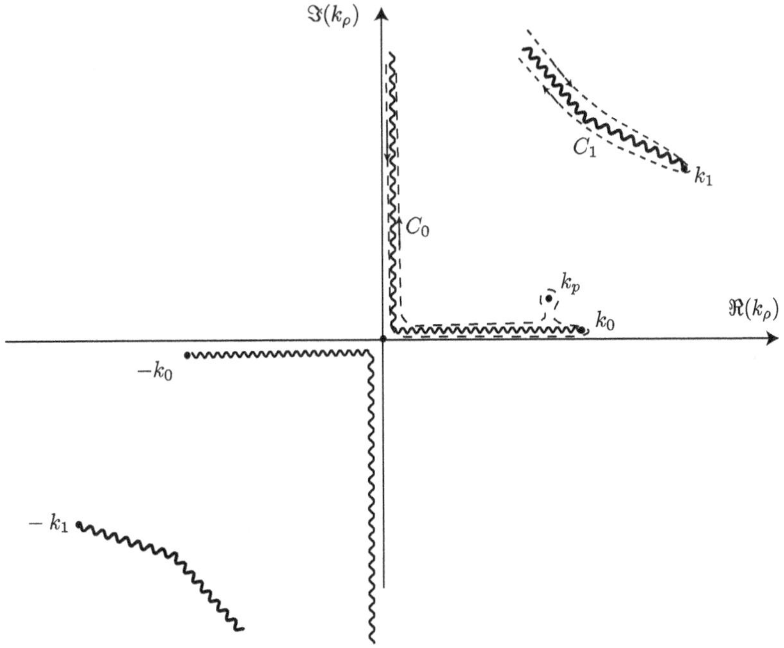


**Figure 16.23.** Sommerfeld branch cuts and the surface wave pole in the complex $k_\rho$-plane. Branch cuts arise from the presence of $k_{z0}$ and $k_{z1}$.

For the purpose of steepest descent evaluation of the integral for large distances, we make the following change of variables: $k_\rho = k_0 \sin\alpha$, $\alpha = \alpha_r + i\alpha_i$, $z + h = R_2 \cos\theta$, $\rho = R_2 \sin\theta$, where $\theta$ is the observation angle made from the dipole with respect to the positive $z$-axis. We then get $k_{z0} = k_0 \cos\alpha$, $k_{z1} = k_0\sqrt{\kappa - \sin^2\alpha}$ and $\sin\alpha_p = \sqrt{\kappa/(\kappa+1)}$ for the location of the poles in the $\alpha$-plane. As remarked previously, for most cases of practical interest, the poles will be located in the $k_\rho$-plane close to $\pm k_0$, which implies that $\alpha_p \approx \pi/2$. Our concern would then be when the saddle point of the exponent is close to $\pi/2$. Using the above change of variables in equation (16.70) and simplifying, we arrive at

$$I = f(\rho)\int_C \frac{\sqrt{\sin\alpha}\,\cos\alpha}{\kappa\cos\alpha + \sqrt{\kappa - \sin^2\alpha}}\, e^{ik_0R_2\cos(\alpha-\theta)} g(\alpha, R_2)\, d\alpha, \qquad (16.73)$$

where the contour of integration is shown in figure 16.24 and the function $g_1(k_\rho, \rho)$ has been renamed as $g(\alpha, R_2)$. As discussed under example 16.4.0.1 (see figure 16.15), the saddle point for the function $q_1(\alpha) = i\cos(\alpha - \theta)$ occurs at $\alpha_s = \theta$ and the SDP is given by $\cos(\alpha_r - \theta)\cosh\alpha_i = 1$. Hence, when the observation angle $\theta$ is close to 90°, the saddle point will be close to a pole of the integrand. Near the saddle point the function $g(\alpha_s, R_2) = H_0^{(1)}(r)\exp(-ir + i\pi/4)\sqrt{\pi r/2}$, $r = k_0R_2\sin^2\theta$. For $r = 1$, $g(\alpha_s, R_2) \approx 0.96 - i0.1 \approx 1$. Hence it is safe to ignore the function $g(\alpha, R_2)$ as long as $R_2/\lambda_0 \geqslant 1/[\pi(1 - \cos 2\theta)]$. For example, for observation points along the ground,

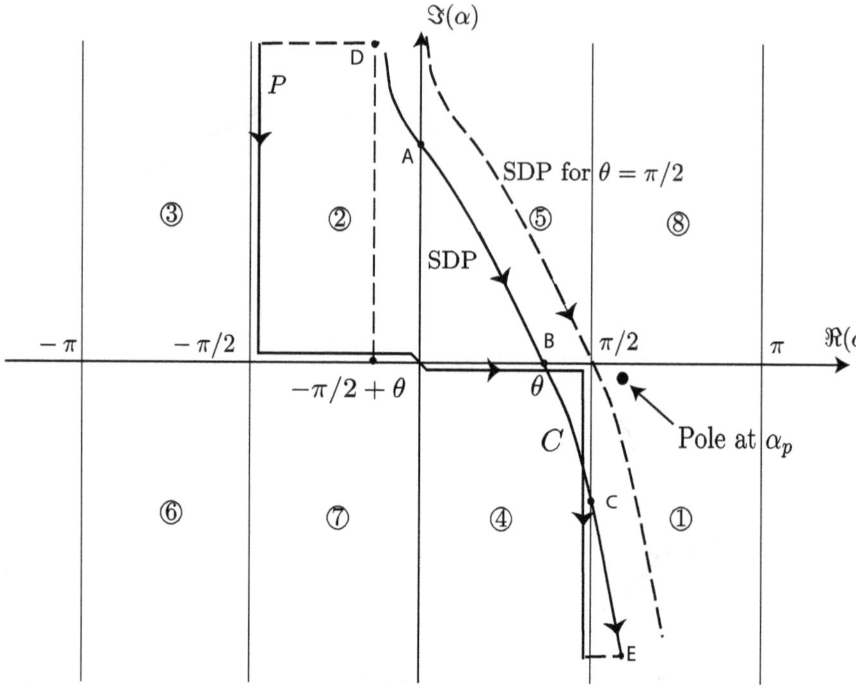


**Figure 16.24.** SDP contour in the $\alpha$-plane for a general $\theta$ and for $\theta = \pi/2$ in the neighborhood of the pole at $\alpha_p$.

$\theta = \pi/2$ and we need $R_2 \geqslant 0.16\lambda_0$. We will assume this to hold and drop the function $g(\alpha, R_2)$ in the subsequent analysis. Consequently,

$$I = f(\rho)\int_C \frac{\sqrt{\sin\alpha}\,\cos\alpha}{\kappa\cos\alpha + \sqrt{\kappa - \sin^2\alpha}}\, e^{ik_0R_2\cos(\alpha-\theta)}\,d\alpha. \tag{16.74}$$

We write $q_1(\alpha) = \mathrm{i}\cos(\alpha - \theta)$ and note that near the saddle point $q_1(\alpha) \approx \mathrm{i} - \mathrm{i}(\alpha - \theta)^2/2$ with $q_1(\alpha_s) = \mathrm{i}$, $q_1'(\alpha_s) = 0$, $q_1'(\alpha_s) = -\mathrm{i}$. To proceed with the steepest-descent evaluation, we make a further change of variable of the form $\cos(\alpha - \theta) = 1 + \mathrm{i}\beta^2/2$ (equivalently $\beta = 2\exp(\mathrm{i}\pi/4)\sin[(\alpha - \theta)/2]$), with $\beta = \beta_r + \mathrm{i}\beta_i$, so that $q_1(\alpha) := q(\beta) = \mathrm{i} - \beta^2/2$. The SDP maps onto the curve $\beta_r\beta_i = 0$, $(\beta_r^2 - \beta_i^2) > 0$. Hence the SDP maps onto the positive real axis in the $\beta$-plane on which $q(\beta) = \mathrm{i} - \beta_r^2/2$. Now $d\alpha/d\beta = \exp(-\mathrm{i}\pi/4)/\cos[(\alpha - \theta)/2] = -2\mathrm{i}/\sqrt{\beta^2 - 4\mathrm{i}}$. Let $g(\beta) = g_1(\beta)/g_2(\beta)$, where

$$g_1(\beta) = \sqrt{\sin\alpha}\,\cos\alpha\frac{d\alpha}{d\beta}\Bigg|_{\cos(\alpha-\theta)=1+\mathrm{i}\beta^2/2}, \tag{16.75}$$

$$g_2(\beta) = \kappa\cos\alpha + \sqrt{\kappa - \sin^2\alpha}\,\Big|_{\cos(\alpha-\theta)=1+\mathrm{i}\beta^2/2}. \tag{16.76}$$

The saddle point in the $\beta$-plane gets mapped to $\beta = 0$ and the pole of the integrand gets mapped to

$$\beta = \beta_p = 2e^{i\pi/4}\sin\left(\frac{\alpha_p - \theta}{2}\right)$$
$$= (1+i)\frac{\left(\sqrt{\kappa+1}-1\right)^{1/2}\cos(\theta/2) - \left(\sqrt{\kappa+1}+1\right)^{1/2}\sin(\theta/2)}{(\kappa+1)^{1/4}}. \tag{16.77}$$

The poles will then lie in the first and third quadrants of the $\beta$-plane. For example, with $\kappa = 12 + i18$ and $\theta = \pi/2$, the poles are located at $\pm\beta_p = \pm(0.203 + i0.065)$. The integral in equation (16.74) for large $k_0R_2$ can now be recast as

$$I = f(\rho)e^{ik_0R_2}\int_{-\infty}^{\infty} g(\beta)e^{-k_0R_2\beta^2/2}\,d\beta. \tag{16.78}$$

Subtracting the pole from the integrand enables one to write

$$g(\beta) = \frac{A}{\beta - \beta_p} + \frac{(\beta - \beta_p)g(\beta) - A}{\beta - \beta_p} \triangleq \frac{A}{\beta - \beta_p} + g_0(\beta). \tag{16.79}$$

The first term is the pole term with residue $A$. The remaining term $g_0(\beta)$ in analytic in the neighborhood of $\beta_p$, but may contain other singularities. Inspection of $d\alpha/d\beta$ reveals that $g_0(\beta)$ will have branch point singularities when $\beta^2 = 4i$ or when $\beta = \pm\sqrt{2}(1+i) := \pm\beta_b$. However, $\beta_b$ is far enough from the pole $\beta_p$ so that a Taylor series expansion of $g_0(\beta)$ about the saddle point $\beta = 0$ will have a reasonably large radius of convergence for treatment by the standard SDP technique. The residue of $g(\beta)$ at $\beta_p$ is

$$A = \frac{g_1(\beta_p)}{\frac{dg_2(\beta_p)}{d\beta}} = \frac{-\sqrt{\sin\alpha_p}\cos\alpha_p}{\kappa\sin\alpha_p + \frac{\sin\alpha_p\cos\alpha_p}{\sqrt{\kappa - \sin^2\alpha_p}}}$$
$$= \frac{\kappa^{3/4}}{(\kappa^2-1)(\kappa+1)^{1/4}} \sim \frac{1}{\kappa^{3/2}},\ |\kappa| \gg 1 \tag{16.80}$$

on using $\sin\alpha_p = \sqrt{\kappa/(\kappa+1)}$, $\cos\alpha_p = 1/\sqrt{\kappa+1}$, $\sqrt{\kappa - \sin^2\alpha_p} = -\kappa\cos\alpha_p$ (from the defining equation for determining the pole locations). As with the standard SDP technique, the first term in the Taylor's series expansion is itself adequate for regular functions such as $g_0(\beta)$. Using $g_0(\beta) \approx g_0(\beta = 0) = g(0) + A/\beta_p$, we obtain

$$I = f(\rho)e^{ik_0R_2}\int_{-\infty}^{\infty}\left[\frac{A}{\beta - \beta_p} + g_0(\beta)\right]e^{-k_0R_2\beta^2/2}\,d\beta$$
$$\sim f(\rho)e^{ik_0R_2}\left[\int_{-\infty}^{\infty}\frac{A}{\beta - \beta_p}\,d\beta + g_0(0)\int_{-\infty}^{\infty}e^{-k_0R_2\beta^2/2}\,d\beta\right] \tag{16.81}$$
$$= I_p + f(\rho)e^{ik_0R_2}\left[\frac{\sqrt{\sin\alpha_s}\cos\alpha_s e^{-i\pi/4}}{\kappa\cos\alpha_s + \sqrt{\kappa - \sin^2\alpha_s}} + \frac{A}{\beta_p}\right]_{\alpha_s=\theta}\sqrt{\frac{2\pi}{k_0R_2}},$$

where $I_p$ represents the contribution from the pole term. It is equal to

$$\begin{aligned} I_p &= Af(\rho)e^{ik_0R_2}\int_{-\infty}^{\infty}\frac{e^{-k_0R_2\beta^2/2}}{\beta-\beta_p}\,d\beta \\ &= Af(\rho)e^{ik_0R_2}i\pi e^{-(k_0R_2/2)\beta_p^2}\operatorname{erfc}\left(-i\sqrt{k_0R_2/2}\,\beta_p\right), \end{aligned} \tag{16.82}$$

where erfc($z$) is the complementary error function defined as

$$\operatorname{erfc}(z) = \frac{2}{\sqrt{\pi}}\int_z^{\infty} e^{-\beta^2}\,d\beta. \tag{16.83}$$

Thus it is seen that the effect of the proximity of a first-order pole near the saddle point is to introduce the extra term $I_e$ given by

$$I_e = Af(\rho)e^{ik_0R_2}\left[i\pi e^{-(k_0R_2/2)\beta_p^2}\operatorname{erfc}\left(-i\sqrt{k_0R_2/2}\,\beta_p\right) + \frac{1}{\beta_p}\sqrt{\frac{2\pi}{k_0R_2}}\right], \tag{16.84}$$

compared to the standard SDP analysis. If $|\beta_p| \gg 1$, one may use the large argument approximation of the complementary error function $e^{z^2}\operatorname{erfc}(z) \sim 1/(\sqrt{\pi}z)$ and obtain $I_e = 0$ as expected.

The term, $I_p$, associated with the pole is valid both for ordinary materials which have $\Re(\kappa) > 0$ as well as for plasmonic materials (metals at optical frequencies, see figure 1.1) which have $\Re(\kappa) < 0$. In the former case the contribution $I_p$ is known as the *Zenneck wave* (ZW) [10] and in the latter as the *surface plasmon polariton* (SPP) [11]. ■■

**Example 16.5.0.2** Sommerfeld half-space problem, $e^{j\omega t}$ time convention, vertical branch cuts.

In example 16.5.0.1, the asymptotic solution was determined using the hyperbolic branch cuts (i.e. Sommerfeld branch cuts). But there are other choices of branch cuts that can also yield other useful asymptotic decompositions, particularly when $k\rho \neq 0$ and $(z + h) < \rho$. Evaluate the asymptotic value of the integral (16.68) using vertical branch cuts and verify that the solution is consistent with that of example 16.5.0.1. (See also Michalski and Mosig [12].)

For the sake of illustration, we will assume the time dependence to be $e^{j\omega t}$ so that familiarity is also gained with branch cuts under this time convention. In this example we shall assume the lower half-space to be either an ordinary or normal medium (real part of permittivity > 0) or a plasmonic medium (real part of permittivity < 0). Relabeling the wavenumbers in the top and bottom half-spaces as $k$ and $k_2$, respectively, the potential $A_z$ in the top region under the $e^{j\omega t}$ time dependence can be written as

$$A_z = \frac{\mu_0 I_0 \ell}{4\pi}\left[\frac{e^{-jkR_1}}{R_1} - \frac{e^{-jkR_2}}{R_2} - j\kappa\int_{-\infty}^{\infty}\frac{k_\rho}{\kappa k_z + k_{z2}}H_0^{(2)}(k_\rho\rho)e^{-jk_z(z+h)}\,dk_\rho\right], \tag{16.85}$$

where $\kappa = \varepsilon_r' - j\varepsilon_r''$ is the relative permittivity of the lower half-space, $k_z = \sqrt{k^2 - k_\rho^2}$, and $k_{z2} = \sqrt{\kappa k^2 - k_\rho^2}$. If the lower half-space is a plasmonic material we write $\varepsilon_r' = -\varepsilon_m$, $\varepsilon_m \gg 1$. For the integral to be convergent we require at the outset that $\Im(k_z) < 0$, $\Im(k_{z2}) < 0$ along the original contour, $\Gamma_0$, shown in figure 16.26. We now focus on the asymptotic evaluation of the integral

$$I = \int_{-\infty}^{\infty} \frac{k_\rho}{\kappa k_z + k_{z2}} H_0^{(2)}(k_\rho \rho) e^{-jk_z(z+h)}\, dk_\rho, \tag{16.86}$$

using vertical branch cuts.

Figure 16.25(a) shows the four-sheeted $k_\rho$ plane with the top sheet (sheet I) being the proper sheet on which $\Im(k_z) < 0$, $\Im(k_{z2}) < 0$. The various sheets are separated by Sommerfeld branch cuts erected at $k$ and $k_2$ in the complex $k_\rho$-plane. Sheets II (red), III (blue), and IV (purple) are all improper (meaning that $e^{-jk_z d}$ and/or $e^{-jk_{z2} d}$ will not decay exponentially for large $k_\rho$ when $d > 0$). Sommerfeld branch cuts are shown as green wavy lines in figure 16.25(b). Note that the Sommerfeld branch cuts appear in the second and fourth quadrants of the complex $k_\rho$-plane for the $e^{j\omega t}$ time-convention (only fourth-quadrant branch cuts are shown here) in contrast to appearing in the first and third quadrants for the $e^{-i\omega t}$ time-convention (see figure 16.23). The values of $k_z$ and $k_{z2}$ on the remaining sheets are indicated in the figure and the connection between various sheets via the Sommerfeld cuts is shown by vertical lines. For instance, when a contour, originally on the top sheet, crosses the Sommerfeld branch cut passing through $k$, it enters sheet II (transition from

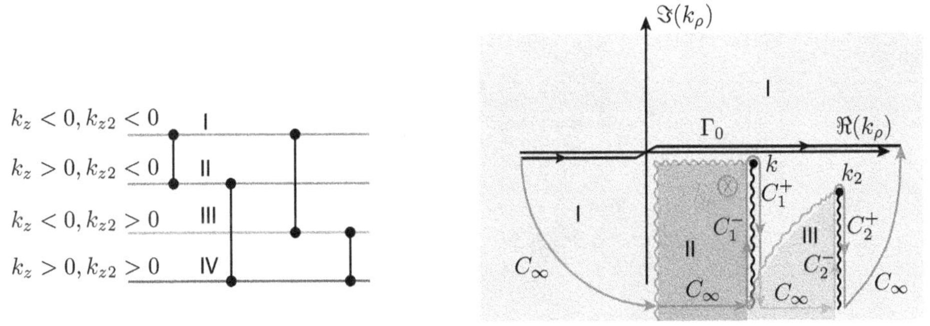


(a) (b)

**Figure 16.25.** Four-sheeted $k_\rho$-plane and the branch cuts. (a) The proper (green: sheet I) and improper (red: sheet II, blue: sheet III, purple: sheet IV) sheets of the $k_\rho$-plane showing the connection between the various sheets. (b) Sommerfeld branch-cuts (the green wavy lines) and the vertical branch cuts (black wavy lines) for a Zenneck wave. Deformation of the original contour $\Gamma_0$ along the vertical branch cuts. After crossing the Sommerfeld cut, the contour enters sheet II and traverses $C_1^-$ of the vertical branch cut, crosses the Sommerfeld cut and traverses $C_1^+$ (green: sheet I plus blue: sheet III below the second Sommerfeld cut). It then traverses $C_2^-$ (blue: sheet III) and finally $C_2^+$ (green: sheet I). The pole at $k_p$ which lies to the left of $k$ and on sheet I (green) is clearly not captured during the deformation. Note that $\Re(k_p) < k_0$, $|k_p| < k_0$ and the integrand vanishes on $C_\infty$ for large radius.

green to red), when it crosses the branch cut passing through $k_2$, it enters sheet III (transition from green to blue), and so on.

As in example 16.5.0.1, setting the denominator of the integrand to zero gives the location of the pole in the $k_\rho$-plane:

$$k_\mathrm{p} = k\sqrt{\frac{\kappa}{\kappa+1}} = \begin{cases} k\sqrt{\dfrac{\varepsilon_\mathrm{r}' - j\varepsilon_\mathrm{r}''}{(\varepsilon_\mathrm{r}'+1) - j\varepsilon_\mathrm{r}''}}, \; 0 < \Re(k_\mathrm{p}) < k, \text{ Zenneck wave pole,} \\ k\sqrt{\dfrac{\varepsilon_\mathrm{m} + j\varepsilon_\mathrm{r}''}{(\varepsilon_\mathrm{m}-1) + j\varepsilon_\mathrm{r}''}}, \; \Re(k_\mathrm{p}) > k, \quad \text{plasmonic wave pole,} \end{cases} \tag{16.87}$$

where we have designated the pole for $\varepsilon_\mathrm{r}' > 0$ as a *Zenneck wave pole* and the pole for $\varepsilon_\mathrm{r}' = -\varepsilon_\mathrm{m} < 0$ as the *plasmonic wave pole*. The Zenneck wave pole lies to the left of the point $k_\rho = k$, whereas the plasmonic wave pole lies to its right, but both of these lie on the top sheet. These are indicated in figures 16.25(b) and 16.26. The relative position of the pole will have implications in the asymptotic analysis as we shall demonstrate shortly.

When the original contour $\Gamma_0$ is deformed around the Sommerfeld branch cuts as in example 16.5.0.1, the entire deformed contour remains on the top sheet and any contribution from the infinite semi-circular arc in the lower half plane is vanishingly

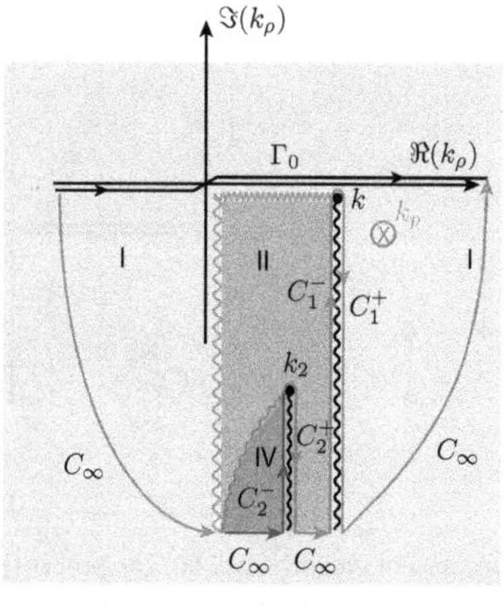


**Figure 16.26.** Sommerfeld branch-cuts (the green wavy lines) and the vertical branch cuts for the plasmonic case. Deformation of the original contour $\Gamma_0$ along the vertical branch cuts. The contour enters sheet IV (purple) after crossing both the Sommerfeld branch cuts and traverses the contours $C_2^-$ (purple: sheet IV) and $C_2^+$ (red: sheet II) of the vertical branch cut at $k_2$. It then traverses the contour $C_1^-$ (red: sheet II) and $C_1^+$ (green: sheet I) of the vertical branch cut at $k$. The pole at $k_\mathrm{p}$ which lies to the right of $k$ and on sheet I is clearly captured during deformation. Note that $|k_2| \gg k_0$, $\Re(k_\mathrm{p}) > k_0$, $|k_p| > k_0$. Integrand vanishes on $C_\infty$ for large radius.

small due to the exponential decay afforded by factors of the form $e^{-jk_z d}$ or $e^{-jk_{z2}d}$, and $H_0^{(2)}(k_\rho\rho)$. Also, the pole is always captured during the deformation. However, when the contour is deformed along vertical branch cuts, a portion of the infinite semi-circular arc enters the improper sheets. For instance, for an ordinary lower medium (see figure 16.25), the contour enters sheet II and traverses $C_\infty$ in the strip $0 < k_\rho < k$ (the red shaded region) and subsequently $C_1^-$ of the vertical branch cut upon crossing the Sommerfeld cut at $k$. It then crosses the Sommerfeld branch cut again and traverses $C_1^+$ (green: sheet I plus blue: sheet III below the second Sommerfeld cut at $k_2$). Lastly, it traverses $C_\infty$ in the blue region of sheet III, the vertical portion $C_2^-$ (blue: sheet III), and finally $C_2^+$ (green: sheet I upon crossing the Sommerfeld branch cut at $k_2$ again). The pole at $k_\mathrm{p}$ which lies to the left of $k$ and on sheet I (green) is clearly not captured during the deformation. Note that on the portion of the infinite semi-circular arcs in the red and blue shaded regions $\Im(k_z)$ and $\Im(k_{z2})$ remain, respectively, finite. It is straightforward to see this. For instance, with $(k_\rho - k) = \kappa_1^2 e^{j\psi_1}$ and $(k_\rho + k) = \kappa_2^2 e^{j\psi_2}$, the vertical wavenumber is equal to $k_z^\mathrm{I} = \kappa_1\kappa_2 e^{j(\psi_1+\psi_2-\pi)/2}$ on sheet I and $k_z^\mathrm{II} = \kappa_1\kappa_2 e^{j(\psi_1+\psi_2+\pi)/2} = -k_z^\mathrm{I}$ on sheet II. Consider now $k_\rho = k_\mathrm{r} - js^2$, $0 < k_\mathrm{r} < k$, $0 < s < \infty$ within the red strip. Then $k_z^\mathrm{II} = (k^2 - (k_\mathrm{r} - js^2)^2)^{1/2} = (k^2 - k_\mathrm{r}^2 + s^4 + 2jk_\mathrm{r}s^2)^{1/2} = \sqrt{k^2 - k_\mathrm{r}^2 + s^4}$ $(1 + 2jk_\mathrm{r}s^2/(k^2 - k_\mathrm{r}^2 + s^4))^{1/2} \sim s^2(1 + jk_\mathrm{r}/s^2) = s^2 + jk_\mathrm{r}$ as $s \to \infty$. Therefore $\Im(k_z^\mathrm{II}) \to k_\mathrm{r} < k$ and remains bounded. Hence a term of the form $e^{jk_z^{II} d}$ will remain finite at $C_\infty$ on sheet II. The presence of a second factor such as $H_0^{(2)}(k_\rho\rho)$ will provide the exponential decay for large $k_\rho\rho$, thereby rendering the contribution from $C_\infty$ to the integral zero even on sheet II. A similar convergence argument holds in the shaded blue region at infinity on sheet III. The only difference between the ordinary medium and a plasmonic medium is that the pole at $k_\mathrm{p}$ is captured in the latter during deformation as it lies to the right of $k$ (see figure 16.26). Note that in the case of a plasmonic medium, $|k_2| \gg k$ and $\arg(k_2) \approx -\pi/2$.

Denoting the vertical contours at the branch cuts at $k$ and $k_2$ as $\Gamma_1$ and $\Gamma_2$ we have

$$I = \int_{\Gamma_1+\Gamma_2} \frac{k_\rho}{\kappa k_z + k_{z2}} H_0^{(2)}(k_\rho\rho) e^{-jk_z(z+h)} \, dk_\rho + 2\pi j R_\mathrm{p}, \tag{16.88}$$

where $R_\mathrm{p} = 0$ for an ordinary medium and

$$\begin{aligned} R_\mathrm{p} &= \lim_{k_\rho \to k_\mathrm{p}} \frac{(k_\rho - k_\mathrm{p})k_\rho}{\kappa k_z + k_{z2}} H_0^{(2)}(k_\rho\rho) e^{-jk_z(z+h)} \\ &= \frac{k_\mathrm{p}}{2(\kappa + 1)} H_0^{(2)}(k_\mathrm{p}\rho) e^{-jk(z+h)/\sqrt{\kappa+1}}, \end{aligned} \tag{16.89}$$

is the residue at $k_\rho = k_\mathrm{p}$ for the plasmonic medium. Note that $\sqrt{\kappa + 1}$ is to have a positive imaginary part so that $k_{z\mathrm{p}} = \sqrt{k^2 - k_\mathrm{p}^2} = k/\sqrt{\kappa + 1}$ will have a negative imaginary part on the top sheet. Consequently, the residue term decays exponentially both in the $z$- and $\rho$-directions and constitutes the surface plasmon polariton.

For most practical cases of interest, the contribution from $\Gamma_2$ will be negligible as it decays at least as rapidly as $e^{-\Im(k_2)\rho}$ for $|k_2| \gg k$ and will thus be ignored in the subsequent analysis. So we consider the approximation

$$I \approx \int_{\Gamma_1} \frac{k_\rho}{\kappa k_z + k_{z2}} H_0^{(2)}(k_\rho \rho) e^{-jk_z(z+h)}\, dk_\rho + 2\pi j R_p =: I_1 + 2\pi j R_p, \tag{16.90}$$

by dropping the contribution from $\Gamma_2$. Let us affect a change of variable by setting $k_\rho = k - js^2$ so that $dk_\rho = -2js ds$. By writing $k_\rho = k_{\rho r} + jk_{\rho i}$ and $s = s_r + js_i$ in terms of real and imaginary parts, we see that $k_{\rho r} = k + 2s_r s_i$ and $k_{\rho i} = s_i^2 - s_r^2$. The images of the original contour $\Gamma_0$ and the right edge, $D_0$ (the red portion of $\Gamma_0$), of $\Gamma_1$ in the complex $s$-plane are shown in figure 16.27. The contour $\Gamma_0$ maps to $s_r = \pm s_i$ and $D_0$ maps to the positive real axis $s_i = 0$, $0 \leqslant s_r < \infty$. Writing $k_\rho - k = |s|^2 e^{-j\pi/2}$, $k_\rho + k = \kappa_2^2 e^{-j\psi_2}$, $-\pi/2 < \psi_2 < 0$ on $D_0$, we obtain $k_z = |s|\kappa_2 e^{-j(\pi/2+\psi_2+\pi)/2} =: k_z^{\mathrm{I}}$ on the right edge (green) of $\Gamma_1$ and $k_z = |s|\kappa_2 e^{-j(\pi/2+\psi_2-\pi)/2} =: k_z^{\mathrm{II}}$ on the left edge (red) of $\Gamma_1$. Note that as $s$ varies over $(0, \infty)$ on $D_0$, $k_z^{\mathrm{I}}$ remains in the third quadrant[1], i.e. $\Re(k_z^{\mathrm{I}}) < 0$ and $\Im(k_z^{\mathrm{I}}) < 0$, and that $k_z^{\mathrm{II}} = -k_z^{\mathrm{I}}$. However, the values of $k_{z2}$ remain the same on both sides of $\Gamma_1$.

Combining the terms on the left and right sides of the vertical branch cut and inserting the factor $1 = e^{jk_\rho\rho}e^{-jk_\rho\rho}$ we obtain

$$\begin{aligned} I_1(\rho, z) &= \int_{D_0} \frac{2k_\rho[\kappa k_z^{\mathrm{I}} \cos(k_z^{\mathrm{I}}(z+h)) + jk_{z2} \sin(k_z^{\mathrm{I}}(z+h))]}{(\kappa^2 - 1)(k_p^2 - k_\rho^2)} \\ &\quad \times H_0^{(2)}(k_\rho \rho) e^{jk_\rho\rho} e^{-jk_\rho\rho}\, dk_\rho \\ &= \int_{s=0}^{\infty} \frac{sF(s)}{(s^2 - s_p^2)} e^{-s^2\rho}\, ds, \end{aligned} \tag{16.91}$$

where

$$F(s; \mathbf{r}) = \left. \frac{-4k_\rho[\kappa k_z^I \cos(k_z^I(z+h)) + jk_{z2} \sin(k_z^I(z+h))] H_0^{(2)}(k_\rho\rho) e^{j(k_\rho - k)\rho}}{(\kappa^2 - 1)(k_p + k_\rho)} \right|_{k_\rho = k - js^2},$$

is a smoothly varying function of $s$ and $k_p = k - js_p^2$. It is clear from equation (16.91) that the path $D_0$ is also a steepest descent path provided that $\rho > (z + h)$ since $\Im(k_z^{\mathrm{I}}) < k$ as already remarked. Furthermore, $s = 0$ is the stationary point. Note that $F(0; \mathbf{r}) = 0$ since $k_z^{\mathrm{I}} = 0$ when $k_\rho = k$. Now $s_p^2 = -jk/(\kappa + 1 + \sqrt{\kappa(\kappa + 1)})$. When $|\kappa| \gg 1$, i.e. in the plasmonic case, $s_p$ is very close to the origin (stationary point). Even in an ordinary medium $s_p$ can remain close to the origin. The function $F(s; \mathbf{r})$ can be expanded as $F(s; \mathbf{r}) = sF'(0; \mathbf{r}) + \frac{s^2}{2}F''(0; \mathbf{r}) + \cdots$. In view of the steepest

---

[1] This is also clear when we express $k_z$ directly in $s$: $k_z^{\mathrm{I}} = -s(s^2 + 2jk)^{1/2}$ and the square root is interpreted as having positive real and imaginary parts.

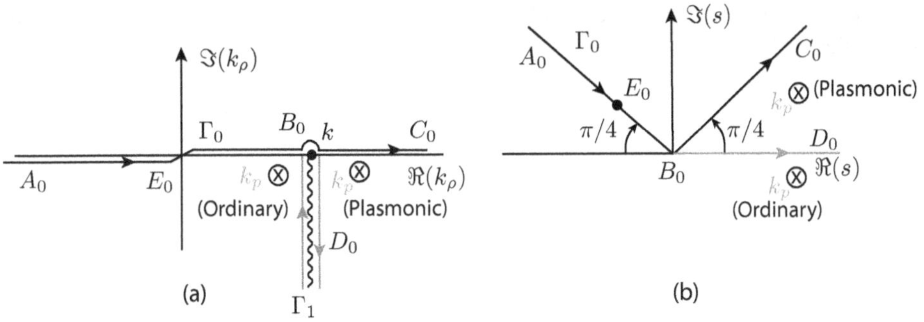


**Figure 16.27.** (a) The original contour, $\Gamma_0$, and vertical cut contour, $\Gamma_1$, at $k$ (red plus green). (b) Image of the vertical cut in the $s$-plane. The Zenneck wave pole at $k_p$ is not captured, whereas the plasmonic pole at $k_p$ is captured when $\Gamma_0$ is deformed to the vertical cut.

descent nature of $D_0$ we only retain the first term in this series and evaluate one of the integrals and cast the other in terms of a special function (the Faddeeva function) to obtain

$$\begin{aligned} I_1(\rho, z) &\approx F'(0; \mathbf{r}) \int_0^\infty \frac{s^2}{s^2 - s_p^2} e^{-s^2\rho}\, ds = F'(0; \mathbf{r}) \\ &\times \left[ \int_0^\infty e^{-s^2\rho}\, ds + s_p^2 \int_0^\infty \frac{e^{-s^2\rho}}{s^2 - s_p^2}\, ds \right] \\ &= F'(0; \mathbf{r}) \left[ \frac{1}{2}\sqrt{\frac{\pi}{\rho}} + \frac{j\pi s_p}{2} \left\{ w(\sqrt{p}) - 2e^{-p}\, \Theta(-\Im(s_p)) \right\} \right], \end{aligned} \tag{16.92}$$

where

$$w(z) = e^{-z^2} \operatorname{erfc}(-jz) = \frac{2z}{\pi j} \int_0^\infty \frac{e^{-t^2}}{t^2 - z^2}\, dt, \quad \Im(z) > 0, \tag{16.93}$$

is the Faddeeva function [13], $\Theta(\cdot)$ is the unit step function and the unitless parameter $p = s_p^2\rho = j(k_p - k)\rho$ is known as the *Sommerfeld numerical distance*. The integral representation of the Faddeeva function in equation (16.93) is valid for $\Im(z) > 0$, but is analytically continued into the entire complex plane by extracting out the pole of the integrand [10] as evident in the occurrence of an extra term involving the unit step function in equation (16.92). In other words, the Faddeeva function has an integral representation

$$w(z) = e^{-z^2} \operatorname{erfc}(-jz) = \frac{2z}{\pi j} \int_0^\infty \frac{e^{-t^2}}{t^2 - z^2}\, dt, \quad \Im(z) > 0 \tag{16.94}$$

$$= \frac{2z}{\pi j} \int_0^\infty \frac{e^{-t^2}}{t^2 - z^2}\, dt + 2e^{-z^2}, \quad \Im(z) < 0, \tag{16.95}$$

and the expression of the asymptotic solution given in equation (16.92) automatically reflects this. Note that

$$F'(0; \mathbf{r}) = \lim_{s\to 0} \frac{F(s; \mathbf{r})}{s} = \frac{4\sqrt{2j}\left(k - k_{\mathrm{p}}\right)}{\sqrt{k}(\kappa - 1)}\left[\kappa + jk(z + h)\sqrt{\kappa - 1}\,\right]H_0^{(2)}(k\rho). \quad (16.96)$$

The complete asymptotic field in the upper half-space is given by

$$A_z \sim \frac{\mu_0 I_0 \ell}{4\pi}\left[\frac{\mathrm{e}^{-jk_0 R_1}}{R_1} - \frac{\mathrm{e}^{-jk_0 R_2}}{R_2} + 2\pi\kappa R_{\mathrm{p}}\Theta(\Im(s_{\mathrm{p}})) - j\kappa I_1\right], \quad (16.97)$$

and is subject to the condition that $\rho \gg (z + h)$. Note that when $z \ll \lambda$, $R_1 \approx R_2$ and the first two terms within the square brackets of equation (16.97) cancel out. The field is then contributed mostly by the integral and residue[2] terms. Figure 16.28 shows a comparison of the magnitude of the integral $I$ computed by (i) numerical integration of equation (16.86) and (ii) asymptotic evaluation using equations (16.89), (16.90), and (16.92). It is seen that excellent agreement is obtained between the two as soon as $\rho$ exceeds $10(z + h)$ for ordinary media and $\rho$ exceeds $(z + h)$ for plasmonic media. Figure 16.29 shows a comparison of the phase of the integrals for ordinary and plasmonic media. Once again a favorable agreement for phase becomes evident for $\rho > 20(z + h)$. ■■

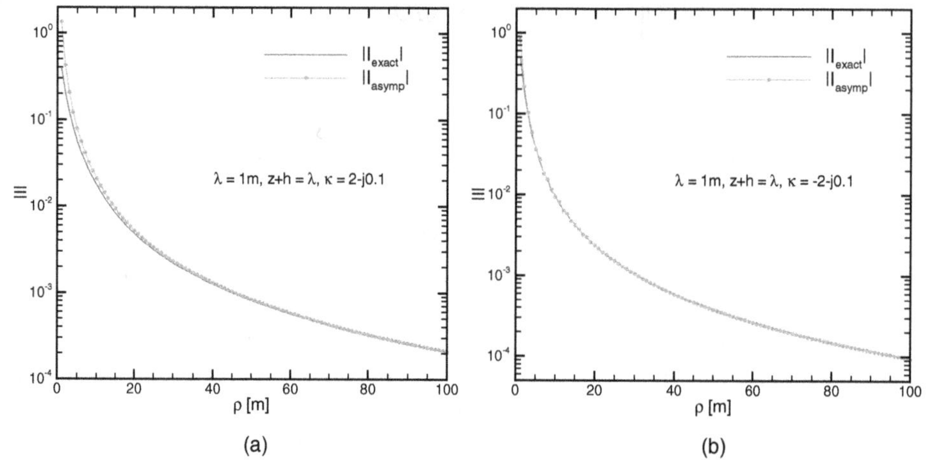


**Figure 16.28.** Numerical integration of equation (16.86) versus asymptotic evaluation using equation (16.90). Magnitude of the integral is plotted versus horizontal distance $\rho/\lambda$ for $\lambda = 1$ m, $z + h = \lambda$. (a) Ordinary medium with $\kappa = 2 - j0.1$. (b) Plasmonic medium with $\kappa = -2 - j0.1$.

[2] Note that the residue term exists only in the plasmonic case where $\Im(s_{\mathrm{p}}) > 0$. This explains the presence of the factor $\Theta(\Im(s_{\mathrm{p}}))$ associated with the residue term in (16.97).

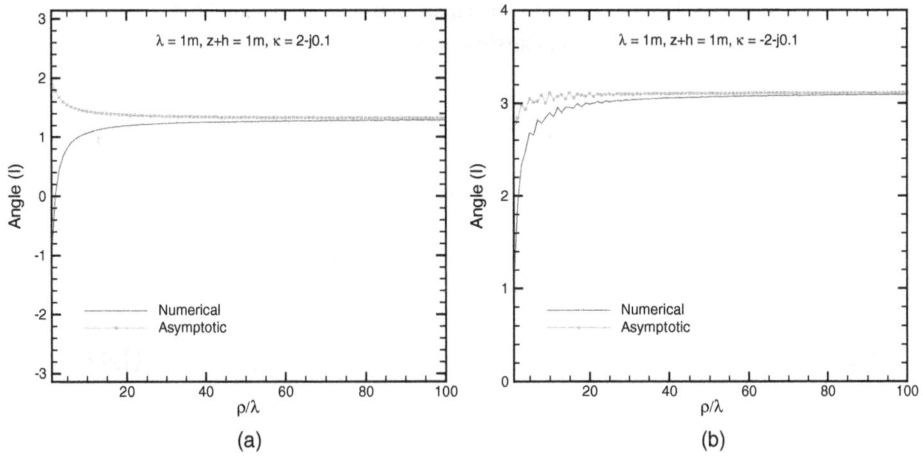


**Figure 16.29.** Numerical integration of equation (16.86) versus asymptotic evaluation using equation (16.90). Phase of the integral is plotted versus horizontal distance $\rho/\lambda$ for $\lambda = 1$ m, $z + h = \lambda$. (a) Ordinary medium with $\kappa = 2 - j0.1$. (b) Plasmonic medium with $\kappa = -2 - j0.1$.

## References


[1] Felsen L B and Marcuwitz N 1994 *Radiation and Scattering of Waves* (Piscataway: NJ: IEEE)

[2] Ablowitz M J and Fokas A S 1997 *Complex Variables: Introduction and Applications* (Cambridge: Cambridge University Press)

[3] Bleistein N and Handelsman R A 1986 *Asymptotic Evaluation of Integrals* (New York: Dover)

[4] Ishimaru A 2017 *Electromagnetic Wave Propagation, Radiation, and Scattering* (Hoboken, NJ: Wiley)

[5] Chew W C 1995 *Waves and Fields in Inhomogeneous Media* (New York: IEEE)

[6] Mittra R and Lee S W 1971 *Analytical Techniques in the Theory of Guided Waves, Macmillan Series in Electrical Science* (New York: Macmillan)

[7] Silver S and Saunders W K 1950 The external field produced by a slot in an infinite cylinder *J. Appl. Phys.* **21** 153–8

[8] Harrington R F 1961 *Time-Harmonic Electromagnetic Fields* (New York: McGraw-Hill)

[9] Balanis C A 2012 *Advanced Engineering Electromagnetics* 2nd edn (New York: Wiley)

[10] Collin R E 2004 Hertzian dipole radiating over a lossy earth or sea: some early and late 20th-century controversies *IEEE Antennas Propag. Mag.* **46**

[11] Kawata S (ed) 2001 *Near-field Optics and Surface Plasmon Polaritons* (New York: Springer)

[12] Michalski K A and Mosig J R 2015 The Sommerfeld half-space problem revisisted: from radio frequencies and Zenneck waves to visible light and Fano modes *J. Electromagn. Waves Appl.* **30** 1–42

[13] Abramowitz M and Stegun I A (ed) 1971 *Handbook of Mathematical Functions with Formulas, Graphs, and Mathematical Tables* (New York: Dover)

**IOP** Publishing

Engineering Electrodynamics
A collection of theorems, principles and field representations
**Ramakrishna Janaswamy**

# Chapter 17

# Covariant formulation of Maxwell's equations

There is an intimate connection between Maxwell's equations and geometry that prompted several workers in the field to view electrodynamics through differential forms [1–3]. The recent emergence of cloaking devices [4] and transformational electromagnetics [5] has provided further impetus to the argument that Maxwell's equations must be understood independent of coordinate systems. The basic philosophy of relativity as expounded by Einstein [6] is that *the general laws of nature are to be expressed by equations which hold good for all system of coordinates, that is, are covariant with respect to any substitutions whatever*. Maxwell's equations remain invariant across different coordinate systems and it is apt to describe them in a covariant form (i.e. tensor form) by casting the various quantities involved as tensors. The main character of a tensor is that the equations of transformation of its components are *linear and homogeneous*. The coefficients of transformation could, however, be non-constant. In this chapter we provide a self-consistent and brief introduction to tensor calculus [6–10] and then demonstrate how to use the background material to cast Maxwell's equations in a tensorial form. It may be mentioned that some of the relativistic concepts such as a geodesics have already been utilized in the high-frequency electromagnetics community (see example 17.1.1.4) without recourse to relativity. After laying the groundwork for tensor analysis and casting Maxwell's equations in a covariant form, we shall derive the equations that are central to transformational electromagnetics and demonstrate the example of a cylindrical cloak.

## 17.1 Preliminaries of tensor calculus

**Definition 17.1.1** *Spacetime*. A four-dimensional space consisting of three space coordinates and one time coordinate.

**Definition 17.1.2** *Event*. A point in spacetime.

doi:10.1088/978-0-7503-1716-0ch17

**Definition 17.1.3.** *Worldline.* A one-dimensional set of events in spacetime.

**Definition 17.1.4.** *Interval in spacetime.* A quadratic form consisting of space and time increments.

**Definition 17.1.5.** *Euclidean space.* A finite-dimensional space that is flat and in which the quadratic form describing the interval is positive definite.

In an Euclidean space it is possible to use Cartesian coordinates to describe points.

**Definition 17.1.6.** *Pseudo-Euclidean space.* The four-dimensional spacetime consisting of ordinary 3D Euclidean space and a fourth coordinate corresponding to time. The resulting quadratic form describing the interval is not necessarily positive definite.

**Definition 17.1.7** *Proper frame.* The frame in which the observed body is at rest.

The length of a rod in such a frame is called the *proper length.* The time interval recorded by a clock attached to the observed body is the *proper time.* The charge density in a frame where charges are at rest is the *proper charge density.*

**Definition 17.1.8.** *Inertial frame.* A frame of reference in which a freely moving body, i.e. a moving body which is not acted upon by external forces, proceeds with constant velocity.

**Definition 17.1.9** *Coordinates of a point P.* A set of variables $x^1, x^2, \ldots, x^N$ in a space, $V_N$, of $N$ dimensions.

**Definition 17.1.10** *Curve.* A curve is defined as the totality of points given by the equations

$$x^r = f^r(u), \quad r = 1, 2, \ldots, N, \tag{17.1}$$

where $u$ is a parameter and $f^r$ are $N$ functions.

**Definition 17.1.11.** *Surface.* A surface is defined as the totality of points given by the equations

$$x^r = f^r(u_1, u_2, \ldots, u_{N-1}), \tag{17.2}$$

where the $u$ are parameters. Equivalently, a surface is represented by a single equation of the form $F(x^1, x^2, \ldots, x^N) = 0$ which divides the space $V_N$ into two parts, one side in which $F > 0$ and the other side in which $F < 0$.

**Definition 17.1.12.** *Range convention.* When a small Latin (or sometimes Greek) suffix (superscript or subscript) occurs unrepeated in a term, it is understood to take all values 1, 2, … , $N$, where $N$ is the number of dimensions of the space.

**Definition 17.1.13** *Summation convention.* When a small Latin or Greek suffix is repeated in a term, summation with respect to (w.r.t.) that suffix is understood, the range of summation being 1, 2, … , $N$.

Suppose there are a new set of coordinates $x'^1, x'^2, \ldots, x'^N$ that are related to the original coordinates $x^1, x^2, \ldots, x^N$ via the $N$ single-valued functions $f^r$:

$$\begin{aligned} x'^r &= f^r(x^1, x^2, \ldots, x^N), \quad r = 1, 2, \ldots, N \\ &=: f^r(x^s), \end{aligned} \tag{17.3}$$

by the range convention. The Jacobian of the transformation is

$$J' = \det\begin{pmatrix} \dfrac{\partial x'^1}{\partial x^1} & \cdots & \dfrac{\partial x'^1}{\partial x^N} \\ \vdots & \cdots & \vdots \\ \dfrac{\partial x'^N}{\partial x^1} & \cdots & \dfrac{\partial x'^N}{\partial x^N} \end{pmatrix} \neq 0 =: \det\left[\frac{\partial x'^r}{\partial x^s}\right], \quad r, s = 1, 2, \ldots N. \tag{17.4}$$

Thus all first order derivatives of the first new variable $x'^1$ w.r.t. the old variables $x^1, x^2, \ldots, x^N$ appear on the first row, all first order derivatives of the second new variable $x'^2$ appear on the second row, and so on. By the chain rule the total differentials in the two coordinates are related as

$$\begin{aligned} \mathrm{d}x'^r &= \sum_{s=1}^{N} \frac{\partial x'^r}{\partial x^s}\mathrm{d}x^s, \quad r = 1, 2, \ldots, N \\ &=: \frac{\partial x'^r}{\partial x^s}\mathrm{d}x^s \end{aligned} \tag{17.5}$$

by the range and summation conventions. The partial derivatives are evaluated at the point $P$. Thus the transformation of the differentials of the coordinates is a linear, homogeneous transformation (i.e. linear and passes through the origin meaning $\mathrm{d}x'^r = 0$ if $\mathrm{d}x^s = 0$), the coefficients being functions of position in $V_N$. All subsequent results in this section will be shown using this summation convention.

**Definition 17.1.14** *Invariant.* A scalar quantity independent of the coordinate system used.

Examples of invariants are total electric charge (unit of charge = electron charge) and action (unit of action = Planck's constant). However, the length and time interval are *not necessarily* invariants. By the chain rule the partial derivatives of an invariant $\psi$ transform according to

$$\frac{\partial \psi}{\partial x'^{r}} = \frac{\partial \psi}{\partial x^{s}} \frac{\partial x^{s}}{\partial x'^{r}}, \tag{17.6}$$

where the partial derivatives of the transformation on the right-hand side of equation (17.6) are now the inverse of those in equation (17.5).

**Definition 17.1.15** *Contravariant tensor of rank one (i.e. contravariant vector).* A set of quantities $T^r$ associated with a point $P$ are said to be the components of a *contravariant tensor of rank one* if they transform according to equation (17.5), i.e.

$$T'^{r} = \frac{\partial x'^{r}}{\partial x^{s}} T^{s}. \tag{17.7}$$

Thus an infinitesimal displacement of equation (17.5) is the prototype of a general contravariant vector. Another example is velocity of a particle. For a curve $x^r = x^r(u)$ defined w.r.t. a parameter $u$, the derivatives $\mathrm{d}x^r/\mathrm{d}u$ are components of a contravariant vector. Note that in a 3D Euclidean space, the contravariant components $T^r$ of an ordinary vector $\mathbf{T}$ are associated with the unitary vectors[1] $\mathbf{a}_r$ such that $\mathbf{T} = T^r \mathbf{a}_r$.

**Definition 17.1.16.** *Covariant tensor of rank one (i.e. covariant vector).* A set of quantities $T_r$ associated with a point $P$ are said to be the components of a *covariant tensor of rank one* if they transform according to equation (17.6), i.e.

$$T'_{r} = \frac{\partial x^{s}}{\partial x'^{r}} T_{s}. \tag{17.8}$$

The gradient of an invariant $\nabla_s \psi = \partial \psi / \partial x^s$ given in equation (17.6) is the prototype of a general covariant vector. An example is the electric field vector at a point. Another example is the propagation vector of a plane wave. Note that in a 3D Euclidean space, the covariant components $T_r$ of an ordinary vector $\mathbf{T}$ are associated with the reciprocal unitary vectors $\mathbf{a}^r$,[2] such that $\mathbf{T} = T_r \mathbf{a}^r$.

---

[1] In 3D Euclidean space, if the position vector of a point is denoted by the vector $\mathbf{r}$, the basis vector $\frac{\partial \mathbf{r}}{\partial x^m}$ is known as a *unitary vector* $\mathbf{a}_m$.

[2] In 3D Euclidean space a basis vector, $\mathbf{a}^n$, known as the *reciprocal unitary vector*, can be uniquely defined such that $\mathbf{a}_m \cdot \mathbf{a}^n = \delta_m^n$.

**Definition 17.1.17.** *Contravariant/covariant tensor of second rank.* A set of quantities $T^{rs}$ (respectively, $T_{rs}$) are said to the components of a contravariant (respectively, covariant) tensor of rank two if they transform according to

$$T^{'rs} = \frac{\partial x^{'r}}{\partial x^m}\frac{\partial x^{'s}}{\partial x^n}T^{mn} \text{ (respectively, } T'_{rs} = \frac{\partial x^m}{\partial x^{'r}}\frac{\partial x^n}{\partial x^{'s}}T_{mn}). \tag{17.9}$$

The definitions can be extended to contravariant or covariant tensors of any rank or even mixed tensors of any rank. Thus if the components of a set of quantities $T^r_{st}$ transform according to

$$T^{'r}_{st} = \frac{\partial x^{'r}}{\partial x^m}\frac{\partial x^n}{\partial x^{'s}}\frac{\partial x^p}{\partial x^{'t}}T^m_{np}, \tag{17.10}$$

then we say that they are the components of *mixed tensor* of third rank, with one contravariant and two covariant suffixes. An example of a mixed tensor is the Kronecker's delta $\delta^r_s$ which is unity if $r = s$ and zero if $r \neq s$. Even though coordinates are not explicitly involved in the definition of Kronecker's delta it is useful to associate it with coordinates to examine how it transforms. That it needs to transform as a mixed tensor is obvious when we write

$$\delta^{'r}_s = \frac{\partial x^{'r}}{\partial x^m}\frac{\partial x^n}{\partial x^{'s}}\delta^m_n = \frac{\partial x^{'r}}{\partial x^m}\frac{\partial x^m}{\partial x^{'s}} = \frac{\partial x^{'r}}{\partial x^{'s}}. \tag{17.11}$$

The right-hand side is non-zero and equal to unity only when $r = s$ thus preserving the numerical value of the Kronecker's delta. We can summarize the results of chain rule using Kronecker's delta as

$$\frac{\partial x^{'r}}{\partial x^m}\frac{\partial x^m}{\partial x^{'s}} = \delta^r_s = \frac{\partial x^r}{\partial x^{'m}}\frac{\partial x^{'m}}{\partial x^s}. \tag{17.12}$$

Consider the mixed tensors $A^i_j = \frac{\partial x^{'i}}{\partial x^j}$ and $B^j_k = \frac{\partial x^j}{\partial x^{'k}}$, which represent the transformation matrices ($i$= row index, $j$= column index in $A^i_j$) for the forward transformation $x^{'i} = f^i(x^j)$ and the inverse transformation $x^j = h^j(x^{'k})$, respectively. The determinants of these matrices are the Jacobians $J'$ and $J$, respectively. In view of equation (17.12) we have $C^i_k = A^i_j B^j_k = \delta^i_k$. Thus we have the identity

$$J'J = \det(A^i_j)\det(B^j_k) = \det(C^i_k) = 1. \tag{17.13}$$

Thus $J$ and $J'$ will both have the same signs and a modulus reciprocal to each other. A transformation with $J = J' = \pm 1$ is known as an *orthogonal transformation.*

**Definition 17.1.18** *Symmetrical and antisymmetrical tensors.* A contravariant or a covariant tensor of second or higher rank is said to be symmetrical (antisymmetrical or skew-symmetric) if two components which are obtained from one another by the interchange of two indices are equal (respectively, equal and of opposite sign).

Symmetry or antisymmetry of a tensor is preserved during coordinate transformations. *An antisymmetrical tensor of rank two in a four-dimensional space (an ordinary 4 × 4 matrix with potentially 16 entries; antisymmetry requires diagonal elements to be zero and off-diagonal element 'ij' to be negative of off-diagonal element 'ji') has only six non-zero components. Thus it can used to represent components of a six-vector.* An example is the electromagnetic field strength tensor which contains three components of electric field and three components of magnetic flux density. An antisymmetrical tensor of third rank has only four non-zero components, while the antisymmetrical tensor of fourth rank has only one non-zero component in a four-dimensional space.

**Definition 17.1.19** *Outer product of tensors.* The outer product of two tensors of rank $m$ and $n$, whether they are of the same type or of different type (that is, contravaraint type or covariant type or mixed type), is another tensor of rank $m + n$ and is written as a juxtaposition of the two tensors where the indices take all possible values.

Thus the outer product of a contravariant tensor $A_{rs}$ and a mixed tensor $B_n^m$ is a mixed tensor $C_{rsn}^m = A_{rs}B_n^m$ of rank four with one contravariant suffix and three covariant suffixes.

**Definition 17.1.20** *Contraction.* If in a mixed tensor we use the same letter for a superscript and a subscript we say that the tensor is *contracted* to result in a tensor which has the tensor character indicated by the remaining suffixes.

Contraction reduces the rank of a mixed tensor by 2. Thus $A_{ms}B_n^m$ has the character of a covariant tensor $T_{sn}$ of second rank. This is because $T_{sn}$ can be regarded as $C_{msn}^m$, which is a contraction of $C_{rsn}^m = A_{rs}B_n^m$ with $r = m$. Along the same lines $A_{rm}B_n^m$ is also a covariant tensor of the second rank. The tensor $A_{ms}B_n^m$ obtained by contracting an outer product is also known as the *inner product* of the tensors $A_{rs}$ and $B_n^m$ w.r.t. the suffixes $r$, $m$.

Note that for a mixed tensor of rank two we have from the transformation rule (17.10) and the result (17.12) that

$$T_r^{'r} = \delta_m^n T_n^m = T_n^n, \tag{17.14}$$

which is a tensor of rank zero and, thus, an invariant.

The following theorem can be used to test whether a quantity qualifies as being a tensor.

**Theorem 17.1.21.** *Quotient rule.* If the contraction of a test quantity having an arbitrary number of subscripts and superscripts with a known tensor results in another tensor, then the quantity under test is a tensor.

As an example suppose that $a_{rs}$ is a set of quantities whose tensor character is under investigation. Let $X^r$ and $Y^r$ be arbitrary contravariant tensors and suppose that $a_{rs}X^rY^s$ is given as an invariant. Then according to the above theorem, $a_{rs}$ are components of a covariant tensor of second rank. This is because an invariant is a tensor of rank zero.

**Definition 17.1.21** *Alternation.* Alternation is a determinant-like operation, denoted by square brackets around the indices and defined by the formulas

$$T_{[\kappa\lambda]} = \frac{1}{2!}(T_{\kappa\lambda} - T_{\lambda\kappa}) \tag{17.15}$$

$$T_{[\kappa\lambda\nu]} = \frac{1}{3!}(T_{\kappa\lambda\nu} + T_{\lambda\nu\kappa} + T_{\nu\kappa\lambda} - T_{\kappa\nu\lambda} - T_{\nu\lambda\kappa} - T_{\lambda\kappa\nu}), \tag{17.16}$$

where even permutations of the suffixes at a single time take a + sign and odd permutations take a − sign.

Alteration applies to a tensor of any rank greater than one.

**Definition 17.1.22** *Mixing.* Mixing is a determinant-like operation, denoted by ordinary brackets around the indices and defined by the formulas

$$T_{(\kappa\lambda)} = \frac{1}{2!}(T_{\kappa\lambda} + T_{\lambda\kappa}) \tag{17.17}$$

$$T_{(\kappa\lambda\nu)} = \frac{1}{3!}(T_{\kappa\lambda\nu} + T_{\lambda\nu\kappa} + T_{\nu\kappa\lambda} - T_{\kappa\nu\lambda} + T_{\nu\lambda\kappa} + T_{\lambda\kappa\nu}), \tag{17.18}$$

where both even permutations and odd permutations take a + sign.

Mixing applies to a tensor of any rank greater than one. Alternation and mixing operations apply to either contra- or covariant indices and will not necessarily result in an invariant.

**Definition 17.1.23.** *Pseudo-tensor.* A quantity whose components transform like the components of a tensor except that they are multiplied by the sign $\alpha = (J/|J|) = (J'/|J'|)$ of the transformation determinants.

Examples are the cross product of two ordinary vectors or the ordinary curl of a vector. The sum of two pseudo-tensors of the same rank is again a pseudo-tensor of equal rank. The outer product of a tensor with a pseudo-tensor is a pseudo-tensor with a rank equal to the sum of the ranks of the two factors in the product. However, the outer product of two pseudo-tensors is a tensor. The operation of contraction can be performed with pseudo-tensors in the same way as with tensors.

**Definition 17.1.24.** *Tensor densities.* A quantity whose components transform like a tensor but are weighted by the absolute Jacobian of the transformation raised to some power not equal to zero.

For instance a weighted mixed tensor density of rank three transforms as

$$B'^{\alpha}_{.\beta\gamma} = |J|^k \frac{\partial x'^{\alpha}}{\partial x^{\ell}} \frac{\partial x^m}{\partial x'^{\beta}} \frac{\partial x^n}{\partial x'^{\gamma}} B^{\ell}_{.mn}, \tag{17.19}$$

where $J = \det(\partial x^r / \partial x'^s)$ is the determinant of the inverse transformation and $k \neq 0$ is known as the weight of the transformation. The dots in the mixed tensors have no significance other than to indicate that a subscript and a superscript are often not written on the same vertical line considering that the suffixes can change positions when a tensor is contracted (an exception to this rule is Kronecker's delta). The dots will be filled by appropriate suffixes when the mixed tensors are multiplied by covariant or contravariant vectors. The dots are also useful sometimes to resolve the ambiguity in the order of indices given that $T^{i}_{.j} = g^{im}T_{mj} = g_{j\ell}T^{i\ell} = T^{i.}_{\;j} \neq T_{j.}^{\;i}$ $= g^{im}T_{jm} = g_{j\ell}T^{\ell i} = T_{j}^{.i}$ in general. (The two mixed tensors are equal only if the corresponding tensor $T_{ij}$ is symmetric in which case $T^{i}_{.j} = T_{j.}^{\;i} = T^{i}_{j.}$.)

**Definition 17.1.25.** *Reflection.* A reflection corresponds to a reversal in the direction of an odd number of coordinate axes.

**Definition 17.1.26** *Permutation symbols $\varepsilon_{mnr}$, $\varepsilon^{mnr}$, i.e. Levi–Civita symbols.* In a space of three dimensions the permutation symbols $\varepsilon_{mnr}$, $\varepsilon^{mnr}$ are defined by the following conditions:

(i) $\varepsilon_{mnr} = 0$, $\varepsilon^{mnr} = 0$ if two of the suffixes coincide.

(ii) $\varepsilon_{mnr} = 1$, $\varepsilon^{mnr} = 1$ if the sequence of numbers *mnr* is the sequence 123, or an even permutation (number of changes in a pair) of it such as 231, 312, etc.

(iii) $\varepsilon_{mnr} = -1$, $\varepsilon^{mnr} = -1$ if the sequence of numbers *mnr* is an odd permutation of the sequence 123 such as 132, 213, etc.

Accordingly we have $\varepsilon^{123} = \varepsilon^{231} = \varepsilon^{312} = 1$ and $\varepsilon^{132} = \varepsilon^{213} = \varepsilon^{321} = -1$. Although these quantities do not involve any coordinate system (such as Kronecker's delta) it is useful to associate them with a coordinate system $x^r$. To preserve their numerical values in any coordinate system the permutation symbols transform according to[3]

$$\varepsilon'_{mnr} = J'\varepsilon_{stu}\frac{\partial x^s}{\partial x'^m}\frac{\partial x^t}{\partial x'^n}\frac{\partial x^u}{\partial x'^r}, \qquad J' = \det\left(\frac{\partial x'^p}{\partial x^q}\right), \tag{17.20}$$

or as

$$\varepsilon'^{mnr} = J\varepsilon^{stu}\frac{\partial x'^m}{\partial x^s}\frac{\partial x'^n}{\partial x^t}\frac{\partial x'^r}{\partial x^u}, \qquad J = \det\left(\frac{\partial x^p}{\partial x'^q}\right) = 1/J'. \tag{17.21}$$

The presence of the Jacobian of the transformation in the transformation equations makes the permutation symbol dependent on the handedness of the coordinate system. Hence it does not have the character of a true tensor which should be independent of the handedness. Consider the *parity-inversion* transformation (or a reflection)

$$x'^1 = -x^1;\; x'^2 = -x^2;\; x'^3 = -x^3, \tag{17.22}$$

so that $\partial x^s/\partial x'^m = -\delta^s_m$ and $J' = -1$. A true covariant tensor $A_{mnr}$ would have transformed under parity inversion as $A'_{mnr} = -A_{stu}\delta^s_m\delta^t_n\delta^u_r$. However, the permutation symbol transforms as $\varepsilon'_{mnr} = \varepsilon_{stu}\delta^s_m\delta^t_n\delta^u_r$. If a transformation preserves the handedness of a coordinate system (examples are translation, rotation, Lorentz transformation) such that the Jacobian is positive, then the permutation tensor behaves like an ordinary tensor. Otherwise it does not. For instance under a transformation $x'^1 = -x^1$, $x'^2 = -x^2$, $x'^3 = x^3$, which corresponds to rotation through 180° about the $x^3$ axis of a Cartesian coordinate system ($x^1$, $x^2$, $x^3$), the permutation symbol $\varepsilon_{mnr}$ does transform like a true tensor. For this reason the permutation symbol is an oriented tensor or a pseudo-tensor under orthogonal transformations (i.e. $|J| = 1$). In a Cartesian coordinate system (where there is no distinction between covariant and contravariant quantities), the properties satisfied by permutation symbol can be expressed as

$$\varepsilon_{mrs}\varepsilon^{mpq} = \delta_r^{\,p}\delta_s^{\,q} - \delta_r^{\,q}\delta_s^{\,p}, \tag{17.23}$$

$$\varepsilon_{jmn}\varepsilon^{imn} = 2\delta^{\,i}_{j}, \tag{17.24}$$

$$\varepsilon^{mnr}\varepsilon_{mnr} = 6. \tag{17.25}$$

The components of the vector cross product (a pseudo-vector or an axial vector) $\mathbf{Z} = \mathbf{a}_i Z^i = \mathbf{a}^i Z_i = \mathbf{X} \times \mathbf{Y}$ of two 3D tensors $\mathbf{X}$, $\mathbf{Y}$ (polar vectors) with $\mathbf{X} = \mathbf{a}_m X^m = \mathbf{a}^m X_m$, $\mathbf{Y} = \mathbf{a}_n Y^n = \mathbf{a}^n Y_n$ can be expressed in terms of the permutation symbol as

[3] Note that from the definition of $\varepsilon_{mnr}$ it follows that $\varepsilon_{stu}\frac{\partial x^s}{\partial x'^1}\frac{\partial x^t}{\partial x'^2}\frac{\partial x^u}{\partial x'^3} = J$.

$$Z_i = \sqrt{\alpha}\,\varepsilon_{imn}X^mY^n \text{or} Z^i = \frac{1}{\sqrt{\alpha}}\varepsilon^{imn}X_mY_n, \tag{17.26}$$

where $\sqrt{\alpha} = \mathbf{a}_1 \cdot (\mathbf{a}_2 \times \mathbf{a}_3)$. In Cartesian coordinates where $\mathbf{a}^i = \mathbf{a}_i$ and $\sqrt{\alpha} = 1$, it can be written as

$$Z^i = \varepsilon^{imn}X_mY_n = Z_i = \varepsilon_{imn}X^mY^n. \tag{17.27}$$

Because the cross product of a vector with itself is zero, we have the identity $\varepsilon^{imn}X_mX_n = 0 = \varepsilon_{imn}X^mX^n$. Using the definition of a cross product, the vector triple product of **X**, **Y**, and **P** can be written as

$$\mathbf{P}\cdot(\mathbf{X}\times\mathbf{Y}) = \frac{1}{\sqrt{\alpha}}\varepsilon^{imn}P_iX_mY_n = \sqrt{\alpha}\,\varepsilon_{imn}P^iX^mY^n. \tag{17.28}$$

It is straightforward to see from the definitions of the cross product and the triple product that $\mathbf{a}_m \times \mathbf{a}_n = \sqrt{\alpha}\,\varepsilon_{imn}\mathbf{a}^i$ , $\mathbf{a}^m \times \mathbf{a}^n = \frac{1}{\sqrt{\alpha}}\varepsilon^{imn}\mathbf{a}_i$ , $\mathbf{a}_i \cdot (\mathbf{a}_m \times \mathbf{a}_n) = \sqrt{\alpha}\,\varepsilon_{imn}$ , and $\mathbf{a}^i \cdot (\mathbf{a}^m \times \mathbf{a}^n) = \frac{1}{\sqrt{\alpha}}\varepsilon^{imn}$. Using the permutation symbol we can associate an anti-symmetric tensor $B_{jk}$ of rank two in a 3D space (which has three non-zero components) with a pseudo-vector $\hat{\mathbf{B}} = (\hat{B}^1, \hat{B}^2, \hat{B}^3)$ by

$$\hat{B}^i = \frac{1}{2}\varepsilon^{ijk}B_{jk}, \tag{17.29}$$

$$\text{i. e.}\quad \hat{\mathbf{B}} = (\hat{B}^1, \hat{B}^2, \hat{B}^3) = (B_{23}, B_{31}, B_{12}). \tag{17.30}$$

A permutation symbol $\varepsilon_{m_1m_2\cdots m_N} = \varepsilon^{m_1m_2\ldots m_N}$ in $N$ dimensions can be defined in an analogous manner. It vanishes if any two suffixes coincide. It is equal to $+1$ or $-1$ according to whether the number of permutations required to transform $m_1m_2\ldots m_N$ into the ordered sequence $12\ldots N$ is, respectively, even or odd. A permutation symbol is skew-symmetric in any pair of suffixes. For instance, $\varepsilon_{mnrs} = -\varepsilon_{mrns} = \varepsilon_{msnr}$.

Like the permutation symbols in 3D, $\varepsilon_{mnpq}$ and $\varepsilon^{mnpq}$ transform as

$$\varepsilon'_{mnpq} = J'\varepsilon_{stuv}\frac{\partial x^s}{\partial x'^m}\frac{\partial x^t}{\partial x'^n}\frac{\partial x^u}{\partial x'^p}\frac{\partial x^v}{\partial x'^q}, \qquad J' = \det\left(\frac{\partial x'^i}{\partial x^j}\right), \tag{17.31}$$

$$\varepsilon'^{mnpq} = J\varepsilon^{stuv}\frac{\partial x'^m}{\partial x^s}\frac{\partial x'^n}{\partial x^t}\frac{\partial x'^p}{\partial x^u}\frac{\partial x'^q}{\partial x^v}, \qquad J = \det\left(\frac{\partial x^i}{\partial x'^j}\right), \tag{17.32}$$

and represent pseudo-tensors under orthogonal transformations. Note that equation (17.31) implies

$$J\varepsilon'_{mnpq} = \varepsilon_{stuv}\frac{\partial x^s}{\partial x'^m}\frac{\partial x^t}{\partial x'^n}\frac{\partial x^u}{\partial x'^p}\frac{\partial x^v}{\partial x'^q}. \tag{17.33}$$

A closely related identity involving a matrix $\{g^{rs}\}$ with determinant $g$ is

$$g\varepsilon^{mnpq} = \varepsilon_{ijkl}g^{im}g^{jn}g^{kp}g^{lq}. \tag{17.34}$$

Additional identities are

$$\varepsilon_{r_1r_2\ldots r_N}\varepsilon^{r_1r_2\ldots r_N} = N!, \tag{17.35}$$

$$\varepsilon_{s_1\ldots s_M\, r_1\ldots r_{N-M}}\varepsilon^{k_1\ldots k_M\, r_1\ldots r_{N-M}} = (N - M)!\delta^{k_1\ldots k_M}_{s_1\ldots s_M}, \tag{17.36}$$

$$\varepsilon_{s_1\ldots s_N}\varepsilon^{k_1\ldots k_N} = \delta^{k_1\ldots k_N}_{s_1\ldots s_N}, \tag{17.37}$$

where

$$\delta^{k_1\ldots k_N}_{s_1\ldots s_N} = \begin{cases} +1, & \text{if } k_1\ldots k_N \text{ are distinct integers selected from the range } 1, 2, \ldots, N \text{ and if } s_1, \ldots, s_N \text{ is an even permutation of } k_1, \ldots, k_N. \\ -1, & \text{if } k_1\ldots k_N \text{ are distinct integers selected from the range } 1, 2, \ldots, N \text{ and if } s_1, \ldots, s_N \text{ is an odd permutation of } k_1, \ldots, k_N. \\ 0, & \text{if any two of } k_1, \ldots, k_N \text{ are equal or if any two of } s_1, \ldots, s_N \text{ are equal, or of the set of numbers } k_1, \ldots, k_N \text{ differs, apart from order, from the set } s_1, \ldots, s_N. \end{cases} \tag{17.38}$$

Using the four-dimensional permutation symbol $\varepsilon_{ijkl}$ we can associate a dual pseudo-tensor $\hat{F}_{ij}$ to an antisymmetric tensor $F^{kl}$ by the relation

$$\hat{F}_{ij} = \frac{1}{2}\varepsilon_{ijkl}F^{kl}, \tag{17.39}$$

that is valid under orthogonal transformations. The inverse relation is obtained by multiplying equation (17.39) by $\frac{1}{2}\varepsilon^{ijmn}$ and utilizing equation (17.36) to result in

$$F^{mn} = \frac{1}{2}\varepsilon^{ijmn}\hat{F}_{ij} = \frac{1}{2}\varepsilon^{mnij}\hat{F}_{ij}. \tag{17.40}$$

### 17.1.1 Riemannian space

**Definition 17.1.27** *Riemannian space.* A space $V_N$ is said to be Riemannian if there is given in it a symmetric, metric covariant tensor $g_{mn}$ of the second rank such that the quadratic form

$$\Phi = ds^2 = g_{mn} dx^m dx^n, \tag{17.41}$$

known as the *metric form* (i.e. fundamental form or linear element or line element or interval) is an invariant.

In 3D space, the quantity $ds$ may be identified with incremental arc-length. Symmetry means $g_{mn} = g_{nm}$, thereby, implying that there are $N(N+1)/2$ independent components in a metric tensor. For instance there are three independent components in 2D space, six independent components in 3D, and so on. If one were to imagine that the coordinates $(x^1, x^2, \dots, x^N)$ are obtained via transformation from original Cartesian coordinates $(z^1, z^2, \dots, z^N)$ in a local subspace in which the metric tensor is $h_{mn}$ then

$$g_{mn} = \frac{\partial z^i}{\partial x^m}\frac{\partial z^j}{\partial x^n} h_{ij}. \tag{17.42}$$

Note that $g_{mn}$ are not necessarily unitless[4], but could have varied units depending on the units of the coordinates $x^j$. The determinant of the metric tensor is denoted by the symbol $g = \det(g_{mn})$. It could be positive or negative and could have any units depending on the coordinate system. It is, however, unitless in Cartesian coordinates in 3D space.

In *Euclidean space* the metric form $\Phi$ is *positive definite*: $\Phi > 0$ and $\Phi = 0$ only if all differentials $dx^m$ vanish. As an example, in spherical coordinates $(r, \theta, \phi)$ in 3D the curvilinear coordinates are $x^1 = r$, $x^2 = \theta$, $x^3 = \phi$ and related to a Cartesian coordinates through $z^1 = x^1 \sin x^2 \cos x^3$, $z^2 = x^1 \sin x^2 \sin x^3$, $z^3 = x^1 \cos x^2$. The associated metric tensor is $g_{11} = 1$, $g_{22} = r^2$, $g_{33} = r^2 \sin^2\theta$ and all other components of $g_{mn}$ are zero. The determinant of the metric tensor is $g = r^4 \sin^2\theta > 0$ and the metric form is $\Phi = (dr)^2 + r^2(d\theta)^2 + r^2 \sin^2\theta(d\phi)^2 > 0$. The unitary vectors are $\mathbf{a}_1 = \hat{\mathbf{r}}$, $\mathbf{a}_2 = r\hat{\boldsymbol{\theta}}$, $\boldsymbol{a}_3 = r\sin\theta\hat{\boldsymbol{\phi}}$, where $\hat{\mathbf{r}}, \hat{\boldsymbol{\theta}}, \hat{\boldsymbol{\phi}}$ are unit vectors along the $r$, $\theta$, and $\phi$ directions.

In a general Riemannian space, the metric form need not be positive definite. As an example $\Phi = (dx^1)^2 - (dx^2)^2$ in 1D-spacetime, which can vanish if $dx^1 = dx^2 \neq 0$. For some displacements $dx^r$ the metric form may be positive and for others it may be negative or zero. If $\Phi = 0$ for $dx^r$ not all zero, the displacement is called a *null displacement*. For any displacement $dx^r$ which is not null, there exists an indicator $\varepsilon$, chosen equal to $+1$ or $-1$ such that $\varepsilon\Phi$ is always positive. The *length* $ds > 0$ of the displacement $dx^r$ is defined to be $ds = \sqrt{\varepsilon\Phi}$.

From the covariant metric tensor $g_{mn}$ we can define a conjugate quantity[5] $g^{mn}$ such that $g_{mr}g^{ms} = \delta_r^s$. Since $g_{mn}$ is symmetric so is $g^{mn}$. By the quotient rule $g^{mn}$ is a contravariant tensor of rank two since $\delta_r^s$ is a mixed tensor. In a Euclidean 3D space

[4] In 3D Euclidean space the covariant metric tensor can be expressed in terms of the unitary vectors as $g_{mn} = \mathbf{a}_m \cdot \mathbf{a}_n$.

[5] In 3D Euclidean space the contravariant metric tensor $g^{mn}$ can be expressed in terms of the reciprocal unitary vectors as $g^{mn} = \mathbf{a}^m \cdot \mathbf{a}^n$.

employing Cartesian coordinates $g_{mn} = h_{mn} = \delta_{mn}$, $h = \det(h_{mn}) = 1$, and there will be no difference between covariant and contravariant components. In a pseudo-Euclidean spacetime employing Cartesian coordinates $x^1 = x$, $x^2 = y$, $x^3 = z$, and the time, $x^4 = ct$, one usually takes for the metric tensor $g_{mn} = -h_{mn} = -\delta_{mn}$, $1 \leqslant m, n, \leqslant 3$, $g_{4m} = h_{4m} = \delta_{4m}$, resulting in $g = \det(h_{mn}) = -1$. Such a four-dimensional system of coordinates with these values of $g_{mn}$ is known as a *Galilean system*.

A key result in any Riemannian space is that

$$g_{mn}g^{mn} = N. \tag{17.43}$$

As an example, in spherical coordinates of a 3D space, $g^{11} = 1, g^{22} = 1/r^2$, $g^{33} = 1/r^2 \sin^2\theta$, and $g_{mn}g^{mn} = 3$. Another important result is

$$g^{mn}\frac{\partial g_{mn}}{\partial x^r} = g_{mn}\frac{\partial g^{mn}}{\partial x^r} = \frac{\partial}{\partial x^r}\ln\mid g\mid = \frac{1}{\mid g\mid}\frac{\partial\mid g\mid}{\partial x^r}, = \frac{2}{\sqrt{\mid g\mid}}\frac{\partial\sqrt{\mid g\mid}}{\partial x^r}. \tag{17.44}$$

Because the metric tensor transforms as

$$g'_{ij} = \frac{\partial x^m}{\partial x'^i}\frac{\partial x^n}{\partial x'^j}g_{mn}, \tag{17.45}$$

its determinant $g$ transforms as

$$g' = \det(g'_{ij}) = \det\left(\frac{\partial x^m}{\partial x'^i}\right)\det\left(\frac{\partial x^n}{\partial x'^j}\right)\det(g_{mn}) = J^2 g, \tag{17.46}$$

where $J = \det\left(\frac{\partial x^m}{\partial x'^i}\right)$ is the Jacobian of the inverse transformation. The volume element $\mathrm{d}v = \prod_{i=1}^{N}\mathrm{d}x^i$ of the Riemannian space transforms as

$$\mathrm{d}v' = J'\mathrm{d}v. \tag{17.47}$$

Combining these two yields $\sqrt{\mid g'\mid}\mathrm{d}v' = \sqrt{\mid g\mid}\mathrm{d}v$ in view of the fact that $JJ' = 1$. Thus the quantity $\sqrt{\mid g\mid}\mathrm{d}v$ is an invariant. If the transformations are orthogonal[6] (such as translation, rotation, Lorentz transformation, etc) such that the absolute Jacobians $|J|$, $|J'|$ are unity, then both the volume element and the determinant are preserved separately.

The metric tensor measures the orthogonality or not of the coordinate system as well as the curvature of the Riemannian space. It is important to keep in mind that the non-diagonal components of a metric tensor could be non-zero when non-orthogonal coordinates are used to describe the geometry of spacetime even when the underlying space is flat. In four-dimensional, flat spacetime, and inertial systems it is always possible to revert to the Galilean values of the metric tensor over all space with $g = -1$. In a non-inertial reference frame (where acceleration is present) in a real spacetime, it is possible to define Galilean values of the metric tensor in an

[6] A linear transformation $z_m' = A_{mn}z_n + B_m$ involving Cartesian coordinates $z_m$ is orthogonal if the coefficients satisfy $A_{mp}A_{mq} = \delta_{pq}$.

infinitesimal volume or in a subspace. After reduction to the diagonal form at any given point, the matrix representation of the metric tensor $g_{mn}$ will have one positive eigenvalue and three negative eigenvalues. Consequently, the product of the eigenvalues, i.e. the determinant of the metric tensor, will remain negative.

In the Riemannian space equipped with the metric tensor $g_{mn}$ and its conjugate $g^{mn}$, complements of covariant and contravariant tensors are easily formed according to $A^{\mu} = g^{\mu\nu}A_{\nu}$, $B_{\mu} = g_{\mu\nu}B^{\nu}$, $A^{\mu\nu} = g^{\mu\alpha}g^{\nu\beta}A_{\alpha\beta}$, $B_{\mu\nu} = g_{\mu\alpha}g_{\nu\beta}B^{\alpha\beta}$. Similarly a *reduced tensor* $B_{\mu\nu}$ associated with $A_{\mu\nu}$ is defined as $B_{\mu\nu} = g_{\mu\nu}g^{\alpha\beta}A_{\alpha\beta}$ and the reduced tensor $B^{\mu\nu}$ associated with $A^{\mu\nu}$ is obtained from $B^{\mu\nu} = g^{\mu\nu}g_{\alpha\beta}A^{\alpha\beta}$.

When equation (17.46) is combined with equation (17.31) we can associate the Levi–Civita symbol $\varepsilon_{mnpq}$ with a covariant quantity

$$\eta_{mnpq} = \pm\sqrt{|\,g\,|}\,\varepsilon_{mnpq}, \tag{17.48}$$

that transforms like a pseudo-tensor under *any coordinate transformations*

$$\eta'_{mnpq} = \alpha\frac{\partial x^{s}}{\partial x'^{m}}\frac{\partial x^{t}}{\partial x'^{n}}\frac{\partial x^{u}}{\partial x'^{p}}\frac{\partial x^{v}}{\partial x'^{q}}\eta_{stuv}, \qquad \alpha = \text{sign}(J'). \tag{17.49}$$

Using equation (17.34), the corresponding contravariant pseudo-tensor has components

$$\eta^{mnpq} = g^{ms}g^{nt}g^{pu}g^{qv}\eta_{stuv} = \pm\sqrt{|\,g\,|}\,g^{ms}g^{nt}g^{pu}g^{qv}\varepsilon_{stuv} = \pm\text{sign}(g)\frac{\varepsilon^{mnpq}}{\sqrt{|\,g\,|}}, \tag{17.50}$$

since the determinant $\det(g^{ij}) = 1/\det(g_{ij}) = g^{-1}$. The choice of signs in equations (17.48) and (17.50) must match. For instance one may take $\eta_{mnpq} = \sqrt{|\,g\,|}\,\varepsilon_{mnpq}$ and $\eta^{mnpq} = \text{sign}(g)\varepsilon^{mnpq}/\sqrt{|\,g\,|}$.

Using the psuedo-vector $\eta_{mnpq}$ we can define a pseudo-tensor dual to an antisymmetrical tensor $F^{ij}$ by

$$\hat{F}_{ij} = \frac{1}{2}\eta_{ijmn}F^{mn} = \frac{1}{2}\sqrt{|\,g\,|}\,\varepsilon_{ijmn}F^{mn}. \tag{17.51}$$

The inverse relation is obtained by multiplying equation (17.51) by $\frac{1}{2}\text{sign}(g)\eta^{ijpq}$ and utilizing equation (17.36) to result in

$$\begin{aligned}\frac{1}{2}\text{sign}(g)\eta^{ijpq}\hat{F}_{ij} &= \frac{1}{4}\text{sign}(g)\eta^{ijpq}\eta_{ijmn}F^{mn} = \frac{1}{2}\delta^{pq}_{mn}F^{mn} = \frac{1}{2}(F^{pq} - F^{qp}) \\ &= F^{pq}.\end{aligned} \tag{17.52}$$

Similarly using the covariant form of the antisymmetric tensor, $F_{ij}$, we define the corresponding contravariant from of the dual pseudo-tensor $\hat{F}^{ij}$

$$\hat{F}^{ij} = \frac{1}{2}\eta^{ijmn}F_{mn} = \frac{1}{2}\text{sign}(g)|\,g\,|^{-1/2}\,\varepsilon^{ijmn}F_{mn}, \tag{17.53}$$

$$F_{ij} = \frac{1}{2}\text{sign}(g)\eta_{ijmn}\hat{F}^{mn} = \frac{1}{2}\text{sign}(g)\sqrt{|\,g\,|}\,\varepsilon_{ijmn}\hat{F}^{mn}. \tag{17.54}$$

Note that $\hat{F}^{ij}\hat{F}_{ij} = \text{sign}(g)F^{mn}F_{mn}$.

**Definition 17.1.28** *Magnitude of a vector in Riemannian space.* The magnitude $X$ of a contravariant vector $X^r$ is the positive real quantity satisfying $X^2 = \varepsilon g_{mn}X^mX^n$ where $\varepsilon$ is the indicator of $X^r$. For a covariant vector $X_r$, the magnitude $X > 0$ is such that $X^2 = \varepsilon g^{mn}X_mX_n$.

**Definition 17.1.29.** *Angle between vectors.* The angle $\theta \in (0, \pi)$ between two contravariant vectors $X^r$ and $Y^r$ having respective magnitudes $X$ and $Y$ is defined from $XY\cos\theta = g_{mn}X^mY^n$.

### *17.1.1.1 Geodesics and Christoffel symbols*

**Definition 17.1.30.** *Geodesic in Riemannian space.* A geodesic joining points $P$ and $Q$ in a Riemannian space $V_N$ is a curve whose length has stationary value w.r.t. arbitrary small variations of the curve, the end points being held constant. In mathematical terms a geodesic satisfies the *variational condition*

$$\delta\int_P^Q \mathrm{d}s = 0, \tag{17.55}$$

where $\mathrm{d}s$ is the elemental displacement.

Taking the differentials of the quadratic form we have

$$\delta\mathrm{d}s^2 = 2\mathrm{d}s\delta\mathrm{d}s = \delta(g_{mn}\mathrm{d}x^m\mathrm{d}x^n) = \mathrm{d}x^m\mathrm{d}x^n\frac{\partial g_{mn}}{\partial x^r}\delta x^r + 2g_{mn}\mathrm{d}x^m\delta(\mathrm{d}x^n). \tag{17.56}$$

Therefore on letting $p^m = \frac{\mathrm{d}x^m}{\mathrm{d}s}$ we obtain

$$\begin{aligned}\delta\int_P^Q \mathrm{d}s &= \int_P^Q\left[\frac{1}{2}p^mp^n\frac{\partial g_{mn}}{\partial x^r}\delta x^r + g_{mn}p^m\frac{\mathrm{d}\delta x^n}{\mathrm{d}s}\right]\mathrm{d}s\\ &= \int_P^Q\left[\frac{1}{2}p^mp^n\frac{\partial g_{mn}}{\partial x^r}\delta x^r - \frac{\mathrm{d}}{\mathrm{d}s}(g_{mn}p^m)\delta x^n\right]\mathrm{d}s\\ &= -\int_P^Q\left[\frac{\mathrm{d}}{\mathrm{d}s}(g_{rm}p^m) - \frac{1}{2}p^mp^n\frac{\partial g_{mn}}{\partial x^r}\right]\delta x^r\mathrm{d}s,\end{aligned} \tag{17.57}$$

on integrating the second term in the first line of the right-hand side by parts and using the end conditions $(\delta x^n)|_P^Q = 0$. Equating the coefficient of the arbitrary variation $\delta x^r$ to zero and noting that

$$p^m \frac{\mathrm{d}g_{rm}}{\mathrm{d}s} = p^m \frac{\partial g_{rm}}{\partial x^n} \frac{\mathrm{d}x^n}{\mathrm{d}s} = p^m \frac{\partial g_{rm}}{\partial x^n} p^n = \frac{1}{2}\left( \frac{\partial g_{rm}}{\partial x^n} + \frac{\partial g_{rn}}{\partial x^m} \right) p^m p^n, \tag{17.58}$$

the equations of a geodesic are obtained as

$$g_{rm} \frac{\mathrm{d}p^m}{\mathrm{d}s} + [mn, r] p^m p^n = 0, \tag{17.59}$$

where

$$[mn, r] = \frac{1}{2}\left( \frac{\partial g_{rm}}{\partial x^n} + \frac{\partial g_{rn}}{\partial x^m} - \frac{\partial g_{mn}}{\partial x^r} \right) = [nm, r], \tag{17.60}$$

is called a *Christoffel symbol of the first kind.* These symbols satisfy the identity $[rm, n] + [rn, m] = \partial g_{mn}/\partial x^r$.

As an example, in cylindrical coordinates with $x^1 = \rho$, $x^2 = \phi$, $x^3 = z$ and the associated metric tensor $g_{11} = 1 = g_{33}$, $g_{22} = \rho^2$, $g_{mn} = 0$, $m \neq n$, the non-zero Christoffel symbols of the first kind are $[12, 2] = -[22, 1] = \rho$. In spherical coordinates ($x^1 = r$, $x^2 = \theta$, $x^3 = \phi$), the non-zero Christoffel symbols of the first kind are $[12, 2] = -[22, 1] = r$, $[13, 3] = -[33, 1] = r \sin^2\theta$, $[23, 3] = -[33, 2] = r^2 \sin\theta \cos\theta$.

The geodesic equations can also be written in terms of the *Christoffel symbol of the second kind*

$$\left\{ {s \atop mn} \right\} = g^{sr}[mn, r] = \left\{ {s \atop nm} \right\}, \tag{17.61}$$

as

$$\frac{\mathrm{d}^2 x^r}{\mathrm{d}s^2} + \left\{ {r \atop mn} \right\} \frac{\mathrm{d}x^m}{\mathrm{d}s} \frac{\mathrm{d}x^n}{\mathrm{d}s} = 0. \tag{17.62}$$

It is clear from equation (17.59) or (17.62) that the geodesic equations are coupled, second order non-linear differential equations in the variables $x^r$. The geodesic is obtained by solving the either the system (17.59) or (17.62) subject to given initial values of $x^r$ and $\mathrm{d}x^r/\mathrm{d}s$. Geodesics are routinely used in ray analysis and for studying antenna interactions on curved bodies [11] among other applications. Also, an understanding of geodesics is extremely useful in the design of electromagnetic cloaks as we shall show in section 17.5.

The Christoffel symbols of the first and second kind are also related as $[mn, r] = g_{rs}\left\{ {s \atop mn} \right\}$. An important result based on the identity (17.44) is

$$\left\{ {n \atop rn} \right\} = \frac{1}{2} g^{mn} \frac{\partial g_{mn}}{\partial x^r} = \frac{\partial}{\partial x^r} \ln \sqrt{|g|} = \frac{1}{\sqrt{|g|}} \frac{\partial}{\partial x^r} \sqrt{|g|}, \tag{17.63}$$

where $g = \det(g_{mn})$. If the metric tensor in a Riemannian space is constant then the Christoffel symbols are both zero.

In cylindrical coordinates the only non-zero Christoffel symbols of the second kind are $\left\{ {1 \atop 22} \right\} = -\rho$, $\left\{ {2 \atop 12} \right\} = \frac{1}{\rho}$. In spherical coordinates the only non-zero Christoffel symbol of the second kind are $\left\{ {1 \atop 22} \right\} = -r$, $\left\{ {1 \atop 33} \right\} = -r\sin^2\theta$, $\left\{ {2 \atop 12} \right\} = \frac{1}{r} = \left\{ {3 \atop 13} \right\}$, $\left\{ {3 \atop 23} \right\} = \cot\theta$, $\left\{ {2 \atop 33} \right\} = -\sin\theta\cos\theta$.

**Example 17.1.1.1.** Geodesic on a cylinder.
As an example we demonstrate the solution of geodesic equations in cylindrical coordinates with the metric form $\mathrm{d}s^2 = (\mathrm{d}\rho)^2 + \rho^2(\mathrm{d}\phi)^2 + (\mathrm{d}z)^2$. Let us assume that the coordinate system is oriented such that the $z$-axis is vertical. The geodesic equations are

$$\frac{\mathrm{d}^2\rho}{\mathrm{d}s^2} - \rho\left(\frac{\mathrm{d}\phi}{\mathrm{d}s}\right)^2 = 0, \tag{17.64}$$

$$\frac{\mathrm{d}^2\phi}{\mathrm{d}s^2} + \frac{2}{\rho}\frac{\mathrm{d}\rho}{\mathrm{d}s}\frac{\mathrm{d}\phi}{\mathrm{d}s} = 0, \tag{17.65}$$

$$\frac{\mathrm{d}^2 z}{\mathrm{d}s^2} = 0. \tag{17.66}$$

Let the initial conditions be $\rho(s=0) = \rho_0$, $\frac{\mathrm{d}\rho}{\mathrm{d}s}(s=0) = b_0$, $\phi(s=0) = \phi_0$, $\rho_0\frac{\mathrm{d}\phi}{\mathrm{d}s}(s=0) = K$, $z(s=0) = z_0$, $\frac{\mathrm{d}z}{\mathrm{d}s}(s=0) = \beta$. Equation (17.66) has the solution $z = \beta s + z_0$, where $\beta$ and $z_0$ are some constants. The constant $\beta$ can be thought as the relative vertical velocity. Multiplication of equation (17.65) with $\rho^2$ suggests that $\rho^2\mathrm{d}\phi/\mathrm{d}s$ is a constant. Evaluating this constant at $s = 0$ we write $\rho^2\mathrm{d}\phi/\mathrm{d}s = \rho_0 K$. Substituting this into equation (17.64) yields $\rho^3\mathrm{d}^2\rho/\mathrm{d}s^2 = \rho_0^2K^2$. We attempt a solution of this equation of the form $\rho^2(s) = \alpha^2s^2 + 2\rho_0 b_0 s + \rho_0^2$. Then we see that $\rho^3\mathrm{d}^2\rho/\mathrm{d}s^2 = \rho_0^2(\alpha^2 - b_0^2)$. Hence the constants are related through $\alpha^2 = K^2 + b_0^2$. Substituting the assumed form of solution for $\rho^2$ into the equation $\rho^2\mathrm{d}\phi/\mathrm{d}s = \rho_0 K$ and integrating after a change of variable $(\alpha^2 s + \rho_0 b_0) = K\rho_0\tan\zeta$ yields the solution for $\phi(s)$. The exact solution of the geodesic equations for $s \geqslant 0$ is then

$$\rho^2(s) = \alpha^2 s^2 + 2\rho_0 b_0 s + \rho_0^2, \tag{17.67}$$

$$\phi(s) = \arctan\left(\frac{\alpha^2 s + \rho_0 b_0}{K\rho_0}\right) + d_0; \quad \alpha^2 = K^2 + b_0^2, \tag{17.68}$$

$$z(s) = \beta s + z_0, \tag{17.69}$$

where the constant $d_0 = \phi_0 - \arctan(b_0/K)$. Finally on imposing the condition that $\mathrm{d}s^2 = (\mathrm{d}\rho)^2 + \rho^2(\mathrm{d}\phi)^2 + (\mathrm{d}z)^2$, we see that $\alpha^2 + \beta^2 = b_0^2 + K^2 + \beta^2 = 1$. To gain more insight into the solution we observe that on combining equations (17.68) and (17.69) $\rho(s)\cos[\phi(s) - d_0] = K\rho_0/\alpha$, which also implies that $x(s)\cos d_0 + y(s)\sin d_0 = K\rho_0/\alpha$ since $x(s) = \rho(s)\cos[\phi(s)]$ and $y(s) = \rho(s)\sin[\phi(s)]$. This is the equation of a straight line in the $xy$-plane. Combined with the solution $z(s) = \beta s + z_0$, the geodesic will be a straight line in the 3D space even though the metric tensor is non-constant. This is because the underlying space is still flat and a non-constant metric tensor does not necessarily imply a curved space. A geodesic can be a straight line even though it may not be apparent from the metric tensor being non-constant. However, a curved geodesic is always associated with a non-constant metric tensor.

We will now outline an approximate procedure that may be useful when the geodesic equations cannot be solved exactly. We illustrate this for the above example for the $(\rho, \phi)$ coordinates. It is convenient to first cast the geodesic equations as a system of first order non-linear equations by defining the state vector $\boldsymbol{y} = [\rho, u = \mathrm{d}\rho/\mathrm{d}s, v = \mathrm{d}\phi/\mathrm{d}s]^T$ and writing the state evolution equations as $\dot{\boldsymbol{y}} = \boldsymbol{f}$, where superscript $T$ indicates matrix transpose, an overhead dot represents $\mathrm{d}/\mathrm{d}s$ and $\boldsymbol{f} = [u, \rho v^2, -2uv/\rho]^T$. To solve the system in the interval $s \in (0, s_0)$, initial conditions $\boldsymbol{y}(s = 0) = \boldsymbol{y}_0$ are needed.

An exact solution of the non-linear state equation with the initial condition $y_0 = [\rho_0, u_0, 0]^T =: \tilde{y}_0$ is $\boldsymbol{y}(s) = [u_0 s + \rho_0, u_0, 0]^T =: \tilde{y}(s)$ or equivalently $\tilde{\rho}(s) = u_0 s + \rho_0$, $\tilde{\phi}(s) = \phi_0$, where $\phi_0 \in (0, 2\pi)$ is some constant. The constant $u_0$ may be thought of as the relative radial velocity. Coupled with $z(s) = \beta s + z_0$, the overall geodesic equation is subject to $\beta^2 + u_0^2 = 1$ relating the two relative velocities to give a consistent metric form. If $u_0 = 0$ the geodesic corresponds to a vertical line emanating from the point $P_0$ with cylindrical coordinates $(\rho_0, \phi_0, z_0)$ on the surface of a cylinder of radius $\rho_0$. If $\beta = 0$, the geodesic corresponds to a radial line in the horizontal plane $z = z_0$ emanating from $P_0$. In general it will be a straight line in the meridian plane $\phi = \phi_0$ making an angle $\arccos(\beta)$ with the $z$-axis and emanating from $P_0$.

Let us now explore the solution with a modified initial condition $y_0 = [\rho_0, 0, \alpha\rho_0^{-1}]^T$, where $\alpha$ is a constant. A non-zero initial velocity in the $\phi$-direction is now specified, but the radial velocity $u_0$ is set to zero. Because a direct solution of the non-linear equations maybe difficult in the general case, linearization of the state equations about the nominal solution $\tilde{y}$ is one good way to approach the problem. We chose the solution with $u_0 = 0$ to be the nominal solution and write $\boldsymbol{y} = \tilde{y} + \mathrm{y}_\delta$, $\mathrm{y}_0 = \tilde{y}_0 + y_\delta(0)$. Consequently we arrive at the linear system $\dot{y}_\delta \simeq \mathbf{A}y_\delta$ where

$$\mathbf{A} = \left.\frac{\partial \boldsymbol{f}}{\partial \boldsymbol{y}}\right|_{y=\tilde{y}} = \begin{pmatrix} 0 & 1 & 0 \\ 0 & 0 & 0 \\ 0 & 0 & 0 \end{pmatrix}; \quad y_\delta(0) = \begin{pmatrix} 0 \\ 0 \\ \beta\rho_0^{-1} \end{pmatrix}. \tag{17.70}$$

The solution of this linear system yields $\rho_\delta(s) = 0$; $\phi_\delta(s) = \alpha\rho_0^{-1}s + \phi_0$. Thus the total solution is $\rho(s) = \rho_0$; $\phi(s) = \alpha\rho_0^{-1}s + \phi_0$; $z(s) = \beta s + z_0$ subject to $\alpha^2 + \beta^2 = 1$ to yield a consistent metric form. We may take $\beta = \cos\theta_0$ and $\alpha = \sin\theta_0$. The geodesic is therefore a helical arc $\ell$ on the surface of a cylinder of radius $\rho_0$

emanating from the point $P_0$ with a helix angle of $\theta_0$ (helix angle = $\arccos(\hat{\mathbf{z}} \cdot \frac{d\ell}{ds})$). For $\theta_0 = 0$, the helix reduces to the vertical line found previously. For $\theta_0 = \pi/2$, the geodesic becomes a circle in the horizontal plane $z - z_0$ on the surface of the cylinder of radius $\rho_0$.

Even though the total solution found with the above initial condition is based on a linear approximation, it turns out to be the exact solution for a geodesic developed in 2D *for motion on the surface of a cylinder of radius* $\rho_0$. This can be easily verified by considering a 2D space with coordinates $x^1 = \phi$, $x^2 = z$ and a metric form $ds^2 = \rho_0^2(dx^1)^2 + (dx^2)^2$. Because *the metric tensor is constant*, the Christoffel symbols of the second kind are all zero and the geodesic equations now become $d^2x^1/ds^2 = 0$, $d^2x^2/ds^2 = 0$. With the initial condition $x^1(0) = \phi_0$, $dx^1(0)/ds = \rho_0^{-1}\sin\theta_0$, $x^2(0) = z_0$, $dx^2(0)/ds = \cos\theta_0$, we obtain a solution $x^1(s) = \sin\theta_0\rho_0^{-1}s + \phi_0$ and $x^2(s) = \cos\theta_0 s + z_0$ for the geodesic on the surface of the cylinder. If the cylinder were to be unrolled onto a plane, the geodesic is seen to a straight line on the surface. Indeed a cylinder is isometric with a Euclidean plane and it is not surprising to see this result. The linear approximation found by considering a geodesic in a 3D space, but with initial conditions imposed such that motion is restricted *to the surface of a cylinder* yields the same result. ■■

**Example 17.1.1.2** Geodesic on a sphere.
As a second example the geodesics on the surface of the sphere of radius $r$ can be obtained by using the usual spherical coordinates $x^1 = \theta$, $x^2 = \phi$ on the surface, where $\theta$ is the polar angle and $\phi$ is the azimuth angle. The metric form is $ds^2 = r^2(dx^1)^2 + r^2\sin\theta^2(dx^2)^2$ with the non-zero metric coefficients $g_{11} = r^2$, $g_{22} = r^2\sin^2\theta$ (also non-constant). The only non-zero Christoffel symbols are $\left\{ {1 \atop 22} \right\} = -\sin\theta\cos\theta$, $\left\{ {2 \atop 12} \right\} = \cot\theta$. Hence the geodesic equations are

$$\frac{d^2\theta}{ds^2} - \sin\theta\cos\theta\left(\frac{d\phi}{ds}\right)^2 = 0 \tag{17.71}$$

$$\frac{d^2\phi}{ds^2} + 2\cot\theta\frac{d\theta}{ds}\frac{d\phi}{ds} = 0 \Longrightarrow \frac{d}{ds}\left(\sin^2\theta\frac{d\phi}{ds}\right) = 0. \tag{17.72}$$

One possible solution to these equations is the longitude $\phi(s) = \phi_0$, $\theta(s) = s/r + \theta_0$, where $(\theta_0, \phi_0)$ are constants. This solution is a great circle on the sphere (intersection of a plane through the origin and the sphere itself). Because a sphere is not isometric with an Euclidean plane, the geodesic curve is not a straight line. Also note that of all the possible latitudes $\theta(s) = \theta_0$, $\phi(s) = s/(r\sin\theta_0) + \phi_0$, only the equator with $\theta_0 = \pi/2$ qualifies as a geodesic. This is also clear from the governing equations. Of course the equator is also a great circle. ■■

**Example 17.1.1.3.** Geodesic on a paraboloid.

Here we look at the more complicated example of a paraboloid, which is a body of revolution of a parabolic curve about the $z$-axis. The tangential coordinates on the surface of a paraboloid, $2zz_0 = \rho^2$, shown in figure 18.4, are $(u, v) = (\rho, \phi)$. The line element and non-zero Christoffel symbols of the second kind are worked out in example 18.1.2.2 and are

$$\mathrm{d}s^2 = \rho^2\mathrm{d}\phi^2 + \left(1 + \frac{\rho^2}{z_0^2}\right)\mathrm{d}\rho^2, \tag{17.73}$$

$$\left\{\begin{matrix}2\\11\end{matrix}\right\} = \frac{-z_0^2\rho}{\rho^2 + z_0^2}, \quad \left\{\begin{matrix}1\\12\end{matrix}\right\} = \frac{1}{\rho}, \quad \left\{\begin{matrix}2\\12\end{matrix}\right\} = \frac{\rho}{\rho^2 + z_0^2}. \tag{17.74}$$

The geodesic equations are

$$\frac{\mathrm{d}^2\phi}{\mathrm{d}s^2} + \frac{2}{\rho}\frac{\mathrm{d}\phi}{\mathrm{d}s}\frac{\mathrm{d}\rho}{\mathrm{d}s} = 0, \tag{17.75}$$

$$\frac{\mathrm{d}^2\rho}{\mathrm{d}s^2} - \frac{z_0^2\rho}{\rho^2 + z_0^2}\left(\frac{\mathrm{d}\phi}{\mathrm{d}s}\right)^2 + \frac{\rho}{\rho^2 + z_0^2}\left(\frac{\mathrm{d}\rho}{\mathrm{d}s}\right)^2 = 0. \tag{17.76}$$

It is clear from these equations that parallels (i.e. $\rho$= constant) on paraboloids cannot be geodesics. This is similar to the situation that latitudes on a sphere cannot be geodesics. However, meridians (i.e. $\phi$ = constant) can be geodesics[7].

On multiplying throughout by $\rho^2$ it is easy to see from equation (17.75) that

$$\frac{\mathrm{d}}{\mathrm{d}s}\left(\rho^2\frac{\mathrm{d}\phi}{\mathrm{d}s}\right) = 0 \Longrightarrow q(s) := \frac{\mathrm{d}\phi(s)}{\mathrm{d}s} = \frac{C_1}{\rho^2(s)} = \frac{C_1}{2z_0z}, \quad C_1 = \text{constant}. \tag{17.77}$$

If $q$ (velocity variable in $\phi$) and $\rho$ are specified at some initial point $s = s_i$ on the geodesic as $q_i$ and $\rho_i$, respectively, then the constant $C_1 = q_i\rho_i^2 = 2q_iz_iz_0$, which could be either positive or negative; $C_1 = 0$ results in meridian geodesics. Substituting equation (17.77) into equation (17.73) gives

$$\left(\frac{\mathrm{d}\rho}{\mathrm{d}s}\right)^2 = \frac{z_0^2}{\rho^2}\frac{\rho^2 - C_1^2}{\rho^2 + z_0^2} \Longrightarrow \frac{\mathrm{d}\rho}{\mathrm{d}s} = \pm\frac{z_0}{\rho}\sqrt{\frac{\rho^2 - C_1^2}{\rho^2 + z_0^2}}. \tag{17.78}$$

The positive sign corresponds to an ascending geodesic on which $\rho(s)$ increases with $s$, while the negative sign corresponds to a descending geodesic on which $\rho(s)$ decreases with $s$. At the initial point $s = s_i$ we note from equation (17.78) that with $\rho_i = \rho(s_i)$

[7] The following theorem is useful in this regard: for a surface of revolution having a parametrization $\mathbf{X}(u, v) = (f(v)\cos u, f(v)\sin u, g(v))$, any meridian is a geodesic. A parallel is a geodesic only at those parameter points$v_0$ where $f'(v_0) = 0$.

$$\left| \frac{d\rho(s_i)}{ds} \right| = z_0 \sqrt{\frac{1 - \rho_i^2 q_i^2}{\rho_i^2 + z_0^2}}. \tag{17.79}$$

Therefore the initial velocity in $\phi$ must satisfy $|q_i| \rho_i < 1$ for non-meridian trajectories (i.e. for $C_1 \neq 0$). This also implies that $|C_1| < \rho_i$. It is easier to determine the arc-length $s$ in terms of $\rho$ by inverting equation (17.78) and integrating. To this end the change of variable $\rho^2 = C_1^2 \cosh^2 t + z_0^2 \sinh^2 t$ permits $\sqrt{\rho^2 - C_1^2} = b \sinh t$, $\sqrt{\rho^2 + z_0^2} = b \cosh t$, $\rho d\rho = b^2 \sinh t \cosh t \, dt$ and $t = \log\left(\left[\sqrt{\rho^2 + z_0^2} + \sqrt{\rho^2 - C_1^2}\right]/b\right)$, where $b^2 = C_1^2 + z_0^2$. Equation (17.78) is then modified to $z_0 \, ds/dt = \pm b^2 \cosh^2 t$, which can be easily integrated. The result is

$$\begin{aligned} s = s_i \pm \frac{1}{2z_0}\Bigg[ & \sqrt{(\rho^2 + z_0^2)(\rho^2 - C_1^2)} - \sqrt{(\rho_i^2 + z_0^2)(\rho_i^2 - C_1^2)} + \\ & + b^2 \log\left( \frac{\sqrt{\rho_2 + z_0^2} + \sqrt{\rho^2 - C_1^2}}{\sqrt{\rho_i^2 + z_0^2} + \sqrt{\rho_i^2 - C_1^2}} \right) \Bigg]. \end{aligned} \tag{17.80}$$

Dividing equation (17.77) by equation (17.78) we obtain, on using the same change of variable as above,

$$\frac{d\phi}{d\rho} = \pm \frac{C_1}{z_0 \rho} \sqrt{\frac{\rho^2 + z_0^2}{\rho^2 - C_1^2}} \Longrightarrow \frac{d\phi}{dt} = \pm \frac{C_1}{z_0}\left[ 1 + \frac{z_0^2}{b^2 \sinh^2 t + C_1^2} \right],$$

and

$$\begin{aligned} \phi(\rho) = \phi(\rho_i) \pm \Bigg[ & \frac{C_1}{z_0} \log\left( \frac{\sqrt{\rho^2 + z_0^2} + \sqrt{\rho^2 + C_1^2}}{\sqrt{\rho_i^2 + z_0^2} + \sqrt{\rho_i^2 + C_1^2}} \right) + \\ & + \cos^{-1}\left( \frac{C_1 \sqrt{\rho^2 + z_0^2}}{b\rho} \right) - \cos^{-1}\left( \frac{C_1 \sqrt{\rho_i^2 + z_0^2}}{b\rho_i} \right) \Bigg], \end{aligned} \tag{17.81}$$

where [12, p 130, identity 2.458-1] has been used in the evaluation of the integral in $t$.

Figure 17.1 shows a sample non-meridian geodesic ascending on the paraboloid $\rho^2 = 2zz_0$ with initial coordinates $z_i = 10$, $\phi(\rho_i) = 0$, an initial velocity $q_i = 1/2z_i$, and $z_0 = 1$. Formula (17.81) is used to generate the numerical results. Clearly, the geodesic intersects several meridians in its ascent. Using equation (17.77) in the formula (18.107), the curvature, $\kappa_p$, and the radius of curvature, $R_p = 1/\kappa_p$, at any point on the geodesic are

$$\kappa_p = -\frac{(C_1^2 + z_0^2)}{(\rho^2 + z_0^2)^{3/2}}; \quad R_p = -\frac{(\rho^2 + z_0^2)^{3/2}}{(C_1^2 + z_0^2)}. \tag{17.82}$$

■■

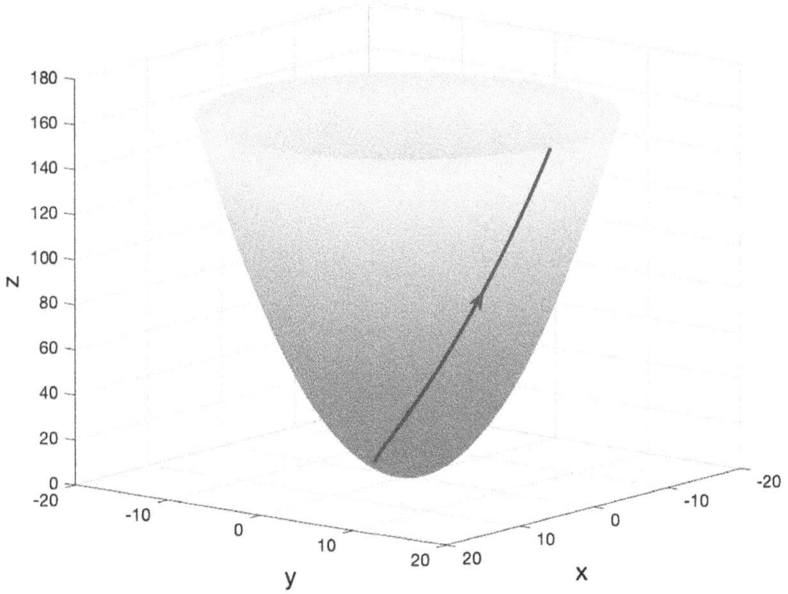


**Figure 17.1.** Geodesic (blue trace) on the paraboloid $\rho^2 = 2zz_0$ initiated at the point $z_i = 10$, $\phi(\rho_i) = 0$ and with initial azimuthal velocity $q_i = 1/2z_i$ and $z_0 = 1$.

**Example 17.1.1.4.** Ray parameter for antennas on convex surfaces.
For conformal antennas located on convex, conducting surfaces, the mutual coupling between the antenna elements at a given frequency is governed predominantly by the geodesic path connecting the two antennas [13]. The central parameter therein is the *generalized Fock parameter* $\xi$ (units = $\mathrm{m}^{-1}$) defined in terms of the ray curvature on the geodesic $\kappa$ and the wavenumber $k$ by

$$\xi(s) = \left(\frac{k}{2}\right)^{1/3} \int_{s_i}^{s} |\,\kappa\,|^{2/3}\ \mathrm{d}s. \tag{17.83}$$

In this example we determine the Fock parameter for the paraboloid of example 17.1.1.3. Substituting the expression for the curvature given in equation (17.82) and the relation (17.78) into equation (17.83), the Fock parameter for the paraboloid is

$$\begin{aligned} \xi &= \left(\frac{k}{2}\right)^{1/3} \int_{\rho_i}^{\rho} |\,\kappa_p\,|^{2/3} \frac{1}{\dfrac{\mathrm{d}\rho}{\mathrm{d}s}}\,\mathrm{d}\rho \\ &= \frac{1}{z_0}\left(\frac{kb}{2}\right)^{1/3} \int_{\rho_i}^{\rho} \frac{\rho\,\mathrm{d}\rho}{\sqrt{\rho^2 + z_0^2}\sqrt{\rho^2 - C_1^2}} = \frac{1}{z_0}\left(\frac{kb}{2}\right)^{1/3} \int_{t_i}^{t} \mathrm{d}t \\ &= \frac{1}{z_0}\left(\frac{kb}{2}\right)^{1/3} \ln\left(\frac{\sqrt{\rho^2 + z_0^2} + \sqrt{\rho^2 - C_1^2}}{\sqrt{\rho_i^2 + z_0^2} + \sqrt{\rho_i^2 - C_1^2}}\right), \end{aligned} \tag{17.84}$$

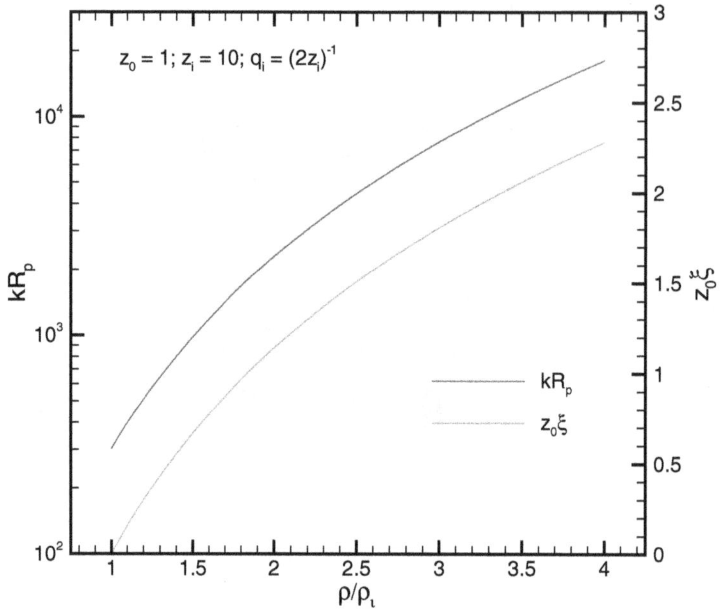


**Figure 17.2.** Normalized radius of curvature $kR_p$ and normalized Fock paramater $z_0\xi$ along the geodesic on a paraboloid $\rho^2 = 2zz_0$.

where the same change of variable from $\rho$ to $t$ as in example 17.1.1.3 has been employed. Figure 17.2 shows a sample plot of the radius of curvature $R_p$ and the generalized Fock parameter $\xi$ along the geodesic. Both of these parameters are shown to increase as the ray ascends the paraboloid starting from an initial $\rho_i = \sqrt{2z_0z_i}$. ■■

*17.1.1.2 Absolute and covariant derivatives*

It can be shown that neither of the Christoffel symbols is a tensor. For instance the first Christoffel symbol transforms as

$$[mn, r]' = [pq, s]\frac{\partial x^p}{\partial x'^m}\frac{\partial x^q}{\partial x'^n}\frac{\partial x^s}{\partial x'^r} + g_{pq}\frac{\partial x^p}{\partial x'^r}\frac{\partial^2 x^q}{\partial x'^m \partial x'^n}, \tag{17.85}$$

which is clearly not a homogeneous transformation as required for a tensor. Similarly, the Christoffel symbol of the second kind transforms as

$$\left\{ {r \atop \mathrm{mn}} \right\}' = \left\{ {s \atop \mathrm{pq}} \right\}\frac{\partial x'^r}{\partial x^s}\frac{\partial x^p}{\partial x'^m}\frac{\partial x^q}{\partial x'^n} + \frac{\partial x'^r}{\partial x^s}\frac{\partial^2 x^s}{\partial x'^m \partial x'^n}, \tag{17.86}$$

which clearly reveals that it does not transform like a tensor. However, in special transformations where the second derivative terms $\partial^2 x^q / \partial x'^m \partial x'^n$ are zero the two symbols behave as mixed tensors of rank three.

Even though the partial derivatives of an invariant, such as those encountered in the gradient given in equation (17.6), form a covariant tensor, the partial derivative of a general tensor is not a tensor. However, by adding certain terms to the derivative we obtain a tensor. This additional terms are encompassed in the definition of an absolute derivative.

**Definition 17.1.31.** *Absolute derivative of a contravariant vector.* Given a curve $x^r = x^r(u)$ the absolute derivative of a contravariant vector $T^r$ is defined as

$$\frac{\delta T^r}{\delta u} = \frac{\mathrm{d}T^r}{\mathrm{d}u} + \left\{ {r \atop \mathrm{mn}} \right\} T^m \frac{\mathrm{d}x^n}{\mathrm{d}u}, \tag{17.87}$$

which is itself a contravariant vector. The quantity $u$ is some parameter, not necessarily the arc-length.

Note that the tensor needs to be specified only along the given curve for the absolute derivative to be defined. Also note that the two constituent parts comprising an absolute derivative are separately not necessarily tensors even though their sum is. In particular, the ordinary derivative of a contravariant tensor is not necessarily a tensor even if d$u$ in an invariant. If the absolute derivative of a contravariant vector $T^r$ is zero along a curve, we say that $T^r$ is *propagated parallelly* along that curve. In this sense the geodesic equation (17.62) which can be expressed in terms of the absolute derivative as

$$\frac{\delta}{\delta s}\left(\frac{\mathrm{d}x^r}{\mathrm{d}s}\right) = 0, \tag{17.88}$$

propagates the unit tangent vector $\mathrm{d}x^r/\mathrm{d}s$ (which is a contravariant vector) along the geodesic. If the tensor is defined in a whole region as opposed to being defined only on a curve we can write

$$\frac{\mathrm{d}T^r}{\mathrm{d}u} = \frac{\partial T^r}{\partial x^n}\frac{\mathrm{d}x^n}{\mathrm{d}u}, \tag{17.89}$$

in the absolute derivative. The common factor $\mathrm{d}x^n/\mathrm{d}u$ (which is a contravariant tensor) can then be factored out from the absolute derivate and the remaining quantity leads us to the definition of a covariant derivative of a tensor.

**Definition 17.1.32.** *Covariant derivative.* The covariant derivative of a contravariant tensor $T^r$ specified in a region is defined as

$$T^r_{\ |n} = \frac{\partial T^r}{\partial x^n} + \left\{ {r \atop \mathrm{mn}} \right\} T^m, \tag{17.90}$$

which is a mixed tensor of rank two.

Other common notation for denoting the covariant derivative is $T^r_{,n}$ but the bar notation is less confusing because the comma notation is also used to indicate partial differentiation. *Note that the two constituent parts comprising a covariant derivative are not separately tensors even though their sum is.* In particular, the partial derivative alone of a contravariant tensor is not necessarily a tensor. Due to the non-vanishing values of the Christoffel symbol the covariant derivative applies to a whole set of quantities unlike the partial derivative which applies only to a single quantity. Also note that the second term on the right-hand side of equation (17.90) only involves a linear combination of the tensor components.

**Definition 17.1.33.** *Absolute derivative of a covariant vector.* The absolute derivative of a covariant vector $T_r$ specified on a curve is defined as

$$\frac{\delta T_r}{\delta u} = \frac{\mathrm{d}T_r}{\mathrm{d}u} - \begin{Bmatrix} m \\ \mathrm{rn} \end{Bmatrix} T_m \frac{\mathrm{d}x^n}{\mathrm{d}u}, \tag{17.91}$$

which is itself a covariant vector.

From this the covariant derivative of a covariant vector $T_r$ given in a whole region is obtained by extracting out the common factor $\mathrm{d}x^n/\mathrm{d}u$:

$$T_{r\,|n} = \frac{\partial T_r}{\partial x^n} - \begin{Bmatrix} m \\ \mathrm{rn} \end{Bmatrix} T_m. \tag{17.92}$$

The result is a covariant tensor of rank two. An immediate consequence of this definition is that

$$\begin{aligned} T_{\mu\,|\nu} - T_{\nu\,|\mu} &= \frac{\partial T_\mu}{\partial x^\nu} - \begin{Bmatrix} r \\ \mu\nu \end{Bmatrix} T_r - \frac{\partial T_\nu}{\partial x^\mu} + \begin{Bmatrix} r \\ \nu\mu \end{Bmatrix} T_r \\ &= \frac{\partial T_\mu}{\partial x^\nu} - \frac{\partial T_\nu}{\partial x^\mu} = -\left(T_{\nu\,|\mu} - T_{\mu\,|\nu}\right), \end{aligned} \tag{17.93}$$

for any pair $(\mu, \nu)$. Thus the difference in the cross covariant derivative of a *covariant tensor* behaves the same way as difference of cross partial derivatives of the tensor. However, this property is not shared by the covariant derivative of a contravariant tensor.

**Definition 17.1.34** *Absolute derivatives of second rank tensors.* The absolute derivative of a contravariant, a covariant, and a mixed tensors of second rank specified on a curve are, respectively, defined as

$$\frac{\delta T^{rs}}{\delta u} = \frac{\mathrm{d}T^{rs}}{\mathrm{d}u} + \begin{Bmatrix} r \\ \mathrm{mn} \end{Bmatrix} T^{ms} \frac{\mathrm{d}x^n}{\mathrm{d}u} + \begin{Bmatrix} s \\ \mathrm{mn} \end{Bmatrix} T^{rm} \frac{\mathrm{d}x^n}{\mathrm{d}u}, \tag{17.94}$$

$$\frac{\delta T_{rs}}{\delta u} = \frac{\mathrm{d}T_{rs}}{\mathrm{d}u} - \begin{Bmatrix} m \\ \mathrm{rn} \end{Bmatrix} T_{ms} \frac{\mathrm{d}x^n}{\mathrm{d}u} - \begin{Bmatrix} m \\ \mathrm{sn} \end{Bmatrix} T_{rm} \frac{\mathrm{d}x^n}{\mathrm{d}u}, \tag{17.95}$$

$$\frac{\delta T^{r}_{.s}}{\delta u} = \frac{\mathrm{d}T^{r}_{.s}}{\mathrm{d}u} + \left\{ {r \atop \mathrm{mn}} \right\} T^{m}_{.s} \frac{\mathrm{d}x^{n}}{\mathrm{d}u} - \left\{ {m \atop \mathrm{sn}} \right\} T^{r}_{.m} \frac{\mathrm{d}x^{n}}{\mathrm{d}u}. \tag{17.96}$$

The covariant derivatives of the second rank tensors specified in a whole region are obtained by extracting out the common factor $\mathrm{d}x^n/\mathrm{d}u$ from the total derivatives and are defined as

$$T^{rs}_{|n} = \frac{\partial T^{rs}}{\partial x^{n}} + \left\{ {r \atop \mathrm{mn}} \right\} T^{ms} + \left\{ {s \atop \mathrm{mn}} \right\} T^{rm}, \tag{17.97}$$

$$T_{rs\,|n} = \frac{\partial T_{rs}}{\partial x^{n}} - \left\{ {m \atop \mathrm{rn}} \right\} T_{ms} - \left\{ {m \atop \mathrm{sn}} \right\} T_{rm}, \tag{17.98}$$

$$T^{r}_{.s\,|n} = \frac{\partial T^{r}_{.s}}{\partial x^{n}} + \left\{ {r \atop \mathrm{mn}} \right\} T^{m}_{.s} - \left\{ {m \atop \mathrm{sn}} \right\} T^{r}_{.m}. \tag{17.99}$$

Equation (17.97) is a third rank mixed tensor with two contravariant indices and one covariant index, equation (17.98) is a third rank covariant tensor and equation (17.99) is a third rank mixed tensor with one contravariant index and two covariant indices.

From equation (17.98) and from the relation between the Christoffel symbols of the first and second kind it can be easily shown that *the covariant derivative of the metric tensor $g_{rs}$ is identically zero*. Also, the covariant derivative of the dual metric tensor $g^{rs}$ is also zero. Thus $g^{rs}_{|t} = 0$, $g_{rs\,|t} = 0$. Another interesting result is

$$\begin{aligned} F_{\mu\nu\,|\sigma} + F_{\nu\sigma\,|\mu} + F_{\sigma\mu\,|\nu} &= \frac{\partial F_{\mu\nu}}{\partial x^{\sigma}} - \left\{ {m \atop \mu\sigma} \right\} F_{m\nu} - \left\{ {m \atop \nu\sigma} \right\} F_{\mu m} + \frac{\partial F_{\nu\sigma}}{\partial x^{\mu}} - \left\{ {m \atop \nu\mu} \right\} F_{m\sigma} \\ &\quad - \left\{ {m \atop \sigma\mu} \right\} F_{\nu m} + \frac{\partial F_{\sigma\mu}}{\partial x^{\nu}} - \left\{ {m \atop \sigma\nu} \right\} F_{m\mu} - \left\{ {m \atop \mu\nu} \right\} F_{\sigma m} \\ &= \frac{\partial F_{\mu\nu}}{\partial x^{\sigma}} + \frac{\partial F_{\nu\sigma}}{\partial x^{\mu}} + \frac{\partial F_{\sigma\mu}}{\partial x^{\nu}} - \left\{ {m \atop \mu\sigma} \right\} (F_{m\nu} + F_{\nu m}) \\ &\quad - \left\{ {m \atop \nu\sigma} \right\} (F_{\mu m} + F_{m\mu}) - \left\{ {m \atop \nu\mu} \right\} (F_{m\sigma} + F_{\sigma m}) \\ &= \frac{\partial F_{\mu\nu}}{\partial x^{\sigma}} + \frac{\partial F_{\nu\sigma}}{\partial x^{\mu}} + \frac{\partial F_{\sigma\mu}}{\partial x^{\nu}}, \text{ provided } F_{\mu\nu} = -F_{\nu\mu}. \end{aligned} \tag{17.100}$$

Hence the cyclical sum of covariant derivatives of a second rank *antisymmetric tensor* is the equal to the cyclical sum of its partial derivatives. This is true irrespective of the dimension and nature of the underlying Riemannian space. The left-hand side of equation (17.100) is an antisymmetric covariant tensor of rank three.

Starting from an arbitrary covariant tensor $T_r$ and successively taking two covariant derivatives it is straightforward to show that

$$T_{r\,|mn} - T_{r\,|nm} = R^{\,s}_{.rmn}T_s, \tag{17.101}$$

where the mixed fourth rank tensor

$$R^{\,s}_{.rmn} = \frac{\partial}{\partial x^m}\left\{ {s \atop rn} \right\} - \frac{\partial}{\partial x^n}\left\{ {s \atop rm} \right\} + \left\{ {p \atop rn} \right\}\left\{ {s \atop pm} \right\} - \left\{ {p \atop rm} \right\}\left\{ {s \atop pn} \right\}, \tag{17.102}$$

in known as the *Riemann–Chistoffel curvature tensor*. Some of its properties are

(i) It is antisymmetric in the last two subscripts: $R^{\,s}_{.rmn} = -R^{\,s}_{.rnm}$

(ii) It is cyclically symmetric in its subscripts: $R^{\,s}_{.rmn} + R^{\,s}_{.mnr} + R^{\,s}_{.nrm} = 0$.

(iii) Contracting it with respect to its last subscript yields the symmetric covariant tensor $R_{rm} = R^{\,n}_{.rmn}$ of rank two:

$$\begin{aligned} R_{rm} &= \frac{\partial}{\partial x^m}\left\{ {n \atop rn} \right\} - \frac{\partial}{\partial x^n}\left\{ {n \atop rm} \right\} + \left\{ {p \atop rn} \right\}\left\{ {n \atop pm} \right\} - \left\{ {p \atop rm} \right\}\left\{ {n \atop pn} \right\} \\ &= \frac{1}{2}\frac{\partial^2(\ln |g|)}{\partial x^r \partial x^m} - \frac{1}{2}\left\{ {p \atop rm} \right\}\frac{\partial}{\partial x^p}\ln |g| - \frac{\partial}{\partial x^n}\left\{ {n \atop rm} \right\} + \left\{ {n \atop pm} \right\}\left\{ {p \atop rn} \right\}. \end{aligned} \tag{17.103}$$

This tensor is known as the *Ricci* tensor. If the choice of coordinates may be made such that $|g| = 1$, then the Ricci tensor simplifies to $\left\{ {n \atop pm} \right\}\left\{ {p \atop rn} \right\} - \frac{\partial}{\partial x^n}\left\{ {n \atop rm} \right\}$. The curvature invariant $R$ is defined in terms of the Ricci tensor as $R = g^{mn}R_{mn} = g^{mn}R^{\,s}_{.mns} = R^{\,s}_{.s}$.

(iv) It satisfies the *Bianchi identity*

$$R^{\,r}_{.smn\,|t} + R^{\,r}_{.snt\,|m} + R^{\,r}_{.stm\,|n} = 0. \tag{17.104}$$

The Riemann–Christoffel tensor and the Ricci tensor measure the intrinsic curvature of the space. They are dependent only on the metric tensor and its derivatives up to second order. If the space is such that there is a coordinate system with reference to which $g_{mn}$ are constants, then all the $R^{\,s}_{.rmn}$ vanish. Such a subspace may be labeled as being *locally flat*. If we choose any new system of coordinates in place of the original ones, the new $g'_{mn}$ will not be constants, but as a consequence of its tensor nature, the transformed components of $R^{\,s}_{.rmn}$ will still vanish in the new system. *The necessary and sufficient condition for the space to be considered as being flat is that the Riemann–Christoffel tensor be zero in a finite subspace of the Riemann space.* So the Euclidean 3D space is flat even after considering that the metric tensor is non-constant and Christoffel symbols can be non-zero (such as occurring when cylindrical or a spherical coordinate system is employed). According to general relativity the properties of a metric tensor in an arbitrary coordinate system in a curved space are indeed induced by matter.

**Definition 17.1.35** *Covariant derivative of a general tensor.* The formula for the covariant derivative of the most general tensor is

$$\begin{aligned}T^{r_1\ldots r_m}_{s_1\ldots s_n}{}_{|p} = {} & \frac{\partial}{\partial x^p}T^{r_1\ldots r_m}_{s_1\ldots s_n} + \left\{ {r_1 \atop qp} \right\} T^{qr_2\ldots r_m}_{s_1\ldots s_n} + \ldots \\ & + \left\{ {r_m \atop qp} \right\} T^{r_1\ldots r_{m-1}q}_{s_1\ldots s_n} - \left\{ {q \atop s_1 p} \right\} T^{r_1\ldots r_m}_{qs_2\ldots s_n} - \ldots \\ & - \left\{ {q \atop s_n p} \right\} T^{r_1\ldots r_m}_{s_1\ldots s_{n-1}q}.\end{aligned} \tag{17.105}$$

The result is a $(m + n + 1)$th rank mixed tensor with $m$ contravariant indices and $(n + 1)$ covariant indices.

A consequence of the fact that the covariant density of the metric tensor $g_{rs}$ or $g^{rs}$ is zero is that

$$T^{n}_{r.m}{}_{|t} = (g_{rs}T^{sn}_{\,..m})_{|t} = g_{rs}T^{sn}_{\,..m}{}_{|t} \tag{17.106}$$

meaning that it is immaterial whether we differentiate the covariant, contravariant, or any one of the mixed representations of a tensor. The corresponding differentiated tensors can be converted to each other via the metric tensor just like the original tensors.

*17.1.1.3 Divergence, curl of tensors*

If the Christoffel symbols vanish in a subspace then the absolute derivative reduces to the ordinary derivative, the covariant derivative reduces to the partial derivative, and each of these is a tensor. Absolute and covariant derivatives follow the same rules as ordinary derivatives for sum and product of quantities. As a special case of the covariant derivative, the divergence of a contravariant tensor $T^r$ is defined by contracting $T^r_{\;|n}$ with respect to the indices $r$ and $n$:

$$\mathrm{Div}(T^r) = T^r_{\;|n=r} = \frac{\partial T^r}{\partial x^r} + \left\{ {r \atop mr} \right\} T^m = \frac{1}{\sqrt{|g|}}\frac{\partial}{\partial x^r}\left(\sqrt{|g|}\,T^r\right), \tag{17.107}$$

where $g = \det(g_{mn})$ is the determinant of the metric tensor and equation (17.63) was used in arriving at the last equality in equation (17.107). For instance, in spherical coordinates in a 3D space $g = r^4 \sin^2\theta$ and one obtains the well known formula

$$\mathrm{Div}(T^r) = \frac{1}{r^2}\frac{\partial(r^2T^1)}{\partial r} + \frac{1}{\sin\theta}\frac{\partial(\sin\theta T^2)}{\partial\theta} + \frac{\partial T^3}{\partial\phi} = \nabla\cdot T^r, \tag{17.108}$$

for the divergence. In an analogous manner, the divergence of an *antisymmetric* contravariant tensor of second order, $T^{rs}$, is obtained by contracting with respect to $s$ and $n$ in equation (17.97). Utilizing equation (17.63) and the symmetry/asymmetry of the Christoffel symbol and the tensor $T^{rs}$ one obtains

$$\mathrm{Div}_s(T^{rs}) = T^{rs}_{\;\;|n=s} = \frac{1}{\sqrt{|g|}}\frac{\partial}{\partial x^s}\left(\sqrt{|g|}\,T^{rs}\right) =: B^r, \tag{17.109}$$

which is seen to be a contravariant tensor of rank one. The index on the divergence operator indicates the index of the second order tensor with respect to which derivative is performed. Note that the divergence is not equal to $\partial T^{rs}/\partial x^s$ unless the determinant of the metric tensor is a constant. Using this definition of divergence it is also easy to see that for an antisymmetric tensor $F_{rs}$

$$g^{sn}F_{rs\,|n} = g_{r\alpha}\mathrm{Div}_\beta(F^{\alpha\beta}). \tag{17.110}$$

Similarly, contracting with respect to the indices $r$ and $n$ in equation (17.99) and taking into consideration equations (17.44), (17.60), (17.61), and (17.63) the divergence of a mixed tensor of the second rank can be obtained as

$$\mathrm{Div}_n(T^n_{.s}) = T^n_{.s\,|n} =: T_s = \frac{1}{\sqrt{|g|}}\frac{\partial}{\partial x^n}\left(\sqrt{|g|}\,T^n_{.s}\right) - \frac{1}{2}\frac{\partial g_{rn}}{\partial x^s}T^{rn}, \tag{17.111}$$

$$= \frac{1}{\sqrt{|g|}}\frac{\partial}{\partial x^n}\left(\sqrt{|g|}\,T^n_{.s}\right) + \frac{1}{2}\frac{\partial g^{rn}}{\partial x^s}T_{rn}, \tag{17.112}$$

provided that the corresponding tensor $T^{rn} = g^{mr}T^n_{.m}$ is *symmetric*.

Another useful property for an antisymmetric contravariant tensor $F^{\mu\nu}$ in a space with a metric tensor $g_{mn}$ is

$$K_\sigma := F_{\sigma\mu}\mathrm{Div}_\nu(F^{\mu\nu}) = F_{\sigma\mu}\frac{1}{\sqrt{|g|}}\frac{\partial}{\partial x^\nu}\left(\sqrt{|g|}\,F^{\mu\nu}\right) = \mathrm{Div}_\nu(T^\nu_{.\sigma}), \tag{17.113}$$

where the mixed tensor $T^\nu_{.\sigma}$ is equal to

$$T^\nu_{.\sigma} = F_{\sigma\mu}F^{\mu\nu} + \frac{1}{4}\delta^\nu_\sigma F^{\alpha\beta}F_{\alpha\beta}. \tag{17.114}$$

This formula reduces to [6, equation (66)] when $|g| = 1$. It is easy to verify that $T^{\nu r} = g^{\sigma r}T^\nu_{.\sigma}$ is symmetric with respect to $r$ and $\nu$.

The components of the tensor curl of a covariant tensor $A_\mu$ are defined via

$$\mathrm{Curl}_{\mu\nu}(A) = \frac{\partial A_\nu}{\partial x^\mu} - \frac{\partial A_\mu}{\partial x^\nu}. \tag{17.115}$$

Comparison with equation (17.93) reveals that the right-hand side is equal to the antisymmetric tensor $B_{\mu\nu} = A_{\nu\,|\mu} - A_{\mu\,|\nu}$. Thus the tensor curl of a covariant tensor is an antisymmetrical covariant tensor of *rank two*[8]. The curl of a covariant antisymmetric tensor $F_{ij}$ of rank two is defined as

$$\mathrm{Curl}_{ijk}(F) = F_{ij\,|k} + F_{jk\,|i} + F_{ki\,|j} = \frac{\partial F_{ij}}{\partial x^k} + \frac{\partial F_{jk}}{\partial x^i} + \frac{\partial F_{ki}}{\partial x^j}, \tag{17.116}$$

[8] In 3D Euclidean space with cyclical indices $(i, \mu, \nu)$ it is common to associate the quantity $A_{\mu|\nu} - A_{\nu|\mu}$ with the unitary vector $\mathbf{a}_i$ and consider a cyclical sum to define the curl.

where the last equality is due to equation (17.100). The resulting tensor of rank three is antisymmetric with respect to any two indices. Thus it vanishes if any two indices are equal. In a four-dimensional space it has only four independent components. If $F$ itself is formed from the curl of a covariant tensor as in equation (17.115), then it is easy to see that Curl(Curl($A$)) = 0. Other identities from ordinary vector analysis also follow such as Curl($\nabla_i\psi$) = 0. Furthermore Div(Div($F$)) = 0 for an antisymmetric tensor $F$.

The covariant derivatives can be used to associate the ordinary curl in 3D space to the components of an antisymmetric tensor using permutation symbols. Consider the curl, $\nabla \times \mathbf{A}$, of a vector $\mathbf{A} = A_i\mathbf{a}^i$ in 3D space with covariant components $A_i$ defined by [14]

$$\begin{aligned}\mathbf{B} = \nabla \times \mathbf{A} = &\frac{1}{\sqrt{|\alpha|}}\left[\left(\frac{\partial A_3}{\partial x^2} - \frac{\partial A_2}{\partial x^3}\right)\mathbf{a}_1 + \left(\frac{\partial A_1}{\partial x^3} - \frac{\partial A_3}{\partial x^1}\right)\mathbf{a}_2\right.\\ &\left.+\left(\frac{\partial A_2}{\partial x^1} - \frac{\partial A_1}{\partial x^2}\right)\mathbf{a}_3\right]\\ =:&\, B^m\mathbf{a}_m,\end{aligned} \tag{17.117}$$

where $\mathbf{a}_m$ and $\mathbf{a}^m$ are the unitary and reciprocal unitary vectors, respectively. The metric tensor corresponding to the space dimensions is denoted by $\alpha_{mn} = \mathbf{a}_m \cdot \mathbf{a}_n$ and the square root of its absolute determinant by $\sqrt{|\alpha|} = |\hat{\mathbf{a}}_1 \times \hat{\mathbf{a}}_2 \cdot \hat{\mathbf{a}}_3| = \sqrt{|\det(\alpha_{mn})|}$. In combination with the permutation symbol we define the quantities $\eta_{mnr} = \varepsilon_{mnr}\sqrt{|\alpha|}$ and $\eta^{mnr} = \varepsilon^{mnr}/\sqrt{|\alpha|}$ that form covariant and contravariant pseudo-tensors, respectively, of rank three. The components of curl of $\mathbf{A}$ can then be represented as

$$B^m = \eta^{mnr}\frac{\partial A_r}{\partial x^n} =: \eta^{mnr}A_{r,n}, \tag{17.118}$$

where a comma preceding a suffix is a short-form notation for the partial derivative with respect to the coordinate represented by that suffix. The observation is that it is the covariant components of a vector that appear in the definition of ordinary curl and the result is a quantity whose components are those of a contravariant pseudo-vector. The pseudo-vector nature of $B^m$ is clear from the presence of the permutation symbols.

By slightly modifying the definition (17.29) (by the inclusion of the factor $\sqrt{|\alpha|}$, which has been assumed to be unity in equation (17.29)) we associate the pseudo-tensor $B^m$ with an antisymmetric covariant tensor $B_{rs}$ via

$$B^m = \frac{1}{2}\eta^{mrs}B_{rs}. \tag{17.119}$$

In view of equation (17.24) this relation is equivalent to

$$B_{rs} = \eta_{rsm}B^m = \eta_{mrs}B^m, \tag{17.120}$$

where $B_{rs} = -B_{sr}$. For the pseudo-tensor $B^m$ in equation (17.118), the skew-symmetric tensor is

$$B_{mn} = \eta_{tmn} B^t = \eta_{tmn} \eta^{trs} A_{s,r} = \varepsilon_{tmn} \varepsilon^{trs} A_{s,r} = (A_{n,m} - A_{m,n}) = A_{n|m} - A_{m|n},$$

which is seen to coincide with the tensor curl definition given in equation (17.115). Thus $B_{mn} = \mathrm{Curl}_{mn}(A)$, where $A = A_i$. In terms of $B^i$ and $A_i$ it has the explicit representation

$$B_{mn} = \sqrt{|\alpha|}\begin{pmatrix} 0 & B^3 & -B^2 \\ -B^3 & 0 & B^1 \\ B^2 & -B^1 & 0 \end{pmatrix} = \begin{pmatrix} 0 & A_{2|1} - A_{1|2} & A_{3|1} - A_{1|3} \\ A_{1|2} - A_{2|1} & 0 & A_{3|2} - A_{2|3} \\ A_{1|3} - A_{3|1} & A_{2|3} - A_{3|2} & 0 \end{pmatrix}. \quad (17.121)$$

In accordance with equation (17.107) the three-dimensional divergence of a vector $\mathbf{A} = \mathbf{a}_i A^i$ is defined as

$$\nabla \cdot \mathbf{A} = \frac{1}{\sqrt{|\alpha|}} \frac{\partial}{\partial x^i}\left(\sqrt{|\alpha|} A^i\right). \quad (17.122)$$

*17.1.1.4 Laplacian and d'Alembertian of invariants*

The Laplacian $\nabla^2\psi$ of an invariant $\psi$ is defined as

$$\nabla^2\psi = (g^{nm}\nabla_m\psi)_{|n} = \left(g^{nm}\frac{\partial\psi}{\partial x^m}\right)_{|n} = g^{nm}\left(\frac{\partial\psi}{\partial x^m}\right)_{|n} = g^{nm}(\nabla_m\psi)_{|n}, \quad (17.123)$$

where the third equality follows in view of $g^{rs}_{\ |t} = 0$ and $\nabla_m$ is an abbreviation for $\partial/\partial x^m$. The result is an invariant since it is the contraction of the second rank contravariant tensor $g^{nm}$ and the second rank covariant tensor $(\nabla_m\psi)_{|n}$. Using equation (17.107) we can write an explicit covariant expression for the d'Alembertian of an invariant as

$$\Box^2\psi = \mathrm{Div}_i(\mathrm{grad}\psi) = \frac{1}{\sqrt{|g|}}\frac{\partial}{\partial x^i}\left(\sqrt{|g|}\, g^{ik}\frac{\partial\psi}{\partial x^k}\right). \quad (17.124)$$

Because the gradient of an invariant produces a covariant tensor and the divergence is defined for a contravariant tensor, the representation of the former is first converted to that of a contravariant tensor in the above before the divergence operator is applied. In summary the operations of gradient and curl increase the rank of a tensor field by one unit, while the divergence operation decreases the rank by one unit.

### 17.1.2 Minkowski space

Minkowski space is a four-dimensional spacetime ($x^1 = x$, $x^2 = y$, $x^3 = z$, $x^4 = ct$), where the space part is the ordinary Euclidean space with Cartesian coordinates $(x, y, z)$ endowed with the metric tensor[9] $h_{mn} = -\delta_{mn}$, $1 \leqslant m, n \leqslant 3$ and the time

[9] We will reserve the symbol $g_{mn}$ to represent a metric tensor in an arbitrary coordinate system, but shall use the symbol $h_{mn}$ to represent it in a Cartesian coordinate system.

part has the metric tensor component $h_{4m} = \delta_{4m}$. The determinant of the metric tensor is $h = -1$. The metric form is $ds^2 = (cdt)^2 - (dx)^2 - (dy)^2 - (dz)^2$, $c$ being the speed of light in free space. Thus the metric form is not necessarily positive definite. A tensor of rank one in this space is called the four-vector. The coordinates $x^i$ are known as *Galilean coordinates* in special relativity, where frames of reference can move at constant velocity relative to each other. The curvature of this space is zero. By writing the metric form as $ds^2 = c^2 d\tau^2$ we define the *proper time* $d\tau$ such that

$$d\tau = \sqrt{1 - \frac{dx^2 + dy^2 + dz^2}{c^2 dt^2}}\, dt = \sqrt{1 - \frac{u^2}{c^2}}\, dt =: \frac{1}{\gamma(u)} dt, \tag{17.125}$$

where $u = |d\mathbf{r}/dt|$ denotes speed and $\gamma(u) > 1$. Thus the proper time is less than the time interval $dt$ measured in a frame that observes the speed $u$. The 'position vector' in this space is designated by $R^\mu = (x^1, x^2, x^3, x^4) = [\mathbf{r}, ct]$, where $\mathbf{r}$ is the ordinary three-vector measured from the origin. The covariant form of this four-vector is $R_\mu = h_{\mu\nu} R^\nu = [-\mathbf{r}, ct]$. The squared magnitude of the position vector is $R^2 = R_\mu R^\mu = -r^2 + c^2 t^2$.

The linear transformation with constant coefficients in this space that preserves the line element is the *Lorentz transformation* $x'^r = (\partial x'^r / \partial x^s) x^s$ with

$$\frac{\partial x'^r}{\partial x^s} = \begin{pmatrix} \gamma & 0 & 0 & -\beta\gamma \\ 0 & 1 & 0 & 0 \\ 0 & 0 & 1 & 0 \\ -\beta\gamma & 0 & 0 & \gamma \end{pmatrix}, \tag{17.126}$$

where $\beta = u/c < 1$ is the relative speed between the two inertial frames $S$, $S'$ involved in the Lorentz transformation and $\gamma = 1/\sqrt{1 - \beta^2} > 1$. Note that the Jacobian $J'$ of the Lorentz transformation is unity. Thus the Lorentz transformation is an orthogonal transformation. The inverse Lorentz transformation $x^r = (\partial x^r / \partial x'^s) x'^s$ has the transformation coefficients

$$\frac{\partial x^r}{\partial x'^s} = \begin{pmatrix} \gamma & 0 & 0 & \beta\gamma \\ 0 & 1 & 0 & 0 \\ 0 & 0 & 1 & 0 \\ \beta\gamma & 0 & 0 & \gamma \end{pmatrix}. \tag{17.127}$$

Because the metric tensor transforms as the latter of equation (17.9), it is easy to verify that $h'_{mn} = h_{mn}$. For instance $h'_{11} = (\frac{\partial x^m}{\partial x'^1})(\frac{\partial x^n}{\partial x'^1}) h_{mn} = (\gamma)(\gamma)(-1) + (\beta\gamma)(\beta\gamma)(+1)$ $= (\beta^2 - 1)\gamma^2 = -1 = h_{11}$. Thus the metric tensor is preserved under the Lorentz transformation. From equation (17.47) and $J' = 1$, we have the equality of volume elements $dv' = dv$ in the four-dimensional space, which further implies that time dilation and length contraction go hand-in-hand under a Lorentz transformation. If $dV$ is the 3D-space volume element in the frame $S$ and $dV'$ is the 3D-space volume element in the moving frame $S'$ then $dv = dV dt = dv' = dV' d\tau \Longrightarrow dV' = \gamma dV$.

For such a Lorentz transformation that operates in a Euclidean space the covariant derivative coincides with partial derivative and the derivative operator is represented by the special covariant symbol

$$\partial_\mu := \frac{\partial}{\partial x^\mu} = \left[\nabla, \frac{1}{c}\frac{\partial}{\partial t}\right], \tag{17.128}$$

where $\nabla$ is the usual gradient operator of 3D space in the Cartesian coordinates. The corresponding contravariant differential symbol $\partial^\mu = h^{\mu\nu}\partial_\nu$ is

$$\partial^\mu := \frac{\partial}{\partial x_\mu} = \left[-\nabla, \frac{1}{c}\frac{\partial}{\partial t}\right]. \tag{17.129}$$

The four-dimensional Laplacian operator or the *d'Alembertian* is represented by the symbol

$$\Box^2 = \partial_\mu \partial^\mu = \left[-\nabla^2 + \frac{1}{c^2}\frac{\partial^2}{\partial t^2}\right], \tag{17.130}$$

which is an invariant operator under the Lorentz transformation. The divergence of a four-vector $A^\mu = [\mathbf{A}, A^4]$, where $\mathbf{A}$ is an ordinary vector in 3D, in Minkowski space is the *invariant*

$$\partial_\mu A^\mu = \nabla \cdot \mathbf{A} + \frac{1}{c}\frac{\partial A^4}{\partial t}. \tag{17.131}$$

Since the displacement $\mathrm{d}x^\mu$ is a tensor and $\mathrm{d}\tau \propto \mathrm{d}s$ is an invariant, the quantity $U^\mu = \mathrm{d}x^\mu/\mathrm{d}\tau$ is also a tensor and is known as the 4-velocity. For an entity traveling with a velocity $\mathbf{u}$ as measured in a reference frame $S$, we have

$$U^\mu = \frac{\mathrm{d}x^\mu}{\mathrm{d}\tau} = \frac{\mathrm{d}x^\mu}{\mathrm{d}t}\frac{\mathrm{d}t}{\mathrm{d}\tau} = \gamma[\mathbf{u}, c], \tag{17.132}$$

in that frame. The corresponding covariant form is $U_\mu = h_{\mu\nu}U^\nu = \gamma[-\mathbf{u}, c]$. Note that $U^\mu U_\mu = \gamma^2[\mathbf{u}, c] \cdot [-\mathbf{u}, c] = c^2$, which is an invariant.

## 17.2 The covariant form of Maxwell's equations in Euclidean pseudo-space

Having laid the groundwork for tensor analysis, we now look at formulating Maxwell's equations in a covariant form by considering Lorentz transformation and Cartesian coordinates. Once expressed in a covariant form the equations remain valid in any other coordinate system as well as under any other transformation.

If $\rho$ and $\rho_0$ denote the electric charge densities in the frames $S$ and $S'$ (frame in which charges are at rest), respectively, then on using the invariance of the number of charges and the fact that the 3D-space volume elements transform as $\mathrm{d}V' = \gamma \mathrm{d}V$, it is easy to see that $\rho = \gamma\rho_0$. The quantity $\rho_0$ is known as the *proper charge density*, which is an invariant. The *current density four-vector* is defined as

$$J^{\mu} = \rho_0 U^{\mu} = \rho_0 \gamma[\mathbf{u}, c] = [\mathscr{J}, \rho c], \tag{17.133}$$

where $\mathscr{J} = \rho\mathbf{u}$ is the ordinary current density vector in a 3D space associated with a moving charge density $\rho$. Thus the current density four-vector encompasses both the ordinary current density vector and the scalar electric charge density into one single tensor quantity $J^{\mu}$, implying that charge density and current density are different aspects of the same entity. The unit of the current density tensor in Cartesian coordinates is A $m^{-1}$. Using equation (17.107) or (17.131) the divergence of the current density four-vector is the invariant $\partial_{\mu} J^{\mu} = \nabla \cdot \mathscr{J} + \frac{\partial \rho}{\partial t}$. But the equation of continuity suggests that this invariant is zero.

It is well known that in free space the scalar electric potential $\psi_e$ and the magnetic vector potential $\mathscr{A}$ satisfy $\Box^2 \psi_e = \rho/\varepsilon_0$ and $\Box^2 \mathscr{A} = \mu_0 \mathscr{J}$. The two equations can be combined by defining the potential four-vector $\Phi^{\mu} = [c\mathscr{A}, \psi_e]$ and writing

$$\Box^2 \Phi^{\mu} = \eta_0 J^{\mu}, \tag{17.134}$$

where $\eta_0 = \sqrt{\mu_0/\varepsilon_0}$ is the intrinsic impedance of vacuum. Thus the scalar and vector potentials are no longer quantities requiring independent description but are different aspects of the same entity $\Phi$. The Lorentz gauge condition $\nabla \cdot \mathscr{A} + 1/c^2 \partial\phi/\partial t = 0$ can de directly represented in terms of the divergence, equation (17.131), of the four-vector as $\partial_{\mu}\Phi^{\mu} = 0$, which is an invariant. The gauge transformation $\psi_e \to \psi_e + \partial\chi/\partial t$, $c\mathscr{A} \to c\mathscr{A} - c\nabla\chi$ can be combined into a single equation $\Phi^{\mu} \to \Phi^{\mu} + c\partial^{\mu}\chi$ using equation (17.129). The gauge function satisfies the wave equation $\Box^2 \chi = 0$.

*Coulomb's law*. It is interesting to see what the ramifications of Coulomb's law are from a tensor viewpoint. Let $(x^j)_1$ and $(x^j)_2$ be the Galilean coordinates of two points and let $R^j = (x^j)_2 - (x^j)_1 =: [\mathbf{r}, ct]$ (corresponding $R_j = [-\mathbf{r}, ct]$) be the displacement between the two points. For a point charge $Q$ (invariant) placed at the source point $(x^j)_1$ the electric potential, $\psi_0$, at the field point $(x^j)_2$ in a proper frame (where is the charge is at rest) is given by Coulomb's law

$$\psi_0 = \frac{Q}{4\pi\varepsilon_0} \frac{1}{r_0}, \tag{17.135}$$

where $r_0$ is the proper vector distance between the source and field points. In accordance with causality, the potential at $(x^j)_2$ is to be measured at the time corresponding to the retardation condition $R^j R_j = -r^2 + c^2 t^2 = 0$ in any frame. In the proper frame $r = r_0$, $ct = r_0$, $R^j = [\mathbf{r}_0, r_0]$, the four-velocity is $U^j = [0, c]$ and the invariant $U^j R_j = c r_0$. Hence the electric potential can be expressed as $\psi_0 = (Q/4\pi\varepsilon_0) \cdot (c/U^j R_j)$. The magnetic vector potential, $\mathscr{A}_0$, in the proper frame is zero. The four-potential in the proper frame is then

$$\Phi = \frac{Q}{4\pi\varepsilon_0} \frac{U^j}{U^j R_j}, \tag{17.136}$$

subject to the condition $R^jR_j = 0$. However, owing to its tensor form this expression is valid in *any inertial frame*. In a frame moving at a velocity **u** relative to the proper frame (or equivalently, in a frame is which the charge is perceived as moving uniformly) we have from equation (17.132) that $U^j = \gamma[\mathbf{u}, c]$. The invariant $U^jR_j = \gamma(-\mathbf{u}\cdot\mathbf{r} + c^2t) = \gamma(-\mathbf{u}\cdot\mathbf{r} + cr)$. The non-zero magnetic vector potential in that frame is then

$$\mathscr{A} = \left(\frac{Q}{4\pi\varepsilon_0 c}\right)\frac{\mathbf{u}}{r - \dfrac{\mathbf{u}\cdot\mathbf{r}}{c}}, \tag{17.137}$$

and the electric scalar potential

$$\psi_e = \left(\frac{Q}{4\pi\varepsilon_0}\right)\frac{1}{r - \dfrac{\mathbf{u}\cdot\mathbf{r}}{c}}. \tag{17.138}$$

The potentials (17.137) and (17.138) are known as Liénard–Wiechert potentials of a uniformly moving charge. Completely electrostatic forces in the proper frame are converted to a combination of electric and magnetic forces in a uniformly moving frame.

The curl of the covariant tensor $\Phi_\mu = [-c\mathscr{A}, \psi_e]$ given by equation (17.115) is used to define a new antisymmetric tensor

$$\mathrm{Curl}_{\mu\nu}(\Phi) = \frac{\partial\Phi_\nu}{\partial x^\mu} - \frac{\partial\Phi_\mu}{\partial x^\nu} = \Phi_{\nu|\mu} - \Phi_{\mu|\nu} =: F_{\mu\nu} = -F_{\nu\mu}, \tag{17.139}$$

having components

$$\begin{aligned} F_{4\nu} &= -\frac{\partial\psi_e}{\partial x^\nu} - \frac{\partial\mathscr{A}_\nu}{\partial t} = \mathscr{E}_\nu; \quad F_{44} = 0 \\ F_{\mu\nu} &= -c\left(\frac{\partial\mathscr{A}_\nu}{\partial x^\mu} - \frac{\partial\mathscr{A}_\mu}{\partial x^\nu}\right), \quad \mu, \nu = 1, 2, 3 \Longrightarrow \end{aligned} \tag{17.140}$$

$$F_{12} = c\left(\frac{\partial\mathscr{A}_1}{\partial x^2} - \frac{\partial\mathscr{A}_2}{\partial x^1}\right) = -c\sqrt{\alpha}\,\mathscr{B}^3 \tag{17.141}$$

$$F_{13} = c\left(\frac{\partial\mathscr{A}_1}{\partial x^3} - \frac{\partial\mathscr{A}_3}{\partial x^1}\right) = c\sqrt{\alpha}\,\mathscr{B}^2 \tag{17.142}$$

$$F_{23} = c\left(\frac{\partial\mathscr{A}_2}{\partial x^3} - \frac{\partial\mathscr{A}_3}{\partial x^2}\right) = -c\sqrt{\alpha}\,\mathscr{B}^1, \tag{17.143}$$

where $\mathscr{E}_\mu$ and $\mathscr{B}^\mu$ are components of the electric field and the magnetic flux density, respectively, and $\alpha > 0$ is the determinant of the spatial part of the metric tensor[10].

[10] The connection between the spatial metric tensor and the tensor for four-dimensional spacetime is shown in equation (17.164).

The quantity $F_{\mu\nu}$ is known as the *electromagnetic field tensor*. It has units of V m$^{-1}$ in Cartesian coordinates and has components

$$F_{\mu\nu} = \begin{pmatrix} 0 & -c\sqrt{\alpha}\,\mathscr{B}^3 & c\sqrt{\alpha}\,\mathscr{B}^2 & -\mathscr{E}_1 \\ c\sqrt{\alpha}\,\mathscr{B}^3 & 0 & -c\sqrt{\alpha}\,\mathscr{B}^1 & -\mathscr{E}_2 \\ -c\sqrt{\alpha}\,\mathscr{B}^2 & c\sqrt{\alpha}\,\mathscr{B}^1 & 0 & -\mathscr{E}_3 \\ \mathscr{E}_1 & \mathscr{E}_2 & \mathscr{E}_3 & 0 \end{pmatrix} = \begin{pmatrix} 0 & -c\mathscr{B}^3 & c\mathscr{B}^2 & -\mathscr{E}_1 \\ c\mathscr{B}^3 & 0 & -c\mathscr{B}^1 & -\mathscr{E}_2 \\ -c\mathscr{B}^2 & c\mathscr{B}^1 & 0 & -\mathscr{E}_3 \\ \mathscr{E}_1 & \mathscr{E}_2 & \mathscr{E}_3 & 0 \end{pmatrix}, \quad (17.144)$$

where $\mu$ is the index for row and $\nu$ is the index for column in this matrix representation and we have used $\alpha = 1$ for the Cartesian coordinates under consideration. Note that because of the occurrence of both covariant and contravariant field components, the various elements of the electromagnetic field tensor need not have the same units in general, although they do so in a Cartesian system. The corresponding contravariant tensor has components

$$F^{\mu\nu} = -F^{\nu\mu} = h^{\mu r}h^{\nu s}F_{rs} = \begin{pmatrix} 0 & -c\sqrt{\alpha}\,\mathscr{B}^3 & c\sqrt{\alpha}\,\mathscr{B}^2 & \mathscr{E}_1 \\ c\sqrt{\alpha}\,\mathscr{B}^3 & 0 & -c\sqrt{\alpha}\,\mathscr{B}^1 & \mathscr{E}_2 \\ -c\sqrt{\alpha}\,\mathscr{B}^2 & c\sqrt{\alpha}\,\mathscr{B}^1 & 0 & \mathscr{E}_3 \\ -\mathscr{E}_1 & -\mathscr{E}_2 & -\mathscr{E}_3 & 0 \end{pmatrix} = \begin{pmatrix} 0 & -c\mathscr{B}^3 & c\mathscr{B}^2 & \mathscr{E}_1 \\ c\mathscr{B}^3 & 0 & -c\mathscr{B}^1 & \mathscr{E}_2 \\ -c\mathscr{B}^2 & c\mathscr{B}^1 & 0 & \mathscr{E}_3 \\ -\mathscr{E}_1 & -\mathscr{E}_2 & -\mathscr{E}_3 & 0 \end{pmatrix}, \quad (17.145)$$

where $\mu$ and $\nu$ are, respectively, the row and column indices. The space part of these tensors (the top left 3 × 3 submatrix) corresponds to the axial-vector component of the EM field (the magnetic flux density) and the time part (bottom right row-column) corresponds to the polar vector component of the EM field (electric field strength). It is seen that space part of $F_{\mu\nu}$ and $F^{\mu\nu}$ remain the same, however, the time part reverses sign between the two. Because the electromagnetic field tensor in equation (17.139) is formed from the curl of a covariant tensor we have from equation (17.116) that

$$\mathrm{Curl}_{\mu\nu\lambda}(F) = \frac{\partial F_{\mu\nu}}{\partial x^\lambda} + \frac{\partial F_{\nu\lambda}}{\partial x^\mu} + \frac{\partial F_{\lambda\mu}}{\partial x^\nu} = 0. \quad (17.146)$$

In view of the antisymmetric nature of $F_{\mu\nu}$, equation (17.146) may also be expressed using the alternation operation as

$$\partial_{[\lambda}F_{\mu\nu]} = 0. \quad (17.147)$$

To examine what equation (17.146) implies, it will be useful to expand its components in Cartesian coordinates:

$$\begin{aligned}&\mu = 2, \nu = 3, \lambda = 4: -\frac{\partial \mathscr{B}^1}{\partial t} - \frac{\partial \mathscr{E}_3}{\partial x^2} + \frac{\partial \mathscr{E}_2}{\partial x^3} = 0\\ &\Longrightarrow (\nabla \times \mathscr{E})_1 = -\frac{\partial \mathscr{B}^1}{\partial t}\end{aligned} \tag{17.148}$$

$$\begin{aligned}&\mu = 3, \nu = 4, \lambda = 1: -\frac{\partial \mathscr{E}_3}{\partial x^1} + \frac{\partial \mathscr{E}_1}{\partial x^3} + \frac{\partial \mathscr{B}^2}{\partial t} = 0\\ &\Longrightarrow (\nabla \times \mathscr{E})_2 = -\frac{\partial \mathscr{B}^2}{\partial t}\end{aligned} \tag{17.149}$$

$$\begin{aligned}&\mu = 4, \nu = 1, \lambda = 2: \frac{\partial \mathscr{E}_1}{\partial x^2} - \frac{\partial \mathscr{B}^3}{\partial t} - \frac{\partial \mathscr{E}_2}{\partial x^1} = 0\\ &\Longrightarrow (\nabla \times \mathscr{E})_3 = -\frac{\partial \mathscr{B}^3}{\partial t}\end{aligned} \tag{17.150}$$

$$\begin{aligned}&\mu = 1, \nu = 2, \lambda = 3: \frac{\partial \mathscr{B}^3}{\partial x^3} + \frac{\partial \mathscr{B}^1}{\partial x^1} + \frac{\partial \mathscr{B}^2}{\partial x^2} = 0\\ &\Longrightarrow \nabla \cdot \mathscr{B} = 0.\end{aligned} \tag{17.151}$$

These equations encompass the two Maxwell's equations $\nabla \times \mathscr{E} = -\frac{\partial \mathscr{B}}{\partial t}$ and $\nabla \cdot \mathscr{B} = 0$. So these two Maxwell's equations are already contained in the electromagnetic field tensor by virtue of its definition. Taking the divergence of $F^{\mu\nu}$ in accordance with equation (17.109) results in $K^\mu = \mathrm{Div}_\nu(F^{\mu\nu})$ with components

$$\begin{aligned}
K^1 &= -c\partial_2 \mathscr{B}^3 + c\partial_3 \mathscr{B}^2 + \frac{1}{c}\partial_t \mathscr{E}_1 = -c(\nabla \times \mathscr{B})_1 + \frac{1}{c}\partial_t \mathscr{E}_1 = -\eta_0 \mathscr{E}_1 = -\eta_0 \mathcal{J}^1\\
K^2 &= c\partial_1 \mathscr{B}^3 - c\partial_3 \mathscr{B}^1 + \frac{1}{c}\partial_t \mathscr{E}_2 = -c(\nabla \times \mathscr{B})_2 + \frac{1}{c}\partial_t \mathscr{E}_2 = -\eta_0 \mathcal{J}_2 = -\eta_0 J^2\\
K^3 &= -c\partial_1 \mathscr{B}^2 + c\partial_2 \mathscr{B}^1 + \frac{1}{c}\partial_t \mathscr{E}_3 = -c(\nabla \times \mathscr{B})_3 + \frac{1}{c}\partial_t \mathscr{E}_3 = -\eta_0 \mathscr{J}_3 = -\eta_0 \mathcal{J}^3\\
K^4 &= -\partial_1 \mathscr{E}_1 - \partial_2 \mathscr{E}_2 - \partial_3 \mathscr{E}_3 = -\nabla \cdot \mathscr{E} = -\frac{\rho}{\varepsilon_0} = -\eta_0 J^4,
\end{aligned}$$

where we have used the Maxwell's equations $\nabla \cdot \mathscr{E} = \rho/\varepsilon_0$ and $c\nabla \times \mathscr{B} = \eta_0 \mathscr{J} + (1/c)\partial \mathscr{E}/\partial t$ with $\eta_0 = \sqrt{\mu_0/\varepsilon_0} =: Y_0^{-1}$ denoting the intrinsic impedance of vacuum[11]. Thus we have the tensor equation

[11] For the time being we ignore the fact that it is the covariant components of a vector that must appear inside the curl operator and that the result of the curl operation produces a contravariant vector. As a matter of fact in Maxwell's equations it is the magnetic field strength $\mathscr{H}$ whose native components are covariant that appears inside the curl operator and the result is proportional to the electric flux density $\mathscr{D}$ whose native components are contravariant. This will become clear later when we define the second electromagnetic tensor $H^{\mu\nu}$ in equation (17.173).

$$\text{Div}_\nu(F^{\mu\nu}) = -\eta_0 J^\mu \Longrightarrow \partial_\nu(Y_0 F^{\mu\nu}) = -J^\mu, \tag{17.152}$$

which combines both of the above Maxwell's equations involving three-vectors into one equation involving the tensors $F^{\mu\nu}$ and $J^\mu$. Multiplying (17.152) by $h_{\alpha\mu}$ and making use of the identity (17.110), this equation may be equivalently specified as

$$h^{\mu n} F_{\alpha\mu|n} = -\eta_0 J_\alpha, \tag{17.153}$$

where we have used $J_\alpha = h_{\alpha\mu} J^\mu = [-\mathscr{J}, \rho c]$.

Thus Maxwell's equations in free space are succinctly described in tensor notation by one of equations (17.152), (17.153), and (17.146), where the various tensor quantities are defined in equations (17.144), (17.145), and (17.133). *Even though the equations have been derived under the Lorentz transformation and Cartesian coordinates, the relations are valid in any Riemannian space, flat or curved by virtue of their tensor nature. The numerical values of the tensor components may, however, change depending on the coordinate system.*

It is possible to re-express the information contained in equation (17.146) in a form similar to equation (17.152) by defining a dual pseudo-electromagnetic field tensor $\hat{F}^{\mu\nu}$ (see equation (17.53)) as

$$\hat{F}^{\mu\nu} = \frac{1}{2}\eta^{\mu\nu\lambda\beta} F_{\lambda\beta} = \frac{1}{\sqrt{|g|}} \begin{pmatrix} 0 & \mathscr{E}_3 & -\mathscr{E}_2 & c\sqrt{\alpha}\,\mathscr{B}^1 \\ -\mathscr{E}_3 & 0 & \mathscr{E}_1 & c\sqrt{\alpha}\,\mathscr{B}^2 \\ \mathscr{E}_2 & -\mathscr{E}_1 & 0 & c\sqrt{\alpha}\,\mathscr{B}^3 \\ -c\sqrt{\alpha}\,\mathscr{B}^1 & -c\sqrt{\alpha}\,\mathscr{B}^2 & -c\sqrt{\alpha}\,\mathscr{B}^3 & 0 \end{pmatrix}, \tag{17.154}$$

where we have chosen to retain the determinant of the metric tensor as a factor outside the matrix (even though it is equal to unity here) so as to use the same expression also in a curvilinear coordinate system later. It is seen that the components of $\hat{F}^{\mu\nu}$ are obtained from $F^{\mu\nu}$ by making the duality transformation $c\mathscr{B}^i \to -\mathscr{E}_i$ and $\mathscr{E}_i \to c\mathscr{B}^i$ in the latter. Taking the divergence of this pseudo-tensor yields $\hat{L}^\mu = \text{Div}_\nu(\hat{F}^{\mu\nu})$ with components

$$\begin{aligned}
\hat{L}^1 &= \partial_2\mathscr{E}_3 - \partial_3\mathscr{E}_2 + \partial_t\mathscr{B}^1 = (\nabla \times \mathscr{E})_1 + \partial_t\mathscr{B}^1 \\
\hat{L}^2 &= -\partial_1\mathscr{E}_3 + \partial_3\mathscr{E}_1 + \partial_t\mathscr{B}^2 = (\nabla \times \mathscr{E})_2 + \partial_t\mathscr{B}^2 \\
\hat{L}^3 &= \partial_1\mathscr{E}_2 - \partial_2\mathscr{E}_1 + \partial_t\mathscr{B}^3 = (\nabla \times \mathscr{E})_3 + \partial_t\mathscr{B}^3 \\
\hat{L}^4 &= c\partial_1\mathscr{B}^1 + c\partial_2\mathscr{B}^2 + c\partial_3\mathscr{B}^3 = c\nabla \cdot \mathscr{B}.
\end{aligned}$$

In view of Maxwell's equations the right-hand sides of all these four equations are zero. Thus we have

$$\text{Div}_\nu(\hat{F}^{\mu\nu}) = 0. \tag{17.155}$$

Equation (17.155) may be used in place of equation (17.146) to describe the relevant Maxwell's equations. By contracting $F_{\mu\nu}$ with $F^{\mu\nu}$ and $\hat{F}^{\mu\nu}$ with respect to both indices we obtain the invariants

$$F^{\mu\nu}F_{\mu\nu} = 2c^2\mathscr{B}\cdot\mathscr{B} - 2\mathscr{E}\cdot\mathscr{E} = 2[(c\mathscr{B})^2 - \mathscr{E}^2] = -\hat{F}^{\mu\nu}\hat{F}_{\mu\nu}. \tag{17.156}$$

$$\hat{F}^{\mu\nu}F_{\mu\nu} = -4c\mathscr{E}\cdot\mathscr{B}. \tag{17.157}$$

For example, these invariants are both zero for a plane or spherical TEM waves propagating in free space. Let us now apply the property (17.113) to the electromagnetic tensor $F^{\mu\nu}$. The left-hand side can be evaluated by utilizing equations (17.152), (17.133), and (17.144) to result in the four-vector

$$F_{\sigma\mu}\mathrm{Div}_\nu(F^{\mu\nu}) = -\eta_0 F_{\sigma\mu}J^\mu = \frac{1}{\varepsilon_0}\left[\mathscr{J}\times\mathscr{B} + \rho\mathscr{E}, \frac{-\mathscr{E}\cdot\mathscr{J}}{c}\right]. \tag{17.158}$$

The tensor $T^{\nu}_{.\sigma}$ appearing on the right-hand side of equation (17.113) and defined in equation (17.114) can be evaluated by utilizing equation (17.156) to result in

$$\begin{aligned}
T^{\nu}_{.\sigma} &= F_{\sigma\mu}F^{\mu\nu} + \frac{1}{4}\delta^{\nu}_{\sigma}F^{\lambda\beta}F_{\lambda\beta} = F_{\sigma\mu}F^{\mu\nu} + \frac{1}{2}\delta^{\nu}_{\sigma}(c^2\mathscr{B}^2 - \mathscr{E}^2) \\
&= \left(\begin{array}{c|c}
\mathscr{E}_m\mathscr{E}^n + c^2\mathscr{B}_m\mathscr{B}^n - \frac{1}{2}\delta^n_m(c^2\mathscr{B}^2 + \mathscr{E}^2) & \begin{matrix}-c(\mathscr{E}\times\mathscr{B})_1\\ -c(\mathscr{E}\times\mathscr{B})_2\\ -c(\mathscr{E}\times\mathscr{B})_3\end{matrix} \\ \hline
c(\mathscr{E}\times\mathscr{B})_1 \; c(\mathscr{E}\times\mathscr{B})_2 \; c(\mathscr{E}\times\mathscr{B})_3 & \frac{1}{2}(\mathscr{E}^2 + c^2\mathscr{B}^2)
\end{array}\right) \\
&= \frac{1}{\varepsilon_0}\left(\begin{array}{c|c}
\mathscr{E}_m\mathscr{D}^n + \mathscr{H}_m\mathscr{B}^n - \frac{1}{2}\delta^n_m\mathscr{E}_\ell\mathscr{D}^\ell - \frac{1}{2}\delta^n_m\mathscr{H}_\ell\mathscr{B}^\ell & \begin{matrix}-cG_1\\ -cG_2\\ -cG_3\end{matrix} \\ \hline
cG_1 \; cG_2 \; cG_3 & \frac{1}{2}(\mathscr{E}\cdot\mathscr{D} + \mathscr{B}\cdot\mathscr{H})
\end{array}\right) \\
&= \frac{1}{\varepsilon_0}\left(\begin{array}{c|c}
\mathscr{T}^{n}_{.m} & \begin{matrix}-cG_1\\ -cG_2\\ -cG_3\end{matrix} \\ \hline
cG_1 \; cG_2 \; cG_3 & \frac{1}{2}(\mathscr{E}\cdot\mathscr{D} + \mathscr{B}\cdot\mathscr{H})
\end{array}\right),
\end{aligned} \tag{17.159}$$

where $\mathscr{G} = \mathscr{D}\times\mathscr{B}$, $\mathscr{T}^{n}_{.m}$ is the momentum tensor (5.62) defined in 3D and $G_i = (\mathscr{D}\times\mathscr{B})_i$ denotes the $i$th component of the vector $(\mathscr{D}\times\mathscr{B})$.[12] Taking the divergence of $T^{\nu}_{.\sigma}$ and equating the corresponding components on the left- and right-hand sides of equation (17.113) gives

[12] In anticipation of converting $\varepsilon_0\mathscr{E}\to\mathscr{D}$ and $\mu_0^{-1}\mathscr{B}\to\mathscr{H}$ we have expressed the term $\mathscr{E}^n$ as a contravariant component and the term $\mathscr{B}_m$ as a covariant component above so that the expressions will involve $\mathscr{D}^n = \varepsilon_0\mathscr{E}^n$ and $\mathscr{H}_m = \mu_0^{-1}\mathscr{B}_m$.

$$\nabla \cdot^2 \mathscr{T} = \rho \mathscr{E} + \mathscr{J} \times \mathscr{B} + \frac{\partial \mathscr{G}}{\partial t}, \tag{17.160}$$

$$-\mathscr{E} \cdot \mathscr{J} = c^2 \nabla \cdot \mathscr{G} + \frac{1}{2}\frac{\partial}{\partial t}(\mathscr{E} \cdot \mathscr{D} + \mathscr{B} \cdot \mathscr{H}), \tag{17.161}$$

which are nothing but laws governing the conservation of momentum and the conservation of energy. As a matter of fact (17.160) coincides with equations (5.64) and (17.161) coincides with equations (5.10). The mixed tensor $T^{\nu}_{.\sigma}$ of equation (17.159) which contains both the momentum components and the energy components is known as the *energy–momentum field tensor*. The space part of this tensor corresponds to Maxwell's stress tensor and the time part to the Poynting vector and energy density. Since the trace of a mixed tensor must be an invariant, we observe that $T^{v}_{v} = 0$ in the case of the energy–momentum field tensor.

## 17.3 Maxwell's equations in an arbitrary spacetime

As stated previously, Maxwell's equations given in a covariant form in equations (17.152) and (17.146) continue to hold in an arbitrary four-dimensional curvilinear system of coordinates ($x^1$, $x^2$, $x^3$, $x^4 = ct$), where the metric tensor $g_{\mu\nu}$ can depart from being Galilean. To generalize the other quantities to a curvilinear system, ordinary derivatives occurring in a Galilean system must be replaced with covariant derivatives. Let us first examine the properties of the metric tensor in a curvilinear coordinate system. In a rest frame with $\mathrm{d}x^1 = 0 = \mathrm{d}x^2 = \mathrm{d}x^3$ and $\mathrm{d}x^4 = c\mathrm{d}t$, $\mathrm{d}s = c\mathrm{d}\tau$, $\tau$ is the proper time, the metric form is $\mathrm{d}s^2 = c^2\mathrm{d}\tau^2 = g_{44}(\mathrm{d}x^4)^2 > 0$. Thus

$$g_{44} > 0, \quad \mathrm{d}\tau = \frac{1}{c}\sqrt{g_{44}}\,\mathrm{d}x^4, \tag{17.162}$$

where $g_{44}$ is a dimensionless quantity. The spatial distance $\mathrm{d}\ell$ between point $A$, having spatial coordinates $x^\lambda$, and point $B$, having spatial coordinates $x^\lambda + \mathrm{d}x^\lambda$, $\lambda = 1, 2, 3$ can be determined by directing a light signal from $B$ to $A$ and then back over the same path and measuring the time interval $\mathrm{d}x^4$ as observed from $B$. The roundtrip distance $2\mathrm{d}\ell$ is equal to the proper time interval $\mathrm{d}\tau = \sqrt{g_{44}}$ $(\mathrm{d}x^4_{B\to A} + \mathrm{d}x^4_{A\to B})/c$ times $c$. Expressing $\mathrm{d}x^4_{B\to A}$ and $\mathrm{d}x^4_{A\to B}$ in terms of the spatial increments for the light signal (for which $\mathrm{d}s = 0$) yields [15]

$$\mathrm{d}\ell^2 = \left(-g_{\mu\nu} + \frac{g_{4\mu}g_{4\nu}}{g_{44}}\right)\mathrm{d}x^\mu \mathrm{d}x^\nu =: \alpha_{\mu\nu}\mathrm{d}x^\mu \mathrm{d}x^\nu, \ 1 \leqslant \mu, \nu \leqslant 3, \tag{17.163}$$

where $\alpha_{\mu\nu}$ is the 3D metric tensor measuring the geometric properties of space and is related to $g_{\mu\nu}$ via

$$\alpha_{\mu\nu} = -g_{\mu\nu} + \frac{g_{4\mu}g_{4\nu}}{g_{44}}, \quad 1 \leqslant \mu, \nu \leqslant 3. \tag{17.164}$$

In general $g_{\mu\nu}$ depend on the time coordinate $x^4$ and it is seen that the space metric $\alpha_{\mu\nu}$ is not equal to the spatial part of the spacetime metric. The only instance where this is possible is when $g_{4\mu} = g_{\mu 4}$ vanish. In a system of coordinates where the latter is true, the time-axis is everywhere orthogonal to the spatial coordinate curves. Such is system is known as *time-orthogonal*. However, using $g^{ij}g_{jk} = \delta^i_k$, it can be easily shown that even in a general coordinate system

$$\alpha^{\mu\nu} = -g^{\mu\nu}, \ 1 \leqslant \mu, \nu \leqslant 3. \tag{17.165}$$

Hence the contravariant three-dimensional metric tensor is the negative of the spatial part of the four-dimensional spacetime contravariant metric tensor. Because the quadratic form in equation (17.163) is positive definite, we must have

$$\alpha_{11} > 0, \quad \det\begin{pmatrix} \alpha_{11} & \alpha_{12} \\ \alpha_{12} & \alpha_{22} \end{pmatrix} > 0, \quad \alpha = \det\begin{pmatrix} \alpha_{11} & \alpha_{12} & \alpha_{13} \\ \alpha_{12} & \alpha_{22} & \alpha_{23} \\ \alpha_{13} & \alpha_{23} & \alpha_{33} \end{pmatrix} > 0, \tag{17.166}$$

which translate to

$$\det\begin{pmatrix} g_{33} & g_{34} \\ g_{34} & g_{44} \end{pmatrix} < 0, \quad \det\begin{pmatrix} g_{22} & g_{23} & g_{24} \\ g_{23} & g_{33} & g_{34} \\ g_{24} & g_{34} & g_{44} \end{pmatrix} > 0, \quad g = \det(g_{\mu\nu}) < 0, \quad g_{ii} < 0, \tag{17.167}$$

for $1 \leqslant i \leqslant 3$. For instance in a Galilean system $g_{11} = g_{22} = g_{33} = -1 < 0$, $g_{44} = 1 > 0$, $g = -1$ and $\alpha_{mn} = \delta_{mn}$, $\alpha = 1$. In general the determinants $\alpha$ and $g$ are related by

$$-g = g_{44}\alpha > 0. \tag{17.168}$$

The definition of the electromagnetic field tensor $F_{\mu\nu}$ in terms of the four-potential $\Phi_\mu$, equation (17.139), needs no modification because it is already in a covariant form. The four-velocity is $U^\mu = \mathrm{d}x^\mu/\mathrm{d}\tau = c\mathrm{d}x^\mu/\sqrt{g_{44}}\mathrm{d}x^4$ and a charge density $\rho$ due to a charge $\mathrm{d}Q$ contained in an elemental volume $\alpha \mathrm{d}x^1\mathrm{d}x^2\mathrm{d}x^3 = \alpha \mathrm{d}V$ is $\rho = \mathrm{d}Q/\alpha\mathrm{d}V$. The four-current density associated with the charge density is then equal to

$$J^\mu = \rho U^\mu = \frac{c\rho}{\sqrt{g_{44}}}\frac{\mathrm{d}x^\mu}{\mathrm{d}x^4}. \tag{17.169}$$

In terms of the four-current density the charge density $\rho$ is simply equal to $\frac{\sqrt{g_{44}}}{c}J^4$. Taking the divergence of $J^\mu$ according to equation (17.107) yields the invariant

$$\frac{1}{\sqrt{|g|}}\frac{\partial}{\partial x^\mu}\left(\sqrt{|g|}J^\mu\right) = 0. \tag{17.170}$$

With the revised form of four-current density given in equation (17.169), Maxwell's equations (17.152) now read

$$\mathrm{Div}_\nu(Y_0F^{\mu\nu}) = -J^\mu \Longrightarrow \partial_\nu\left(Y_0\sqrt{|g|}\,F^{\mu\nu}\right) = j^\mu, \tag{17.171}$$

where $j^\mu = -\sqrt{|g|}\,J^\mu = -c\sqrt{\alpha}\,\rho\mathrm{d}x^\mu/\mathrm{d}x^4$ and we have used $g = -g_{44}\alpha$. With reference to equation (17.144) the electromagnetic field tensor $F_{\mu\nu}$ is still defined in terms of field components according to (see equation (17.120))

$$F_{4\nu} = \mathscr{E}_\nu,\ F_{\mu\nu} = -cB_{\mu\nu} = -c\eta_{\mu\nu\lambda}\mathscr{B}^\lambda,\ 1 \leqslant \mu, \nu, \lambda \leqslant 3. \tag{17.172}$$

## 17.4 Covariant form of Maxwell's equations in stationary matter

We assume the constitutive relations $\mathscr{D} = \varepsilon_r\varepsilon_0\mathscr{E}$, $\mathscr{B} = \mu_0\mu_r\mathscr{H}$ and $\mathscr{J} = \sigma_r Y_0\mathscr{E}$ in a frame $S_0$ in which the electromagnetic matter is at rest. Here $\varepsilon_r$, $\mu_r$, and $\sigma_r$ are, respectively, the relative permittivity, the relative permeability, and the relative conductivity of the material as measured by a local observer in the proper coordinate system. In addition to the tensor $F^{\mu\nu}$ of equation (17.145) we define a second antisymmetric electromagnetic tensor $H^{\mu\nu}$ in terms of the dual field components $(c\sqrt{\alpha}\,\mathscr{D}^i, \mathscr{H}_i)$ motivated by the transformation $H^{\mu\nu} = Y_0\sqrt{|g|}\,F^{\mu\nu}$ [16], with the space part of $F^{\mu\nu}$ expressed in an axial vector (in the native covariant components for the magnetic field strength $\mathscr{H}$) and the time part expressed in polar vector (in the native contravariant components for the electric flux density $\mathscr{D}$) to result in

$$H^{\mu\nu} = \begin{pmatrix} 0 & -\mathscr{H}_3 & \mathscr{H}_2 & c\sqrt{\alpha}\,\mathscr{D}^1 \\ \mathscr{H}_3 & 0 & -\mathscr{H}_1 & c\sqrt{\alpha}\,\mathscr{D}^2 \\ -\mathscr{H}_2 & \mathscr{H}_1 & 0 & c\sqrt{\alpha}\,\mathscr{D}^3 \\ -c\sqrt{\alpha}\,\mathscr{D}^1 & -c\sqrt{\alpha}\,\mathscr{D}^2 & -c\sqrt{\alpha}\,\mathscr{D}^3 & 0 \end{pmatrix}. \tag{17.173}$$

Note that this particular definition of $H^{\mu\nu}$ is unique in relation to the definition of $F^{\mu\nu}$ in equation (17.145) in the sense that it is suggested directly by the tensor form of Maxwell's equations in equation (17.171). In a Cartesian coordinates it will have units [A m$^{-1}$]. The corresponding covariant tensor is (using $\alpha = 1$)

$$H_{\mu\nu} = \begin{pmatrix} 0 & -\mathscr{H}_3 & \mathscr{H}_2 & -c\sqrt{\alpha}\,\mathscr{D}^1 \\ \mathscr{H}_3 & 0 & -\mathscr{H}_1 & -c\sqrt{\alpha}\,\mathscr{D}^2 \\ -\mathscr{H}_2 & \mathscr{H}_1 & 0 & -c\sqrt{\alpha}\,\mathscr{D}^3 \\ c\sqrt{\alpha}\,\mathscr{D}^1 & c\sqrt{\alpha}\,\mathscr{D}^2 & c\sqrt{\alpha}\,\mathscr{D}^3 & 0 \end{pmatrix}. \tag{17.174}$$

Using this second electromagnetic field tensor, Maxwell's equations (17.171) can be re-expressed as

$$\partial_\nu(H^{\mu\nu}) = j^\mu. \tag{17.175}$$

One may introduce a dual pseudo-tensor, $\hat{H}^{\mu\nu}$, dual to equation (17.173) along the lines of equation (17.154) by defining

$$\hat{H}^{\mu\nu} = \frac{1}{2}\eta^{\mu\nu\alpha\beta}H_{\alpha\beta} = \frac{1}{\sqrt{|g|}}\begin{pmatrix} 0 & c\sqrt{\alpha}\,\mathscr{D}^3 & -c\sqrt{\alpha}\,\mathscr{D}^2 & \mathscr{H}_1 \\ -c\sqrt{\alpha}\,\mathscr{D}^3 & 0 & c\sqrt{\alpha}\,\mathscr{D}^1 & \mathscr{H}_2 \\ c\sqrt{\alpha}\,\mathscr{D}^2 & -c\sqrt{\alpha}\,\mathscr{D}^1 & 0 & \mathscr{H}_3 \\ -\mathscr{H}_1 & -\mathscr{H}_2 & -\mathscr{H}_3 & 0 \end{pmatrix}. \tag{17.176}$$

Using the tensors $F_{\mu\nu}$, $H_{\mu\nu}$, $\hat{F}_{\mu\nu}$, and $\hat{H}_{\mu\nu}$ it is straightforward to extend the constitutive relations to *uniform motion* in a frame $S$ by writing [8, 9]

$$U^{\mu}H_{\mu\nu} = \varepsilon_r Y_0 U^{\mu}F_{\mu\nu}, \tag{17.177}$$

$$Y_0 U_{\mu}\hat{F}^{\mu\nu} = \mu_r U_{\mu}\hat{H}^{\mu\nu}, \tag{17.178}$$

$$\eta_0 J_{\beta}U^{\beta}U^{\lambda} - c^2\eta_0 J^{\lambda} = c\sigma_r U_{\gamma}F^{\gamma\lambda}, \tag{17.179}$$

where $U^{\lambda}$ are components of the velocity of the matter at the point of interest. In the proper frame $U^{\lambda} = [0, c] = U_{\lambda}$. Equation (17.177) reduces to $H_{4\nu} = \varepsilon_r Y_0 F_{4\nu}$ $\Longrightarrow \mathscr{D} = \varepsilon_r\varepsilon_0\mathscr{E}$. Equation (17.178) reduces to $Y_0\hat{F}^{4\nu} = \mu_r\hat{H}^{4\nu} \Longrightarrow \mathscr{B} = \mu_r\mu_0\mathscr{H}$. Equation (17.179) reduces to $-\eta_0 J^{\lambda} = \sigma_r F^{4\lambda}$, $1 \leqslant \lambda \leqslant 3 \Longrightarrow \mathscr{J}_{\lambda} = \sigma_r Y_0\mathscr{E}_{\lambda}$. Hence these expressions for the constitutive relations, being valid in one set of coordinates, are valid in all sets of coordinates owing to their tensor character.

## 17.5 Transformational electromagnetics

It is interesting to explore the relations arising between $\mathscr{D}$ and the pair $(\mathscr{E}, \mathscr{H})$ and between $\mathscr{B}$ and the pair $(\mathscr{E}, \mathscr{H})$ that are induced purely by the metric tensor of a curvilinear coordinate system. In a flat space in a vacuum where the metric tensor takes its Galilean values, we have the well known isotropic relations $\mathscr{D} = \varepsilon_0\mathscr{E}$ and $\mathscr{B} = \mu_0\mathscr{H}$. Let us see what the tensor relation between $F_{\mu\nu}$ and $H^{\mu\nu}$ described in a curvilinear coordinate system suggests for the constitutive relations.

Lowering the indices of $F^{\mu\nu}$ in the definition $H^{\mu\nu} = Y_0\sqrt{|g|}\,F^{\mu\nu}$ we have

$$\begin{aligned} F_{\mu\nu} &= \eta_0\,|g|^{-\frac{1}{2}}\,g_{\mu\gamma}g_{\nu\beta}H^{\gamma\beta} \\ &= -\eta_0\,|g|^{-\frac{1}{2}}\,g_{\mu\gamma}g_{\nu\beta}H^{\beta\gamma} = -\eta_0\,|g|^{-\frac{1}{2}}\,g_{\mu\beta}g_{\nu\gamma}H^{\gamma\beta} \\ &= \frac{1}{2}\eta_0\,|g|^{-\frac{1}{2}}\,(g_{\mu\gamma}g_{\nu\beta} - g_{\mu\beta}g_{\nu\gamma})H^{\gamma\beta} \end{aligned} \tag{17.180}$$

$$=: \frac{1}{2}\chi_{\mu\nu\gamma\beta}H^{\gamma\beta}, \tag{17.181}$$

where the quantity

$$\chi_{\mu\nu\gamma\beta} = \eta_0\,|g|^{-\frac{1}{2}}(g_{\mu\gamma}g_{\nu\beta} - g_{\mu\beta}g_{\nu\gamma}), \tag{17.182}$$

is a *tensor density* of rank four and of weight −1 that describes the linear constitutive relations in arbitrary Riemannian space[13]. The density property of the tensor $\chi_{\mu\nu\gamma\beta}$ is clear due to the presence of the scalar factor $|g|^{-\frac{1}{2}}$ in its definition in equation (17.182). This factor transforms as $|g'|^{-\frac{1}{2}} = |J'|\,|g|^{-\frac{1}{2}}$ (see equation (17.46)) under coordinate transformations, indicating that it is a scalar density of weight −1. From the symmetry properties of $H^{\gamma\beta}$ and $F_{\mu\nu}$ it is easy to see that the constitutive tensor $\chi_{\mu\nu\gamma\beta}$ is antisymmetric in the two index pairs $(\mu, \nu)$ and $(\gamma, \beta)$ and symmetric with respect to the interchange of the pair $(\mu, \nu)$ with $(\gamma, \beta)$:

$$\chi_{\mu\nu\gamma\beta} = -\chi_{\nu\mu\gamma\beta} = -\chi_{\mu\nu\beta\gamma} = \chi_{\gamma\beta\mu\nu}. \tag{17.183}$$

The constitutive tensor can be represented by a 6 × 6 symmetric matrix (6 × 7/2 = 21 independent elements) [16]. Note that the tensor constitutive relation (17.180) between the two electromagnetic tensors $F_{\mu\nu}$ and $H^{\gamma\beta}$ suggests that the relationship between the quantities in the pair $(\mathscr{D}, \mathscr{E})$ is no longer separate from the relationship between the quantities in the pair $(\mathscr{B}, \mathscr{H})$. The coupling between the two pairs is automatically included by the tensor $\chi_{\mu\nu\gamma\beta}$. Specializing equation (17.180) to $\mu = 4$ gives

$$\begin{aligned} Y_0\sqrt{|g|}\,F_{4\nu} &= (g_{4m}g_{\nu n} - g_{4n}g_{\nu m})H^{mn} \\ &= (g_{41}g_{\nu 2} - g_{42}g_{\nu 1})H^{12} + (g_{41}g_{\nu 3} - g_{43}g_{\nu 1})H^{13} + (g_{41}g_{\nu 4} - g_{44}g_{\nu 1})H^{14} \\ &\quad + (g_{42}g_{\nu 3} - g_{43}g_{\nu 2})H^{23} + (g_{42}g_{\nu 4} - g_{44}g_{\nu 2})H^{24} + (g_{43}g_{\nu 4} - g_{44}g_{\nu 3})H^{34}. \end{aligned}$$

Substituting the values of $F_{4\nu}$ and $H^{mn}$, evaluating for $1 \leqslant \nu, \tau, \beta, \lambda \leqslant 3$ and utilizing the properties of the three-dimensional permutation symbol and the definition (17.163) we obtain

$$\begin{aligned} \varepsilon_0\frac{\sqrt{|g|}}{g_{44}}\mathscr{E}_\nu &= \frac{1}{g_{44}}(g_{4\lambda}g_{\nu 4} - g_{44}g_{\nu\lambda})\sqrt{\alpha}\,\mathscr{D}^\lambda + \frac{1}{c}\frac{1}{g_{44}}\varepsilon^{\tau\beta\lambda}g_{4\beta}g_{\nu\tau}\mathscr{H}_\lambda \\ &= \sqrt{\alpha}\,\alpha_{\nu\lambda}\mathscr{D}^\lambda + \frac{1}{c}\frac{1}{g_{44}}\varepsilon^{\tau\beta\lambda}g_{4\beta}g_{\nu\tau}\mathscr{H}_\lambda, \end{aligned} \tag{17.184}$$

where equation (17.164) has been used relating $\alpha_{\lambda\nu}$ and $g_{\lambda\nu}$. Multiplying equation (17.184) with $\alpha^{\gamma\nu} = -g^{\gamma\nu}$, defining the 3D vector component $g_\beta = -g_{4\beta}/g_{44}$ and utilizing $\sqrt{|g|} = \sqrt{\alpha g_{44}}$, $\alpha_{\nu\lambda}\alpha^{\gamma\nu} = \delta^\gamma_\lambda$, $g^{\nu\gamma}g_{\nu\tau} = \delta^\gamma_\tau$, and $\alpha^{\gamma\nu}\mathscr{E}_\nu = \mathscr{E}^\gamma$, we obtain the bianisotropic relation

$$\sqrt{\alpha}\,\mathscr{D}^\gamma = \varepsilon_0\sqrt{\frac{\alpha}{g_{44}}}\,\alpha^{\gamma\nu}\mathscr{E}_\nu - \frac{1}{c}\varepsilon^{\gamma\beta\lambda}g_\beta\mathscr{H}_\lambda, \tag{17.185}$$

---

[13] The magnetic flux density $\mathscr{B}$, the electric flux density $\mathscr{D}$, the four-current density $J^\mu$, and the second electromagnetic tensor $H^{\mu\nu}$ are all regarded as tensor densities of weight +1, while the electric field $\mathscr{E}$, the magnetic field $\mathscr{H}$ and the first electromagnetic tensor $F_{\mu\nu}$ are all regarded as ordinary tensors with weight 0 [16]. Since $H^{\mu\nu}$ has weight +1 and $F_{\gamma\beta}$ has weight 0, the quantity $\chi_{\mu\nu\gamma\beta}$ relating the two has weight −1.

$$=:\varepsilon_0\zeta_0^{\gamma\nu}\mathscr{E}_\nu + \varepsilon_0\eta_0\xi_0^{\gamma\lambda}\mathscr{H}_\lambda, \tag{17.186}$$

where

$$\zeta_0^{\gamma\nu} = \sqrt{\frac{\alpha}{g_{44}}}\alpha^{\gamma\nu} =: \varepsilon_{\rm er}\sqrt{\alpha}\,\alpha^{\gamma\nu} = -\varepsilon_{\rm er}\sqrt{\alpha}\,g^{\gamma\nu}, \text{ and } \xi_0^{\gamma\lambda} = -\varepsilon^{\gamma\beta\lambda}g_\beta, \tag{17.187}$$

are proportional to the relative permittivity tensor (symmetric) and relative magnetoelectric tensor (antisymmetric), respectively. Note that $\zeta_0^{\gamma\nu}$ is positive definite (since $\alpha^{\gamma\nu}$ is positive definite), but $\xi_0^{\gamma\lambda}$ is not necessarily so. In a vector form equation (17.186) may be written as $\mathscr{D} = \varepsilon_0\varepsilon_{\rm er}\mathscr{E} - \varepsilon_0\eta_0(\mathbf{g}\times\mathscr{H})$ since $\varepsilon^{\gamma\beta\lambda}g_\beta\mathscr{H}_\lambda = \sqrt{\alpha}(\mathbf{g}\times\mathscr{H})^\gamma$ (see equation (17.26)). In an identical manner working with the dual pseudo-tensors $\hat{F}_{ij}$ and $\hat{H}^{mn}$ and the corresponding relation $\hat{F}_{ij} = g_{im}g_{jn}\frac{\hat{H}^{mn}}{Y_0\sqrt{|g|}}$ we arrive at the dual relation

$$\sqrt{\alpha}\,\mathscr{B}^\gamma = \mu_0\zeta_0^{\gamma\nu}\mathscr{H}_\nu - \mu_0 Y_0\xi_0^{\gamma\lambda}\mathscr{E}_\lambda, \tag{17.188}$$

where the relative permeability $\mu_{\rm er} = \sqrt{1/g_{44}}$ is seen to be the same as the relative permittivity $\varepsilon_{\rm er}$ in equation (17.187). Equation (17.188) may be written in vector form as $\mathscr{B} = \mu_0\mu_{\rm er}\mathscr{H} + \mu_0\eta_0^{-1}(\mathbf{g}\times\mathscr{E})$. In a Galilean system $\alpha = 1$, $\zeta_0^{\gamma\nu} = \delta^{\gamma\nu}$ and $\xi_0^{\gamma\nu} = 0$ and one recovers the free-space relations $\mathscr{D}^\nu = \varepsilon_0\mathscr{E}^\nu$ and $\mathscr{B}^\nu = \mu_0\mathscr{H}^\nu$.

On observing equations (17.186) and (17.188) we may say that, with respect to its effect on the electromagnetic field, a non-Galilean metric tensor plays the role of an effective medium with relative permittivity and permeability $\varepsilon_{\rm er} = \mu_{\rm er} = \sqrt{1/g_{44}}$. In addition, the vector component $g_\beta$ of the metric tensor introduces bianisotropy through the magnetoelectric quantity $\xi_0^{\gamma\lambda}$. Note that the quantities $\zeta_0^{\gamma\nu}$, $\xi_0^{\gamma\nu}$ are of rank two in a three-dimensional space and are dependent only on the properties of the metric tensor of the underlying curvilinear coordinates. In the case of time-orthogonal systems (where $g_{44} = 1$, $g_{\beta4} = 0$, $g_\beta = 0$, $|g| = \alpha$), the quantity $\xi_0^{\gamma\lambda}$ vanishes and the medium reduces to a uniaxial one.

In view of the fact that $\alpha^{\gamma\nu} = \mathbf{a}^\gamma\cdot\mathbf{a}^\nu$, where $\mathbf{a}^\gamma$ are reciprocal unitary vectors, and $\sqrt{\alpha} = \mathbf{a}_1\cdot(\mathbf{a}_2\times\mathbf{a}_3)$, the quantity $\zeta^{\gamma\nu}$, in general, has the matrix representation

$$\zeta^{\gamma\nu} = \frac{1}{\sqrt{g_{44}}}[\mathbf{a}_1\cdot(\mathbf{a}_2\times\mathbf{a}_3)]\begin{bmatrix}\mathbf{a}^1\cdot\mathbf{a}^1 & \mathbf{a}^1\cdot\mathbf{a}^2 & \mathbf{a}^1\cdot\mathbf{a}^3\\ \mathbf{a}^2\cdot\mathbf{a}^1 & \mathbf{a}^2\cdot\mathbf{a}^2 & \mathbf{a}^2\cdot\mathbf{a}^3\\ \mathbf{a}^3\cdot\mathbf{a}^1 & \mathbf{a}^3\cdot\mathbf{a}^2 & \mathbf{a}^3\cdot\mathbf{a}^3\end{bmatrix}, \tag{17.189}$$

where $\gamma$ is the row index and $\nu$ is the column index. Equation (17.189) with $g_{44}$ set equal to unity is identical to that given in [4, equation (13), accompanying online material], where the authors express the tensor in terms of the normalized unitary and reciprocal unitary vectors $\mathbf{u}_i = \mathbf{a}_i/Q_i$, $\mathbf{u}^i = \mathbf{a}^iQ_i$, where $Q_i = \sqrt{\mathbf{a}_i\cdot\mathbf{a}_i}$.

Let us now try to draw an equivalence between two spaces:

(i) A flat space in which the free-space constitutive relations $\mathscr{D}_\nu = \varepsilon_0\mathscr{E}_\nu$, $\mathscr{B}_\nu = \mu_0\mathscr{H}_\nu$ hold. This space is regarded as a virtual space and is labeled as an *electromagnetic space* [17]. In this space the spatial contravariant metric tensor is $\alpha_0^{mn} = \delta^{mn}$ and the geodesic for a plane wave is a straight line. Furthermore the electromagnetic space is time-orthogonal with $g_{44} = 1$, $g_\beta = 0$.

(ii) A *physical space* endowed with a space metric tensor $\alpha_{mn}$ (corresponding determinant $\alpha = \det(\alpha_{mn}) > 0$) and filled with a bianisotropic material (possibly inhomogeneous). The material is chosen such that the relative permittivity tensor of such a medium is equal to its relative permeability tensor, both being proportional to a quantity $\zeta^{\lambda\nu}$. Furthermore, the material provides additional coupling between the electric and magnetic fields via a magneto-dielectric quantity $\xi^{\lambda\nu}$. The relationship between the electric and magnetic fields in such a space is

$$\sqrt{\alpha}\,\mathscr{D}^\lambda = \varepsilon_0\zeta^{\lambda\nu}\mathscr{E}_\nu + \varepsilon_0\eta_0\xi^{\lambda\nu}\mathscr{H}_\nu; \qquad \sqrt{\alpha}\,\mathscr{B}^\lambda = \mu_0\zeta^{\lambda\nu}\mathscr{H}_\nu - \varepsilon_0 Y_0\xi^{\lambda\nu}\mathscr{E}_\nu. \tag{17.190}$$

In this physical space, the geodesic for propagating waves is not a straight, but a curved line.

Because a relationship such as equation (17.190) can also be induced purely by the properties of a metric tensor as is evident from examining equations (17.186) and (17.188), we connect the two spaces by imagining a *time-invariant* coordinate transformation between the electromagnetic space with coordinates $(x_0^1, x_0^2, x_0^3, x_0^4 = ct)$ and a physical space with coordinates $(x^1, x^2, x^3, x^4 = ct)$. Under this transformation the physical space will remain time-orthogonal with $g_{44} = 1$, $g_\beta = 0$. There is then a one-to-one correspondence between the geodesics in the two spaces. For example, if a geodesic in the electromagnetic space avoids a certain region $V_0$, then the corresponding geodesic in the physical space will avoid the mapped region $V_0 \mapsto V$. This idea is used in the design of cloaking devices as well as many other exotic man-made devices and materials [4, 17, 18]. It remains to determine the components of the permittivity and magneto-dielectric tensors to accomplish this equivalence. We define the following transformational quantities:

$$A_{\nu_0}^{\lambda} := \frac{\partial x^\lambda}{\partial x_0^\nu}; \quad A_\lambda^{\nu_0} := \frac{\partial x_0^\nu}{\partial x^\lambda}, \quad 1 \leqslant \nu, \lambda \leqslant 3. \tag{17.191}$$

Under the coordinate transformation, the spatial metric tensor $\alpha_0^{mn}$ is transformed to a spatial metric tensor $\tau^{mn}$ with components

$$\tau^{mn} = A_{m_0}^{m} A_{n_0}^{n} \delta^{m_0 n_0} = \sum_{m_0=1}^{3} A_{m_0}^{m} A_{m_0}^{n}. \tag{17.192}$$

The determinant of the metric tensor $\alpha_0 = \det(\alpha_{0mn}) = [\det(\alpha_0^{mn})]^{-1}$ is transformed to the determinant $\tau = \det(\tau_{mn}) = [\det(\tau^{mn})]^{-1}$. Then in accordance with equation (17.186) we set

$$\zeta^{mn} = \sqrt{\tau}\,\tau^{mn}; \quad \xi^{mn} = 0. \tag{17.193}$$

This is a quantity that resulted entirely from the coordinate transformation of an original electromagnetic space. In combination with equation (17.190) we now define the relative permittivity tensor $\varepsilon_r^{\lambda\nu}$ (equal to the relative permeability tensor $\mu_r^{\lambda\nu}$) in the physical space as

$$\varepsilon_{\mathrm{r}}^{\lambda\nu} = \mu_{\mathrm{r}}^{\lambda\nu} = \frac{1}{\sqrt{\alpha}}\zeta^{\lambda\nu} = \sqrt{\frac{\tau}{\alpha}}\,\tau^{\lambda\nu}; \qquad \varepsilon_{\mathrm{r}\nu}^{\ \lambda} = \alpha_{\nu k}\,\varepsilon_{\mathrm{r}}^{\lambda k}, \tag{17.194}$$

while the relative magneto-dielectric tensor $\xi_{\mathrm{r}}^{\lambda\nu} = 0$. Note that the subscript 'r' here is not a tensor index, but rather denotes the word 'relative'. Thus a physical space filled with an anisotropic material with permittivity tensor given by equation (17.194) will have a curved geodesic corresponding to the straight line geodesic in the electromagnetic space.

### 17.5.1 Design of a cylindrical cloak

Suppose we want to shield an object that is contained within a cylinder of radius $r_1$ and the cloaking region (anisotropic region) is to be confined within a radius of $r_2$. We let $x_0^1 = x$, $x_0^2 = y$, $x_0^3 = z$, where $(x, y, z)$ are the Cartesian coordinates of a point with corresponding cylindrical coordinates ($\rho = \sqrt{x^2 + y^2}$, $\phi = \arctan(y/x)$, $z$) be points in the electromagnetic space. We want all the geodesics in the physical space to circumvent the cylindrical region $0 \leqslant \rho' \leqslant r_1$ as shown in figure 17.3. The coordinates of the physical space are $x^1 = \rho'$, $x^2 = \phi'$, $x^3 = z'$. The desired transformation and the inverse transformation to accomplish cloaking are

$$\rho' = r_1 + \rho\Delta r; \qquad x = \frac{(\rho' - r_1)}{\Delta r}\cos\phi' \tag{17.195}$$

$$\phi' = \phi; \qquad y = \frac{(\rho' - r_1)}{\Delta r}\sin\phi' \tag{17.196}$$

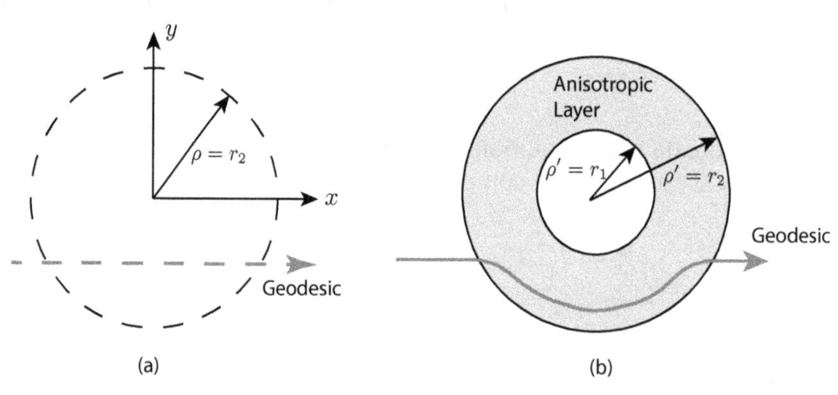


**Figure 17.3.** Mapping of a geodesic (red line) from electromagnetic space (a) to physical space (b). The shaded region contains anisotropic material with relative permittivity and permeability equal to $\varepsilon_{rn}^{m}$. Interior of circle of radius $\rho' = r_1$ in physical space has no counterpart in the electromagnetic space and can be utilized to shield objects.

$$z' = z; \qquad z = z', \tag{17.197}$$

in the region $0 \leqslant \rho \leqslant r_2$ and $\rho' = \rho$, $\phi' = \phi$, $z' = z$ in the region $r_2 \leqslant \rho < \infty$, where $\Delta r = (r_2 - r_1)/r_2$. The transformation maps the entire electromagnetic space to the region $r_1 \leqslant \rho' < \infty$, $0 \leqslant \phi' \leqslant 2\pi$, $-\infty < z' < \infty$ of the physical space. In particular, the region $0 \leqslant \rho' < r_1$ of the physical space has no counterpart in the electromagnetic space and is thus not visited by any geodesic. We have $\partial\rho'/\partial z = 0 = \partial\phi'/\partial z = \partial z'/\partial x = \partial z'/\partial y$, $\partial z'/\partial z = 1$ and

$$\frac{\partial\rho'}{\partial x} = \frac{\partial\rho'}{\partial\rho}\frac{\partial\rho}{\partial x} + \frac{\partial\rho'}{\partial\phi}\frac{\partial\phi}{\partial x} = \Delta r\cos\phi; \qquad \frac{\partial\rho'}{\partial y} = \frac{\partial\rho'}{\partial\rho}\frac{\partial\rho}{\partial y} + \frac{\partial\rho'}{\partial\phi}\frac{\partial\phi}{\partial y} = \Delta r\sin\phi$$

$$\frac{\partial\phi'}{\partial x} = \frac{\partial\phi'}{\partial\rho}\frac{\partial\rho}{\partial x} + \frac{\partial\phi'}{\partial\phi}\frac{\partial\phi}{\partial x} = -\frac{\sin\phi}{\rho}; \qquad \frac{\partial\phi'}{\partial y} = \frac{\partial\phi'}{\partial\rho}\frac{\partial\rho}{\partial y} + \frac{\partial\phi'}{\partial\phi}\frac{\partial\phi}{\partial y} = \frac{\cos\phi}{\rho}.$$

Thus

$$A^{\lambda}_{\nu_0} = \begin{pmatrix} \Delta r\cos\phi & \Delta r\sin\phi & 0 \\ -\dfrac{\sin\phi}{\rho} & \dfrac{\cos\phi}{\rho} & 0 \\ 0 & 0 & 1 \end{pmatrix}, \tag{17.198}$$

where $\lambda$ is the index for row and $\nu_0$ is the index for the column. The contravariant spatial metric tensor $\tau^{mn}$ is

$$\tau^{mn} = A^{m}_{m_0}A^{n}_{n_0}\delta^{m_0n_0} = \begin{pmatrix} (\Delta r)^2 & 0 & 0 \\ 0 & \dfrac{1}{\rho^2} & 0 \\ 0 & 0 & 1 \end{pmatrix}, \tag{17.199}$$

where $m$ is the index for row and $n$ is the index for the column and $\tau = \rho^2/(\Delta r)^2$. Note that if $\Delta r = 1$, the tensor $\tau^{mn}$ and the determinant $\tau$ would coincide with their values in ordinary cylindrical coordinates. In the physical space now filled with an anisotropic material the unitary vectors are $\mathbf{a}_1 = \hat{\boldsymbol{\rho}}'$, $\mathbf{a}_2 = \rho'\hat{\boldsymbol{\phi}}'$, $\mathbf{a}_3 = \hat{\mathbf{z}}$ on employing cylindrical coordinates. The metric tensor is

$$\alpha_{mn} = \begin{pmatrix} 1 & 0 & 0 \\ 0 & \rho'^2 & 0 \\ 0 & 0 & 1 \end{pmatrix}, \tag{17.200}$$

where $m$ is the index for row and $n$ is the index for the column. The determinant of the metric tensor is $\alpha = \rho'^2$. Finally, the relative permittivity (or permeability) tensor (17.194) to realize the cloak is given by

$$\varepsilon r_n^m = \sqrt{\frac{\tau}{\alpha}}\,\tau^{m\nu}\alpha_{\nu n} = \begin{pmatrix} \frac{\rho\Delta r}{\rho'} & 0 & 0 \\ 0 & \frac{\rho'}{\rho\Delta r} & 0 \\ 0 & 0 & \frac{\rho}{\rho'}\frac{1}{\Delta r} \end{pmatrix}, \quad r_1 \leqslant \rho' \leqslant r_2, \tag{17.201}$$

where $\mathscr{D}_n = \varepsilon_0\varepsilon_{rn}^{\ m}\mathscr{E}_m$. In other words the components of electric flux density and electric field as related as $\mathscr{D}_1 = \mathbf{a}_1 \cdot \mathscr{D} = \varepsilon_0(\frac{\rho\Delta r}{\rho'})\mathscr{E}_1$, $\mathscr{D}_2 = \mathbf{a}_2 \cdot \mathscr{D} = \varepsilon_0(\frac{\rho'}{\rho\Delta r})\mathscr{E}_2$, $\mathscr{D}_3 = \mathbf{a}_3 \cdot \mathscr{D} = \varepsilon_0(\frac{\rho}{\rho'\Delta r})\mathscr{E}_3$ in the region $r_1 \leqslant \rho' \leqslant r_2$. If $\Delta r = 1$, $\varepsilon_{rn}^{\ m}$ reduces to an identity matrix.

In a space with metric tensor $\tau_{mn}$ (non-constant values), the only non-zero Christoffel symbols of the first kind are $[21, 2] = [12, 2] = -[22, 1] = \rho/\Delta r = (\rho' - r_1)/(\Delta r)^2$. According to equation (17.59) the geodesic equations within the anisotropic layer having a metric tensor $\tau_{mn}$ are

$$\frac{\mathrm{d}^2x^1}{\mathrm{d}s^2} - \rho\Delta r\left(\frac{\mathrm{d}x^2}{\mathrm{d}s}\right)^2 = 0, \tag{17.202}$$

$$\frac{\mathrm{d}^2x^2}{\mathrm{d}s^2} + \frac{2}{\rho\Delta r}\left(\frac{\mathrm{d}x^1}{\mathrm{d}s}\right)\left(\frac{\mathrm{d}x^2}{\mathrm{d}s}\right) = 0, \tag{17.203}$$

$$\frac{\mathrm{d}^2x^3}{\mathrm{d}s^2} = 0, \tag{17.204}$$

where $s$ is the arc-length parameter. Although it is possible to solve these equations directly in a manner similar to solving equations (17.64)–(17.66), it is much easier to first solve the geodesic equations in the electromagnetic space and then transform the final solution to the physical space[14]. If $(x_{0p}, y_{0p}, z_{0p})$ and $(x_{0q}, y_{0q}, z_{0q})$ are the initial and final coordinates of two points $P$ and $Q$ in the electromagnetic space and if $s_q = \sqrt{(x_{0q} - x_{0p})^2 + (y_{0q} - y_{0p})^2 + (z_{0q} - z_{0p})^2}$ denotes the straight line distance between the two, and if $m_x = (x_{0q} - x_{0p})/s_q$, $m_y = (y_{0q} - y_{0p})/s_q$, $m_z = (z_{0q} - z_{0p})/s_q$ denote the slopes of that straight line along the $x$-, $y$-, and $z$-directions, then $x_0(s) = m_x s + x_{0p}$, $y_0(s) = m_y s + y_{0p}$, $z_0(s) = m_z s + z_{0p}$ for $0 \leqslant s \leqslant s_q$ and

$$\rho'(s) = x^1(s) = r_1 + \Delta r\sqrt{[x_0(s)]^2 + [y_0(s)]^2}, \tag{17.205}$$

[14] In the trivial case of $\Delta r = 1$ (which implies that $r_1 = 0$), equations (17.202)–(17.204) are similar to the equations (17.64)–(17.66), which have already been solved exactly. This results in a straight line geodesic. Indeed it is this non-unity value of $\Delta r$ that renders the space to be curved.

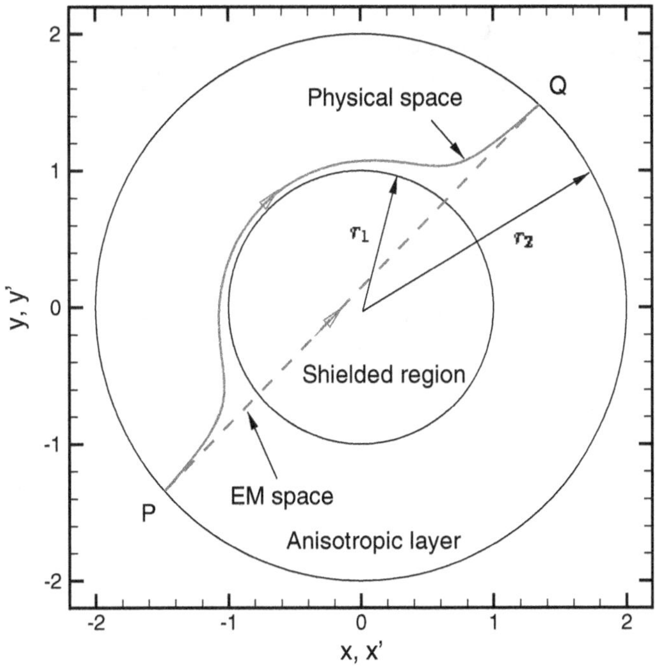


**Figure 17.4.** Geodesic in the electromagnetic space (dashed red line) and in the physical space (solid red line) in the presence of metric-induced anisotropic layer.

$$\phi'(s) = x^2(s) = \arctan\left(\frac{y_0(s)}{x_0(s)}\right), \tag{17.206}$$

$$z'(s) = x^3(s) = z_0(s), \tag{17.207}$$

where $m_x^2 + m_y^2 + m_z^2 = 1$. Figure 17.4 shows the geodesics in the electromagnetic and physical spaces for the points $P$ and $Q$ taken arbitrarily to lie on the bounding circle $\rho' = r_2 = \rho$ of the anisotropic layer. Clearly the path taken by an electromagnetic wave in the physical space is a curved line and avoids the interior of $\rho' = r_1$, where an object to be hidden can be located. This is in contrast to the geodesic in the electromagnetic space, which is a straight line connecting $P$ and $Q$. Note that conclusions drawn in this example are based on an exact analysis and not subject to any ray-optic approximations.

### 17.5.2 General transformational equations

By making use of some vector and dyadic identities[15], it is possible to come up with the transformation equations for the medium constants in the special case where the transformation matrix is constant [19] as the following example shows.

[15] A dyad $A$ is a quantity that is formed by placing two vectors **a** and **b** in juxtaposition $A = \mathbf{ab} \neq \mathbf{ba}$, where the ordering the two vectors is important. For an extensive treatment of dyadics, see the text by Lindell [19].

**Example 17.5.2.1** Derive the transformational equations for a general bianisotropic medium under constant affine transformation $\mathbf{r}' = A\mathbf{r}$, where the components of matrix $A$ are *constant*[16].

The transformation matrix $A$ moves every point $\mathbf{r}$ to a different point $\mathbf{r}'$. An example of such a transformation is reflection about a fixed plane of every point. Another example is uniform stretching or squeezing in a certain direction. Let us assume that the transformation is one-to-one so that the matrix $A$ is non-singular. Let the transformed field vectors be denoted with a prime so that the transformed electromagnetic field is $\mathbf{E}'(\mathbf{r}) = T_e\mathbf{E}(\mathbf{r}')$, $\mathbf{H}'(\mathbf{r}) = T_h\mathbf{H}(\mathbf{r}')$, and so on. That is, the transformation takes the original electric field value at the transformed point $\mathbf{r}'$ and further modifies it by transforming it by the matrix $T_e$ to the new value of electric field $\mathbf{E}'(\mathbf{r})$ associated with the point $\mathbf{r}$. The matrices $T_e$ and $T_h$ depend on $A$ and must be chosen so as to preserve Maxwell's equations. By imposing the dyadic relation $\nabla'\mathbf{r}' = \nabla\mathbf{r}$ and using the notation $A^{-1T} := (A^{-1})^T = (A^T)^{-1}$, the gradient operator gets transformed as $\nabla' = A^{-1T}\nabla$, which implies that $\nabla = A^T\nabla'$. Using this it can be shown that the curl operation gets transformed as [19, p 114]

$$\nabla \times \mathbf{E}'(\mathbf{r}) = (\det A)A^{-1}\nabla' \times (A^{-1T}T_e\mathbf{E}(\mathbf{r}')). \tag{17.208}$$

Thus the time-harmonic Maxwell's curl equations (1.36) and (1.37) are transformed to

$$\nabla' \times [A^{-1T}T_e\mathbf{E}(\mathbf{r}')] = -j\omega\frac{A}{\det A}\mathbf{B}'(\mathbf{r}) - \frac{A}{\det A}\mathbf{M}'(\mathbf{r}) \tag{17.209}$$

$$\nabla' \times [A^{-1T}T_h\mathbf{H}(\mathbf{r}')] = j\omega\frac{A}{\det A}\mathbf{D}'(\mathbf{r}) + \frac{A}{\det A}\mathbf{J}'(\mathbf{r}). \tag{17.210}$$

When these equations are compared with the Maxwell's equations in the transformed space,

$$\nabla' \times \mathbf{E}(\mathbf{r}') = -j\omega\mathbf{B}(\mathbf{r}') - \mathbf{M}(\mathbf{r}') \tag{17.211}$$

$$\nabla' \times \mathbf{H}(\mathbf{r}') = j\omega\mathbf{D}(\mathbf{r}') + \mathbf{J}(\mathbf{r}'), \tag{17.212}$$

one arrives at $T_e = \alpha A^T$ and $T_h = \beta A^T$, where $\alpha$ and $\beta$ are scalar constants. By requiring that the transformation $A^{-1}$ reverts all quantities back to their original values, the constants $\alpha$ and $\beta$ are subject to $\alpha^2 = 1 = \beta^2$. Thus $\alpha$ and $\beta$ are $\pm 1$. The various fields and sources are thus transformed as

$$\mathbf{E}'(\mathbf{r}) = \alpha A^T\mathbf{E}(\mathbf{r}'); \quad \mathbf{H}'(\mathbf{r}) = \beta A^T\mathbf{H}(\mathbf{r}'), \tag{17.213}$$

$$\mathbf{B}'(\mathbf{r}) = \alpha(\det A)A^{-1}\mathbf{B}(\mathbf{r}'); \quad \mathbf{D}'(\mathbf{r}) = \beta(\det A)A^{-1}\mathbf{D}(\mathbf{r}'), \tag{17.214}$$

$$\mathbf{M}'(\mathbf{r}) = \alpha(\det A)A^{-1}\mathbf{M}(\mathbf{r}'); \quad \mathbf{J}'(\mathbf{r}) = \beta(\det A)A^{-1}\mathbf{J}(\mathbf{r}'). \tag{17.215}$$

[16] To be consistent with the notation in this section, we drop the double bars in matrices or dyadics $\bar{\bar{A}}$.

If one starts from the constitutive relation

$$\mathbf{D} = \varepsilon\mathbf{E} + \xi\mathbf{H}, \tag{17.216}$$

(i) multiplies both sides with $\beta \det(A)A^{-1}$, (ii) replaces the lhs with $\mathbf{D}'$, (iii) expresses the fields $\mathbf{E}$ and $\mathbf{H}$ on the rhs in terms of $\mathbf{E}'$ and $\mathbf{H}'$ using equation (17.213), and (iv) compares the resulting expression with the transformed constitutive relations

$$\mathbf{D}' = \varepsilon'\mathbf{E}' + \xi'\mathbf{H}', \tag{17.217}$$

then one arrives at the following relations between the transformed parameters and the original parameters

$$\varepsilon' = \alpha\beta \det(A)A^{-1}\varepsilon A^{-1T}; \quad \xi' = \det(A)A^{-1}\xi A^{-1T}. \tag{17.218}$$

Similarly,

$$\mu' = \alpha\beta \det(A)A^{-1}\mu A^{-1T}; \quad \zeta' = \det(A)A^{-1}\zeta A^{-1T}. \tag{17.219}$$

For example, if one starts off with an isotropic medium with scalars $\varepsilon$ and $\mu$, then transformation converts the medium to an anisotropic medium with parameters

$$\varepsilon' = \alpha\beta\varepsilon \det(A)(A^T A)^{-1}; \quad \mu' = \alpha\beta\mu \det(A)(A^T A)^{-1} \Longrightarrow \mu'(\varepsilon')^{-1} = \frac{\mu}{\varepsilon}I, \tag{17.220}$$

where $I$ is an identity matrix. So $\varepsilon'$ and $\mu'$ are proportional to the same symmetric matrix $(A^T A)^{-1}$. Conversely, if one starts off with a special anisotropic medium wherein the constitutive parameters are multiples of the same symmetric matrix $S$ such that $\varepsilon = \varepsilon_0 S$, $\mu = \mu_0 S$, and if one chooses the transformation matrix dictated by the relation $AA^T = S$, then the medium is transformed to an isotropic one with parameters $\mu' = \mu_0\alpha\beta \det(A)$, $\varepsilon' = \varepsilon_0\alpha\beta \det(A)$. The transformed medium has an intrinsic impedance matched to that of free space with $\mu'/\varepsilon' = \mu_0/\varepsilon_0$, but with a refractive index $\pm|\det(A)|$. ■■

## References

[1] Deschamps G A 1981 Electromagnetics and differential forms *Proc. IEEE* **69** 676–96
[2] Warnick K F, Selfridge R H and Arnold D V 1997 Teaching electromagnetic field theory using differential forms *IEEE Trans. Educ.* **40** 53–68
[3] Russer P 2005 The geometry of electrodynamics *Eur. Microwave J.* **1** 3–16
[4] Pendry J B, Schurig D and Smith D R 2006 Controlling electromagnetic fields *Science* **312** 1780–2
[5] Kwon D-H and Werner D H 2010 Transformation electromagnetics: an overview of the theory and applications *IEEE Antennas Propag. Mag.* **52** 24–46
[6] Einstein A 1923 *The Principle of Relativity: A Collection of Original Memoirs on the Special and General Theory of Relativity* ed W Perrett and G B Jeffery (New York: Dover)
[7] Synge J L and Schild A 1978 *Tensor Calculus* (New York: Dover)
[8] Tolman R C 1967 *Relativity Thermodynamics and Cosmology* (New York: Dover)
[9] Møller C 1972 *The Theory of Relativity* 2nd edn (Oxford: Clarendon)

[10] Carroll S M *Spacetime and Geometry: An Introduction to General Relativity* (Cambridge: Cambridge University Press)

[11] Jha R M and Wiesbeck W 1995 The geodesic constant method: a novel approach to analytical surface-ray tracing on convex conducting bodies *IEEE Antennas Propag. Mag.* **37** 28–38

[12] Gradshteyn I S and Ryzhik I M 2015 *Tables of Integrals, Series and Products* 8th edn (New York: Academic)

[13] Pathak P H and Wang N 1981 Ray analysis of mutual coupling between antennas on a convex surface *IEEE Trans. Antennas Propag.* **29** 911–22

[14] Stratton J A 1941 *Electromagnetic Theory* (New York: McGraw-Hill)

[15] Landau L D and Lifshitz E M 1975 *The Classical Theory of Fields* 4th edn (Burlington, MA: Elsevier)

[16] Post E J 1997 *Formal Structure of Electromagnetics* (New York: Dover)

[17] Leonhardt U and Philbin T G 2006 General relativity in electrical engineering *New J. Phys.* **8** 2–18

[18] Werner D H and Kwon D-H (ed) 2014 *Transformational Electromagnetics and Metamaterials* (New York: Springer)

[19] Lindell I V 1995 *Methods for Electromagnetic Field Analysis* (IEEE Press Series on Electromagnetic Wave Theory) (New York: IEEE)

**IOP** Publishing

Engineering Electrodynamics
A collection of theorems, principles and field representations
**Ramakrishna Janaswamy**

# Chapter 18

# Maxwell's equations in the sense of distributions

Traditionally Maxwell's equations are presented in a differential or integral form and the supplemental information on boundary and initial conditions is provided separately from additional assumptions. At a boundary between two different material media, the fields are no longer continuous and the boundary conditions serve the purpose of sewing the fields in the two regions together to provide consistency. In this regard it is desirable to have a field formulation where the boundary conditions follow directly from Maxwell's equations without the need for additional assumptions. Such a formulation is possible if one postulates at the outset that *Maxwell's equations are valid everywhere in the sense of distributions*. The distribution formulation for electromagnetic theory was carried out in [1, 2] and [3] and a detailed account is given in the books [4, 5]. A good description of general distribution theory is given in [6–8] and [9]. The main essence of distribution theory is that instead of characterizing a function $f(x)$ by means of its value at each point $x$, we view it as a *functional*, $\langle f, \varphi\rangle$, on some class $\mathcal{D}$ of test functions $\varphi$. The process is analogous to characterizing a differentiable, real-valued function $f(x)$ in $0 \leq x \leq 2\pi$ by means of its complex Fourier coefficients $2\pi c_n = \int_0^{2\pi} f(x)\exp(jnx)\mathrm{d}x$ rather than by its point values in $0 \leq x \leq 2\pi$. The advantage of the distribution viewpoint is that it allows for larger kinds of functions to be described rigorously. These ideas are elaborated below by first providing precise definitions and then developing the appropriate distributional identities. The arguments of the various distributions considered here are space coordinates. At the end we will demonstrate the generality and versatility of the distributional approach by deriving the classical boundary conditions for fields and deriving some newer boundary conditions for artificial surfaces (metasurfaces) and for the potentials.

## 18.1 Preliminaries of distributions

**Definition 18.1.1.** *Support.* The support of a function $f(\mathbf{r})$ is the closure of the set of points in the $n$-dimensional Euclidean space $R_n$ on which $f(\mathbf{r}) \neq 0$.

 

**Definition 18.1.2** *Test function.* A test function $\varphi(\mathbf{r})$ is a real continuous function of finite support endowed with continuous derivatives of all orders. A collection $\varphi_n(\mathbf{r})$, $n = 1, 2, \ldots$ of these functions forms a space $\mathcal{D}$.

An example of a test function is

$$\varphi(\mathbf{r}; \mathbf{r}_0) = \begin{cases} \exp\left(\dfrac{-|\mathbf{r}_0|^2}{|\mathbf{r}_0|^2 - |\mathbf{r}|^2}\right) & \text{for } |\mathbf{r}| < |\mathbf{r}|_0 \\ 0 & \text{for } |\mathbf{r}| \geqslant \mathbf{r}_0, \end{cases} \tag{18.1}$$

having a support $(0 \leq |\mathbf{r}| \leq |\mathbf{r}_0|)$. A possible sequence of valid test functions is $\varphi_n = n^{-1}\varphi(\mathbf{r}; \mathbf{r}_0)$, $n = 1, 2, \ldots$, which converges in $C^{\infty}_{(0,\, \mathbf{r}_0)}$, the space of real continuous functions with continuous derivatives of all orders and with bounded support $(0 \leq |\mathbf{r}| \leq |\mathbf{r}_0|)$. A counter-example is $\varphi(\mathbf{r}) = \mathbf{r} \cdot \mathbf{r}$ as it does not have finite support even though it is infinitely differentiable. An invalid sequence of test functions is $n^{-1}\varphi(n^{-1}\mathbf{r}; \mathbf{r}_0)$ as the set has no common bounded support. In the following we denote by $V$ the maximum support of the sequence of test functions and their derivatives in 3D space and by $\Sigma$ the surface of $V$ on which $\varphi_n$ vanishes.

**Definition 18.1.3.** *Linear functional on $\mathcal{D}$.* We say that $z$ is a linear functional on $\mathcal{D}$ if there exists a rule which associates a complex number denoted by $\langle z, \varphi\rangle$ for every member $\varphi$ belonging to $\mathcal{D}$ such that

$$\langle z, \alpha_1\varphi_1 + \alpha_2\varphi_2\rangle = \alpha_1\langle z, \varphi_1\rangle + \alpha_2\langle z, \varphi_2\rangle, \tag{18.2}$$

for real $\alpha_1, \alpha_2$. The functional is continuous if, $\varphi_n \to \varphi$ as $n \to \infty$ implies $\langle z, \varphi_n\rangle \to \langle z, \varphi\rangle$.

The linearity property (18.2) is also valid as is for complex-valued testing functions $\varphi$ and complex $\alpha_1, \alpha_2$. Note that the rhs of the linearity relation (18.2) does not contain complex conjugates of $\alpha_1$ and $\alpha_2$. The quantity $\langle z, \varphi\rangle$ may be thought of as the *action of z on $\varphi$*.

**Definition 18.1.4.** *Distribution.* A distribution $z$ (i.e. a generalized function) is a continuous linear functional defined on $\mathcal{D}$. The number $\langle z, \varphi\rangle$ is the action of $z$ on $\varphi$. Often the functional $\langle z, \varphi\rangle$ is simply denoted by the letter $z$ omitting the reference to the test function $\varphi$. A locally integrable function $z(\mathbf{r})$ (i.e. a function which is integrable over a compact set[1]), hereby, labeled as the generating function also defines a distribution $z$ by the relation

[1] A function $f(x)$ is said to be locally integrable if $\int_I |f(x)|\, dx$ exists for every finite interval $I$. The class of locally integrable functions includes all piecewise continuous functions and other functions such as $1/x^{1-\epsilon}$, where $\epsilon > 0$.

$$\langle z, \varphi \rangle = \iiint_{-\infty}^{\infty} z(\mathbf{r})\varphi(\mathbf{r})\,d\mathbf{r}. \tag{18.3}$$

Given this we say the distribution $z$ has a function representation $z(\mathbf{r})$ when we wish to call attention to the existence of a function by means of which the distribution functional may be computed by means of the integral (18.3). Not all distributions may have a corresponding function representation. Those distributions that can be generated by a locally integrable function through equation (18.3) are known as *regular*, while all other distributions are known as *singular*. The symbols $z$, $z(x)$, $\langle z, \varphi \rangle$ are often used interchangeably. If $z$ is singular, the representation (18.3) is purely symbolic and $z(\mathbf{r})$ is known as a *generalized function* corresponding to the distribution $z$.

As an example the integral

$$\int_0^{\infty} \varphi(x)\,dx = \int_{-\infty}^{\infty} \Theta(x)\varphi(x)\,dx, \tag{18.4}$$

may be used to define the distribution $\Theta = \langle \Theta, \varphi \rangle$ when the generating function $\Theta(x)$ is a unit step function. A second example is the 3D Dirac distribution

$$\langle \delta_{\mathbf{r}_0}, \varphi \rangle = \varphi(\mathbf{r}_0) = \iiint_{-\infty}^{\infty} \delta(\mathbf{r} - \mathbf{r}_0)\varphi(\mathbf{r})\,d\mathbf{r}. \tag{18.5}$$

The quantity $\delta(\mathbf{r} - \mathbf{r}_0)$ may be identified with the generating function for the Dirac distribution $\delta_{\mathbf{r}_0}$ and is commonly referred to as the Dirac distribution or Dirac delta function[2].

What is meant by a distribution is not simply the collection of the functional values, but rather the collection of its effect on the class of test functions. The effect must be independent of any specific test function even though the numerical value of the functional will depend on it. *A vector distribution is a vector each of whose components is a distribution*. The functional of a vector distribution $\mathbf{A} = \sum_{i=1}^{3} \hat{\mathbf{x}}_i A_i$ on a space $\mathscr{D}$ of vector test functions $\boldsymbol{\phi} = \sum_{i=1}^{3} \hat{\mathbf{x}}_i \varphi_i$, where $a_i$ and $\varphi_i$, $i = 1, 2, 3$ are the Cartesian components with respect to the unit vectors $\hat{\mathbf{x}}_i$, is defined as

$$\langle \mathbf{A}, \boldsymbol{\phi} \rangle = \sum_{i=1}^{3} \langle A_i, \varphi_i \rangle. \tag{18.6}$$

The result is a scalar. It is useful to think of distributions as a limit function in a sequence of well-behaved ordinary functions, known as good functions.

[2] On strict mathematical grounds $\delta(\mathbf{r})$ is not a function although it may thought of as the limit of a sequence of functions. For that reason some authors feel that the Dirac point distribution does not have a function representation.

**Definition 18.1.5** *Good functions.* A good function is one which is infinitely differentiable and is such that all its derivatives go to zero at infinity faster than $|\mathbf{r}|^{-N}$ for any integer $N > 0$.

An example is $\exp(-|\mathbf{r}|^2/2\sigma^2)$, $\sigma$ real. Note that good functions need not have finite support unlike test functions.

**Definition 18.1.6.** *Regular sequence.* A sequence of good functions $s_n(\mathbf{r})$ is called regular if for any test function $\varphi$ the limit

$$\lim_{n\to\infty} \iiint_V s_n(\mathbf{r})\varphi(\mathbf{r})\, d\mathbf{r}, \tag{18.7}$$

exists.

An example is the sequence $s_n(\mathbf{r}) = (n/\sqrt{2\pi})\exp(-n^2|\mathbf{r}|^2/2)$, $n = 1, 2, \ldots$. In fact the Dirac distribution $\delta(\mathbf{r})$ can be thought of as the limit of this sequence as $n \to \infty$.

**Definition 18.1.7.** *Order of a 1D distribution.* A distribution is said to be of order $k$ if it can be expressed as the $(k + 1)$th derivative of a regular distribution but cannot be expressed as a derivative of lower order of a regular distribution. Mathematically, the order $k$ of a distribution $z$ is the smallest value of integer $k$ such that

$$|\langle z, \varphi\rangle| \leq C_0\, \|\varphi\|_k, \tag{18.8}$$

where

$$\|\varphi\|_p = \max_{\alpha<x<\beta}\left(|\varphi|, |\varphi'|, \ldots, |\varphi^{(p)}|\right), \quad p = 0, 1, \ldots, \tag{18.9}$$

$C_0$ is a positive constant and $\alpha \leq x \leq \beta$ is the support of the test function.

For example using test functions having a support over $-a \leq x \leq a$ one has for the Dirac distribution that

$$|\langle \delta, \varphi\rangle| = |\varphi(0)| \leq \max_{|x|\leq a} |\varphi(x)|. \tag{18.10}$$

Therefore the order of a delta distribution[3] is $k = 0$. Two theorems are important for finite-ordered distributions.

**Theorem 18.1.1.** L Schwartz.
*Let $f$ be a distribution of order $k$ in an interval $(\alpha, \beta)$. Then there exists a function $F(x) \in C(\alpha, \beta)$ such that $f = F^{(k+2)}(x)$ in the sense of distributions. The function $F(x)$ is not unique.*

[3] The definition of order is not consistent in the literature but we have used the definition given in [8] and [5].

For example the Dirac distribution, which is of order $k = 0$ can be represented as $\delta(x) = F''(x)$, $F \in C[-a, a]$. Candidate functions for $F(x)$ are

$$F(x) = \frac{1}{2}|x|$$

or

$$F(x) = \begin{cases} 0, & x < 0 \\ x, & x > 0 \end{cases}.$$

**Theorem 18.1.2.** Theorem on $k$th order point distribution.
*Let $f$ be a distribution of order $k$ defined on the real line $R$. If the support of $f$ consists only of the point $x = 0$, then it is of the following form:*

$$f = \sum_{j=0}^{k} c_j \delta^{(j)}(x), \tag{18.11}$$

*where $c_j$ are real or complex-valued constants.*

For a proof see [5]. A direct consequence of theorem 18.1.2 is the following.

**Corollary 18.1.1** *If $g(x)$ is infinitely differentiable for $x < x_0$ and $x > x_0$ and if $g(x)$ and all its derivatives have left-hand and right-hand limits, then for $m \geqslant 1$*

$$g^{(m)} = \{g^{(m)}\} + \sum_{j=1}^{m} \Delta[g^{(m-j)}]\delta^{(j-1)}(x - x_0), \tag{18.12}$$

*where $\{g^{(j)}\}$ denotes the $j$th derivative defined at all points but $x = x_0$ and $\Delta[g^{(j)}]$ denotes the difference between the right-hand limit and the left-hand limit of the $j$th derivative of $g(x)$ at $x = x_0$.*

For example, if $g(x) = \exp(-|x|)$

$$g' = \frac{\mathrm{d}}{\mathrm{d}x}\mathrm{e}^{-|x|} = \mp\mathrm{e}^{-|x|},\ x \gtrless 0 \tag{18.13}$$

$$g'' = \frac{\mathrm{d}^2}{\mathrm{d}x^2}\mathrm{e}^{-|x|} = \mathrm{e}^{-|x|} - 2\delta(x) = g - 2\delta(x) \tag{18.14}$$

$$g^{(3)} = \frac{\mathrm{d}^3}{\mathrm{d}x^3}\mathrm{e}^{-|x|} = \mp\mathrm{e}^{-|x|} - 2\delta'(x), \tag{18.15}$$

since $\Delta[g'] = -2$, $\Delta[g''] = 0$, and $\Delta[g^{(3)}] = -2$. In particular we have the relation $g'' - g = -2\delta(x)$.

**Definition 18.1.8** *Heaviside jump distribution.* Consider a closed volume $V_k$ bounded by a surface $\Sigma_k$. The Heaviside distribution, $\Theta_k(\mathbf{r})$, is that distribution whose functional representation is

$$\Theta_k(\mathbf{r}) = \begin{cases} 1, & \text{if } \mathbf{r} \text{ is inside } V_k \\ 0, & \text{if } \mathbf{r} \text{ is outside } V_k \end{cases}, \tag{18.16}$$

and defines the functional

$$\langle \Theta_k, \varphi \rangle = \iiint_{V_k} \varphi(\mathbf{r}) \, \mathrm{d}v. \tag{18.17}$$

The $k$ in the above definition has no significance other than to be a place holder for an integer index to describe multiple regions $V_1$, $V_2$, ... in later development.

**Definition 18.1.9.** *Product of a distribution and a good function.* The product of a distribution $\psi$ and a good function $g$ is defined via

$$\langle \psi g, \varphi \rangle = \langle \psi, g\varphi \rangle. \tag{18.18}$$

This definition makes sense because $g\varphi$ also belongs to the space of test functions. Note that the product of two distributions cannot always be defined; for instance, $\delta^2(x) = \delta(x)\delta(x)$ cannot be defined as a distribution.

**Definition 18.1.10.** *Dirac surface distribution.* The Dirac surface distribution $\delta_S$ (units of $\mathrm{m}^{-1}$) concentrated on the surface $S$ is defined as

$$\langle \delta_S, \varphi \rangle = \iiint_V \delta_S \varphi \, \mathrm{d}v = \iint_S \varphi(\mathbf{r}) \, \mathrm{d}s. \tag{18.19}$$

On a surface with coordinates $(u_1, u_2)$ a surface charge density of the form $\rho_s(u_1, u_2)$ may be represented by the volume density $\rho = \rho_s \delta_S$. The corresponding functional pertaining to the single-layer distribution $\rho$ is

$$\langle \rho_s \delta_S, \varphi \rangle = \iint_S \rho_s \varphi \, \mathrm{d}s. \tag{18.20}$$

Similarly, a surface electric current density $\mathbf{J}_s$ (which can have both a normal and tangential component) can be represented as a volume density $\mathbf{J} = \mathbf{J}_s \delta_S$.

**Definition 18.1.11.** *Dirac line distribution.* The Dirac line distribution $\delta_C$ (units of $\mathrm{m}^{-2}$) concentrated on a curve $C$ is defined as

$$\langle \delta_C, \varphi \rangle = \iiint_V \delta_C \varphi \, \mathrm{d}v = \int_C \varphi(\mathbf{r}) \, \mathrm{d}\ell. \tag{18.21}$$

A distribution of line charge density $\rho_c$ on a curve $C$ may be represented by the volume density $\rho = \rho_c \delta_C$ and the corresponding functional

$$\langle \rho_c \delta_C, \varphi \rangle = \int_C \rho_c \varphi \, \mathrm{d}\ell. \tag{18.22}$$

**Theorem 18.1.3** Transformation of Dirac distribution under coordinate transformations, [10, p 293].
*Let curvilinear coordinates $\xi_1, \xi_2, \dots, \xi_n$ be obtained from Cartesian coordinates $x_1, x_2, \dots, x_n$ by the transformation law*

$$x_i = u_i(\xi_1, \xi_2, \dots, \xi_n), \quad i = 1, \dots, n. \tag{18.23}$$

*Let a point $P$ with coordinates $\mathbf{r}' = (\alpha_1, \alpha_2, \dots, \alpha_n)$ in the x-coordinate system have the coordinates $(\beta_1, \beta_2, \dots, \beta_n)$ in the $\xi$-coordinate system. Then*

$$\delta(\mathbf{r} - \mathbf{r}') = \delta(x_1 - \alpha_1) \cdots \delta(x_n - \alpha_n) = |J|^{-1} \delta(\xi_1 - \beta_1) \cdots \delta(\xi_n - \beta_n), \tag{18.24}$$

*where $J = \det(\partial x_i / \partial \xi_j) \neq 0$ is the Jacobian of the transformation.*

A point at which $J = 0$ is called a singular point of the transformation. At a singular point some coordinates are either multi-valued or take no determinate value. These are known as *ignorable coordinates.* For example in a spherical coordinate system in 3D the coordinates $(\theta, \phi)$ are both ignorable at the origin. Let $\xi_{r+1}, \dots, \xi_n$ be ignorable at a point P. Then

$$\delta(\mathbf{r} - \mathbf{r}') = J_r^{-1} \delta(\xi - \xi_1) \cdots \delta(\xi - \xi_r), \tag{18.25}$$

where

$$J_r = \int_{\xi_n} \cdots \int_{\xi_{r+1}} J \, \mathrm{d}\xi_{r+1} \dots \mathrm{d}\xi_n. \tag{18.26}$$

For example, in spherical coordinates in 3D, $J = r^2 \sin\theta$ vanishes at $r = 0$ and $\theta$, $\phi$ are both ignorable. Then $J_1 = 4\pi r^2$ which yields

$$\delta(\mathbf{r}) = \delta(x)\delta(y)\delta(z) = \delta(r)/4\pi r^2. \tag{18.27}$$

On the other hand, only $\phi$ is ignorable on the $z$-axis where $\theta = 0$. In that case $J_2 = 2\pi r^2 \sin\theta$ and

$$\delta(x)\delta(y)\delta(z - z_1) = \delta(r - z_1)\delta(\theta)/(2\pi r^2 \sin\theta). \tag{18.28}$$

In both of these results it is assumed that $\int_0^\infty \delta(r)\,\mathrm{d}r = 1$ and $\int_0^\pi \delta(\theta)\,\mathrm{d}\theta = 1$. However, it is more common in the literature to assume $\int_0^\infty \delta(r)\,\mathrm{d}r = 1/2$ and $\int_0^\pi \delta(\theta)\,\mathrm{d}\theta = 1/2$ [11, p 839], [12, p 542] because the range of the variables $r$ and $\theta$ is regarded as non-negative. In such a case it is more prudent to take

$$\delta(\mathbf{r}) = \frac{1}{2\pi r^2}\delta(r); \quad \delta(x)\delta(y)\delta(z - z_1) = \frac{2}{\pi r^2 \sin\theta}\delta(r - z_1)\delta(\theta). \tag{18.29}$$

It may be noted that either of the definitions is correct as long as it is used consistently [10, p 154].

### 18.1.1 Derivatives of distributions

**Definition 18.1.12.** *Partial derivative of distribution.* The partial derivative, $\partial\rho/\partial x_i$, of a distribution $\rho$ is defined as

$$\left\langle \frac{\partial\rho}{\partial x_i}, \varphi \right\rangle = \iiint_V \frac{\partial\rho}{\partial x_i}\varphi\,\mathrm{d}v = -\iiint_V \rho\frac{\partial\varphi}{\partial x_i}\,\mathrm{d}v = \left\langle \rho, -\frac{\partial\phi}{\partial x_i} \right\rangle. \tag{18.30}$$

The definition is suggested by integration by parts and the fact that $\varphi$ vanishes on the boundary $\Sigma$ of its support. It is evident that whenever a distribution is differentiated, the differentiation may be transferred to the test function. Since the test functions are infinitely differentiable, *every distribution may be differentiated any number of times.* Indeed let $k_1, k_2, \ldots, k_n$ be non-negative integers such that $|k| = \sum_i k_i$ and define the partial derivative operator

$$D^k = \frac{\partial^{|k|}}{\partial x_1^{k_1}\partial x_2^{k_2}\cdots\partial x_n^{k_n}}, \tag{18.31}$$

with the understanding that if any $k_i$ is zero, the differentiation with respect to that variable is omitted. Then the $|k|$ derivative of the distribution $\rho$ is defined from

$$\langle D^k\rho, \varphi\rangle = (-1)^{|k|}\langle \rho, D^k\varphi\rangle. \tag{18.32}$$

**Definition 18.1.13** *Differential operation on distribution.* For an arbitrary linear differential operator of order $p$ in $n$ variables

$$\mathscr{L} = \sum_{|k|\le p} A_k(\mathbf{r})D^k, \tag{18.33}$$

where the coefficients $A_k(\mathbf{r})$ are infinitely differentiable real functions, the functional relation

$$\langle \mathscr{L}\rho, \varphi \rangle = \langle \rho, \mathscr{L}^{\dagger}\varphi \rangle, \tag{18.34}$$

defines the distribution $\mathscr{L}\rho$, where the operator $\mathscr{L}^{\dagger}$ defining

$$\mathscr{L}^{\dagger}\psi = \sum_{|k|\le p} (-1)^{|k|} D^{k}(A_k\psi), \tag{18.35}$$

is the formal adjoint operator of $\mathscr{L}$.

The action of $\mathscr{L}\rho$ on a $\varphi$ is equivalent to the action of $\rho$ on the test function $\mathscr{L}^{\dagger}\varphi$. As an example in ID, if $\mathscr{L}$ is the differential operator

$$\mathscr{L} = A_n(x)\frac{\mathrm{d}^n}{\mathrm{d}x^n} + \cdots + A_1(x)\frac{\mathrm{d}}{\mathrm{d}x} + A_0(x),$$

where all the coefficients are infinitely differentiable, then

$$\mathscr{L}^{\dagger}\varphi = \sum_{k=0}^{n}(-1)^k\frac{\mathrm{d}^k}{\mathrm{d}x^k}(A_k\varphi).$$

From the above it is clear that the operator $\mathcal{L}$ will be *formally self-adjoint* if has constant coefficients and contains *partial derivatives only of even order*.

Some useful formulas involving the Dirac distribution in 1D assuming that $\alpha(x)$ is differentiable are

$$\delta(ax) = \frac{1}{|a|}\delta(x), \quad a \neq 0, \tag{18.36}$$

$$\langle \delta^{(m)}, \varphi \rangle = (-1)^m \varphi^{(m)}(0) \quad (m\text{th derivative of Dirac distribution}), \tag{18.37}$$

$$\alpha(x)\delta'(x) = \alpha(0)\delta'(x) - \alpha'(0)\delta(x), \tag{18.38}$$

$$x\delta^{(m)}(x) = -m\delta^{(m-1)}(x); \quad x^n\delta^{(m)}(x) = 0, \quad n > m+1, \tag{18.39}$$

$$\frac{\mathrm{d}}{\mathrm{d}x}\delta(\alpha(x)) = \alpha'(x)\delta'(\alpha(x)). \tag{18.40}$$

**Definition 18.1.14** *Gradient of distribution.* The gradient, $\nabla\psi$, of a scalar distribution $\psi$ is defined as

$$\langle \nabla\psi, \varphi \rangle = \iiint_V (\nabla\psi)\varphi \, \mathrm{d}v = -\iiint_V \psi\nabla\varphi \, \mathrm{d}v. \tag{18.41}$$

An equivalent definition is

$$\langle \nabla\psi, \boldsymbol{\phi} \rangle = -\langle \psi, \nabla\cdot\boldsymbol{\phi} \rangle = -\sum_{i=1}^{3}\iiint_V \psi\frac{\partial\varphi_i}{\partial x_i}\, \mathrm{d}v. \tag{18.42}$$

The functional in equation (18.41) is a scalar, while that in equation (18.42) is a vector. Either of these two definitions for a distribution occurring at the end-points will yield the same properties for the quantity $\nabla\psi$. Note that both of these definitions can be treated as special cases of equation (18.34).

**Definition 18.1.15** *Partial derivative operator for Dirac surface distribution.* The partial derivative operator, $\partial_a$ acting on the Dirac surface distribution $\delta_S$ in a direction $\hat{\mathbf{a}}$ is defined as

$$\langle \partial_a \delta_S, \varphi \rangle = -\iint_S \frac{\partial \varphi}{\partial a}\, \mathrm{d}s. \tag{18.43}$$

When $\hat{\mathbf{a}}$ coincides with the unit normal $\hat{\mathbf{n}}$ to S, and $\tau(u_1, u_2)$ is a surface density then the functional pertaining to the *double-layer distribution* $(-\tau\partial_n\delta_S)$ is

$$\langle -\tau\partial_n\delta_S, \varphi \rangle = \iint_S \tau\frac{\partial \varphi}{\partial n}\, \mathrm{d}s. \tag{18.44}$$

Similarly, a double layer of surface current density $\mathbf{c}_s(u_1, u_2)$ (units of A) may be represented as the volume density $\mathbf{J} = -\mathbf{c}_s\partial_n\delta_S$. Note that $\partial_n\delta_S \neq \partial\delta_S/\partial n$ for a non-planar surface $S$. We will later see the relation between the two.

**Definition 18.1.16.** *Gradient of Dirac surface distribution.* The gradient $\nabla\delta_S$ of the Dirac surface distribution is defined as

$$\langle \nabla\delta_S, \varphi \rangle = -\iint_S \nabla\varphi\, \mathrm{d}s. \tag{18.45}$$

For a surface density distribution $\tau(u_1, u_2)$, the corresponding volume distribution is $\rho = \tau\delta_S$ for which the gradient from equation (18.41) is

$$\langle \nabla(\tau\delta_S), \varphi \rangle = -\iint_S \tau\nabla\varphi\, \mathrm{d}s. \tag{18.46}$$

**Definition 18.1.17** *Normal derivative of Dirac surface distribution.* The normal derivative of the Dirac surface distribution is defined via

$$\begin{aligned}\left\langle \frac{\partial\delta_S}{\partial n}, \varphi \right\rangle &= \langle \hat{\mathbf{n}}\cdot\nabla\delta_S, \varphi\rangle = \sum_{i=1}^{n}\left\langle n_i\frac{\partial\delta_S}{\partial x_i}, \varphi\right\rangle \\ &= \sum_{i=1}^{n}\left\langle \frac{\delta_S}{\partial x_i}, n_i\varphi\right\rangle = -\left\langle \delta_S, \frac{\partial(n_i\varphi)}{\partial x_i}\right\rangle,\end{aligned} \tag{18.47}$$

where $n_i$ are the direction cosines in the $i$th coordinate direction in an $n$-dimensional space.

Note that since $\nabla \cdot \hat{\mathbf{n}} = -2J_c$, where $J_c$ is the mean curvature and $\sum \partial(n_i\varphi)/\partial x_i = \sum(n_i \partial\varphi/\partial x_i + \varphi \partial n_i/\partial x_i) = \hat{\mathbf{n}} \cdot \nabla\varphi + \varphi\nabla \cdot \hat{\mathbf{n}} = \hat{\mathbf{n}} \cdot \nabla\varphi - 2J_c\varphi$, we have

$$\begin{aligned}\left\langle \frac{\partial \delta_S}{\partial n}, \varphi \right\rangle &= -\left\langle \delta_S, \frac{\partial \varphi}{\partial n} - 2J_c\varphi \right\rangle \\ &= \langle \partial_n \delta_S, \varphi\rangle + \langle 2J_c\delta_S, \varphi\rangle.\end{aligned} \tag{18.48}$$

Thus

$$\partial_n \delta_S = \frac{\partial \delta_S}{\partial n} - 2J_c\delta_S \neq \frac{\partial \delta_S}{\partial n}, \tag{18.49}$$

unless $J_c = 0$. The result (18.49) is also derived in [4, p 208] using an entirely physical approach.

**Definition 18.1.18.** *Surface divergence of tangential surface distribution.* The surface divergence, $\nabla_s \cdot \mathbf{J}_s$, of a tangential surface distribution $\mathbf{J}_s$ on $S$ is defined as

$$\langle \nabla_s \cdot \mathbf{J}_s, \varphi \rangle = -\iint_S \mathbf{J}_s \cdot \nabla_s\varphi \, \mathrm{d}s. \tag{18.50}$$

If the surface distribution is discontinuous on a surface $S$ with a bounding curve $C$ (see figure 18.1), then

$$\begin{aligned}\langle \nabla_s \cdot \mathbf{J}_s, \varphi \rangle &= -\iint_S \{\mathbf{J}_s\} \cdot \nabla_s\varphi \, \mathrm{d}s \\ &= \iint_S \{\varphi\nabla_s \cdot \mathbf{J}_s\} \, \mathrm{d}s - \iint_S \{\nabla_s \cdot (\mathbf{J}_s\varphi)\} \, \mathrm{d}s \\ &= \iint_S \varphi\{\nabla_s \cdot \mathbf{J}_s\} \, \mathrm{d}s - \oint_C \hat{\mathbf{u}}_m \cdot \mathbf{J}_s\varphi \, \mathrm{d}\ell \\ &= \iint_S (\{\nabla_s \cdot \mathbf{J}_s\} - \hat{\mathbf{u}}_m \cdot \mathbf{J}_s \, \delta_C)\varphi \, \mathrm{d}s,\end{aligned} \tag{18.51}$$

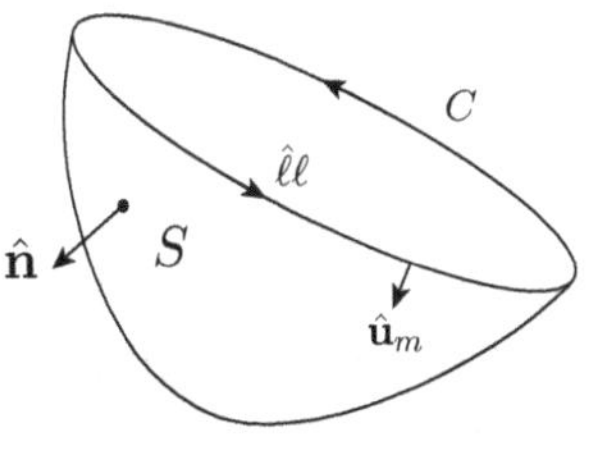


**Figure 18.1.** Open surface $S$ bounded by curve $C$. The normal to the surface $\hat{\mathbf{n}}$ and the tangential vectors $(\hat{\ell}, \hat{\mathbf{u}}_m)$ satisfy $\hat{\ell} = \hat{\mathbf{n}} \times \hat{\mathbf{u}}_m$.

where a quantity within the curly brackets $\{\cdot\}$ is defined everywhere on $S$ except $C$ (i.e. in regions where the distribution has a functional representation). The function representation is found by evaluating the expression inside the brackets as a function instead of as a distribution. The identity $\nabla_s \cdot (\varphi \mathbf{J}_s) = \varphi \nabla \cdot \mathbf{J}_s + \mathbf{J}_s \cdot \nabla \varphi$ was used in the second step of the above derivation and the surface Gauss divergence theorem (see equation (B.18)) in the third step. Since equation (18.51) is true for any test function $\varphi$ we can write

$$\nabla_s \cdot \mathbf{J}_s = \{\nabla_s \cdot \mathbf{J}_s\} - \hat{\mathbf{u}}_m \cdot \mathbf{J}_s \, \delta_C. \tag{18.52}$$

which expresses the surface divergence of the discontinuous surface distribution $\mathbf{J}_s$ as the sum of its divergence everywhere except on $C$ and a singular contribution that depends on the normal component of $\mathbf{J}_s$ on $C$. If the surface current density $\mathbf{J}_s$ is to be associated with a volume current density $\mathbf{J}$, then the correct way to express its divergence is $\nabla \cdot \mathbf{J} = \{\nabla_s \cdot \mathbf{J}_s\} \, \delta_S - \hat{\mathbf{u}}_m \cdot \mathbf{J}_s \, \delta_C$. The presence of $\delta_S$ is needed to account for the fact that the function $\{\mathbf{J}_s\}$ is concentrated in $S$.

**Definition 18.1.19.** *Divergence of a distribution.* The divergence, $\nabla \cdot \mathbf{A}$, of a vector distribution $\mathbf{A}$ is defined as

$$\langle \nabla \cdot \mathbf{A}, \varphi \rangle = - \iiint_V \mathbf{A} \cdot \nabla \varphi \, \mathrm{d}v. \tag{18.53}$$

**Definition 18.1.20.** *Curl of a distribution.* The curl, $\nabla \times \mathbf{A}$, of a vector distribution $\mathbf{A}$ is defined in either of two equivalent ways

$$\langle \nabla \times \mathbf{A}, \varphi \rangle = \iiint_V \mathbf{A} \times \nabla \varphi \, \mathrm{d}v, \tag{18.54}$$

$$\langle \nabla \times \mathbf{A}, \boldsymbol{\phi} \rangle = \iiint_V \mathbf{A} \cdot \nabla \times \boldsymbol{\phi} \, \mathrm{d}v. \tag{18.55}$$

**Definition 18.1.21** *Curl–curl of a distribution.* The curl–curl, $\nabla \times \nabla \times \mathbf{A}$, of a vector distribution $\mathbf{A}$ is defined in either of two equivalent ways

$$\langle \nabla \times \nabla \times \mathbf{A}, \varphi \rangle = - \iiint_V \mathbf{A} \nabla^2 \varphi \, \mathrm{d}v + \iiint_V \mathbf{A} \cdot \nabla \nabla \varphi \, \mathrm{d}v, \tag{18.56}$$

$$\langle \nabla \times \nabla \times \mathbf{A}, \boldsymbol{\phi} \rangle = \iiint_V \mathbf{A} \cdot \nabla \times \nabla \times \boldsymbol{\phi} \, \mathrm{d}v. \tag{18.57}$$

**Definition 18.1.22** *Grad–div of a distribution.* The grad–div, $\nabla \nabla \cdot \mathbf{A}$, of a vector distribution $\mathbf{A}$ is defined in either of the two equivalent ways

$$\langle \nabla\nabla \cdot \mathbf{A}, \varphi \rangle = \iiint_V \mathbf{A} \cdot \nabla\nabla\varphi \, \mathrm{d}v, \tag{18.58}$$

$$\langle \nabla\nabla \cdot \mathbf{A}, \boldsymbol{\phi} \rangle = \iiint_V \mathbf{A} \cdot \nabla\nabla \cdot \boldsymbol{\phi} \, \mathrm{d}v. \tag{18.59}$$

In the case of Maxwell's equations, each field or source variable will be treated as a distribution. The fields are continuous with continuous gradients except on a regular surface $S$ (i.e. a Lyapunov surface, see definition B.1.1). In the following we assume that $S$ intersects the volume $V = V_1 \cup V_2$ as shown in figure 18.2 and divides it into two subvolumes $V_1$ and $V_2$. Let $\hat{\mathbf{n}}$ be the unit normal normal oriented arbitrarily on $S$, but inwardly into the volume $V$ on $\Sigma$. A quantity inside the curly brackets $\{\cdot\}$ represents the distribution anywhere in $V$ except on $S$ (i.e. in regions where the distribution can have function representation). Using the definition of divergence of $\nabla \cdot \mathbf{A}$ in equation (18.53) we obtain

$$\begin{aligned}
\langle \nabla \cdot A, \varphi \rangle &= -\iiint_V \{A\} \cdot \nabla\varphi \, \mathrm{d}v \\
&= \iiint_V \{\varphi\nabla \cdot A - \nabla \cdot (\varphi A)\} \, \mathrm{d}v \\
&= \iiint_V \varphi\{\nabla \cdot A\} \, \mathrm{d}v - \oiint_\Sigma \varphi\hat{n} \cdot A \, \mathrm{d}s + \iint_S \varphi(\hat{n} \cdot A_2 - \hat{n} \cdot A_1) \, \mathrm{d}s \\
&= \iiint_V \varphi\{\nabla \cdot A\} \, \mathrm{d}v + \iint_S \varphi\hat{n} \cdot \Delta[A] \, \mathrm{d}s \\
&= \iiint_V \varphi\{\nabla \cdot A\} \, \mathrm{d}v + \iiint_V \varphi\hat{n} \cdot \Delta[A]\delta_S \, \mathrm{d}v,
\end{aligned}$$

where $\Delta[\cdot]$ denotes the increment of the argument function in crossing the surface $S$ in the direction of $\hat{\mathbf{n}}$ (for instance, $\Delta[\mathbf{A}] = \mathbf{A}_2 - \mathbf{A}_1$). The identity $\nabla \cdot (\varphi\mathbf{A}) = \varphi\nabla \cdot \mathbf{A} + \mathbf{A} \cdot \nabla\varphi$ was used in the second step of the above derivation and the Gauss divergence theorem (see equation (B.4)) applied separately to each volume $V_k$ in the third step. We thus have

$$\langle \nabla \cdot \mathbf{A}, \varphi \rangle = \langle \{\nabla \cdot \mathbf{A}\} + \hat{\mathbf{n}} \cdot \Delta[\mathbf{A}]\delta_S, \varphi \rangle.$$

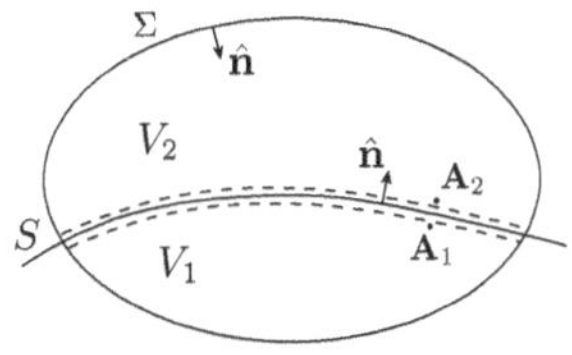


**Figure 18.2.** Surface, $S$, of discontinuity intersecting the support $V = V_1 \cup V_2$ of $\varphi$. The field takes a value $\mathbf{A}_2$ just above the surface $S$ in the direction of $\hat{\mathbf{n}}$ and $\mathbf{A}_1$ just below. The incremental changes of the field across $S$ are $\Delta[\hat{\mathbf{n}} \cdot \mathbf{A}] = \hat{\mathbf{n}} \cdot \mathbf{A}_2 - \hat{\mathbf{n}} \cdot \mathbf{A}_1$ and $\Delta[\hat{\mathbf{n}} \times \mathbf{A}] = \hat{\mathbf{n}} \times \mathbf{A}_2 - \hat{\mathbf{n}} \times \mathbf{A}_1$.

Since this is true for all test functions we have the relation

$$\nabla \cdot \mathbf{A} = \{\nabla \cdot \mathbf{A}\} + \hat{\mathbf{n}} \cdot \Delta[\mathbf{A}]\delta_S, \tag{18.60}$$

which expresses the divergence of the vector distribution $\mathbf{A}$ with discontinuity on $S$ as the sum of its divergence everywhere except on $S$ and the incremental change, $\hat{\mathbf{n}} \cdot \Delta[\mathbf{A}]$, in the normal component of $\mathbf{A}$ across the surface $S$. As an example, for a constant radially directed current $I_0$ flowing within a sphere $V_1$ of radius $r_1$, the current density distribution is $\mathbf{J}(\mathbf{r}) = \hat{\mathbf{r}}\frac{I_0}{4\pi r^2}\Theta_1(\mathbf{r})$, where $\Theta_1$ is the Heaviside jump distribution in $V_1$. The divergence of the current density in the sense of distribution is

$$\nabla \cdot \mathbf{J} = \hat{\mathbf{r}} \cdot (\mathbf{J}(r = r_1 +) - \mathbf{J}(r = r_1 -))\delta(r - r_1) = -\frac{I_0}{4\pi r_1^2}\delta(r - r_1), \tag{18.61}$$

where $\delta_{S_1} = \delta(r - r_1)$. From the equation of continuity (1.39) we know this to be equal to $-j\omega q_{\mathrm{ev}}$ for time-harmonic excitation, with $q_{\mathrm{ev}}$ being the volume charge density. We then arrive at the result that a constant, radially flowing current inside a sphere implies negative volume charge density distributed on the outer spherical shell.

Similarly applying the definition of curl in equation (18.54) we obtain

$$\begin{aligned}
\langle \nabla \times A, \varphi \rangle &= \iiint_V \{A\} \times \nabla\varphi \, \mathrm{d}v \\
&= \iiint_V \{\varphi \nabla \times A - \nabla \times (\varphi A)\} \, \mathrm{d}v \\
&= \iiint_V \varphi\{\nabla \times A\} \, \mathrm{d}v - \oint\!\!\!\oint_\Sigma \varphi \hat{n} \times A \, \mathrm{d}s + \iint_S \varphi(\hat{n} \times A_2 - \hat{n} \times A_1) \, \mathrm{d}s \\
&= \iiint_V \varphi\{\nabla \times A\} \, \mathrm{d}v + \iint_S \varphi \hat{n} \times \Delta[A] \, \mathrm{d}s \\
&= \iiint_V \varphi\{\nabla \times A\} \, \mathrm{d}v + \iiint_V \varphi \hat{n} \times \Delta[A]\delta_S \, \mathrm{d}v.
\end{aligned}$$

The identity $\nabla \times (\varphi\mathbf{A}) = \varphi\nabla \times \mathbf{A} - \mathbf{A} \times \nabla\varphi$ was used in the second step of the above derivation and the curl theorem (see equation (B.4)) applied separately to each volume $V_k$ in the third step. We thus have

$$\langle \nabla \times \mathbf{A}, \varphi \rangle = \langle \{\nabla \times \mathbf{A}\} + \hat{\mathbf{n}} \times \Delta[\mathbf{A}]\delta_S, \varphi \rangle,$$

which implies that

$$\nabla \times \mathbf{A} = \{\nabla \times \mathbf{A}\} + \hat{\mathbf{n}} \times \Delta[\mathbf{A}]\delta_S. \tag{18.62}$$

Thus the curl of the vector distribution $\mathbf{A}$ with a discontinuity on $S$ is the sum of its curl everywhere except on $S$ and the incremental change, $\hat{\mathbf{n}} \times \Delta[\mathbf{A}]$, in the tangential component of $\mathbf{A}$ across the surface $S$. By applying equation (18.62) successively we obtain the second identity related to curl:

$$\nabla \times \nabla \times \mathbf{A} = \{\nabla \times \nabla \times \mathbf{A}\} + \hat{\mathbf{n}} \times \Delta[\nabla \times \mathbf{A}]\, \delta_S + \nabla \times (\hat{\mathbf{n}} \times \Delta[\mathbf{A}]\delta_S). \quad (18.63)$$

To complete the rhs of equation (18.63) we would first need an expression for the curl of the tangential vector distribution contained in the last term. It has been worked out in [4, p 52], [12, p 501] and the result is

$$\begin{aligned}\nabla \times (\hat{\mathbf{n}} \times \Delta[\mathbf{A}]\delta_S) = &-(\boldsymbol{\pi} \cdot \Delta[\mathbf{A}] + \hat{\mathbf{n}}\nabla_s \cdot (\hat{\mathbf{n}} \times (\hat{\mathbf{n}} \times \Delta[\mathbf{A}])))\delta_S \\ &+ \hat{\mathbf{n}} \times (\hat{\mathbf{n}} \times \Delta[\mathbf{A}])\partial_n\delta_S,\end{aligned} \quad (18.64)$$

where

$$\boldsymbol{\pi} = \frac{1}{R_1}\hat{\mathbf{u}}\hat{\mathbf{u}} + \frac{1}{R_2}\hat{\mathbf{v}}\hat{\mathbf{v}}, \quad (18.65)$$

is the *curvature dyadic* on the surface $S$ with respect to the orthogonal tangential coordinates $(\hat{\mathbf{u}}u, \hat{\mathbf{v}}v)$ and $R_1$ and $R_2$ are the principal radii of curvature along $\hat{\mathbf{u}}$ and $\hat{\mathbf{v}}$ directions, respectively. If the surface $S$ is planar, then $\boldsymbol{\pi} = 0$. Using equation (18.64) in equation (18.63) we arrive at

$$\begin{aligned}\nabla \times \nabla \times \mathbf{A} = &\{\nabla \times \nabla \times \mathbf{A}\} + \hat{\mathbf{n}} \times \Delta[\nabla \times \mathbf{A}]\delta_S + \boldsymbol{\pi} \cdot \hat{\mathbf{n}} \times (\hat{\mathbf{n}} \times \Delta[\mathbf{A}])\delta_S \\ &- \hat{\mathbf{n}}\, \nabla_s \cdot (\hat{\mathbf{n}} \times (\hat{\mathbf{n}} \times \Delta[\mathbf{A}]))\delta_S + \hat{\mathbf{n}} \times (\hat{\mathbf{n}} \times \Delta[\mathbf{A}])\partial_n\delta_S.\end{aligned} \quad (18.66)$$

The coefficient of $\delta_S$ in equation (18.66) corresponds to the single-layer potential, while the coefficient of $-\partial_n\delta_S$ corresponds to the double-layer potential.

As an example on the distributional aspects of curl, consider an azimuthally directed current flowing within a sphere $V_1$ of radius $r_1$ such that the total current carried in any meridian plane is equal to $I_0$. Using the current density expression $\mathbf{J} = \hat{\boldsymbol{\phi}}\frac{I_0}{\pi r_1 r}\Theta_1(\mathbf{r})$, where $\Theta_1$ is the Heaviside jump distribution in $V_1$, we obtain for the curl and curl–curl of current density in the sense of distributions to be

$$\nabla \times \mathbf{J} = \hat{\mathbf{r}}\frac{I_0 \cot\theta}{\pi r^2 r_1}\Theta_1(\mathbf{r}) + \hat{\boldsymbol{\theta}}\frac{I_0}{\pi r_1^2}\delta(r - r_1), \quad (18.67)$$

$$\nabla \times \nabla \times \mathbf{J} = \hat{\boldsymbol{\phi}}\frac{I_0}{\pi r_1 r \rho^2}\Theta_1(\mathbf{r}) + \hat{\boldsymbol{\phi}}\frac{I_0}{\pi r_1^2}\left(\delta'(r - r_1) + \frac{1}{r_1}\delta(r - r_1)\right), \quad (18.68)$$

where $\rho = r\sin\theta$ is the radial distance in cylindrical coordinates. Since $\nabla \times \mathbf{J}$ is often associated with an equivalent magnetic current for the non-static case, one can say that the azimuthal electric current density of the type considered in this example is equivalent to a bulk radial magnetic current density proportional to the first term on the rhs of equation (18.67) plus a tangential surface magnetic current proportional to the second term.

Finally, using the definition of gradient in equation (18.41), we see that

$$
\begin{aligned}
\langle \nabla\psi, \varphi\rangle &= -\iiint_V \{\psi\}\nabla\varphi \,\mathrm{d}v \\
&= \iiint_V \{\varphi\nabla\psi - \nabla(\psi\varphi)\}\,\mathrm{d}v \\
&= \iiint_V \varphi\{\nabla\psi\}\,\mathrm{d}v + \iint_S \hat{\mathbf{n}}(\psi_2 - \psi_1)\varphi\,\mathrm{d}s \\
&= \iiint_V (\{\nabla\psi\} + \hat{\mathbf{n}}\Delta[\psi]\delta_S)\varphi\,\mathrm{d}v = \langle\{\nabla\psi\} + \hat{\mathbf{n}}\Delta[\psi]\delta_S\rangle,
\end{aligned}
$$

which implies that

$$
\nabla\psi = \{\nabla\psi\} + \hat{\mathbf{n}}\Delta[\psi]\delta_S. \tag{18.69}
$$

Gradient theorem (see equation (B.7)) was used separately in each region $V_k$ in the third step to reduce the volume integral to a surface integral. Thus the presence of discontinuities in the distribution produce an additional normal component proportional to the jump $[\psi]$ in the expression for the gradient. As an example for a Heaviside jump distribution $\Theta_k(\mathbf{r})$ with a support $V_k$ and boundary $S_k$ we obtain

$$
\nabla\Theta_k = \hat{\mathbf{n}}\,\delta_{S_k}, \tag{18.70}
$$

in the sense of distributions.

We can take the divergence of equation (18.69) and apply equation (18.60) to result in

$$
\begin{aligned}
\nabla^2\psi &= \{\nabla^2\psi\} + \hat{\mathbf{n}}\cdot\nabla(\Delta[\psi])\delta_S + \nabla\cdot(\hat{\mathbf{n}}\Delta[\psi]\delta_S) \\
&= \{\nabla^2\psi\} + \Delta\left[\frac{\partial\psi}{\partial n}\right]\delta_S + \Delta[\psi]\partial_n\delta_S.
\end{aligned} \tag{18.71}
$$

We can take the gradient of equation (18.60) and apply equation (18.41) to result in

$$
\begin{aligned}
\nabla\nabla\cdot\mathbf{A} &= \{\nabla\nabla\cdot\mathbf{A}\} + \nabla(\Delta[\hat{\mathbf{n}}\cdot\mathbf{A}]\delta_S) + \hat{\mathbf{n}}\Delta[\nabla\cdot\mathbf{A}]\delta_S \\
&= \{\nabla\nabla\cdot\mathbf{A}\} + \Delta[\nabla(\hat{\mathbf{n}}\cdot\mathbf{A}) + \hat{\mathbf{n}}\nabla\cdot\mathbf{A}]\delta_S + \Delta[\hat{\mathbf{n}}\cdot\mathbf{A}]\nabla\delta_S.
\end{aligned} \tag{18.72}
$$

Other identities for vector distributions may be established in a similar manner. For example, the operator $\mathcal{L}_{dc} = \nabla\cdot\nabla\times$ acting on a vector $\mathbf{B} = \sum_i \hat{\mathbf{a}}_i B_i$ is equal to zero. This is also true for distributions when one recognizes that $\mathcal{L}_{dc}$ may be represented in terms of the permutation symbols and second order derivatives as (see equations (17.118) and (17.107))

$$
\mathcal{L}_{dc}\mathbf{B} := \nabla\cdot\nabla\times\mathbf{B} = \sum_{ijk}\epsilon_{ijk}\frac{\partial^2 B_i}{\partial x_j\partial x_k}. \tag{18.73}
$$

The operator $\mathcal{L}_{dc}$ is formally self-adjoint and is equal to the null operator since $\epsilon_{ikj} = -\epsilon_{ijk}$.

### 18.1.2 Theorem on scalar, vector functions with surface discontinuities

The results derived in equations (18.60), (18.62), (18.69), (18.71), (18.72), and (18.66) can be summarized in the form of the following theorem.

**Theorem 18.1.4.** Scalar and vector functions with surface discontinuities.
*Let* $\psi$ *be a bounded scalar function and* $\mathbf{A}$ *be a bounded vector function, both with continuous first derivatives in a volume* $V$ *except on a regular surface* $S$ *that divides* $V$ *into two subvolumes* $V_1$ *and* $V_2$ *(see figure 18.2). Let the corresponding functions in the volume* $V_i$ *be* $\psi_i$, $\mathbf{A}_i$, $i = 1, 2$. *Then the distributions* $\psi(\mathbf{r}) = \psi_1(\mathbf{r})\Theta_1(\mathbf{r}) + \psi_2(\mathbf{r})\Theta_2(\mathbf{r})$, $\mathbf{A}(\mathbf{r}) = \mathbf{A}_1(\mathbf{r})\Theta_1(\mathbf{r}) + \mathbf{A}_2(\mathbf{r})\Theta_2(\mathbf{r})$ *have the properties*

$$\nabla\psi = \{\nabla\psi\} + \hat{\mathbf{n}}\Delta[\psi]\delta_S, \tag{18.74}$$

$$\nabla^2\psi = \{\nabla^2\psi\} + \Delta\left[\frac{\partial\psi}{\partial n}\right]\delta_S + \Delta[\psi]\partial_n\delta_S, \tag{18.75}$$

$$\nabla\cdot\mathbf{A} = \{\nabla\cdot\mathbf{A}\} + \Delta[\hat{\mathbf{n}}\cdot\mathbf{A}]\delta_S, \tag{18.76}$$

$$\nabla\times\mathbf{A} = \{\nabla\times\mathbf{A}\} + \Delta[\hat{\mathbf{n}}\times\mathbf{A}]\delta_S, \tag{18.77}$$

$$\begin{aligned}\nabla\nabla\cdot\mathbf{A} &= \{\nabla\nabla\cdot\mathbf{A}\} + \Delta[\nabla(\hat{\mathbf{n}}\cdot\mathbf{A}) + \hat{\mathbf{n}}\nabla\cdot\mathbf{A}]\delta_S + \Delta[\hat{\mathbf{n}}\cdot\mathbf{A}]\nabla\delta_S, \\ \nabla\times\nabla\times\mathbf{A} &= \{\nabla\times\nabla\times\mathbf{A}\} + \Delta[\hat{\mathbf{n}}\times\nabla\times\mathbf{A}]\delta_S \qquad (18.78)\\ &\quad + \boldsymbol{\pi}\cdot\hat{\mathbf{n}}\times(\Delta[\hat{\mathbf{n}}\times\mathbf{A}])\delta_S \\ &\quad - \hat{\mathbf{n}}\,\nabla_s\cdot(\hat{\mathbf{n}}\times(\Delta[\hat{\mathbf{n}}\times\mathbf{A}]))\delta_S + \hat{\mathbf{n}}\times(\Delta[\hat{\mathbf{n}}\times\mathbf{A}])\partial_n\delta_S. \qquad (18.79)\end{aligned}$$

*where* $\Delta[g]$ *denotes the jump in the function g in crossing the surface S in the direction of unit normal* $\hat{\mathbf{n}}$ *(i.e.* $\Delta[g] = g_2 - g_1$*),* $\delta_S$ *is the surface Dirac distribution,* $\partial_n\delta_S$ *is the normal derivative operator on the surface Dirac distribution and* $\pi$ *is the curvature dyadic given in equation (18.65).*

The following observations can be made from the results of this theorem:

1. The gradient operation on a scalar distribution extracts its jump discontinuity across the boundary.
2. The Laplacian operation on a scalar distribution extracts both its jump as well as the jump in its normal derivative across the boundary.
3. The divergence operation on a vector distribution extracts the jump discontinuity in its normal component across the boundary.
4. The curl operation on a vector distribution extracts the jump discontinuity in its tangential component across the boundary.

It is important to keep in mind that $\delta_S$ is not the same as $\delta(w)$, where $w$ is a coordinate normal (not necessarily a spatial coordinate) to $S$. However, the two are related as we show below. Let the surface $S$, dividing the volume $V$ into two parts $V_1$ and $V_2$ be described by the equation $w(\mathbf{r}) = 0$ and let $(u, v, w)$ be a local orthogonal coordinate system erected at a point on it. We assume that the mapping from $(x, y, z)$ to $(u, v, w)$ is one-to-one so that the Jacobian of the transformation does not vanish. Let the unit normal be directed from region 1 to region 2 as shown in figure 18.3(a). Consider a scalar distribution $\psi$ which has a jump discontinuity on $S$ and has differentiable parts $\psi_i$ in volumes $V_i$, $i = 1, 2$. Then the distribution $\psi = \psi_1(1 - \Theta_2(w)) + \psi_2\Theta_2(w)$, where $\Theta_2(w)$ is a one-dimensional Heaviside jump distribution in the variable $w$ taking a value 1 for $w > 0$ and 0 for $w < 0$, has a gradient according to equation (18.74) as

$$\nabla\psi = \{\nabla\psi\} + \hat{\mathbf{n}}\,\Delta[\psi]\delta_S. \tag{18.80}$$

A second way to compute the gradient is to use the formula for the gradient of a product of two scalar functions to result in

$$\begin{aligned}\nabla\psi &= (1 - \Theta_2(w))\nabla\psi_1 + \Theta_2(w)\nabla\psi_2 + \Delta[\psi]\nabla\Theta_2(w)\\ &= \{\nabla\psi\} + \Delta[\psi]\frac{\mathrm{d}\Theta_2}{\mathrm{d}w}\nabla w\\ &= \{\nabla\psi\} + \hat{\mathbf{n}}\,\Delta[\psi]|\,\nabla w\,|\,\delta(w).\end{aligned} \tag{18.81}$$

Comparing equations (18.80) and (18.81) we see that

$$\delta_S = |\nabla w|\,\delta(w) = |\nabla w|\,\delta(\xi\,|\nabla w|) = \delta(\xi), \tag{18.82}$$

where $\xi = w(x, y, z)/|\nabla w|$ is the *spatial* coordinate along the normal to $S$ and the last equality in equation (18.82) follows from the property (18.36) of the 1D delta distribution.

**Example 18.1.2.1** Specific surface Dirac distributions.
On a sphere of radius $r_1$ we have $w = r - r_1 = \sqrt{x^2 + y^2 + z^2} - r_1 = 0$. Then $|\nabla w| = 1$, $\xi = r - r_1$ and $\delta_S = \delta(w) = \delta(\xi)$.

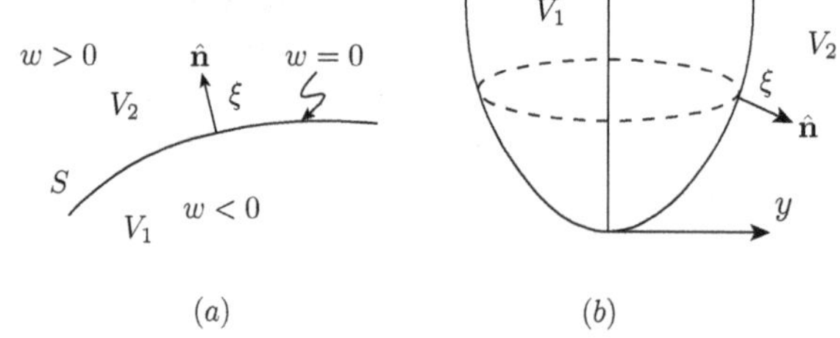


**Figure 18.3.** Spatial coordinate $\xi$ along the normal to (a) surface $S$ on which $w = 0$ separating $V_1$ from $V_2$ and (b) paraboloid $2zz_0 = x^2 + y^2$ with apex at the origin.

On a paraboloid shown in figure 18.3(b), the surface $S$ is described by $w = x^2 + y^2 - 2zz_0 = \rho^2 - 2zz_0 = 0$, $z_0 \neq 0$, where $\rho = \sqrt{x^2 + y^2}$ is the usual distance from the $z$-axis in cylindrical coordinates. The gradient on the surface is $\nabla w = 2(\hat{\boldsymbol{\rho}}\rho - \hat{\mathbf{z}}z_0)$, $|\nabla w| = 2R := 2\sqrt{\rho^2 + z_0^2}$ and the unit normal on the surface is $\hat{\mathbf{n}} = (\hat{\boldsymbol{\rho}}\rho - \hat{\mathbf{z}}z_0)/R$. The distance coordinate along the unit normal is $\xi = w/|\nabla w| = (\rho^2 - 2zz_0)/2R$ and $\delta_S = \delta(\xi) = 2R\delta(w) \neq \delta(w)$. ■■

We have already seen in equation (18.49) that in a distributional sense

$$\partial_n \delta_S = \frac{\partial \delta_S}{\partial n} - 2J_c\delta_S, \tag{18.83}$$

where $J_c$ is the mean curvature of $S$ at the point under consideration. The mean curvature vanishes for a flat surface. When associated with a double layer of density $\tau$, equation (18.83) suggests that

$$-\tau\partial_n\delta_S = -\tau\frac{\partial \delta_S}{\partial n} + 2J_c\tau\delta_S. \tag{18.84}$$

In other words, the volume charge density of a double layer (charge neutrality is maintained) given by the lhs is equal to (i) the density produced by depositing equal and opposite density $\rho_s$ on both sides of the double layer of height $h$ such that $\tau = \lim_{h\to 0} h\rho_s$ (the first term on the rhs; charge neutrality across the layer is not maintained by this term alone because the elemental surface area changes in going from the bottom surface to the top surface of the double layer due to non-zero curvature) plus (ii) the additional charge density $2J_c\tau\delta_S$ placed in the middle of the double layer required to maintain charge neutrality of the equi-density layer (the second term on the rhs). Another useful result that is evident from equation (18.82) on denoting $\nabla w = |\nabla w|\hat{\mathbf{n}} = \zeta^{-1}\hat{\mathbf{n}}$, $\zeta^{-1} = |\nabla w| = \partial w/\partial n$ is that

$$\frac{\partial \delta_S}{\partial n} = \hat{\mathbf{n}} \cdot \nabla\delta_S = \frac{1}{\zeta^2}\left(\delta'(w) - \frac{\partial \zeta}{\partial n}\delta(w)\right), \tag{18.85}$$

where a prime on the Dirac distribution $\delta(w)$ signifies d/d$w$. Using $\nabla \cdot \hat{\mathbf{n}} = -2J_c$ and $\zeta^{-2}\partial\zeta/\partial n = -\nabla^2 w - 2\zeta^{-1}J_c$, we also cast (18.85) as

$$\frac{\partial \delta_S}{\partial n} = \zeta^{-2}\delta'(w) + (\nabla^2 w + 2\zeta^{-1}J_c)\delta(w). \tag{18.86}$$

A double layer of density $\tau$ will be represented in the $w$ coordinate system as

$$\begin{aligned} -\tau\frac{\partial \delta_S}{\partial n} + 2J_c\tau\delta_S &= -\frac{\tau}{\zeta^2}\delta'(w) - \tau(\nabla^2 w + 2\zeta^{-1}J_c)\delta(w) + 2J_c\tau\delta_S \\ &= -\frac{\tau}{\zeta^2}\delta'(w) - \tau\nabla^2 w\delta(w). \end{aligned} \tag{18.87}$$

Once again $\tau$ appears as a coefficient of not only $\delta'(w)$ but also of $\delta(w)$. If both single-layer and double-layer potentials are present with densities $\mu$ and $\tau$, respectively, then the distribution is written as

$$\mu\delta_S - \tau\partial_n\delta_S = \left(\frac{\mu}{\zeta} - \tau\nabla^2 w\right)\delta(w) - \frac{\tau}{\zeta^2}\delta'(w). \tag{18.88}$$

In particular, the coefficient of $\delta(w)$ cannot be identified with the single-layer potential as it contains contribution from both the single-layer and the double-layer potentials.

Theorem 18.1.2 allows any electromagnetic field distribution $\mathbf{F}$ of order $k$ with field singularities on a surface $S$ defined by $w(x, y, z) = 0$ to be expressed as

$$\mathbf{F}(u, v, w) = \{\mathbf{F}(u, v, w)\} + \sum_{m=0}^{k} \mathbf{F}_m(u, v)\delta^{(m)}(w), \tag{18.89}$$

where $\mathbf{F}_m(u, v)$ are functions defined on the surface $S$ with coordinates $(u, v)$ of the triplet coordinate system $(u, v, w)$.

**Example 18.1.2.2** Surface theory of an elliptic paraboloid.
We shall now look at the details of various quantities associated with equation (18.83) by means of an example surface. The end goal is to elucidate the various steps involved in the determination of the surface and normal coordinates that lead to the expressions for the mean curvature for a general surface.

Consider the elliptic paraboloid $2z_0z = \rho^2 =: v^2$ shown in figure 18.4, where $(\rho, \phi, z)$ are the usual cylindrical coordinates. Let the surface coordinates be $(u, v)$. The rectangular coordinates $(x, y, z)$ of a point on the surface in terms of the two surface coordinates $(u, v)$ are

$$\mathbf{X} = (x, y, z) = \left(v\cos u,\ v\sin u,\ \frac{v^2}{2z_0}\right). \tag{18.90}$$

It is evident that $v = \rho$ and $u = \phi$, so that $u$ is the coordinate along the parallels ($z$ = constant) and $v$ is the coordinate along the meridians ($\phi$ = constant). The

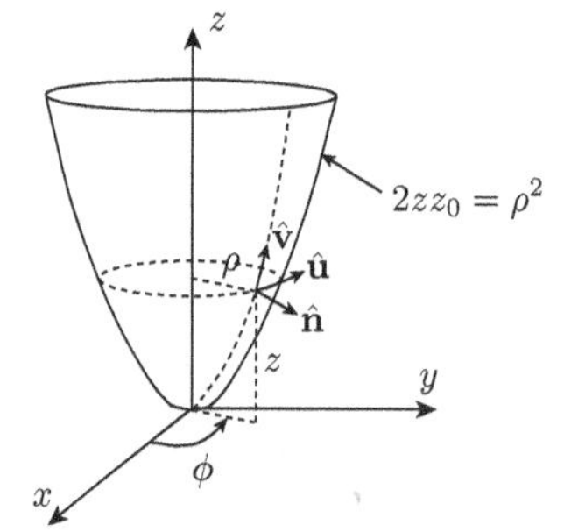


**Figure 18.4.** Paraboloid $2zz_0 = \rho^2$ with apex at the origin.

parameter $z_0$ controls the waist width of the paraboloid at any height (waist $\propto\sqrt{z_0}$). Let $\mathbf{X}_p$ and $\mathbf{X}_{pq}$, respectively, denote the first partial derivative of $\mathbf{X}$ w.r.t. the variable $p$ and the second partial derivative of $\mathbf{X}$ w.r.t. the variables $p$, $q$. Then

$$\mathbf{X}_u = (-v \sin u, v \cos u, 0) = v\hat{\boldsymbol{\phi}}, \tag{18.91}$$

$$\mathbf{X}_v = (\cos u, \sin u, \frac{v}{z_0}) = \left(\hat{\rho} + \frac{v}{z_0}\hat{\mathbf{z}}\right) \tag{18.92}$$

$$\mathbf{X}_{uu} = (-v \cos u, -v \sin u, 0) = -v\hat{\rho}, \tag{18.93}$$

$$\mathbf{X}_{vv} = (0, 0, \frac{1}{z_0}) = \frac{1}{z_0}\hat{\mathbf{z}}, \tag{18.94}$$

$$\mathbf{X}_{uv} = X_{vu} = (-\sin u, \cos u, 0) = \hat{\boldsymbol{\phi}}. \tag{18.95}$$

Since the coordinates $u$ and $v$ are perpendicular in this example we observe that $\mathbf{X}_u \cdot \mathbf{X}_v = 0$. The unit vectors in the tangential plane, $(\mathbf{u}, \mathbf{v})$, and the unit normal, $\hat{\mathbf{n}}$, to the surface are [13, p 62]

$$\hat{\mathbf{u}} = \frac{\mathbf{X}_u}{|\mathbf{X}_u|} = \hat{\boldsymbol{\phi}} \tag{18.96}$$

$$\hat{\mathbf{v}} = \frac{\mathbf{X}_v}{|\mathbf{X}_v|} = \frac{1}{\sqrt{v^2 + z_0^2}}(\hat{\rho} z_0 + \hat{\mathbf{z}} v) \tag{18.97}$$

$$\hat{\mathbf{n}} = \frac{\mathbf{X}_u \times \mathbf{X}_v}{|\mathbf{X}_u \times \mathbf{X}_v|} = \frac{1}{\sqrt{v^2 + z_0^2}}(\hat{\rho} v - \hat{\mathbf{z}} z_0). \tag{18.98}$$

Note that the expression for $\hat{\mathbf{n}}$ derived here coincides with the result derived in example 18.1.2.1 by a different means. It is easy to verify that $(\mathbf{u}, \mathbf{v}, \mathbf{n})$ form an orthogonal right-handed system. Furthermore

$$\hat{\rho} = \frac{z_0\hat{\mathbf{v}} + v\hat{\mathbf{n}}}{\sqrt{v^2 + z_0^2}}; \quad \hat{\mathbf{z}} = \frac{v\hat{\mathbf{v}} - z_0\hat{\mathbf{n}}}{\sqrt{v^2 + z_0^2}}, \tag{18.99}$$

so that the various second partial derivatives in equations (18.93)–(18.95) contain linear combinations of both surface tangential vectors and the normal vector. The first and second fundamental forms of the surface for the general case are [13, p 59, 75]

$$\text{I: } \mathrm{d}s^2 = \mathrm{d}\mathbf{X} \cdot \mathrm{d}\mathbf{X} = E\mathrm{d}u^2 + 2F\mathrm{d}u\mathrm{d}v + G\mathrm{d}v^2 \tag{18.100}$$

$$\text{II: } (\mathbf{X}_{uu}\mathrm{d}u^2 + 2\mathbf{X}_{uv}\mathrm{d}u\mathrm{d}v + \mathbf{X}_{vv}\mathrm{d}v^2) \cdot \hat{\mathbf{n}} = (L\mathrm{d}u^2 + 2M\mathrm{d}u\mathrm{d}v + N\mathrm{d}v^2), \tag{18.101}$$

where $s$ is the arc-length parameter. The quantities $E$, $F$, and $G$ also constitute components of the metric tensor (see equation (17.41)) on the surface with $g_{11} = E$, $g_{12} = F$, $g_{22} = G$. Here the coefficients of the first and second forms are

$$E = \mathbf{X}_u \cdot \mathbf{X}_u = v^2; \quad F = \mathbf{X}_u \cdot \mathbf{X}_v = 0; \quad G = \mathbf{X}_v \cdot \mathbf{X}_v = \frac{1}{z_0^2}(v^2 + z_0^2), \tag{18.102}$$

$$\begin{aligned} L &= \mathbf{X}_{uu} \cdot \hat{\mathbf{n}} = -\frac{v^2}{\sqrt{v^2 + z_0^2}}; \quad M = \mathbf{X}_{uv} \cdot \hat{\mathbf{n}} = 0; \\ N &= \mathbf{X}_{vv} \cdot \hat{\mathbf{n}} = -\frac{1}{\sqrt{v^2 + z_0^2}}. \end{aligned} \tag{18.103}$$

The curvature, $\kappa(s)$, along any curve $\mathbf{X}(s) = \mathbf{X}(u(s), v(s))$ is defined by [14, p 55] $\kappa(s) = \frac{\mathrm{II}}{\mathrm{I}} = |\ddot{\mathbf{X}}|$, where an over dot signifies derivative w.r.t. $s$. The extremum values of $\kappa$ in two special directions (*lines of curvature*) are the *principal curvatures*. Because $F = 0 = M$, the lines of curvature in this example are simply the parametric lines[4] $\mathbf{X}_u$ and $\mathbf{X}_v$; a result that remains true whenever $F = 0 = M$ [13, p 81]. As a consequence $(\hat{\mathbf{u}}, \hat{\mathbf{v}})$ themselves are the directions of principal curvature. The principal curvatures are

$$\kappa_1 = \frac{L}{E} = -\frac{1}{\sqrt{v^2 + z_0^2}} < 0; \quad \kappa_2 = \frac{N}{G} = -\frac{z_0^2}{(v^2 + z_0^2)^{3/2}} < 0. \tag{18.104}$$

Finally the mean curvature is

$$J_c = -\frac{\nabla \cdot \hat{\mathbf{n}}}{2} = \frac{\kappa_1 + \kappa_2}{2} = -\frac{1}{\sqrt{v^2 + z_0^2}} \left( \frac{\frac{1}{2}v^2 + z_0^2}{v^2 + z_0^2} \right) < 0, \tag{18.105}$$

and the Gaussian curvature $\kappa_G = \kappa_1 \kappa_2$ is

$$\kappa_G = \frac{z_0^2}{(v^2 + z_0^2)^2} > 0. \tag{18.106}$$

At the apex of the paraboloid, the surface is rotationally symmetric with the normal vector $\hat{\mathbf{n}} = -\hat{\mathbf{z}}$ and the curvatures $\kappa_1 = \kappa_2 = J_c = -z_0^{-1}$.

For the special case of $F = 0 = M$, the curvature along any curve in the direction $\mathrm{d}u/\mathrm{d}v$ is

[4] A *parametric line* is a curve along which only one coordinate changes.

$$\kappa = \frac{\mathrm{II}}{\mathrm{I}} = L\left(\frac{du}{ds}\right)^2 + N\left(\frac{dv}{ds}\right)^2 = \left(L - \frac{NE}{G}\right)\left(\frac{du}{ds}\right)^2 + \frac{N}{G}$$
$$= (\kappa_1 - \kappa_2)E\left(\frac{du}{ds}\right)^2 + \kappa_2. \tag{18.107}$$

Using the values of $\kappa_1$, $\kappa_2$, and $E$ from equations (18.104) and (18.102), the curvature of any curve on the paraboloid is

$$\kappa = -\frac{1}{(z_0^2 + v^2)^{3/2}}\left[z_0^2 + v^4\left(\frac{du}{ds}\right)^2\right]. \tag{18.108}$$

For instance, on a parallel $du/ds = v^{-1}$ and the curvature is $\kappa = -1/(z_0^2 + v^2)^{1/2} = \kappa_1$. On the meridian $du/ds = 0$ and the curvature is $\kappa = -z_0^2/(z_0^2 + v^2)^{3/2} = \kappa_2$.

The paraboloid geometry also serves as a good example for the calculation of Christoffel symbols that are useful in the determination of the geodesic on the surface. Christoffel symbols of the first kind (see equation (17.60)) can be directly calculated from the derivatives of the parameter coordinates using the alternate definition [14, p 135] $[ik, \ell] = \mathbf{X}_{ik} \cdot \mathbf{X}_m$, $i, k, \ell, m = 1, 2$, where the index 1 is associated with $u$ and 2 with $v$. The only non-zero Christoffel symbols of the first kind are

$$-[11, 2] = [12, 1] = [21, 1] = v; \quad [22, 2] = \frac{v}{z_0^2}. \tag{18.109}$$

The non-zero Christoffel symbols of the second kind, $\left\{ {\ell \atop ik} \right\} = [ik, \ell]/g_{\ell\ell}$, are $\left\{ {2 \atop 11} \right\} = -z_0^2 v/(v^2 + z_0^2)$, $\left\{ {1 \atop 12} \right\} = 1/v$, and $\left\{ {2 \atop 22} \right\} = v/(v^2 + z_0^2)$. The Christoffel symbols of the second kind are related to the parameters of the second fundamental form via the linear system

$$\begin{bmatrix} \mathbf{X}_{uu} \\ \mathbf{X}_{uv} \\ \mathbf{X}_{vv} \end{bmatrix} = \begin{bmatrix} \left\{ {1 \atop 11} \right\} & \left\{ {2 \atop 11} \right\} & L \\ \left\{ {1 \atop 12} \right\} & \left\{ {2 \atop 11} \right\} & M \\ \left\{ {1 \atop 22} \right\} & \left\{ {2 \atop 22} \right\} & N \end{bmatrix} \begin{bmatrix} \mathbf{X}_u \\ \mathbf{X}_v \\ \hat{\mathbf{n}} \end{bmatrix}, \tag{18.110}$$

and to the Christoffel symbols of the first kind via equation (17.61). ■■

**Definition 18.1.23.** *Distribution* $r^{-1}$. Let $r$ be the distance from a point in space. Construct a spherical coordinate system with coordinates $(r, \theta, \phi)$ about that point. Then the distribution $1/r$ is defined by

[4] A *parametric line* is a curve along which only one coordinate changes.

$$\langle r^{-1}, \varphi \rangle = \lim_{\epsilon \to 0} \int_{r=\epsilon}^{\infty} \oiint \frac{\varphi(r)}{r} \, \mathrm{d}v. \tag{18.111}$$

For a regular density function $\tau(\mathbf{r})$, the distribution $\tau/r$ is defined by

$$\langle \tau r^{-1}, \varphi \rangle = \lim_{\epsilon \to 0} \int_{r=\epsilon}^{\infty} \oiint \frac{\tau(r)\varphi(r)}{r} \, \mathrm{d}v. \tag{18.112}$$

Similarly for a regular vector density function **F**

$$\langle F/r, \varphi \rangle = \lim_{\epsilon \to 0} \int_{r=\epsilon}^{\infty} \oiint \frac{F(r)\varphi(r)}{r} \, \mathrm{d}v. \tag{18.113}$$

The following identities related to the distribution $r^{-1}$ can be readily established from its definition and the techniques employed in section 10.1.1:

$$\nabla(\tau/r) = \{\nabla(\tau/r)\} \tag{18.114}$$

$$\nabla \times (\mathbf{F}/r) = \{\nabla \times (\mathbf{F}/r)\} \tag{18.115}$$

$$\nabla \cdot (\mathbf{F}/r) = \{\nabla \cdot (\mathbf{F}/r)\} \tag{18.116}$$

$$\nabla \times \nabla \times (\mathbf{F}/r) = \{\nabla \times \nabla \times (\mathbf{F}/r)\} + \frac{8\pi}{3}\mathbf{F}(0)\delta(\mathbf{r}) \tag{18.117}$$

$$\nabla^2(\tau/r) = \{\nabla^2(\tau/r)\} - 4\pi\tau(0)\delta(\mathbf{r}). \tag{18.118}$$

Several other identities based on the distribution $r^{-1}$ are proved and listed in [1]. Of special importance is the distribution $\delta(r)/r$ which behaves similar to $-\delta'(r)$ [11, p 839]. Hence in a distributional sense

$$\frac{\delta(r)}{r} = -\delta'(r). \tag{18.119}$$

## 18.2 Derivation of boundary conditions using distributions

As stated previously the advantage of distribution theory is that it includes the singular behavior of various fields. In particular the behavior of fields near material boundaries is already contained in Maxwell's equations if one treats them as being true in the sense of distributions. No additional assumptions need be made to derive boundary conditions. We illustrate this generality of the distributional approach by considering two examples.

### 18.2.1 Classical boundary conditions

The first example shows how to derive traditional boundary conditions across material interfaces using the theory of distributions.

**Example 18.2.1.1.** Classical boundary conditions for **E**, **D**, **H**, **B**.
In this example we show the derivation of the classical boundary conditions for the time-harmonic field variables **E, D, B, H** directly from the distributional representations.

With reference to figure 18.2 let the electric charge density, $q_e$, the electric current density, **J**, the magnetic charge density $q_m$, and the magnetic current density **M** in a volume $V$ be expressed as

$$q_{ev} = \{q_{ev}\} + q_{es}\,\delta_S, \quad \mathbf{J} = \{\mathbf{J}\} + \mathbf{J}_s\delta_S, \tag{18.120}$$

$$q_{mv} = \{q_{mv}\} + q_{ms}\,\delta_S, \quad \mathbf{M} = \{\mathbf{M}\} + \mathbf{M}_s\delta_S, \tag{18.121}$$

where $(q_{es}, q_{ms})$ and $(\mathbf{J}_s, \mathbf{M}_s)$ are the densities on the surface of discontinuity $S$. Then on using the results (18.76) and (18.77) of theorem 18.1.4 together with Maxwell's equations (1.36) and (1.37) we obtain

$$\{\nabla \times \mathbf{E}\} + \Delta[\hat{\mathbf{n}} \times \mathbf{E}]\,\delta_S = -j\omega\mu\mathbf{H} - \{\mathbf{M}\} - \mathbf{M}_s\delta_S, \tag{18.122}$$

$$\{\nabla \cdot \mathbf{D}\} + \Delta[\hat{\mathbf{n}} \cdot \mathbf{D}]\,\delta_S = \{q_{ev}\} + q_{es}\,\delta_S, \tag{18.123}$$

$$\{\nabla \times \mathbf{H}\} + \Delta[\hat{\mathbf{n}} \times \mathbf{H}]\,\delta_S = j\omega\epsilon\mathbf{E} + \{\mathbf{J}\} + \mathbf{J}_s\,\delta_S, \tag{18.124}$$

$$\{\nabla \cdot \mathbf{B}\} + \Delta[\hat{\mathbf{n}} \cdot \mathbf{B}]\,\delta_S = \{q_{mv}\} + q_{ms}\,\delta_S. \tag{18.125}$$

Equating the coefficients of $\delta_S$ on both sides we obtain the well known boundary conditions (1.63) and (1.64):

$$\Delta[\hat{\mathbf{n}} \times \mathbf{E}] = -\mathbf{M}_s, \tag{18.126}$$

$$\Delta[\hat{\mathbf{n}} \cdot \mathbf{D}] = q_{es}, \tag{18.127}$$

$$\Delta[\hat{\mathbf{n}} \times \mathbf{H}] = \mathbf{J}_s, \tag{18.128}$$

$$\Delta[\hat{\mathbf{n}} \cdot \mathbf{B}] = q_{ms}. \tag{18.129}$$

Doing a similar analysis on the equation of continuity $\nabla \cdot \mathbf{J} + j\omega q_{ev} = 0$ yields $\Delta[\hat{\mathbf{n}} \cdot \mathbf{J}] = -j\omega q_{es}$ which coincides with the latter of (B.24) since $\Delta[\hat{\mathbf{n}} \cdot \mathbf{J}] = \nabla_s \cdot \mathbf{J}_s$.

■■

### 18.2.2 Boundary conditions including GSTCs on interfaces with single-layer, double-layer densities

The next example shows how to derive boundary conditions in the presence of singular densities present on the interface. The boundary conditions so derived also encompass the sheet transition conditions that are relevant to metasurfaces [15].

**Example 18.2.2.1** Boundary conditions for more general interfaces.
To demonstrate the versatility of the distribution approach let us now determine the boundary conditions to be satisfied by the various time-harmonic fields when there exist both a single and a double-layer charge and current distributions on a surface $S$ (figure 18.6). Maxwell's equations in the material medium are[5]

$$\begin{aligned} &\nabla \times \mathbf{E} = -j\omega \mathbf{B} - \mathbf{K}; \ \mathbf{B} = \mu_0 \mathbf{H} + \mathbf{M} =: \mu_0 \mu_r \mathbf{H} \\ &\nabla \times \mathbf{H} = j\omega \mathbf{D} + \mathbf{J}; \ \ \mathbf{D} = \epsilon_0 \mathbf{E} + \mathbf{P} =: \epsilon_0 \epsilon_r \mathbf{E} \\ &\nabla \cdot \mathbf{B} = q^{\mathrm{m}}; \qquad \quad \nabla \cdot \mathbf{D} = q^{\mathrm{e}}, \end{aligned} \tag{18.130}$$

where $\mathbf{J}$ and $\mathbf{K}$ denote the electric and magnetic volume current density, $\mathbf{P}$ and $\mathbf{M}$ denote the electric and magnetic polarization in the medium and $(q^{\mathrm{e}}, q^{\mathrm{m}})$ are the electric and magnetic charge densities. The relative permittivity, $\epsilon_r$, and permeability, $\mu_r$, could be tensors and complex-valued. The equations of continuity

$$\nabla \cdot \mathbf{J} = -j\omega q^{\mathrm{e}}; \quad \nabla \cdot \mathbf{K} = -j\omega q^{\mathrm{m}}, \tag{18.131}$$

are already contained in the set (18.130). Let $w(x, y, z) = 0$ be surface $S$ across which the various fields could be discontinuous. The unit normal to the surface $\hat{\mathbf{n}} = \zeta \nabla w$ points into the medium with $w > 0$ (see figure 18.5), where $\zeta = |\nabla w|^{-1}$. A local Cartesian coordinate system $(u, v, n)$ can be erected on $S$, which can change orientation from point to point on the surface. We will now determine the boundary conditions for the fields under a much more general setting than those given in equations (18.126)–(18.129). In particular the field boundary conditions will allow for non-zero normal components of electric and magnetic polarizations as well as double-layer densities for the charge and current distributions to exist on the interface. In order to derive the general boundary conditions for the fields we follow the approach outlined in [5]. To this end we treat every field $\mathbf{F}$ as a distribution of order one (extension to higher orders is straightforward) and express (see theorem 18.1.2)

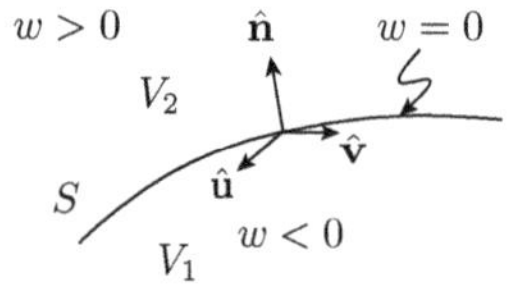


**Figure 18.5.** Surface $S$ on which $w = 0$ separating $V_1$ from $V_2$. The unit normal $\hat{\mathbf{n}} = \hat{\mathbf{u}} \times \hat{\mathbf{v}}$ points from $V_1$ to $V_2$.

[5] We are using a slightly different notation on placing the indices on the charge distributions on superscripts instead of subscripts. We reserve the subscripts for defining distributional quantities.

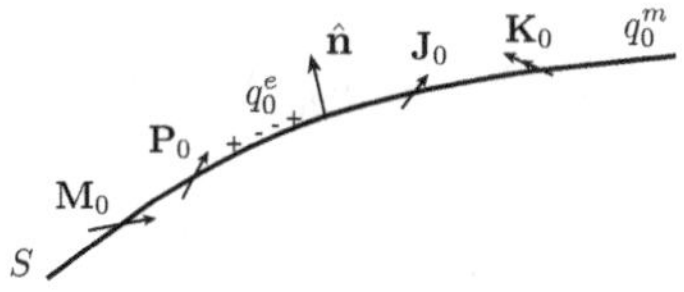


**Figure 18.6.** Surface $S$ on which impressed and induced single-layer electric and magnetic sources exist.

$$\mathbf{F} = \Theta_1\mathbf{F}_1 + \Theta_2\mathbf{F}_2 + \mathbf{F}_0\delta(w) + \mathbf{F}_1\delta'(w) =: \widetilde{\mathbf{F}}+\mathbf{F}_0\,\delta(w) + \mathbf{F}_1\delta'(w). \tag{18.132}$$

Here $\Theta_k$ is the Heaviside jump distribution in the volume $V_k$, $\widetilde{\mathbf{F}}$ is the regular part[6] which is defined everywhere except on $S$, and $\mathbf{F}_0$ and $\mathbf{F}_1$ are possible single-layer and double-layer density distributions[7] (functions of $(u, v)$ coordinates) that exist only on $S$. The singular distributions can have both tangential and normal parts. If one imposes the requirement that $\mathbf{F}$ be finite everywhere, then both $\mathbf{F}_0$ and $\mathbf{F}_1$ are taken as zero. Similarly, we express a scalar function $\psi$ as the distribution $\psi = \widetilde{\psi}+\psi_0\,\delta(w)$ $+\psi_1\,\delta'(w)$. We note the following properties

$$\nabla(\delta(w)) = \zeta^{-1}\hat{\mathbf{n}}\delta'(w) \tag{18.133}$$

$$\nabla\times(\delta(w)\mathbf{F}_0) = \delta(w)\nabla\times\mathbf{F}_0 + \zeta^{-1}\delta'(w)\,\hat{\mathbf{n}}\times\mathbf{F}_0 \tag{18.134}$$

$$\nabla\cdot(\delta(w)\mathbf{F}_0) = \delta(w)\nabla\cdot\mathbf{F}_0 + \zeta^{-1}\delta'(w)\,\hat{\mathbf{n}}\cdot\mathbf{F}_0 \tag{18.135}$$

$$\nabla(\delta(w)\psi_0) = \delta(w)\nabla\psi_0 + \zeta^{-1}\delta'(w)\,\hat{\mathbf{n}}\,\psi_0 \tag{18.136}$$

$$\nabla(\delta'(w)\psi_0) = \delta'(w)\nabla\psi_0 + \zeta^{-1}\delta''(w)\,\hat{\mathbf{n}}\,\psi_0. \tag{18.137}$$

On applying these properties together with the results from theorem 18.1.4 to every equation in (18.130) and (18.131), noting that $\delta_S = \zeta^{-1}\delta(w)$, and matching the coefficients of $\delta(w)$, $\delta'(w)$, and $\delta''(w)$ on both side of the equation we obtain the desired boundary conditions. To elaborate we obtain the boundary and auxiliary conditions

$$\Delta[\hat{\mathbf{n}}\times\mathbf{E}] = -\zeta(\nabla\times\mathbf{E}_0 + j\omega\mathbf{B}_0 + \mathbf{K}_0);\quad \hat{\mathbf{n}}\times\mathbf{E}_1 = 0, \tag{18.138}$$

$$\Delta[\hat{\mathbf{n}}\times\mathbf{H}] = -\zeta(\nabla\times\mathbf{H}_0 - j\omega\mathbf{D}_0 - \mathbf{J}_0);\quad \hat{\mathbf{n}}\times\mathbf{H}_1 = 0, \tag{18.139}$$

$$\Delta[\hat{\mathbf{n}}\cdot\mathbf{B}] = \zeta(q_0^{\mathrm{m}} - \nabla\cdot\mathbf{B}_0);\quad \hat{\mathbf{n}}\cdot\mathbf{B}_1 = 0, \tag{18.140}$$

$$\Delta[\hat{\mathbf{n}}\cdot\mathbf{D}] = \zeta(q_0^{\mathrm{e}} - \nabla\cdot\mathbf{D}_0);\quad \hat{\mathbf{n}}\cdot\mathbf{D}_1 = 0, \tag{18.141}$$

$$\Delta[\hat{\mathbf{n}}\cdot\mathbf{J}] = -\zeta(\nabla\cdot\mathbf{J}_0 + j\omega q_0^{\mathrm{e}});\quad \hat{\mathbf{n}}\cdot\mathbf{J}_1 = 0, \tag{18.142}$$

[6] A tilde is used here to denote the regular part instead of the notation $\{\cdot\}$ used in the previous sections.

[7] Note that $\mathbf{F}_0$ and $\mathbf{F}_1$ are the coefficients of the Dirac distributions $\delta(w)$ and $\delta'(w)$, respectively, but not of the surface distributions $\delta_S$ and $\partial\delta_S/\partial n$.

$$\Delta[\hat{\mathbf{n}} \cdot \mathbf{K}] = -\zeta(\nabla \cdot \mathbf{K}_0 + j\omega q_0^{\mathrm{m}}); \quad \hat{\mathbf{n}} \cdot \mathbf{K}_1 = 0, \tag{18.143}$$

$$\nabla \times \mathbf{E}_1 = -\zeta^{-1}\hat{\mathbf{n}} \times \mathbf{E}_0 - j\omega\mathbf{B}_1 - \mathbf{K}_1, \tag{18.144}$$

$$\nabla \times \mathbf{H}_1 = -\zeta^{-1}\hat{\mathbf{n}} \times \mathbf{H}_0 + j\omega\mathbf{D}_1 + \mathbf{J}_1, \tag{18.145}$$

$$\nabla \cdot \mathbf{B}_1 = -\zeta^{-1}\hat{\mathbf{n}} \cdot \mathbf{B}_0 + q_1^{\mathrm{m}}, \tag{18.146}$$

$$\nabla \cdot \mathbf{D}_1 = -\zeta^{-1}\hat{\mathbf{n}} \cdot \mathbf{D}_0 + q_1^{\mathrm{e}}, \tag{18.147}$$

$$\nabla \cdot \mathbf{J}_1 = -\zeta^{-1}\hat{\mathbf{n}} \cdot \mathbf{J}_0 - j\omega q_1^{\mathrm{e}} \tag{18.148}$$

$$\nabla \cdot \mathbf{K}_1 = -\zeta^{-1}\hat{\mathbf{n}} \cdot \mathbf{K}_0 - j\omega q_1^{\mathrm{m}}. \tag{18.149}$$

The goal here is to express $\mathbf{E}_k$ and $\mathbf{H}_k$ in terms of all $\mathbf{P}_{\mathrm{m}}$ and $\mathbf{M}_{\mathrm{m}}$ and then obtain the boundary conditions for $\Delta[\hat{\mathbf{n}} \times \mathbf{E}]$, $\Delta[\hat{\mathbf{n}} \times \mathbf{H}]$, $\Delta[\hat{\mathbf{n}} \cdot \mathbf{E}]$, and $\Delta[\hat{\mathbf{n}} \cdot \mathbf{H}]$. We list here the key steps:

1. Equations (18.138)–(18.143) suggest that $\mathbf{E}_1$ and $\mathbf{H}_1$ are purely normal vectors, while $\mathbf{B}_1$, $\mathbf{D}_1$, $\mathbf{J}_1$, and $\mathbf{K}_1$ are all purely tangential vectors. From $\hat{\mathbf{n}} \cdot \mathbf{D}_1 = 0$ we obtain $\hat{\mathbf{n}} \cdot \mathbf{E}_1 = -(1/\epsilon_0)\hat{\mathbf{n}} \cdot \mathbf{P}_1 =: -P_{1n}/\epsilon_0$. Hence $\mathbf{E}_1 = -\hat{\mathbf{n}}P_{1n}/\epsilon_0$, where $P_{1n}$ is the normal component of $\mathbf{P}_1$, i.e. $\mathbf{P}_1 = \mathbf{P}_{1t} + \hat{\mathbf{n}}P_{1n}$, where $\mathbf{P}_{1t}$ is the projection of $\mathbf{P}_1$ on the tangent plane to the surface $S$. Similarly, $\mathbf{H}_1 = -\hat{\mathbf{n}}M_{1n}/\mu_0$, where $\mathbf{M}_1 = \mathbf{M}_{1t} + \hat{\mathbf{n}}M_{1n}$. We also have $\mathbf{B}_1 = \mathbf{M}_{1t}$, $\mathbf{D}_1 = \mathbf{P}_{1t}$, $\mathbf{J}_1 = \mathbf{J}_{1t}$, $\mathbf{K}_1 = \mathbf{K}_{1t}$.
2. In equation (18.144) the quantity $\epsilon_0\nabla \times \mathbf{E}_1 = -\nabla \times (\hat{\mathbf{n}}P_{1n}) = -\nabla \times (\zeta P_{1n}\nabla w) = -\nabla(\zeta P_{1n}) \times \nabla w = \zeta^{-1}\hat{\mathbf{n}} \times \nabla(\zeta P_{1n})$, which shows that its normal component is zero. Similarly $\mu_0\nabla \times \mathbf{H}_1 = \zeta^{-1}\hat{\mathbf{n}} \times \nabla(\zeta M_{1n})$ in equation (18.145) is a tangential vector. Substituting these in equations (18.144) and (18.145) yields the expressions

$$\epsilon_0\hat{\mathbf{n}} \times \mathbf{E}_0 = -\hat{\mathbf{n}} \times \nabla(\zeta P_{1n}) - \zeta\epsilon_0(j\omega\mathbf{M}_{1t} + \mathbf{K}_{1t}), \tag{18.150}$$

$$\mu_0\hat{\mathbf{n}} \times \mathbf{H}_0 = -\hat{\mathbf{n}} \times \nabla(\zeta M_{1n}) + \zeta\mu_0(j\omega\mathbf{P}_{1t} + \mathbf{J}_{1t}), \tag{18.151}$$

   for the tangential components of $\mathbf{E}_0$ and $\mathbf{H}_0$. From these the tangential components of $\mathbf{D}_0$ and $\mathbf{B}_0$ are obtained as $\hat{\mathbf{n}} \times \mathbf{D}_0 = \epsilon_0\hat{\mathbf{n}} \times \mathbf{E}_0 + \hat{\mathbf{n}} \times \mathbf{P}_{0t}$ and $\hat{\mathbf{n}} \times \mathbf{B}_0 = \mu_0\hat{\mathbf{n}} \times \mathbf{H}_0 + \hat{\mathbf{n}} \times \mathbf{M}_{0t}$.
3. From equations (18.147) and (18.146) we obtain the expressions

$$\epsilon_0\hat{\mathbf{n}} \cdot \mathbf{E}_0 = \zeta q_1^{\mathrm{e}} - P_{0n} - \zeta\nabla \cdot \mathbf{P}_{1t}, \tag{18.152}$$

$$\mu_0\hat{\mathbf{n}} \cdot \mathbf{H}_0 = \zeta q_1^{\mathrm{m}} - M_{0n} - \zeta\nabla \cdot \mathbf{M}_{1t}, \tag{18.153}$$

   for the normal components of $\mathbf{E}_0$ and $\mathbf{H}_0$. The normal components of $\mathbf{D}_0$ and $\mathbf{B}_0$ are $\hat{\mathbf{n}} \cdot \mathbf{D}_0 = \zeta(q_1^{\mathrm{e}} - \nabla \cdot \mathbf{P}_{1t})$ and $\hat{\mathbf{n}} \cdot \mathbf{B}_0 = \zeta(q_1^{\mathrm{m}} - \nabla \cdot \mathbf{M}_{1t})$.
4. Substituting the results from items (i)–(iii) into equations (18.138)–(18.141) yields the desired boundary conditions.

*Special case 1.* Single-layer fields and sources only. Consider first the special case where only single-layer impressed and induced sources exist, but all double-layer fields and sources vanish (all singular components with a subscript 1 vanish[8]) (figure 18.6). Then $\hat{\mathbf{n}} \times \mathbf{E}_0 = 0 = \hat{\mathbf{n}} \times \mathbf{H}_0$, $\hat{\mathbf{n}} \cdot \mathbf{B}_0 = 0 = \hat{\mathbf{n}} \cdot \mathbf{D}_0 = \hat{\mathbf{n}} \cdot \mathbf{J}_0 = \hat{\mathbf{n}} \cdot \mathbf{K}_0$. Furthermore, $\epsilon_0 \mathbf{E}_0 = -\hat{\mathbf{n}} P_{0n}$, $\mu_0 \mathbf{H}_0 = -\hat{\mathbf{n}} M_{0n}$, $\epsilon_0 \nabla \times \mathbf{E}_0 = \zeta^{-1} \hat{\mathbf{n}} \times \nabla(\zeta P_{0n})$, and $\mu_0 \nabla \times \mathbf{H}_0 = \zeta^{-1} \hat{\mathbf{n}} \times \nabla(\zeta M_{0n})$. Relabeling the tangential vectors $\mathbf{J}_0 = \mathbf{J}_{0t}$, $\mathbf{K}_0 = \mathbf{K}_{0t}$, $\mathbf{B}_0 = \mathbf{M}_{0t}$, $\mathbf{D}_0 = \mathbf{P}_{0t}$, we can now express the boundary conditions (18.138) and (18.139) as

$$\Delta[\hat{\mathbf{n}} \times \mathbf{E}] = -\frac{1}{\epsilon_0} \hat{\mathbf{n}} \times \nabla(\zeta P_{0n}) - j\omega\zeta \mathbf{M}_{0t} - \zeta \mathbf{K}_{0t}, \quad (18.154)$$

$$\Delta[\hat{\mathbf{n}} \times \mathbf{H}] = -\frac{1}{\mu_0} \hat{\mathbf{n}} \times \nabla(\zeta M_{0n}) + j\omega\zeta \mathbf{P}_{0t} + \zeta \mathbf{J}_{0t}, \quad (18.155)$$

which are little more general than equations (18.126) and (18.128). Note that the surface currents ($\mathbf{J}_s$, $\mathbf{K}_s$) there are related to the surface currents ($\mathbf{J}_{0t}$, $\mathbf{K}_{0t}$) here via $\mathbf{J}_s = \zeta \mathbf{J}_{0t}$ and $\mathbf{K}_s = \zeta \mathbf{K}_{0t}$, the differences arising due to the use of different delta function representations, i.e. $\delta_S$ versus $\delta(w)$. It is interesting to note from equation (18.154) that the tangential electric field can be discontinuous across an interface even in the absence of magnetic current $\mathbf{K}_{0t}$ and the magnetization $\mathbf{M}_{0t}$ when there is a dipole layer of density $-\hat{\mathbf{n}} \times \nabla(\zeta P_{0n})/\epsilon_0$ on the interface. Along the same lines, the tangential magnetic field can be discontinuous across an interface even in the absence of surface electric current $\mathbf{J}_{0t}$ and polarization $\mathbf{P}_{0t}$ if there is a dipole layer of density $-\hat{\mathbf{n}} \times \nabla(\zeta M_{0n})/\mu_0$ on the interface. The boundary conditions on the normal components of flux densities given in equations (18.140) and (18.141) reduce to

$$\zeta^{-1}\Delta[\hat{\mathbf{n}} \cdot \mathbf{B}] = -\nabla \cdot \mathbf{M}_{0t} + q_0^{\mathrm{m}}, \quad (18.156)$$

$$\zeta^{-1}\Delta[\hat{\mathbf{n}} \cdot \mathbf{D}] = -\nabla \cdot \mathbf{P}_{0t} + q_0^{\mathrm{e}}, \quad (18.157)$$

which suggest that the normal components of **B** and **D** can be discontinuous across an interface even in the absence of surface charge densities provided that the polarization and magnetization vectors have non-vanishing divergence at the surface. The jumps in the current densities given in equations (18.142) and (18.143) are now given by

$$\zeta^{-1}\Delta[\hat{\mathbf{n}} \cdot \mathbf{J}] = -(\nabla \cdot \mathbf{J}_{0t} + j\omega q_0^{\mathrm{e}}), \quad (18.158)$$

$$\zeta^{-1}\Delta[\hat{\mathbf{n}} \cdot \mathbf{K}] = -(\nabla \cdot \mathbf{K}_{0t} + j\omega q_0^{\mathrm{m}}). \quad (18.159)$$

When the interface is planar, $\zeta = 1$ and equations (18.154)–(18.157) are referred to in the literature as the *generalized sheet transition conditions* (GSTCs) when applied to

[8] Setting the coefficient of $\delta(w)$ does not necessarily mean that the single-layer vanishes. However, setting the coefficient of $\delta'(w)$ to zero does imply that the double layer vanishes, see equation 18.88.

a *metasurface*[9] [15, p 39]. In practical applications, it is the electromagnetic properties of the metasurface that determine the surface electric and magnetic polarizations. For instance, in one methodology the polarizations are expressed in terms of the averaged electric and magnetic fields as [16]

$$\mathbf{P}_0 = \epsilon_0(\bar{\bar{\chi}}_e \mathbf{E}_{av} + \eta_0 \bar{\bar{\xi}} \mathbf{H}_{av}); \quad \mathbf{M}_0 = \mu_0\left(\eta_0^{-1} \bar{\bar{\zeta}} \mathbf{E}_{av} + \bar{\bar{\chi}}_m \mathbf{H}_{av}\right), \tag{18.160}$$

where $\mathbf{E}_{av} = (\mathbf{E}^i + \mathbf{E}^r + \mathbf{E}^t)/2$, $\mathbf{H}_{av} = (\mathbf{H}^i + \mathbf{H}^r + \mathbf{H}^t)/2$ are the unit-cell-averaged total (incident plus reflected plus transmitted) electric and magnetic fields on the metasurface and $\bar{\bar{\chi}}_e, \bar{\bar{\xi}}, \bar{\bar{\zeta}}, \bar{\bar{\chi}}_m$ are the unitless surface susceptibility tensors (or matrices). These tensors depend on the structure of the inclusions that make up the periodic cells, but are independent of excitations. The various tensors can be determined by considering the problem of scattering by incident plane waves and relating them to the reflection and transmission coefficients [17].

*Special case 2.* Single and double electric layers in a vacuum ($\mathbf{P} = 0 = \mathbf{M} = \mathbf{K}$, $q^m = 0$).

The medium in this case is homogeneous, but there are double-layer densities ($q_1^e$, $\mathbf{J}_1$) in addition to the single-layer densities ($q_0^e$, $\mathbf{J}_0$) on the surface $S$. The currents $\mathbf{J}_0$ and $\mathbf{J}_1$ need not be parallel to the surface, see figure 18.7(b). In this case we have from equations (18.150) and (18.153) that $\hat{\mathbf{n}} \times \mathbf{E}_0 = 0$, $\hat{\mathbf{n}} \cdot \mathbf{H}_0 = 0$. These together with $\epsilon_0 \hat{\mathbf{n}} \cdot \mathbf{E}_0 = \zeta q_1^e$ from equation (18.152) and $\hat{\mathbf{n}} \times \mathbf{H}_0 = \zeta \mathbf{J}_{1t}$ from equation (18.151) give

$$\epsilon_0 \mathbf{E} = \hat{\mathbf{n}} \zeta q_1^e = \mathbf{D}_0; \quad \mathbf{H}_0 = -\zeta \hat{\mathbf{n}} \times \mathbf{J}_1 = \mu_0^{-1} \mathbf{B}_0. \tag{18.161}$$

The boundary conditions are then

$$\Delta[\hat{\mathbf{n}} \times \mathbf{E}] = -\frac{\zeta}{\epsilon_0} \nabla \times \left(\hat{\mathbf{n}} \zeta q_1^e\right) + j\omega\mu_0 \zeta^2 (\hat{\mathbf{n}} \times \mathbf{J}_1), \tag{18.162}$$

$$\Delta[\hat{\mathbf{n}} \times \mathbf{H}] = \zeta \nabla \times (\zeta \hat{\mathbf{n}} \times \mathbf{J}_1) + j\omega\epsilon_0 \hat{\mathbf{n}} \zeta^2 q_1^e + \zeta \mathbf{J}_0, \tag{18.163}$$

$$\Delta[\hat{\mathbf{n}} \cdot \mathbf{B}] = \mu_0 \zeta \nabla \cdot (\zeta \hat{\mathbf{n}} \times \mathbf{J}_1), \tag{18.164}$$

$$\Delta[\hat{\mathbf{n}} \cdot \mathbf{D}] = \zeta q_0^e - \zeta \nabla \cdot (\hat{\mathbf{n}} \zeta q_1^e). \tag{18.165}$$

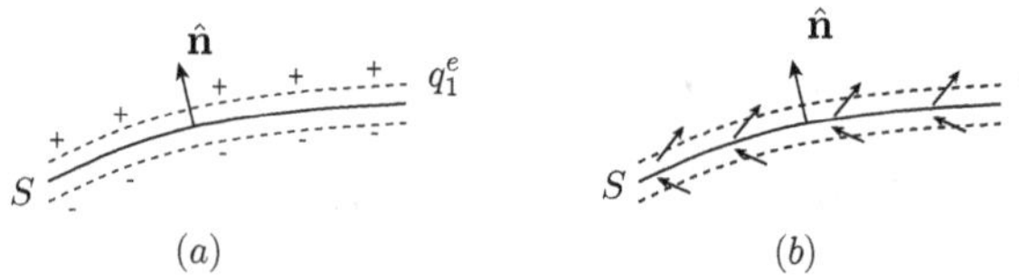


**Figure 18.7.** Double-layer distributions on $S$. (a) Charge distribution. (b) Current distribution.

[9] A metasurface is any periodic two-dimensional structure whose thickness and periodicity are small compared to wavelength in the surrounding media.

Hence neither the tangential electric field nor the normal magnetic flux density are continuous across a boundary in a homogeneous medium if the double-layer charges and currents ($q_1^{\mathrm{e}}$, $\mathbf{J}_1$) are non-zero on the boundary. ■

### 18.2.3 Boundary conditions for potentials

There has been some interest lately to derive integral equations directly in terms of potentials rather than the fields. This case seems to be particularly attractive at low frequencies where traditional formulations suffer from ill-conditioning [18]. To derive the required integral equations directly involving the potentials, one would need to translate the boundary conditions available for fields to those on potentials. We show in the next two examples how distributions theory can be used to rigorously derive boundary conditions associated with potentials such as the Lorenz potentials and the Hertzian potentials.

**Example 18.2.3.1** Boundary conditions for Lorenz potentials.
To further demonstrate the versatility of the distribution approach we extend the results of example 18.2.2.1 and determine the boundary conditions to be satisfied by the Lorenz potentials ($\mathbf{A}$, $\psi^{\mathrm{e}}$) for the time-harmonic case. All fields and sources here will be assumed to be distributions of order zero and all results under special case 1 of example 18.2.2.1 are applicable. Additional results can be derived from the new relations between fields and potentials. Writing the magnetic charge density $q^{\mathrm{m}}$ as $(-\nabla \cdot \mathbf{K})/j\omega$, the relations between the various fields and potentials in a general medium (inhomogeneous as well as anisotropic) are

$$\mathbf{B} + \mathbf{K}/j\omega = \nabla \times \mathbf{A};\ \mu_0\mathbf{H} = \mu_{\mathrm{r}}^{-1}\left(\nabla \times \mathbf{A} - \frac{\mathbf{K}}{j\omega}\right) \qquad (18.166)$$
$$\mathbf{E} = -j\omega\mathbf{A} - \nabla\psi^{\mathrm{e}};\ \ \mathbf{D} = -j\omega\epsilon_0\mathbf{G} - \epsilon_0\boldsymbol{\phi}^{\mathrm{e}}.$$

The relative permittivity $\epsilon_{\mathrm{r}}$ and permeability $\mu_{\mathrm{r}}$ could in general be tensors and complex-valued. Equations (18.166) have to be supplemented by the generalized Lorenz gauge condition (3.86)

$$\nabla \cdot \mathbf{G} = -j\omega\mu_0\epsilon_0\psi^{\mathrm{e}}, \qquad (18.167)$$

in order to decouple the second order differential equations for $\mathbf{A}$ and $\psi^{\mathrm{e}}$. The second order differential equation for $\psi^{\mathrm{e}}$ under the generalized Lorenz gauge condition is given in equation (3.88) and is

$$\nabla \cdot \boldsymbol{\phi}^{\mathrm{e}} + k_0^2\psi^{\mathrm{e}} = -\frac{q^{\mathrm{e}}}{\epsilon_0}, \qquad (18.168)$$

where $k_0 = \omega/c$ is the free-space wavenumber. The second Maxwell's equation $\nabla \times \mathbf{H} = j\omega\mathbf{D} + \mathbf{J}$ could be cast in terms of the potentials as

$$\nabla \times \mathbf{H} + j\omega\epsilon_0\boldsymbol{\phi}^e - \omega^2\epsilon_0\mathbf{G} = \mathbf{J}. \tag{18.169}$$

The procedure for deriving the remaining jump relations is the same as in the previous example. For instance, substituting $\mathbf{A} = \widetilde{\mathbf{A}} + \mathbf{A}_0\delta(w)$, $\mathbf{E} = \widetilde{\mathbf{E}} + \mathbf{E}_0\delta(w)$, and $\psi^e = \widetilde{\psi}^e + \psi_0^e\delta(w)$ into $\nabla\psi^e = -j\omega\mathbf{A} - \mathbf{E}$, using $\nabla(\psi_0^e\delta(w)) = \delta(w)\nabla\psi_0^e + \zeta^{-1}\hat{\mathbf{n}}\delta'(w)\psi_0^e$, the relation (18.74) and equating the coefficients of $\delta(w)$ and $\delta'(w)$ on both sides we obtain the conditions

$$\psi_0^e = 0, \tag{18.170}$$

$$\Delta[\psi^e] = -j\omega\zeta A_{0n} - \zeta E_{0n}, \tag{18.171}$$

where $A_{0n} = \hat{\mathbf{n}} \cdot \mathbf{A}_0$ is the normal component of the singular surface potential $\mathbf{A}_0$ and $\epsilon_0 E_{0n} = \epsilon_0\hat{\mathbf{n}} \cdot \mathbf{E} = -P_{0n}$ from the previous example. Under the assumption of finite $\mathbf{A}$ and $P_{0n}$, equation (18.171) reduces to $\Delta[\psi^e] = 0$ and the electric potential remains continuous across the boundary.

From the defining equation $\nabla \times \mathbf{A} = \mathbf{B} + \mathbf{K}/j\omega$ we obtain the condition $\hat{\mathbf{n}} \times \mathbf{A}_0 = 0$ (so that $\mathbf{A}_0 = \hat{\mathbf{n}}A_{0n}$) and the general boundary condition on the tangential component of $\mathbf{A}$

$$\Delta[\hat{\mathbf{n}} \times \mathbf{A}] = \hat{\mathbf{n}} \times \nabla(\zeta A_{0n}) + \zeta\mathbf{M}_{0t} + \frac{\zeta}{j\omega}\mathbf{K}_{0t}. \tag{18.172}$$

From the jump condition for $\mathbf{H}$ in equation (18.155) or directly from equation (18.169) we obtain the general boundary condition on the tangential component of $\mu_r^{-1}\nabla \times \mathbf{A}$

$$\begin{aligned}\Delta\left[\hat{\mathbf{n}} \times \left(\mu_r^{-1}\nabla \times \mathbf{A}\right)\right] = {} & \frac{1}{j\omega}\Delta[\hat{\mathbf{n}} \times \mu_r^{-1}\mathbf{K}] \\ & + j\omega\mu_0\zeta\mathbf{P}_{0t} + \mu_0\zeta\mathbf{J}_{0t} - \hat{\mathbf{n}} \times \nabla(\zeta M_{0n}).\end{aligned} \tag{18.173}$$

From the Lorenz condition (18.167) we obtain the condition $\hat{\mathbf{n}} \cdot \mathbf{G}_0 = 0$ and the boundary condition

$$\Delta[\hat{\mathbf{n}} \cdot \mathbf{G}] = -\zeta\nabla \cdot \mathbf{G}_0, \tag{18.174}$$

upon substituting $\psi_0^e = 0$. Finally from the differential equation (18.168) we obtain the condition $\hat{\mathbf{n}} \cdot \boldsymbol{\phi}_0^e = 0$ and the boundary condition

$$\Delta[\hat{\mathbf{n}} \cdot \boldsymbol{\phi}_0^e] = -\zeta\nabla \cdot \boldsymbol{\phi}_0^e - \frac{\zeta}{\epsilon_0}q_0^e. \tag{18.175}$$

The boundary conditions in equations (18.171)–(18.175) are valid under general conditions. However, if the electric polarization, magnetic polarization, and magnetic vector potential remain finite at the boundary, then the corresponding singular parts vanish and the boundary conditions are simplified to

$$\Delta[\psi^e] = 0, \mathbf{r} \in S, \tag{18.176}$$

$$\Delta[\hat{\mathbf{n}} \cdot (\epsilon_r \nabla \psi^e] = -\frac{1}{\epsilon_0} q_{es}, \ \mathbf{r} \in S, \tag{18.177}$$

$$\Delta[\hat{\mathbf{n}} \times \mathbf{A}] = \frac{1}{j\omega} \mathbf{K}_s, \ \mathbf{r} \in S, \tag{18.178}$$

$$\Delta[\hat{\mathbf{n}} \times \mu_r^{-1} \nabla \times \mathbf{A}] = \frac{1}{j\omega} \Delta[\hat{\mathbf{n}} \times \mu_r^{-1} \mathbf{K}] + \mu_0 \mathbf{J}_s, \ \mathbf{r} \in S, \tag{18.179}$$

$$\Delta[\hat{\mathbf{n}} \cdot (\epsilon_r \mathbf{A})] = 0, \ \mathbf{r} \in S, \tag{18.180}$$

where $q_{es} = \zeta q_0^e$, $\mathbf{K}_s = \zeta \mathbf{K}_{0t}$, and $\mathbf{J}_s = \zeta \mathbf{J}_{0t}$ are the surface charge, surface tangential magnetic current, and surface tangential electric current densities, respectively. Special forms of equations (18.176)–(18.180) under the assumption of no surface charges and surface currents have been given in [18] based on heuristic arguments. ■■

**Example 18.2.3.2** *Boundary conditions for Hertzian potentials*. In this example we work out the required boundary conditions on the Hertzian potentials $\mathbf{\Pi}^e$, $\mathbf{\Pi}^m$ for the time-harmonic case[10]. The differential equations satisfied by the Hertzian potentials are given in equations (3.92) and (3.93) in which the sources are the stream potentials $\mathbf{Q}^e$, $\mathbf{R}^m$, $\mathbf{R}^e$, and $\mathbf{Q}^m$, which are in turn dependent on the true current densities ($\mathbf{J}$, $\mathbf{K}$) and the charge densities ($q^e$, $q^m$). With no loss in generality we choose the consistent pairs ($\mathbf{Q}^e = \mathbf{J}/j\omega$, $\mathbf{R}^e = 0$) and ($\mathbf{R}^m = \mathbf{K}/j\omega$, $\mathbf{Q}^m = 0$) assuming non-static case. On defining $\mathbf{T}^h = j\omega\epsilon_0\mu_r^{-1}\nabla \times \mathbf{\Pi}^e = (jk_0/\eta_0)\mu_r^{-1}\nabla \times \mathbf{\Pi}^e$, $\mathbf{T}^e = -j\omega\mu_0\epsilon_r^{-1}\nabla \times \mathbf{\Pi}^m = -jk_0\eta_0\epsilon_r^{-1}\nabla \times \mathbf{\Pi}^m$, $k_0 = \omega\sqrt{\mu_0\epsilon_0}$, and $\eta_0 = \sqrt{\mu_0/\epsilon_0}$, equations (3.92) and (3.93) can be rewritten as

$$-\frac{jk_0^3}{\eta_0} \epsilon_r \mathbf{\Pi}^e + \nabla \times \mathbf{T}^h = \mathbf{J}, \tag{18.181}$$

$$jk_0^3 \eta_0 \mu_r \mathbf{\Pi}^m + \nabla \times \mathbf{T}^e = -\mathbf{K}. \tag{18.182}$$

Taking a divergence on both sides leads to

$$k_0^2 \nabla \cdot (\epsilon_0 \epsilon_r \mathbf{\Pi}^e) = q^e, \tag{18.183}$$

$$k_0^2 \nabla \cdot (\mu_0 \mu_r \mathbf{\Pi}^m) = q^m. \tag{18.184}$$

From equations (3.95) and (3.96), the equations for the field flux densities in terms of the Hertzian potentials and the sources are

[10] As in example 18.2.3.1 we use the suffixes *e* and *m* as superscripts instead of subscripts.

$$c\mathbf{B} = -\frac{\mathbf{K}}{jk_0} + jk_0\nabla \times \mathbf{\Pi}^{\rm e} - \frac{1}{jk_0}\nabla \times \mathbf{T}^{\rm e}, \tag{18.185}$$

$$c\mathbf{D} = -\frac{\mathbf{J}}{jk_0} - jk_0\nabla \times \mathbf{\Pi}^{\rm m} + \frac{1}{jk_0}\nabla \times \mathbf{T}^{\rm h}. \tag{18.186}$$

Assuming that all field and potential variables are distributions of order zero we obtain the following conditions from equations (18.181)–(18.186):

$$\hat{\mathbf{n}} \times \mathbf{T}_0^{\rm h} = 0;\ \zeta^{-1}\Delta[\hat{\mathbf{n}} \times \mathbf{T}^{\rm h}] = \mathbf{J}_0 - \nabla \times \mathbf{T}_0^{\rm h} + j\omega\epsilon_0 k_0^2(\epsilon_{\rm r}\mathbf{\Pi}^{\rm e})_0,$$
$$\hat{\mathbf{n}} \times \mathbf{T}_0^{\rm e} = 0;\ \zeta^{-1}\Delta[\hat{\mathbf{n}} \times \mathbf{T}^{\rm e}] = -\mathbf{K}_0 - \nabla \times \mathbf{T}_0^{\rm e} - j\omega\mu_0 k_0^2\left(\mu_{\rm r}\mathbf{\Pi}^{\rm m}\right)_0,$$
$$\hat{\mathbf{n}} \cdot (\epsilon_{\rm r}\mathbf{\Pi}^{\rm e})_0 = 0;\ \zeta^{-1}\Delta[\hat{\mathbf{n}} \cdot (\epsilon_{\rm r}\mathbf{\Pi}^{\rm e})] = \frac{q_0^{\rm e}}{\epsilon_0 k_0^2} - \nabla \cdot (\epsilon_{\rm r}\mathbf{\Pi}^{\rm e})_0,$$
$$\hat{\mathbf{n}} \cdot (\mu_{\rm r}\mathbf{\Pi}^{\rm m})_0 = 0;\ \zeta^{-1}\Delta[\hat{\mathbf{n}} \cdot (\mu_{\rm r}\mathbf{\Pi}^{\rm m})] = \frac{q_0^{\rm m}}{\mu_0 k_0^2} - \nabla \cdot (\mu_{\rm r}\mathbf{\Pi}^{\rm m})_0,$$
$$\hat{\mathbf{n}} \times \mathbf{\Pi}_0^{\rm e} = 0;\ \zeta^{-1}\Delta\left[\hat{\mathbf{n}} \times \left(jk_0\mathbf{\Pi}^{\rm e} - \frac{1}{jk_0}\mathbf{T}^{\rm e}\right)\right]$$
$$= c\mathbf{B}_0 + \frac{\mathbf{K}_0}{jk_0} - \nabla \times \left(jk_0\mathbf{\Pi}_0^{\rm e} - \frac{1}{jk_0}\mathbf{T}_0^{\rm e}\right),$$
$$\hat{\mathbf{n}} \times \mathbf{\Pi}_0^{\rm m} = 0;\ \zeta^{-1}\Delta\left[\hat{\mathbf{n}} \times \left(-jk_0\mathbf{\Pi}^{\rm m} + \frac{1}{jk_0}\mathbf{T}^{\rm h}\right)\right]$$
$$= c\mathbf{D}_0 + \frac{\mathbf{J}_0}{jk_0} + \nabla \times \left(jk_0\mathbf{\Pi}_0^{\rm m} - \frac{1}{jk_0}\mathbf{T}_0^{\rm h}\right).$$

If the fields, Hertzian potentials and the material parameters are assumed to be finite at the boundary, then the singular parts of all variables vanish and one obtains the simplified boundary conditions after incorporating the first two equations above into the last two

$$\Delta[\hat{\mathbf{n}} \times \mathbf{T}^{\rm h}] = \mathbf{J}_{\rm s}, \tag{18.187}$$

$$\Delta[\hat{\mathbf{n}} \times \mathbf{T}^{\rm e}] = -\mathbf{K}_{\rm s}, \tag{18.188}$$

$$k_0^2\epsilon_0\Delta[\hat{\mathbf{n}} \cdot (\epsilon_{\rm r}\mathbf{\Pi}^{\rm e})] = q_{\rm es}, \tag{18.189}$$

$$k_0^2\mu_0\Delta[\hat{\mathbf{n}} \cdot (\mu_{\rm r}\mathbf{\Pi}^{\rm m})] = q_{\rm ms}, \tag{18.190}$$

$$\Delta[\hat{\mathbf{n}} \times \mathbf{\Pi}^{\rm e}] = 0, \tag{18.191}$$

$$\Delta[\hat{\mathbf{n}} \times \mathbf{\Pi}^{\rm m}] = 0. \tag{18.192}$$

Note that the boundary conditions on $\mathbf{\Pi}^{\rm e}$ are dependent only on surface electric sources, while the boundary conditions on $\mathbf{\Pi}^{\rm m}$ are dependent only on surface magnetic sources. In the absence of surface charges and currents, the normal components of $\epsilon_{\rm r}\mathbf{\Pi}^{\rm e}$ and $\mu_{\rm r}\mathbf{\Pi}^{\rm m}$ and the tangential components of $\mathbf{\Pi}^{\rm e}$, $\mathbf{\Pi}^{\rm m}$, $\mathbf{T}^{\rm h}$ and

$\mathbf{T}^e$ are continuous across the boundary. From the above analysis it is also clear that $k_0^2\mathbf{\Pi}^e$ and $\mathbf{T}^e$ behave like the electric field, whereas $k_0^2\mathbf{\Pi}^m$ and $\mathbf{T}^h$ behave like the magnetic field. In fact all formulations and integral equations available for (**E**, **H**) can be translated to ($\mathbf{\Pi}^e$, $\mathbf{\Pi}^m$) with simple change of variables. ■■

## References


[1] Gagnon R J 1970 *Am. J. Phys.* **38** 879–91

[2] Idemen M 1973 The Maxwell's equations in the sense of distributions *IEEE Trans. Antennas Propag.* **AP-21** 736–8

[3] Idemen M 1990 Universal boundary relations of the electromagnetic field *J. Phys. Soc. Jpn* **59** 71–80

[4] van Bladel J 1991 *Singular Electromagnetic Fields and Sources* (The IEEE Series on Electromagnetic Wave Theory) (Piscataway: NJ: IEEE)

[5] Idemen M 2011 *Discontinuities in the Electromagnetic Field* (Hoboken, NJ: Wiley)

[6] Gel'fand I M and Shilov G E 1968 Spaces of fundamental and generalized functions *Generalized Functions* vol 2 (New York: Academic)

[7] Stakgold I 1967 *Boundary Value Problems of Mathematical Physics* vol 1 (New York: Macmillan)

[8] Zemanian A H 1965 *Distribution Theory and Transform Analysis* (New York: McGraw-Hill)

[9] Kanwal R P 2004 *Generalized Functions: Theory and Applications* 3rd edn (New York: Springer)

[10] Friedman B 1956 *Principles and Techniques of Applied Mathematics* (New York: Wiley)

[11] Morse P M and Feshbach H 1953 *Methods of Theoretical Physics* (New York: McGraw-Hill)

[12] van Bladel J 1985 *Electromagnetic Fields* (New York: Hemisphere)

[13] Struik D J 1961 *Lectures on Classical Differential Geometry* 2nd edn (Reading, MA: Addison-Wesley)

[14] Stoker J J 1969 *Differential Geometry* (New York: Wiley)

[15] Yang F and Rahmat-Samii Y 2019 Analytical modeling of electromagnetic surfaces *Surface Electromagnetics with Applications in Antennam Microwave, and Optical Engineering* ed V Asadchy, A Diaz-Rubio, D-H Kwon and S Tretyakov (New York: Cambridge University Press) pp 30–65 ch 2

[16] Kuester E F, Mohamed M A, Piket-May M and Holloway C L 2003 Averaged transition conditions for electromagnetic fields at a metafilm *IEEE Trans. Antennas Propag.* **51** 2641–51

[17] Yang F and Rahmat-Samii Y 2019 Using generalized sheet transition conditions (GTSCs) in the analysis of metasurfaces *Surface Electromagnetics with Applications in Antenna, Microwave, and Optical Engineering* ed C L Holloway and E F Kuester (New York: Cambridge University Press) pp 66–123 ch 3

[18] Chew W C 2014 Vector potential electromagnetics with generalized gauge for inhomogeneous media: formulation *Prog. Electromagn. Res.* **149** 69–84

**IOP** Publishing

Engineering Electrodynamics
A collection of theorems, principles and field representations
**Ramakrishna Janaswamy**

# Chapter 19

# Stochastic representations of wave phenomena

The techniques discussed thus far in this book have all been deterministic. By introducing some sort of artificial randomness into a physical problem, alternative solution techniques become available for the solution of wave and Helmholtz equations. For instance, the propagation of electromagnetic waves in specialized media such as plasmonic media [1] and the determination of the cut-off frequency of the lowest order mode in a dielectric waveguide [2] can both be handled efficiently by random walk techniques. As another example, it is possible to relate the solution of a telegraph equation (lossy wave equation) to the solution of a wave equation by randomizing time [3]. In all these cases, the solution is obtained by taking at the end an expectation of an appropriately formulated stochastic process. The branch of mathematics that deals with the operations on stochastic processes is known as stochastic calculus and an excellent introductory resource in this regard is the book by Shreve [4]. In this chapter we introduce the relevant concepts for stochastic techniques and arrive at some useful representations such as the Feynman–Kac representation for the solution of Helmholtz or wave type equations. Important topics such as Brownian motion, martingales, the Markov process, Itô integrals, and the Itô–Doeblin formula are all discussed in a self-consistent manner.

## 19.1 Preliminaries of stochastic calculus

**Definition 19.1.1** *Borel σ-algebra* [4, p 3]. The $\sigma$-algebra obtained by beginning with a closed interval [0, 1] and adding everything else necessary in order to have a $\sigma$-algebra is called the *Borel σ-algebra* of subsets of [0, 1] and is denoted as $\mathscr{B}[0, 1]$. The sets in this $\sigma$-algebra are called *Borel sets*. The probability measure of closed intervals $[a, b]$ is given by the formula $\mathbb{P}[a, b] = b - a$, $0 \le a \le b \le 1$.

 

**Definition 19.1.2.** $\mathcal{F}$*-measurable* [5, p 8]. Given the probability space $(\Omega, \mathcal{F}, \mathbb{P})$, a function $Y: \Omega \to R^n$ is called $\mathcal{F}$*-measurable* if

$$Y^{-1}(U) := \{\omega \in \Omega;\ Y(\omega) \in U\} \in \mathcal{F}$$

for all open sets $U \in R^n$.

**Definition 19.1.3** *Borel-measurable function* [4, p 21]. Let $f(x)$ be a real-valued function defined on $\mathbb{R}$. If for every Borel subset $\mathscr{B}$ of $\mathbb{R}$ the set $\{x; f(x) \in \mathscr{B}\}$ is also a Borel subset of $\mathbb{R}$, then $f$ is said to be *Borel-measurable*. Every continuous and piecewise continuous function is Borel-measurable. The concept of measurability and $\sigma$-algebra can be understood as follows. A random experiment is performed and an outcome $\omega$ is determined, but the value of $\omega$ is not revealed. Instead, for each set in the $\sigma$-algebra $\mathcal{F}$, we are told whether $\omega$ is in the set. The more sets there are in $\mathcal{F}$, the more information this provides. Measurability states that the information in $\mathcal{F}$ is enough to determine the value of the random variable $Y(\omega)$ even though it may not be enough to determine the value $\omega$ of the outcome of the random experiment.

**Definition 19.1.4** *Indicator function* [4]. The *indicator function* (random function) of a set $A$ is defined by

$$\mathbb{I}_A(\omega) = \begin{cases} 1 & \text{if } \omega \in A, \\ 0 & \text{if } \omega \notin A. \end{cases} \tag{19.1}$$

**Definition 19.1.5.** *Expectation* [4, p 17]. The *expectation* of a random variable $X$ is defined by the *Lebesgue* integral

$$\mathbb{E}\, X = \int_\Omega X\, \mathrm{d}\mathbb{P} = \int_{R^n} x\, \mathrm{d}F_n(x),$$

assuming that $X$ is integrable, i.e. if

$$\mathbb{E}\, |X| = \int_\Omega |X(\omega)| \mathrm{d}\mathbb{P}(\omega) < \infty,$$

or if $X \geq 0$ a.s. In the latter case $\mathbb{E}\, X$ might be $\infty$. Note

$$\int_A X(\omega)\, \mathrm{d}\mathbb{P}(\omega) = \int_\Omega \mathbb{I}_A(\omega) X(\omega)\, \mathrm{d}\mathbb{P}(\omega) \quad \text{for all } A \in \mathcal{F}.$$

If $\varphi$ is a convex, real-valued function (it looks like a cup) defined on the real axis and if $\mathbb{E}\, X < \infty$, then we have the *Jensen's inequality*

$$\varphi(\mathbb{E}\, X) \leq \mathbb{E}\, \varphi(X).$$

**Theorem 19.1.1.** Iterated conditioning [4, p 70].
*If* $\mathcal{H}$ *is a sub-*$\sigma$*-algebra of* $\mathcal{G}$ *(*$\mathcal{H}$ *contains less information than* $\mathcal{G}$, $\mathcal{H} \subseteq \mathcal{G}$*) and* $X$ *is an integrable random variable (namely,* $\int |X(\omega)| \mathrm{d}\mathbb{P}(\omega) < \infty$*), then*

$$\mathbb{E}\,[\mathbb{E}\,[X|\mathcal{G}]|\mathcal{G}] = \mathbb{E}\,[X|\mathcal{H}]. \tag{19.2}$$

**Lemma 19.1.2** Independence lemma [4, p 73].
*Let* $(\Omega, \mathcal{F}, \mathbb{P})$ *be a probability space, and let* $\mathcal{G}$ *be a sub-*$\sigma$*-algebra of* $\mathcal{F}$. *Suppose the random variables* $X_1, \ldots, X_K$ *are* $\mathcal{G}$*-measurable and the random variables* $Y_1, \ldots, Y_L$ *are independent of* $\mathcal{G}$. *Let* $f(x_1, \ldots, x_K; y_1, \ldots, y_L)$ *be a function of the dummy variables* $x_1, \ldots, x_K$ *and* $y_1, \ldots, y_L$ *and define*

$$g(x_1, \ldots, x_K) = \mathbb{E}\,f(x_1, \ldots, x_K;\, Y_1, \ldots, Y_L). \tag{19.3}$$

*Then*

$$\mathbb{E}\,f(X_1, \ldots, X_K;\, Y_1, \ldots, Y_L|\mathcal{G}) = g(X_1, \ldots, X_K). \tag{19.4}$$

What the lemma says is that since the information in $\mathcal{G}$ is sufficient to determine the values of $X_1, \ldots, X_K$, we should hold these random variables constant when estimating $f(X_1, \ldots, X_K; Y_1, \ldots, Y_L)$. The other random variables $Y_1, \ldots, Y_L$ are independent of $\mathcal{G}$ and so we should integrate them out without regard to the information in $\mathcal{G}$. This step is contained in equation (19.3). We obtain an estimate that depends on the values of $X_1, \ldots, X_K$ and to capture this fact we replace the dummy variables $x_1, \ldots, x_K$ in step (19.3) by the random variables $X_1, \ldots, X_K$ in step (19.4).

**Theorem 19.1.3.** Lebesgue dominated convergence theorem [4, p 27].
*Let* $\{f_n(x)\}$, $n = 1, 2, \ldots$ *be a sequence of real-valued Borel-measurable functions on* $\mathbb{R}$ *converging a.s. to a function* $f(x)$. *If there is another non-negative function* $g(x)$ *such that* $\int_{\mathbb{R}} |g(x)|\, \mathrm{d}x < \infty$ *and* $|f_n(x)| \leq g(x)$ *for a.a.* $n$ *then*

$$\lim_{n\to\infty} \int_{\mathbb{R}} f_n(x)\, \mathrm{d}x = \int_{\mathbb{R}} f(x)\, \mathrm{d}x.$$

The Lebesgue dominated convergence theorem applies to averages as well as conditional averages.

## 19.2 Stochastic processes and Brownian motion

**Definition 19.2.1.** *Stochastic process* [6]. A *stochastic process* is a family of random variables $\{X(t)\}$ defined on a probability space $(\Omega, \mathcal{F}, \mathbb{P})$, where $t$ varies in a real interval $I$ ($I$ is open, closed, or half-closed).

For a fixed sample point $\omega \in \Omega$, the function $t \to X(t, \omega)$[1] is called a *sample path* or realization of the process. If for a.a. $\omega$ the sample paths are continuous functions for all $t \in I$, then we say that the stochastic process is *continuous.* A stochastic process $\{X(t),\ t \in I\}$ is said to be *continuous in probability* if for any $s \in I$ and $\varepsilon > 0$,

$$\mathbb{P}[|X(t) - X(s)| > \varepsilon] \to 0 \quad \text{if} \quad t \in I, \quad t \to s.$$

There is a very important reason to include the $\sigma$-fields $\mathcal{F}$ in the study of stochastic processes, and that is to keep track of information. The temporal features of a stochastic process suggests a flow of time, in which, at every moment $t \geq 0$, we can talk about a *past, present*, and *future* and can ask how much an observer of the process knows about it at present, as compared to how much he knew at some point in the past or will know at some point in the future. We equip our sample space $(\Omega, \mathcal{F})$ with a *filtration*, i.e. a non-decreasing family $\{\mathcal{F}_t;\ t \geq 0\}$ of sub $\sigma$-fields of $\mathcal{F}$: $\mathcal{F}_s \subseteq \mathcal{F}_t \subseteq \mathcal{F}$ for $0 \leq s < t < \infty$.

We define $\mathcal{F}_{t^-} = \sigma(\bigcup_{s<t}\mathcal{F}_s)$ to be the $\sigma$-field of events strictly prior to $t > 0$ and $\mathcal{F}_{t^+} = \bigcap_{\varepsilon>0}\mathcal{F}_{t+\varepsilon}$ to be the $\sigma$-field of events immediately after $t \geq 0$. We say that the filtration is *right*-(left-)*continuous* if $\mathcal{F}_t = \mathcal{F}_{t^+}$ (respectively, $\mathcal{F}_t = \mathcal{F}_{t^-}$) holds for every $t \geq 0$.

**Definition 19.2.2.** *Separable stochastic process* [6, p 6]. An $n$-dimensional stochastic process $\{X(t),\ t \in I\}$ is called *separable* if there exists a countable sequence $T = \{t_j\}$ that is a dense subset of $I$ and a subset $N$ of $\Omega$ with $\mathbb{P}(N) = 0$ such that if $\omega \notin N$,

$$\{X(t, \omega) \in F \text{ for all } t \in J\} = \{X(t_j, \omega) \in F \text{ for all } t_j \in J\},$$

for any open subset $J$ of $I$ and for any closed subset $F$ of $R^n$. The set $T$ is called a *set of separability* or simply a *separant.* In other words, a separable stochastic process can be well described probabilistically using a countably infinite sequence of time samples.

**Definition 19.2.3.** *Random time* [5, p 5]. A *random time* $T$ is an $\mathcal{F}$-measurable random variable with values in $[0, \infty)$.

**Definition 19.2.4.** *Martingale* [7]. A stochastic process $\{X(t),\ t \in I\}$ is called a *martingale* if $\mathbb{E}\ |X(t)| < \infty$ for all $t \in I$, and

$$\mathbb{E}\ \{X(t)|X(\tau),\ \tau \leq s,\ \tau \in I\} = X(s) \quad \text{for all} \quad s, t \in I, \quad s \leq t.$$

In other words, *the expected value of a martingale $X(t)$ for all times subsequent to $s$ is equal to the known value $X(s)$.* A martingale has a tendency to neither rise nor fall, but will instead vary with equal probability around the mean value $X(s, \omega)$ for any

[1] The argument $\omega$ is not always indicated in the process $X(t, \omega)$.

given sample path. In a sense a martingale represents an abstract presentation of a 'fair game'. An important energy property of martingales for $s < t$ is that [8, p 35]

$$\mathbb{E}\,[X(t) - X(s)]^2 = \mathbb{E}\,X^2(t) - \mathbb{E}\,X^2(s).$$

**Definition 19.2.5** *Stopping time* [6]. Let $X(t)$, $\alpha \leq t \leq \beta$ be a stochastic process. (If $\beta \to \infty$, we take $\alpha \leq t < \infty$.) A finite-valued random variable $\tau$ is called a *stopping time* with respect to the process $X(t)$ if $\alpha \leq \tau \leq \beta$ and if, for any $\alpha \leq t \leq \beta$, the set $\{\tau \leq t\}$ belongs to $\mathcal{F}_t := \mathcal{F}[X(\lambda), \alpha \leq \lambda \leq t]$.

Stated alternatively, a random time $\tau$ is a stopping time of the filtration $\mathcal{F}_t$, if the event $\{\tau \leq t\}$ belongs to $\mathcal{F}_t$ for every $\alpha \leq t \leq \beta$.

Stopping times describe the instants at which certain random events occur. At the name implies, a random process should be able to stop or be frozen for all times posterior to the stopping time. At any time $t$ one can observe only the events of the probability space $(\Omega, \mathcal{F}_t, \mathbb{P})$. If $\tau$ is a random time then at time $t$ one cannot observe events over the whole $\tau$. One can observe events only over the random variable $\tau \wedge t$. By definition $\tau$ is a stopping time if $\tau \wedge t$ is an $(\Omega, \mathcal{F}_t, \mathbb{P})$-random variable for all $t$. As an example the *first* time $X(t)$ is zero is a valid stopping time. On the other hand the *last* time it is zero is not a stopping time because one will never know when the last zero crossing will be for any realization. The instant at which $X(t)$ reaches its maximum value on a certain interval is also not a stopping time. (Try putting a limit order to sell a stock holding at the time it reaches a maximum value over one year. The order can never be executed!) If $T$ and $S$ are stopping times, then so are $T \wedge S$, $T \vee S$, $T + S$. If $\tau$ is a stopping time and $\theta$ is a positive constant, the $\tau + \theta$ is a stopping time.

**Definition 19.2.6.** *pth variation* [5, p 32]. Let $X = \{X_t;\, 0 \leq t < \infty\}$ be a stochastic process. Fix $t > 0$ and let $\Pi = \{0 = t_0 \leq t_1 \leq t_2 \leq \cdots \leq t_m = t\}$ be a partition of $[0, t]$. The *pth variation* $(p > 0)$ *of X over the partition* $\Pi$ is

$$V_t^{(p)}(\Pi) = \sum_{k=0}^{m-1} \left|X_{t_{k+1}} - X_{t_k}\right|^p.$$

Define the *mesh* of the partition $\Pi$ as $\|\Pi\| = \max_{0 \leq k \leq m-1} |t_{k+1} - t_k|$. If $V_t^{(p)}(\Pi)$ converges in some sense as $\|\Pi\| \to 0$, the limit $V_t^{(p)}$ is called the $p$th variation of $X$ on $[0, t]$. Of particular importance is the *quadratic variation*, $[X, X](t)$, which applies with $p = 2$. Note that the $p$th variation will depend on the sample path and is hence a random variable.

If a function $f(t)$, $t > 0$ has a continuous first derivative, application of mean value theorem will reveal that the first variation $V_t^{(1)} = \int_0^t |f'(\tau)|\,d\tau$ and the quadratic variation of $f$ will be zero. An example of a continuous function with non-zero quadratic variation is a Brownian motion, which we discuss next.

**Definition 19.2.7** *Brownian motion* [9, p 1]. Let $(\Omega, \mathcal{F}, \mathbb{P})$ be a probability space. For each $\omega \in \Omega$, suppose there is a continuous function $W(t)$, $t \geq 0$ that satisfies $W(0) = W_0$. An $n$-dimensional *Brownian motion* $W(t)$ starting at $t = 0$ with the value $W(0) := W_0$ is a stochastic process with the following properties:

(a) *Translational invariance.* $W(t) - W_0$ is independent of $W_0$ and has the same distribution as a Brownian motion with $W_0 = 0$.

(b) *Independence of coordinates.* If $W_0 = 0$, then $\{W^1(t), t \geq 0\}$, $\{W^2(t), t \geq 0\}$, ..., $\{W^n(t), t \geq 0\}$ are independent one-dimensional Brownian motions starting at 0.

(c) *Independent increments.* For all $0 = t_0 < t_1 < \cdots < t_m = t$ the increments for any coordinate component

$$W(t_1) - W_0,\ W(t_2) - W(t_1),\ \ldots,\ W(t_m) - W(t_{m-1}),$$

are independent and each of these increments is *Gaussian* distributed with

$$\begin{aligned} &\mathbb{E}\left[W(t_{i+1}) - W(t_i)\right] = 0 \\ &\mathbb{E}\left[W(t_{i+1}) - W(t_i)\right]^2 = t_{i+1} - t_i, \end{aligned} \tag{19.5}$$

See figure 19.1.

A consequence of item (c) is that $\mathbb{E}\left[W(t)W(s)\right] = t \wedge s$. Another consequence is that *Brownian motion is a martingale*. This can be easily shown. For $0 \leq s \leq t$,

$$\begin{aligned} \mathbb{E}\left[W(t)|\mathcal{F}(s)\right] &= \mathbb{E}\left[W(t) - W(s) + W(s)|\mathcal{F}(s)\right] \\ &= \mathbb{E}\left[W(t) - W(s)|\mathcal{F}(s)\right] + \mathbb{E}\left[W(s)|\mathcal{F}(s)\right] \\ &= \mathbb{E}\left[W(t) - W(s)\right] + W(s) \\ &= W(s). \end{aligned}$$

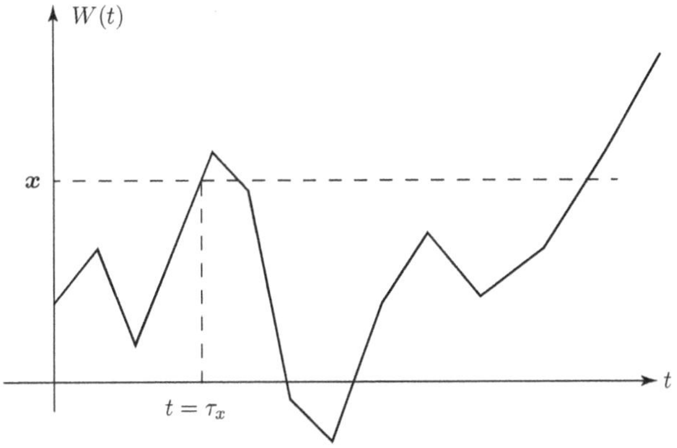


**Figure 19.1.** Sample waveform corresponding to a Brownian motion $W(t)$. As the intervals between the successive points of slope discontinuity shrinks to zero, the waveform $W(t)$ will become infinitely rough, but still remains continuous. The time $\tau_x$ at which $W(t) = x$ for the *first time* is an example of a valid stopping time $\tau$. For times prior to $\tau_x$ we can take $\tau = \infty$. For times posterior to $\tau_x$ we can take $\tau = \tau_x$.

From item (c) it is also easy to show that the expectation of the quadratic variation of a Brownian motion $\mathbb{E}\,[W,\ W](t) = \sum_{i=0}^{m-1} t_{i+1} - t_i = t$ for $t \geq 0$ a.s. It is also straightforward to show using item (c), definition 19.2.7 that the variance of the random variable $[W(t_{i+1}) - W(t_i)]^2$ is

$$\begin{aligned} \mathrm{Var}[W(t_{i+1}) - W(t_i)]^2 &= \mathbb{E}\left[\left(\{W(t_{i+1}) - W(t_i)\}^2 - (t_{(i+1)} - t_i)\right)^2\right] \\ &= 2(t_{i+1} - t_i)^2. \end{aligned} \tag{19.6}$$

Therefore the variance of the quadratic variation of a Brownian motion is

$$\begin{aligned} \mathrm{Var}[W,\ W](t) &= 2\sum_{i=0}^{m-1}(t_{i+1} - t_i)^2 \leq 2\,\|\,\Pi\,\|\sum_{i=0}^{m-1}(t_{i+1} - t_i) \\ &= 2t\,\|\,\Pi\,\| \to 0 \text{as}\ \|\,\Pi\,\| \to 0. \end{aligned}$$

We then conclude that for a Brownian motion the quadratic variation

$$\lim_{\|\Pi\| \to 0}[W,\ W](t) = t, \text{ a.s.} \tag{19.7}$$

and we say that Brownian motion accumulates quadratic variation at the rate of one per unit time. Recall that *a.s.* means that there can be some paths of the Brownian motion for which the assertion $[W,\ W](t) = t$ is not true. However, the set of all such paths has zero probability. Even though a Brownian motion is continuous, it is everywhere non-differentiable. It is this rich property that endows it a special place in stochastic analysis.

Owing to equations (19.5) and (19.6) it is meaningful to argue that when $t_{i+1} - t_i$ is small, $(t_{i+1} - t_i)^2 \ll (t_{i+1} - t_i)$ and the quantity $[W(t_{i+1}) - W(t_i)]^2$, although random, is with high probability near its mean value of $t_{i+1} - t_i$. Hence

$$[W(t_{i+1}) - W(t_i)]^2 \approx t_{i+1} - t_i.$$

For small increments $t_{i+1} - t_i = \delta t \to 0$ we write this as $\delta W(t)\delta W(t) = \delta t$. For the partition $\Pi = \{0 = t_0,\ t_1,\ \ldots,\ t_m = t\}$ we also have the results

$$\lim_{\|\Pi\| \to 0}\sum_{i=0}^{m-1}[W(t_{i+1}) - W(t_i)](t_{i+1} - t_i) = 0$$

$$\lim_{\|\Pi\| \to 0}\sum_{i=0}^{m-1}(t_{i+1} - t_i)^2 = 0.$$

For small $t_{i+1} - t_i = \delta t$ we may express these as $\delta W(t)\delta t = 0$ and $\delta t\delta t = 0$. To summarize we have the following useful *limit* relations when dealing with a Brownian motion

$$\mathrm{d}t\mathrm{d}t = 0, \tag{19.8}$$

$$\mathrm{d}W\mathrm{d}t = 0, \tag{19.9}$$

$$\mathrm{d}W^i(s)\mathrm{d}W^j(t) = \delta_i^j\ \delta(t - s)\ \mathrm{d}t\mathrm{d}s \quad \text{(inside integrals)}, \tag{19.10}$$

$$\mathrm{d}W(t)\mathrm{d}W(t) := \mathrm{d}[W, W](t) = \mathrm{d}t \quad \text{(for differentials)}. \tag{19.11}$$

In equation (19.10), the superscripts $i$, $j$ refer to independent realizations of the Brownian motion, $\delta_i^j$ is Kronecker's delta and $\delta(t)$ is the delta function.

**Definition 19.2.8** *Nonanticipative process* [6, p 55]. Let $W(t)$ be a Brownian motion and let $\mathcal{F}_t$, $t \geq 0$ be an associated filtration. A stochastic process $f(t)$ defined for $0 \leq \alpha \leq t \leq \beta < \infty$ is called *nonanticipative* function with respect to $\mathcal{F}_t$

(i) $f(t)$ is a separable process;

(ii) $f(t)$ is a measurable process, i.e. the function $(t, \omega) \to f(t, \omega)$ from $[\alpha, \beta] \times \Omega$ into $\mathbb{R}$ is measurable (observable probabilities can be assigned to the events in the set);

(iii) for each $t \in [\alpha, \beta]$, $f(t)$ is $\mathcal{F}_t$ measurable. In other words $f(t)$ is adapted to $\mathcal{F}_t$.

We denote by $L_w^p[\alpha, \beta]$, $(1 \leq p < \infty)$ the class of all nonanticipative functions $f(t)$ satisfying

$$\mathbb{P}\left\{\int_\alpha^\beta |f(t)|^p\ \mathrm{d}t < \infty\right\} = 1,$$

and by $M_w^p[\alpha, \beta]$ the subset of $L_w^p[\alpha, \beta]$ consisting of all functions $f$ with

$$\mathbb{E} \int_\alpha^\beta |f(t)|^p\ \mathrm{d}t < \infty.$$

In a sense the term *nonanticipative* with respect to $\mathcal{F}_t$ implies that the processes $f(t)$ and $W(t + \varepsilon)$, $\varepsilon > 0$ are statistically independent.

## 19.3 Itô integral and Itô–Doeblin formula

**Definition 19.3.1.** *Itô integral* [4, p 134]. Let $W(t)$, $t > 0$ be a Brownian motion on a probability space $(\Omega, \mathcal{F}, \mathbb{P})$ and let $\mathcal{F}_t$, $t > 0$ be a filtration for this Brownian motion. Note that $\mathcal{F}(W(t + \lambda) - W(t), \lambda \geq 0)$ is independent of $\mathcal{F}_t$ for all $t > 0$ in view of item (c), definition 19.2.7.

Let $\Delta(t)$, $0 \leq t \leq T$ be an *adapted*[2] stochastic process that satisfies

$$\mathbb{E}\int_0^T \Delta^2(t)\,\mathrm{d}t < \infty.$$

We consider a partition $\Pi = \{0 = u_0 < u_1 < u_2 < \ldots < u_m = t\}$ of the interval $(0, t)$ and approximate $\Delta(u)$ by a staircase function $\Delta_m(u)$ that is *right*-continuous: $\Delta_m(u) = \Delta(u_i)$, $u_i \leq u < u_{i+1}$, $0 \leq i \leq m - 1$. Then *Itô integral* (also known as the *Itô–Stieltjes integral*) is defined as

$$\begin{aligned} I_\Delta(t) := \int_0^t \Delta(u)\,\mathrm{d}W(u) &= \lim_{m\to\infty}\int_0^t \Delta_m(u)\,\mathrm{d}W(u) \\ &= \lim_{m\to\infty}\int_0^t \Delta(u_i)[W(u_{i+1}) - W(u_i)],\ 0 \leq t \leq T, \end{aligned} \tag{19.12}$$

The process $\Delta(t)$ is known as the *integrand* and the process $W(t)$ is the *integrator*. If an integrator $X$ is right-continuous and has finite quadratic variation and the integrand $Y$ is regular for any time interval $[0, b]$, then the Itô integral $\int_a^b Y\,\mathrm{d}X$ exists and the sum as defined in equation (19.12) for the special case of $X(t) = W(t)$ is convergent [8].

The Itô integral has the following properties which can all be established using items (b), (c), definition 19.2.7, and (19.8)–(19.11)

(a) $I_\Delta(t)$ is a martingale,

(b) $\mathbb{E}\, I_\Delta^2(t) = \mathbb{E}\int_0^t \Delta^2(u)\,\mathrm{d}u$ (Itô isometry),

(c) $[I_\Delta, I_\Delta](t) = \int_0^t \Delta^2(u)\,\mathrm{d}u$ (quadratic variation of Itô integral accumulated up to time $t$),

(d) If $\Delta(t) \in L_w^2[0, T]$ and $\varepsilon > 0$, $N > 0$, then

$$\mathbb{P}\left\{\sup_{0\leq t\leq T} |I_\Delta(t)| > \varepsilon\right\} \leq \mathbb{P}\left\{\int_0^T \Delta^2(t)\,\mathrm{d}t > N\right\} + \frac{N}{\varepsilon^2}.$$

It is very important to keep in mind the right-continuity nature of the process $\Delta_m(t)$, which renders it nonanticipative (see definition 19.2.8), see figure 19.2. We now look at a very instructive example to bring out the intricacies of the Itô integral. Consider

[2] Meaning we require $\Delta(t)$ to be $\mathcal{F}_t$ measurable for each $t \geq 0$. In other words, the information available at time $t$ about $W(t)$ is sufficient to evaluate $\Delta(t)$ at that time. When we are standing at time 0 and $t$ is strictly positive, $\Delta(t)$ is unknown to us as it is a random variable. When we get to time $t$, we have sufficient information to evaluate $\Delta(t)$; its randomness has been resolved. As an example $\Delta(t)$ could be taken to be $W(t)$.

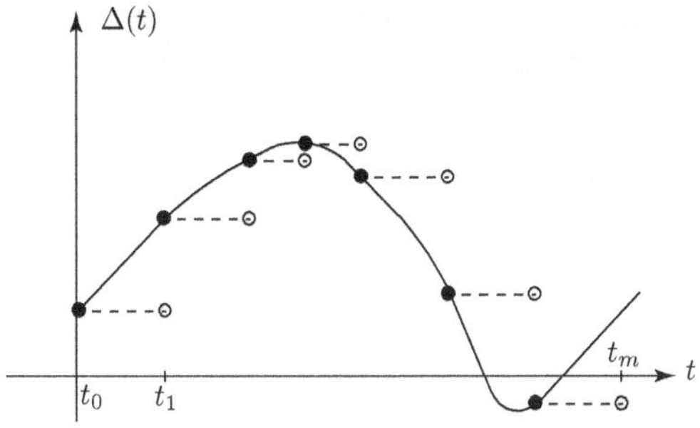


**Figure 19.2.** The integrand $\Delta(t)$ and its right-continuous approximation (shown by the dashed lines) $\Delta_m(t)$. The function $\Delta_m(t)$ is defined at the left end of any sub-interval $(t_i, t_{i+1})$ (the black dot in the figure) and is held constant until the end of that sub-interval. $\Delta_m(t)$ is right-continuous.

$$
\begin{aligned}
I_W(t) = \int_0^t W(u)\,\mathrm{d}W(u) &= \lim_{m\to\infty}\sum_{i=0}^{m-1} W(u_i)[W(u_{i+1}) - W(u_i)] \\
&= \frac{1}{2}W^2(u_m) - \frac{1}{2}\lim_{m\to\infty}\sum_{i=0}^{m-1}[W(u_{i+1}) - W(u_i)]^2 \\
&= \frac{1}{2}W^2(t) - \frac{1}{2}[W, W](t) = \frac{1}{2}W^2(t) - \frac{1}{2}t.
\end{aligned}
\tag{19.13}
$$

$I_W(t)$ is a martingale with an expectation of zero, since $\mathbb{E}\ W^2(t) = t$. We contrast (19.13) with the result obtained from ordinary calculus for a differential function $g(t)$ with $g(0) = 0$:

$$
\int_0^t g(u)\,\mathrm{d}g(u) = \frac{1}{2}\int_0^t \mathrm{d}[g^2(u)] = \frac{1}{2}g^2(t). \tag{19.14}
$$

The extra term $-\frac{1}{2}t$ in equation (19.13) comes from the non-zero quadratic variation of Brownian motion and from using the right-continuous definition of $\Delta_m(t)$ in equation (19.12). Without the extra term, $I_W(t)$ will not be a martingale. The differential form of Itô integral is $\mathrm{d}I_\Delta(t) = \Delta(t)\,\mathrm{d}W(t)$.

**Theorem 19.3.1.** Martingale property [6, p 72].
*If $f \in M_w^2[0, \tau]$ and $\tau$ is a stopping time with respect to $\mathcal{F}_t$, $0 \le \tau \le T$, i.e. $\{\tau \le t\} \in \mathcal{F}_t$ for all $0 \le t \le T$, then the process*

$$
I_f(t) = \int_0^{\tau\wedge t} f(s)\,\mathrm{d}W(s), \quad 0 \le t \le T,
$$

*is a martingale with expectation equals zero.*

**Definition 19.3.2** *Markov process* [4, p 74]. Let $(\Omega, \mathcal{F}, \mathbb{P})$ be a probability space. Let $T$ be a fixed positive number, and $\mathcal{F}_t$, $0 \le t \le T$ be a filtration of the sub-$\sigma$-algebras of $\mathcal{F}$. Consider an adapted stochastic process $X(t), 0 \leq t \leq T$. If for all $0 \leq s \leq t \leq T$ and for *every non-negative*, Borel-measurable function $f(t, x)$, we have

$$\mathbb{E}\,[f(t, X(t))|\mathcal{F}_s] = f(s, X(s)),$$

then $X(t)$ is a *Markov process*.

**Definition 19.3.3.** *Itô process* [4, 6]. Let $W(t)$ be a Brownian motion and let $\mathcal{F}_t$, $t \geq 0$ be an associated filtration. An *Itô process* is a stochastic process of the form

$$X(t) = X(0) + \int_0^t \Delta(u)\,\mathrm{d}W(u) + \int_0^t \Theta(u)\,du, \tag{19.15}$$

where $X(0)$ is non-random and $\Delta(u)$ and $\Theta(u)$ are adapted stochastic processes.

The quadratic variation of the Itô process is

$$[X, X](t) = \int_0^t \Delta^2(u)\,\mathrm{d}u. \tag{19.16}$$

The differential forms of Itô process and of its quadratic variation are

$$\mathrm{d}X(t) = \Delta(t)\,\mathrm{d}W(t) + \Theta(t)\,\mathrm{d}t, \tag{19.17}$$

$$\begin{aligned} \mathrm{d}X(t)\mathrm{d}X(t) := \mathrm{d}[X, X](t) &= \Delta^2(t)\mathrm{d}W(t)\mathrm{d}W(t) \\ &\quad + 2\Delta(t)\Theta(t)\mathrm{d}W(t)\mathrm{d}t + \Theta^2(t)\mathrm{d}t\mathrm{d}t, \\ &= \Delta^2(t)\,\mathrm{d}t, \end{aligned} \tag{19.18}$$

upon using the limit relations (19.8)–(19.11). The coefficient of $\mathrm{d}W(t)$ in equation (19.17) is known as the diffusion component and the coefficient of $\mathrm{d}t$ in equation (19.17) is known as the drift component. Note that the quadratic variation of the Itô process (19.16) is entirely due to the diffusion term and is a non-decreasing function of the time variable. Equations of the form (19.17) involving stochastic functions are known as stochastic differential equations (SDE). Note also that the Itô process is a nonanticipative function. Furthermore it is a continuous process. Hence it belongs to $L_w^\infty[0, T]$ (since $|X(t)| < \infty$).

**Theorem 19.3.2** Itô–Doeblin formula for an Itô process [4, p 146].
*Let $X(t)$, $t \geq 0$ be an Itô process as described in equation (19.15) or (19.17) and let $f(t, x)$ be a function for which the partial derivatives $f_t(t, x), f_x(t, x), f_{xx}(t, x)$ are defined and continuous. Then for for every $t > 0$,*

$$\begin{aligned} f(t, X(t)) = f(0, X(0)) &+ \int_0^t f_t(u, X(u))\,\mathrm{d}u \\ &+ \int_0^t f_x(u, X(u))\,\mathrm{d}X(u) + \frac{1}{2}\int_0^t f_{xx}(u, X(u))\Delta^2(u)\,\mathrm{d}u. \end{aligned} \tag{19.19}$$

This can be easily proved by resorting to Taylor's series expansion up to second order and considering the partition $\Pi = \{0 = t_0 < t_1 < t_2 < \cdots < t_m = t\}$ of $(0, t)$. We are interested in the difference between $f(t, X(t))$ and $f(0, X(0))$. This change in $f(t, X(t))$ over $(0, t)$ can be written as the sum of changes in $f(t, X(t))$ over each of the subintervals $[t_i, t_{i+1}]$:

$$
\begin{aligned}
f(t, X(t)) - f(0, X(0)) = & \sum_{i=0}^{m-1} f_t(t_i, X(t_i))(t_{i+1} - t_i) + \sum_{i=0}^{m-1} f_x(t_i, X(t_i))(X(t_{i+1}) - X(t_i)) \\
& + \frac{1}{2}\sum_{i=0}^{m-1} f_{xx}(t_i, X(t_i))(X(t_{i+1}) - X(t_i))^2 + \\
& \frac{1}{2}\sum_{i=0}^{m-1} f_{tx}(t_i, X(t_i))(t_{i+1} - t_i)(X(t_{i+1}) - X(t_i) \\
& + \frac{1}{2}\sum_{i=0}^{m-1} f_{tt}(t_i, X(t_i))(t_{i+1} - t_i)^2 + \cdots.
\end{aligned}
$$

Taking the limit as $m \to \infty$ and using equations (19.8)–(19.11) we obtain the desired result in equation (19.19). The differential form of equation (19.19) for the total derivative of $f$ is the stochastic differential equation:

$$
\mathrm{d}f(t, X(t)) = f_t(t, X(t))\mathrm{d}t + f_x(t, X(t))\mathrm{d}X(t) + \frac{1}{2}f_{xx}(t, X(t))\Delta^2(t)\mathrm{d}t. \quad (19.20)
$$

Itô's formula is the stochastic generalization of the fundamental theorem of calculus. The important difference with respect to ordinary calculus is the presence of the second order term $f_{xx}(\cdot, \cdot)$ in equations (19.19) and (19.20), which arises entirely from the non-zero quadratic variation of the Brownian process and the assumed *right*-continuity of $\Delta_m(t)$. It is this connection between the second derivative of a function and quadratic variation of a Brownian motion that forms the foundation of the various stochastic representations of the solutions of partial differential equations having second order space derivatives. Equation (19.19) also remains valid when $t$ is replaced by any random variable $\tau$, $0 \leq \tau \leq T$. If, in particular, $\tau$ is a stopping time, when taking expectation and using theorem 19.3.1 we find that

$$
\mathbb{E}f(\tau, X(\tau)) - \mathbb{E}f(0, X(0)) = \mathbb{E}\int_0^\tau (Lf)(t, X(t))\,\mathrm{d}t, \quad (19.21)
$$

where

$$
Lf = f_t + \Theta f_x + \frac{1}{2}\Delta^2 f_{xx}, \quad (19.22)
$$

provided $\Delta(t)f_x(t, X(t))$ belongs to $M_w^2[0, T]$ and $(Lf)(t, X(t))$ belongs to $M_x^1[0, T]$.

**Theorem 19.3.3** Multi-dimensional Itô–Doeblin formula [6, p 90].
*Let $u(t, \boldsymbol{x})$ be a continuous function in $(t, \boldsymbol{x}) \in [0, \infty) \times R^n$ together with its partial derivatives $u_t$, $u_{x_i}$, $u_{x_i x_j}$. Let $\boldsymbol{\xi}(t)$ be an n-dimensional Itô process having an SDE*

$$\mathrm{d}\boldsymbol{\xi}(t) = \boldsymbol{a}(t)\, \mathrm{d}t + \boldsymbol{B}(t)\, \mathrm{d}\boldsymbol{w}(t),$$

*where $\boldsymbol{w}(t) = [w_1(t), \ldots, w_m(t)]'$ is the multi-dimensional Brownian vector, $\boldsymbol{a} = [a_1(t), \ldots, a_n(t)]'$ and $\boldsymbol{B} = \{b_{ij}(t)\}$, $1 \le i \le n$, $1 \le j \le m$, belong element-wise to $L^1_W[0, T]$ and $L^2_W[0, T]$, respectively. Then $u(t, \boldsymbol{\xi}(t))$ satisfies the SDE*

$$\mathrm{d}u(t, \boldsymbol{\xi}(t)) = \left[ u_t(t, \boldsymbol{\xi}(t)) + \sum_{i=1}^{n} u_{x_i}(t, \boldsymbol{\xi}(t)) a_i(t) + \frac{1}{2} \sum_{l=1}^{m} \sum_{i,j=1}^{n} u_{x_i x_j}(t, \boldsymbol{\xi}(t)) b_{il}(t) b_{jl}(t) \right] \mathrm{d}t$$
$$+ \sum_{l=1}^{m} \sum_{i=1}^{n} u_{x_i}(t, \boldsymbol{\xi}(t)) b_{il}(t) \mathrm{d}w_l(t). \tag{19.23}$$

Let

$$A = \boldsymbol{B}\boldsymbol{B}',$$
$$Lu = \frac{1}{2} \sum_{i,j=1}^{n} A_{ij}(t) \frac{\partial^2 u}{\partial x_i \partial x_j} + \sum_{i=1}^{n} a_i(t) \frac{\partial u}{\partial x_i} + \frac{\partial u}{\partial t}.$$

Then equation (19.23) can be expressed in the form

$$\mathrm{d}u(t, \boldsymbol{\xi}(t)) = Lu(t, \boldsymbol{\xi}(t))\, \mathrm{d}t + u_x(t, \boldsymbol{\xi}(t)) \cdot \boldsymbol{A}(t) \mathrm{d}\boldsymbol{w}(t). \tag{19.24}$$

The integral form of equation (19.23) or (19.24) is valid for any random time $\tau$, $0 \le \tau \le T$. Indeed

$$u(\tau, \boldsymbol{\xi}(\tau)) - u(0, \boldsymbol{\xi}(0) = \int_0^{\tau} (Lu)(s, \boldsymbol{\xi}(s))\, \mathrm{d}s + \int_0^{\tau} u_x(s, \boldsymbol{\xi}(s)) \cdot \boldsymbol{A}(s) \mathrm{d}\boldsymbol{w}(s). \tag{19.25}$$

If $\tau$ is any stopping time and $Lu$, $u_x \cdot \boldsymbol{A}$ are in $M^1_w[0, T]$ and $M^2_w[0, T]$, respectively, then taking an expectation we obtain

$$\mathbb{E}\, u(\tau, \boldsymbol{\xi}(\tau)) - \mathbb{E}\, u(0, \boldsymbol{\xi}(0)) = \mathbb{E} \int_0^{\tau} Lu(s, \boldsymbol{\xi}(s))\, \mathrm{d}s. \tag{19.26}$$

Note that for any adapted stochastic process $X(t)$, $\mathbb{E}\, [\gamma(t, X(t))\mathrm{d}W(t)] = 0$ because $\mathrm{d}W(t)$ is treated *as occurring in advance* of any change in the process $\gamma(t, X(t))$ and the mean value of $\mathrm{d}W(t)$ is zero. This is a consequence of the fact that the integrand is evaluated at the left end point of each subdivision $(t_i, t_{i+1})$ in the definition of Itô integral. The result of this choice for the value of the integrand is to destroy the usual time-reversal symmetry that may be contained in the solution of ordinary differential equations. Equation (19.26) is known as *Dynkin's formula* [5]. By considering the function $u(t, x)\mathrm{e}^{-\alpha t}$, $\alpha \ge 0$, we can an extension of the Dynkin's formula

$$\mathbb{E}\,[u(\tau,\,\xi(\tau))e^{-\alpha\tau}] - \mathbb{E}\,u(0,\,\xi(0)) = \mathbb{E}\int_0^{\tau} e^{-\alpha s}(L-\alpha)u(s,\,\xi(s))\,\mathrm{d}s. \tag{19.27}$$

**Corollary 19.3.3.1** Itô product rule [4, p 168].
*Let $X(t)$ and $Y(t)$ be two 1D Itô processes. Then*

$$\mathrm{d}(X(t)Y(t)) = X(t)\mathrm{d}Y(t) + Y(t)\mathrm{d}X(t) + \mathrm{d}X(t)\mathrm{d}Y(t). \tag{19.28}$$

*In particular if* $\mathrm{d}X(t) = a_x(t)\mathrm{d}t + b_{xx}(t)\mathrm{d}W_x(t) + b_{xy}(t)\mathrm{d}W_x$, $\mathrm{d}Y(t) = a_y(t)\mathrm{d}t + b_{yx}(t)\mathrm{d}W_x(t) + b_{yy}(t)\mathrm{d}W_y(t)$, *where $W_x(t)$ and $W_y(t)$ are independent Brownian motions then*

$$\mathrm{d}(X(t)Y(t)) = X(t)\mathrm{d}Y(t) + Y(t)\mathrm{d}X(t) + [b_{xx}(t)b_{yx}(t) + b_{xy(t)}b_{yy}(t)]\mathrm{d}t.$$

**Lemma 19.3.4** [4, p 268].
*Let $X(u)$ be a solution to the stochastic differential equation (SDE)*

$$\mathrm{d}X(u) = \beta(u,\,X(u))\mathrm{d}u + \gamma(u,\,X(u))\mathrm{d}W(u) \tag{19.29}$$

*consisting of the drift coefficient $\beta(u,\,x)$ and the diffusion coefficient $\gamma(u,\,x)$ with an initial condition given at time 0. Let $h(y)$ be a Borel-measurable function. Fix $T > 0$ and let $t \in [0,\,T]$ be given. Define the function*

$$g(t,\,x) = \mathbb{E}\,^t_x h(X(T)) \quad \text{(assumed } < \infty \text{ for all } t,\,x), \tag{19.30}$$

*where $\mathbb{E}\,^t_x$ denotes the expectation operator given $X(t) = x$. Then the stochastic process*

$$g(t,\,X(t)), \quad 0 \le t \le T, \tag{19.31}$$

*obtained by replacing the dummy variable $x$ by $X(t)$ in equation (19.30) satisfies*

$$\mathbb{E}\,[h(X(T))|\mathcal{F}_t] = g(t,\,X(t)), \quad 0 \le t \le T, \tag{19.32}$$

*and is a martingale.*

If we do not have an explicit formula for the distribution of $X(T)$, we could compute $g(t,\,x)$ numerically by beginning at $X(t) = x$ and simulating the SDE (19.29). This would give one realization of $X(T)$ corresponding to one $\omega$. The process is repeated many times and the average of the random variable $h(X(T))$ computed over all realizations to obtain $g(t,\,x)$. The conditional average given $t$ and $x$ is represented by the notation $\mathbb{E}\,^t_x$. A physical picture is in order at this point. Imagine $X(t)$ to represent the trajectories of particles in spacetime, all released from an initial location specified at $t = 0$ and governed by the SDE (19.29). The particle locations at the exit time $t = T$ are given by the random variable $X(T)$. The random variable $h(X(T))$ corresponds to observed particle locations at $t = T$ after having passed

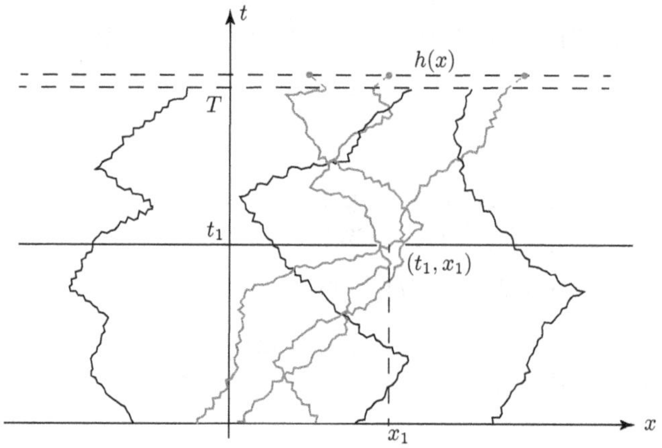


**Figure 19.3.** Itô processes initiated at $t = 0$ and arriving at the exit time $T$. The red trajectories all transit through the spacetime slit at $(x_1, t_1)$. A spatial filter $h(x)$ erected at $t = T$ instantly maps the trajectories to the dots. The expectation operator $\mathbb{E}\,{}^{t_1}_{x_1}$ corresponds to obtaining an average location of the red dots.

through a continuous spatial filter $h(x)$. The averaged filtered location of those particle trajectories (red trajectories in figure 19.3) which pass through an intermediate spacetime split at $(t_1, x_1)$ and arriving at the exit time $T$ is given by $g(t_1, x_1)$.

Equation (19.32) follows from the independence lemma 19.1.2. Using this result we have for $0 \le s \le t \le T$ that

$$\begin{aligned}\mathbb{E}\,[g(t, X(t))|\mathcal{F}_s] &= \mathbb{E}\,[\mathbb{E}\,[h(X(T))|\mathcal{F}_t]|\mathcal{F}_s] \\ &= \mathbb{E}\,[h(X(T))|\mathcal{F}_s] \text{ (using theorem 19.1.1 since } \mathcal{F}_s \subseteq \mathcal{F}_t) \\ &= g(s, X(s)).\end{aligned} \tag{19.33}$$

Hence $g(t, X(t))$ is a martingale. The expected value of a martingale is fixed, and hence the coefficient of $\mathrm{d}t$ term in the differential $\mathrm{d}g(t, X(t))$ must be zero. This leads directly to the Feynman–Kac theorem, which is considered next.

## 19.4 Solution of PDEs by stochastic technique, Feynman–Kac formulas

**Theorem 19.4.1.** Feynman–Kac [4, p 268].
*Consider the stochastic differential equation (19.29). Let $h(y)$ be a Borel-measurable function. Fix $T > 0$ and let $t \in [0, T]$ be given. Define the functions*

$$g(t, x) = \mathbb{E}\,{}^t_x h(X(T)) \quad (\mathbb{E}\,{}^t_x|h(X(T))| < \infty \text{ for all } t, x),$$
$$f(t, x) = \mathbb{E}\,{}^t_x[\mathrm{e}^{-r(T-t)}h(X(T))] \quad (r \ge 0, \mathbb{E}\,{}^t_x|h(X(T))| < \infty \text{ for all } t, x)$$

*where $X(T)$ is the solution of SDE (19.29) defined for $T \ge t$ with initial condition $X(t) = x$. Then $g(t, x)$ satisfies the parabolic partial differential equation*

$$\frac{1}{2}\gamma^2(t, x)g_{xx}(t, x) + \beta(t, x)g_x(t, x) + g_t(t, x) = 0, \tag{19.34}$$

*and the terminal condition*

$$g(T, x) = h(x) \text{ for all } x,$$

*and* $f(t, x)$ *satisfies the parabolic partial differential equation*

$$\frac{1}{2}\gamma^2(t, x)f_{xx}(t, x) + \beta(t, x)f_x(t, x) + f_t(t, x) - rf(t, x) = 0, \tag{19.35}$$

*and the terminal condition*

$$f(T, x) = h(x) \text{ for all } x.$$

According to the Itô–Doeblin formula we have that

$$\begin{aligned} \mathrm{d}g(t, X(t)) &= g_t\,\mathrm{d}t + g_x\,\mathrm{d}X + \tfrac{1}{2}g_{xx}\,\mathrm{d}X\mathrm{d}X \\ &= \left[g_t + \beta g_x + \tfrac{1}{2}\gamma^2 g_{xx}\right]\mathrm{d}t + \gamma g_x\,\mathrm{d}W. \end{aligned}$$

Since $g(t, X(t))$ is a martingale per lemma 19.3.4, we equate the coefficient of $\mathrm{d}t$ to zero to obtain (since $\mathbb{E}\,\gamma g_x\,\mathrm{d}W = \mathbb{E}\,(\gamma g_x\,\mathrm{d}W|\mathcal{F}_t) = \mathbb{E}\,(\gamma g_x)|\mathcal{F}_t)\mathbb{E}\,(\mathrm{d}W) = 0$)

$$g_t(t, X(t)) + \beta(t, X(t))g_x(t, X(t)) + \frac{1}{2}\gamma^2(t, X(t))g_{xx}(t, X(t)) = 0,$$

along every path of $X(t)$. Equation (19.34) follows for every point $(t, x)$ that can be reached by $(t, X(t))$. From the definition of $g(t, x)$ we have $g(T, x) = \mathbb{E}\,_x^T h(X(T))= h(x)$. The proof for $f(t, x)$ can be carried out in a similar fashion by considering the martingale $\mathrm{e}^{-rt}f(t, X(t))$. Equations (19.34) and (19.35) are also known as *Kolmogorov equations* or *backward parabolic equations* [6, p 124].

**Theorem 19.4.2** Principal eigenvalue of elliptic systems [10, p 373].
*Let* $\boldsymbol{\xi}(t)$ *be an n-dimensional Itô process having an SDE*

$$\mathrm{d}\boldsymbol{\xi}(t) = \boldsymbol{b}(\boldsymbol{\xi}(t))\,\mathrm{d}t + \boldsymbol{\sigma}(\boldsymbol{\xi}(t))\,\mathrm{d}\boldsymbol{w}(t), \tag{19.36}$$

*where* $\boldsymbol{w}(t) = [w_1(t), \ldots, w_n(t)]'$ *is the multi-dimensional Brownian vector,* $\boldsymbol{b} = [b_1(t), \ldots, b_n(t)]'$ *and* $\boldsymbol{\sigma} = \{\sigma_{ij}(t)\}$, $1 \le i, j \le n$, *belong element-wise to* $L_W^1[0, T]$ *and* $L_W^2[0, T]$, *respectively. Set* $\boldsymbol{A} = \boldsymbol{\sigma\sigma}'$ *and let*

$$\begin{aligned} |\boldsymbol{A}(\boldsymbol{x})| \le M, &\quad |\boldsymbol{A}(\boldsymbol{x}) - \boldsymbol{A}(\boldsymbol{y})| \le M|\boldsymbol{x} - \boldsymbol{y}|^\alpha \\ |\boldsymbol{b}(\boldsymbol{x})| \le M, &\quad |\boldsymbol{b}(\boldsymbol{x}) - \boldsymbol{b}(\boldsymbol{y})| \le M|\boldsymbol{x} - \boldsymbol{y}|^\alpha, \end{aligned}$$

*for all* $\boldsymbol{x}$, $\boldsymbol{y}$ *in* $R^n$, *where* $M$, $\alpha$ *are positive constants,* $0 < \alpha < 1$, *and the notation* $|\cdot|$ *refers to a matrix or vector element-wise absolute sum. Furthermore let* $\boldsymbol{\xi}' \boldsymbol{A}(\boldsymbol{x})\boldsymbol{\xi} \geq \mu|\boldsymbol{\xi}|^2$, $\mu > 0$ *for all* $\boldsymbol{x}$ *and* $\boldsymbol{\xi}$ *in* $R^n$, *i.e.* $\boldsymbol{A}$ *is a positive definite matrix. Set*

$$Lu = \frac{1}{2}\sum_{i,j=1}^{n} A_{ij}(t)\frac{\partial^2 u}{\partial x_i \partial x_j} + \sum_{i=1}^{n} b_i(t)\frac{\partial u}{\partial x_i}. \tag{19.37}$$

*Let* $D$ *be a bounded domain in* $R^n$ *with* $C^2$ *boundary* $\partial D$ *and consider the eigenvalue problem*

$$\begin{aligned} -Lu &= \lambda u \quad \text{in } D,\\ u &= 0 \quad \text{on } \partial D. \end{aligned}$$

*If* $A_{ij} = \delta_i^j$ *and* $\boldsymbol{b}$ *is the gradient of a function, then* $L$ *will be self-adjoint in a suitable space. Even if* $L$ *is not self-adjoint there exists at least one positive eigenvalue. Let* $\lambda_0$ *be the smallest positive eigenvalue. Then*

$$\lambda_0 = \sup\left\{\Lambda \geq 0;\ \sup_{x\in D} \mathbb{E}_x\, \mathrm{e}^{\Lambda\tau_x} < \infty\right\}, \tag{19.38}$$

*where* $\tau_x$

$$\tau_x = \inf\{t > 0;\ \boldsymbol{\xi}(t) \notin D,\ \boldsymbol{\xi}(0) = \boldsymbol{x} \in D\}, \tag{19.39}$$

*is the first exit time*[3] *from* $D$ *of the solution of equation (19.36) with the initial condition* $\boldsymbol{\xi}(0) = \boldsymbol{x}$, *and* $\mathbb{E}_x$ *is the expectation operator conditioned on* $\boldsymbol{x}$.

**Theorem 19.4.3.** Stochastic representation of solutions of elliptic equations [6, p 145], [10, p 375].
*Suppose* $c(\boldsymbol{x}) < 0$ *and consider the Dirichlet problem*

$$Lu + cu = -f \quad \text{in } D \quad (f \text{ Hölder continuous in } \bar{D} = D \cup \partial D) \tag{19.40}$$

$$u = \psi \quad \text{on } \partial D \quad (\psi \text{ continuous on } \partial D), \tag{19.41}$$

*where* $L$ *is defined in equation (19.37). Define*

$$\gamma_c(\tau) = \exp\left[\int_0^{\tau} c(\boldsymbol{\xi}(t))\,\mathrm{d}t\right]. \tag{19.42}$$

*Then*

$$u(\boldsymbol{x}) = \mathbb{E}_x[\psi(\boldsymbol{\xi}(\tau_x))\gamma_c(\tau_x)] + \mathbb{E}_x\left[\int_0^{\tau_x} f(\boldsymbol{\xi}(s))\gamma_c(s)\,\mathrm{d}s\right], \tag{19.43}$$

[3] *It is easy to show using the Dynkin's formula (19.26) and specific examples of the function u that the first exit time for Brownian motion is finite for finite domains D.*

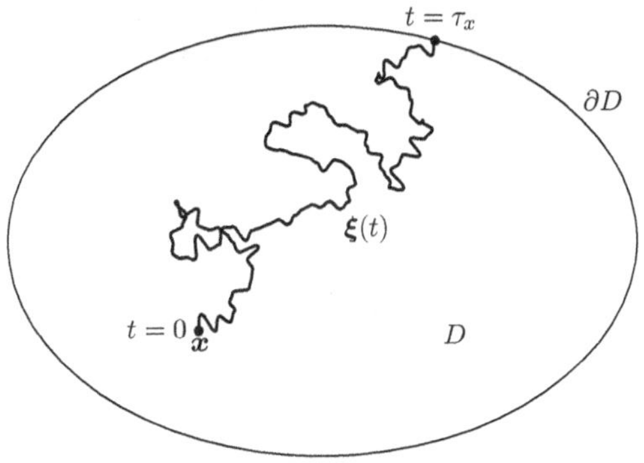


**Figure 19.4.** Itô process initialized at $\boldsymbol{x}$ and hitting the boundary of $D$ for the first time at $t = \tau_X$.

*where $\boldsymbol{\xi}(t)$ is the solution of the SDE (19.36) with initial condition $\boldsymbol{\xi}(0) = \boldsymbol{x}$ and $\tau_X$ is the first exit time of that trajectory from D (figure 19.4).*

The representation (19.43) continues to be valid for $c(\boldsymbol{x})$ positive provided that $\sup_{x \in D} c(\boldsymbol{x}) < \lambda_0$, which assures that $\mathbb{E}\, \gamma_c(\tau_X) < \infty$. Recently Janaswamy [1] used the representation (19.43) to determine electromagnetic fields involving plasmonic materials, which have $\Re(c) < 0$.

**Theorem 19.4.4** Stochastic representation of solutions of parabolic equations [11, p 133]. *Let $u(t, \boldsymbol{x})$ be a solution of*

$$\frac{\partial u(t, \boldsymbol{x})}{\partial t} = L\, u(t, \boldsymbol{x}) + c(t, \boldsymbol{x})u(t, \boldsymbol{x}),\ t > 0,\ \boldsymbol{x} \in D, \tag{19.44}$$

$$u(0, \boldsymbol{x}) = f(\boldsymbol{x}), \quad u(t, \boldsymbol{x}) = g(t, \boldsymbol{x}) \text{ for } \boldsymbol{x} \in \partial D, \tag{19.45}$$

*which is continuous and bounded on $\bar{Q} = [0, T] \times \bar{D}$, where D is a bounded domain with $C^2$ boundary $\partial D$ and L is defined in equation (19.37). Let $\boldsymbol{\xi}$ be the solution of* $\mathrm{d}\boldsymbol{\xi}^t(s) = \boldsymbol{\sigma}(\lambda_s^t, \boldsymbol{\xi}^t(s))\mathrm{d}\boldsymbol{w}(s) + \boldsymbol{b}(\lambda_s^t, \boldsymbol{\xi}^t(s))\mathrm{d}s$, $\boldsymbol{\xi}^t(0) = \boldsymbol{x}$ *where* $\lambda_s^t = t - s$. *Let* $\tau_x^t = \inf\{s\colon \boldsymbol{\xi}^t(s) \in \partial D\}$, $\tau_t = (t \wedge \tau_x^t)$. *Let $\boldsymbol{A}$, $\boldsymbol{b}$ be uniformly Lipschitz continuous in* $Q = [0, T) \times D$, $S = [0, T) \times \partial D$. *Let c be uniformly Hölder continuous in* $(\boldsymbol{x}, t) \in \bar{Q}$, *f continuous on* $\bar{D}$, *and g continuous on* $\bar{S}$ *and* $g(\boldsymbol{x}, t) = f(\boldsymbol{x})$. *Then $u(t, x)$ has the stochastic representation*

$$\begin{aligned} u(t, \boldsymbol{x}) = {} & \mathbb{E}_x\left\{f(\boldsymbol{\xi}^t(t))\mathbb{1}_{\tau_t = t} \exp\left[\int_0^t c(\lambda_s^t, \boldsymbol{\xi}^t(s))\, ds\right]\right\} \\ & + \mathbb{E}_x\left\{g(\tau_x^t, \boldsymbol{\xi}^t(\tau_x^t))\mathbb{1}_{\tau_t \neq t} \exp\left[\int_0^{\tau_x^t} c(\lambda_s^t, \boldsymbol{\xi}^t(s))\, ds\right]\right\}, \end{aligned} \tag{19.46}$$

*where $\mathbb{1}_{\tau_t = t}$ is the indicator function of the set $\{\omega\colon \tau_x^t = t\}$ and $\mathbb{1}_{\tau_t \neq t} = 1 - \mathbb{1}_{\tau_t = t}$. If there are no boundaries, we take $g(t, \boldsymbol{x}) = 0$. Furthermore $\tau_x^t \to \infty$ and hence $\tau_t = t$. We obtain the representation*

$$u(t, \boldsymbol{x}) = \mathbb{E}_x\left\{f(\boldsymbol{\xi}^t(t))\exp\left[\int_0^t c(\lambda_s^t, \boldsymbol{\xi}^t(s))\,\mathrm{d}s\right]\right\}. \tag{19.47}$$

In reference [2], Galdi *et al* applied formula (19.47) to determine the lowest order cut-off frequency of a dielectric waveguide. Mikhailov [12] also discusses the solution of Helmholtz equation for interior problems under the first resonance by stochastic techniques.

## References

[1] Janaswamy R 2017 Field determination near plasmonic structures by the Feynman–Kac stochastic representation *IEEE Antennas Wirel. Propag. Lett.* **16** 1643–6

[2] Galdi V, Pierro V and Pinto I M 1997 Cutoff frequency and dominant eigenfunction computation in complex dielectric geometries via Donsker–Kac formula and Monte Carlo method *Electromagnetics* **17** 1–14

[3] Janaswamy R 2013 On random time and the relation between wave and telegraph equations *IEEE Trans. Antennas Propag.* **61** 2735–44

[4] Shreve S E 2004 *Stochastic Calculus for Finance II: Continuous-Time Models* (New York: Springer)

[5] Øksendal B 2003 *Stochastic Differential Equations: An Introduction with Applications* (New York: Springer)

[6] Friedman A 1975 *Stochastic Differential Equations and Applications* vol 1 (New York: Academic)

[7] Arnold L 1974 *Stochastic Differential Equations: Theory and Applications* (New York: Wiley)

[8] Medvegyev P 2007 *Stochastic Integration Theory* (Oxford: Oxford University Press)

[9] Durrett R 1996 *Stochastic Calculus: A Practical Approach* (Boca Raton, FL: CRC Press)

[10] Friedman A 1976 *Stochastic Differential Equations and Applications* vol 2 (New York: Academic)

[11] Freidlin M 1985 *Functional Integration and Partial Differential Equations* (Princeton, NJ: Princeton University Press)

[12] Mikhailov G A 1995 The Monte-Carlo methods for solving vector and stochastic Helmholtz equations *Siberian Math. J.* **36** 517–25

**IOP** Publishing

Engineering Electrodynamics
A collection of theorems, principles and field representations
**Ramakrishna Janaswamy**

# Appendix A

# Complex variable theory

## A.1 Preliminaries

### A.1.1 Analyticity

A complex valued function[1] $f(z) = u(z) + \mathrm{i}v(z)$ in the complex variable $z = x + \mathrm{i}y$ with real $x$, $y$, $u$, $v$ is said to be *analytic* (or holomorphic or regular) at a point $z_0$ if it is single-valued and has a derivative $f'(z)$ in the neighborhood of $z_0$. Let $z_\delta = \delta e^{\mathrm{i}\theta}$ be any point on a small circle of radius $\delta$ and whose center is at $z = z_0$. The derivative of the function at a point $z_0$ is defined as the limit

$$f'(z_0) = \lim_{\delta \to 0} \frac{f(z_0 + z_\delta) - f(z_0)}{\delta}. \tag{A.1}$$

Thus the limit must be independent of the angle $\theta$ as $\delta \to 0$. By successively considering $\theta = 0$ and $\theta = \pi/2$, the following necessary and sufficient conditions for a function to be analytic at the point $z = z_0$ are established:

$$\frac{\partial u}{\partial x} = \frac{\partial v}{\partial y}, \tag{A.2}$$

$$\frac{\partial u}{\partial y} = -\frac{\partial v}{\partial x}. \tag{A.3}$$

These equations are known as the *Cauchy–Riemann* (C–R) conditions [1, p 67]. The two equations may be combined into a single one and be expressed as

$$\mathrm{i}\frac{\partial f}{\partial x} = \frac{\partial f}{\partial y}. \tag{A.4}$$

[1] In this chapter the symbols 'i' and 'j' are used interchangeably to denote $\sqrt{-1}$.

doi:10.1088/978-0-7503-1716-0ch20

An analytic function will also have continuous second derivatives at $z_0$ with respect to (w.r.t) $x$ and $y$. By taking second derivatives of $u$ and $v$ and making use of the C–R conditions, it can be readily seen that

$$\frac{\partial^2 u}{\partial x^2} + \frac{\partial^2 u}{\partial y^2} = 0, \qquad \frac{\partial^2 v}{\partial x^2} + \frac{\partial^2 v}{\partial y^2} = 0. \tag{A.5}$$

Thus the real and imaginary parts of an analytic function satisfy Laplace's equation.

### A.1.2 Singularities

For every function except a constant, there are one or more points in the $z$-plane at which it ceases to be analytic. These exceptional points are called the *singularities* of the function. At all other points the function $f(z)$ is said to be analytic. For example, if $f(z) = 1/(z - z_0)$, then $\mathrm{d}f/\mathrm{d}z = -1/(z - z_0)^2$. Consequently, both the function and its first derivative are infinite at $z = z_0$ and $f(z)$ fails to be analytic there. Another example is $f(z) = (z - z_0)^\alpha$ with $-1 < \alpha < 1$. Here the function fails to be single-valued at $z = z_0$; if $z$ is made to go around a circle with center at $z = z_0$, multiple values are obtained after $z$ traverses one complete circle.

#### *A.1.2.1 Types of singularities*

There are three kinds of singularities in the complex plane: (i) poles or non-essential singularities, (ii) essential singularities, and (iii) branch points. If the function $f(z)$ has an expansion of the type

$$f(z) = \sum_{n=0}^{\infty} a_n(z - z_0)^n + \sum_{n=1}^{m} \frac{b_n}{(z - z_0)^n}, \tag{A.6}$$

at the point $z = z_0$, then it is said to have a pole of order $m$. The coefficient $b_1$ is known as the *residue* of the pole. It is evaluated using the expression

$$b_1 = \frac{1}{(m-1)!}\frac{\mathrm{d}^{m-1}}{\mathrm{d}z^{m-1}}[(z - z_0)^m f(z)]_{z=z_0}. \tag{A.7}$$

If the series of negative powers about $z = z_0$ in equation (A.6) does not terminate (i.e. if $m \to \infty$) then the point $z = z_0$ is called the *essential* singularity of $f(z)$.

A point $z_0$ about which the function $w = f(z)$ becomes multi-valued is known as a *branch point*. The order of the branch point is the number of complete rotations in the $z$-plane that would be needed to define the function $w$ uniquely. For example, the point $z = 0$ is a branch point of order 2 of the $f(z) = \sqrt{z}$. The point $z = 0$ is a branch point of the function $f(z) = z^{p/q}$, of order $q$, where $p$, $q$ with $p \neq q$ are integers with no common factors. The point $z = 0$ is a branch point of order infinity of the function $f(z) = \ln z$. Branch points always occur in pairs. The point $z = \infty$ is also a branch point of $f(z) = \sqrt{z}$ as can be easily seen by making the transformation $\zeta = 1/z$ and looking at the transformed function $g(\zeta) = f(1/z) = \sqrt{\zeta}$. A curve that connects two branch points is known as a *branch cut*. At a branch point, a particular physical point in the complex $z$-plane will map to several points in the $w$-plane and there is ambiguity in the inverse mapping from the $w$-plane to the $z$-plane. To make the function $w$ single-valued, the complex $z$-plane is duplicated as many times as the order of the branch point and each such duplicated plane is known as a *Riemann sheet*. The Riemann sheets are connected with each other by means of the branch cuts. Each branch of the multi-valued function $w$ is then thought of as being produced by values of $z$ that lie on a particular sheet. As the point $z$ is traversed on a Riemann sheet without passing through a branch cut, it continues to define one particular branch of a multi-valued function $w$. In the example, $w = \sqrt{z}$, a branch cut can be chosen to run from $z = 0$ to $z = -\infty$ along the negative real axis. Writing in the polar form $z = r\,\exp(\mathrm{i}\phi)$ and restricting $-\pi < \phi < \pi$, one sheet will define $w = w_1 = \sqrt{r}\,\exp(\mathrm{i}\phi/2)$. On this sheet, the real part of $w$ will always remain positive. The second sheet will define $w = w_2 = \sqrt{r}\,\exp(\mathrm{i}\phi/2)$ with $\phi$ taking on values $\pi < \phi < 3\pi$. On the second sheet, the real part of $w$ will always remain negative. The branch cut in this instance is defined such that $\Re(w) = 0$. Note that merely defining $w_1 = +\sqrt{z}$ and $w_2 = -\sqrt{z}$ without restricting the argument (angle $\phi$) as above will not define a single-valued function, even though such a process works well for the square root of real numbers. Restricting the argument is tantamount to making the function single-valued by defining branch cuts. Other choices of branch cuts are equally valid. For example, for $w = \sqrt{z}$, a second choice of the branch cut is along the positive real axis, which is defined by $\Im(w) = 0$.

If the only singularities of a function are poles in a finite plane, then the function is called a *meromorphic* function.

## A.2 Theorems from complex analysis

If a function $f(z)$ is analytic within and upon a closed contour $C$ and if $z_0$ is an interior point, then

$$\oint_C f(z)\,\mathrm{d}z = 0 \quad \text{(Cauchy's theorem).} \tag{A.8}$$

$$\frac{1}{2\pi i}\oint_C \frac{f(z)}{(z - z_0)^m}\,\mathrm{d}z = \frac{1}{(m-1)!}\frac{\mathrm{d}^{m-1}}{\mathrm{d}z^{m-1}}f(z_0) \quad \text{(residue theorem).} \tag{A.9}$$

For the special case of $m = 1$, the residue theorem is known by the name *Cauchy's integral formula.*

*Liouville's theorem* states that the only function that is analytic everywhere and is bounded at infinity is a constant.

Suppose a function $f(z)$ has a branch cut on the real axis from $a$ to $b$. Let $D_\varepsilon(x) = f(x + \mathrm{i}\varepsilon) - f(x - \mathrm{i}\varepsilon)$, $a \le x \le b$, $\varepsilon = 0^+$. If $f(z)$ has no other singularities and vanishes at infinity, then it can be shown from the Cauchy's integral formula that

$$f(z) = \frac{1}{2\pi \mathrm{i}}\int_a^b \frac{D_\varepsilon(x)}{(x - z)}\,\mathrm{d}x \quad \text{(discontinuity theorem).} \tag{A.10}$$

If $|a|$ or $|b| \to \infty$, the integral converges only when $D_\varepsilon(x)$ decays at least algebraically at $|x| = \infty$. The behavior of the function near the end points is governed by:

$$f(z \simeq a) = \mathcal{O}(\ln(z - a)), \quad \text{if } D_\varepsilon(x \to a) = D(a) \tag{A.11}$$

$$= \mathcal{O}((z - a)^{-\alpha}) \quad \text{if } D_\varepsilon(x \to a) = (x - a)^{-\alpha},\ 0 < \alpha < 1 \tag{A.12}$$

$$f(z \simeq b) = \mathcal{O}(\ln(b - z)), \quad \text{if } D_\varepsilon(x \to b) = D(b) \tag{A.13}$$

$$= \mathcal{O}((b - z)^{-\beta}), \quad \text{if } D_\varepsilon(x \to b) = (b - x)^{-\beta},\ 0 < \beta < 1. \tag{A.14}$$

**Example A.2.0.1.** Let $D_\varepsilon(x) = 2x^{-1/2}$ on the branch cut $0 \leq x < \infty$. Note that $|D_\varepsilon(x)| \to 0$ as $x = b \to \infty$. We then have

$$f(z) = \frac{2}{2\pi i}\int_0^\infty \frac{1}{\sqrt{x}(x - z)}\,dx = \frac{1}{i\pi}\int_0^\infty \frac{2dt}{t^2 - z}$$

$$= \frac{1}{i\pi}\int_0^\infty \frac{1}{\sqrt{z}}\left[\frac{1}{t - \sqrt{z}} - \frac{1}{t + \sqrt{z}}\right]dt = \frac{1}{i\pi\sqrt{z}}\left[\int_0^\infty \frac{dt}{t - \sqrt{z}} - \int_0^{-\infty} \frac{-dt}{-t + \sqrt{z}}\right]$$

$$= \frac{1}{\sqrt{z}}\frac{1}{i\pi}\int_{-\infty}^\infty \frac{dt}{t - \sqrt{z}} = \frac{1}{\sqrt{z}}.$$

■■

**Theorem A.2.1** Argument principle [2, p 259].

*Let $f(z)$ be a meromorphic function defined inside and outside and on a simple closed curve C, with no zeros or poles on C. Then*

$$I = \frac{1}{2\pi \mathrm{i}}\oint_C \frac{f'(z)}{f(z)}\,\mathrm{d}z = N - P = \frac{1}{2\pi}[\arg f(z)]_C,$$

*where N and P are the number of zeros and poles, respectively, of $f(z)$ inside C; where a multiple zero or pole is counted according to its multiplicity, and where* $\arg f(z)$ *is the argument of $f(z)$ and* $[\arg f(z)]_C$ *denotes the change in the argument of $f(z)$ over C.*

**Example A.2.0.2.** Determine the number of zeros located inside the first quadrant if the function

$$f(z) = z^3 + 1, \quad z = x + \mathrm{i}y.$$

Consider the contour $C$ shown in figure A1. Since $f(z)$ is a polynomial, it is analytic inside $C$ and $P = 0$. At any point $\arg f(z) = \phi = \tan^{-1}\left(\frac{\Im[f(z)]}{\Re[f(z)]}\right)$. On the circular arc $|z| = R$,

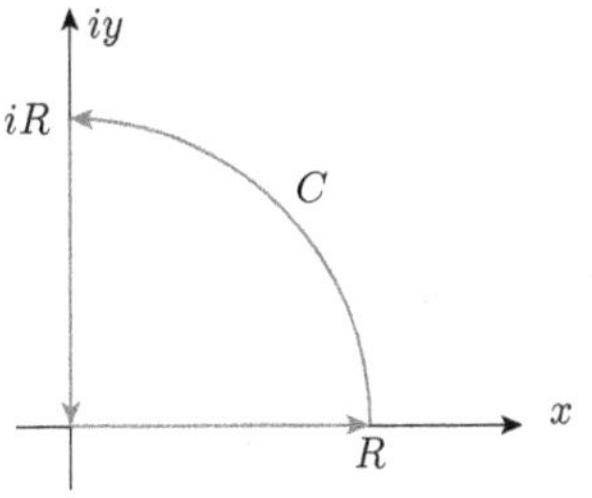


**Figure A1.** Contour in the first quadrant.

$$z = Re^{i\theta}, \quad f(z) = R^3 e^{i3\theta}\left(1 + \frac{1}{R^3 e^{i3\theta}}\right).$$

On the real axis $z = x$, $f(z) = x^3 + 1$. At $x = R$, $y = 0$, and $\theta = 0$. For large $R$ on the circular arc, $\arg f(z) = 3\theta$. At $z = iR$, $\theta = \pi/2$; hence $\arg f(z) = 3\pi/2$. Along the imaginary axis $z = iy$, $f(z) = -iy^3 + 1$. As $y$ traverses from $y \to \infty$ to $y = \varepsilon = 0+$, $\tan\phi$ goes from $\tan^{-1}(-\infty) = 3\pi/2$ to $\tan^{-1}\left(\frac{-\varepsilon}{1}\right) = 2\pi$. Then the argument principle gives

$$N - P = N = \frac{1}{2\pi}[0 + (3\pi/2 - 0) + (2\pi - 3\pi/2)] = 1.$$

Indeed the three zeros of $f(z)$ are $z_1 = e^{i\pi/3}$, $z_2 = e^{i\pi}$, $z_3 = e^{5i\pi/3}$, out of which $z_1$ lies in the first quadrant. ■■

As a corollary to the argument principle, we have Rouché's theorem.

**Theorem A.2.2** Rouché's theorem.
*Let $f(z)$ and $g(z)$ be analytic on and inside a simple closed contour C. If $|f(z)| > |g(z)|$ on C, then $f(z)$ and $f(z) + g(z)$ have the same number of zeros inside the contour C.*

As an example if $f(z) = 10$, $g(z) = z^8 - 4z^3$ and $C = C_1$ is the unit circle $|z| = 1$ then $|g(z)| \le |z|^8 + 4|z|^3 = 5 < |f(z)|$ on $C_1$. The polynomial $P(z) = f(z) + g(z) = z^8 - 4z^3 + 10$ has no roots inside $C_1$. On the other hand if $f(z) = z^8$, $g(z) = -4z^3 + 10$ and $C = C_2$ is a circle of radius 2: $|z| = 2$, then $|f(z)| = 2^8 = 256$, $|g(z)| \le 4|z|^3 + 10 = 42$ on $C_2$, so $|f(z)| > |g(z)|$. Hence the number of roots of $P(z) = f(z) + g(z) = z^8 - 4z^3 + 10$ is the same as the number of roots of $f(z) = z^8$ inside $C_2$. Since $f(z)$ has eight roots inside $C_2$ (essentially all at the origin), $P(z)$ will also have eight roots inside $C_2$. But since no roots of $P(z)$ lie inside $C_1$, they all lie in the region $1 < |z| < 2$.

**Lemma A.2.3.** Jordan's lemma [3, p 255], [2, p 222].

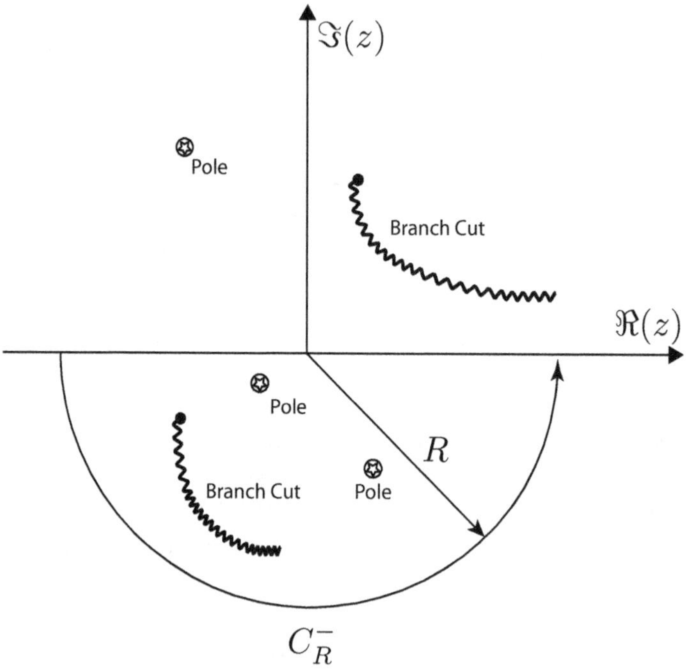


**Figure A2.** A large semicircular arc $C_R^-$ in the lower half of the $z$-plane.

Suppose that on the semicircular arc $C_R^-$ in the lower half of the complex $z$-plane we have $|f(z)| \to 0$ uniformly as $R \to \infty$ (i.e. $|f(z)| \leq K_R$, where $K_R$ depends only on $R$ but not on $\arg(z)$, and $K_R \to 0$ as $R \to \infty$) (figure A2). Then

$$\lim_{R\to\infty} \int_{C_R^-} e^{-jkz} f(z)\, dz = 0 \quad (k > 0), \tag{A.15}$$

where the half circle on the integral implies that it is performed on the large semicircular arc in the lower half plane.

**Lemma A.2.4** Watson's lemma [4, p 389], [2, p 490].
Consider the integral

$$I(z) = \int_0^\infty f(t) \mathrm{e}^{-zt}\, \mathrm{d}t, \qquad z \in \mathbb{C}. \tag{A.16}$$

Assume that $f(t)$ has the asymptotic series expansion

$$f(t) \sim \sum_{n=0}^{\infty} a_n t^{\alpha_n}, \qquad t \to 0^+; \quad \alpha_0 > -1. \tag{A.17}$$

(a) If $f(t) = \mathcal{O}(\mathrm{e}^{\beta t})$ as $t \to \infty$ and $f(t)$ is continuous in $(0, \infty)$, then

$$I(z) \sim \frac{1}{z}\sum_{n=0}^{\infty} a_n \frac{\Gamma(\alpha_n + 1)}{z^{\alpha_n}} \text{as } z \to \infty, \tag{A.18}$$

$$\text{for } -\frac{\pi}{2} < \arg(z - \beta) < \frac{\pi}{2}.$$

(b) If $f(t) = \mathcal{O}(e^{\beta t})$ as $t \to \infty$ and $f(t)$ is analytic in $\phi_1 \leq \arg(t) \leq \phi_2$, then $I(z)$ has the asymptotic behavior given in equation (A.18) for $z$ in

$$-\frac{\pi}{2} - \phi_2 < \arg(z - \beta) < \frac{\pi}{2} - \phi_1.$$

Jordan's lemma and Watson's lemma find extensive use in electromagnetic field analysis. For example uses in the analysis of Green's functions and in the determination of asymptotic fields, see [5, 6].

## A.3 Integral transforms

**Theorem A.3.1.** Analytic properties of functions defined by integrals [1, p 99], [7, p 11]. *Let $g(s, \zeta) = f(\zeta)h(s, \zeta)$ satisfy the conditions*

(i) *$h(s, \zeta)$ is a continuous function of the complex variables $s = \alpha + j\omega$ and $\zeta$ where $s$ lies inside a region $R$ and $\zeta$ lies on a contour $\Gamma$. The contour $\Gamma$ could possible extend to infinity in which case $\zeta$ lies on any bounded part of $\Gamma$.*

(ii) *$h(s, \zeta)$ is a regular function (i.e. analytic at every point in the domain of interest) of $s$ in $R$ for every $\zeta$ on any bounded part of $\Gamma$.*

(iii) *$f(\zeta)$ has only a finite number of discontinuities on $\Gamma$ and a finite number of maxima and minima on any finite part of $\Gamma$.*

(iv) *$f(\zeta)$ is bounded except at a finite number of points. If $\zeta_0$ is such a point so that $g(s, \zeta) \to \infty$ as $\zeta \to \zeta_0$, then*

$$G(s) = \int_{\Gamma} g(s, \zeta)\, \mathrm{d}\zeta = \lim_{\delta \to 0} \int_{\Gamma - \delta} g(s, \zeta)\, \mathrm{d}\zeta, \tag{A.19}$$

*exists where the notation $(\Gamma - \delta)$ denotes the contour $\Gamma$ apart from a small length surrounding $\zeta_0$ and* $\lim(\delta \to 0)$ *denotes the limit as this excluded length tends to zero. The limit must be approached uniformly when $s$ lies in any closed domain $R' \subset R$.*

(v) *In the case where $\Gamma$ extends to infinity, the integral defining $G(s)$ must be uniformly convergent when $s$ lies in any close domain $R' \subset R$.*

*Then $G(s)$ is a regular function of $s$ in $R$. Moreover,*

$$\frac{\mathrm{d}G(s)}{\mathrm{d}s} = \int_{\Gamma} f(\zeta)\frac{\partial h}{\partial s}\, \mathrm{d}\zeta \tag{A.20}$$

*and similarly for higher derivatives.*

For the analytic function $G(s = \alpha + j\omega)$ we have from the Cauchy–Riemann conditions that

$$\frac{\partial G}{\partial \alpha} = -j\frac{\partial G}{\partial \omega}. \tag{A.21}$$

**Theorem A.3.2.** Bilateral Laplace transform [7, p 23].
*Let $f(t)$ be a function of the real variable t such that $|f(t)| \le A\,\mathrm{e}^{\alpha_- t}$ as $t \to +\infty$ and $|f(t)| \le B\,\mathrm{e}^{\alpha_+ t}$ as $t \to -\infty$ with $\alpha_- < \alpha_+$. Define*

$$F(s) = \int_{-\infty}^{\infty} f(t)\mathrm{e}^{-st}\,\mathrm{d}t, \quad s = \alpha + j\omega. \tag{A.22}$$

*Then with $f(t)$ subject to conditions (iii) and (iv) of theorem (A.3.1), $F(s)$ is an analytic function of s, regular in the vertical strip $\alpha_- < \alpha < \alpha_+$ (see figure A3) and*

$$f(t) = \frac{1}{2\pi j}\int_{\alpha_0 - j\infty}^{\alpha_0 + j\infty} F(s)\mathrm{e}^{st}\,\mathrm{d}s, \tag{A.23}$$

*for any $\alpha_0$ in $\alpha_- < \alpha_0 < \alpha_+$.*

For piecewise continuous function we have the following result [8, p 244].

**Theorem A.3.3** *If $f(t)$ is absolutely integrable in every finite interval and if the integral*

$$F(s) = \int_{-\infty}^{\infty} f(t)\mathrm{e}^{-st}\,\mathrm{d}t, \quad s = \alpha + j\omega, \tag{A.24}$$

*converges in the strip $\alpha_- < \alpha < \alpha_+$, then the inversion integral*

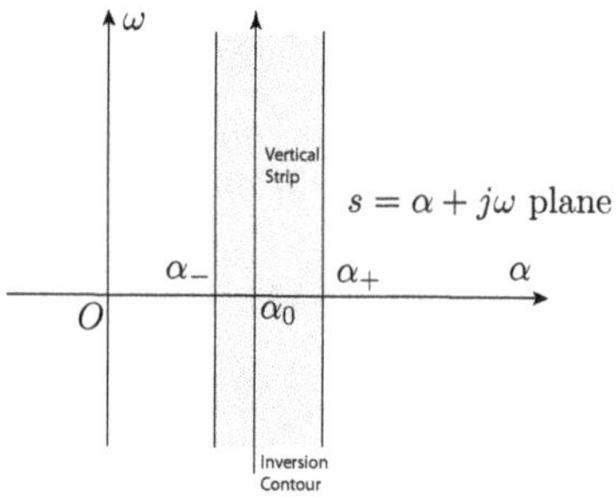


**Figure A3.** Inversion contour for the bilateral Laplace transform.

$$\frac{1}{2\pi j}\int_{\alpha_0 - j\infty}^{\alpha_0 + j\infty} F(s)e^{st}\,ds, \quad \alpha_- < \alpha_0 < \alpha_+, \tag{A.25}$$

*converges to* $\frac{1}{2}[f(t+)+f(t-)]$ *for any t where* $f(t+)$ *and* $f(t-)$ *exist.*

## References


[1] Titchmarsh E C 1939 *The Theory of Functions* 2nd edn (New York: Oxford University Press)

[2] Ablowitz M J and Fokas A S 1997 *Complex Variables: Introduction and Applications* (Cambridge: Cambridge University Press)

[3] Henrici P 1974 *Applied and Computational Complex Analysis* vol I (New York: Wiley)

[4] Henrici P 1977 *Applied and Computational Complex Analysis* vol II (New York: Wiley)

[5] Janaswamy R 2018 Consistency requirements for integral representations of Green's functions —part I *IEEE Trans. Antennas Propag.* **66** 4060–8

[6] Janaswamy R 2018 Consistency requirements for integral representations of Green's functions —part II: an erroneous representation *IEEE Trans. Antennas Propag.* **66** 4069–76

[7] Noble B 1988 *Methods Based on the Wiener–Hopf Technique for the Solution of Partial Differential Equations* (New York: Chelsea)

[8] Widder D V 1941 *The Laplace Transform* (Princeton, NJ: Princeton University Press)

# Appendix B

## Vector analysis

In this appendix we list several definitions and results from vector analysis that are of relevance to electromagnetic theory.

### B.1 Preliminaries

**Definition B.1.1.** *Lyapunov surface.* A surface $S$ is a Lyapunov surface [1, p 202] if the following conditions are satisfied:

(i) At each point of the surface it has a tangent plane and hence a well-defined normal.

(ii) If $M_1$ and $M_2$ are two points on the surface, and $\hat{\mathbf{n}}_1$ and $\hat{\mathbf{n}}_2$ are unit normals to the surface $S$ at these points and if $r$ is the distance between these points, then

$$|\,\hat{\mathbf{n}}_1 - \hat{\mathbf{n}}_2\,| \leqslant AR^{\lambda}, \tag{B.1}$$

where $A$ and $0 < \lambda \leq 1$ are constants.

(iii) About any point $P_0$ of the surface a sphere can be described with radius $d$ ($d$ independent of $P_0$) so that the section $\sigma$ of the surface which falls within the sphere meets normal $\hat{\mathbf{n}}_0$ to the surface $S$ at the point $P_0$ no more than once.

(iv) The solid angle $\omega_0$, which any part $\sigma$ of the surface subtends at an arbitrary point $P_0$ is bounded:

$$|\omega_0| \leq K. \tag{B.2}$$

**Definition B.1.2** *Exterior domain.* An exterior domain $\mathscr{D}$ is a domain which consists of all points lying a closed bounded surface $S$.

**Definition B.1.3.** *Hölder continuity.* A function $f(\boldsymbol{r})$, continuous for $|\boldsymbol{r}| \leq r_0$, is said to satisfy Hölder continuity if there are three positive numbers $\alpha$, $A$, and $\tau < r_0$ such that for all $|\boldsymbol{r}_1|$, $|\boldsymbol{r}_2| \leq \tau$, the relation

doi:10.1088/978-0-7503-1716-0ch21

$$|f(\boldsymbol{r}_2) - f(\boldsymbol{r}_1)| \leq A\,|\boldsymbol{r}_2 - \boldsymbol{r}_1|^{\alpha}, \tag{B.3}$$

holds.

**Definition B.1.4** *Functions of class $C^k$.* A function $f$ is said to be of (differentiability) class $C^k$ if the derivatives $f', f'', \dots, f^{(k)}$ exist and are continuous (the continuity is implied by differentiability for all the derivatives except for $f^{(k)}$).
The following integral theorems from vector calculus are relevant:

$$\iiint_V \nabla\cdot\mathbf{A}\,\mathrm{d}v = \oiint_S \hat{\mathbf{n}}\cdot\mathbf{A}\,\mathrm{d}s, \quad \text{(Gauss' divergence theorem)} \tag{B.4}$$

$$\iiint_V \nabla\times\mathbf{A}\,\mathrm{d}v = \oiint_S \hat{\mathbf{n}}\times\mathbf{A}\,\mathrm{d}s, \quad \text{(curl theorem)} \tag{B.5}$$

$$\oiint_S \hat{\mathbf{n}}\times\mathbf{F}\,\mathrm{d}s = \oiint_B \mathbf{r}\,(\hat{\mathbf{n}}\cdot\nabla\times\mathbf{F})\,\mathrm{d}s, \quad \text{(curl lemma)} \tag{B.6}$$

$$\iiint_V \nabla\psi\,\mathrm{d}v = \oiint_S \hat{\mathbf{n}}\psi\,\mathrm{d}s, \quad \text{(gradient theorem)} \tag{B.7}$$

$$\iint_S \hat{\mathbf{n}}\times\nabla\psi\,\mathrm{d}s = \iint_S \hat{\mathbf{n}}\times\nabla_s\psi\,\mathrm{d}s = \oint_C \psi\,\mathrm{d}\ell, \quad \text{(gradient lemma)} \tag{B.8}$$

$$\iint_S \nabla\times\mathbf{A}\cdot\hat{\mathbf{n}}\,\mathrm{d}s = \oint_C \mathbf{A}\cdot\mathrm{d}\ell, \quad \text{(Stokes's theorem)} \tag{B.9}$$

$$\iiint_V u\nabla^2 g + \nabla u\cdot\nabla g\,\mathrm{d}v = \oiint_S u\frac{\partial g}{\partial n}\,\mathrm{d}s, \quad \text{(Green's first identity)} \tag{B.10}$$

$$\iiint_V (u\nabla^2 g - g\nabla^2 u)\,\mathrm{d}v = \oiint_S \left(u\frac{\partial g}{\partial n} - g\frac{\partial u}{\partial n}\right)\mathrm{d}s, \quad \text{(Green's second identity)} \tag{B.11}$$

$$\begin{aligned}&\iiint_V \left[\psi\nabla\times\nabla\times\mathbf{A} + \mathbf{A}\nabla^2\psi + (\nabla\cdot\mathbf{A})\nabla\psi\right]\mathrm{d}v \quad \text{(Scalar-vector Green's theorem)}\\ &= \oiint_S \left[(\hat{\mathbf{n}}\times\nabla\times\mathbf{A})\psi + (\hat{\mathbf{n}}\times\mathbf{A})\times\nabla\psi + (\hat{\mathbf{n}}\cdot\mathbf{A})\nabla\psi\right]\mathrm{d}s.\end{aligned} \tag{B.12}$$

The surface $S$ in the gradient lemma and Stokes' theorem are open surfaces with a bounding rim $C$. The surface gradient operator in the gradient lemma is $\nabla_s = \nabla - \hat{\mathbf{n}}\frac{\partial}{\partial n}$. The functions in the Green's first identity are assumed to be continuous in $V$ plus $S$ with continuous first derivatives in $V$. Further $u$ is assumed to have continuous first derivative in $V$ plus $S$ and continuous second derivative in $V$. In Green's second

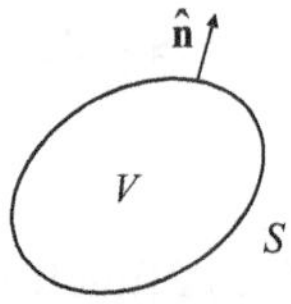


**Figure B.1.** Geometry for the various volume related identities.

identity we assume continuity of the first derivative of both $u$ and $g$ in $V$ plus $S$ and continuity of the second derivative in $V$. Vectors versions of Green's identities are also possible. If **P** and **Q** are vector functions of position that have continuous first and second derivatives, application of the Gauss' divergence theorem to a volume $V$ bounded by a closed surface $S$ with an outward unit normal $\hat{\mathbf{n}}$ (see figure B.1) gives

$$\iiint_V \nabla \cdot (\mathbf{P} \times \nabla \times \mathbf{Q})\, \mathrm{d}v = \oiint_S (\mathbf{P} \times \nabla \times \mathbf{Q}) \cdot \hat{\mathbf{n}}\, \mathrm{d}s. \tag{B.13}$$

But $\nabla \cdot (\mathbf{P} \times \nabla \times \mathbf{Q}) = (\nabla \times \mathbf{Q}) \cdot (\nabla \times \mathbf{P}) - \mathbf{P} \cdot (\nabla \times \nabla \times \mathbf{Q})$. Substituting this above leads to the *vector Green's first identity*:

$$\iiint_V (\nabla \times \mathbf{Q}) \cdot (\nabla \times \mathbf{P})\, \mathrm{d}v - \iiint_V \mathbf{P} \cdot (\nabla \times \nabla \times \mathbf{Q})\, \mathrm{d}v = \oiint_S (\mathbf{P} \times \nabla \times \mathbf{Q}) \cdot \hat{\mathbf{n}}\, \mathrm{d}s. \tag{B.14}$$

Interchanging **P** and **Q** and subtracting from equation (B.14) leads to the *vector Green's second identity*:

$$\iiint_V (\mathbf{Q} \cdot \nabla \times \nabla \times \mathbf{P} - \mathbf{P} \cdot \nabla \times \nabla \times \mathbf{Q})\, \mathrm{d}v = \oiint_S (\mathbf{P} \times \nabla \times \mathbf{Q} - \mathbf{Q} \times \nabla \times \mathbf{P}) \cdot \hat{\mathbf{n}}\, \mathrm{d}s. \tag{B.15}$$

### B.1.1 Surface divergence of tangential vectors

Let **u** be a tangential vector[1] on a surface $S$. Draw a small closed curve $C_\delta$ on $S$ and let $\hat{\nu}$ denote an outward drawn unit vector on the curve $C_\delta$ which is perpendicular to both $C_\delta$ and the unit normal $\hat{\mathbf{n}}$ drawn at a point inside the curve $c_\delta$. Let $S_\delta$ be the portion of $S$ enclosed by $C_\delta$ and let $A_\delta$ be its surface area (figure B.2)

$$A_\delta = \oiint_{S_\delta} \mathrm{d}s. \tag{B.16}$$

The surface divergence $\nabla_s \cdot$ of **u** at the point of the surface at the normal is defined by

$$\nabla_s \cdot \mathbf{u} \triangleq \lim_{C_\delta \to 0} \frac{1}{A_\delta} \int_{C_\delta} \mathbf{u} \cdot \hat{\nu}\, \mathrm{d}\ell. \tag{B.17}$$

[1] The surface divergence can also be defined for an arbitrary vector in 3D. See [2].

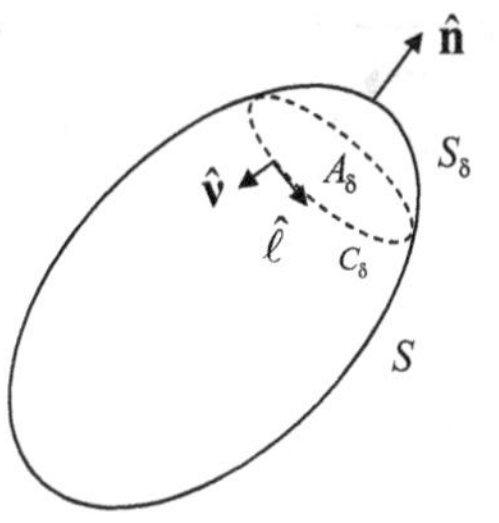


**Figure B.2.** Definition of surface divergence.

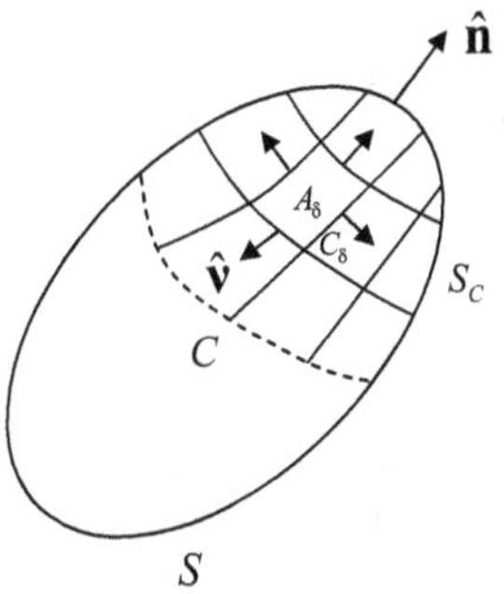


**Figure B.3.** Surface divergence theorem.

For an arbitrary curve $C$ on $S$, the portion $S_C$ of the surface $S$ enclosed by $C$ may be split up into small regions $C_\delta$ as shown in figure B.3. Using $A_\delta \nabla_s \cdot \mathbf{u} \approx \int_{C_\delta} \mathbf{u} \cdot \boldsymbol{\nu}\, \mathrm{d}\ell$ in each subregion and adding contributions from all regions we obtain the surface counterpart of Gauss' divergence theorem:

$$\iint_{S_C} \nabla_s \cdot \mathbf{u}\, \mathrm{d}s = \oint_C \mathbf{u} \cdot \boldsymbol{\nu}\, \mathrm{d}\ell \quad \text{(surface divergence theorem).} \tag{B.18}$$

If $C$ is allowed to shrink to a point such that $S_C$ becomes $S$, then the closed surface integral vanishes:

$$\oiint_S \nabla_s \cdot \mathbf{u}\, \mathrm{d}s = 0. \tag{B.19}$$

**Example B.1.1.1.** Relation between surface currents and charges.
We use the definition given above for the surface divergence and establish the relation between surface currents and surface charges in electromagnetic fields.

Consider a patch shown in figure B.4. The unit vectors $\hat{\ell}$, $\hat{\mathbf{n}}$, $\hat{\boldsymbol{\nu}}$ form a right-handed system so that $\hat{\ell} = \hat{\mathbf{n}} \times \hat{\boldsymbol{\nu}}$ and $\mathrm{d}\boldsymbol{\ell} = \hat{\mathbf{n}} \times \hat{\boldsymbol{\nu}}\, \mathrm{d}\ell$. For $\boldsymbol{u} = \hat{\mathbf{n}} \times \mathbf{E}$, we obtain

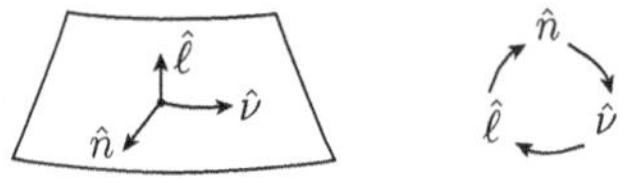

**Figure B.4.** Surface divergence of magnetic current density.

$$\begin{aligned}\nabla_s \cdot (\hat{\mathbf{n}} \times \mathbf{E}) &= \lim_{C_\delta \to 0} \frac{1}{A_\delta} \int_{C_\delta} (\hat{\mathbf{n}} \times \mathbf{E}) \cdot \hat{\nu}\, d\ell = \lim_{C_\delta \to 0} \frac{1}{A_\delta} \int_{C_\delta} (\hat{\nu} \times \hat{\mathbf{n}}) \cdot \mathbf{E}\, d\ell \\ &= -\lim_{C_\delta \to 0} \frac{1}{A_\delta} \int_{C_\delta} \mathbf{E} \cdot d\ell = -\lim_{C_\delta \to 0} \frac{1}{A_\delta} \oiint_{S_\delta} (\nabla \times \mathbf{E}) \cdot \hat{\mathbf{n}}\, ds \\ &= \lim_{C_\delta \to 0} \frac{1}{A_\delta} \oiint_{S_\delta} [j\omega\mu \mathbf{H} + \mathbf{M}] \cdot \hat{\mathbf{n}}\, ds \\ &= j\omega\mu \hat{\mathbf{n}} \cdot \mathbf{H} + (\hat{\mathbf{n}} \cdot \mathbf{M}), \end{aligned} \tag{B.20}$$

$$= j\omega\mu\hat{\mathbf{n}} \cdot \mathbf{H} \qquad \text{(source free region)}, \tag{B.21}$$

where the last equality in the second line follows from Stokes' theorem and the last equality in equation (B.20) follows from equation (B.16) and the first Maxwell curl equation (1.36). Similarly,

$$\nabla_s \cdot (\hat{\mathbf{n}} \times \mathbf{H}) = -j\omega\varepsilon\hat{\mathbf{n}} \cdot \mathbf{E} - (\hat{\mathbf{n}} \cdot \mathbf{J}), \tag{B.22}$$

$$= -j\omega\varepsilon\hat{\mathbf{n}} \cdot \mathbf{E} \qquad \text{(source free region)}. \tag{B.23}$$

Denoting $\mathbf{J}_s = \hat{\mathbf{n}} \times \mathbf{H}$, $\mathbf{M}_s = -\hat{\mathbf{n}} \times \mathbf{E}$, $q_{es} = \varepsilon\hat{\mathbf{n}} \cdot \mathbf{E}$, and $q_{ms} = \mu\hat{\mathbf{n}} \cdot \mathbf{H}$, we can also express equations (B.21) and (B.23) as

$$\nabla_s \cdot \mathbf{M}_s = -j\omega q_{ms}, \quad \nabla_s \cdot \mathbf{J}_s = -j\omega q_{es}, \tag{B.24}$$

which are the surface counterparts of the equations of continuity (1.39). Hence the normal component of one field component (**E** or **H**) can be related to the surface divergence of tangential part of the other field component (**H** or **E**, respectively). ■■

*B.1.1.1 Surface divergence in general orthogonal coordinates*

Using the definition in equation (B.17), it is easy to derive an expression for the surface divergence in an orthogonal coordinate system. Let $\xi_1$ and $\xi_2$ be the coordinate lines along the unit vectors $\hat{\mathbf{a}}_1$ and $\hat{\mathbf{a}}_2$ on a surface, figure B.5. The arc lengths along $\hat{\mathbf{a}}_1$ and $\hat{\mathbf{a}}_2$ are, respectively, $d\ell_1 = h_1 d\xi_1$ and $d\ell_2 = h_2 d\xi_2$, where $h_1$ and $h_2$ are scale factors for $\xi_1$ and $\xi_2$. An arbitrary tangential vector **u** can be expressed as $\mathbf{u} = \hat{\mathbf{a}}_1 u_1 + \hat{\mathbf{a}}_2 u_2$, where $u_i = \hat{\mathbf{a}}_i \cdot \mathbf{u}$, $i = 1, 2$. Consider the curvilinear patch shown in figure. The elemental area of the patch is $A_\delta = d\ell_1 d\ell_2 = h_1 h_2 d\xi_1 d\xi_2$. The coordinates of the center of the

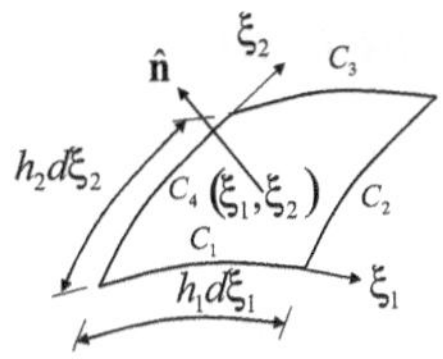


**Figure B.5.** Surface divergence in an orthogonal coordinate system.

patch are $(\xi_1, \xi_2)$. The contribution to the right-hand side (rhs) of equation (B.17) resulting from sides $C_1$ and $C_2$ are

$$\int_{C_1+C_3} \mathbf{u} \cdot \hat{\nu}\, \mathrm{d}\ell = \int_{C_3} \mathbf{u} \cdot \hat{\mathbf{a}}_2\, \mathrm{d}\ell_1 - \int_{C_1} \mathbf{u} \cdot \hat{\mathbf{a}}_2\, \mathrm{d}\ell_1 = \int_{C_3} u_2 h_1\, \mathrm{d}\xi_1 - \int_{C_1} u_2 h_1\, \mathrm{d}\xi_1.$$

On side $C_3$, a two term approximation of the Taylor series expansion gives

$$u_2 h_1 \approx u_2(\xi_1, \xi_2) h_1(\xi_1, \xi_2) + \frac{1}{2}\frac{\partial}{\partial \xi_2}[u_2(\xi_1, \xi_2) h_1(\xi_1, \xi_2)]\, \mathrm{d}\xi_2.$$

Similarly, on side $C_1$,

$$u_2 h_1 \approx u_2(\xi_1, \xi_2) h_1(\xi_1, \xi_2) - \frac{1}{2}\frac{\partial}{\partial \xi_2}[u_2(\xi_1, \xi_2) h_1(\xi_1, \xi_2)]\, \mathrm{d}\xi_2.$$

Therefore

$$\int_{C_1+C_3} \mathbf{u} \cdot \hat{\nu}\, \mathrm{d}\ell = \frac{\partial}{\partial \xi_2}(h_1 u_2)\, \mathrm{d}\xi_1 \mathrm{d}\xi_2.$$

Similarly,

$$\int_{C_2+C_4} \mathbf{u} \cdot \hat{\nu}\, \mathrm{d}\ell = \frac{\partial}{\partial \xi_1}(h_2 u_1)\, \mathrm{d}\xi_1 \mathrm{d}\xi_2.$$

Substituting all of these into equation (B.17), we arrive at

$$\nabla_s \cdot \mathbf{u} = \frac{1}{h_1 h_2}\left[\frac{\partial}{\partial \xi_1}(u_1 h_2) + \frac{\partial}{\partial \xi_2}(u_2 h_1)\right]. \tag{B.25}$$

For example, on the curved surface of a circular cylinder with its axis along the $z$-axis, $\xi_1 = \phi$, $h_1 = \rho$, $\xi_2 = z$, $h_2 = 1$ and we obtain

$$\nabla_s \cdot \mathbf{u} = \frac{1}{\rho}\left[\frac{\partial u_\phi}{\partial \phi} + \frac{\partial}{\partial z}(\rho u_z)\right] = \frac{1}{\rho}\frac{\partial u_\phi}{\partial \phi} + \frac{\partial u_z}{\partial z}, \tag{B.26}$$

where $(\rho, \phi, z)$ are the usual cylindrical coordinates.

## B.2 Theorems from potential theory

**Lemma B.2.1.** *Let a function* $\rho(\mathbf{r})$ *be continuous on the regular, closed and smooth surface F. Then*

$$U(\mathbf{r}) = \oint\!\!\!\oint_F \frac{\rho(\mathbf{r}')}{|\mathbf{r} - \mathbf{r}'|} \, \mathrm{d}s',$$

*is continuous everywhere.*

**Lemma B.2.2.** *Let* $\mathcal{J}$ *and* $\nabla_s \cdot \mathcal{J}$ *be continuous on the closed twice continuously differentiable surface F. Then*

$$\oint\!\!\!\oint_F \nabla_s \cdot \mathcal{J} \, \mathrm{d}s = 0.$$

*Moreover, with a twice continuously differentiable function U we have*

$$\oint\!\!\!\oint_F U \nabla_s \cdot \mathcal{J} \, \mathrm{d}s + \oint\!\!\!\oint_F \mathcal{J} \nabla_s \cdot U \, \mathrm{d}s = 0.$$

**Theorem B.2.3.** Continuity of scalar potential, [3, p 179].
*Suppose a function* $\rho$ *is defined on the closed and smooth surface F, and assume that there are three positive constants A,* $\alpha$*, and* $\gamma$ *such that for* $\mathbf{r}$ *and* $\mathbf{r}'$ *on F satisfying* $|\mathbf{r} - \mathbf{r}'| \leq \gamma$

$$|\rho(\mathbf{r}) - \rho(\mathbf{r}')| \leq A\,|\mathbf{r} - \mathbf{r}'|^{\alpha}, \quad 0 < \alpha < 2.$$

*Let* $G_{\mathrm{i}}$ *be the interior of F and* $G_{\mathrm{e}}$ *be exterior to F. Then the vector field*

$$\oint\!\!\!\oint_F \rho(\mathbf{r}') \nabla' \frac{1}{|\mathbf{r} - \mathbf{r}'|} \, \mathrm{d}s',$$

*is continuous in* $G_{\mathrm{i}} + F$ *and in* $G_{\mathrm{e}} + F$.

**Theorem B.2.4.** Jump in scalar potential [3, pp 193, 205].
*Suppose the function* $\rho$ *defined on the surface F satisfies the conditions of theorem B.2.3. Let* $H(\mathbf{r}')$ *denote the mean curvature at any point* $\mathbf{r}'$ *of F and the unit normal* $\hat{\mathbf{n}}$ *point from the interior to the exterior. Let* $F_i$ $(F_e)$ *denote the integral F as a point is approached from the interior (exterior). Then*

$$\begin{aligned}\oint\!\!\!\oint_{F_i} \rho(\mathbf{r}') \nabla' \frac{1}{|\mathbf{r} - \mathbf{r}'|} \, \mathrm{d}s' &= -2\pi\rho(\mathbf{r})\hat{\mathbf{n}}(\mathbf{r}) + \oint\!\!\!\oint_F (\rho(\mathbf{r}') - \rho(\mathbf{r})) \nabla' \frac{1}{|\mathbf{r} - \mathbf{r}'|} \, \mathrm{d}s' \\ &\quad + \rho(\mathbf{r}) \oint\!\!\!\oint_F \hat{\mathbf{n}}(\mathbf{r}') \left[ \frac{\partial}{\partial n'} \frac{1}{|\mathbf{r} - \mathbf{r}'|} - \frac{H(\mathbf{r}')}{|\mathbf{r} - \mathbf{r}'|} \right] \mathrm{d}s',\end{aligned} \tag{B.27}$$

*and*

$$\oiint_{F_e} \rho(\mathbf{r}')\nabla'\frac{1}{|\mathbf{r}-\mathbf{r}'|}\,\mathrm{d}s' - \oiint_{F_i} \rho(\mathbf{r}')\nabla'\frac{1}{|\mathbf{r}-\mathbf{r}'|}\,\mathrm{d}s' = 4\pi\rho(\mathbf{r})\hat{\mathbf{n}}(\mathbf{r}),$$

$$\oiint_{F_e} \rho(\mathbf{r}')\nabla'\psi\,\mathrm{d}s' - \oiint_{F_i} \rho(\mathbf{r}')\nabla'\psi\,\mathrm{d}s' = \rho(\mathbf{r})\hat{\mathbf{n}}(\mathbf{r}),$$

*for a point* $\mathbf{r}$ *on F, where* $\psi = \frac{e^{-jk|\mathbf{r}-\mathbf{r}'|}}{4\pi|\mathbf{r}-\mathbf{r}'|}$ *is the free-space Green's function.*

**Theorem B.2.5** Jump in vector potential [3, pp 194, 205].
*Let a surface field* $\mathscr{J}$ *be continuous on a closed surface F. Let the unit normal* $\hat{\mathbf{n}}$ *point from the interior to the exterior of F. Let* $F_i$ *(*$F_e$*) denote the integral F as a point is approached from the interior (exterior). Then*

$$\hat{\mathbf{n}}(\mathbf{r}) \times \oiint_{F_i} \mathscr{J}(\mathbf{r}') \times \nabla'\frac{1}{|\mathbf{r}-\mathbf{r}'|}\,\mathrm{d}s' = -2\pi\mathscr{J}(\mathbf{r}) + \hat{\mathbf{n}}(\mathbf{r}) \times \oiint_{F} \mathscr{J}(\mathbf{r}') \times \nabla'\frac{1}{|\mathbf{r}-\mathbf{r}'|}\,\mathrm{d}s',$$

$$\hat{\mathbf{n}}(\mathbf{r}) \times \oiint_{F_i} \mathscr{J}(\mathbf{r}') \times \nabla'\psi\,\mathrm{d}s' = -\tfrac{1}{2}\mathscr{J}(\mathbf{r}) + \hat{\mathbf{n}}(\mathbf{r}) \times \oiint_{F} \mathscr{J}(\mathbf{r}') \times \nabla'\psi\,\mathrm{d}s',$$

*and*

$$\hat{\mathbf{n}}(\mathbf{r}) \times \oiint_{F_e} \mathscr{J}(\mathbf{r}') \times \nabla'\frac{1}{|\mathbf{r}-\mathbf{r}'|}\,\mathrm{d}s' = 2\pi\mathscr{J}(\mathbf{r}) + \hat{\mathbf{n}}(\mathbf{r}) \times \oiint_{F} \mathscr{J}(\mathbf{r}') \times \nabla'\frac{1}{|\mathbf{r}-\mathbf{r}'|}\,\mathrm{d}s',$$

$$\hat{\mathbf{n}}(\mathbf{r}) \times \oiint_{F_e} \mathscr{J}(\mathbf{r}') \times \nabla'\psi\,\mathrm{d}s' = \tfrac{1}{2}\mathscr{J}(\mathbf{r}) + \hat{\mathbf{n}}(\mathbf{r}) \times \oiint_{F} \mathscr{J}(\mathbf{r}') \times \nabla'\psi\,\mathrm{d}s',$$

*for a point* $\mathbf{r}$ *on F, where* $\psi = \frac{e^{-jk|\mathbf{r}-\mathbf{r}'|}}{4\pi|\mathbf{r}-\mathbf{r}'|}$ *is the free-space Green's function.*

**Theorem B.2.6.** Helmholtz theorem [4, p 93].
*Any vector* $\mathbf{C}(\mathbf{r})$ *whose value vanishes at least as* $\frac{1}{r}$ *at infinity and whose source densities (i.e.* $\nabla \cdot \mathbf{C}$, $\nabla \times \mathbf{C}$*) vanish at least as* $\frac{1}{r^2}$ *at infinity may be uniquely expressed as the sum of a solenoidal field (i.e. zero divergence) and irrotational field (i.e. zero curl), i.e. if* $r|\mathbf{C}(\mathbf{r})| < \infty$, $r^2|\nabla \cdot \mathbf{C}(\mathbf{r})| < \infty$, *and* $r^2|\nabla \times \mathbf{C}(\mathbf{r})| < \infty$ *for all r, then*

$$\mathbf{C} = -\nabla\Phi + \nabla \times \mathbf{A}, \tag{B.28}$$

*where* $\Phi(\mathbf{r})$ *and* $\mathbf{A}(\mathbf{r})$ *are known as the scalar potential and the vector potential of* $\mathbf{C}(\mathbf{r})$*, respectively.*

*Proof.* Let the source densities $\nabla \cdot \mathbf{C} = \psi$ and $\nabla \times \mathbf{C} = \mathbf{B}$ be specified. Note that $\nabla \times \mathbf{B} = 0$. From the vector identity $\nabla \times \nabla \times \mathbf{C} = \nabla\nabla \cdot \mathbf{C} - \nabla^2\mathbf{C}$ we have

$$\begin{aligned}\nabla^2\mathbf{C} &= \nabla\nabla \cdot \mathbf{C} - \nabla \times \nabla \times \mathbf{C}\\ &= \nabla\psi - \nabla \times \mathbf{B} =: -\mathbf{F}.\end{aligned} \tag{B.29}$$

Each rectangular component of **C** satisfies the Poisson's equation $\nabla^2 C_i = -F_i$, where $i = x$ or $y$ or $z$ with a solution[2]

$$C_i(\mathbf{r}) = \frac{1}{4\pi} \iiint\limits_V \frac{F_i(\mathbf{r}')}{|\mathbf{r} - \mathbf{r}'|} \, dv'.$$

Combining the three components and using equation (B.29) we can write

$$\mathbf{C}(\mathbf{r}) = \frac{1}{4\pi} \iiint\limits_V \frac{-\nabla'\psi(\mathbf{r}')}{|\mathbf{r} - \mathbf{r}'|} \, dv' + \frac{1}{4\pi} \iiint\limits_V \frac{\nabla' \times \mathbf{B}(\mathbf{r}')}{|\mathbf{r} - \mathbf{r}'|} \, dv'. \tag{B.30}$$

Now

$$\begin{aligned} \nabla'\left(\frac{\psi(\mathbf{r}')}{|\mathbf{r} - \mathbf{r}'|}\right) &= \frac{\nabla'\psi(\mathbf{r}')}{|\mathbf{r} - \mathbf{r}'|} + \psi(\mathbf{r}')\nabla'\frac{1}{|\mathbf{r} - \mathbf{r}'|} \\ &= \frac{\nabla'\psi(\mathbf{r}')}{|\mathbf{r} - \mathbf{r}'|} - \nabla\frac{\psi(\mathbf{r}')}{|\mathbf{r} - \mathbf{r}'|}. \end{aligned} \tag{B.31}$$

Using equation (B.31) and the gradient theorem (B.7) into the first integral in equation (B.30) we obtain

$$\begin{aligned} \iiint\limits_V \frac{-\nabla'\psi(\mathbf{r}')}{|\mathbf{r} - \mathbf{r}'|} \, dv' &= -\nabla \iiint\limits_V \frac{\psi(\mathbf{r}')}{|\mathbf{r} - \mathbf{r}'|} \, dv' - \oiint\limits_S \frac{\psi(\mathbf{r}')}{|\mathbf{r} - \mathbf{r}'|} \hat{\mathbf{n}}' \, ds' \\ &= -\nabla \iiint\limits_V \frac{\psi(\mathbf{r}')}{|\mathbf{r} - \mathbf{r}'|} \, dv', \end{aligned} \tag{B.32}$$

because the surface integral vanishes in the limit as $r \to \infty$ due to the hypothesis. Likewise

$$\begin{aligned} \nabla' \times \left(\frac{\mathbf{B}(\mathbf{r}')}{|\mathbf{r} - \mathbf{r}'|}\right) &= \frac{\nabla' \times \mathbf{B}(\mathbf{r}')}{|\mathbf{r} - \mathbf{r}'|} - \mathbf{B}(\mathbf{r}') \times \nabla'\frac{1}{|\mathbf{r} - \mathbf{r}'|} \\ &= \frac{\nabla' \times \mathbf{B}(\mathbf{r}')}{|\mathbf{r} - \mathbf{r}'|} - \nabla \times \left(\frac{\mathbf{B}(\mathbf{r}')}{|\mathbf{r} - \mathbf{r}'|}\right). \end{aligned} \tag{B.33}$$

Therefore

$$\begin{aligned} \iiint\limits_V \frac{\nabla' \times \mathbf{B}(\mathbf{r}')}{|\mathbf{r} - \mathbf{r}'|} \, dv' &= \iiint \nabla' \times \left(\frac{\mathbf{B}(\mathbf{r}')}{|\mathbf{r} - \mathbf{r}'|}\right) dv' + \iiint\limits_V \nabla \times \left(\frac{\mathbf{B}(\mathbf{r}')}{|\mathbf{r} - \mathbf{r}'|}\right) dv' \\ &= \oiint\limits_S \hat{\mathbf{n}}' \times \frac{\mathbf{B}(\mathbf{r}')}{|\mathbf{r} - \mathbf{r}'|} \, ds' + \nabla \times \iiint\limits_V \frac{\mathbf{B}(\mathbf{r}')}{|\mathbf{r} - \mathbf{r}'|} \, dv' \\ &= \nabla \times \iiint\limits_V \frac{\mathbf{B}(\mathbf{r}')}{|\mathbf{r} - \mathbf{r}'|} \, dv', \end{aligned} \tag{B.34}$$

[2] A unique solution to the Poisson's equation is not guaranteed unless the conditions at infinity are imposed.

where the surface integral in the second line arises out of the curl theorem (**B**.5) and subsequently vanishes in the limit as $r \to \infty$ because of the hypothesis. We then have from equation (B.30) that

$$\begin{aligned} \mathbf{C}(\mathbf{r}) &= \frac{1}{4\pi}\iiint_V \frac{-\nabla'\psi(\mathbf{r}')}{|\mathbf{r}-\mathbf{r}'|}\,\mathrm{d}v' + \frac{1}{4\pi}\iiint_V \frac{\nabla'\times\mathbf{B}(\mathbf{r}')}{|\mathbf{r}-\mathbf{r}'|}\,\mathrm{d}v' \\ &= -\nabla\frac{1}{4\pi}\iiint_V \frac{\psi(\mathbf{r}')}{|\mathbf{r}-\mathbf{r}'|}\,\mathrm{d}v' + \nabla\times\frac{1}{4\pi}\iiint_V \frac{\mathbf{B}(\mathbf{r}')}{|\mathbf{r}-\mathbf{r}'|}\,\mathrm{d}v' \\ &=: -\nabla\Phi + \nabla\times\mathbf{A}, \end{aligned} \tag{B.35}$$

where

$$\Phi = \frac{1}{4\pi}\iiint_V \frac{\psi(\mathbf{r}')}{|\mathbf{r}-\mathbf{r}'|}\,\mathrm{d}v', \tag{B.36}$$

$$\mathbf{A} = \frac{1}{4\pi}\iiint_V \frac{\mathbf{B}(\mathbf{r}')}{|\mathbf{r}-\mathbf{r}'|}\,\mathrm{d}v'. \tag{B.37}$$

■

If the vector $\mathbf{C}(\mathbf{r})$ has surface discontinuities one can still use the above form of Helmholtz theorem if one regards it as a distribution. If the relations (18.75) and (18.76) from theorem 18.1.4 are used to separate out the surface discontinuities then one arrives at the following more useful statement of the Helmholtz theorem for sources with discontinuities.

**Theorem B.2.7** Helmholtz theorem for fields with surface discontinuities.
*If a vector* $\mathbf{C}(\mathbf{r})$ *has a support confined to a volume* $V$ *with surface* $S$ *and outward normal* $\hat{\mathbf{n}}$ *and is zero outside* $V$ *then it has the unique decomposition given in equation (B.28) where*

$$\Phi = \frac{1}{4\pi}\iiint_V \frac{\{\nabla'\cdot\mathbf{C}(\mathbf{r}')\}}{|\mathbf{r}-\mathbf{r}'|}\,\mathrm{d}v' - \frac{1}{4\pi}\oiint_S \frac{\hat{\mathbf{n}}'\cdot\mathbf{C}(\mathbf{r}')}{|\mathbf{r}-\mathbf{r}'|}\,\mathrm{d}s', \tag{B.38}$$

$$\mathbf{A} = \frac{1}{4\pi}\iiint_V \frac{\{\nabla'\times\mathbf{C}(\mathbf{r}')\}}{|\mathbf{r}-\mathbf{r}'|}\,\mathrm{d}v' - \frac{1}{4\pi}\oiint_S \frac{\hat{\mathbf{n}}'\times\mathbf{C}(\mathbf{r}')}{|\mathbf{r}-\mathbf{r}'|}, \tag{B.39}$$

*where terms in the curly brackets* $\{\cdot\}$ *denote the regular part defined everywhere except on the surface S.*

**Example B.2.0.1** Decomposition of a constant column of current.
To demonstrate the application of Helmholtz theorem, consider the example of a $\hat{\mathbf{z}}$-directed constant current of dipole moment $p$ filling a cylindrical pillbox of height $2\ell$ and radius $\ell$ centered around the origin

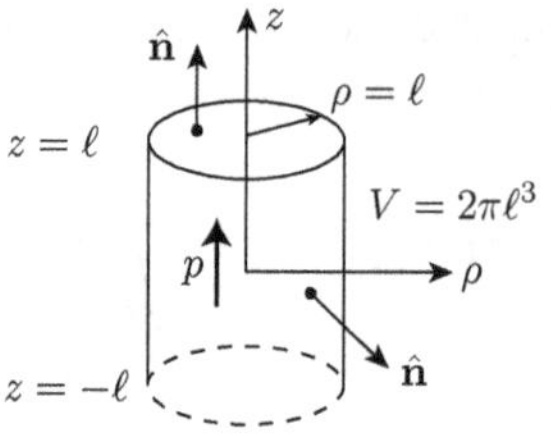


**Figure B.6.** Constant current in a cylindrical volume.

$$\mathbf{J} = \begin{cases} \hat{\mathbf{z}}\dfrac{p}{V}, & 0 < \rho \leq \ell;\ |z| < \ell \\ 0, & \text{otherwise} \end{cases}, \tag{B.40}$$

where $V = 2\pi\ell^3$ is the volume of the pillbox, see figure B.6. For $V \to 0$ the current represents a VED with density $\hat{\mathbf{z}}p\delta(\mathbf{r} - 0)$. Theorem B.2.7 is applicable in this case with $\{\nabla \cdot \mathbf{J}\} = 0$, $\{\nabla \times \mathbf{J}\} = 0$. Also $\hat{\mathbf{n}} \cdot \mathbf{J} = \pm p/V$ at $z = \pm\ell$ and $\hat{\mathbf{n}} \times \mathbf{J} = -\hat{\boldsymbol{\phi}}p/V$. Then with $R_{\substack{1\\2}} = \sqrt{\rho^2 + \rho'^2 - 2\rho\rho' \cos(\phi - \phi') + (z \mp \ell)^2}$, the scalar potential is equal to

$$\begin{aligned}\Phi(r, \theta, \phi) &= -\frac{p}{4\pi V}\int_0^{2\pi}\int_{\rho'=0}^{\ell}\left(\frac{1}{R_1} - \frac{1}{R_2}\right)\rho'\,\mathrm{d}\rho'\mathrm{d}\phi' \\ &= -\frac{pz\ell}{\pi V}\int_0^{2\pi}\int_0^{\ell}\frac{1}{R_1R_2(R_2 + R_1)}\rho'\,\mathrm{d}\rho'\mathrm{d}\phi' \\ &\to -\frac{pz\ell}{\pi V}\int_0^{2\pi}\int_0^{\ell}\frac{1}{2r^3}\rho'\,\mathrm{d}\rho'\mathrm{d}\phi' \\ &= -\frac{pz}{4\pi r^3} = -\frac{p}{4\pi}\frac{\cos\theta}{r^2},\quad r \neq 0,\end{aligned} \tag{B.41}$$

in the limit as $V \to 0$, since $R_1$, $R_2 \to r$ in that case.

Furthermore, using $|\mathbf{r} - \mathbf{r}'| = \sqrt{r^2 + r'^2 - 2r[\ell \sin\theta\cos(\phi - \phi') + z'\cos\theta]}$, $-\hat{\boldsymbol{\phi}}' = \hat{\mathbf{x}}\cos\phi' - \hat{\mathbf{y}}\sin\phi'$, the vector potential in the limit as $V \to 0$ is given by

$$\begin{aligned}\mathbf{A}(r, \theta, \phi) &= -\frac{p\ell}{4\pi V}\int_{z'=-\ell}^{\ell}\int_0^{2\pi}\frac{(\hat{\mathbf{x}}\cos\phi' - \hat{\mathbf{y}}\sin\phi')\,\mathrm{d}\phi'\mathrm{d}z'}{\sqrt{r^2 + r'^2 - 2r[\ell\sin\theta\cos(\phi - \phi') + z'\cos\theta]}}, \\ &\to -\frac{p\ell}{4\pi r V}\int_{-\ell}^{\ell}\int_0^{2\pi}\left[1 + \frac{r'^2}{2r^2} + \frac{z'\cos\theta}{r} + \frac{\ell\sin\theta\cos(\phi - \phi')}{r}\right] \\ &\qquad (\hat{\mathbf{x}}\cos\phi' - \hat{\mathbf{y}}\sin\phi')\,\mathrm{d}\phi'\mathrm{d}z' \\ &= -\frac{p\ell^3\sin\theta}{2\pi r^2 V}\int_0^{2\pi}\cos(\phi - \phi')[\hat{\mathbf{x}}\cos\phi' - \hat{\mathbf{y}}\sin\phi']\,\mathrm{d}\phi' \\ &= \frac{p\sin\theta\hat{\boldsymbol{\phi}}}{4\pi r^2},\quad r \neq 0,\end{aligned} \tag{B.42}$$

where binomial approximation has been used in the second step. It is seen that despite the current density being of finite support, the constituent scalar and vector potentials both have infinite support. It is easy to verify that $\nabla\Phi = \nabla \times \mathbf{A}$ so that $\mathbf{J} = -\nabla\Phi + \nabla \times \mathbf{A} = 0$ for $r \neq 0$ in the limit as $V \to 0$. ■■

We state here without proof some recent generalizations to the Helmholtz theorem valid for causal fields [5].

**Definition B.2.1.** *Causal vector field.* A time-dependent field $\mathbf{F}(\mathbf{r}; t)$ of finite support in time is known as a causal vector field if its sources (i.e. its curl, its divergence, and its time-derivative, expressed here in primed coordinates) are evaluated at a retarded time $t' = t - |\mathbf{r} - \mathbf{r}'|/c$, where $c$ is the speed of light in free-space.

**Theorem B.2.8.** Causal Helmholtz theorem.
*Any causal vector field* $\mathbf{F}$ *that behaves as* $o(r^{-1})$ *at* $r \to \infty$ *can be expanded as*

$$\mathbf{F} = -\nabla \iiint_V \frac{[\nabla' \cdot \mathbf{F}]}{4\pi R} \mathrm{d}v' + \nabla \times \iiint_V \frac{[\nabla' \times \mathbf{F}]}{4\pi R} \mathrm{d}v' + \frac{1}{c^2}\frac{\partial}{\partial t} \iiint_V \frac{[\partial \mathbf{F}/\partial t]}{4\pi R} \mathrm{d}v', \tag{B.43}$$

*where quantities within the square brackets* $[\cdot]$ *are evaluated at the retarded time* $t' = t - R/c$ *and* $R = |\mathbf{r} - \mathbf{r}'|$.

**Theorem B.2.9** Causal Helmholtz theorem for antisymmetric tensors.
*Any antisymmetric tensor that decays sufficiently rapidly at infinity can be expressed as*

$$F^{\mu\nu} = \partial^{\mu\nu}_{..\lambda} \int G_a \partial'_\alpha F^{\alpha\lambda}(x')\, \mathrm{d}^4x' - \hat{\partial}^{\mu\nu}_{..\lambda} \int G_a \partial'_\alpha \hat{F}^{\alpha\lambda}(x')\, \mathrm{d}^4x', \tag{B.44}$$

*where* $\partial^{\mu\nu}_{..\lambda} = \delta^\nu_\lambda \partial^\mu - \delta^\mu_\lambda \partial^\nu$ *is the curl operator (see equation* (17.115)), $\hat{\partial}^{\mu\nu}_{\lambda} = \frac{1}{2}\varepsilon^{\mu\nu\alpha\beta}\partial_{\alpha\beta\lambda}$, $\varepsilon^{\mu\nu\alpha\beta}$ *is the four-dimensional Levi–Civita symbol,* $\partial^\mu = \partial/\partial x^\mu$, $\partial_\mu = \partial/\partial x_\mu$, $\hat{F}^{\mu\nu}$ *is the dual of* $F^{\mu\nu}$ *(see equation* (17.40)*),* $G_a = \delta(t' - t + R/c)/(4\pi cR)$ *is the Green's function of the d'Alembert's equation* $\partial_\mu\partial^\mu G_a = \delta^4(x - x')$, $\mathrm{d}^4x'$ *is the volume element in spacetime and the integrals are taken over all spacetime.*

## References

[1] Sobolev S L 1964 *Partial Differential Equations of Mathematical Physics* (New York: Pergamon)

[2] van Bladel J 1991 *Singular Electromagnetic Fields and Sources* (The IEEE Series on Electromagnetic Wave Theory) (Piscataway, NJ: IEEE)

[3] Müller C 1969 *Foundations of the Mathematical Theory of Electromagnetic Waves* (New York: Springer)
[4] Javid M and Brown P M 1963 *Field Analysis and Electromagnetics* (New York: McGraw-Hill)
[5] Heras R 2016 The Helmholtz theorem and retarded fields *Eur. J. Phys.* **37** 1–11

# Appendix C

## Bessel functions

### C.1 Bessel differential equation

Consider the *Bessel equation*

$$z^2\frac{\mathrm{d}^2\mathscr{C}}{\mathrm{d}z^2} + z\frac{\mathrm{d}\mathscr{C}}{\mathrm{d}z} + (z^2 - \nu^2)\mathscr{C} = 0, \tag{C.1}$$

which has a regular singularity at $z = 0$ and an irregular singularity at $z \to \infty$. The solutions of the Bessel equation are denoted by $J_\nu(z)$, $Y_\nu(z)$, which are termed as Bessel function of the first kind and second kind of order $\nu$, respectively, or by $H_\nu^{(1)}(z)$, $H_\nu^{(2)}(z)$, which are termed as Hankel functions of the first and second kind of order $\nu$, respectively. For positive $\nu$, the functions $J_\nu(z)$ are all well-behaved at the origin, while the functions $Y_\nu(z)$, $H_\nu^{(1)}(z)$, $H_\nu^{(2)}(z)$ are all singular at $z = 0$. The relation between these functions are

$$H_\nu^{\substack{(1)\\(2)}}(z) = J_\nu(z) \pm \mathrm{i}\,Y_\nu(z). \tag{C.2}$$

References [1, 2] contain many identities and integrals involving Bessel functions. Some of these relevant to wave propagation are listed below.

### C.2 General properties of Bessel functions

For *n*th order linear differential equations, the Wronskian is a quantity that depends on the determinant of a coefficient matrix encountered in the 'variation of parameters' method of solution [3, p 18, 65]. The Wronskian between two differential functions $f$ and $g$ is defined as $\mathscr{W}(f, g) = fg' - gf'$. It measures the linear dependence/independence of the solutions of the homogeneous differential equation. For linear independence the Wronskian must be non-zero; otherwise the solutions will be linearly dependent. For a second order equation of the type

doi:10.1088/978-0-7503-1716-0ch22

$$\frac{\mathrm{d}}{\mathrm{d}z}\left[p(z)\frac{\mathrm{d}y}{\mathrm{d}z}\right] = q(z)y, \tag{C.3}$$

the Wronskian, $\mathscr{W}(y_1, y_2)$, involving any two solutions $y_1(z)$ and $y_2(z)$ is equal to $\mathscr{W}(y_1, y_2) = \frac{C}{p(z)}$, where $C$ is a constant, which may be evaluated by using $y_1(z)$ and $y_2(z)$ at any convenient point $z$. The Wronskian is explicitly independent of the function $q(z)$. For the Bessel equation $p(z) = z$, $q(z) = \nu^2/z - z$. The Wronskians of the Bessel equation are

$$\mathscr{W}\{J_\nu(z),\ Y_\nu(z)\} = \frac{2}{\pi z}, \tag{C.4}$$

$$\mathscr{W}\left\{H_\nu^{(1)}(z),\ H_\nu^{(2)}(z)\right\} = \frac{4}{\mathrm{i}\pi z}. \tag{C.5}$$

Using equation (C.4), the Wronskian for $J_\nu(z)$, $H_\nu^{(2)}(z)$ can be written as

$$\mathscr{W}\{J_\nu(z),\ H_\nu^{(2)}(z)\} = \frac{2}{\mathrm{i}\pi z}, \tag{C.6}$$

It is seen the above Wronskians are all independent of the order $\nu$ and are non-zero. Hence $H_\nu^{(1)}(z)$ and $H_\nu^{(2)}(z)$ are linearly independent solutions of the Bessel equation as are $J_\nu(z)$ and $Y_\nu(z)$.

### C.2.1 Recurrence relations

The recurrence relations satisfied by any of $\mathscr{C}_\nu(z) = J_\nu(z),\ Y_\nu(z),\ H_\nu^{(1)}(z),\ H_\nu^{(2)}(z)$ are

$$\mathscr{C}_{\nu-1}(z) + \mathscr{C}_{\nu+1}(z) = \frac{2\nu}{z}\mathscr{C}_\nu(z), \tag{C.7}$$

$$\mathscr{C}_{\nu-1}(z) - \mathscr{C}_{\nu+1}(z) = 2\frac{\mathrm{d}}{\mathrm{d}z}\mathscr{C}_\nu(z). \tag{C.8}$$

### C.2.2 Small argument approximations

For small argument $z$ and for fixed $\nu \geq 0$, the various functions behave as

$$J_\nu(z) \sim \left(\frac{z}{2}\right)^\nu \frac{1}{\Gamma(\nu+1)}, \quad J_\nu'(z) \sim \left(\frac{z}{2}\right)^{\nu-1} \frac{1}{2\Gamma(\nu)}, \tag{C.9}$$

$$Y_\nu(z) \approx -\mathrm{i}H_\nu^{(1)}(z) \approx \mathrm{i}H_\nu^{(2)}(z) \sim -\left(\frac{z}{2}\right)^{-\nu}\frac{\Gamma(\nu)}{\pi}, \quad \nu > 0, \tag{C.10}$$

$$Y_0(z) \sim \frac{2}{\pi}\left[\ln\left(\frac{z}{2}\right) + \gamma\right], \quad \gamma = 0.577\,215\,664\ldots, \tag{C.11}$$

$$J_\nu(z_1)H_\nu^{(2)}(z_2) = \frac{\mathrm{i}}{\nu\pi}\left(\frac{z_1}{z_2}\right)^\nu, \quad |z_1|, |z_2| \ll 1, \nu > 0. \tag{C.12}$$

### C.2.3 Analytical continuation and negative orders

$$J_\nu(z\mathrm{e}^{\mathrm{i}m\pi}) = \mathrm{e}^{\mathrm{i}m\pi\nu}J_\nu(z), \tag{C.13}$$

$$Y_\nu(z\mathrm{e}^{im\pi}) = \mathrm{e}^{-\mathrm{i}m\pi\nu}Y_\nu(z) + 2\mathrm{i}\sin(m\nu\pi)\cot(\nu\pi)J_\nu(z), \tag{C.14}$$

$$J_\nu(z^*) = [J_\nu(z)]^*, \quad Y_\nu(z^*) = [Y_\nu(z)]^*, \tag{C.15}$$

$$J_{-\nu}(z) = \cos(\nu\pi)J_\nu(z) - \sin(\nu\pi)Y_\nu(z), \tag{C.16}$$

$$Y_{-\nu}(z) = \cos(\nu\pi)Y_\nu(z) + \sin(\nu\pi)J_\nu(z). \tag{C.17}$$

where $m$ is an integer and the superscript * denotes complex conjugation.

$$H_\nu^{(1)}(z\mathrm{e}^{\mathrm{i}\pi}) = -\mathrm{e}^{\mathrm{i}\pi\nu}H_\nu^{(2)}(z), \tag{C.18}$$

$$H_\nu^{(2)}(z\mathrm{e}^{-\mathrm{i}\pi}) = -\mathrm{e}^{\mathrm{i}\pi\nu}H_\nu^{(1)}(z), \tag{C.19}$$

$$\overset{(1)}{H_\nu^{(2)}}(z^*) = \left[\overset{(2)}{H_\nu^{(1)}}(z)\right]^*, \tag{C.20}$$

$$H_{-\nu}^{(1)}(z) = \mathrm{e}^{\mathrm{i}\nu\pi}H_\nu^{(1)}(z), \tag{C.21}$$

$$H_{-\nu}^{(2)}(z) = \mathrm{e}^{-\mathrm{i}\nu\pi}H_\nu^{(2)}(z). \tag{C.22}$$

On combining equations (C.4) and (C.16), it is seen that

$$\mathscr{W}\{J_\nu(z), J_{-\nu}(z)\} = -\frac{2}{\pi z}\sin(\nu\pi), \tag{C.23}$$

which is non-zero if $\nu$ is not an integer or zero. Hence, $J_\nu(z)$ and $J_{-\nu}(z)$ constitute an independent set for $\nu \neq 0, 1, 2, \ldots$. Also note that $J_{-\nu}(z)$ for $\nu > 0$ and $\nu \neq 0, 1, 2, \ldots$ is singular at $z = 0$.

### C.2.4 Large argument approximations

When $\nu$ is fixed and $|z| \to \infty$

$$J_\nu(z) \sim \sqrt{\frac{2}{\pi z}}\cos\left(z - \frac{\nu\pi}{2} - \frac{\pi}{4}\right), \quad |\arg(z)| < \pi, \tag{C.24}$$

$$Y_\nu(z) \sim \sqrt{\frac{2}{\pi z}} \sin\left(z - \frac{\nu\pi}{2} - \frac{\pi}{4}\right), \quad |\arg(z)| < \pi, \tag{C.25}$$

$$H_\nu^{(1)}(z) \sim \sqrt{\frac{2}{\mathrm{i}\pi z}}\ \mathrm{i}^{-\nu}\mathrm{e}^{\mathrm{i}z}, \ -\pi < \arg(z) < 2\pi, \tag{C.26}$$

$$H_\nu^{(2)}(z) \sim \sqrt{\frac{2\mathrm{i}}{\pi z}}\ \mathrm{i}^{\nu}\mathrm{e}^{-\mathrm{i}z}, \ -2\pi < \arg(z) < \pi, \tag{C.27}$$

$$J_\nu{}'(z) \sim -\sqrt{\frac{2}{\pi z}} \sin\left(z - \frac{\nu\pi}{2} - \frac{\pi}{4}\right), \quad |\arg(z)| < \pi, \tag{C.28}$$

$$H_\nu^{(2)'}(z) \sim -\mathrm{i}\sqrt{\frac{2\mathrm{i}}{\pi z}}\ \mathrm{i}^{\nu}\mathrm{e}^{-\mathrm{i}z}, \ -2\pi < \arg(z) < \pi. \tag{C.29}$$

### C.2.5 Large order approximations

For $z$ fixed and $\nu \to \infty$

$$J_\nu(z) \sim \frac{1}{\sqrt{2\nu\pi}}\left(\frac{ez}{2\nu}\right)^\nu \to 0, \tag{C.30}$$

$$Y_\nu(z) \approx -\mathrm{i}H_\nu^{(1)}(z) \approx \mathrm{i}H_\nu^{(2)}(z) \sim -\sqrt{\frac{2}{\nu\pi}}\left(\frac{2\nu}{ez}\right)^\nu \to \infty, \tag{C.31}$$

$$J_\nu(z_1)\, Y_\nu(z_2) \approx -\mathrm{i}J_\nu(z_1)H_\nu^{(1)}(z_2) \approx \mathrm{i}J_\nu(z_1)H_\nu^{(2)}(z_2) \sim -\frac{1}{\nu\pi}\left(\frac{z_1}{z_2}\right)^\nu. \tag{C.32}$$

### C.2.6 Integral relations

$$J_\nu(z) = \frac{(-\mathrm{i})^\nu}{\pi}\int_0^\pi \mathrm{e}^{\mathrm{i}z\cos\theta}\cos\nu\theta\ \mathrm{d}\theta, \tag{C.33}$$

$$\underbrace{\mathscr{L}[J_\nu(x)] := \int_0^\infty J_\nu(x)\mathrm{e}^{-sx}\ \mathrm{d}x}_{\text{Laplace Transform}} = \frac{1}{\sqrt{(s^2+1)}(s+\sqrt{s^2+1})^\nu}, \qquad \Re(\nu) > -1 \tag{C.34}$$

$$\underbrace{\mathscr{M}[J_\nu(x)] := \int_0^\infty x^{s-1}J_\nu(x)\ \mathrm{d}x}_{\text{Mellin Transform}} = 2^{s-1}\frac{\Gamma\left(\frac{\nu+s}{2}\right)}{\Gamma\left(1+\frac{\nu-s}{2}\right)}, \ -\Re(\nu) < \Re(s) < \frac{3}{2} \tag{C.35}$$

$$\mathscr{M}[Y_\nu(x)] := \int_0^\infty x^{s-1}Y_\nu(x)\,\mathrm{d}x = 2^{s-1}\frac{\cot\left[(\nu - s)\dfrac{\pi}{2}\right]\Gamma\left(\dfrac{\nu+s}{2}\right)}{\Gamma\left(1+\dfrac{\nu-s}{2}\right)}, \quad |\Re(\nu)| < \Re(s) < \frac{3}{2}. \tag{C.36}$$

### C.2.7 Addition theorems

$$\mathscr{C}_\nu(u+v) = \sum_{m=-\infty}^{\infty} \mathscr{C}_{\nu\mp m}(u)J_m(v), \qquad |v| < |u|, \tag{C.37}$$

where $\mathscr{C}_\nu(z) = J_\nu(z),\ Y_\nu(z),\ H_\nu^{(1)}(z),\ H_\nu^{(2)}(z)$. The restriction $|v| < |u|$ is unnecessary when $\mathscr{C} = J$ and $\nu$ is an integer or zero.

For $\boldsymbol{\rho} = (\rho, \phi)$, $\boldsymbol{\rho}' = (\rho', \phi')$, and $\lambda > 0$

$$H_0^{\substack{(1)\\(2)}}(\lambda\,|\boldsymbol{\rho} - \boldsymbol{\rho}'|) = \sum_{m=-\infty}^{\infty} J_m(\lambda\rho_<)H_m^{\substack{(1)\\(2)}}(\lambda\rho_>)\mathrm{e}^{\mathrm{i}m(\phi-\phi')}, \tag{C.38}$$

where $\rho_< = \min(\rho, \rho')$, $\rho_> = \max(\rho, \rho')$ and $|\boldsymbol{\rho} - \boldsymbol{\rho}'| = \sqrt{\rho^2 + \rho'^2 - 2\rho\rho'\cos(\phi - \phi')}$. Equation (C.38) also implies that

$$J_0(\lambda\,|\boldsymbol{\rho} - \boldsymbol{\rho}'|) = \sum_{m=-\infty}^{\infty} J_m(\lambda\rho_<)J_m(\lambda\rho_>)\mathrm{e}^{\mathrm{i}m(\phi-\phi')}, \tag{C.39}$$

$$Y_0(\lambda\,|\boldsymbol{\rho} - \boldsymbol{\rho}'|) = \sum_{m=-\infty}^{\infty} J_m(\lambda\rho_<)Y_m(\lambda\rho_>)\mathrm{e}^{\mathrm{i}m(\phi-\phi')}. \tag{C.40}$$

Expressing $\boldsymbol{\rho} = \rho\mathrm{e}^{\mathrm{i}\phi} = x + \mathrm{i}y$, $\boldsymbol{\rho}' = \rho'\mathrm{e}^{\mathrm{i}\phi'} = x' + \mathrm{i}y'$, $R\mathrm{e}^{\mathrm{i}\Omega} = \rho\mathrm{e}^{\mathrm{i}\phi} - \rho'\mathrm{e}^{\mathrm{i}\phi'} = |\boldsymbol{\rho} - \boldsymbol{\rho}'|\mathrm{e}^{\mathrm{i}\Omega}$, we have the following generalization of the additional theorem (C.38) applicable to a cylindrical harmonic (figure C1)

$$\mathscr{C}_\nu(\lambda|\boldsymbol{\rho} - \boldsymbol{\rho}'|)\mathrm{e}^{\mathrm{i}\nu(\Omega-\phi')} = \begin{cases} \displaystyle\sum_{m=-\infty}^{\infty} J_m(\lambda\rho)\mathscr{C}_{m-\nu}(\lambda\rho')\mathrm{e}^{\mathrm{i}m(\phi-\phi')}, & \rho < \rho', \\ \displaystyle\sum_{m=-\infty}^{\infty} J_{m-\nu}(\lambda\rho)\mathscr{C}_m(\lambda\rho')\mathrm{e}^{\mathrm{i}m(\phi-\phi')}, & \rho > \rho'. \end{cases} \tag{C.41}$$

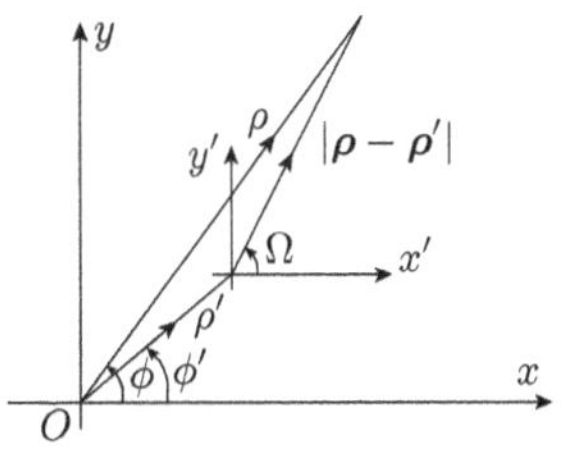


**Figure C1.** Parameters associated with the addition theorem.

### C.2.8 Multiplicative theorem

$$\mathscr{C}_\nu(\lambda z) = \lambda^{\pm\nu} \sum_{m=0}^{\infty} (\mp 1)^m \frac{(\lambda^2 - 1)^m \left(\frac{z}{2}\right)^m}{m!} \mathscr{C}_{\nu\pm m}(z), \quad |\lambda^2 - 1| < 1. \tag{C.42}$$

If $\mathscr{C} = J$ and the upper signs are taken, the restriction on $\lambda$ is unnecessary.

### C.2.9 Upper bounds

Let $z = x + \mathrm{i}y$

$$|J_\nu(x)| \leq 1, \quad \nu \geq 0 \tag{C.43}$$

$$|J_\nu(x)| \leq \frac{1}{\sqrt{2}}, \quad \nu \geq 1 \tag{C.44}$$

$$|J_\nu(z)| \leq \frac{\left|\frac{1}{2}z\right|^\nu e^{|y|}}{\Gamma(\nu + 1)}, \quad \nu \geq -\frac{1}{2} \tag{C.45}$$

$$|J_n(nz)| \leq \left| \frac{z^n \exp\left(n\sqrt{1 - z^2}\right)}{\left(1 + \sqrt{1 - z^2}\right)^n} \right| \tag{C.46}$$

$$0 < J_\nu(\nu) < \frac{2^{\frac{1}{3}}}{3^{\frac{2}{3}}\Gamma\left(\frac{2}{3}\right)\nu^{\frac{1}{3}}}, \quad \nu > 0. \tag{C.47}$$

### C.2.10 Relation to modified Bessel functions

$$J_\nu(ze^{\mathrm{i}\pi/2}) = e^{\mathrm{i}\pi\nu/2} I_\nu(z), \quad (-\pi < \arg(z) \leq \pi/2) \tag{C.48}$$

$$J_\nu(ze^{-\mathrm{i}3\pi/2}) = e^{-\mathrm{i}3\pi\nu/2} I_\nu(z), \quad (\pi/2 < \arg(z) \leq \pi) \tag{C.49}$$

$$H_\nu^{(2)}(ze^{-\mathrm{i}\pi/2}) = \frac{2\mathrm{i}}{\pi} e^{\mathrm{i}\nu\pi/2} K_\nu(z), \quad (-\pi/2 < \arg(z) \leq \pi) \tag{C.50}$$

$$H_\nu^{(1)}(ze^{\mathrm{i}\pi/2}) = -\frac{2\mathrm{i}}{\pi} e^{-\mathrm{i}\nu\pi/2} K_\nu(z), \quad (-\pi < \arg(z) \leq \pi/2) \tag{C.51}$$

$$I_\nu(z_1)K_\nu(z_2) \sim \frac{e^{(z_1 - z_2)}}{2\sqrt{z_1 z_2}}, \quad |z_1|, |z_2| \gg \nu \tag{C.52}$$

$$I_\nu(z_1)K_\nu(z_2) \sim \frac{1}{2\nu}\left(\frac{z_1}{z_2}\right)^\nu, \quad |z_1|, |z_2| \ll 1. \tag{C.53}$$

**Table C1.** $p$th zero $\chi_{np}$ of $J_n(x)$.

| $n$ | $p = 1$ | $p = 2$ | $p = 3$ | $p = 4$ |
|---|---|---|---|---|
| 0 | 2.405 | 5.520 | 8.654 | 11.792 |
| 1 | 3.832 | 7.016 | 10.173 | 13.324 |
| 2 | 5.136 | 8.417 | 11.620 | 14.796 |
| 3 | 6.380 | 9.761 | 13.015 | 16.223 |
| 4 | 7.588 | 11.065 | 14.373 | 17.616 |

**Table C2.** $p$th zero $\chi'_{np}$ of $J'_n(x)$.

| $n$ | $p = 1$ | $p = 2$ | $p = 3$ | $p = 4$ |
|---|---|---|---|---|
| 0 | 3.831 | 7.016 | 10.173 | 13.324 |
| 1 | 1.841 | 5.331 | 9.536 | 11.706 |
| 2 | 3.054 | 6.706 | 9.970 | 13.170 |
| 3 | 4.201 | 8.015 | 11.346 | 14.586 |
| 4 | 5.317 | 9.283 | 12.682 | 15.964 |

### C.2.11 Zeros of $J_n(x)$ and $J'_n(x)$

See tables C1 and C2 for the $p$th zero of Bessel function of the first kind and its derivative.

### C.2.12 Useful integrals involving Bessel functions

In the integrals below $\mathscr{C}$ and $\mathscr{Z}$ could be any one of $J$, $Y$, $H^{(1)}$, $H^{(2)}$.

$$\int_0^\infty \rho J_n(k_\rho \rho) J_n(k'_\rho \rho)\, \mathrm{d}\rho = \frac{\delta(k_\rho - k'_\rho)}{k'_\rho}, \tag{C.54}$$

$$\int x^{-\nu}\, \mathscr{C}_{\nu+1}(x)\, \mathrm{d}x = -x^{-\nu}\, \mathscr{C}_\nu(x), \tag{C.55}$$

$$\int x^{\nu+1}\, \mathscr{C}_\nu(x)\, \mathrm{d}x = x^{\nu+1} \mathscr{C}_{\nu+1}(x), \tag{C.56}$$

$$\int x\mathscr{Z}_p(ax)\mathscr{C}_p(ax)\,\mathrm{d}x = \frac{x^4}{4}\Big[2\mathscr{Z}_p(ax)\mathscr{C}_p(ax) - \mathscr{Z}_{p-1}(ax)\mathscr{C}_{p+1}(ax) - \mathscr{Z}_{p+1}(ax)\mathscr{C}_{p-1}(ax)\Big], \tag{C.57}$$

$$\int x\mathscr{Z}_p(ax)\mathscr{C}_p(bx)\,\mathrm{d}x = \frac{ax\mathscr{Z}_{p+1}(ax)\mathscr{C}_p(bx) - bx\mathscr{Z}_p(ax)\mathscr{C}_{p+1}(bx)}{a^2 - b^2}, \tag{C.58}$$

$$= \frac{bx\mathscr{Z}_p(ax)\mathscr{C}_{p-1}(bx) - ax\mathscr{Z}_{p-1}(ax)\mathscr{C}_p(bx)}{a^2 - b^2}, \tag{C.59}$$

$$\int x\mathscr{Z}_p^2(ax)\,\mathrm{d}x = \frac{x^2}{2}\Big[\mathscr{Z}_p^2(ax) - \mathscr{Z}_{p-1}(ax)\mathscr{Z}_{p+1}(ax)\Big], \tag{C.60}$$

$$\int x\mathscr{Z}_0(x)\,\mathrm{d}x = x\mathscr{Z}_1(x), \tag{C.61}$$

$$\int \mathscr{Z}_1(x)\,\mathrm{d}x = -\mathscr{Z}_0(x), \tag{C.62}$$

$$\int \frac{1}{z^2}\,\hat{\mathscr{Z}}_p(z)\,\hat{\mathscr{C}}_q(z)\,\mathrm{d}z = \frac{\hat{\mathscr{C}}_q\hat{\mathscr{Z}}'_p(z) - \hat{\mathscr{Z}}_p\hat{\mathscr{C}}'_q(z)}{(p-q)(p+q+1)}, \tag{C.63}$$

$$\int_{-\infty}^{\infty} F(k_z)H_n^{(2)}(k_\rho\rho)\mathrm{e}^{jk_z z}\,\mathrm{d}k_z \sim 2\frac{\mathrm{e}^{-jkr}}{r}j^{n+1}F(-k\cos\theta),\ r\to\infty, \tag{C.64}$$

where $k_z^2 + k_\rho^2 = k^2$, $z = r\cos\theta$, $\rho = r\sin\theta$.

$$\int x^{-1}\,\mathscr{Z}_p(ax)\mathscr{C}_q(ax)\,\mathrm{d}x = ax\frac{\mathscr{Z}_p(ax)\mathscr{C}_{q+1}(ax) - \mathscr{Z}_{p+1}(ax)\mathscr{C}_q(ax)}{p^2 - q^2} + \frac{\mathscr{Z}_p(ax)\mathscr{C}_q(ax)}{p+q}, \tag{C.65}$$

$$= ax\frac{\mathscr{Z}_{p-1}(ax)\mathscr{C}_q(ax) - \mathscr{Z}_p(ax)\mathscr{C}_{q-1}(ax)}{p^2 - q^2} - \frac{\mathscr{Z}_p(ax)\mathscr{C}_q(ax)}{p+q}, \tag{C.66}$$

$$= ax\frac{\mathscr{C}_q(ax)\mathscr{Z}'_p(ax) - \mathscr{Z}_p(ax)\mathscr{C}'_q(ax)}{p^2 - q^2}. \tag{C.67}$$

$$\frac{e^{\pm ik|\mathbf{r}-\mathbf{r}'|}}{|\mathbf{r}-\mathbf{r}'|} = \frac{1}{2i}\int_{-\infty}^{\infty} H_0^{\substack{(1)\\(2)}}(k_\rho|\boldsymbol{\rho}-\boldsymbol{\rho}'|)e^{ik_z(z-z')}\,dk_z \tag{C.68}$$

$$=\frac{1}{2i}\int_{-\infty}^{\infty}\sum_{n=-\infty}^{\infty} J_n(k_\rho \rho_<)H_n^{\substack{(1)\\(2)}}(k_\rho \rho_>)e^{i\,n(\phi-\phi')}e^{ik_z(z-z')}\,dk_z, \tag{C.69}$$

where $k_\rho^2 + k_z^2 = k^2$, $\rho_< = \min(\rho, \rho')$, $\rho_> = \max(\rho, \rho')$, and $|\boldsymbol{r} - \boldsymbol{r}'|^2 = |\boldsymbol{\rho} - \boldsymbol{\rho}'|^2 + (z - z')^2$.

The functions with hats in equation (C.63) are the Spherical Ricatti-Bessl functions (see section C.3.8).

### C.2.13 Series of Bessel functions

$$e^{iz\cos\varphi} = \sum_{n=0}^{\infty}(2n+1)i^n j_n(z)P_n(\cos\varphi), \tag{C.70}$$

$$= \sum_{n=-\infty}^{\infty} i^n J_n(z)e^{in\varphi} = J_0(z) + 2\sum_{n=1}^{\infty} i^n J_n(z)\cos n\varphi, \tag{C.71}$$

$$e^{\pm iz\sin\varphi} = J_0(z) + 2\sum_{n=1}^{\infty} J_{2n}(z)\cos 2n\varphi \pm 2i\sum_{n=0}^{\infty} J_{2n+1}(z)\sin(2n+1)\varphi. \tag{C.72}$$

## C.3 Spherical Bessel functions

The differential equation satisfied by the spherical Bessel functions is

$$z^2\frac{d^2R}{dz^2} + 2z\frac{dR}{dz} + [z^2 - n(n+1)]R = 0, \quad n = 0, \pm 1, \pm 2, \ldots. \tag{C.73}$$

Its solutions are denoted by $b_n(z)$, which are related to the cylindrical Bessel functions through the relationship

$$b_n(z) = \sqrt{\frac{\pi}{2z}}\,\mathscr{C}_{n+\frac{1}{2}}(z), \tag{C.74}$$

where $\mathscr{C}$ is any of $J$, $Y$, $H^{(1)}$, $H^{(2)}$. Thus

$$j_n(z) = \sqrt{\frac{\pi}{2z}}\,J_{n+\frac{1}{2}}(z) \tag{C.75}$$

$$y_n(z) = \sqrt{\frac{\pi}{2z}}\,Y_{n+\frac{1}{2}}(z) \tag{C.76}$$

$$h_n^{(1)}(z) = \sqrt{\frac{\pi}{2z}}\,H_{n+\frac{1}{2}}^{(1)}(z) = j_n(z) + iy_n(z) \tag{C.77}$$

$$h_n^{(2)}(z) = \sqrt{\frac{\pi}{2z}}\, H_{n+\frac{1}{2}}^{(2)}(z) = j_n(z) - \mathrm{i}y_n(z). \tag{C.78}$$

The function $p(z)$ for the spherical Bessel functions is $p(z) = z^2$ and Wronskians are

$$\mathscr{W}\{j_n(z), y_n(z)\} = \frac{1}{z^2}, \tag{C.79}$$

$$\mathscr{W}\left\{h_n^{(1)}(z), h_n^{(2)}(z)\right\} = \frac{2}{\mathrm{i}z^2}, \tag{C.80}$$

$$\mathscr{W}\left\{j_n(z), h_n^{(2)}(z)\right\} = \frac{1}{\mathrm{i}z^2}. \tag{C.81}$$

Furthermore,

$$y_n(z) = (-1)^{n+1} j_{-n-1}(z), \quad n = 0, \pm 1, \pm 2, \ldots \tag{C.82}$$

$$h_{-n-1}^{(1)}(z) = \mathrm{i}(-1)^n h_n^{(1)}(z), \quad n = 0, \pm 1, \pm 2, \ldots \tag{C.83}$$

$$h_{-n-1}^{(2)}(z) = -\mathrm{i}(-1)^n h_n^{(2)}(z), \quad n = 0, \pm 1, \pm 2, \ldots. \tag{C.84}$$

For complex values of the argument we have

$$j_n(z\mathrm{e}^{\mathrm{i}\pi/2}) = \mathrm{e}^{\mathrm{i}n\pi/2}\mathrm{i}_n(z), \; -\pi < \arg(z) \le \pi/2 \tag{C.85}$$

$$y_n(z\mathrm{e}^{\mathrm{i}\pi/2}) = \mathrm{e}^{-\mathrm{i}3n\pi/2}k_n(z), \; -\pi < \arg(z) \le \pi/2, \tag{C.86}$$

where $\mathrm{i}_n(z)$ and $k_n(z)$ are the *modified spherical Bessel functions.*

### C.3.1 Relation to elementary functions

The spherical Bessel functions can be expressed in a finite series of elementary functions as

$$j_n(z) = z^{-1}\left[\sin\left(z - \frac{1}{2}n\pi\right)P\left(n + \frac{1}{2}, z\right) + \cos\left(z - \frac{1}{2}n\pi\right)Q\left(n + \frac{1}{2}, z\right)\right] \tag{C.87}$$

$$y_n(z) = -z^{-1}\left[\cos\left(z - \frac{1}{2}n\pi\right)P\left(n + \frac{1}{2}, z\right) - \sin\left(z - \frac{1}{2}n\pi\right)Q\left(n + \frac{1}{2}, z\right)\right] \tag{C.88}$$

$$h_n^{(1)}(z) = z^{-1}\left[-\mathrm{i}P\left(n + \frac{1}{2}, z\right) + Q\left(n + \frac{1}{2}, z\right)\right]\mathrm{e}^{\mathrm{i}\left(z - \frac{1}{2}n\pi\right)} \tag{C.89}$$

$$h_n^{(2)}(z) = z^{-1}\left[\mathrm{i}P\left(n + \frac{1}{2}, z\right) + Q\left(n + \frac{1}{2}, z\right)\right]\mathrm{e}^{-\mathrm{i}\left(z - \frac{1}{2}n\pi\right)} \tag{C.90}$$

where

$$P\left(n+\frac{1}{2},z\right)=\sum_{k=0}^{[\frac{1}{2}n]}(-1)^k(n+\frac{1}{2},2k)(2z)^{-2k} \tag{C.91}$$

$$Q\left(n+\frac{1}{2},z\right)=\sum_{k=0}^{[\frac{1}{2}(n-1)]}(-1)^k\left(n+\frac{1}{2},2k+1\right)(2z)^{-2k-1} \tag{C.92}$$

$$\left(n+\frac{1}{2},k\right)=\frac{(n+k)!}{k!(n-k)!}. \tag{C.93}$$

Some of the lower order functions are

$$j_0(z)=\frac{\sin z}{z},\quad y_0(z)=-j_{-1}(z)=-\frac{\cos z}{z} \tag{C.94}$$

$$j_1(z)=\frac{1}{z}\left(\frac{\sin z}{z}-\cos z\right),\quad y_1(z)=j_{-2}(z)=-\frac{1}{z}\left(\frac{\cos z}{z}+\sin z\right) \tag{C.95}$$

$$j_2(z)=\left(\frac{3}{z^3}-\frac{1}{z}\right)\sin z-\frac{3}{z^2}\cos z \tag{C.96}$$

$$y_2(z)=-j_{-3}(z)=-\left(\frac{3}{z^3}-\frac{1}{z}\right)\cos z-\frac{3}{z^2}\sin z \tag{C.97}$$

$$h_0^{(1)}(z)=\frac{e^{iz}}{iz},\quad h_0^{(2)}(z)=-\frac{e^{-iz}}{iz} \tag{C.98}$$

$$h_1^{(1)}(z)=\frac{e^{iz}}{z}\left(\frac{1}{iz}-1\right),\quad h_1^{(2)}(z)=-\frac{e^{-iz}}{z}\left(\frac{1}{iz}+1\right) \tag{C.99}$$

$$h_2^{(1)}(z)=-\frac{e^{iz}}{z}\left[i\left(\frac{3}{z^2}-1\right)+\frac{3}{z}\right],\quad h_2^{(2)}(z)=\frac{e^{-iz}}{z}\left[i\left(\frac{3}{z^2}-1\right)-\frac{3}{z}\right]. \tag{C.100}$$

### C.3.2 Small argument approximations

For a fixed $n$ and $z \to 0$

$$j_n(z)\sim\frac{2^n n! z^n}{(2n+1)!}=\frac{z^n}{1.3.5.\ \ldots(2n+1)} \tag{C.101}$$

$$y_n(z)\approx -ih_n^{(1)}(z)\approx ih_n^{(2)}(z)\sim-\frac{1.3.5\ldots(2n-1)}{z^{n+1}}. \tag{C.102}$$

### C.3.3 Analytical continuation

$$j_n(z\mathrm{e}^{\mathrm{i}m\pi}) = \mathrm{e}^{\mathrm{i}mn\pi} j_n(z) \tag{C.103}$$

$$y_n(z\mathrm{e}^{\mathrm{i}m\pi}) = (-1)^m \mathrm{e}^{\mathrm{i}mn\pi} y_n(z) \tag{C.104}$$

$$h_n^{(1)}(z\mathrm{e}^{\mathrm{i}\pi(2m+1)}) = (-1)^n h_n^{(2)}(z) \tag{C.105}$$

$$h_n^{(2)}(z\mathrm{e}^{\mathrm{i}\pi(2m+1)}) = (-1)^n h_n^{(1)}(z) \tag{C.106}$$

$$h_n^{(\ell)}(z\mathrm{e}^{\mathrm{i}2m\pi}) = h_n^{(\ell)}(z), \quad \ell = 1, 2;\ m, n = 0, 1, 2, .... \tag{C.107}$$

### C.3.4 Recurrence relations

$$b_{n-1}(z) + b_{n+1}(z) = (2n+1)\frac{b_n(z)}{z} \tag{C.108}$$

$$nb_{n-1}(z) - (n+1)b_{n+1}(z) = (2n+1)\frac{\mathrm{d}}{\mathrm{d}z}b_n(z) \tag{C.109}$$

$$zb_n'(z) = nb_n(z) - zb_{n+1}(z) = -(n+1)b_n(z) + b_{n-1}(z). \tag{C.110}$$

### C.3.5 Addition theorems

With $r$, $\rho$, $\xi$, $\theta$, $\lambda$ arbitrary complex and $R = \sqrt{r^2 + \rho^2 - 2r\rho\cos\xi}$

$$\frac{\sin\lambda R}{\lambda R} = \sum_{n=0}^{\infty}(2n+1)j_n(\lambda r)j_n(\lambda\rho)P_n(\cos\xi), \tag{C.111}$$

$$-\frac{\cos\lambda R}{\lambda R} = \sum_{n=0}^{\infty}(2n+1)j_n(\lambda r)y_n(\lambda\rho)P_n(\cos\xi), \quad |r\mathrm{e}^{\pm\mathrm{i}\xi}| < |\rho|, \tag{C.112}$$

$$h_0^{(2)}(\lambda R) = \sum_{n=0}^{\infty}(2n+1)j_n(\lambda r)h_n^{(2)}(\lambda\rho)P_n(\cos\xi), \quad |r\mathrm{e}^{\pm\mathrm{i}\xi}| < |\rho|, \tag{C.113}$$

$$= \sum_{n=0}^{\infty}(2n+1)j_n(\lambda r_<)h_n^{(2)}(\lambda r_>)P_n(\cos\xi), \quad r_{\lessgtr} = \frac{\min}{\max}(r, \rho), \tag{C.114}$$

$$\mathrm{e}^{\mathrm{i}\lambda r\cos\theta} = \sum_{n=0}^{\infty}(2n+1)j_n(\lambda r)\mathrm{i}^n P_n(\cos\theta), \tag{C.115}$$

where $P_n(z)$ is the Legendre polynomial of the first kind of degree $n$. In equations (C.111)–(C.113), if $r$ and $\rho$ denote spherical distances of position vectors $(r, \theta_1, \phi_1)$ and $(\rho, \theta_2, \phi_2)$, then $\xi$ is the angle between these position vectors with $\cos\xi = \cos\theta_1\cos\theta_2 + \sin\theta_1\sin\theta_2\cos(\phi_1 - \phi_2)$.

### C.3.6 Cross products

$$j_n(z)y_{n-1}(z) - j_{n-1}(z)y_n(z) = z^{-2} \tag{C.116}$$

$$j_{n+1}(z)y_{n-1}(z) - j_{n-1}(z)y_{n+1}(z) = (2n+1)z^{-3}. \tag{C.117}$$

### C.3.7 Some useful series

$$\sum_{n=0}^{\infty}(2n+1)j_n^2(z) = 1 \tag{C.118}$$

$$\sum_{n=0}^{\infty}(-1)^n(2n+1)j_n^2(z) = \frac{\sin 2z}{2z} \tag{C.119}$$

$$\sum_{n=0}^{\infty}j_n^2(z) = \frac{\mathrm{Si}(2z)}{2z} \tag{C.120}$$

$$\sum_{n=0}^{\infty}n(n+1)(2n+1)j_n^2(z) = \frac{2z^2}{3}. \tag{C.121}$$

### C.3.8 Spherical Ricatti–Bessel functions

In spherical coordinates, the radial components of the vector potentials, $A_r$, $F_r$ do not satisfy the Helmholtz equation. However, they are related to the scalar potential function $\psi$ that satisfies the scalar Helmholtz equation through $A_r, F_r \propto z\psi$. The vector potential components satisfy the differential equation

$$z^2\frac{d^2\hat{R}}{dz^2} + [z^2 - \nu(\nu+1)]\hat{R} = 0, \tag{C.122}$$

which is different from equation (C.73). The function $p(z)$ for the Ricatti–Bessel equation (C.122) is $p(z) = 1$. Hence the Wronskian between two independent solutions will be independent of $z$. The solutions of equation (C.122) are denoted by $\hat{\mathscr{C}}_\nu(z)$ and are related to the spherical Bessel functions $c_\nu(z)$ and the cylindrical Bessel functions $\mathscr{C}_\nu(z)$ through

$$\hat{\mathscr{C}}_\nu(z) = z\, c_\nu(z) = \sqrt{\frac{\pi z}{2}}\, \mathscr{C}_{\nu+\frac{1}{2}}(z), \tag{C.123}$$

and

$$\frac{\mathrm{d}}{\mathrm{d}z}\hat{\mathscr{C}}_\nu(z) = \sqrt{\frac{\pi z}{2}}\frac{\mathrm{d}}{\mathrm{d}z}\mathscr{C}_{\nu+\frac{1}{2}}(z) + \frac{1}{2z}\hat{\mathscr{C}}_\nu(z). \tag{C.124}$$

Consequently, the singularities present at the origin in $\hat{Y}_\nu(z)$, $\hat{H}_\nu^{(1)}(z)$, and $\hat{H}_\nu^{(2)}(z)$ are weakened by the presence of the multiplicative factor $z$. In particular, $\hat{Y}_0(z)$ is no longer singular, but simply equals $-\cos z$ and can appear in the expansion of functions regular at the origin. The recurrence relations for the spherical Ricatti–Bessel functions are

$$\hat{\mathscr{C}}_{\nu-1}(z) + \hat{\mathscr{C}}_{\nu+1}(z) = \frac{2\nu+1}{z}\hat{\mathscr{C}}_\nu(z), \tag{C.125}$$

$$\hat{\mathscr{C}}_{\nu-1}(z) - \hat{\mathscr{C}}_{\nu+1}(z) = 2\frac{\mathrm{d}}{\mathrm{d}z}\hat{\mathscr{C}}_\nu(z) - \frac{1}{z}\hat{\mathscr{C}}_\nu(z), \tag{C.126}$$

$$(\nu+1)\hat{\mathscr{C}}_{\nu-1}(z) - \nu\hat{\mathscr{C}}_{\nu+1}(z) = (2\nu+1)\frac{\mathrm{d}}{\mathrm{d}z}\hat{\mathscr{C}}_\nu(z). \tag{C.127}$$

From these it is easy to derive $z\hat{\mathscr{C}}'_\nu(z) = (\nu+1)\hat{\mathscr{C}}_\nu(z) - z\hat{\mathscr{C}}_{\nu+1}(z) = z\hat{\mathscr{C}}_{\nu-1}(z) - \nu\hat{\mathscr{C}}_\nu(z)$. Note also that

$$\frac{\mathrm{d}^2\hat{\mathscr{C}}_\nu}{\mathrm{d}z^2} + \hat{\mathscr{C}}_\nu = \frac{\nu(\nu+1)}{z^2}\hat{\mathscr{C}}_\nu, \tag{C.128}$$

for all spherical Ricatti–Bessel functions. Equations (C.123) and (C.23) imply that the Wronskian

$$\mathscr{W}[\hat{J}_\nu(z), \hat{Y}_\nu(z)] = z^2\mathscr{W}[j_\nu(z), y_\nu(z)] = 1, \tag{C.129}$$

which is independent of $z$ as remarked earlier. Furthermore,

$$\mathscr{W}[\hat{J}_\nu(z), \hat{H}_\nu^{(2)}(z)] = -\mathrm{i}, \tag{C.130}$$

$$\mathscr{W}[\hat{H}_\nu^{(1)}(z), \hat{H}_\nu^{(2)}(z)] = -2\mathrm{i}. \tag{C.131}$$

From the independence of the Wronskian on the argument it follows that

$$\hat{Y}_\nu''(z)\hat{J}_\nu(z) - \hat{J}_\nu''(z)\hat{Y}_\nu(z) = 0. \tag{C.132}$$

Other Wronskians can be written down in a similar fashion.

*C.3.8.1 Exact series form and lower order functions*

$$\hat{H}_n^{(1)}(z) = \mathrm{i}^{-(n+1)}\mathrm{e}^{\mathrm{i}z}\sum_{k=0}^{n}(-1)^k\frac{(n+k)!}{k!(n-k)!}\frac{1}{(2\mathrm{i}z)^k} \tag{C.133}$$

$$\hat{H}_n^{(2)}(z) = \mathrm{i}^{(n+1)}\mathrm{e}^{-\mathrm{i}z}\sum_{k=0}^{n}\frac{(n+k)!}{k!(n-k)!}\frac{1}{(2\mathrm{i}z)^k} \tag{C.134}$$

$$\hat{H}_0^{\substack{(1)\\(2)}}(z) = \mp\mathrm{i}\mathrm{e}^{\pm\mathrm{i}z} \tag{C.135}$$

$$\hat{J}_0(z) = \Re\left[\hat{H}_0^{(2)}(z)\right] = \sin z \tag{C.136}$$

$$\hat{Y}_0(z) = -\Im\left[\hat{H}_0^{(2)}(z)\right] = -\cos z. \tag{C.137}$$

Note that all zeroth order Ricatti–Bessel functions are finite at the origin. It is also worth noting that for integer orders, the spherical Bessel functions and the Ricatti–Bessel functions are free of any branch point singularities in the complex $z$-plane. This is in contrast to the cylindrical Hankel functions of integer order, which have a branch point singularity at $z = 0$.

A useful indefinite integral that follows from (C.60) is that

$$2\int\hat{\mathscr{C}}_\nu^2(x)\,\mathrm{d}x = x\left[\hat{\mathscr{C}}_\nu^2(x) - \hat{\mathscr{C}}_{\nu-1}(x)\hat{\mathscr{C}}_{\nu+1}(x)\right], \tag{C.138}$$

$$= x\left[\hat{\mathscr{C}}_\nu^{\prime 2}(x) + \left(1 - \frac{\nu(\nu+1)}{x^2}\right)\hat{\mathscr{C}}_\nu^2(x)\right] - \hat{\mathscr{C}}_\nu(x)\hat{\mathscr{C}}_\nu'(x), \tag{C.139}$$

which remains valid for $\hat{\mathscr{C}}_\nu = \hat{J}_\nu, \hat{Y}_\nu, \hat{H}_\nu^{(1)}, \hat{H}_\nu^{(2)}$.

*C.3.8.2 Small argument approximations*
For fixed order $n$ and small argument $|z| \ll n$, the following asymptotic forms hold

$$\hat{H}_n^{(2)}(z) \sim \mathrm{i}\frac{(2n)!}{n!2^n z^n} =: \mathrm{i}k_n z^{-n}, \quad k_n = \frac{(2n)!}{n!2^n}, \tag{C.140}$$

$$\frac{\hat{H}_n^{(2)\prime}(z)}{\hat{H}_n^{(2)}(z)} \sim -\frac{n}{z}\left(1 - \frac{z^2}{n(2n-1)}\right); \quad \hat{H}_n^{(2)\prime}(z) \sim -\mathrm{i}nk_n z^{-(n+1)}. \tag{C.141}$$

*C.3.8.3 Large order approximations*
For fixed argument $z$ and large orders $\nu \gg |z|$, the following approximations hold

$$\hat{J}_\nu(z) \sim \frac{1}{\sqrt{2e}}\left(\frac{ez}{2\nu+1}\right)^{\nu+1}; \quad \hat{J}'_\nu(z) \sim \frac{\nu}{z}\hat{J}_\nu(z), \tag{C.142}$$

$$\hat{Y}_\nu(z) \sim -\sqrt{\frac{2}{e}}\left(\frac{2\nu+1}{ez}\right)^{\nu}; \quad \hat{Y}'_\nu(z) \sim -\frac{\nu}{z}\hat{Y}_\nu(z), \tag{C.143}$$

$$\hat{J}_\nu(z)\hat{Y}_\nu(z) \sim -\frac{z}{2\nu+1}, \tag{C.144}$$

$$\hat{H}_\nu^{(2)'}(z) \sim \frac{\nu}{z}\hat{H}_\nu^{(1)}(z) \sim \frac{\nu}{z}\left[\hat{J}_\nu(z) - \mathrm{i}\left(\frac{z}{2\nu+1}\right)\frac{1}{\hat{J}_\nu(z)}\right] \sim -\frac{\mathrm{i}\nu}{2\nu+1}\frac{1}{\hat{J}_\nu(z)}. \tag{C.145}$$

*C.3.8.4 Inequalities*

$$\Re\left(\hat{H}_\nu^{(2)}(x)\hat{H}_\nu^{(1)'}(x)\right) = \hat{J}_\nu(x)\hat{J}'_\nu(x) + \hat{Y}_\nu(x)\hat{Y}'_\nu(x) < 0, \quad x > 0, \nu > 0, \tag{C.146}$$

$$\Im\left(\hat{H}_\nu^{(2)}(x)\hat{H}_\nu^{(1)'}(x)\right) = 1 > 0, \tag{C.147}$$

$$2 + \frac{\nu(\nu+1)}{x^2}\left|\hat{H}_\nu^{(2)}(x)\right|^2 > \left|\hat{H}_\nu^{(2)'}(x)\right|^2 + \left|\hat{H}_\nu^{(2)}(x)\right|^2, \; x > 0, \nu > 0. \tag{C.148}$$

*C.3.8.5 Large argument approximations*

For a fixed order $\nu$ and large argument $|z| \gg \nu$ or $|x| \gg \nu$, $x$ real, the following asymptotic forms hold

$$\hat{J}_\nu(z) \sim \sin(z - \frac{\nu\pi}{2}) + \frac{\nu(\nu+1)}{z}\cos(z - \frac{\nu\pi}{2}) + \mathcal{O}\left(\frac{1}{z^2}\right), \tag{C.149}$$

$$-\hat{Y}_\nu(z) \sim \cos(z - \frac{\nu\pi}{2}) - \frac{\nu(\nu+1)}{z}\sin(z - \frac{\nu\pi}{2}) + \mathcal{O}\left(\frac{1}{z^2}\right), \tag{C.150}$$

$$\hat{H}_\nu^{(2)}(z) \sim \mathrm{i}^{\nu+1}\mathrm{e}^{-\mathrm{i}z}\left(1 + \frac{\nu(\nu+1)}{2\mathrm{i}z}\right) + \mathcal{O}\left(\frac{1}{z^2}\right), \tag{C.151}$$

$$\hat{H}_\nu^{(1)}(z) \sim \mathrm{i}^{-(\nu+1)}\mathrm{e}^{\mathrm{i}z}\left(1 - \frac{\nu(\nu+1)}{2\mathrm{i}z}\right) + \mathcal{O}\left(\frac{1}{z^2}\right), \tag{C.152}$$

$$\left|\hat{H}_\nu^{(1)}(x)\right|^2 = \left|\hat{H}_\nu^{(2)}(x)\right|^2 = \hat{J}_\nu^{\,2}(x) + \hat{Y}_\nu^{\,2}(x) \sim 1 + \frac{\nu(\nu+1)}{2x^2}, \tag{C.153}$$

$$\hat{J}_\nu'(x)\hat{J}_\nu(x) + \hat{Y}_\nu'(x)\hat{Y}(x) \sim -\frac{\nu(\nu+1)}{2x^3}, \tag{C.154}$$

$$\hat{J}_\nu'(z) \sim -\ \hat{Y}_\nu(z) + \mathcal{O}\left(\frac{1}{z^2}\right) \tag{C.155}$$

$$\hat{Y}_\nu'(z) \sim \hat{J}_\nu(z) + \mathcal{O}\left(\frac{1}{z^2}\right) \tag{C.156}$$

$$\left|\hat{H}_\nu^{(1)'}(x)\right|^2 = \left|\hat{H}_\nu^{(2)'}(x)\right|^2 \sim \hat{J}_\nu'^2(x) + \hat{Y}_\nu'^2(x) \sim 1 - \frac{\nu(\nu+1) - 0.5}{2x^2}. \tag{C.157}$$

## C.4 Plots for Bessel functions

Figure C2 and C3 show plots of some lower order cylindrical and spherical Bessel functions of the first and second kind, respectively, with respect to the real argument for fixed integer orders. Figure C4 shows the variation of the Bessel functions with respect to the order for a fixed real and positive argument. Note that $|Y_\nu(z)|$ and $|y_\nu(z)|$ start to take very large values, whereas $|J_\nu(z)|$ and $|j_\nu(z)|$ start to diminish as the positive order exceeds the argument (figures C5 and C6).

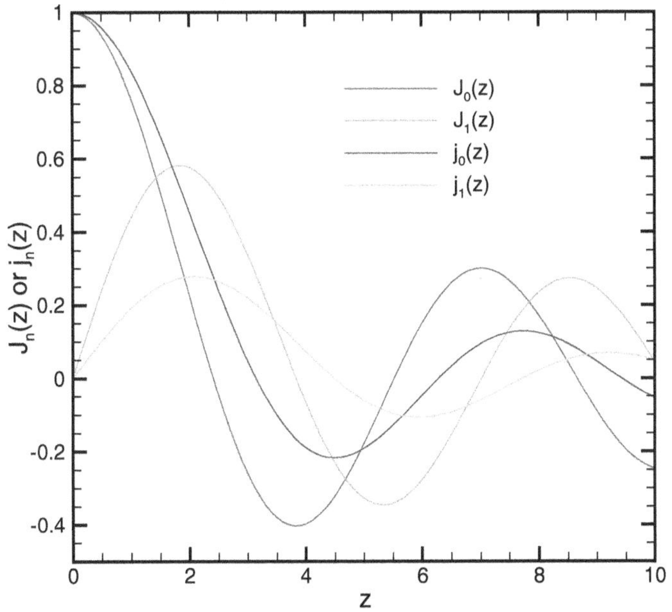


**Figure C2.** Cylindrical and spherical Bessel functions of the first kind.

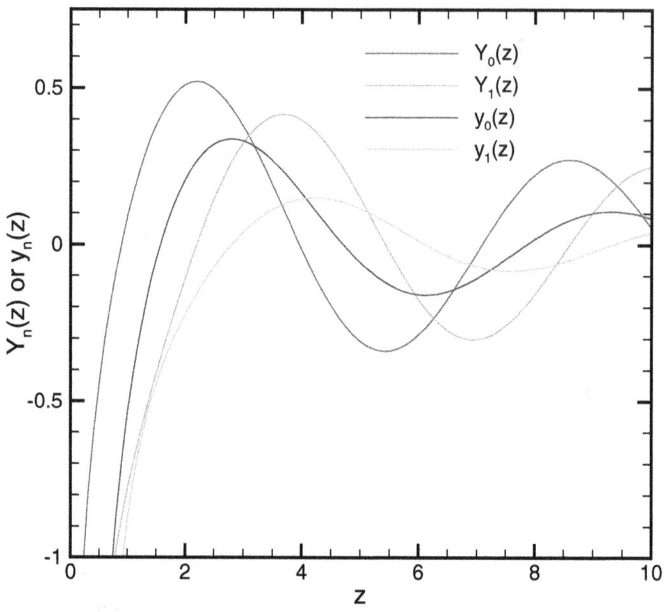


**Figure C3.** Cylindrical and spherical Bessel functions of the second kind.

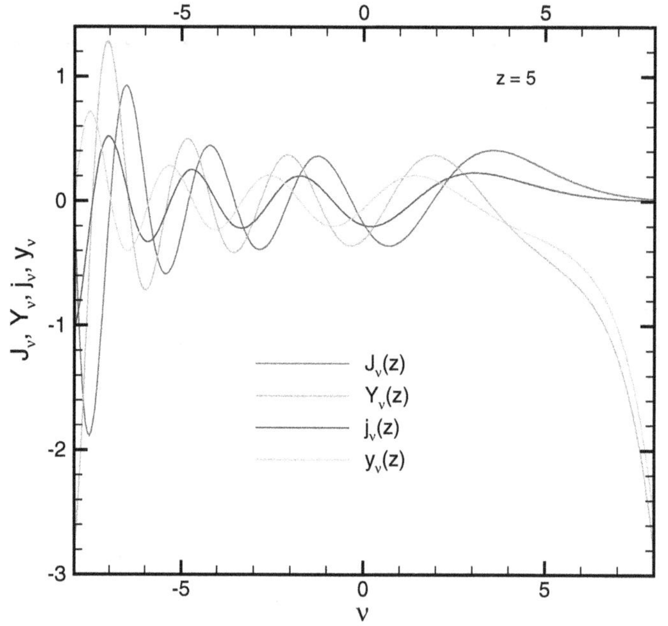


**Figure C4.** Variation of cylindrical and spherical Bessel functions with respect to order.

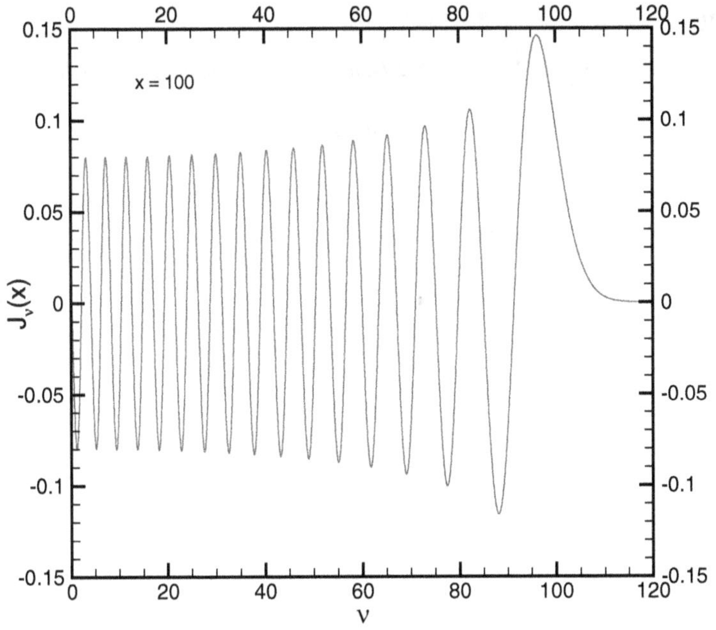


**Figure C5.** Variation of $J_\nu(x)$ with respect to order $\nu$ for $x = 100$. For orders exceeding the argument, the Bessel function decays super exponentially.

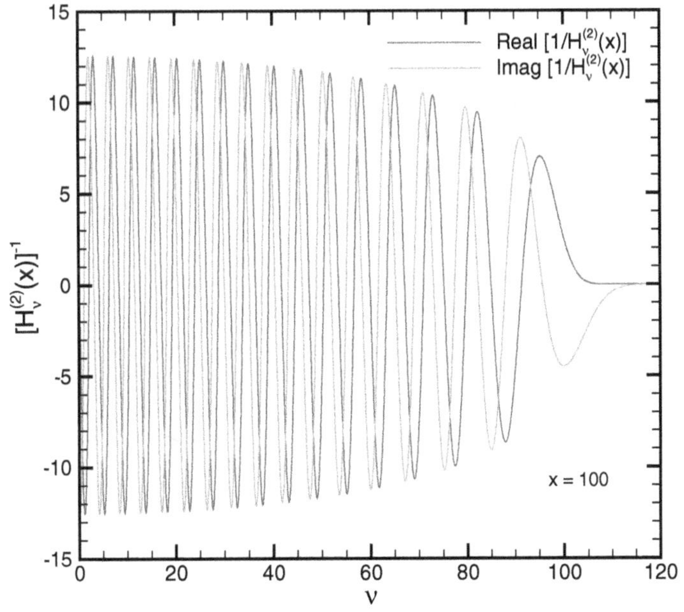


**Figure C6.** Variation of $[H_\nu^{(2)}(x)]^{-1}$ with respect to order $\nu$ for $x = 100$. For orders exceeding the argument, the reciprocal of the Hankel function decays super exponentially.

## References

[1] Abramowitz M and Stegun I A (ed) 1972 *Handbook of Mathematical Functions with Formulas, Graphs, and Mathematical Tables* (New York: Dover)
[2] Gradshteyn I S and Ryzhik I M 2015 *Tables of Integrals, Series and Products* 8th edn (New York: Academic)
[3] Hairer E, Nørsett S P and Wanner G 1986 *Solving Ordinary Differential Equations I: Non-stiff Problems* (New York: Springer)

**IOP** Publishing

Engineering Electrodynamics
A collection of theorems, principles and field representations
**Ramakrishna Janaswamy**

# Appendix D

# Associated Legendre functions and Legendre polynomials

## D.1 Legendre differential equation

Consider the *associated Legendre equation*

$$\frac{1}{\sin\theta}\frac{\mathrm{d}}{\mathrm{d}\theta}\left(\sin\theta\frac{\mathrm{d}\Theta}{\mathrm{d}\theta}\right)+\left[\nu(\nu+1)-\frac{\mu^2}{\sin^2\theta}\right]\Theta=0, \tag{D.1}$$

where $\mu$ and $\nu$ are not necessarily integers. This can be put in an equivalent form by using the substitution $u=\cos\theta$, $\mathrm{d}/\mathrm{d}\theta=(\mathrm{d}u/\mathrm{d}\theta)(\mathrm{d}/\mathrm{d}u)=-\sin\theta(\mathrm{d}/\mathrm{d}u)$ to result in

$$(1-u^2)\frac{\mathrm{d}^2\Theta}{\mathrm{d}u^2}-2u\frac{\mathrm{d}\Theta}{\mathrm{d}u}+\left[\nu(\nu+1)-\frac{\mu^2}{1-u^2}\right]\Theta=0. \tag{D.2}$$

Although we are mostly interested in the solution of this equation on the real axis in the interval $u\in[-1,1]$, many of the relations shown below are also valid for $u=x+iy=z$ complex. If a formula is only valid for real arguments, we show the argument as $x$.

It is clear from equation (D.2) that the differential equation remains unchanged if any one of the following transformations is made: $u\to-u$, or $\mu\to-\mu$, or $\nu\to-(\nu+1)$. Thus, if $w(u)$ is a solution of equation (D.2), so is $w(-u)$, although they need not be linearly dependent on each other. Initially, we focus on the following range of variable and parameter values: $\Re(u)\geq0$, $\Re(\mu)\geq0$, and $\Re(\nu)\geq-\frac{1}{2}$, which will lie at the intersections of the variable or parameter values and $\Re(\cdot)$ denotes 'real part of'. In order to understand the properties of the solution in the complex plane, we express equation (D.2) in the following standard form by dividing throughout by $(1-u^2)$:

doi:10.1088/978-0-7503-1716-0ch23

$$\frac{d^2\Theta}{du^2} + f(u)\frac{d\Theta}{du} + g(u)\Theta = 0, \tag{D.3}$$

where

$$f(u) = \frac{2u}{u^2 - 1} = \frac{1}{u + 1} + \frac{1}{u - 1}, \tag{D.4}$$

$$\begin{aligned} g(u) &= -\left[\frac{\nu(\nu + 1)}{u^2 - 1} + \frac{\mu^2}{(u^2 - 1)^2}\right] \\ &= \left[\frac{\nu(\nu + 1)}{2} - \left(\frac{\mu}{2}\right)^2\right]\left(\frac{1}{u + 1} - \frac{1}{u - 1}\right) - \left(\frac{\mu}{2}\right)^2\left[\frac{1}{(u + 1)^2} + \frac{1}{(u - 1)^2}\right]. \end{aligned} \tag{D.5}$$

Thus, $u = \pm 1$ are regular singularities[1] of (D.3). More specifically, $(u \pm 1)f(u) = 1 =: f_0$ and $(u \pm 1)^2 g(u) = -(\mu/2)^2 =: g_0$ at $u = \pm 1$. In addition, the point at infinity is also a regular singularity. This can be seen by making the transformation $u = 1/z$, $d/du = -z^2 d/dz$ in equation (D.3) and simplifying to yield

$$\frac{d^2\Theta}{dz^2} + p(z)\frac{d\Theta}{dz} + q(z)\Theta = 0, \tag{D.6}$$

where $p(z) = f(z)$ and

$$q(z) = -\left[\frac{\nu(\nu + 1)}{z^2(1 - z^2)} + \frac{\mu^2}{(1 - z^2)^2}\right], \tag{D.7}$$

and observing the behavior at $z = 0$. Clearly, $zp(z) = 0 =: p_0$ and $z^2 q(z) = -\nu(\nu + 1) =: q_0$ at $z = 0$. Thus the points $u = \pm 1, \infty$ are the three regular singular points of the associated Legendre equation. A fundamental pair of solutions of equation (D.3) may be developed with respect to each singular point and the solution itself can be singular at these singular points. The nature of singularity (i.e. whether a pole or a branch point) is dictated by the solution to the *indicial equations* defined relative to the singularity points:

$$u = \pm 1: \alpha(\alpha - 1) + f_0\alpha + g_0 = 0 \Longrightarrow \alpha_{1,2} = -\mu/2, \mu/2 \tag{D.8}$$

$$u = \infty: \alpha(\alpha - 1) + p_0\alpha + q_0 = 0 \Longrightarrow \alpha_{1,2} = -\nu, \nu + 1. \tag{D.9}$$

The solutions for $\alpha$ are known as *exponents* or *indices*. The solution behaves as $u^{\alpha_j}, j = 1, 2$ in the neighborhood of the singularity. Thus the behavior (i.e. growth or decay) at $u = \pm 1$ is dictated only by the *order* $\mu$ and behavior at infinity is dictated only by the *degree* $\nu$. Hence if $\alpha_j$ is a negative integer, the solution has a pole at the singularity, and when $\alpha_j$ is non-integral, it has a branch point. If the exponents do not differ by an integer (i.e. if $\mu$ is not an integer or if $2\nu + 1$ is not an integer) then

[1] A point $u = u_0$ is said to be a *regular singularity* [1, 2] if either $f(u)$ or $g(u)$ is not analytic at $u_0$, but *both* $(u - u_0)f(u)$ and $(u - u_0)^2 g(u)$ are analytic there.

one of the two solutions in the fundamental pair has a branch point at the corresponding singularity.

### D.1.1 Independent solutions

The two independent solutions of equation (D.3) are the *associated Legendre functions* of the first and second kind, $P_\nu^\mu(u)$ and $Q_\nu^\mu(u)$, respectively of *order* $\mu$ and *degree* $\nu$. For the range of variable and parameter values of interest here, i.e. $\Re(\nu) \geq -\frac{1}{2}$, $\Re(\mu) \geq 0$, and $u \geq 0$, it is shown in [2] that $P_\nu^{-\mu}(u)$ and $Q_\nu^\mu(u)$, which are developed with respect to the singularity at $u = 1$ with exponent $= \mu/2$ and $u = \infty$ with exponent $= (\nu + 1)$, respectively, comprise a *numerically satisfactory* pair of solutions. For arbitrary $\nu$ and $\mu$, the associated Legendre functions are related to the *hypergeometric function*, $F(\alpha, \beta; \gamma; z)$, through the gamma function, $\Gamma(z)$, as follows [3]:

$$P_\nu^{-\mu}(u) = \frac{1}{\Gamma(1+\mu)}\left(\frac{u-1}{u+1}\right)^{\frac{\mu}{2}} F\left(\nu+1, -\nu; 1+\mu; \frac{1-u}{2}\right), \tag{D.10}$$

$$\begin{aligned} e^{-i\mu\pi} Q_\nu^\mu(u) = &\frac{\Gamma(\mu+\nu+1)\Gamma(-\mu)}{2\Gamma(1+\nu-\mu)}\left(\frac{u-1}{u+1}\right)^{\frac{\mu}{2}} F\left(\nu+1, -\nu; 1+\mu; \frac{1-u}{2}\right) \\ &+ \frac{\Gamma(\mu)}{2}\left(\frac{u+1}{u-1}\right)^{\frac{\mu}{2}} F\left(\nu+1, -\nu; 1-\mu; \frac{1-u}{2}\right), \quad [\mu \neq 0,] \end{aligned} \tag{D.11}$$

$$\begin{aligned} = &\frac{\pi^{\frac{1}{2}} 2^\mu e^{\pm i\frac{\pi}{2}(\mu-\nu)}}{(u^2-1)^{\frac{\mu}{2}}}\left\{ \mp i \frac{\Gamma\left(\frac{1+\nu+\mu}{2}\right)}{2\,\Gamma\left(1+\frac{\nu-\mu}{2}\right)} F\left(-\frac{\nu+\mu}{2}, \frac{1+\nu-\mu}{2}; \frac{1}{2}; u^2\right)\right. \\ &\left. + \frac{u\,\Gamma\left(1+\frac{\nu+\mu}{2}\right)}{\Gamma\left(\frac{1+\nu-\mu}{2}\right)} F\left(\frac{1-\nu-\mu}{2}, 1+\frac{\nu-\mu}{2}; \frac{3}{2}; u^2\right)\right\}, \\ &[|u| < 1, \Im(u) \gtrless 0], \end{aligned} \tag{D.12}$$

where the representations (D.10) and (D.11) are valid for $|1 - u| < 2$ and

$$F(\alpha, \beta; \gamma; z) = \frac{\Gamma(\gamma)}{\Gamma(\alpha)\Gamma(\beta)} \sum_{k=0}^{\infty} \frac{\Gamma(\alpha+k)\Gamma(\beta+k)}{\Gamma(\gamma+k)} \frac{z^k}{k!}, \quad [|z| \leq 1], \tag{D.13}$$

$$= (1-z)^{-\alpha} F\left(\alpha, \gamma-\beta; \gamma; \frac{z}{z-1}\right), \tag{D.14}$$

$$=(1-z)^{-\beta}F\left(\beta, \gamma-\alpha; \gamma; \frac{z}{z-1}\right), \tag{D.15}$$

$$=(1-z)^{\gamma-\alpha-\beta}F(\gamma-\alpha, \gamma-\beta; \gamma; z), \tag{D.16}$$

$$F(\alpha, \beta; \gamma; 1) = \frac{\Gamma(\gamma)\Gamma(\gamma-\alpha-\beta)}{\Gamma(\gamma-\alpha)\Gamma(\gamma-\beta)}, \quad [\gamma \neq 0, -1, -2, \dots, \Re(\gamma-\alpha-\beta) > 0]. \tag{D.17}$$

If the order $\mu$ is zero, the associated Legendre equation reduces to the *Legendre equation*

$$(1-u^2)\frac{d^2\Theta}{du^2} - 2u\frac{d\Theta}{du} + \nu(\nu+1)\Theta = 0, \tag{D.18}$$

and the solutions are simply labeled as $P_\nu(u)$ and $Q_\nu(u)$ and referred to as *Legendre functions.*

In equations (D.10)–(D.17), it has been assumed that a branch cut lies along the line $(-\infty, 1)$. Accordingly, we assign $-\pi < \arg(u) < \pi$ when computing functions of the form $(u \pm 1)^\alpha$, for non-integer $\alpha$. In particular, $(u-1)^\alpha = (1-u)^\alpha e^{i\alpha\pi}$. When $u$ is a real number lying on the interval $[-1, 1]$, so that $(u = x = \cos\theta)$, we take the following functions as linearly independent solutions of the associated Legendre equation:

$$P_\nu^\mu(x) = \frac{1}{2}\left[e^{\frac{1}{2}i\mu\pi}P_\nu^\mu(\cos\theta + i0) + e^{-\frac{1}{2}i\mu\pi}P_\nu^\mu(\cos\theta - i0)\right] \tag{D.19}$$

$$=\frac{1}{\Gamma(1-\mu)}\left(\frac{1+x}{1-x}\right)^{\frac{1}{2}\mu}F\left(-\nu, \nu+1; 1-\mu; \frac{1-x}{2}\right), \tag{D.20}$$

$$Q_\nu^\mu(x) = \frac{e^{-i\mu\pi}}{2}\left[e^{-\frac{1}{2}i\mu\pi}Q_\nu^\mu(x+i0) + e^{\frac{1}{2}i\mu\pi}Q_\nu^\mu(x-i0)\right] \tag{D.21}$$

$$\begin{aligned}=&\frac{\pi}{2\sin[(\nu+\mu)\pi]\Gamma(1-\mu)}\\&\left[\cos[(\nu+\mu)\pi]\left(\frac{1+x}{1-x}\right)^{\frac{1}{2}\mu}F\left(-\nu, \nu+1; 1-\mu; \frac{1-x}{2}\right)\right.\\&\left.-\left(\frac{1-x}{1+x}\right)^{\frac{1}{2}\mu}F\left(-\nu, \nu+1; 1-\mu; \frac{1+x}{2}\right)\right],\end{aligned} \tag{D.22}$$

$$=\frac{\pi}{2\sin\mu\pi}\left[\cos\mu\pi P_\nu^\mu(x) - \frac{\Gamma(\nu+\mu+1)}{\Gamma(\nu-\mu+1)}P_\nu^{-\mu}(x)\right] \tag{D.23}$$

$$Q_\nu(x) = \frac{\pi}{2\sin\nu\pi}[\cos\nu\pi P_\nu(x) - P_\nu(-x)]. \tag{D.24}$$

Formulas for $P_\nu^\mu(x)$ and $Q_\nu^\mu(x)$ are obtained with the replacement of $(u-1)$ by $(1-x)e^{\pm i\pi}$, $(u^2-1)$ by $(1-x^2)e^{\pm i\pi}$, $(u+1)$ by $(x+1)$ for $u = x \pm i0$.

Functions with negative arguments are related to the functions with positive arguments via

$$P_\nu^\mu(-u) = e^{\mp i\nu\pi}P_\nu^\mu(u) - \frac{2}{\pi}e^{-i\mu\pi}\sin[(\mu+\nu)\pi]Q_\nu^\mu(u), \qquad \Im(u)\gtrless 0, \tag{D.25}$$

$$P_\nu^\mu(-x) = \cos[(\nu+\mu)\pi]P_\nu^\mu(x) - \frac{2}{\pi}\sin[(\nu+\mu)\pi]Q_\nu^\mu(x), \tag{D.26}$$

$$Q_\nu^\mu(-u) = -e^{\pm i\nu\pi}Q_\nu^\mu(u), \qquad \Im(u)\gtrless 0, \tag{D.27}$$

$$Q_\nu^\mu(-x) = -\cos[(\nu+\mu)\pi]Q_\nu^\mu(x) - \frac{\pi}{2}\sin[(\nu+\mu)\pi)]P_\nu^\mu(x), \tag{D.28}$$

where $\Im(\cdot)$ denotes 'imaginary part of'. Note that if $\nu + \mu$ is not equal to an integer or zero, $P_\nu^\mu(-u)$ is independent of $P_\nu^\mu(u)$ as it contains a part of the function of the second kind. Hence $P_\nu^\mu(u)$ and $P_\nu^\mu(-u)$ comprise a linearly independent set of solutions of the associated Legendre equation in that case. Functions with negative and positive indices are related as

$$P_\nu^\mu(u) = \frac{\Gamma(\nu+\mu+1)}{\Gamma(\nu-\mu+1)}P_\nu^{-\mu}(u) + \frac{2e^{-i\pi\mu}\sin(\mu\pi)}{\pi}Q_\nu^\mu(u), \tag{D.29}$$

$$P_{-(\nu+1)}^\mu(u) = P_\nu^\mu(u), \tag{D.30}$$

$$Q_\nu^{-\mu}(u) = \frac{e^{-i2\pi\mu}\Gamma(\nu-\mu+1)}{\Gamma(\nu+\mu+1)}Q_\nu^\mu(u), \tag{D.31}$$

$$Q_{-(\nu+1)}^\mu(u) = -\pi e^{i\mu\pi}\cos(\nu\pi)P_\nu^\mu(u) + \frac{\sin\pi(\nu+\mu)}{\sin\pi(\nu-\mu)}Q_\nu^\mu(u). \tag{D.32}$$

### D.1.2 Functions with integer orders

If the order $\mu$ takes integer values, which we denote by $m$, some useful relations are

$$P_\nu^m(u) = (u^2-1)^{\frac{m}{2}}\frac{d^m}{du^m}P_\nu(u), \tag{D.33}$$

$$Q_\nu^m(u) = (u^2-1)^{\frac{m}{2}}\frac{d^m}{du^m}Q_\nu(u), \tag{D.34}$$

$$P_\nu^m(x) = (-1)^m(1-x^2)^{\frac{m}{2}}\frac{d^m}{dx^m}P_\nu(x), \tag{D.35}$$

$$Q_\nu^m(x) = (-1)^m(1 - x^2)^{\frac{m}{2}}\frac{\mathrm{d}^m}{\mathrm{d}x^m}Q_\nu(x), \tag{D.36}$$

$$P_\nu^m(x) = (-1)^m\frac{\Gamma(\nu + m + 1)}{\Gamma(\nu - m + 1)}P_\nu^{-m}(x). \tag{D.37}$$

If both $\nu$ and $\mu$ are integers, the functions are denoted by $P_n^m(u)$ and $Q_n^m(u)$ and referred to as *Legendre polynomials* of the first and second kind, respectively.

### D.1.3 Wronskians

For $n$th order linear differential equations, the Wronskian is a quantity that depends on the determinant of a coefficient matrix encountered in the 'variation of parameters' method of solution [4, p 18, 65]. The Wronskian between two differential functions $f$ and $g$ is defined as $\mathscr{W}(f, g) = fg' - gf'$. It measures the linear dependence/independence of the solutions of the homogeneous differential equation. For linear independence, the Wronskian must be non-zero; otherwise the solutions will be linearly dependent. For a second-order equation of the type

$$\frac{\mathrm{d}}{\mathrm{d}u}\left[\alpha(u)\frac{\mathrm{d}y}{\mathrm{d}u}\right] = \beta(u)y, \tag{D.38}$$

the Wronskian, $\mathscr{W}(y_1, y_2)$, involving the two solutions $y_1(u)$ and $y_2(u)$ is equal to $\mathscr{W}(y_1, y_2) = C/\alpha(u)$, where $C$ is a constant which may be evaluated by using $y_1$ and $y_2$ at any convenient point $u$. The Wronskian is explicitly independent of the function $\beta(u)$. For the associated Legendre equation (D.2), $\alpha(u) = 1 - u^2$, whereas for the form shown in equation (D.1) in variable $\theta$, the function $\alpha(\theta) = \sin\theta$. The Wronskians of associated Legendre equation are

$$\mathscr{W}\left\{P_\nu^\mu(u), Q_\nu^\mu(u)\right\} = \frac{\mathrm{e}^{\mathrm{i}\mu\pi}\,2^{2\mu}\,\Gamma\left(\frac{\nu+\mu+2}{2}\right)\Gamma\left(\frac{\nu+\mu+1}{2}\right)}{(1-u^2)\,\Gamma\left(\frac{\nu-\mu+2}{2}\right)\Gamma\left(\frac{\nu-\mu+1}{2}\right)}, \tag{D.39}$$

$$\mathscr{W}\left\{P_\nu^{-\mu}(\cos\theta), P_\nu^{-\mu}(-\cos\theta)\right\} = -\frac{2\mathrm{e}^{\mathrm{i}\mu\pi}}{\sin\theta\,\Gamma(\mu-\nu)\Gamma(\nu+\mu+1)} \tag{D.40}$$

$$\mathscr{W}\left\{P_\nu^{-\mu}(u), Q_\nu^\mu(u)\right\} = \frac{\mathrm{e}^{\mathrm{i}\pi\mu}}{(1-u^2)}, \tag{D.41}$$

$$\mathscr{W}\left\{P_\nu^{-\mu}(u), P_\nu^\mu(u)\right\} = \frac{2\sin(\mu\pi)}{\pi(1-u^2)}. \tag{D.42}$$

Hence $P_\nu^{-\mu}(u)$ and $P_\nu^\mu(u)$ are linearly independent solutions of the associated Legendre equation as long as $\mu$ is non-integer or zero. Using equation (D.39), the Wronskian for the Legendre equation can be written as

$$\mathscr{W}\{P_\nu(u),\, Q_\nu(u)\} = \frac{1}{(1-u^2)}, \tag{D.43}$$

a quantity seen to be independent of $\nu$. Hence, $P_\nu(u)$ and $Q_\nu(u)$ are always linearly independent of each other in any finite plane. References [3, 5] contain many identities and integrals involving Legendre functions. Some of those relevant to wave propagation are listed below.

### D.1.4 Recurrence relations

The recurrence relations satisfied by both $P_\nu^\mu(u)$ and $Q_\nu^\mu(u)$ are

$$\mathscr{C}_\nu^{\mu+1}(u) = (u^2-1)^{-\frac{1}{2}}\{(\nu-\mu)u\mathscr{C}_\nu^\mu(u) - (\nu+\mu)\mathscr{C}_{\nu-1}^\mu(u)\}, \tag{D.44}$$

$$(u^2-1)\frac{\mathrm{d}\mathscr{C}_\nu^\mu(u)}{\mathrm{d}u} = (\nu+\mu)(\nu-\mu+1)\sqrt{u^2-1}\,\mathscr{C}_\nu^{\mu-1}(u) - \mu u\mathscr{C}_\nu^\mu(u), \tag{D.45}$$

$$(\nu-\mu+1)\mathscr{C}_{\nu+1}^\mu(u) = (2\nu+1)u\mathscr{C}_\nu^\mu(u) - (\nu+\mu)\mathscr{C}_{\nu-1}^\mu(u), \tag{D.46}$$

$$(u^2-1)\frac{\mathrm{d}\mathscr{C}_\nu^\mu(u)}{\mathrm{d}u} = \nu u\mathscr{C}_\nu^\mu(u) - (\nu+\mu)\mathscr{C}_{\nu-1}^\mu(u), \tag{D.47}$$

$$\mathscr{C}_{\nu+1}^\mu(u) = \mathscr{C}_{\nu-1}^\mu(u) + (2\nu+1)\sqrt{u^2-1}\,\mathscr{C}_\nu^{\mu-1}(u), \tag{D.48}$$

$$\frac{\mathrm{d}}{\mathrm{d}\theta}P_n(\cos\theta) = \frac{n}{\sin\theta}[\cos\theta P_n(\cos\theta) - P_{n-1}(\cos\theta)] = P_n^1(\cos\theta), \tag{D.49}$$

$$= \frac{1}{\sin\theta}\frac{n(n+1)}{2n+1}[P_{n+1}(\cos\theta) - P_{n-1}(\cos\theta)] \tag{D.50}$$

$$(2n+1)\cos\theta P_n(\cos\theta) = (n+1)P_{n+1}(\cos\theta) + nP_{n-1}(\cos\theta), \tag{D.51}$$

where $\mathscr{C}_\nu^\mu = \{P_\nu^\mu,\, Q_\nu^\mu\}$.

### D.1.5 Integral representations

Useful integral representations for non-integer and integer values of order and degree are

$$P_\nu^\mu(u) = \frac{2^\mu}{(u^2-1)^{\frac{\mu}{2}}\sqrt{\pi}\,\Gamma\left(\frac{1}{2}-\mu\right)}\int_{-1}^{1}\frac{(u+t\sqrt{u^2-1})^{\nu+\mu}}{(1-t^2)^{\mu+\frac{1}{2}}}\,\mathrm{d}t, \tag{D.52}$$
$$[2\,\Re(\mu) < 1,\ |\arg(u\pm 1)| < \pi],$$

$$P_\nu^m(u) = \frac{(-1)^m \nu(\nu-1)\ldots(\nu-m+1)}{\pi} \int_0^\pi \frac{\cos m\alpha}{[u+\sqrt{u^2-1}\cos\alpha]^{\nu+1}}\,\mathrm{d}\alpha, \tag{D.53}$$

$$[|\arg(u)| < \pi/2],$$

$$P_n^m(\cos\theta) = \frac{(-1)^m\sqrt{2}\,(n+m)!}{\Gamma(m+\frac{1}{2})\sqrt{\pi}\,(n-m)!}\frac{1}{\sin^m\theta}\int_0^\theta (\cos\alpha-\cos\theta)^{m-\frac{1}{2}}\cos(n+\frac{1}{2})\alpha\,\mathrm{d}\alpha, \tag{D.54}$$

$$P_\nu(\cos\theta) = \frac{\sqrt{2}}{\pi}\int_0^\theta \frac{\cos\left(\nu+\frac{1}{2}\right)\alpha}{\sqrt{\cos\alpha-\cos\theta}}\,\mathrm{d}\alpha, \tag{D.55}$$

$$Q_\nu^\mu(u) = \frac{\mathrm{e}^{\mathrm{i}\mu\pi}\Gamma(\mu+\nu+1)}{2^{\nu+1}\Gamma(\nu+1)(u^2-1)^{\frac{\mu}{2}}}\int_{-1}^1 \frac{(1-t^2)^\nu}{(u-t)^{\nu+\mu+1}}\,\mathrm{d}t, \tag{D.56}$$

$$[\Re(\mu+\nu+1) > 0,\ \Re(\mu+1) > 0,\ |\arg(u\pm 1)| < \pi],$$

$$Q_n^m(u) = (-1)^m\frac{(n+m)!}{n!}\int_0^\zeta (u-\sqrt{u^2-1}\cosh\alpha)^n\cosh(m\alpha)\,\mathrm{d}\alpha$$

where

$$\zeta = \frac{1}{2}\ln\left(\frac{u+1}{u-1}\right) = \coth^{-1}(u) \tag{D.57}$$

$$Q_\nu(u) = \int_0^\infty \frac{1}{\left(u+\sqrt{u^2-1}\cosh\alpha\right)^{\nu+1}}\,\mathrm{d}\alpha, \qquad [\Re(\nu) > -1], \tag{D.58}$$

$$Q_n(u) = \frac{1}{2^{n+1}}\int_{-1}^1 \frac{(1-t^2)^n}{(u-t)^{n+1}}\,\mathrm{d}t = \frac{1}{2}\int_{-1}^1 \frac{P_n(t)}{u-t}\,\mathrm{d}t, \quad [|\arg(u-1)| < \pi]. \tag{D.59}$$

### D.1.6 General properties of associated Legendre functions

Some general properties of the solutions are:

1. For $\mu$ and $\nu$ non-integers, the functions $Q_\nu^\mu(u)$ are singular at $u = \pm 1$. This is evident from equation (D.11). For $\Re(\mu) > 0$, $P_\nu^{-\mu}(u)$ is singular at $u = -1$, but finite at $u = 1$. This is evident from the equation (D.10) on recognizing that $F(\alpha, \beta; \gamma; z)$ is finite at $z = 1$ for the conditions satisfied in equation (D.17). However, $P_\nu^\mu(u)$ for $\mu > 0$ is singular at both $u = \pm 1$ unless $\mu$ is an integer or equals zero. This is clear from equation (D.29). In the latter case,

$P_\nu^\mu(u)$ is singular only at $u = -1$ (see subsection D.1.11). For $\nu - \mu \neq 0, 1, 2, \ldots$, $P_\nu^{-\mu}(\cos\theta)$ and $P_\nu^{-\mu}(-\cos\theta)$ are linearly independent.

2. For $\nu$ non-integer or zero, $P_\nu^m(+1)$ are non-singular, but $Q_\nu^m(u)$ are singular at $u = \pm 1$. This is evident from the representations (D.53) and (D.58) together with equation (D.34). Also, $P_\nu^m(-1 \pm i0)$ is singular as is evident from equation (D.25). Hence $P_\nu^m(\cos\theta)$ is regular at $\theta = 0$, but singular at $\theta = \pi$. Conversely, $P_\nu^m(-\cos\theta)$ is singular at $\theta = 0$, but regular at $\theta = \pi$. For $\nu$ a non-integer, $P_\nu^m(\cos\theta)$ and $P_\nu^m(-\cos\theta)$ constitute a linearly independent set of solutions. Note that even though $Q_\nu^m(\cos\theta)$ and $P_\nu^m(\cos\theta)$ are both singular at $\theta = \pi$, certain combinations of these functions can be finite at $\theta = \pi$. This is because $\cos(\nu\pi)P_\nu^m(\cos\theta) - (2/\pi)\sin(\nu\pi)Q_\nu^m(\cos\theta) = (-1)^m P_\nu^m(-\cos\theta)$ by equation (D.26), which is finite on the $\theta = \pi$ axis when $\nu$ is a non-integer or zero. *Hence when the domain of interest includes $\theta = \pi$ axis, it is incorrect to outright ignore $Q_\nu^m(\cos\theta)$ from function expansion.*
3. For $\nu = n$, an integer, but $\mu$ not an integer or zero, the functions $P_n^\mu(u)$ for $\mu > 0$ and $Q_n^\mu(u)$ are both still singular. This is evident from equations (D.52) and (D.25) and equations (D.58) and (D.27), respectively.
4. If both $\nu$ and $\mu$ are equal to integers $n$ and $m$, respectively, then $P_n^m(u)$ are finite, but $Q_n^m(u)$ remain infinite at $u = \pm 1$.
5. Note that if $x \in (-1, 1)$ then the functions $P_\nu^{\pm\mu}(x)$ and $Q_\nu^\mu(x)$ as defined in equations (D.20) and (D.22) are purely real. In certain problems it might be advantageous to express any solution of the Legendre equation in terms of $P_\nu^{-\mu}(x)$ and $P_\nu^{-\mu}(-x)$ rather than expressing it in terms of $P_\nu^{-\mu}(x)$ and $Q_\nu^\mu(x)$, when $\nu$ is a non-integer or zero.
6. *Zeros of $P_\nu^\mu(x)$ for $x \in (-1, 1)$*: Let $E(z)$ denote the greatest integer less than $z$ when $z > 1$ and equal to zero for all other values of $z$. Let $\nu = n + \alpha$, $\mu = m + \beta$, where $0 < \alpha, \beta < 1$ and $m, n$ are integers. The number of zeros of $P_\mu^\nu(x)$within the real interval $(-1, 1)$ is $E(\nu - \mu + 1)$ when $\mu < 0, \nu + \frac{1}{2} \geq 0$. If $\mu > 0, \nu + \frac{1}{2} \geq 0$, it is $E(\nu - \mu + 1)$ if $n \geq m$, $\alpha > \beta$, or if $n < m$ and $m - n$ is odd; it is $E(\nu - \mu + 1) + 1$ if $n \geq m$, $\alpha \leq \beta$, or if $n < m$ and $m - n$ is even. If $\nu$ is a positive integer, the number of zeros is $E(\nu - |\mu| + 1)$. If $\mu$ is real and positive and $x$ is fixed in the interval $(-1, 1)$, $P_\nu^{-\mu}(x)$ as a function of $\nu$ has no complex zeros. When $\mu$ is positive, $P_\nu^\mu(u)$ as a function of $\nu$ has at most $2E(\mu)$ complex zeros.
7. Zeros of $Q_\nu^\mu(x)$ for $x \in (-1, 1)$: The number of zeros of $Q_\nu^\mu(\cos\theta)$ is $E(\nu - \mu + 1) + k$ where $k$ may have the values $-1, 0, 1, 2$ and must be such that the number is even or odd according as $\cos\nu\pi$ and $\cos\mu\pi$ have opposite signs or the same signs. When $n$ is a positive integer, the number of zeros of $Q_n(\cos\theta)$ is $\nu + 1$. Furthermore, the function $Q_n(u)$ has no zeros, real or complex outside of the branch cut $(-\infty, 1)$.
8. For a fixed $\theta$, $P_\nu(\cos\theta) - P_\nu(-\cos\theta)$ as a function of $\nu$ will have infinite roots. The first few zero crossings are shown later in figure D8.

### D.1.7 Addition theorem

*Let $z$, $z_1$, $z_2$, and $\phi$ be real or complex numbers such that $z = z_1 z_2 - (z_1^2 - 1)^{1/2}(z_2^2 - 1)^{1/2}\cos\phi$, the branches of the square roots such that $\Re(z^2 - 1)^{1/2}$ being positive for $\Re(z) > 0$ with branch cut along the interval* $[-1, 1]$. *Then*

$$P_n(z) = P_n(z_1)P_n(z_2) + 2\sum_{m=1}^{n}(-1)^m\frac{(n-m)!}{(n+m)!}P_n^m(z_1)P_n^m(z_2)\cos(m\phi). \tag{D.60}$$

### D.1.8 Derivatives with respect to order

$$\begin{aligned}\frac{\partial P_\nu^\mu(x)}{\mathrm{d}\nu} &= \frac{1}{\Gamma(\mu+1)}\left(\frac{1-x}{1+x}\right)^{\frac{1}{2}\mu}\sum_{n=1}^{\infty}\frac{(-\nu)(1-\nu)\ldots(n-1-\nu)(\nu+1)(\nu+2)\ldots(\nu+n)}{(\mu+1)(\mu+2)\ldots(\mu+n)1\cdot 2\ldots n}\\ &\times[\psi(\nu+n+1) - \psi(\nu-n+1)]\left(\frac{1-x}{2}\right)^n,\\ &\Re(\mu) > -1; \quad \nu \neq 0, \pm1, \pm2, \ldots,\end{aligned} \tag{D.61}$$

where

$$\psi(z) = \frac{\mathrm{d}\ln\Gamma(z)}{\mathrm{d}z} = \frac{\Gamma'(z)}{\Gamma(z)}, \tag{D.62}$$

is the Euler psi function and $\Gamma(z)$ is the gamma function

$$\Gamma(z) = \int_0^\infty \mathrm{e}^{-t}t^{z-1}\,\mathrm{d}t; \quad \Re(z) > 0, \tag{D.63}$$

with the recurrence property

$$\Gamma(z+1) = z\Gamma(z), \tag{D.64}$$

and the large argument property (Stirling's approximation)

$$\ln[\Gamma(z)] \sim \left(z - \frac{1}{2}\right)\ln z - z + \frac{1}{2}\ln(2\pi); \quad |z| \gg 1, -\pi < \arg z < \pi. \tag{D.65}$$

$$\left[\frac{\partial P_\nu(\cos\theta)}{\mathrm{d}\nu}\right]_{\nu=0} = 2\ln\cos\frac{\theta}{2}, \tag{D.66}$$

$$\left[\frac{\partial P_\nu^{-1}(\cos\theta)}{\mathrm{d}\nu}\right]_{\nu=0} = -\tan\frac{\theta}{2} - 2\cot\frac{\theta}{2}\cdot\ln\cos\frac{\theta}{2}, \tag{D.67}$$

$$\left[\frac{\partial P_\nu^{-1}(\cos\theta)}{\mathrm{d}\nu}\right]_{\nu=1} = -\frac{1}{2}\tan\frac{\theta}{2}\cdot\sin^2\frac{\theta}{2} + \sin\theta\cdot\ln\cos\frac{\theta}{2}. \qquad \text{(D.68)}$$

### D.1.9 Inequalities

$$\sin^{\mu+\frac{1}{2}}\theta\left|P_\nu^{\pm\mu}(\cos\theta)\right| < \frac{2}{\sqrt{\nu\pi}}\frac{\Gamma(\nu\pm\mu+1)}{\Gamma(\nu+1)} \qquad \text{(D.69)}$$

$$\sin^{\mu+\frac{1}{2}}\theta\left|Q_\nu^{\pm\mu}(\cos\theta)\right| < \sqrt{\frac{\pi}{\nu}}\frac{\Gamma(\nu\pm\mu+1)}{\Gamma(\nu+1)}. \qquad \text{(D.70)}$$

### D.1.10 Special values of indices

$$(1-x^2)\frac{\mathrm{d}P_n}{\mathrm{d}x} = \frac{n(n+1)}{2n+1}[P_{n-1}(x) - P_{n+1}(x)] \qquad \text{(D.71)}$$

$$P_0^\mu(\cos\theta) = \frac{1}{\Gamma(1-\mu)}\cot^\mu\left(\frac{\theta}{2}\right), \qquad \text{(D.72)}$$

$$-\nu(\nu+1)P_\nu^{-1}(\cos\theta) = \frac{\mathrm{d}P_\nu(\cos\theta)}{\mathrm{d}\,\theta} = P_\nu^1(\cos\theta), \qquad \text{(D.73)}$$

$$P_\nu^{\frac{1}{2}}(\cos\theta) = \sqrt{\frac{2}{\pi\sin\theta}}\cos\left(\nu+\frac{1}{2}\right)\theta, \qquad \text{(D.74)}$$

$$Q_\nu^{\frac{1}{2}}(\cos\theta) = -\sqrt{\frac{\pi}{2\sin\theta}}\sin\left(\nu+\frac{1}{2}\right)\theta, \qquad \text{(D.75)}$$

$$P_n^m(u) \equiv 0, \qquad \text{for} m > n, \qquad \text{(D.76)}$$

$$P_n(u) = \frac{1}{2^n n!}\frac{\mathrm{d}^n(u^2-1)^n}{\mathrm{d}u^n}, \qquad |P_n(x)| \le 1,\ x\in[-1,1], \qquad \text{(D.77)}$$

$$P_\nu^m(x) = (-1)^m(1-x^2)^{\frac{m}{2}}\frac{\mathrm{d}^m}{\mathrm{d}x^m}P_\nu(x), \qquad \text{(D.78)}$$

$$Q_\nu^m(u) = (u^2-1)^{\frac{m}{2}}\frac{\mathrm{d}^m}{\mathrm{d}u^m}Q_\nu(u), \qquad \text{(D.79)}$$

$$Q_n(x) = P_n(x)Q_0(x) - V_{n-1}(x), \text{ where} \qquad \text{(D.80)}$$

$$Q_0(x) = \frac{1}{2}\ln\left(\frac{1+x}{1-x}\right), \text{ and} \tag{D.81}$$

$$V_{n-1}(x) = \sum_{k=1}^{n} \frac{1}{k} P_{k-1}(x) P_{n-k}(x),\ V_{-1}(x) = 0, \tag{D.82}$$

$$P_0(u) = 1, \quad P_1(u) = u, \quad P_2(u) = \frac{1}{2}(3u^2 - 1), \tag{D.83}$$

$$P_0(\cos\theta) = 1, \quad P_1(\cos\theta) = \cos\theta, \quad P_2(\cos\theta) = \frac{1}{4}(3\cos 2\,\theta + 1), \tag{D.84}$$

$$Q_0(u) = \frac{1}{2}\ln\left(\frac{u+1}{u-1}\right),\ u > 1,\ Q_0(\cos\theta) = \frac{1}{2}\ln\left(\frac{1+\cos\theta}{1-\cos\theta}\right), \tag{D.85}$$

$$Q_1(u) = uQ_0(u) - 1,\ Q_2(u) = P_2(u)Q_0(u) - \frac{3u}{2}, \tag{D.86}$$

$$P_1^1(\cos\theta) = -\sin\theta,\ P_2^1(\cos\theta) = -\frac{3}{2}\sin 2\,\theta,\ P_2^2(\cos\theta) = 3\sin^2\theta, \tag{D.87}$$

$$P_3^1(\cos\theta) = -\frac{3}{8}(\sin\theta + 5\sin 3\,\theta), \tag{D.88}$$

$$P_3^2(\cos\theta) = 15\sin^2\theta\cos\theta,\ P_3^3(\cos\theta) = -15\sin^3\theta, \tag{D.89}$$

$$P_\nu^{-\mu}(\cos\theta) \sim \sqrt{\frac{2}{\pi\sin\theta}}\frac{\Gamma(\nu-\mu+1)}{\Gamma\left(\nu+\frac{3}{2}\right)}\cos\left[\left(\nu+\frac{1}{2}\right)\theta - \frac{\pi}{4} - \frac{\mu\pi}{2}\right]e^{i\mu\pi/2}, \tag{D.90}$$

$[|\nu| \gg |\mu|,\ |\arg\nu| < \pi,\ |\nu|\sin\theta \gg 1]$

$$P_n^m(\cos\theta) \sim n^m\sqrt{\frac{2}{\pi n\sin\theta}}\cos\left[\left(n+\frac{1}{2}\right)\theta + \frac{m\pi}{2} - \frac{\pi}{4}\right],\ m > 0,\ n \gg 1, \tag{D.91}$$

$$P_n^{-m}(\cos\theta) = (-1)^m\frac{(n-m)!}{(n+m)!}P_n^m(\cos\theta). \tag{D.92}$$

### D.1.11 Specific values of argument

$$P_\nu^{-\mu}(u) \sim \frac{(u-1)^{\mu/2}}{2^{\mu/2}\Gamma(\mu+1)},\ [u \to 1,\ \mu \neq -1, -2,, 3, \ldots] \tag{D.93}$$

$$P_\nu(-u) \sim \cos\nu\pi + \frac{\sin\nu\pi}{\pi}\ln(1-u),\ [u \to 1,\ \nu \neq -1, -2, \ldots] \tag{D.94}$$

$$\frac{dP_\nu^{-\mu}(u)}{du} \sim \frac{(u-1)^{\mu/2-1}}{2^{\mu/2+1}\Gamma(\mu)},\ [u \to 1,\ \mu \neq 0, -1, -2, -3, \ldots] \tag{D.95}$$

$$P_\nu^{-\mu}(u) \sim \frac{\Gamma(\nu+\frac{1}{2})}{\pi^{1/2}\Gamma(\mu+\nu+1)}(2u)^\nu,\ [u \to \infty] \tag{D.96}$$

$$Q_\nu^\mu(u) \sim \frac{\sqrt{\pi}e^{i\mu\pi}\Gamma(\nu+\mu+1)}{2^{\nu+1}\Gamma(\nu+\frac{3}{2})u^{\nu+1}},\ [u \to \infty,\ \nu \neq -3/2, -5/2, -7/2, \ldots] \tag{D.97}$$

$$Q_\nu^\mu(u) \sim \frac{2^{\mu/2-1}e^{i\mu\pi}\Gamma(\mu)}{(u-1)^{\mu/2}},\ [u \to 1,\ \Re(\mu) > 0,\ \nu+\mu \neq -1, -2, -3, \ldots] \tag{D.98}$$

$$P_{-1/2}^{-\mu}(u) \sim \frac{\ln(u)}{\Gamma(\mu+\frac{1}{2})}\left(\frac{2}{\pi u}\right)^{1/2},\quad [u \to \infty,\ \mu \neq -1/2, -3/2, -5/2, \ldots] \tag{D.99}$$

$$Q_\nu^0(u) \sim -\frac{\ln(u-1)}{2},\ [u \to 1,\ \nu \neq -1, -2, -3, \ldots] \tag{D.100}$$

$$P_\nu^\mu(0) = \frac{2^\mu}{\sqrt{\pi}}\cos\left(\frac{\pi}{2}(\nu+\mu)\right)\frac{\Gamma\left(\frac{\nu+\mu+1}{2}\right)}{\Gamma\left(\frac{\nu-\mu}{2}+1\right)}, \tag{D.101}$$

$$Q_\nu^\mu(0) = -2^{\mu-1}\sqrt{\pi}\sin\left(\frac{\pi}{2}(\nu+\mu)\right)\frac{\Gamma\left(\frac{\nu+\mu+1}{2}\right)}{\Gamma\left(\frac{\nu-\mu}{2}+1\right)} \tag{D.102}$$

$$\frac{d}{du}P_\nu^\mu(0) = \frac{2^{\mu+1}}{\sqrt{\pi}}\sin\left(\frac{\pi}{2}(\nu+\mu)\right)\frac{\Gamma\left(\frac{\nu+\mu+2}{2}\right)}{\Gamma\left(\frac{\nu-\mu+1}{2}\right)} \tag{D.103}$$

$$\frac{d}{du}Q_\nu^\mu(0) = 2^\mu\sqrt{\pi}\cos\left(\frac{\pi}{2}(\nu+\mu)\right)\frac{\Gamma\left(\frac{\nu+\mu+2}{2}\right)}{\Gamma\left(\frac{\nu-\mu+1}{2}\right)} \tag{D.104}$$

$$P_\nu(1) = 1, \tag{D.105}$$

$$P_n(-1) = (-1)^n \tag{D.106}$$

$$P_n^m(\pm 1) = 0, \quad m > 0 \tag{D.107}$$

$$\frac{\mathrm{d}P_n(\cos\theta)}{\mathrm{d}\,\theta} \sim -\,n\tan\left(\frac{\theta}{2}\right) \text{ for } \theta \sim 0 \tag{D.108}$$

$$\frac{\mathrm{d}P_n(\cos\theta)}{\mathrm{d}\,\theta} \sim (-1)^n n\cot\left(\frac{\theta}{2}\right) \text{ for } \theta \sim \pi \tag{D.109}$$

$$\lim_{\theta\to 0}\frac{\mathrm{d}}{\mathrm{d}\,\theta}P_n(\cos\theta) = 0 \tag{D.110}$$

$$\lim_{\theta\to 0}\frac{1}{\sin\theta}\frac{\mathrm{d}}{\mathrm{d}\,\theta}P_n(\cos\theta) = -\frac{n}{2} \tag{D.111}$$

$$\lim_{\theta\to 0}\frac{mP_n^m(\cos\theta)}{\sin\theta} = \lim_{\theta\to 0}\frac{\mathrm{d}}{\mathrm{d}\,\theta}P_n^m(\cos\theta) = \begin{cases} 0, & m \neq 1 \\ -\dfrac{n(n+1)}{2}, & m = 1 \end{cases} \tag{D.112}$$

$$\frac{\mathrm{d}P_n}{\mathrm{d}\,\theta}(\cos\theta) = -\sqrt{1-x^2}\,\frac{\mathrm{d}P_n}{\mathrm{d}x}(x) \tag{D.113}$$

$$P_\nu^\mu(\cos\theta) \sim \left[\left(\nu + \frac{1}{2}\right)\cos\frac{\theta}{2}\right]^\mu J_{-\mu}\left(2\nu\sin\frac{\theta}{2}\right), \quad \mu \geq 0,\ \nu \gg 1,\ \theta \ll 1. \tag{D.114}$$

### D.1.12 Generating function for integer order and degree

With $R^2 = (x - x')^2 + (y - y')^2 + (z - z')^2 = r^2 + r'^2 - 2rr'\cos\zeta$, $\cos\zeta = \cos\theta\cos\theta' + \sin\theta\sin\theta'\cos(\phi - \phi')$, $r_> = \max(r, r')$, $r_< = \min(r, r')$, the generating function of spherical harmonics is

$$\begin{aligned} \frac{1}{R} &= \frac{1}{r_>}\sum_{n=0}^{\infty}\left(\frac{r_<}{r_>}\right)^n P_n(\cos\zeta) \\ &= \frac{1}{r_>}\sum_{n=0}^{\infty}\sum_{m=-n}^{n}\frac{(n-m)!}{(n+m)!}\left(\frac{r_<}{r_>}\right)^n P_n^m(\cos\theta)P_n^m(\cos\theta')\mathrm{e}^{jm(\phi-\phi')}. \end{aligned} \tag{D.115}$$

### D.1.13 Useful integrals involving Legendre functions

Denoting the Kronecker's delta by $\delta_n^k$, we have

$$\int_0^\pi P_n^m(\cos\theta)P_k^m(\cos\theta)\sin\theta\,\mathrm{d}\,\theta = \int_{-1}^1 P_n^m(x)P_k^m(x)\,\mathrm{d}x = \frac{2\delta_n^k}{2n+1}\frac{(n+m)!}{(n-m)!}, \tag{D.116}$$

$$\int_0^\pi P_k^m(\cos\theta)Q_n^m(\cos\theta)\sin\theta\,\mathrm{d}\,\theta = \int_{-1}^1 Q_n^m(x)P_k^m(x)\,\mathrm{d}x = (-1)^m\frac{1-(-1)^{n+k}}{(k-n)(k+n+1)}\frac{(n+m)!}{(n-m)!}, \tag{D.117}$$

$$\int_0^\pi P_n(\cos\theta)\mathrm{e}^{\mathrm{i}\alpha\cos\theta}\sin\theta\,\mathrm{d}\,\theta = \int_{-1}^1 P_n(x)\mathrm{e}^{\mathrm{i}x\alpha}\,\mathrm{d}x = 2\mathrm{i}^n j_n(\alpha) = \mathrm{i}^n\sqrt{\frac{2\pi}{\alpha}}J_{n+\frac{1}{2}}(\alpha), \tag{D.118}$$

$$\int_{-1}^1 P_n^m(x)P_k^m(x)\frac{\mathrm{d}x}{1-x^2} = \int_0^\pi \frac{1}{\sin\theta}P_n^m(\cos\theta)\,P_k^m(\cos\theta)\,\mathrm{d}\,\theta = \frac{\delta_n^k}{m}\frac{(n+m)!}{(n-m)!}, \quad [m \neq 0], \tag{D.119}$$

$$\int_{-1}^1\left[(1-x^2)\frac{\mathrm{d}P_n^m(x)}{\mathrm{d}x}\frac{\mathrm{d}P_k^m(x)}{\mathrm{d}x} + \frac{m^2}{1-x^2}P_n^m(x)P_k^m(x)\right]\mathrm{d}x = \frac{2\delta_n^k\,n(n+1)}{2n+1}\frac{(n+m)!}{(n-m)!}, \tag{D.120}$$

$$\int_0^\pi \sin\theta\,\frac{\mathrm{d}}{\mathrm{d}\,\theta}P_n(\cos\theta)\,\frac{\mathrm{d}}{\mathrm{d}\,\theta}P_k(\cos\theta)\,\mathrm{d}\,\theta = \frac{2n(n+1)}{2n+1}\delta_n^k, \tag{D.121}$$

$$\int_{-1}^1 (1+x)^\sigma P_\nu(x)\,\mathrm{d}x = \frac{2^{\sigma+1}[\Gamma(\sigma+1)]^2}{\Gamma(\sigma+\nu+2)\Gamma(\sigma-\nu+1)}, \; \Re(\sigma) > 1 \tag{D.122}$$

$$\int_{\theta_1}^{\theta_2} \sin\theta\;\Theta_\nu(\theta)\bar{\Theta}_\nu(\theta)\,\mathrm{d}\,\theta = \left\{\frac{\sin\theta}{2\nu+1}\left[\frac{\partial\bar{\Theta}_\nu}{\partial\nu}\frac{\partial\Theta_\nu}{\partial\,\theta} - \Theta_\nu\frac{\partial^2\bar{\Theta}_\nu}{\partial\,\theta\partial\nu}\right]\right\}_{\theta_1}^{\theta_2}, \tag{D.123}$$

where $\Theta_\nu(\theta)$, $\bar{\Theta}(\theta)$ could be linear combinations of $P_\nu(\cos\theta)$ or $Q_\nu(\cos\theta)$. For instance, with $\bar{\Theta}_\nu(\theta) = P_\nu(\cos\theta) = \Theta_\nu(\theta)$, $\theta_1 = 0$, $\theta_2 = \pi$ and using equation (D.66), $\mathrm{d}P_\nu(\cos\theta)/\mathrm{d}\,\theta = 0$ for $\nu = 0$, we obtain from equation (D.123)

$$\int_0^\pi \sin\theta\;\Theta_0(\theta)\bar{\Theta}_0(\theta)\,\mathrm{d}\,\theta = 2 \tag{D.124}$$

which agrees with the result of formula (D.116).

$$\int_{\theta_1}^{\theta_2} \frac{\partial P_\nu}{\partial \theta}(\cos\theta)\frac{\partial P_\mu}{\partial \theta}(\cos\theta)\sin\theta \, \mathrm{d}\,\theta = \frac{\sin\theta}{(\mu-\nu)(\mu+\nu+1)}\times \left[\mu(\mu+1)P_\mu(\cos\theta)\frac{\partial P_\nu}{\partial\theta}(\cos\theta) - \nu(\nu+1)P_\nu(\cos\theta)\frac{\partial P_\mu}{\partial\theta}(\cos\theta)\right]_{\theta_1}^{\theta_2}, \ \mu \neq \nu$$

$$= \sin\theta\left[P_\mu(\cos\theta)\frac{\partial P_\mu}{\partial\theta}(\cos\theta) - \frac{\mu(\mu+1)}{2\mu+1}\left(\frac{\partial P_\mu}{\partial\mu}(\cos\theta)\frac{\partial P_\mu}{\partial\theta}(\cos\theta) - P_\mu(\cos\theta)\frac{\partial^2 P_\mu}{\partial\mu\partial\theta}(\cos\theta)\right)\right]_{\theta_1}^{\theta_2}, \ \mu = \nu$$

$$\int_0^\pi \frac{\mathrm{d}}{\mathrm{d}\,\theta}\left[P_n^m(\cos\theta)P_q^m(\cos\theta)\right]\mathrm{d}\,\theta = 0, \tag{D.125}$$

$$\int_0^\pi \left[\frac{\mathrm{d}P_n^m(\cos\theta)}{\mathrm{d}\,\theta}\frac{\mathrm{d}P_q^m(\cos\theta)}{\mathrm{d}\,\theta} + \frac{m^2}{\sin^2\theta}P_n^m(\cos\theta)P_q^m(\cos\theta)\right]\sin\theta \, \mathrm{d}\,\theta = \delta_n^q \frac{2n(n+1)}{2n+1}\frac{(n+m)!}{(n-m)!}. \tag{D.126}$$

If $Y_n^m(\theta, \phi) = \mathrm{e}^{\mathrm{i}m\phi}P_n^m(\cos\theta)$ and $S_1$ is a unit sphere with $\alpha \in (0, \pi)$, $\beta \in (0, 2\pi)$, then

$$j_n(kr)Y_n^m(\theta, \phi) = \frac{(-i)^n}{4\pi}\oint\!\!\oint_{S_1} Y_n^m(\alpha, \beta)e^{ikr(\sin\theta\sin\alpha\cos(\beta-\phi)+\cos\theta\cos\alpha)}\sin\alpha \, d\alpha \, d\beta. \tag{D.126}$$

### D.1.14 Series representation of Legendre polynomials

$$P_n(z) = 2^{-n}\sum_{k=0}^{\lfloor \frac{n}{2} \rfloor}\frac{(-1)^k[2(n-k)]!}{k!(n-k)!(n-2k)!}z^{n-2k}, \tag{D.128}$$

$$= \sum_{k=0}^{n}\frac{(-1)^k(n+k)!2^{-(k+1)}}{(k!)^2(n-k)!}[(1-z)^k + (-1)^n(1+z)^k]. \tag{D.129}$$

### D.1.15 Plots for associated Legendre functions

Figure D1–D5 show plots of $P_\nu^\mu(x)$, $P_\nu^{-\mu}(x)$, and $Q_\nu^\mu(x)$ for various values of $\mu$ and $\nu$ as functions of the argument $x$. Notice that for $|\mu| < 1$, $|\nu| < 1$, the functions do not oscillate in the range $x \in (-1, 1)$. For a given fractional degree $\nu$, the functions become more singular as the order $\mu$ is increased. This can be seen from comparing

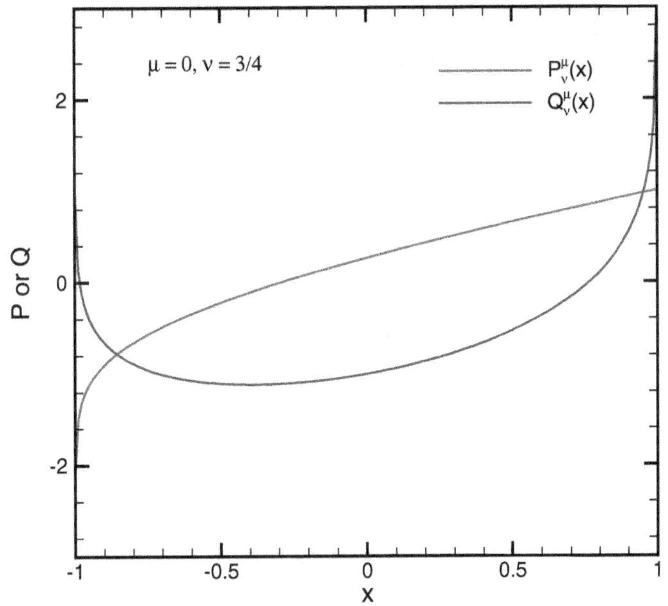


**Figure D1.** Associated Legendre functions for fractional degree and zero order. The function $P_\nu(u)$ is singular only at $u = -1$, but $Q_\nu(u)$ is singular at both $u = \pm 1$.

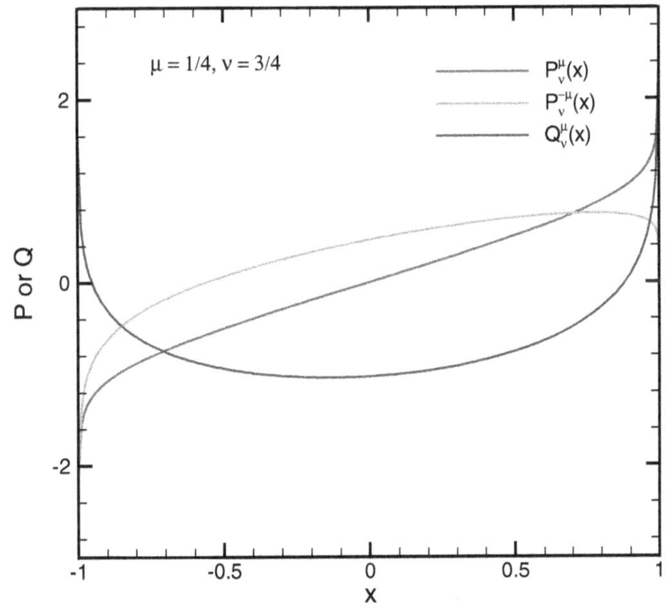


**Figure D2.** Associated Legendre functions for fractional degree and order with $|\mu| < 1$. Note that $P_\nu^{-\mu}(x)$ is regular at $x = 1$, but $P_\nu^{\mu}(x)$ is singular both at $x = \pm 1$.

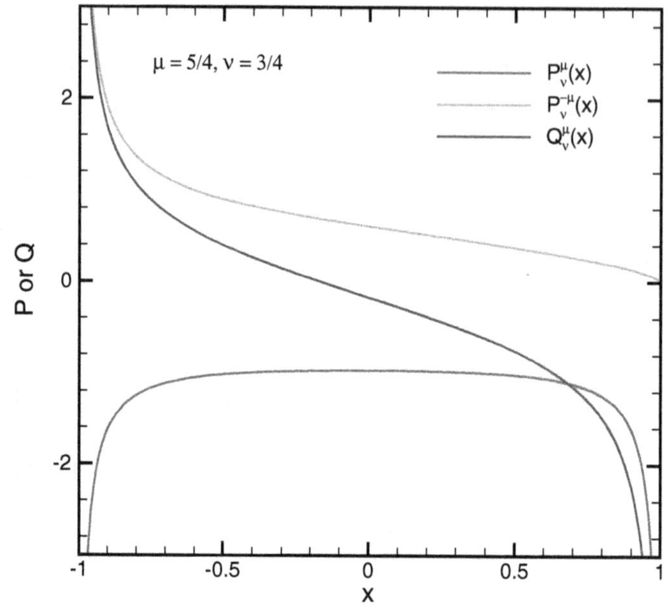


**Figure D3.** Associated Legendre functions for fractional degree and order with $|\mu| > 1$. The singularity of $P_\nu^{-\mu}(x)$ at $x = -1$ becomes stronger as $\mu$ increases.

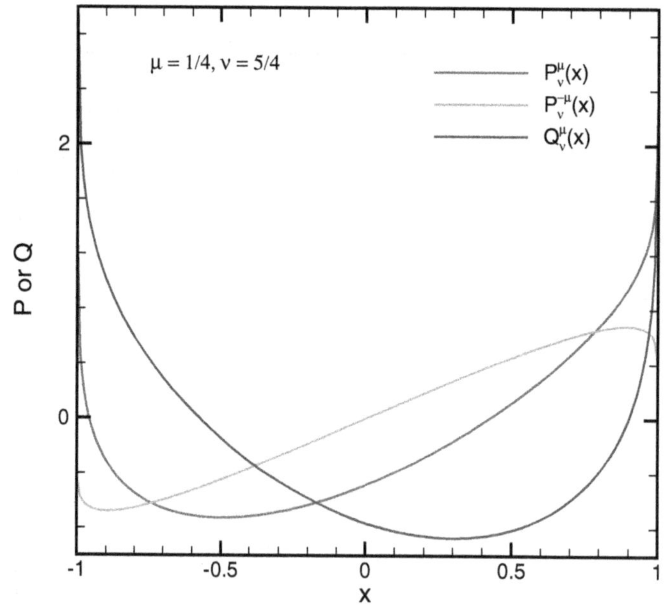


**Figure D4.** Associated Legendre functions for fractional order and degree with $|\nu| > 1$.

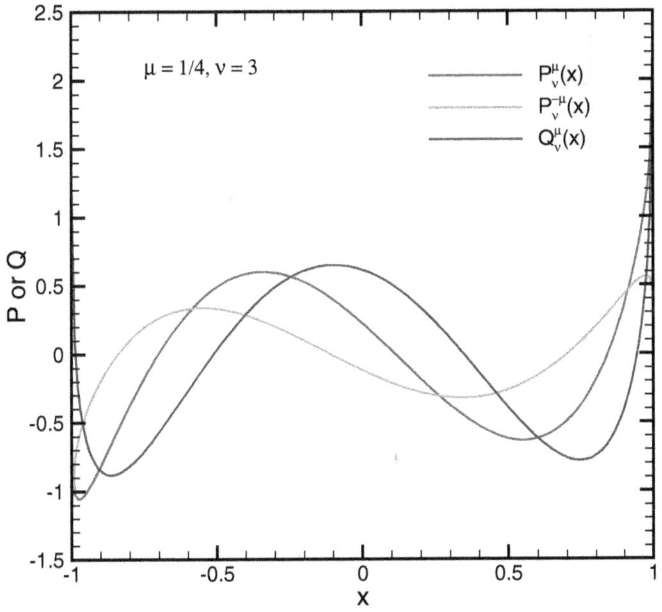


**Figure D5.** Associated Legendre functions for fractional order and integer degree with $\nu > 1$.

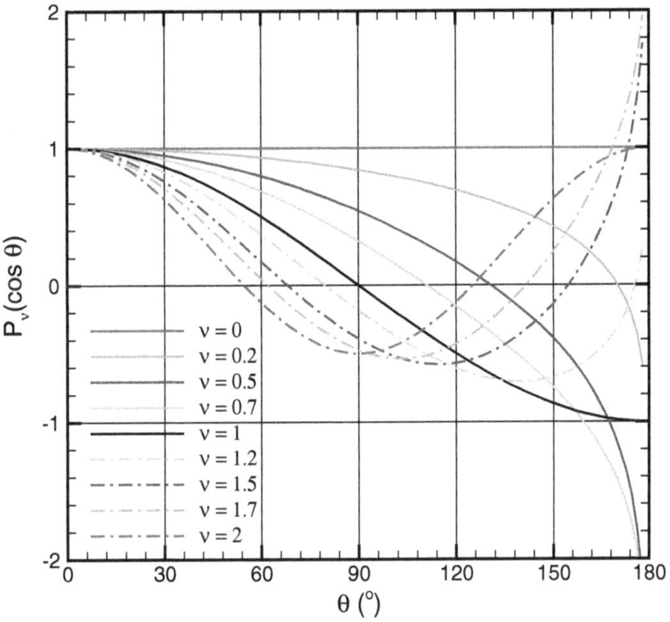


**Figure D6.** Legendre functions of the first kind for fractional degree with $0 \le \nu \le 2$.

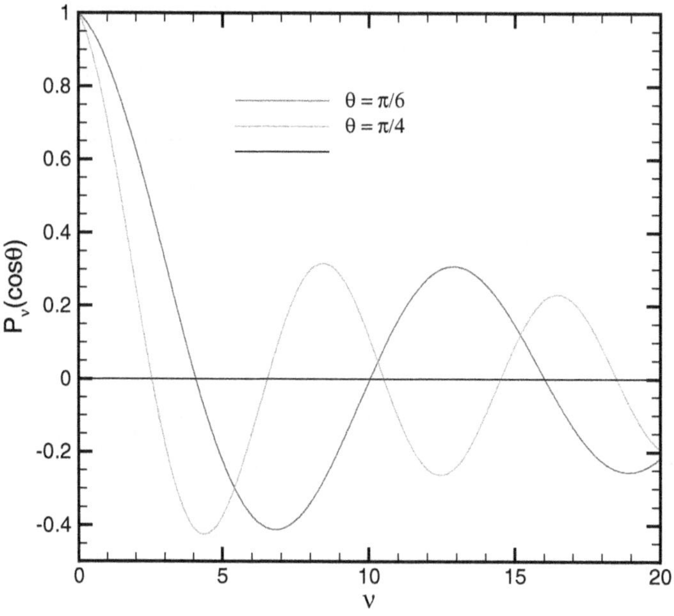


**Figure D7.** Legendre functions of the first kind versus fractional degree $\nu \in 0 \leq \nu \leq 20$.

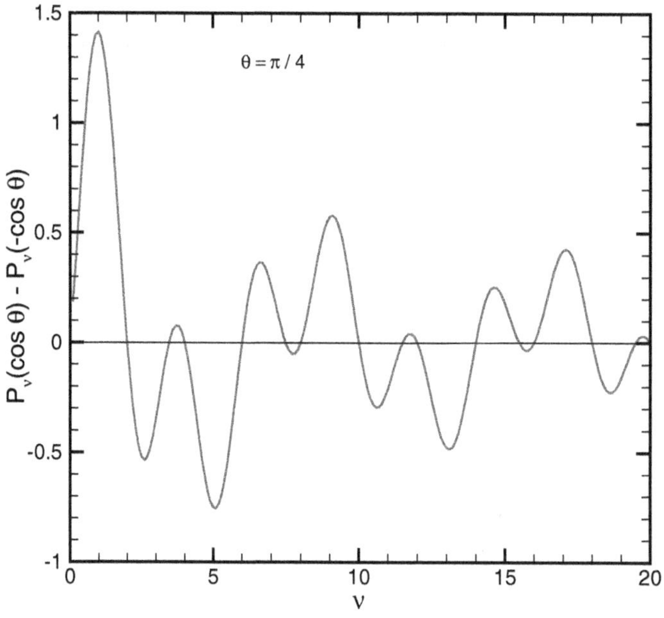


**Figure D8.** Variation of $L_\nu^-(\cos\theta) = P_\nu(\cos\theta) - P_\nu(-\cos\theta)$ versus the degree $\nu$ for a fixed argument $\theta = \pi/4$ showing zero crossings.

figures D1 and D3. However, for large degrees, the functions oscillate as can be seen from figure D5. Figure D6 shows plots of $P_\nu(\cos\theta)$ versus $\theta$ with $\nu$ in the range $0 \le \nu \le 2$. Figure D7 shows plots of $P_\nu(\cos\theta)$ versus $\nu$ for $\theta = \pi/6$, $\theta = \pi/4$.

In some applications, where the region of interest consists of the space between co-axial cones, functions such as $L_\nu^{\pm}(\cos\theta) = P_\nu(\cos\theta) \pm P_\nu(-\cos\theta)$ appear and one is interested in knowing the roots of $L_\nu^{\pm}$ as a function of $\nu$ for a fixed $\theta$. Figure D.8 shows the distribution of zeros of $L_\nu^{-}(\cos\theta)$ for $\theta = \pi/4$.

## References

[1] Ince E L 1944 *Ordinary Differential Equations* (New York: Dover)

[2] Olver F W J 1997 *Asymptotics and Special Functions* (Natick, MA: A K Peters)

[3] Abramowitz M and Stegun I A (ed) 1972 *Handbook of Mathematical Functions with Formulas, Graphs, and Mathematical Tables* (New York: Dover)

[4] Hairer E, Nørsett S P and Wanner G 1986 *Solving Ordinary Differential Equations I: Non-stiff Problems* (New York: Springer)

[5] Gradshteyn I S and Ryzhik I M 2015 Tables of integrals *Series and Products* 8th edn (New York: Academic)

www.ingramcontent.com/pod-product-compliance
Lightning Source LLC
Chambersburg PA
CBHW082117210326
41599CB00031B/5795